# the Elements

| III | IV | V | VI | VII | O |
|---|---|---|---|---|---|
| | | | | | 2<br>**2**<br>**He**<br>4.0026 |
| 2 3<br>**5**<br>**B**<br>10.811 | 2 4<br>**6**<br>**C**<br>12.01115 | 2 5<br>**7**<br>**N**<br>14.0067 | 2 6<br>**8**<br>**O**<br>15.9994 | 2 7<br>**9**<br>**F**<br>18.9984 | 2 8<br>**10**<br>**Ne**<br>20.183 |
| 2 8 3<br>**13**<br>**Al**<br>26.9815 | 2 8 4<br>**14**<br>**Si**<br>28.086 | 2 8 5<br>**15**<br>**P**<br>30.9738 | 2 8 6<br>**16**<br>**S**<br>32.064 | 2 8 7<br>**17**<br>**Cl**<br>35.453 | 2 8 8<br>**18**<br>**Ar**<br>39.948 |

| | | | III | IV | V | VI | VII | O |
|---|---|---|---|---|---|---|---|---|
| 28 16 2<br>**28**<br>**Ni**<br>58.71 | 28 18 1<br>**29**<br>**Cu**<br>63.546 | 28 18 2<br>**30**<br>**Zn**<br>65.37 | 2 8 18 3<br>**31**<br>**Ga**<br>69.72 | 2 8 18 4<br>**32**<br>**Ge**<br>72.59 | 2 8 18 5<br>**33**<br>**As**<br>74.9216 | 2 8 18 6<br>**34**<br>**Se**<br>78.96 | 2 8 18 7<br>**35**<br>**Br**<br>79.904 | 2 8 18 8<br>**36**<br>**Kr**<br>83.80 |
| 2 8 18 18<br>**46**<br>**Pd**<br>106.4 | 2 8 18 18 1<br>**47**<br>**Ag**<br>107.868 | 2 8 18 18 2<br>**48**<br>**Cd**<br>112.40 | 2 8 18 18 3<br>**49**<br>**In**<br>114.82 | 2 8 18 18 4<br>**50**<br>**Sn**<br>118.69 | 2 8 18 18 5<br>**51**<br>**Sb**<br>121.75 | 2 8 18 18 6<br>**52**<br>**Te**<br>127.60 | 2 8 18 18 7<br>**53**<br>**I**<br>126.9044 | 2 8 18 18 8<br>**54**<br>**Xe**<br>131.30 |
| 2 8 18 32 17 1<br>**78**<br>**Pt**<br>195.09 | 2 8 18 32 18 1<br>**79**<br>**Au**<br>196.967 | 2 8 18 32 18 2<br>**80**<br>**Hg**<br>200.59 | 2 8 18 32 18 3<br>**81**<br>**Tl**<br>204.37 | 2 8 18 32 18 4<br>**82**<br>**Pb**<br>207.19 | 2 8 18 32 18 5<br>**83**<br>**Bi**<br>208.980 | 2 8 18 32 18 6<br>**84**<br>**Po**<br>(210) | 2 8 18 32 18 7<br>**85**<br>**At**<br>(210) | 2 8 18 32 18 8<br>**86**<br>**Rn**<br>(222) |

| | | | | | | | | |
|---|---|---|---|---|---|---|---|---|
| 2 8 18 25 8 2<br>**63**<br>**Eu**<br>151.96 | 2 8 18 25 9 2<br>**64**<br>**Gd**<br>157.25 | 2 8 18 27 8 2<br>**65**<br>**Tb**<br>158.924 | 2 8 18 28 8 2<br>**66**<br>**Dy**<br>162.50 | 2 8 18 29 8 2<br>**67**<br>**Ho**<br>164.930 | 2 8 18 30 8 2<br>**68**<br>**Er**<br>167.26 | 2 8 18 31 8 2<br>**69**<br>**Tm**<br>168.934 | 2 8 18 32 8 2<br>**70**<br>**Yb**<br>173.04 | 2 8 18 32 9 2<br>**71**<br>**Lu**<br>174.97 |

| | | | | | | | | |
|---|---|---|---|---|---|---|---|---|
| 2 8 18 32 24 9 2<br>**95**<br>**Am**<br>(243) | 2 8 18 32 25 9 2<br>**96**<br>**Cm**<br>(247) | 2 8 18 32 26 9 2<br>**97**<br>**Bk**<br>(247) | 2 8 18 32 28 9 2<br>**98**<br>**Cf**<br>(249) | 2 8 18 32 29 9 2<br>**99**<br>**Es**<br>(254) | 2 8 18 32 30 9 2<br>**100**<br>**Fm**<br>(253) | 2 8 18 32 30 9 2<br>**101**<br>**Md**<br>(256) | 2 8 18 32 31 9 2<br>**102**<br>**No**<br>(254?) | 2 8 18 32 32 9 2<br>**103**<br>**Lw†**<br>(257) |

Atomic weights are based on carbon-12;
values in parentheses are for the most stable or the most familiar isotope.
† Symbol is unofficial.

# INTRODUCTION TO ORGANIC AND BIOLOGICAL CHEMISTRY

SAMIA SARKAR

# INTRODUCTION TO
# Organic and Biological
# Chemistry

**JOHN  R.  HOLUM,  PH.D.**
Professor of Chemistry, Augsburg College

**JOHN  WILEY  &  SONS,  INC.**
New York  ·  London  ·  Sydney  ·  Toronto

Library of Congress Catalog Card Number: 69-19093
SBN 471  40851  4
Printed in the United States of America

To Mary,

and to our children,

Elizabeth

Ann

Kathryn

# PREFACE

Particularly in regard to content, the short course in organic chemistry needs a change; this book is my suggestion for a different selection of material. The one-term or one-semester course is taken by few who intend to become professional organic chemists, which means that their goals are different from those of students in the full-year course and that they will respond to different motivations. For most of them the short course in organic chemistry is their last opportunity for formal study in the field, and they rightfully want to learn what will be useful to them in their future occupations. Many will become medical technologists or nurses. Some are going into dietetics, nutrition, physical therapy, and other paramedical fields. In some schools most are entering various vocations in the agricultural sciences. Future teachers of high school biology and chemistry take the course, as well as many with goals defined only as a career in biology. For all these students the problem what can be obtained from coal tar may be of some interest, but they would much rather learn (or should want to know) something about the chemical inventory and reactions in living cells. A great many of the reactions traditionally included in the short terminal course obviously cannot occur in any living system and are promptly forgotten after the final examination because the student knows that they are irrelevant to anything he will ever do. In a new short organic course the selection of material must surely be oriented toward the life sciences, and for this reason the theme of this textbook is the molecular basis of life.

I would be the last to claim that the molecular basis of life is all there is to life. I am not a hard determinist. But life does have a molecular basis. One of the difficulties is that the molecules have, even for many organic chemists, a mind-numbing complexity. Their structures fill the page like the desiccated remains of some microcosmic alphabet soup. What should the teacher do about them? Rote memorization is not the answer in this course, no more than memorizing whole maps is prescribed for students learning to read maps. Learning the map signs is enough, for then any map, however complicated, can be read. The map signs for biochemically important compounds are functional groups, types of reagents, and knowledge of some fundamental principles that lead us to expect particular events when certain reagents and conditions are presented to certain functional groups.

If the biochemical orientation is accepted for the brief organic course, organic reaction mechanisms must also be included, not as things in themselves but rather as tools for illuminating biochemical reactions at the molecular level of life. Although the molecules in cells are often very complicated, most of their changes can be broken down into a succession of minor molecular surgeries. A molecule of water (or something like water) is inserted or removed; the pieces of the element hydrogen are put in or taken out; carbon dioxide breaks off or is affixed; ammonia (or something like ammonia) adds or is eliminated; a proton is transferred from one place to another; opposite charges draw molecules or parts of molecules together; covalent bonds are made and broken. Sometimes, in more complicated reactions, a large molecule seemingly breaks in half or two large molecules join. But acetal formation is easy to learn among simple molecules; the aldol condensation makes sense, as does the reverse aldol condensation; the Claisen ester condensation, and its reverse, are quite similar; and, after being introduced to the ways of a carbonium ion, what is more natural than to expect it to attack a double bond. Events in cells are of course catalyzed by enzymes, not acids, bases, or powdered metals. The pH of the medium is very close to neutral, the temperature is mild, and the pressure is essentially atmospheric. But protons are transferred, and so too are hydride ions. Partial opposite charges attract each other. Steric factors operate. The fundamental laws of physics and chemistry are not suspended within a living cell—at least so we ardently believe. An *in vitro* aldol condensation may be much more of a jerky process than one taking place *in vivo*—we do not actually know—but the end result is the same and knowledge of the *in vitro* mechanism helps to make the *in vivo* event plausible. In trying to make biochemistry come alive in this text, most of the essentials of modern organic chemistry are exploited.

A few of the many dramatic illustrations of the molecular basis of life are developed as natural consequences of fundamental laws. How does the food we eat and the air we breathe become translated into the flexing of a muscle? The R. E. Davies theory, presented for the first time in any such textbook, is one exciting proposal. To be alive an organism must exchange with its environment materials, energy, and information. How are materials broken down and reorganized? How is energy tapped, stored, and used? How is information obtained, used, and transmitted? What are some of the fearful consequences of seemingly minor molecular disturbances? Students in the short course in organic chemistry can become excited over these questions. As they study them they will learn a great deal of both organic chemistry and biochemistry.

All that I have said in the Preface thus far could also be said about the first-year terminal course in chemistry for those not intending to become professional chemists. But the first-year student usually requires a background in basic principles of chemistry (and some physics). I have written another text that begins with these principles (*Principles of Physical, Organic, and Biological Chemistry,* John Wiley and Sons, New York, 1969). In the prepublication stage it was suggested that the last two-thirds of the text be made available for the short course in organic chemistry. This text, an expanded version of the last two-thirds of the longer book (it contains more metabolic pathways and more traditional organic chemistry), is the result.

Dr. Arne Langsjoen (Gustavus Adolphus College) and Dr. A. H. Blatt (Queens College) both read the entire manuscript and made many very valuable suggestions and important corrections. I am deeply grateful to them. What mistakes may remain are all mine. Please send me a note when you find them.

John R. Holum

Augsburg College
Minneapolis, Minnesota 55404
January 1969

# CONTENTS

# CHAPTER ONE

# Introduction to Organic Chemistry. Alkanes

## CONCERNING THINGS ORGANIC

Living organisms possess a capacity to synthesize a great number and variety of substances. The list of organic materials extends far beyond the obvious and most important example, food, to include petroleum, dyes, fibers, drugs, and natural plastics such as gums and rubber. Crude precursory attempts at chemical technology can be seen in the efforts of early people to isolate, purify, and improve these products. Some of these efforts involved little more than using heat on natural substances. When wood is heated without burning it, wood alcohol is obtained as well as acetone and pyroligneous acid, now known to contain acetic acid.

Heat may also cause charring and decomposition. Therefore methods that could be used when heat alone was unproductive were developed. Procedures for extracting substances from barks, roots, leaves, and fruit yielded many drugs and dyes. Making tea and coffee is a simple extraction with water the solvent. Modern and comparable extractions in industry use substances such as benzene, ether, light petroleums, and carbon tetrachloride as well as water as common solvents.

The assumed or actual curative powers of natural extracts were but one reason why increasing numbers of people were drawn to the investigation of organic compounds. Dyes, perfumes, flavoring agents, and spices were lures too. Some investigations were undertaken simply because, like Mount Everest to mountain climbers, they were there to be conquered. Until the early part of the nineteenth century, very few scientists saw much hope of making the realm of the organic into a science. One of the frustrations that hounded the early workers was the difficulty of purifying organic substances, which often decompose or deteriorate during or after efforts to isolate them. It was difficult to obtain evidence that these substances even obeyed the same chemical laws that held among minerals, for example, the law of definite proportions and the law of multiple proportions. The elementary composition of organic substances did not become clear until the beginning of the nineteenth century when Antoine Lavoisier demonstrated that carbon and hydrogen are essential elements in organic substances. It was soon discovered that oxygen, nitrogen, phosphorus, and sulfur also occur commonly in organic compounds. Better procedures for purification as well as better quantitative techniques soon established the fact that organic substances do indeed obey the laws of chemical composition. Yet they remained mysterious. The mineral kingdom includes carbonates and water, but in the hands of the early investigators these sources of the elements carbon and hydrogen could not be used to synthesize organic compounds in the test tube.

**Vitalism.** Chemists came to believe that they could never make organic substances from minerals. They could burn them, transform them into other products, and analyze them, but to make them seemed in the nature of things beyond the prowess and imagination of man. The ultimate and only source of organic substances appeared to be organisms, plant and animal, and the belief grew that living systems possessed a special force acting independently of other forces—a special force that could be passed from one living thing to another in the normal processes of reproduction and growth. The force was likened to an invisible flame burning steadily within living organisms. From this source of mysterious energy—energy unlike that available from other sources—nature derived the power to put together soil nutrients, water, and carbon dioxide into sugars, proteins, fibers, drugs, dyes, perfumes, spices, and a host of other organic substances. The gulf between organic and inorganic compounds seemed to be unbridgeable. Only an accidental event could have closed the gap, and it occurred to one Friedrich Wöhler in 1828.

**Wöhler's Synthesis of Urea.** Wöhler had been working with cyanogen, $C_2N_2$, and ammonia, both of which obey the known laws of inorganic chemistry and both of which were not at the time considered to be organic. Cyanogen reacts with water to produce cyanic acid and hydrocyanic acid:

$$C_2N_2 + H_2O \longrightarrow HCNO + HCN$$

$$\underset{\text{Cyanic acid}}{} \quad \underset{\text{Hydrocyanic acid}}{}$$

When ammonia is added, ammonium ions and cyanate ions form:

$$NH_3 + HCNO \longrightarrow NH_4^+ + CNO^-$$

$$\underset{\text{Cyanate ion}}{}$$

Wöhler prepared a solution containing these ions, and he hoped to obtain crystalline ammonium cyanate by heating it to drive off the water. He did obtain a crystalline, white solid, but it possessed none of the properties of the expected salt. Instead, its physical and chemical properties were those of an entirely different compound,

$$\underset{\text{Urea}}{NH_2-\overset{\overset{\displaystyle O}{\|}}{C}-NH_2}$$

urea. Urea had been isolated from urine half a century earlier. We now know that it is an end product of the metabolism of proteins, being made in the liver and removed from circulation at the kidneys. Although relatively simple, it is plainly of organic origin.

Wöhler's synthesis was checked by other scientists and rechecked by Wöhler himself. By 1830 distinguished scientists from different parts of Europe hailed his brilliant work. Still, the victory was not entirely complete. It was possible to argue that the sources of both the ammonia and the cyanogen were animal bones—dead bones, to be sure, but bones nevertheless. Thus the "vital force" might still linger in the substances used by Wöhler. This argument was taken so seriously that neither Wöhler nor his contemporaries trumpeted the overthrow of the vital force theory. It was not until 1844 when Kolbe synthesized acetic acid, $CH_3CO_2H$, from completely inorganic compounds that vitalism was decisively disproved. The synthesis of numerous other organic compounds without the aid of living organisms[1] soon followed.

---

Adolf Wilhelm Hermann Kolbe (1818–1884). A student of Wöhler's and an assistant to Bunsen, Kolbe is best remembered among organic chemists for his synthesis of salicylic acid.

---

Wöhler's and Kolbe's work constituted a scientific advance of immense proportions. The fields of synthetic dyes, drugs, and polymers stand as monuments to it. Because the vital force theory no longer inhibits attempts to synthesize organic substances artificially, many not known in nature are continually being made as organic chemists seek for further information about products of possible commercial or medicinal value as well as for a deeper understanding of the underlying laws that govern the behavior of all matter.

Organic chemistry has reached such a stage of development that we can study it in some depth as though living organisms, which gave the field its name and its start, did not exist. This procedure has its advantages because the organic chemicals found in living things are usually quite complicated. Since our theme is the molecular basis of life, we may not ignore them indefinitely, of course. However, by delaying a full-fledged investigation of them until we have found our bearings among simpler molecules, the final phase of our study will be much easier. The complicated molecules associated with living processes have one simplifying feature to help us—they can be understood rather well in terms of only a small number of key molecular "parts." In fact, our study of organic chemistry will evolve around the following question: What are the simplest and most useful "elements" in terms of which bio-

[1] There is irony here and George Wald has pointed it out. Wöhler's and Kolbe's experiments, as well as the work of any organic chemist, involve a very important living agency, the chemist himself. What Wöhler demonstrated was that chemists can make organic compounds externally as well as internally. There is no turning back, however. The vital force theory is defunct as far as chemists are concerned. (George Wald, *Scientific American*, August 1954, page 48.)

logical chemicals and their reactions may be understood? These "elements" are molecular parts or *functional groups* that occur in biological chemicals, and we shall study them first among simple substances. This focus on things biochemical will not be to the exclusion of all other aspects. In addition to its primary relevance to the processes of life and death and of health and disease, organic chemistry has a special significance in our technological age. Its important role in the worlds of petroleum products, of drugs and dyes, and of synthetic fibers and plastics cannot be ignored.

## CARBON'S UNIQUENESS

Organic chemistry is the branch of chemistry that deals with the compounds of carbon. Although other elements are present, carbon is given first rank because its atoms provide the skeletons for organic molecules. Carbon is unique among the 103 elements in that its atoms can bond to each other successively many, many times by means of strong covalent bonds. Carbon can do this at the same time that it uses its other valences in bonds to atoms of other elements ranging through the entire group of nonmetallic elements (e.g., hydrogen, oxygen, and fluorine) and even some metallic elements (e.g., lead). Thus both ethane, **1**, and hexafluoroethane, **2**, are known. The carbon "skeleton" in these two compounds consists of only two carbons. In principle, there is no limit to the length of a chain of carbons, and the common household plastic, polyethylene, consists of molecules with carbon chains hundreds of nuclei long.

$$H-\underset{\underset{H}{|}}{\overset{\overset{H}{|}}{C}}-\underset{\underset{H}{|}}{\overset{\overset{H}{|}}{C}}-H \qquad F-\underset{\underset{F}{|}}{\overset{\overset{F}{|}}{C}}-\underset{\underset{F}{|}}{\overset{\overset{F}{|}}{C}}-F$$

1            2
Ethane      Hexafluoroethane

## STRUCTURAL ORGANIC CHEMISTRY

**Geometry of Carbon Compounds.** The bonds in organic compounds are almost exclusively electron pair or covalent bonds. The covalence number of carbon is four, with but very few exceptions (e.g., CO, carbon monoxide). When a particular carbon atom is bonded to four groups, its four single bonds are directed toward the corners of a regular tetrahedron (Figure 1.1). All four hydrogen atoms in methane are equivalent. If one were to be replaced by, say, a chlorine atom, it would not matter which was taken. Only one monochloro derivative of methane exists—methyl chloride, **3**. Only a symmetrical structure for methane can account for this fact. The tetrahedral orientation of the hydrogens in methane is not the only one, however, that is symmetrical. A square-planar structure, **6**, would also have four equivalent hydrogens and would also lead to only one monochloro derivative, **7**. The three hydrogens in **7**, however, are not equivalent. Two are closer to the chlorine nucleus

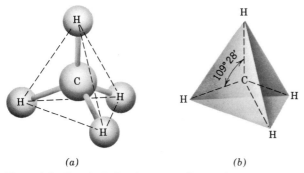

(a)                                                    (b)

**Figure 1.1**  Tetrahedral carbon atom. Part *a* shows a common ball and stick model of methane with dotted lines added to bring out the tetrahedron. A different perspective is given in
part *b*.

and its surrounding large electron cloud than the other. If the square-planar structure for methyl chloride were correct, then from it we should be able to prepare two
possible dichloro derivatives, **8** and **9.** But only one is known, **4,** and only one trich-
loro derivative, **5,** is known. These facts and many others, including results of
X-ray studies, confirm that the carbon in methane and its simple derivatives has its
four bonds directed not to the corners of a square but to the corners of a tetrahedron.
The bond angle in methane, the angle between the lines joining any two hydrogens
to the carbon, is 109°28′, precisely the angle that would be calculated from the
solid geometry of a regular tetrahedron.

In virtually all compounds of carbon, wherever a carbon has four single bonds
going away from it, the geometry at that point will generally be close to tetrahedral.
Moreover, groups that are joined by a single bond can rotate with respect to each
other about that bond. Given the tetravalence of carbon and its tetrahedral geometry,
structures **10** and **11** would seem to be two legitimate ways of orienting the nuclei
in the compound known as 1-chloropropane. If **10** and **11** were each rigid structures,
they would be molecules of different substances. However, only one 1-chloropro-
pane is known. A sample of this liquid presumably contains molecules such as **10**

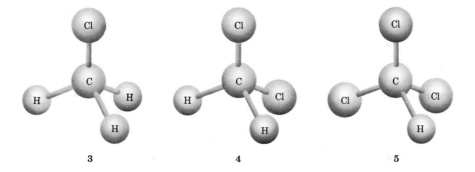

3                                    4                                    5

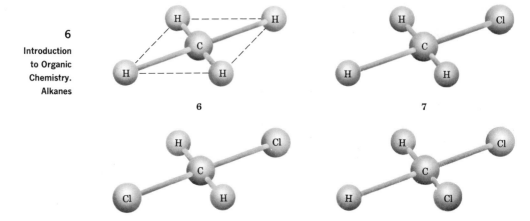

6

7

8

9

and **11** and all other possible forms differing only in the *relative* orientations of the large $CH_3$—$CH_2$— group and the chlorine. If we assume that these two groups are able to rotate about the bond drawn as a heavy line in **10** and **11**, the nonexistence of these two as separate compounds can be understood.

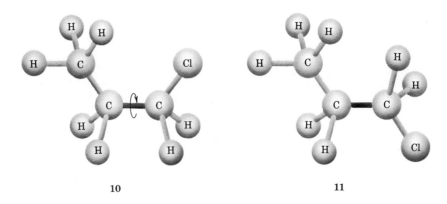

10

11

In summary, we have discussed three essential facts about carbon compounds. Carbon is tetravalent. When only single bonds are involved, it is tetrahedral. Groups attached by single bonds can rotate with respect to each other. We turn next to a theory of bonding in carbon compounds that will unify these facts.

**Bonding in Carbon Compounds.** The electronic configuration of carbon is $1s^2 2s^2 2p_x 2p_y$, which in terms of an energy level diagram would be represented as follows:

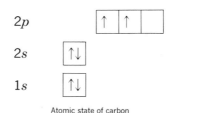

Atomic state of carbon

In the valence shell there are only two unpaired electrons. Only two covalent bonds rather than four would therefore be possible. The empty $p_z$ orbital might serve in the formation of a coordinate covalent bond by accepting a share in a pair of electrons from some Lewis base. But this would give only one more bond for a total of three. Of course, the pair of $2s$ electrons might be used in donating a share to some Lewis acid, some electron-poor species. Then the total number of bonds would be four—two ordinary covalent bonds involving the $p_x$ and $p_y$ orbitals, two coordinate covalent bonds involving the empty $p_z$ orbital as an acceptor and the filled $2s$ orbital as a donor. Although this arrangement might account for the tetravalence of carbon, it would not account for the geometry of its compounds.

The most prevalent current theory accounting for both the tetravalence and the geometry of carbon in its compounds holds that, although the electronic configuration we have used for carbon may be true for its atomic state, in its compounds these valence shell orbitals are no longer present. To have four equivalent *bonds* from a carbon, it is postulated that four equivalent *orbitals* are available for overlapping with orbitals from the other groups attached to the carbon. To understand how four such equivalent orbitals can arise, the following model is used.

One of the $2s$ electrons is envisioned as promoted to the empty $2p_z$ orbital. In the bookkeeping of the energy changes, this costs energy. In eventually forming four equivalent bonds, the cost is more than repaid. The new state of carbon—it may be called a "promoted" state—will be as follows:

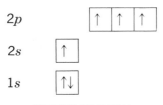

"Promoted" state of carbon

We now have four available orbitals for forming four covalent bonds, but these would not produce the final geometry of a molecule such as methane.

---

Exercise 1.1    What geometry would result if the orbitals of the "promoted" state were used to form bonds to hydrogen?

---

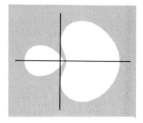

**Figure 1.2**   Cross section of an $sp^3$ hybrid orbital.

From the promoted state it is postulated that the four orbitals—one $s$ orbital and three $p$ orbitals—become reorganized into four identical orbitals. This reorganization or "mixing" is called *hybridization*. The new orbitals are *hybrid orbitals,* hybrids of the $s$ and the three $p$ orbitals. Since there are ultimately several kinds of hybrid orbitals, they must be given names, and whenever they are made from one $s$ and three $p$ orbitals, they are called $sp^3$ hybrid orbitals ("$s$–$p$–three"). The new energy level diagram for carbon in this *valence state* is as follows:

$$2(sp^3) \qquad \boxed{\uparrow}\,\boxed{\uparrow}\,\boxed{\uparrow}\,\boxed{\uparrow}$$

$$1s \qquad \boxed{\uparrow\downarrow}$$

The shape of an $sp^3$ hybrid orbital is indicated in Figure 1.2. It has some $s$ character —the intersection of the axes is enclosed by the orbital—and some $p$ character— there are two lobes. Figure 1.3 shows how these four equivalent orbitals are arranged in relation to each other. The axes of these orbitals point toward the corners of a regular tetrahedron.

Figure 1.4 illustrates how the bonds of methane are formed. Each bond from carbon to a hydrogen is the result of the overlap of an $sp^3$ orbital from carbon and the $1s$ orbital from hydrogen. Each new molecular orbital has cylindrical symmetry about the bonding axis, and such bonds are labeled sigma bonds ($\sigma$-bonds).

The bonding in ethane, $C_2H_6$, is illustrated in Figures 1.5 and 1.6, and two models of the molecule are shown in Figure 1.7. With this molecule we may return

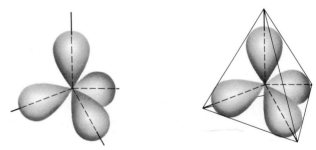

**Figure 1.3**   The four equivalent $sp^3$ hybrid orbitals of carbon are positioned so that their axes point to the corners of a regular tetrahedron. Only the large lobes of each $sp^3$ orbital are indicated in this figure.

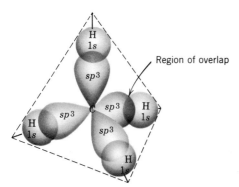

Region of overlap

H
1s

sp3

sp3 C sp3 H
1s

H
1s

sp3

H
1s

**Figure 1.4** Bonds in methane. Each carbon-to-hydrogen covalent bond is thought of as forming from the overlap of an $sp^3$ hybrid orbital from carbon and a $1s$ orbital from hydrogen. Because there is symmetry about the bonding axis, each bond is a $\sigma$-bond. The principle of maximum overlap ensures that the nucleus of each hydrogen is at the corner of the regular tetrahedron outlined by the dotted lines.

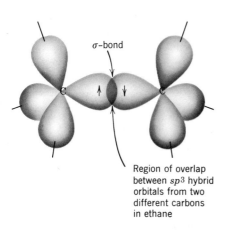

$\sigma$–bond

C ↑ ↓ C

Region of overlap between $sp^3$ hybrid orbitals from two different carbons in ethane

**Figure 1.5** The carbon-carbon single bond in ethane. The oppositely pointing arrows symbolize that the new molecular orbital contains two electrons of opposite spin. The region of maximum electron density for these two falls between the two carbon nuclei and is the principal factor contributing to the bond between them. (The exact locations of the arrows shown in this figure are not significant.) The remaining six $sp^3$ hybrid orbitals are used in ethane to overlap with $1s$ orbitals from six hydrogens as shown in Figure 1.6.

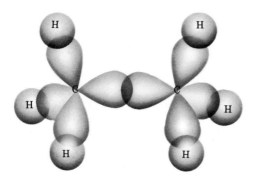

H

H

C

C

H

H

H

H

**Figure 1.6** The overlap of atomic orbitals to form the single bonds in ethane, all of which are $\sigma$-bonds.

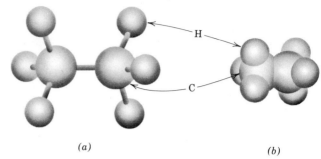

*(a)*

*(b)*

**Figure 1.7**  Two models for ethane. (*a*) Ball and stick model. (*b*) Scale model designed to indicate relative volumes occupied by parts of the molecule.

to the question how two groups attached by a single bond can rotate with respect to each other. In ethane, can the two $CH_3$ groups (methyl groups) rotate with respect to each other about the bond joining them? If such rotation required significant amounts of energy, the answer would be no, for the barrier to free rotation is fundamentally an energy barrier. But in ethane the overlap between the two orbitals forming the bond in question is not increased or decreased in any significant way by such rotation, and for this reason it is allowed. (That the spinning we have been describing actually happens is evident from spectral studies which we cannot discuss.) We do not imply that all possible orientations in ethane are energetically equivalent, for they are not and Figure 1.8 shows why. On the left, in the *a–a'''* series, the orientation is such that the six hydrogens are staggered. On the right, in the *b–b'''* series, the orientation has the six hydrogens mutually eclipsed; looking down the carbon-carbon bond axis, we see that a hydrogen in front eclipses one in back. In this eclipsed arrangement the hydrogens on the different carbons approach each other the most closely, and the repulsion between their electron clouds as well as between their nuclei will be slightly greater than in the staggered orientation. Consequently, the staggered form will be the most stable of all possible orientations. At ordinary temperatures, however, these hydrogen-to-hydrogen repulsions are so small that, although they may inhibit free rotation, they cannot prevent it. The physical and chemical properties of ethane are therefore the result of the average effects of all possible rotational arrangements.

**ISOMERISM**

When Wöhler tried to make ammonium cyanate (molecular formula, $CH_4N_2O$; structural formula, $NH_4^+CNO^-$) and obtained, instead, urea (molecular formula,

$$NH_2-\overset{\overset{\displaystyle O}{\|}}{C}-NH_2$$

$CH_4N_2O$; structural formula, $NH_2-\overset{\overset{\displaystyle O}{\|}}{C}-NH_2$), he added to the small but growing list of substances that were obviously different while still having identical molecular formulas. In 1832 Jöns Jacob Berzelius proposed that such substances be called *isomeric,* from the Greek *isos,* equal, and *meros,* part. He further suggested that the difference between compounds having identical molecular formulas must lie in a different arrangement of their parts, a prediction later confirmed.

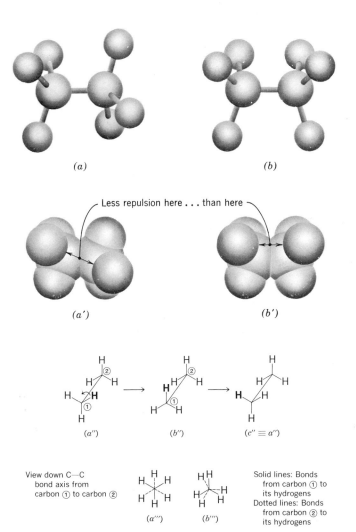

*(a)*

*(b)*

Less repulsion here . . . than here

*(a')*

*(b')*

*(a'')*

*(b'')*

*(c'' ≡ a'')*

View down C—C
bond axis from
carbon ① to carbon ②

*(a''')*

*(b''')*

Solid lines: Bonds
from carbon ① to
its hydrogens
Dotted lines: Bonds
from carbon ② to
its hydrogens

**Figure 1.8** Several representations of the staggered and the eclipsed conformations of ethane. In the *a* to *a'''* series are ways of describing the staggered form; in the *b* to *b'''* series are those for the eclipsed form. In parts *a'''* and *b''* are shown head-on views looking straight down the carbon-carbon axis. The dotted lines are the bonds from the carbon *behind* the front carbon to its hydrogens. In *b'''* a little twist is retained to show that all hydrogens are present, but this figure indicates the origin of the word "eclipsed." The front hydrogens in *b'''* eclipse the rear ones in the manner of a solar eclipse.

Note that in *a'* hydrogen nuclei with attendant electron clouds are farther apart, all around, than in *b'*. Electron clouds and nuclei tend to be as far apart as possible within the limits of the bonds present, and the fact that they are closer in *b'* than in *a'* is the origin of the greater internal energy of the eclipsed form. The situation in *b'* is not so serious, however, that the electron clouds cannot pass by each other during a rotation about the central carbon-carbon bond. Thus at ordinary temperatures "free rotation" is said to exist about the single bonds in ethane.

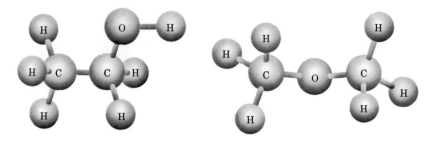

Ethyl alcohol                                               Methyl ether

**Figure 1.9** Isomers of $C_2H_6O$, ethyl alcohol and methyl ether, in ball and stick models. Each isomer has the same molecular formula, but they differ in the sequences in which these atoms are joined together.

Consider the molecular formula $C_2H_6O$. *Within the rules of valence,* how many structures are possible? In how many ways within the rules of valance can the nuclei of two carbons, six hydrogens, and one oxygen be arranged? Trial and error will show just two, given as ball and stick models in Figure 1.9. Substances consisting of each of these molecules are known, ethyl alcohol and methyl ether. Table 1.1 summarizes some of their characteristic properties.

Compounds that are isomeric are said to be *isomers* of each other, and the phenomenon itself is called *isomerism. For two compounds to be related as isomers, they must have identical molecular formulas but different structures.* Experimentally, two isomers must have identical formula weights, identical analytical results for the percentages of each element, but be detectably different in at least one physical or chemical property. The structures must be such that simple rotations about single

**Table 1.1   Properties of Two Isomers: Ethyl Alcohol and Methyl Ether**

| Property | Ethyl Alcohol | Methyl Ether* |
|---|---|---|
| Molecular formula | $C_2H_6O$ | $C_2H_6O$ |
| Boiling point | 78.5°C | −24°C |
| Melting point | −117°C | −138.5°C |
| Density | 0.789 g/cc | 2 g/liter |
| Solubility in water | completely soluble in all proportions | slightly soluble |
| Action of metallic sodium | vigorous reaction, hydrogen evolved | no reaction |
| Structural formula | H H<br>\| \|<br>H—C—C—O—H<br>\| \|<br>H H | H H<br>\| \|<br>H—C—O—C—H<br>\| \|<br>H H |
| Condensed structural formula | $CH_3CH_2OH$ | $CH_3OCH_3$ |

* Sometimes called dimethyl ether.

bonds will not convert one into the other. Although staggered and eclipsed forms of ethane that do not have the same relative locations of the nuclei or the same internal energies can be drawn, they do not correspond to different, isolable compounds. Ethane has no isomers. Free rotation assures this.

**Condensed Structural Formulas.** In writing the structure of ethane as $CH_3CH_3$ instead of as

$$H-\underset{\underset{H}{|}}{\overset{\overset{H}{|}}{C}}-\underset{\underset{H}{|}}{\overset{\overset{H}{|}}{C}}-H$$

considerable space and time is saved. With large structures the saving is more obvious, and we shall use such condensed structural formulas exclusively. Whenever three hydrogens are attached to the same carbon, we write the formula $CH_3-$. If two hydrogens are attached to the same carbon, and only two, the formula is written as $-CH_2-$; $-\overset{|}{C}H$ is used if the carbon has only one hydrogen. All covalent bonds are "understood" in a symbol such as $CH_3CH_3$, but frequently they are included, for example, $CH_3-CH_3$. Other conventions for writing condensed structures will be introduced as needed in subsequent chapters.

## HOW ORGANIC CHEMISTRY IS ORGANIZED AS A SUBJECT

The hundreds of thousands of known organic compounds can be classified into relatively few groups on the basis of similarities in structure. Just as zoologists classify animals into families having structural likenesses, so chemists classify organic compounds according to structural similarities. And just as animals with similar structures behave in many of the same ways, so members of organic structural families exhibit many of the same chemical properties. One of the important classification schemes in organic chemistry is by structure. Our study will be organized according to this scheme. Table 1.2 shows several important classes of organic compounds. We start our study of the families of organic compounds with the alkanes, a subgroup of the hydrocarbons.

# SATURATED HYDROCARBONS. ALKANES

## FAMILIES OF HYDROCARBONS

A study of the hydrocarbons is the most convenient starting point for our study of other families. As the name implies, molecules in this family are derived from the two elements carbon and hydrogen. Only covalent bonds occur, and depending on whether various types of multiple bonds are present or absent, a hydrocarbon may be a member of one of several subfamilies outlined in Figure 1.10.

The alkanes, whose molecules consist of only carbon and hydrogen and only single bonds, are the simplest, and they furnish the basis for systems of naming members of many other families.

**Table 1.2   Some Important Classes of Organic Compounds**

| Class | Characteristic Structural Feature |
|---|---|
| Hydrocarbons | Composed only of carbon and hydrogen, with many subclasses when single, double, or triple bonds are present<br>Alkanes: all single bonds, e.g., $CH_3CH_3$<br>Alkenes: at least one double bond, e.g., $CH_2{=}CH_2$<br>Alkynes: at least one triple bond, e.g., $H{-}C{\equiv}C{-}H$<br>Aromatic: at least one benzenoid ring system, e.g., |
| Alcohols | Contain the $-OH$ group as in $CH_3CH_2{-}OH$ |
| Ethers | Contain $C{-}O{-}C$ system as in $CH_3{-}O{-}CH_3$ |
| Aldehydes | Contain the $-\overset{\displaystyle O}{\overset{\|}{C}}{-}H$ group as in $CH_3{-}\overset{\displaystyle O}{\overset{\|}{C}}{-}H$ |
| Ketones | Contain the $C{-}\overset{\displaystyle O}{\overset{\|}{C}}{-}C$ group as in $CH_3{-}\overset{\displaystyle O}{\overset{\|}{C}}{-}CH_3$ |
| Carboxylic acids | Contain the $-\overset{\displaystyle O}{\overset{\|}{C}}{-}OH$ system as in $CH_3{-}\overset{\displaystyle O}{\overset{\|}{C}}{-}OH$ |
| Esters | Contain the $-\overset{\displaystyle O}{\overset{\|}{C}}{-}O{-}C$ system as in $CH_3{-}\overset{\displaystyle O}{\overset{\|}{C}}{-}O{-}CH_2CH_3$ |
| Amines | Contain trivalent nitrogen, with all covalent bonds single bonds as in $CH_3{-}NH_2$ |
| Amides | Contain the grouping $-\overset{\displaystyle O}{\overset{\|}{C}}{-}\overset{\displaystyle}{\overset{\|}{N}}{-}$ as in $CH_3{-}\overset{\displaystyle O}{\overset{\|}{C}}{-}NH_2$ |

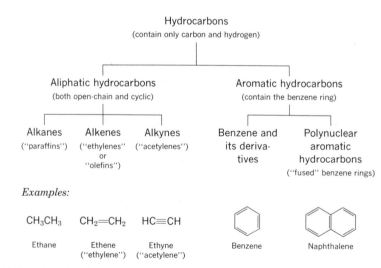

**Figure 1.10**   Hydrocarbon families.

# STRUCTURAL FEATURES OF ALKANES

**Saturated Compounds.** Any molecule of whatever family in which only single bonds occur is said to be *saturated*. Any molecule with one or more double or triple bonds is *unsaturated*.

**Homologous Series.** The structures and important physical properties of the ten smallest "straight chain" members of the alkanes are given in Table 1.3. Straight chain means only that the carbon nuclei follow one another as links in a chain, for example, C—C—C—C—C—C. In contrast, branched chain means that carbon nuclei are bonded as branches to the chain, for example, C—C—C—C—C—C.

(with two C branches below)

These expressions do not describe the geometry of the molecule. With free rotation about single bonds, these chains may flex, coil, and otherwise become kinky.

The series in Table 1.3 is said to be a homologous series because members differ from each other in a consistent, regular way. Here each member differs from the one just before or after it by one $CH_2$ unit. In chemical terminology butane, for example, is "the next higher *homolog* of propane."

**Isomerism among the Alkanes.** Among the alkanes, from butane and on to higher homologs, isomerism is possible. Table 1.4 lists several examples. There are two isomers of formula $C_4H_{10}$, three of formula $C_5H_{12}$, and five of formula $C_6H_{14}$. As the homologous series is ascended, the number of possible isomers approaches astronomical figures. There are 75 possible isomers of decane, $C_{10}H_{22}$, and an estimated $6.25 \times 10^{13}$ possible isomers of $C_{40}H_{82}$. Not all possible isomers have actually been prepared in the pure state and studied, for no useful purpose would be served and time would not permit. The occurrence of isomerism again emphasizes how limited is the information in a molecular formula. Only with a structural formula can the uniqueness of a molecule be understood and correlated with its properties.

**Table 1.3    Straight-Chain Alkanes**

| Number of Carbon Atoms | Name | Molecular Formula* | Condensed Structural Formula | Bp (°C at atmospheric pressure) | Mp (°C) | Density (in g/cc at 20°C) |
|---|---|---|---|---|---|---|
| 1 | Methane | $CH_4$ | $CH_4$ | −161.5 | | |
| 2 | Ethane | $C_2H_6$ | $CH_3CH_3$ | −88.6 | | |
| 3 | Propane | $C_3H_8$ | $CH_3CH_2CH_3$ | −42.1 | | |
| 4 | Butane | $C_4H_{10}$ | $CH_3CH_2CH_2CH_3$ | −0.5 | −138.4 | |
| 5 | Pentane | $C_5H_{12}$ | $CH_3CH_2CH_2CH_2CH_3$ | 36.1 | −129.7 | 0.626 |
| 6 | Hexane | $C_6H_{14}$ | $CH_3CH_2CH_2CH_2CH_2CH_3$ | 68.7 | −95.3 | 0.659 |
| 7 | Heptane | $C_7H_{16}$ | $CH_3CH_2CH_2CH_2CH_2CH_2CH_3$ | 98.4 | −90.6 | 0.684 |
| 8 | Octane | $C_8H_{18}$ | $CH_3CH_2CH_2CH_2CH_2CH_2CH_2CH_3$ | 125.7 | −56.8 | 0.703 |
| 9 | Nonane | $C_9H_{20}$ | $CH_3CH_2CH_2CH_2CH_2CH_2CH_2CH_2CH_3$ | 150.8 | −53.5 | 0.718 |
| 10 | Decane | $C_{10}H_{22}$ | $CH_3CH_2CH_2CH_2CH_2CH_2CH_2CH_2CH_2CH_3$ | 174.1 | −29.7 | 0.730 |

* The molecular formulas of the open-chain alkanes fit the general formula $C_nH_{2n+2}$, where $n$ is the number of carbons in the molecule.

## NOMENCLATURE

The earliest known organic compounds were named after their source, for example, formic acid (Latin *formica,* ants) can be made by grinding ants with water and distilling the result. Hundreds of compounds were named after their sources, but the system becomes impossibly difficult to extend to all compounds. These common names, however, are still used, and the beginning student is faced with the necessity of learning some of them. In addition, to be able to read and talk about more complicated structures without common names, the beginner must learn rules for formal or systematic nomenclature. A third system, the derived system of nomenclature, applies in a few situations.

**Table 1.4  Properties of Isomeric Alkanes**

| Family | Common Name (except where noted) | Structure | Bp (°C at atmospheric pressure) | Mp (°C) | Density (in g/cc) |
|---|---|---|---|---|---|
| Butane isomers | *n*-butane | $CH_3CH_2CH_2CH_3$ | −0.5 | −138.4 | 0.622 (−20°C) |
| | isobutane | $CH_3\overset{\vert}{C}HCH_3$ $CH_3$ | −11.7 | −159.6 | 0.604 (−20°C) |
| Pentane isomers | *n*-pentane | $CH_3CH_2CH_2CH_2CH_3$ | 36.1 | −129.7 | 0.626 (20°C) |
| | isopentane | $CH_3CHCH_2CH_3$ $CH_3$ | 27.9 | −159.9 | 0.620 |
| | neopentane | $CH_3$ $CH_3CCH_3$ $CH_3$ | 9.5 | −16.6 | 0.591 |
| Hexane isomers | *n*-hexane | $CH_3CH_2CH_2CH_2CH_2CH_3$ | 68.7 | −95.3 | 0.659 |
| | 3-methylpentane (no common name) | $CH_3$ $CH_3CH_2CHCH_2CH_3$ | 63.3 | | 0.664 |
| | isohexane | $CH_3$ $CH_3CHCH_2CH_2CH_3$ | 60.3 | −153.7 | 0.653 |
| | 2,3-dimethylbutane (no common name) | $CH_3$ $CH_3CHCHCH_3$ $CH_3$ | 58.0 | −128.5 | 0.662 |
| | neohexane | $CH_3$ $CH_3CCH_2CH_3$ $CH_3$ | 49.7 | −99.9 | 0.649 |

Octane isomers, $C_8H_{18}$—total of 18
Decane isomers, $C_{10}H_{22}$—total of 75
Eicosane isomers, $C_{20}H_{42}$—total of 366,319
Tetracontane isomers, $C_{40}H_{82}$—estimated total of $6.25 \times 10^{13}$ isomers

Organic nomenclature is a rather large field of study in its own right, but it must not be confused with organic *chemistry*. Nomenclature is a necessary part of the chemistry, but it is no substitute for it, contrary to a curious opinion held by some whose only experience with organic chemistry has been with naming compounds. We shall not make a thorough study of nomenclature, but we must cover enough of the subject to be able to examine the physical and chemical properties of organic substances.

**Common Names of Alkanes.** The straight-chain isomers are designated the *normal* isomers, and their common names include *n-* for normal. Examples are *n*-butane, *n*-octane, but not *n*-ethane because ethane has no isomer. In Table 1.3 butane and all subsequent names would be common if *n-* were placed before each. As they stand, they are formal names, to be described later. Except where noted, the names in Table 1.4 are also common names. The names of all the normal alkanes through *n*-decane as well as the names of the isomers through the five-carbon series must be learned. The total carbon content of the alkane should be associated with the prefix portion of its name. Thus "eth-" is a word part for a two-carbon unit; "but-" signifies four carbons whether in *n*-butane or isobutane. The "-ane" ending is characteristic of all alkanes in any system of nomenclature. We shall turn next to the formal system, the Geneva system or International Union of Pure and Applied Chemistry (IUPAC) system, but the student is advised that common names will be used almost exclusively in this text.

**Formal Names. Geneva System. IUPAC Rules for Alkanes.** Representatives from chemical societies all over the world, meeting as the International Union of Pure and Applied Chemistry, with sessions held usually in Geneva, have recommended adoption of the following rules for naming alkanes.

1. The general name for saturated hydrocarbons is *alkane.*

2. The names of the straight-chain members of the alkanes are those listed in Table 1.3. (The designation *n-* is not included. Names going beyond the ten-carbon alkanes are, of course, available, but we shall not need them.)

3. For branched-chain alkanes, base the root of the name on the alkane that corresponds to the longest continuous (i.e., unbranched) chain of carbons in the molecule. For example, in the compound

$$CH_2-CH_2$$
$$CH_2-CH-CH_2 \quad CH_3$$
$$CH_3 \quad CH_3$$

which when "straightened" is

$$CH_3$$
$$CH_3CH_2CHCH_2CH_2CH_2CH_3$$

the longest continuous chain totals seven carbons. The *last* part of the complete name for this compound will therefore be *heptane.* We next learn how to specify the location of the small $CH_3-$ branch.

4. To locate branches, assign a number to each carbon of the longest continuous chain. Begin at whichever end of the chain will result in the smaller number or set of numbers for the carbon(s) holding branches. In our example,

$$
\begin{array}{c}
CH_3 \\
| \\
CH_3CH_2CHCH_2CH_2CH_2CH_3 \\
\begin{array}{ccccccc} 1 & 2 & 3 & 4 & 5 & 6 & 7 \end{array}
\end{array}
$$

If this chain had been numbered from right to left, the carbon holding the branch would have the number 5. Having located the branch(es), we must now be able to name it (them).

5. If a side chain or branch consists only of carbon and hydrogen linked with single bonds, it is called an *alkyl group;* "alkyl" comes from changing the "-ane" ending of "alkane" to "-yl." This change is the key to making up names for alkyl groups, and Table 1.5 lists the names and structures for the most common ones. Their structures must be learned so well that they can be recognized written backward, forward, and upside down. Each is derived (on paper) by taking an alkane and removing a hydrogen to leave an unused bond represented by a line. The site of this unused bond must be clearly known, for it is at this point that the group is attached to the chain. In learning the names and structures of the alkyl groups in Table 1.5, again the best advice is to associate the total number of carbons in the group with the prefix portion of the name of its "parent" alkane. The table shows, for example, four butyl groups, two related to *n*-butane and two to isobutane. All are butyl groups because each has four carbons. To distinguish them from each other, additional word parts are tacked on. The words "secondary" and "tertiary" (abbreviated *sec-* and *tert-* or simply *t-*) denote the condition of the carbon in the group having the unused bond. In the *sec*-butyl group this carbon is directly attached to *two* other carbons and is therefore classified as a *secondary* carbon. In the *t*-butyl group this carbon has direct bonds to three other carbons and is classified as a *tertiary* carbon. When a carbon is attached directly to but one other carbon, it is classified as a *primary* carbon. These classifications of carbons should be learned. In our study we shall on only two occasions use the names for the classes to make names for specific groups. Among the butyl groups, one is *sec*-butyl (a name) and one is *t*-butyl (a name). The other two are primary (a class), and the class name therefore cannot be used to name either, for it would be ambiguous to do so. But one is a straight chain with the unused bond at the *end;* it is therefore called the *n*-butyl group (or, simply, butyl). The other is called the isobutyl group.

Having learned rules for locating and naming side chains, we next consider situations in which identical groups are located on the same carbon of the main chain.

6. Whenever two identical groups are attached at the same place, numbers are supplied for each group. For example,

$$
\begin{array}{ll}
\begin{array}{c}
CH_3 \\
| \\
CH_3CCH_2CH_2CH_2CH_3 \\
| \\
CH_3
\end{array}
&
\begin{array}{l}
\text{Correct name: 2,2-dimethylhexane} \\
\text{Incorrect: 2-dimethylhexane} \\
\text{2,2-methylhexane}
\end{array}
\end{array}
$$

**Table 1.5 Alkyl Groups—Names and Structures**

| Parent Alkane | Structure of Parent Alkane | Structure of Alkyl Group from Alkane | Name of Alkyl Group |
|---|---|---|---|
| Methane | $CH_4$ | $CH_3—$ | methyl |
| Ethane | $CH_3CH_3$ | $CH_3CH_2—$ | ethyl |
| Propane | $CH_3CH_2CH_3$ | $CH_3CH_2CH_2—$ | $n$-propyl |
| | | $\begin{matrix} CH_3 \\ \\ CH_3 \end{matrix}\!\!\!\!\!\searrow\!\!\!\!\!\nearrow CH—$ | isopropyl |
| $n$-Butane | $CH_3CH_2CH_2CH_3$ | $CH_3CH_2CH_2CH_2—$ | $n$-butyl |
| | | $\overset{\quad CH_3}{\underset{\quad |}{CH_3CH_2CH—}}$ | secondary butyl (sec-butyl) |
| Isobutane | $\overset{\ CH_3}{\underset{\ |}{CH_3CHCH_3}}$ | $\overset{CH_3}{\underset{|}{CH_3CHCH_2—}}$ | isobutyl |
| | | $\overset{CH_3}{\underset{CH_3}{\overset{|}{\underset{|}{CH_3C—}}}}$ | tertiary butyl (t-butyl) |

Any normal alkane: If the "free valence" extends from the *end* of the unbranched chain, change the "-ane" ending of the alkane to "-yl"; e.g., $CH_3CH_2CH_2CH_2CH_2CH_2CH_2—$ is $n$-heptyl

| Any alkane in general: R—H | R— | alkyl |
|---|---|---|

7. Whenever two or more different groups are affixed to a chain, two ways are acceptable for organizing all the name parts into the final name. The last word part is always the name of the alkane corresponding to the longest chain.

($a$) The word parts can be ordered by increasing complexity of side chains (e.g., in order of increasing carbon content):

methyl, ethyl, propyl, isopropyl, butyl, isobutyl, *sec*-butyl, *t*-butyl, etc.

($b$) They can be listed in simple alphabetical order. (This corresponds to the indexing system of *Chemical Abstracts*, a publication of the American Chemical Society.)

In this text we shall not be particular about the matter of order. For the following compound both names given are acceptable.

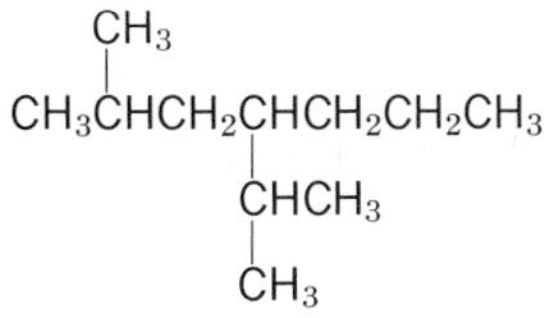

$$\overset{\qquad CH_3}{\underset{\qquad |}{CH_3CHCH_2CHCH_2CH_2CH_3}} \quad \begin{matrix} \text{2-methyl-4-isopropylheptane} \\ \text{or} \quad \text{4-isopropyl-2-methylheptane} \end{matrix}$$

with side chain $\underset{|}{CHCH_3}$ and $CH_3$

Note carefully the use of hyphens and commas in organizing the parts of names. Hypens always separate numbers from word parts, and commas always separate numbers. The intent is to make the final name one word.

8. The formal names for several of the more important nonalkyl substituents are as follows:

| | | | |
|---|---|---|---|
| —F | fluoro | —NO$_2$ | nitro |
| —Cl | chloro | —NH$_2$ | amino[2] |
| —Br | bromo | —OH | hydroxy[2] |
| —I | iodo | | |

Several examples of compounds correctly named according to these rules are given below. In parentheses are shown some common ways in which incorrect names are often devised. As an exercise, describe how each incorrect name violates one or more of the rules.

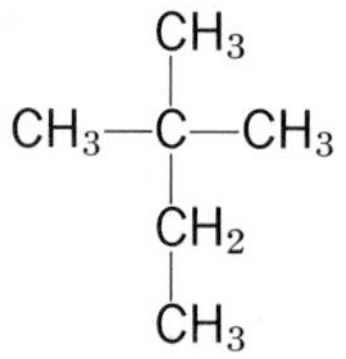

$$CH_3-\underset{\underset{\displaystyle CH_3}{\overset{\displaystyle \underset{|}{CH_2}}{|}}}{\overset{\displaystyle \overset{CH_3}{|}}{C}}-CH_3$$

2,2-Dimethylbutane
*Not* 2-methyl-2-ethylpropane

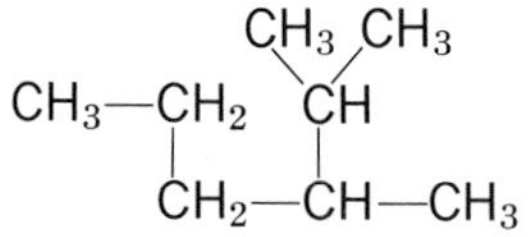

$$CH_3-CH_2\ \underset{\underset{\displaystyle CH_2-CH-CH_3}{|}}{\overset{\displaystyle \overset{CH_3\ \ CH_3}{|\ \ \ \ |}}{CH}}$$

2,3-Dimethylhexane
*Not* 2-isopropylpentane

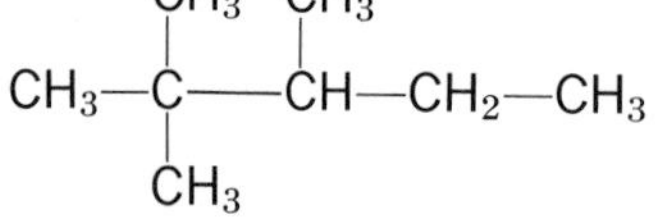

$$CH_3-\underset{\underset{\displaystyle CH_3}{|}}{\overset{\displaystyle \overset{CH_3}{|}}{C}}-\overset{\displaystyle \overset{CH_3}{|}}{CH}-CH_2-CH_3$$

2,2,3-Trimethylpentane
*Not* 2,3-trimethylpentane
*Not* 2-t-butylbutane

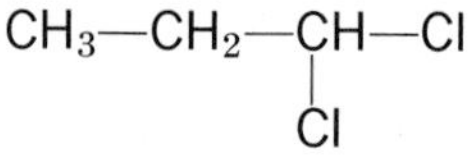

$$CH_3-CH_2-\underset{\underset{\displaystyle Cl}{|}}{\overset{\displaystyle \overset{}{\ }}{CH}}-Cl$$

1,1-Dichloropropane
*Not* 3,3-dichloropropane
*Not* 3,3-chloropropane
*Not* 1-dichloropropane
*Not* 1,1-chloropropane

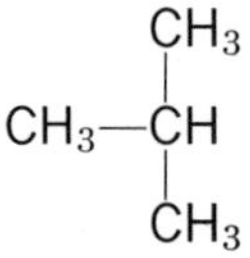

$$CH_3-\overset{\displaystyle \overset{CH_3}{|}}{\underset{\underset{\displaystyle CH_3}{|}}{CH}}$$

2-Methylpropane
*Not* 1,1-dimethylethane
*Not* isobutane, which is its
*common* name

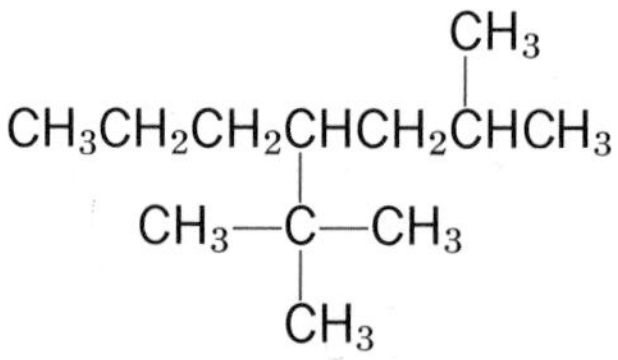

$$CH_3CH_2CH_2\underset{\ }{CH}CH_2\overset{\displaystyle \overset{CH_3}{|}}{CH}CH_3$$
$$CH_3-\underset{\underset{\displaystyle CH_3}{|}}{\overset{\displaystyle \overset{CH_3}{|}}{C}}-CH_3$$

2-Methyl-4-t-butylheptane
*Not* 4-t-butyl-6-methylheptane
*But* 4-t-butyl-2-methylheptane is acceptable

---

[2] These two are used only in special circumstances. As we shall see in later chapters, amino compounds are usually named as amines and hydroxy compounds as alcohols, with special IUPAC rules.

Exercise 1.2   Write the structures (condensed) of each of the following.
  (a)  1-bromo-2-nitropentane
  (b)  2,2,3,3,4,4-hexamethyl-5-isopropyloctane
  (c)  2,2-diiodo-3-methyl-4-isopropyl-5-sec-butyl-6-t-butylnonane
  (d)  1-chloro-1-bromo-2-methylpropane
  (e)  4,4-di-sec-butyldecane

Exercise 1.3   Write IUPAC names for each of the following.

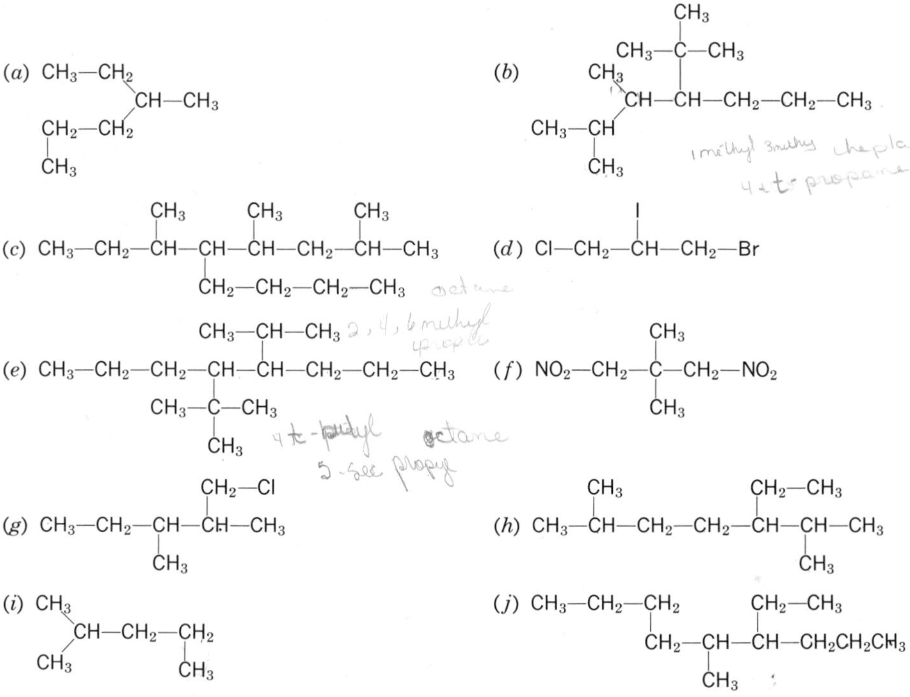

(a) 
$$CH_3-CH_2$$
$$\quad\quad CH-CH_3$$
$$CH_2-CH_2$$
$$CH_3$$

(b) 
$$CH_3$$
$$CH_3-C-CH_3$$
$$CH_3 \quad\quad |$$
$$\quad CH-CH-CH_2-CH_2-CH_3$$
$$CH_3-CH$$
$$\quad\quad CH_3$$

*1 methyl 3methyl chepane*
*4 et propane*

(c) 
$$CH_3 \quad\quad CH_3 \quad\quad CH_3$$
$$CH_3-CH_2-CH-CH-CH-CH_2-CH-CH_3$$
$$\quad\quad\quad CH_2-CH_2-CH_2-CH_3 \quad octane$$

*2,4,6 methyl*
*4 ex el*

(d) 
$$I$$
$$Cl-CH_2-CH-CH_2-Br$$

(e) 
$$CH_3-CH-CH_3$$
$$CH_3-CH_2-CH_2-CH-CH-CH_2-CH_2-CH_3$$
$$\quad\quad\quad CH_3-C-CH_3$$
$$\quad\quad\quad\quad CH_3$$

*4 t-butyl  octane*
*5 sec propyl*

(f) 
$$CH_3$$
$$NO_2-CH_2-C-CH_2-NO_2$$
$$CH_3$$

(g) 
$$CH_2-Cl$$
$$CH_3-CH_2-CH-CH-CH_3$$
$$\quad\quad\quad CH_3$$

(h) 
$$CH_3 \quad\quad CH_2-CH_3$$
$$CH_3-CH-CH_2-CH_2-CH-CH-CH_3$$
$$\quad\quad\quad\quad\quad\quad CH_3$$

(i) 
$$CH_3$$
$$\quad CH-CH_2-CH_2$$
$$CH_3 \quad\quad CH_3$$

(j) 
$$CH_3-CH_2-CH_2 \quad\quad CH_2-CH_3$$
$$\quad\quad CH_2-CH-CH-CH_2CH_2CH_3$$
$$\quad\quad\quad\quad CH_3$$

## PHYSICAL PROPERTIES

**Polar Molecules.** When the force of gravity acts on an object, it acts on all parts of it. The net result of this force, however, is calculated by treating the object as though all its mass were concentrated at one point and the force of gravity acted on just that one particular point, which is called the center of gravity. The subatomic particles, the nuclei and electrons, that make up a molecule have both masses and electric charges. A molecule has both a center of gravity and two centers of electric charge—a center of positive-charge density and a center of negative-charge density. If these two centers do not coincide, the molecule is said to possess a *dipole,* that is, to have two electric poles, somewhat like the two magnetic poles of a magnet. A molecule with an electric dipole is called a *polar molecule.*

In a symmetrical molecule such as hydrogen, H—H, the centers of density of positive and negative charge coincide at a point halfway between the two nuclei, making the molecule nonpolar. A nonpolar molecule can be compared to an unmagnetized nail that has no attraction for another unmagnetized nail. Hydrogen molecules have almost nonexistent forces of attraction between them. Because its molecules have little cause to stick together, hydrogen is a gas under ordinary pressures until it is cooled to $-253°C$.

The element fluorine, F—F, similarly consists of symmetrical, nonpolar molecules and remains a gas under ordinary pressures until cooled to a temperature of $-188°C$. It boils at a higher temperature than hydrogen principally because its molecules are much heavier, and only at a higher temperature do they have sufficient average kinetic energy to exist in the gaseous state.

Hydrogen fluoride, H—F, whose molecules are intermediate in mass between those of hydrogen and fluorine, boils at $19°C$, a much higher boiling point than those of either hydrogen or fluorine. Hydrogen fluoride molecules obviously must have forces of attraction between them. They are unsymmetrical, and the center of positive-charge density is not at the same point as the center of negative-charge density. The hydrogen fluoride molecule is polar, symbolized by placing a $\delta+$ and a $\delta-$ at appropriate places near its structure.

$$\overset{\delta+ \quad \delta-}{\text{H—F}}$$

Hydrogen fluoride
(a polar molecule)

(The Greek lowercase delta, $\delta$, is a commonly used symbol for the word "partial" or "fractional.") The positive-charge density is toward the hydrogen end of this molecule, the negative-charge density in the other end. The magnitudes of the fractional charges are equal, although their algebraic sum is zero. Thus the molecule is electrically neutral but polar. Just as two or more magnets can stick together if they are properly oriented (and are not being vigorously tumbled), so can two or more polar molecules. If their motions are not too violent, polar molecules can aggregate into crystalline arrays or in the looser condition of the liquid state. The degree of molecular polarity, the molecular mass, the temperature, and the external pressure determine the physical state. To understand the physical properties of an organic compound and how they are related to molecular structure, we should be able to predict which molecules will be polar. The concept of electronegativity provides us with the most help in making these predictions.

**Electronegativity.** When two different atoms are joined by a covalent bond, one of the atoms usually exerts a stronger attraction for the pair of bonding electrons than the other. Such an atom is said to be more *electronegative* than the other. For example, fluorine is more electronegative than hydrogen and is, in fact, the most electronegative of all the elements, with oxygen ranking second. The ability of an atom (or a group) to attract to itself electron density from a bonding pair of electrons is called the *inductive effect,* since it induces partial charges. Fluorine is said to have an electron-withdrawing inductive effect. The electronegativities of several elements important to our study are in the following order:

$$F > O > N > Cl \approx Br > C > H$$

== decreasing electronegativity ⟹

Since fluorine and hydrogen are widely separated in this series, the bond between them in hydrogen fluoride, H—F, should be very polar. In other words, the magnitudes of the partial charges, $\delta+$ and $\delta-$, should be large, although they are of course still fractions of a unit of charge. Since chlorine and hydrogen are closer to each other in relative electronegativity, the bond joining them in hydrogen chloride, H—Cl, should be less polar, and forces of attraction between hydrogen chloride molecules should be weaker. They are. Even though hydrogen chloride molecules are heavier than those of hydrogen fluoride, hydrogen chloride boils at $-85°C$, over 100° below the boiling point of hydrogen fluoride.

When a molecule has three or more atoms and therefore two or more covalent bonds, its polarity is the resultant of individual bond polarities. (In vector terminology, the resultant is the vector sum of the individual bond polarities.) Since oxygen is more electronegative than carbon, a carbon-oxygen bond should have a partial positive charge at carbon and a partial negative charge at oxygen. Carbon dioxide, O=C=O, however, is a linear molecule; its bonds, although polar, are disposed in exactly opposite directions, giving the molecule zero polarity.

Because oxygen and hydrogen are widely separated in relative electronegativity, an oxygen-hydrogen bond should be very polar and it is. The two oxygen-hydrogen bonds in a water molecule are not colinear, and therefore the polarity of water is high.

$$
\begin{array}{cc}
\delta+ & \delta+ \\
H & H \\
& O \\
& \delta-
\end{array}
$$

Water

We expect a molecule to be polar whenever it contains polar bonds whose polarities do not cancel, and we expect a bond to be polar whenever it joins atoms of different electronegativities. In hydrocarbons the only bonds present join a carbon either to another carbon or to a hydrogen. Since carbon and hydrogen differ only slightly in relative electronegativity, a carbon-hydrogen bond will be only weakly polar. Hydrocarbons are therefore relatively nonpolar substances.

**Physical Properties of Alkanes.** Molecules of methane (formula weight 16) and water (formula weight 18) have nearly the same masses. Yet methane, being nonpolar, boils at $-161.5°C$. The higher the alkane is in the homologous series of normal alkanes (Table 1.3), the higher its boiling point, melting point, and density. The alkanes as well as all hydrocarbons are insoluble in water, a very polar solvent, and soluble in the relatively nonpolar solvents that include some common hydrocarbons or mixtures of hydrocarbons (benzene, gasoline, ligroine, and petroleum ether) and halogen derivatives (carbon tetrachloride, chloroform, and dichloromethane). The rule of thumb is that "likes dissolve likes," where "likes" refers to polarities. How this rule works will be discussed in Chapter 4.

What we are seeking here are general correlations between structure and property. Just as we have now learned to associate alkanes with low water solubility, we shall in the future be able to predict that *any molecule* of *any family* substantially *alkane-like* will also have this property. The alkanes are not directly involved in the chemical reactions that occur in living systems, but one of the three great classes of foods, the lipids, fats and oils, consist of molecules that are largely alkane-like. Fats and oils do not dissolve in water as the sugars do, for example, yet the great aqueous transport system of the body, the bloodstream, must somehow handle them. Before we can see how this is done, we must appreciate the root of the problem. We begin here our study of "map signs," of functional or nonfunctional groups, by which we later can "read" or predict properties of complicated systems. We have learned thus far that a molecule that is largely alkane-like will probably be insoluble in water, unless another "map sign" in the molecule, to be learned later, overrules this prediction.

We shall now examine chemical properties associated not just with the alkanes but with the *portions* of other types of molecules that are alkane-like. If we organize our study with this more general approach, the various families will not seem so isolated. One of our ultimate goals is to study some examples of the molecular basis for several key processes in living systems. We are working our way toward this end by examining simpler but related systems.

## CHEMICAL PROPERTIES

Alkanes are as a class chemically very unreactive. Carbon-to-carbon single bonds are extremely strong and very difficult to attack chemically; carbon-to-hydrogen bonds, *as they occur in an alkane environment,* are also exceptionally resistant to chemical attack by most laboratory reagents. Under normal circumstances, for example, at or near room temperature or below, concentrated sulfuric acid, sodium metal, and strong alkalies do not affect alkanes. It is thus understandable that alkanes were originally named paraffins (Latin *parum affinus,* little affinity).

**Combustion.** Oxygen is one of a few chemicals that will attack alkanes, but only at elevated temperatures. Combustion is their most useful chemical reaction. The most important product, of course, is heat. Natural gas is largely methane; propane is a common home-heating and cooking fuel known as bottle gas, or Pyrofax; heptane and octane, especially highly branched isomers of them, are found in gasoline. The twelve- to eighteen-carbon alkanes are found in kerosene, jet fuel, and tractor fuel. Diesel oil, fuel oil, and gas oil consist roughly of the twelve- to eighteen-carbon alkanes. The higher alkanes, those containing more than twenty carbons, are the principal components in refined mineral oils, lubricating oils and greases, and paraffin wax. Even asphalt and tar for roads and roofing contain large percentages of hydrocarbons.

The combustion of propane proceeds according to the following equation:

$$CH_3CH_2CH_3 + 5O_2 \longrightarrow 3CO_2 + 4H_2O \quad \Delta H = -531 \text{ kcal/mole}$$

In the absence of enough oxygen, partial combustion may take place and produce carbon monoxide and even carbon. The former is a dangerous poison often present in exhaust fumes of cars. The latter may form deposits on motor pistons and in cylinder heads.

**Chlorination.** The chlorination of alkanes is one of the very few other reactions they undergo. It also illustrates two important and characteristic reactions that occur in several other families of organic compounds, *substitution* reactions and *free-radical* reactions. A substitution reaction is one in which a group or atom already on an organic molecule is replaced (substituted) by another group. The term free radical is applied to the *intermediate* that chemists believe forms as the reaction proceeds from the reactants to the products. First, let us look at the overall steps in the chlorination of methane as an example.

When a mixture of methane and chlorine is either heated or exposed to ultraviolet radiation, the following reaction takes place:

$$CH_4 + Cl_2 \longrightarrow CH_3Cl + HCl$$

Methyl chloride
(chloromethane, bp $-24°C$)

What happens in the mixture, however, is not as simple as this. In the first place, we must keep in mind that coefficients in balanced equations (all are one in this equation) may stand for molecules or for moles, depending on the context. Imagine a mixture of one mole of methane and one mole of chlorine, about $6 \times 10^{23}$ molecules of each. What happens after a few molecules of methyl chloride form? They appear in an atmosphere still very rich in molecules of chlorine, and it is entirely probable that newly formed methyl chloride molecules will be attacked by chlorine molecules. After all, billions of collisions constantly occur *at random* (or nearly so) in the gaseous mixture. Thus newly forming product molecules (methyl chloride) can be expected to compete with the still abundant methane molecules for chlorine. In successful collisions the reaction is the following:

$$CH_3Cl + Cl_2 \longrightarrow CH_2Cl_2 + HCl$$

Methylene chloride
(dichloromethane, bp $40°C$)

Now we have another species that can compete with methane (as well as with still unchanged methyl chloride) for chlorine, and a third "competing" reaction takes place:

$$CH_2Cl_2 + Cl_2 \longrightarrow CHCl_3 + HCl$$

Chloroform
(trichloromethane, bp $61°C$)

This new species possesses one remaining carbon-to-hydrogen bond, and still another competing reaction is possible:

$$CHCl_3 + Cl_2 \longrightarrow CCl_4 + HCl$$

Carbon tetrachloride
(tetrachloromethane, bp $77°C$)

This is the end of it; carbon tetrachloride cannot be further altered by chlorine. In summary, a mixture of methane and chlorine of a 1:1 mole ratio will react, if activated by heat or ultraviolet light, to produce a mixture of all the possible chlorination products. Each reaction is a *substitution* reaction, for hydrogen attached to carbon has been substituted by chlorine.

---

Exercise 1.4   How can we adjust the initial conditions (1:1 mole ratio of reactants) to maximize the formation of methyl chloride and keep production of higher chlorination products at a minimum? (*Hint.* How can a chemist adjust the initial ratio to increase considerably the probability that chlorine molecules will encounter *unchanged methane* molecules rather than newly formed molecules of methyl chloride?) It is not practical to remove the methyl chloride as it forms, and no matter what is done, higher chlorinated products will form. The problem is to minimize these.

Exercise 1.5   How can the initial conditions be adjusted to maximize the production of carbon tetrachloride?

---

Having seen *what* happens, we shall now study *how* it happens in order to deepen our insight into how organic reactions can occur. Chemists call the answer to this "How?" question the *mechanism* of a reaction. In a sense, it has to do with the arrow of an equation. Just how are reactant molecules converted into final products?

**Free-Radical Chain Reactions. Mechanism of Chlorination.**

Step 1. *Initiation. The function of heat or ultraviolet light.* Energy (from heat or ultraviolet light) is absorbed by a chlorine molecule to split it into two chlorine atoms:

$$:\!\ddot{C}l\!:\!\ddot{C}l\!: + \text{energy} \longrightarrow 2:\!\ddot{C}l\cdot$$

<table>
<tr><td></td><td>Heat or<br>ultraviolet light</td><td>Chlorine<br>atom</td></tr>
</table>

The chlorine atoms formed are electrically neutral, but each possesses seven electrons in the outside shell. Six of these are paired and the seventh is unpaired. An unpaired electron is sometimes called an *odd electron* to indicate that it is not one

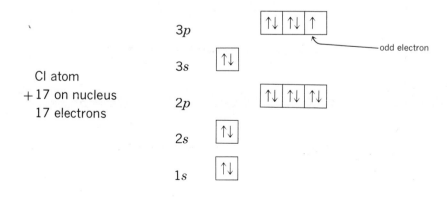

Cl atom
+17 on nucleus
17 electrons

of those evenly paired with another. Any atom or group of atoms that has an odd electron is called a *free radical*.

Step 2. *The fate of the highly active chlorine atom.* The trace amounts of chlorine atoms form in an atmosphere rich in molecules of both chlorine and methane. Collisions with both can and undoubtedly do occur, but it is the collision between a chlorine atom and a methane molecule that interests us.

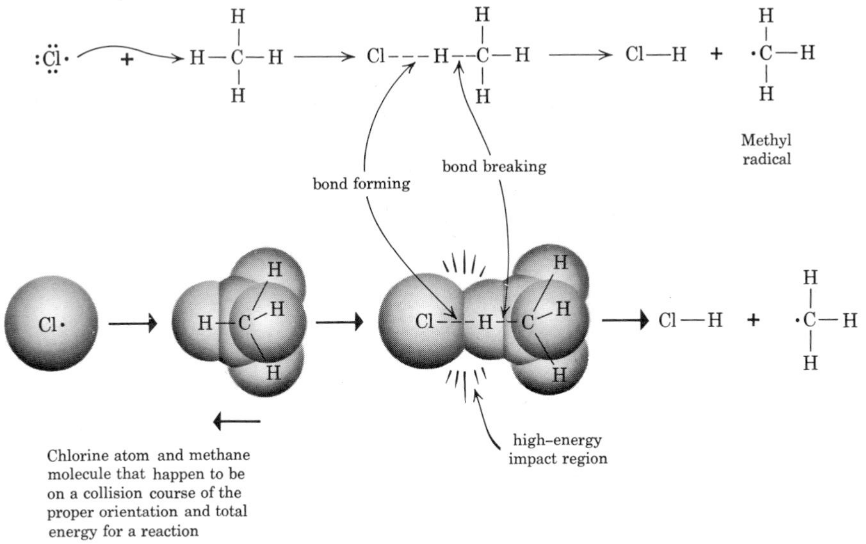

Chlorine atom and methane molecule that happen to be on a collision course of the proper orientation and total energy for a reaction

high–energy impact region

In the impact region the chlorine atom will have penetrated into the electron cloud about the hydrogen nucleus. If this penetration is deep enough, the hydrogen nucleus and the chlorine nucleus are separated by only the normal bond distance in H—Cl. The hydrogen nucleus can then be considered as much bound to the chlorine as to the carbon. Hence, if conditions during the collision are just right (in terms not only of the energy but also of the orientation of the collision), when the particles bounce away, the hydrogen nucleus may depart with the chlorine as a molecule of hydrogen chloride. Left behind will be an electrically neutral $CH_3 \cdot$ species known as a *methyl radical*. Lacking an octet for its carbon, it is now a high-energy species.

Step 3. *The fate of the methyl radical.* Because the methyl radical forms in an environment rich in molecules of both methane and chlorine, it will very likely collide with one of them. (The chlorine *atoms* produced in step 1 are too few for a direct encounter between a chlorine atom and a methyl radical to be very probable.) Collisions between the methyl radical and methane do not lead to any new species:

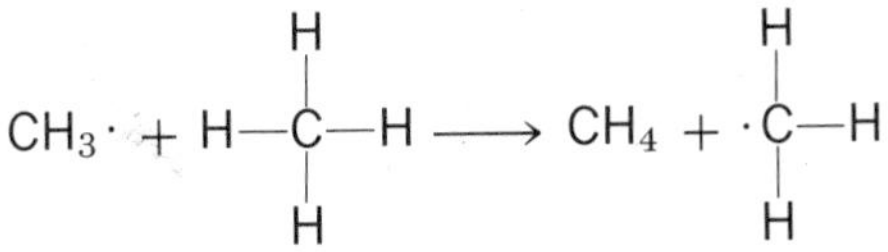

Collisions between methyl radicals and chlorine molecules, however, are significant. If enough energy is present at impact, and if the collision is properly lined up, the following reaction occurs:

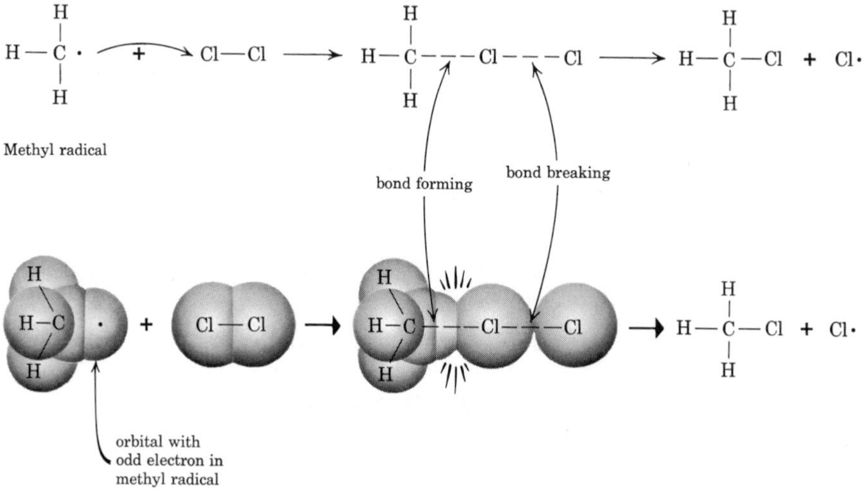

$$H-\underset{\underset{H}{|}}{\overset{\overset{H}{|}}{C}} \cdot \quad + \quad Cl-Cl \longrightarrow H-\underset{\underset{H}{|}}{\overset{\overset{H}{|}}{C}}---Cl---Cl \longrightarrow H-\underset{\underset{H}{|}}{\overset{\overset{H}{|}}{C}}-Cl \quad + \quad Cl\cdot$$

Methyl radical

bond forming          bond breaking

orbital with
odd electron in
methyl radical

Step 3 therefore produces one molecule of the second product, methyl chloride. Equally important, however, is the new chlorine atom that is released. Heretofore we relied on step 1 to produce this, and step 1 was necessary to initiate the reaction. Once it has started, however, and after step 3 has occurred, another step 2 can take place. Step 3 can then of course occur again, with the net result that once initiated by heat or ultraviolet light the reaction sustains itself. Steps 2 and 3 repeat themselves time and time again in a cyclical chain process. For this reason the chlorination of methane is called a free-radical *chain* reaction.

The situation, however, is not quite this simple. Although a chlorine atom will only very rarely collide with a methyl radical,

$$Cl\cdot + \cdot CH_3 \longrightarrow Cl:CH_3 \quad \text{or} \quad CH_3Cl, \text{ methyl chloride}$$

if this does occur, a chain is terminated. Collision between two chlorine atoms or between two methyl radicals will also terminate a chain:

$$CH_3\cdot + \cdot CH_3 \longrightarrow CH_3-CH_3$$

It is estimated that steps 2 and 3 repeat themselves in a cyclic manner about 5000 times before a random, chain-terminating reaction takes place. Since step 1 initially gives two chlorine atoms, it starts two chains. The absorption of only one "bundle" of light energy (quantum or photon) therefore initiates the production of about 10,000 molecules of methyl chloride.

We have described the mechanism for the chlorination of methane in considerable detail because free-radical chain reactions constitute one of the very important types of organic reactions. They are especially important to the plastics industry, for some plastics are produced by a chain process. Photosynthesis, an important process occurring in green plants, *starts* with the absorption of light energy, but from this point on other mechanisms take over. The combustion of gasoline, as well as any combustion process, involves radical chain reactions.

In summary, the steps in a free-radical chain reaction, as illustrated by the chlorination of methane, are as follows:

Step 1. *Chain initiation:*

$$Cl_2 \xrightarrow[\text{light}]{\text{heat or ultraviolet}} 2Cl \cdot$$

Steps 2 and 3. *Chain propagation:*

Step 2.      $Cl \cdot + CH_4 \longrightarrow HCl + \cdot CH_3$
Step 3.      $\cdot CH_3 + Cl_2 \longrightarrow CH_3Cl + Cl \cdot$
Then steps 2, 3, 2, 3, 2, 3, etc.

Step 4. *Chain termination* (several possible ways shown):

$$Cl \cdot \ + \cdot CH_3 \longrightarrow CH_3Cl$$
or $\quad Cl \cdot \ + Cl \cdot \longrightarrow Cl_2$
or $\quad CH_3 \cdot + \cdot CH_3 \longrightarrow CH_3CH_3$    (ethane)

**Chlorination of Higher Homologs of Methane.** The hydrogens in ethane form an equivalent set, which means that all members of the set (hydrogens in this case) are the same with respect to being replaced by some other group. No matter which of the six hydrogens on ethane is replaced by chlorine, only one monochloroethane is possible:

$$CH_3CH_3 + Cl_2 \longrightarrow CH_3CH_2Cl$$

Ethyl chloride
(chloroethane, bp 13°C)

Although ethyl chloride has no isomer, if a second chloro group is put into ethyl chloride, a mixture of isomeric dichloroethanes forms:

$$CH_3CH_2Cl + Cl_2 \xrightarrow[\text{ultraviolet light}]{\text{heat or}} ClCH_2CH_2Cl$$

1,2-Dichloroethane
(ethylene chloride, bp 84°C)

$$+ \qquad\qquad + \ HCl$$
$$CH_3CHCl_2$$

1,1-Dichloroethane
(ethylidene chloride, bp 57°C)

This equation is not balanced and tells us only that the action of macroscopic samples of chlorine on ethyl chloride produces a mixture of the two isomers shown. How much of each forms depends on the precise experimental conditions, and the ratio of isomers must be measured experimentally. (We shall frequently use such unbalanced "reaction sequences" to describe general results.)

The chlorination of propane will produce two isomeric monochloropropanes

(again, an unbalanced "reaction sequence" is shown),

$$CH_3CH_2CH_3 + Cl_2 \longrightarrow CH_3CH_2CH_2Cl + CH_3\underset{\underset{Cl}{|}}{C}HCH_3 + HCl$$

n-Propyl chloride       Isopropyl chloride
(bp 47°C)          (bp 35°C)

because propane, unlike ethane, has two *different* sets of equivalent hydrogens, one set of six (from the two $CH_3$ groups) and one set of two (from the middle $CH_2$ group).

---

Exercise 1.6   (*a*) How many monochloro derivatives of *n*-butane are possible? Give both the common and the IUPAC names. *Hint.* How many different sets of equivalent hydrogens does a *n*-butane molecule have?

(*b*) How many monochloro derivatives of isobutane are possible? Name them according to the common and the IUPAC systems.

---

The multiplicity of products emanating from the halogenation of alkanes reduces the usefulness of this reaction for laboratory syntheses which usually have as their goal the production of a pure compound. Isomer mixtures are often difficult to separate with laboratory equipment. Industrial scientists, however, have solved such problems, and the haloalkanes are important commercial substances. Furthermore, in some industrial processes that do not require pure compounds, the mixtures can be used.

## PREPARATION OF ALKANES

In addition to natural sources such as crude oil, several laboratory methods for making pure alkanes have been developed. These syntheses require the conversion of some compound in another family to an alkane. We shall defer our discussion of these syntheses until they are encountered in the reactions of other families. Moreover, these methods are of little relevance to our ultimate goal, study of the molecular basis of life. The following syntheses of alkanes are listed for reference purposes only.

**The catalytic hydrogenation of alkenes and alkynes** (cf. p. 50):

$$CH_2{=}CH_2 + H_2 \xrightarrow[\text{heat, pressure}]{\text{Pt or Pd}} CH_3CH_3$$

Ethylene                 Ethane

$$HC{\equiv}CH + 2H_2 \xrightarrow[\text{heat, pressure}]{\text{Pt or Pd}} CH_3CH_3$$

Acetylene                Ethane

**The reduction of alkyl halides:**

$$R{-}X + Mg \xrightarrow{\text{ether}} R{-}Mg{-}X \xrightarrow{\text{H}^+} R{-}H + Mg^{2+} + X^-$$

Any alkyl halide          An alkane

Other reagents, such as zinc and acetic acid will also convert some alkyl halides to alkanes.

**The Wurtz reaction.** In general:

$$RX + RX + 2Na \xrightarrow[\text{heat}]{} R{-}R + 2NaX$$

Two molecules of an alkyl halide

An alkane with carbon skeleton twice as long as in the original R—X

Specific example:

$$2CH_3CH_2CH_2CH_2Br + 2Na \xrightarrow[\text{heat}]{}$$

n-Butyl bromide

$$CH_3CH_2CH_2CH_2CH_2CH_2CH_2CH_3 + 2NaBr$$

n-Octane

---

Charles Adolphe Wurtz (1817–1884) discovered this reaction. Naming the more important reactions in organic chemistry after their discoverers was a rather common practice until recently. The designation Wurtz reaction serves as a highly abbreviated "symbol" for all the basic features of the general reaction. In oral or written communication all the pertinent facts, for example, types of reactants, other conditions, and types of products, are indicated in the one brief expression. As we encounter other named reactions in our study, the usefulness of this symbolism will become more and more apparent. Fewer reactions are now named after their discoverers only because learning so many became a burden and because rather large teams of scientists are frequently involved in new discoveries, making it difficult to establish priority.

---

# CYCLOALKANES

A carbon skeleton can take the form of a closed ring as well as of an open chain. The simplest of such cyclic compounds are called the cylcloalkanes or cycloparaffins. The following examples are illustrative.

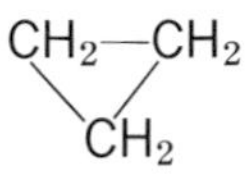

Cyclopropane 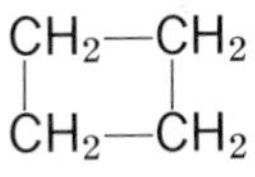      Cyclobutane 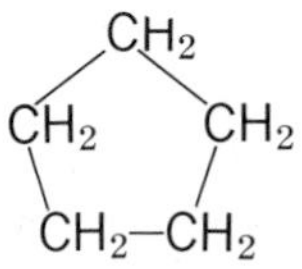      Cyclopentane 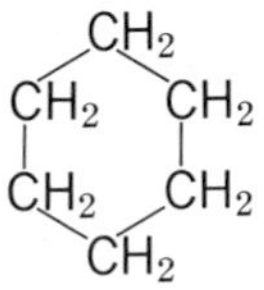      Cyclohexane

Rings of higher carbon content (over thirty) are also known. Since the molecules in these compounds have only single bonds, they generally exhibit the same types of physical and chemical properties as the open-chain alkanes. Cyclopropane and cyclobutane are exceptions.

**Ring Strain.** As written, the C-C-C bond angle for cyclopropane should be that of the internal angle of an equilateral triangle, namely 60°. But the normal bond angle for tetravalent carbon is much larger than that, 109°28′. The internal angle in cyclobutane should be 90°, also smaller than the angle for open-chain systems. The internal angle of a regular pentagon (cf. cyclopentane) is 108°, very close to 109°28′.

We learned earlier in this chapter that the internal energy of a system involving tetravalent carbon is minimized when bond angles are 109°28′. Maximum overlap between bonding orbitals is achieved with this geometry. Any deviation from this angle must therefore imply departure from maximum overlap (Figure 1.11), which in turn means weaker bonds. The bonds in cyclopropane are weaker, and so to a lesser extent are those in cyclobutane. Cyclopentane, however, in which deviation from the normal bond angle is very slight, is normal. The relative reactivities of cyclopropane, cyclobutane, and cyclopentane reflect these considerations. To illustrate,

$$
\begin{array}{c}
\underset{\text{Cyclopropane}}{\overset{\displaystyle CH_2-CH_2}{\underset{\displaystyle CH_2}{\diagdown\diagup}}} + H_2 \xrightarrow[80°C]{Ni} \underset{\text{Propane}}{\overset{\displaystyle CH_2CH_2CH_2}{\underset{\displaystyle H \qquad H}{|\qquad\qquad|}}}
\end{array}
$$

$$
\begin{array}{c}
\underset{\text{Cyclobutane}}{\overset{\displaystyle CH_2-CH_2}{\underset{\displaystyle CH_2-CH_2}{|\qquad\quad|}}} + H_2 \xrightarrow[200°C]{Ni} \underset{n\text{-Butane}}{\overset{\displaystyle CH_2CH_2CH_2CH_2}{\underset{\displaystyle H \qquad\qquad H}{|\qquad\qquad\qquad|}}}
\end{array}
$$

$$
\begin{array}{c}
\underset{\text{Cyclopentane}}{\overset{\displaystyle CH_2}{\underset{\displaystyle CH_2-CH_2}{\overset{\diagup\quad\diagdown}{CH_2 \qquad CH_2}}}} + H_2 \xrightarrow[\text{heat}]{Ni} \text{no reaction}
\end{array}
$$

The temperatures at which the rings for cyclopropane and cyclobutane open are revealing. The three-membered ring is quite easily opened because its carbon-to-carbon bonds are weak and because the cyclopropane ring is said to be subject to internal strain. The cyclobutane ring is less strained than that of cyclopropane, however, and a higher temperature is required to open it. Cyclopentane has essentially no ring strain and undergoes no useful ring-opening reactions.

The internal angle of a regular hexagon is 120°, and this is 10°32′ larger than the normal bond angle for tetravalent carbon. This departure might lead us to predict that cyclohexane should be subject to internal strain similar to but a bit less than that in cyclobutane. The evidence indicates, however, that bonds in cyclohexane are just as strong as those in cylcopentane, certainly far stronger than those in

The $sp^3$–$sp^3$ overlap to form
a C — C bond is maximized
at a bond angle of 109° 28′.

The $sp^3$–$sp^3$ overlap is poorer
here when the bond angle
deviates much from 109° 28′.

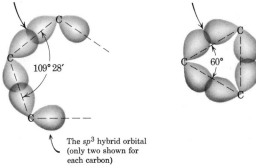

109° 28′

60°

The $sp^3$ hybrid orbital
(only two shown for
each carbon)

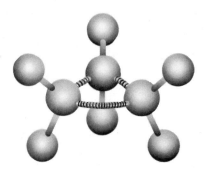

Ball and stick model of
cyclopropane in which
springs must be used
to represent the C — C
bonds in the ring

**Figure 1.11**    The ring strain in cyclopropane is attributed to relatively poor overlap between $sp^3$
hybrid orbitals of the carbons.

cyclobutane or cyclopropane. The explanation is that the cyclohexane ring is not
flat as in a normal hexagon. It is twisted out of the plane, and two conformations
having normal bond angles are shown in Figure 1.12. Important structural features
of cyclohexane are discussed in the legend for this figure. In similar fashion, the
rings with seven carbons or more are also nonplanar, and they are subject to no internal strain. Therefore they all have the normal alkane-like chemical and physical
properties.

Six-membered and five-membered ring systems are abundant in the molecules
of living things. The six-membered ring is especially important in carbohydrate
chemistry where even such features as the preference for a chair form over a boat
form are highly significant.

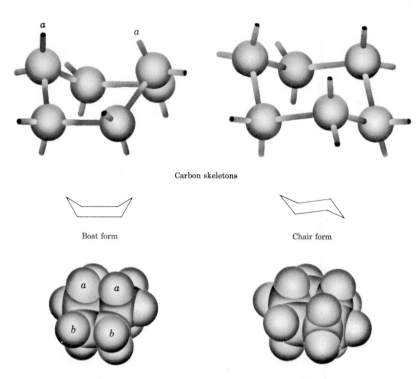

Carbon skeletons

Boat form　　　　　　　　　　　　　　　　Chair form

Scale models

**Figure 1.12**　Conformations of the cyclohexane ring. In both the boat and the chair form, bond angles in the ring are normal (109°28'). The chair form has less internal energy than the boat form, however, and is therefore much more abundant in a sample of cyclohexane. The boat form brings hydrogens from opposite sides of the ring quite close to each other (see those marked *a*), and their electron clouds tend to repel each other. Moreover, the electron clouds marked *b* eclipse each other. These interactions involving groups not bonded directly to each other are called nonbonded repulsions. The hydrogens in the chair form are all staggered.

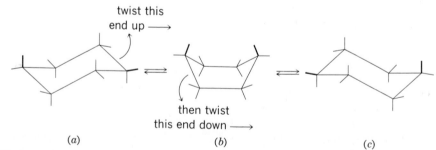

twist this
end up ⟶

then twist
this end down ⟶

(*a*)　　　　　　　　　　　　(*b*)　　　　　　　　　　　　(*c*)

**Figure 1.13**　The chair form of cyclohexane (*a*) can "flip-flop" to another chair form (*c*) and back again by means of the boat form (*b*). A dynamic equilibrium exists. Interconversion occurs rapidly enough at ordinary temperatures to make it expedient simply to treat the cyclohexane system as a flat hexagon.

Cyclopropane  Cyclobutane  Cyclopentane  Cyclohexane

1,2-Dimethylcyclohexane

**Figure 1.14** Symbolism and nomenclature in cycloalkane systems. At each corner in the geometric symbols a carbon is understood to be present. The lines from corner to corner are carbon-carbon bonds. Unless otherwise indicated, hydrogens are assumed to be present at the corner carbons in sufficient number to fill out carbon's tetravalence. When substituents occupy corners on the ring, the ring positions are numbered in a direction and from a beginning point that together yield the set of smallest numbers possible. Alkyl groups are usually given smaller numbers than halogens.

**Interconversion of Chair Forms.** In cyclic systems rotations about the single bonds that are part of the ring itself are severely limited. Because ball and stick models illustrate this restriction best, the student should examine any that are available to see this restricted rotation for himself.

In cyclohexane a small amount of movement is possible; one chair form can interconvert via a boat form into another equivalent but not identical chair form, as illustrated in Figure 1.13. This ready interconvertibility makes it possible to use simple flat hexagons to represent cyclohexane, as though it were a planar molecule.

**Condensed Symbols for Cyclic Systems.** Chemists usually represent cycloalkane systems with simple geometric forms such as those shown in Figure 1.14. Their symbolism is described in the legend.

## GEOMETRICAL ISOMERISM. "CIS-TRANS" ISOMERISM

Two different 1,2-dimethylcyclopropanes (Figure 1.15) are known. It is the lack of free rotation about single bonds in ring systems that makes their formation possible.

This kind of isomerism is known as *geometric isomerism*. The difference between these isomers lies not so much *where* on a chain or skeleton substituents are located but rather in their geometric orientation. Because two different compounds cannot have the same name, chemists call the isomer in which the substituents project in generally opposite directions the *trans* isomer. In the *cis* isomer the sub-

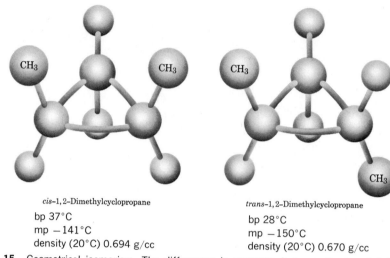

*cis*–1,2–Dimethylcyclopropane
bp 37°C
mp −141°C
density (20°C) 0.694 g/cc

*trans*–1,2–Dimethylcyclopropane
bp 28°C
mp −150°C
density (20°C) 0.670 g/cc

**Figure 1.15** Geometrical isomerism. The differences in geometry between the two 1,2-di-methylcyclopropanes are manifested in differences in observable properties.

stituents generally protrude in the same direction. Other examples of cis-trans isomerism are

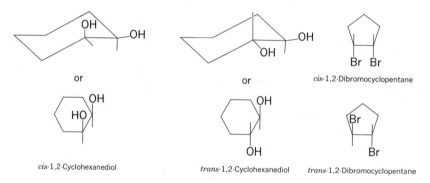

or

or

*cis*-1,2-Dibromocyclopentane

*cis*-1,2-Cyclohexanediol

*trans*-1,2-Cyclohexanediol

*trans*-1,2-Dibromocyclopentane

Geometrical isomerism is but one type of a general kind of isomerism, *stereo-isomerism* (Greek *stereos,* solid, meaning here three-dimensional shape). One other type, optical isomerism, we shall discuss later. In general, stereoisomerism has rather profound implications for the chemical events in living organisms. An interesting example is the difference between two geometric isomers of trimedlure, a sex attractant for the Mediterranean fruit fly. The trans isomer is a more powerful attractant than the cis isomer. Traps baited with *trans*-trimedlure and a rapid-acting insecticide have been set in Florida since 1957 to eradicate this pest, which annually used to cause severe damage to Florida's fruit crops. Although the exact chemistry of how the attractant works is not known, molecular geometry is as critical a factor as molecular structure. A likely explanation will be discussed in Chapter 11.

cis

trans

Trimedlure isomers

# REFERENCES AND ANNOTATED READING LIST

## BOOKS

O. T. Benfey. *From Vital Force to Structural Formulas*. Houghton Mifflin Company, Boston, 1964. The historical development of the structural theory in organic chemistry is presented with many interesting excerpts (in translation) from original works.

F. Wöhler. "On the Artificial Production of Urea," in *Readings in the Literature of Science*, edited by W. C. Dampier and M. Dampier. Harper Torchbook TB512, Harper and Row, New York, 1959, page 211. This is a translation of the complete article appearing in *Annalen der Physik und Chemie*, 1828, in which Wöhler reported the first synthesis of an organic compound from substances of mineral origin.

J. Read. *A Direct Entry to Organic Chemistry*. Harper Torchbook TB523, Harper and Row, New York, 1960. Chapter 3, "Mapping the Molecule: Differences Between Inorganic and Organic Compounds," has a good section on the phenomenon of isomerism.

B. Jaffe. *Crucibles: The Story of Chemistry*. Premier Reprint s49, Fawcett World Library, New York, 1957. Chapter 8 tells the story of Wöhler in a popularized style.

N. L. Allinger and J. Allinger. *Structures of Organic Molecules*. Prentice-Hall, Englewood Cliffs, N.J., 1965. Structures, the kinds of evidence for them, and stereoisomerism are discussed. (Paperback.)

R. Stewart. *The Investigation of Organic Reactions*. Prentice-Hall, Englewood Cliffs, N.J., 1966. The first chapter reviews principles of free energy and equilibria, especially as they apply to organic reactions. Types of intermediates, including free radicals, are discussed in the second chapter.

K. Mislow. *Introduction to Sterochemistry*. W. A. Benjamin, New York, 1966. Designed for students who have taken or who are enrolled in a beginning organic chemistry course, this book introduces all the types of stereoisomerism. (Paperback.)

R. Breslow. *Organic Reaction Mechanisms*. W. A. Benjamin, New York, 1966. This book supplements and goes beyond our own discussion of reaction mechanisms. (Paperback.)

O. Runquist. *A Programmed Review of Organic Chemistry, Nomenclature*. Burgess Publishing Company, Minneapolis, Minn., 1965. Nomenclature is easily self-taught, and this book is designed for that.

O. Runquist. *A Programmed Review of Organic Chemistry. Reactions I* (1965) and *Reactions II* (1966). Burgess Publishing Company, Minneapolis, Minn. These two self-help guides are especially designed for the study of reaction mechanisms.

O. T. Benfey. *The Names and Structures of Organic Compounds*. John Wiley and Sons, New York, 1966. This is a programmed learning book on organic nomenclature. (Paperback.)

**ARTICLES**

C. D. Hurd. "The General Philosophy of Organic Nomenclature." *Journal of Chemical Educa-tion,* Vol. 38, 1961, page 43. Some of the reasons behind the rules of nomenclature are discussed, and a short bibliography to systems of naming compounds is given.

## PROBLEMS AND EXERCISES

1. Check each of the following structures to see whether within the rules of covalence they represent possible compounds.

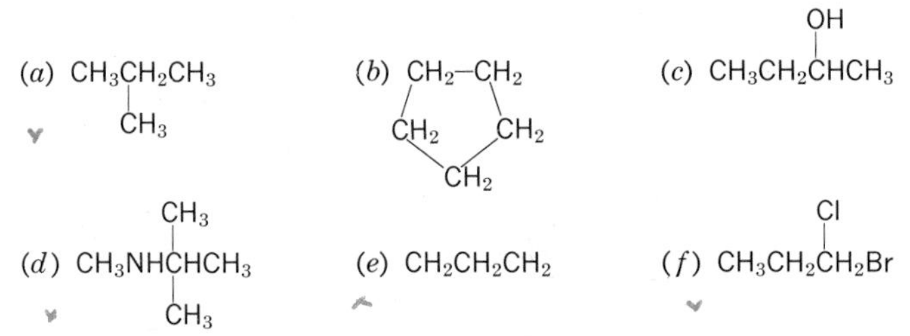

    (a) $CH_3CH_2CH_3$
      $\mathrm{CH_3}$

    (b) $CH_2-CH_2$
      $CH_2 \quad CH_2$
        $CH_2$

    (c) $CH_3CH_2CHCH_3$ (with OH on third carbon)

    (d) $CH_3NHCHCH_3$
      $\mathrm{CH_3}$

    (e) $CH_2CH_2CH_2$

    (f) $CH_3CH_2CH_2Br$ (with Cl on second carbon)

2. Convert each of the following structural formulas into a condensed structure. Straighten out the longest continuous carbon chain so that it is written on one line. (Cyclic structures, of course, are exceptions.)

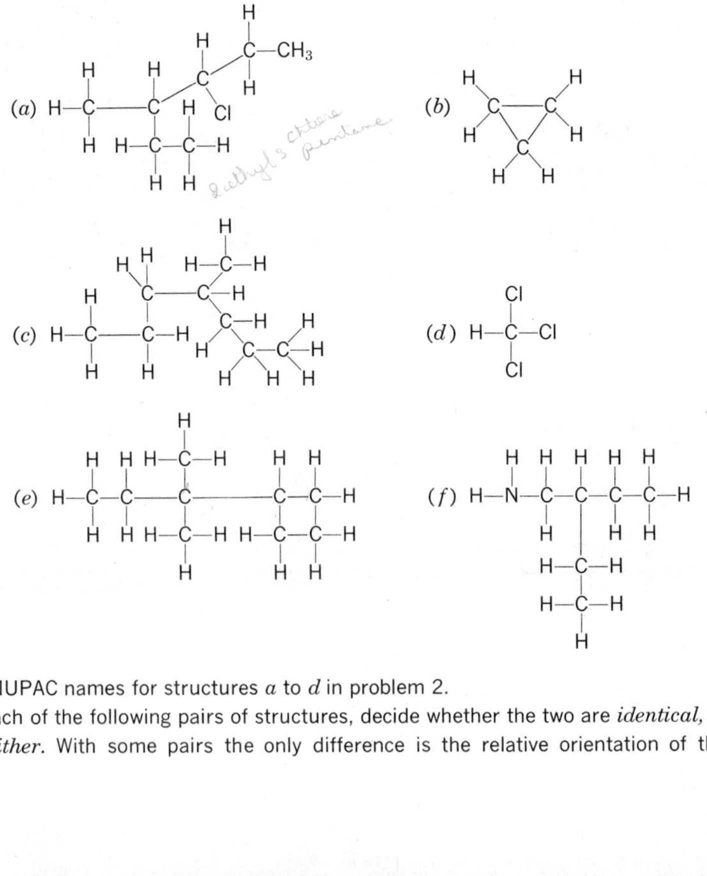

3. Write IUPAC names for structures *a* to *d* in problem 2.
4. For each of the following pairs of structures, decide whether the two are *identical, isomers,* or *neither.* With some pairs the only difference is the relative orientation of the same

structure in space (i.e., on paper). Others differ only in relative rotational position within the molecule. Assume free rotation about all single bonds except those that are part of a cyclic system.

(a)
$$\overset{\displaystyle CH_2-CH_2}{\underset{}{|}} \\ CH_3-CH_2 \quad CH_3$$

$$CH_3-CH_2 \\ \overset{}{\underset{}{|}} \\ CH_2-CH_2-CH_3$$

(b) $CH_3-O-CH_2CH_3$      $CH_3CH_2-O-CH_3$

(c)

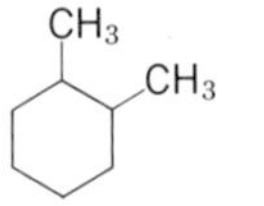

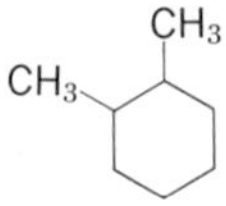

(d)

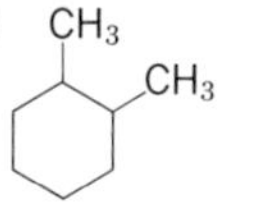

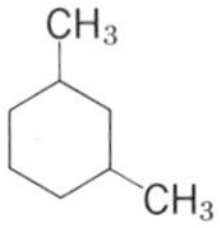

(e) $H-O-CH_2CH_2CH_2CH_3$

$$\overset{\displaystyle CH_3}{\underset{}{|}} \\ CH_3CHCH_2OH$$

(f)
$$\overset{\displaystyle CH_3}{\underset{}{|}} \\ CH_3CHCH_2CH_2Br$$

$$\overset{\displaystyle CH_3}{\underset{}{|}} \\ CH_3CHCH_2CH_2CH_2Br$$

(g)
$$CH_3CHCH_2CH_2CH_2 \\ \overset{}{\underset{}{|}} \qquad \overset{}{\underset{}{|}} \\ CH_3CCH_3 \qquad CHCH_3 \\ \overset{}{\underset{}{|}} \qquad \overset{}{\underset{}{|}} \\ CH_3 \qquad CH_3$$

$$\overset{\displaystyle CH_3}{\underset{}{|}} \qquad\qquad \overset{\displaystyle CH_3}{\underset{}{|}} \\ CH_3CHCH_2CH_2CH_2CHC-CH_3 \\ \overset{}{\underset{}{|}} \quad \overset{}{\underset{}{|}} \\ CH_3 \quad CH_3$$

(h) $CH_3CH=CHCH_3$      $CH_2=CHCH_2CH_3$

5. Write condensed structures for all the isomeric monochloro derivatives of 2,2,3-trimethylpentane.
6. Ethyl chloride may be formed in the following reaction,

$$Cl_2 + CH_3-CH_3 \xrightarrow[\text{ultraviolet light}]{\text{heat or}} CH_3-CH_2-Cl + H-Cl$$

Ethyl chloride

which is known to be a free-radical chain mechanism. Write equations for the steps in this mechanism as they apply to this reaction.
7. If the common name of chloromethane is *methyl chloride* and if the common name for chloroethane is *ethyl chloride,* what are the common names for the following?

(a)
$$\overset{\displaystyle CH_3}{\underset{}{|}} \\ CH_3C-Cl \\ \overset{}{\underset{}{|}} \\ CH_3$$

(b)
$$CH_3CHCH_3 \\ \overset{}{\underset{}{|}} \\ Cl$$

(c)
$$\overset{\displaystyle CH_3}{\underset{}{|}} \\ Cl-CH \\ \overset{}{\underset{}{\diagdown}} \\ CH_3$$

(d)
$$\overset{\displaystyle CH_3}{\underset{}{|}} \\ CH_3CHCH_2Cl$$

(e)
$$CH_3 \\ \overset{}{\underset{}{|}} \\ CH_2 \\ \overset{}{\underset{}{|}} \\ CH_2 \\ \overset{}{\underset{}{|}} \\ Cl$$

(f)
$$CH_3CHCH_2CH_3 \\ \overset{}{\underset{}{|}} \\ Cl$$

$$(g) \quad \underset{\underset{\text{CH}_3}{|}}{\overset{\overset{\text{CH}_3}{|}}{\text{CH}_2}} - \underset{}{\overset{\overset{\text{Cl}}{|}}{\text{CH}}}$$

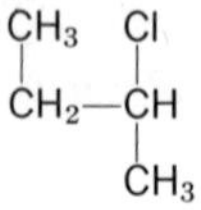

$$(h) \quad \text{CH}_3 \underset{\underset{\text{CH}_3}{|}}{\overset{\overset{\text{Cl}}{|}}{\text{C}}} \text{CH}_3$$

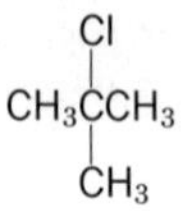

8. From the standpoint of the development of organic chemistry, what was the significance of the vital force theory?

# CHAPTER TWO

# Unsaturated Aliphatic Hydrocarbons

## ALKENES

### THE CARBON-CARBON DOUBLE BOND

The carbon-carbon double bond, which occurs in a wide variety of natural products, is a site of unsaturation in a molecule. Many of the lipids, the fats and oils, are unsaturated, as anyone who sees advertisements for "polyunsaturated cooking oils" knows. Several of the simpler alkenes are listed in Table 2.1.

The first member of the alkene family, ethylene, has the molecular formula $C_2H_4$. Within the rules of covalence it can have only one structure, and this must

Table 2.1   Properties of Some 1-Alkenes

| Name (IUPAC) | Structure | Bp (°C) | Mp (°C) | Density (in g/cc at 10°C) |
|---|---|---|---|---|
| Ethene | $CH_2{=}CH_2$ | −104 | −169 | — |
| Propene | $CH_2{=}CHCH_3$ | −48 | −185 | — |
| 1-Butene | $CH_2{=}CHCH_2CH_3$ | −6 | −185 | — |
| 1-Pentene | $CH_2{=}CHCH_2CH_2CH_3$ | +30 | −165 | 0.641 |
| 1-Hexene | $CH_2{=}CHCH_2CH_2CH_2CH_3$ | 64 | −140 | 0.673 |
| 1-Heptene | $CH_2{=}CHCH_2CH_2CH_2CH_2CH_3$ | 94 | −119 | 0.697 |
| 1-Octene | $CH_2{=}CHCH_2CH_2CH_2CH_2CH_2CH_3$ | 121 | −102 | 0.715 |
| 1-Nonene | $CH_2{=}CHCH_2CH_2CH_2CH_2CH_2CH_2CH_3$ | 147 | −81 | 0.729 |
| 1-Decene | $CH_2{=}CHCH_2CH_2CH_2CH_2CH_2CH_2CH_2CH_3$ | 171 | −66 | 0.741 |
| Cyclopentene |  | 44 | −135 | 0.722 |
| Cyclohexene |  | 83 | −104 | 0.811 |

include a carbon-carbon double bond. Long before modern theories of chemical bonds were developed, it was known that one of the two bonds in the double bond

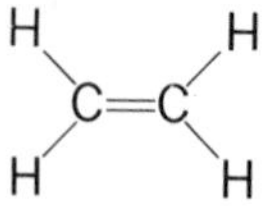

Ethylene

is quite reactive, especially toward acids. To understand the properties of this system, we must see what modern theory has to say.

$sp^2$ **Hybridization.** When carbon is attached to four other atoms or groups, it utilizes $sp^3$ hybrid orbitals. In a double bond, however, each carbon is attached to only three other groups and, in terms of the theory most commonly used by organic chemists, carbon uses different hybrid orbitals. We start with carbon in its so-called atomic state:

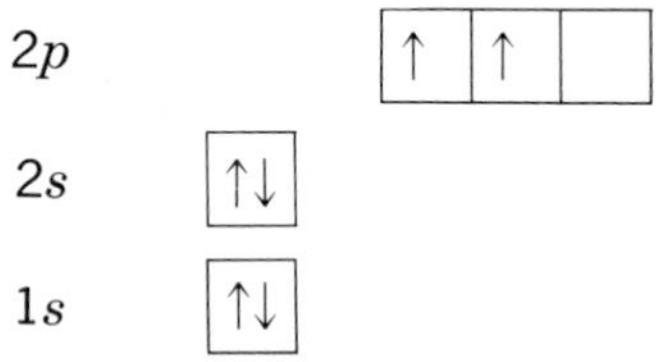

$2p$  ↑ ↑ ☐

$2s$  ↑↓

$1s$  ↑↓

Again, as before, we envision the promotion of a $2s$ electron into the empty $2p_z$ orbital to give the "promoted state":

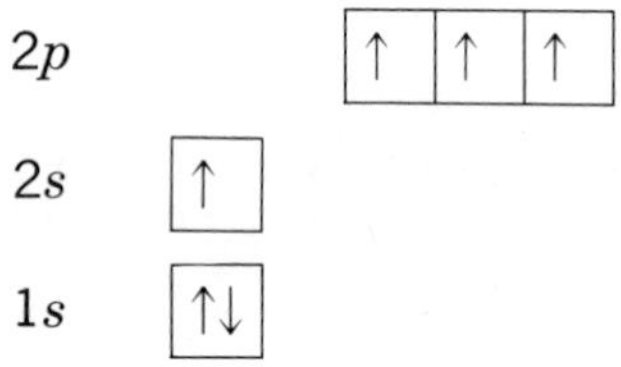

$2p$  ↑ ↑ ↑

$2s$  ↑

$1s$  ↑↓

At this point we follow a different course than that used for $sp^3$ hybridization in which all four orbitals at the second level were "mixed." Instead we "mix" the $2s$ and just two of the $2p$ orbitals. Hence the designation $sp^2$ ($s$–$p$–two) hybridization. The third $2p$ orbital is left unhybridized. Thus in this theory the valence state for carbon when it is bonded to only three groups is, in terms of an energy level diagram,

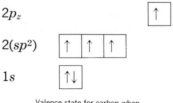

$2p_z$  ↑

$2(sp^2)$  ↑ ↑ ↑

$1s$  ↑↓

Valence state for carbon when
bonded to just three groups

There are three new hybrid orbitals at the second level plus one unhybridized $p$ orbital. The shapes of these new $sp^2$ hybrid orbitals are approximately those shown in Figure 2.1, each with both $s$ character and $p$ character. The arrangement of minimum energy for these orbitals, each with one electron, is shown in Figure 2.2.

**The Pi Bond.** Now that we have the orbital arrangement believed to exist for a carbon in a double bond, we shall assemble in Figures 2.3 and 2.4 two such carbons plus four hydrogens to show the final structure of ethylene. Of special importance is the overlap between the two unhybridized $p_z$ orbitals (Figure 2.4). If this overlap occurs, the energy is also lowered and the two "halves" can have only one orientation in relation to each other. Any rotation about the carbon-carbon bond would break the overlap, but at room temperature the system does not have enough energy for this rupture. For all practical purposes there is no free rotation about a carbon-carbon double bond. Hence geometrical isomerism will be possible for ethylenic hydrocarbons.

One of the two bonds in a double bond is a sigma bond (or $\sigma$-bond). The overlap between an $sp^2$ orbital of one carbon and an $sp^2$ orbital of the other gives a new molecular orbital with symmetry about the bonding axis. The second bond results from the overlap of two $p$ orbitals and does not have this symmetry. Having two lobes of identical shape, it resembles a $p$ orbital and is called a pi bond (or $\pi$-bond). Its two sausage-shaped lobes together constitute one molecular orbital, and somewhere in this region are two electrons whose electric charges help to hold the two carbon nuclei together. The two electrons in the sigma bond plus the two in the pi bond provide a stronger attractive force for the carbon nuclei, which are closer together than in ethane. The carbon-carbon distance in ethane is 1.54 Å; in ethylene it is 1.33 Å.

The electrons in the pi bond are called the pi electrons, and they are not located as close to the skeleton as the sigma electrons are. The overlap in the pi bond is not as effective as that in the sigma bond, with the net result that the pi electrons are capable of reacting with electron-poor species (Lewis acids, including protons). As we shall see, *the key to the chemistry of the double bond is the availability of its pi electrons.*

## ISOMERISM AMONG THE ALKENES

Isomers are possible in the alkene family whenever the double bond can be located differently between carbons or whenever the groups attached to the double bond permit cis-trans relations. Four isomers that have the formula $C_4H_8$ and are alkenes are known. Their structures are

| $CH_2$=$CHCH_2CH_3$ | | | |
|---|---|---|---|
| 1-Butene (α-butylene) | cis-2-Butene (cis-β-butylene) | trans-2-Butene (trans-β-butylene) | 2-Methyl-1-propene (isobutylene) |

In 1-butene the double bond is between the first and the second carbon. In both of the 2-butenes it is between the second and third. If there were free rotation about a

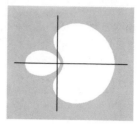

**Figure 2.1**   Cross section of an $sp^2$ hybrid orbital.

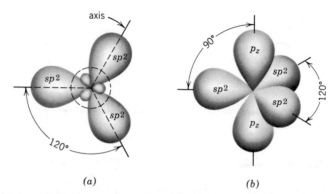

$(a)$                                    $(b)$

**Figure 2.2**   Bonding orbitals available from carbon when it is bonded to just three other groups. ($a$) Top view in which the axes of the $sp^2$ orbitals all lie in the plane of the paper, forming angles of 120°. The circular dashed line locates the $p_z$ orbital that is perpendicular to the plane. ($b$) Perspective view, showing the larger lobes of the $sp^2$ orbitals and the two lobes of the unhybridized $p_z$ orbital.

double bond, 2-butene, like $n$-butane, could have no geometric isomers. But two 2-butenes are known. Their physical properties are listed in Table 2.2, which is included only to demonstrate that differences in structure do mean differences in observable properties. The six isomeric pentenes are also shown in Table 2.2.

**Table 2.2   Physical Properties of the Butenes and Pentenes**

| Name | Bp (°C) | Mp (°C) | Density (in g/cc) |
|---|---|---|---|
| Butenes |  |  |  |
| 2-Methyl-1-propene | −6.90 | −140.4 | 0.640 (−20°C) |
| 1-Butene | −6.26 | −185 | 0.641 |
| *trans*-2-Butene | +0.88 | −105.6 | 0.649 |
| *cis*-2-Butene | +3.72 | −138.9 | 0.667 |
| Pentenes |  |  |  |
| 1-Pentene | 30.0 | −165.2 | 0.641 (+20°C) |
| *cis*-2-Pentene | 36.9 | −151.4 | 0.656 |
| *trans*-2-Pentene | 36.4 | −140.2 | 0.648 |
| 2-Methyl-1-butene | 31.2 | −137.5 | 0.650 |
| 3-Methyl-2-butene | 20.1 | −168.5 | 0.627 |
| 2-Methyl-2-butene | 38.6 | −133.8 | 0.662 |

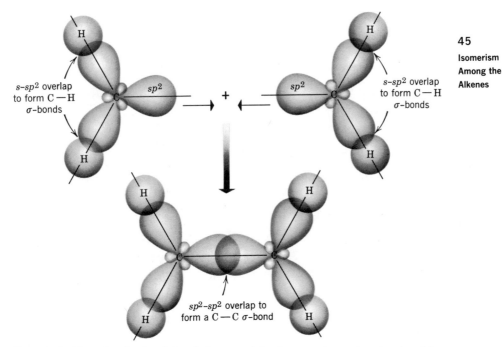

**Figure 2.3** The σ-bond network in ethylene consists of one carbon-carbon bond and four carbon-hydrogen bonds. The $p_z$ orbitals are not shown, but their lobes lie above and below the plane of the page with their axes piercing the paper at each carbon.

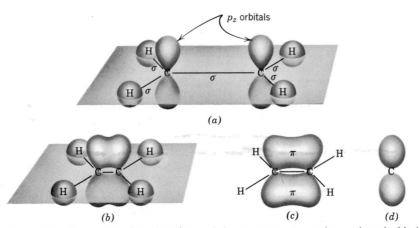

**Figure 2.4** The nature of the bonding and the geometry at a carbon-carbon double bond, according to one common theory. The solid lines labeled σ are described in Figure 2.3.

(*a*) This is the hypothetical situation before the $p_z$ orbitals overlap. (*b*) Maximum overlap of the $p_z$ orbitals can occur only when the six nuclei shown here lie in the same plane. (*c*) The double bond in ethylene consists of one σ-bond (the heavy line) and one π-bond. The π-bond has the appearance of two sausages above and below the plane. (*d*) A cross-sectional view of the π-bond looking down the carbon-carbon axis shows its two lobes.

**Exercise 2.1**   Identify each of the following compounds by name.

(a) CH₃   H
$$\text{C}$$
$$\text{C}$$
H   CH₃

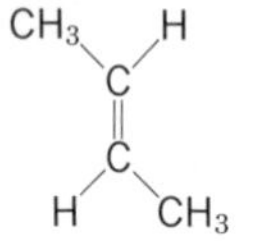

(b) $CH_3CH_2CH{=}CH_2$

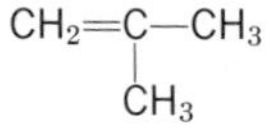

(c) $CH_2{=}C{-}CH_3$
    $\overset{|}{C}H_3$

(d) H      H
$$\text{C}{=}\text{C}$$
CH₃   CH₃

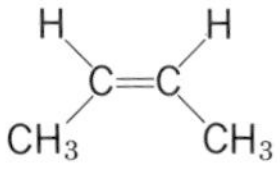

**Exercise 2.2**   Examine each of the following structures, write trial structures for geometric isomers, and determine whether or not cis-trans isomerism is possible.

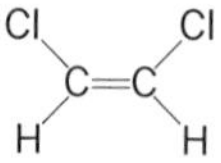

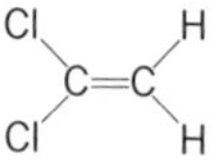

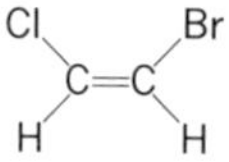

(a) $CH_3{-}CH_2$   H
C=C
H   H

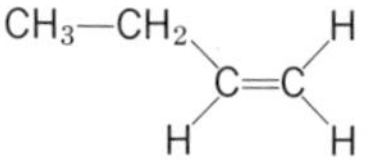

(b) Cl   Cl
C=C
H   H

(c) Cl   H
C=C
Cl   H

(d) Cl   Br
C=C
H   H

(e) Cl   H
C=C
Br   H

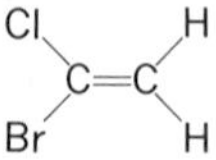

(f) $CH_3CH_2CH{=}CHCH_2CH_3$

(g) $CH_3CH_2$   H
C=C
$CH_3CH_2$   CH₃

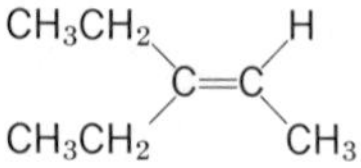

**Exercise 2.3**   Which of the following statements is true concerning the conditions under which cis-trans isomerism in simple alkenes is possible? Check these statements against the structures in exercise 2.2.

(a) Geometric isomerism *can* exist if *either* carbon of the double bond has two identical groups attached to it.

(b) Geometric isomerism *cannot* exist if *either* carbon of the double bond has two identical groups attached to it.

In general, physical properties of isomeric alkenes are different but the chemical properties, which depend more on the presence of a double bond than on its precise location, are quite similar. What differences occur, aside from actual products formed, have to do more with the *rates* of reactions than with the *types* of reactions. For this reason, unless attention to cis-trans isomerism is important, alkenes may be represented in simplified, one-line structures. For example, 2-butene may be written $CH_3CH{=}CHCH_3$. (Note that the double bond is never "understood" in condensed structures as single bonds are. It is always written into even the most condensed structures.)

## NOMENCLATURE

**Common Names.** The common names of alkenes have the same ending, "-ylene." An alkene with the carbon skeleton of the three-carbon alkane, propane, is called "propylene." The "prop-" prefix indicates a total of three carbons; the

"-ylene" suffix places it in the ethylene or alkene family. The various butylenes (cf. p. 43) are differentiated arbitrarily by the prefixes $\alpha$, $\beta$, and "iso-." In the homologous series the common system loses most of its practicality above the butylenes, and the IUPAC system is used.[1]

**IUPAC Names.** The IUPAC rules for naming the alkenes are as follows.

1. The characteristic name ending is "-ene."
2. The longest continuous chain *that contains the double bond* is selected. This chain is named by selecting the alkane with the *identical* chain length and changing the suffix in its name from "-ane" to "-ene."
3. The chain is numbered to give the *first* carbon of the double bond the lowest possible number (e.g., $CH_3CH_2CH{=}CH_2$ is 1-butene, not 1,2-butene and not 3-butene).
4. The locations of the groups attached to the main chain are identified by numbers. The word parts are then assembled in a manner analogous to that in alkane nomenclature.

The following examples illustrate correct application of these rules. Study them, noting especially the placing of commas and hyphens. Because rule 3 gives precedence to the double bond, rather than to the side chains, positions bearing side chains will sometimes have larger numbers than they would if an alternate numbering were used. Common names are given in parentheses.

$$CH_2{=}CH_2 \qquad CH_3CH{=}CH_2 \qquad CH_3CH{=}CHCH_2CH_3$$

Ethene[2]　　　　　　Propene[2]　　　　　2-Pentene (*cis* or *trans*)
("ethylene")　　　　("propylene")　　　　　　("$\beta$-amylene")

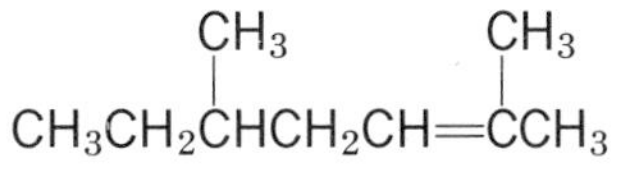

$$\underset{\text{2,5-Dimethyl-2-heptene}}{CH_3CH_2\overset{\displaystyle CH_3}{\overset{|}{C}}HCH_2CH{=}\overset{\displaystyle CH_3}{\overset{|}{C}}CH_3}$$

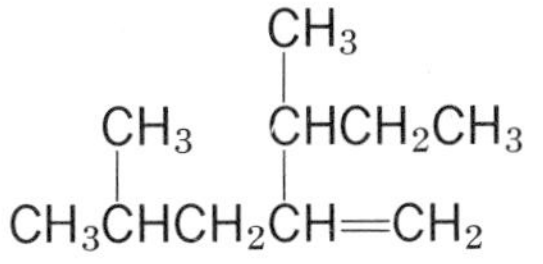

$$\underset{\text{2-Isobutyl-3-methyl-1-pentene}}{CH_3\overset{\displaystyle CH_3}{\overset{|}{C}}HCH_2\overset{\displaystyle \overset{CH_3}{\overset{|}{C}}HCH_2CH_3}{\overset{|}{C}}H{=}CH_2}$$

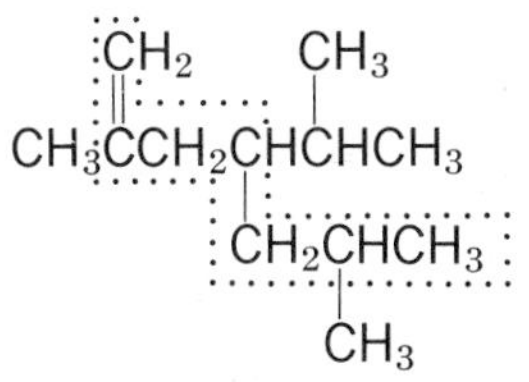

4-Isopropyl-2,6-dimethyl-1-heptene
(dotted lines enclose the longest chain having the double bond)

Some important unsaturated groups that occur many times in organic systems have common names, for example,

---

[1] A third system of nomenclature is known and used, but we shall not employ it in this text.
[2] Whenever the double bond cannot be located differently to make an isomer, its position need not be designated by a number.

$$CH_2{=}CH{-} \qquad\qquad CH_2{=}CHCH_2{-}$$

Vinyl                          Allyl

Thus the compound $CH_2{=}CH{-}Cl$ is nearly always called *vinyl chloride*, although its formal name is chloroethene; the compound $CH_2{=}CHCH_2{-}Br$ is almost always called *allyl bromide,* although its formal name is 3-bromo-1-propene.

---

**Exercise 2.4**   Write condensed structures for each of the following.

(a) 4-methyl-2-pentene          (b) 3-*n*-propyl-1-heptene
(c) 3,3-dimethyl-4-chloro-1-butene          (d) 2,3-dimethyl-2-butene
(e) allyl iodide          (f) vinyl bromide

**Exercise 2.5**   Write IUPAC names and the common names if they have already been mentioned for each of the following. Use cis or trans designations if they are applicable.

(a) $CH_3$   $CH_3$
        C
        ‖
       $CH_2$

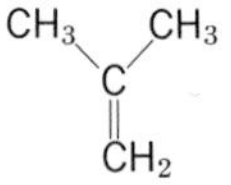

(b) $CH_3{-}CH{-}CH_2{-}C{-}CH_2{-}CH{-}CH_3$ with $CH_3$ groups above on the first CH and last CH, and below the central C a $C$ bearing $CH_3$ and $CH_2{-}CH_3$

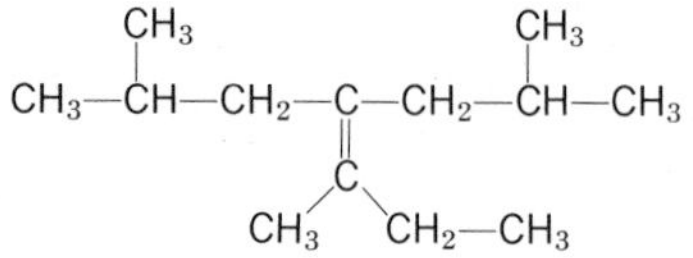

(c) $CH_3{-}CH{=}CH{-}Cl$

(d) $Br{-}CH_2{-}CH{=}CH_2$

(e) $CH_3{-}CH{-}CH_2{-}CH{=}CH_2$ with a $CH_2{-}CH_3$ branch on the second carbon

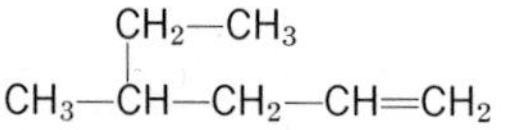

(f) If (cyclohexene ring) is called cyclohexene what is its *full* structure? What is the structure of cyclopentene?

(g) If (ring with $CH_3$ at position 3, numbered 1–6) is 3-methylcyclohexene,[3] what is (ring with $CH_3$ and $CH_3$) ?

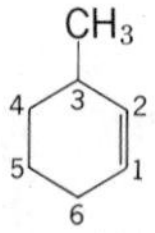

---

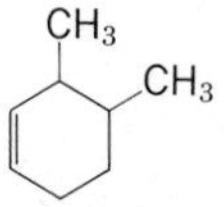

## PHYSICAL PROPERTIES OF ALKENES

The alkenes, as a class, closely resemble the alkanes in physical properties; this generalization is the most useful one we can carry forward to our study of other families. Alkene molecules, which in the higher homologs are mostly alkane-like anyway, are relatively nonpolar. Alkenes are insoluble in water, but they dissolve in the typical nonpolar solvents such as benzene, ether, carbon tetrachloride, chloroform, and ligroin (a mixture of liquid alkanes).

## CHEMICAL PROPERTIES OF ALKENES

There are basically two sites in alkenes that are attacked by chemicals. One is the double bond itself, the other the position on the chain directly attached to

---

[3] The double bond in an acyclic system is *always* regarded as *starting from* position 1. Numbering then proceeds through the double bond around the ring in the direction that will give the lower set of numbers to the substituents.

the double bond, the *allylic position*. In this chapter we study reactions at the double bond itself.

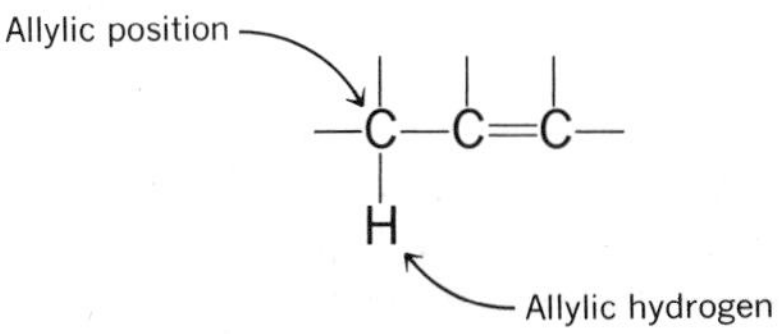

Allylic position —

Allylic hydrogen

**Addition Reactions.** In marked contrast to the alkanes, alkenes undergo many reactions, nearly all involving attack at the double bond. In further contrast to alkanes, these reactions are not substitutions. They are a new type called addition reactions, that is, two molecules come together and react—that is, join—without loss of any fragment from either. (We recall that chlorination of methane, a substitution reaction, produces a fragment, H—Cl, with the hydrogen coming from the alkane.) The following are typical addition reactions of alkenes.

1. **Halogenation,** the addition of chlorine or bromine (only these two halogens will add to a double bond). In general:

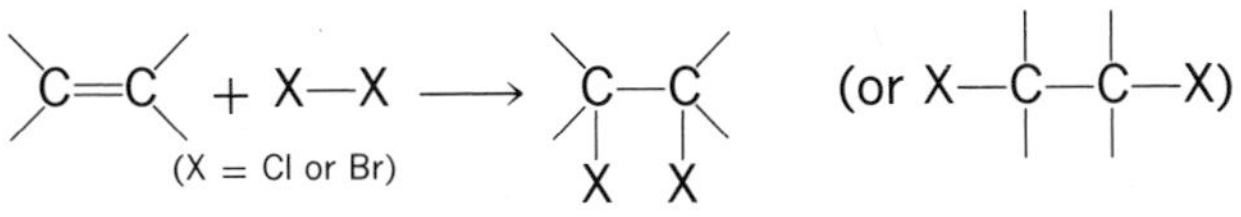

$$\text{C=C} + \text{X—X} \longrightarrow \text{C—C} \quad \text{(or X—C—C—X)}$$
$$\overset{|}{\underset{X}{}} \overset{|}{\underset{X}{}}$$

(X = Cl or Br)

Specific examples:

$$CH_3\text{—}CH\text{=}CH\text{—}CH_3 + Cl_2 \longrightarrow CH_3\text{—}\underset{\underset{Cl}{|}}{CH}\text{—}\underset{\underset{Cl}{|}}{CH}\text{—}CH_3$$

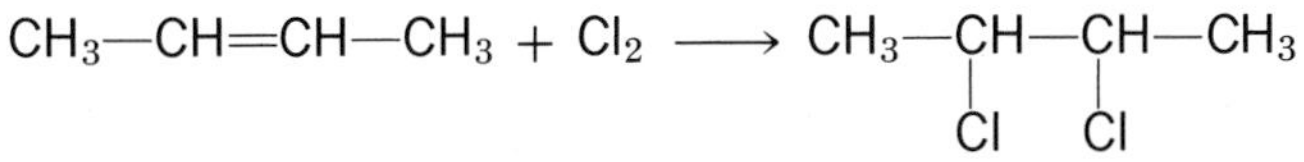

2-Butene            2,3-Dichlorobutane (81%)[4]

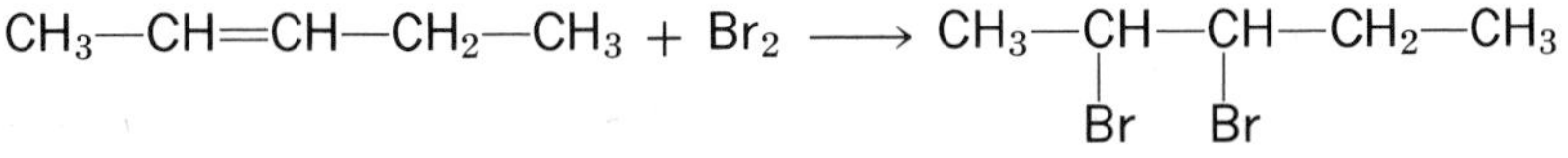

$$CH_3\text{—}CH\text{=}CH\text{—}CH_2\text{—}CH_3 + Br_2 \longrightarrow CH_3\text{—}\underset{\underset{Br}{|}}{CH}\text{—}\underset{\underset{Br}{|}}{CH}\text{—}CH_2\text{—}CH_3$$

2-Pentene            2,3-Dibromopentane (94%)

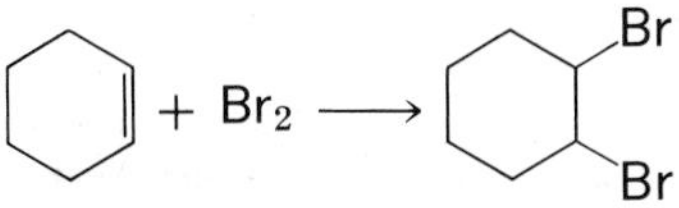

Cyclohexene      1,2-Dibromocyclohexane (100%)

[4] Throughout the chapters on organic chemistry, most reactions are shown with a figure in parentheses representing the mole percent yield. Side reactions are almost inevitable in organic synthesis. Moreover, in isolating the product, losses occur. Many reactions do not go to completion; unchanged starting materials remain and must be separated from the product. The finally isolated, pure product is nearly always less, on a mole basis, than would be expected if every mole of starting material forms the desired product. The percentages cited in nearly all examples in this text are yields actually reported and then tabulated in *Synthetic Organic Chemistry* by R. B. Wagner and H. D. Zook (John Wiley and Sons, New York, 1953).

Usually this reaction is carried out at room temperature, or below, and the reagents are dissolved in some inert solvent such as carbon tetrachloride. If chlorine is used, the gas is bubbled into the solution. Bromine, a corrosive liquid that must be handled with great care, is usually added as a dilute solution in carbon tetrachloride. This solution of bromine added to an unknown provides a convenient test tube test for the presence of a double bond, for the dibromo compounds that form are colorless or pale yellow, in contrast to the deep-brown and almost opaque bromine solution. If an unknown hydrocarbon believed to be either an alkene or an alkane is tested, it could react immediately at room temperature, with the characteristic bromine color disappearing rapidly. If, simultaneously, fumes of hydrogen bromide gas are *not* evolved, the unknown is probably an alkene, for hydrogen bromide does form if bromine reacts with an alkane by a substitution process.

2. **Hydrogenation,** the addition of the elements of hydrogen, H—H. In general:

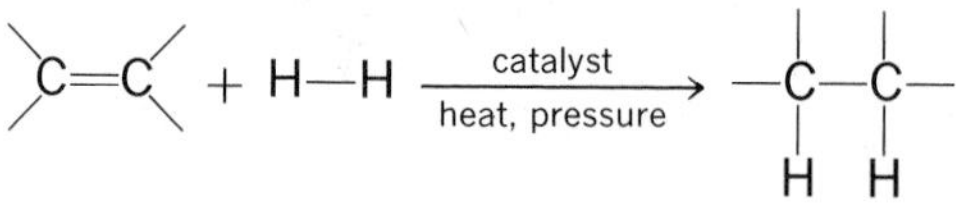

$$\text{C}=\text{C} + \text{H}-\text{H} \xrightarrow[\text{heat, pressure}]{\text{catalyst}} -\overset{\displaystyle |}{\underset{\displaystyle \text{H}}{\text{C}}}-\overset{\displaystyle |}{\underset{\displaystyle \text{H}}{\text{C}}}-$$

Specific examples:[5]

$$\text{CH}_2{=}\text{CH}_2 + \text{H}_2 \xrightarrow[\text{heat, pressure}]{\text{Ni}} \text{CH}_3-\text{CH}_3$$

Ethylene                           Ethane

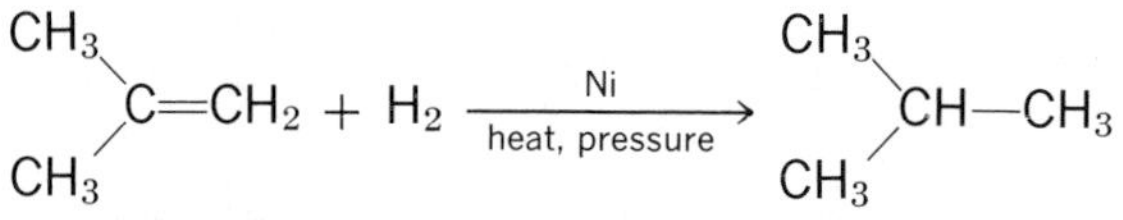

$$\begin{array}{c}\text{CH}_3 \\ \phantom{xx}\diagdown \\ \phantom{xxx}\text{C}{=}\text{CH}_2 \\ \phantom{xx}\diagup \\ \text{CH}_3\end{array} + \text{H}_2 \xrightarrow[\text{heat, pressure}]{\text{Ni}} \begin{array}{c}\text{CH}_3 \\ \phantom{xx}\diagdown \\ \phantom{xxx}\text{CH}-\text{CH}_3 \\ \phantom{xx}\diagup \\ \text{CH}_3\end{array}$$

Isobutylene                          Isobutane

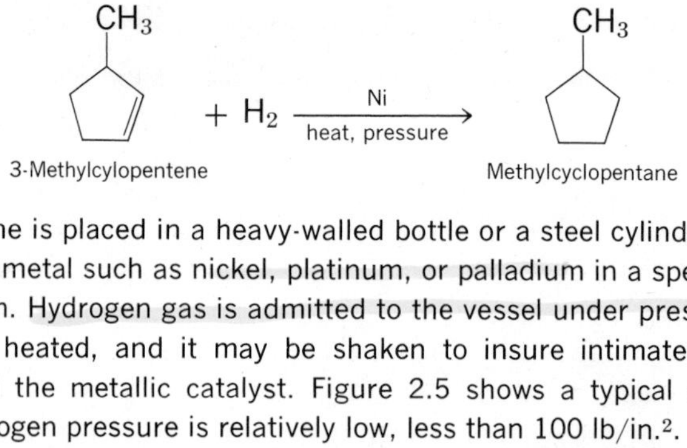

3-Methylcyclopentene     + H$_2$ $\xrightarrow[\text{heat, pressure}]{\text{Ni}}$     Methylcyclopentane

The alkene is placed in a heavy-walled bottle or a steel cylinder together with the catalyst, a metal such as nickel, platinum, or palladium in a specially prepared, powdered form. Hydrogen gas is admitted to the vessel under pressure; the bottle is sometimes heated, and it may be shaken to insure intimate contact of the reactants with the metallic catalyst. Figure 2.5 shows a typical apparatus used when the hydrogen pressure is relatively low, less than 100 lb/in.$^2$.

One important commercial application of this reaction is the hydrogenation of vegetable oils. Their molecules contain alkene links which, although the number per molecule varies, qualify as "polyunsaturated." Animal fats, structurally, are almost

---

[5] The examples have been made up to illustrate the reaction. Reported examples involve more complicated alkenes than those shown here.

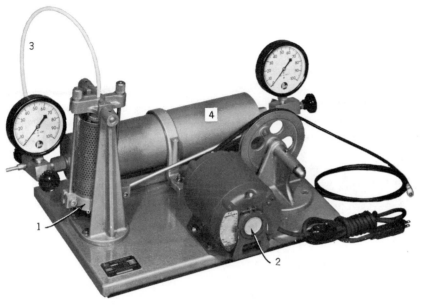

**Figure 2.5** Typical low-pressure hydrogenation apparatus. The unsaturated compound to-gether with a catalyst and an inert solvent, is placed in a thick-walled glass bottle positioned in a heavy-mesh, protective metal screen and clamped on a rocker, 1. Rocking action, powered by the motor, 2, mixes the contents of the flask while hydrogen gas under pressure is led in by a hose, 3, from a storage tank, 4. (Courtesy of Parr Instrument Company.)

identical with vegetable oils, but they have fewer double bonds per molecule. By controlled hydrogenation of some of the double bonds in vegetable oils, they become, like the animal fats, solids at room temperature. This reaction is discussed further in Chapter 9.

3. **Addition of hydrogen chloride or hydrogen bromide.** Molecules of hydrogen or of any of the halogens are symmetrical, but those of hydrogen chloride or hydrogen bromide are not. If a molecule of either acid adds itself to an unsymmetrically substituted double bond, two directions for the addition are possible. Propylene, for example, may react with hydrogen chloride to give either $n$-propyl chloride or isopropyl chloride or a mixture of both:

$$CH_3\!-\!CH\!=\!CH_2 + H\!-\!Cl \longrightarrow CH_3\!-\!\underset{H}{\overset{\phantom{x}}{C}}H\!-\!\underset{Cl}{\overset{\phantom{x}}{C}}H_2 \quad \text{(or } CH_3CH_2CH_2Cl)$$

Propylene               $n$-Propyl chloride

$$\text{or} \quad CH_3\!-\!CH\!=\!CH_2 + H\!-\!Cl \longrightarrow CH_3\!-\!\underset{Cl}{\overset{\phantom{x}}{C}}H\!-\!\underset{H}{\overset{\phantom{x}}{C}}H_2 \quad \text{(or } CH_3\underset{Cl}{\overset{\phantom{x}}{C}}HCH_3)$$

Isopropyl chloride

In either case, the double bond becomes a single bond, and the two parts of the hydrogen chloride molecule become attached to the carbons at each end of the former double bond. The product that actually forms is largely isopropyl chloride. Essentially no *n*-propyl chloride is produced.

The several other examples of this reaction to be shown display a consistency first noted by a Russian chemist, Markovnikov.[6] According to Markovnikov's rule, when an unsymmetrical reagent, H—G, adds across an unsymmetrical carbon-carbon double bond of a simple alkene, the hydrogen of H—G attaches to the carbon of the double bond that is already bonded to the greater number of hydrogens. ("Them that has, gets.") For purposes of this rule, "unsymmetrical carbon-carbon double bond" means one in which the carbons bear unequal numbers of hydrogens.

Specific examples:

$$\underset{\text{Ethylene}}{CH_2{=}CH_2} + H{-}Br \longrightarrow \underset{\text{Ethyl bromide}}{CH_3CH_2Br} \qquad \text{(only possible product)}$$

$$\underset{\text{Isobutylene}}{CH_3{-}\underset{\underset{CH_3}{|}}{C}{=}CH_2} + H{-}Cl \longrightarrow \underset{t\text{-Butyl chloride}}{CH_3{-}\overset{\overset{Cl}{|}}{\underset{\underset{CH_3}{|}}{C}}{-}CH_3} \qquad \left(\text{not } CH_3{-}\underset{\underset{CH_3}{|}}{CH}{-}CH_2{-}Cl\right)$$

$$\underset{\text{2-Butene}}{CH_3{-}CH{=}CH{-}CH_3} + H{-}Cl \longrightarrow \underset{sec\text{-Butyl chloride}}{CH_3{-}CH_2{-}\underset{\underset{Cl}{|}}{CH}{-}CH_3} \qquad \text{(only possible product)}$$

$$\underset{\text{2-Pentene}}{CH_3{-}CH_2{-}CH{=}CH{-}CH_3} + H{-}Br \longrightarrow \underset{\text{3-Bromopentane}}{CH_3{-}CH_2{-}\underset{\underset{Br}{|}}{CH}{-}CH_2{-}CH_3} + \underset{\text{2-Bromopentane}}{CH_3{-}CH_2{-}CH_2{-}\underset{\underset{Br}{|}}{CH}{-}CH_3}$$

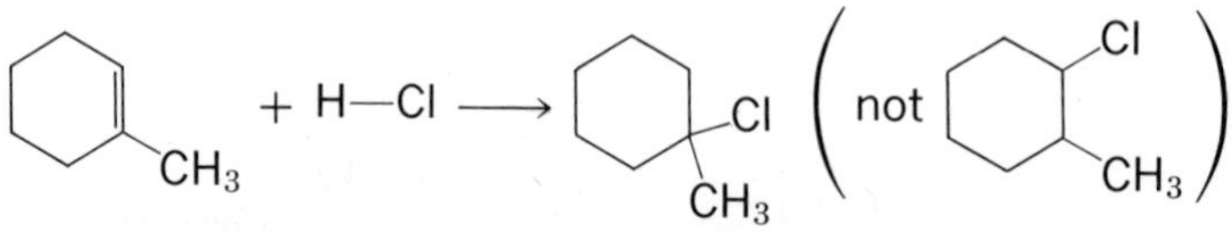


1-Methylcyclohexene 1-Chloro-1-methylcyclohexane

[6] Vladimir Vasil'evich Markovnikov (1838–1904).

The reaction is carried out by bubbling the hydrogen halide in gaseous form into the alkene, which may or may not be dissolved in some solvent inert to the reagents (e.g., carbon tetrachloride).[7]

For preparing alkyl chlorides and bromides, this reaction is superior to direct halogenation of an alkane. The latter gives mixtures of mono-, di-, and higher halogenated materials, and the reaction also produces mixtures of isomers.

Alkyl chlorides and bromides (and iodides) are extremely important intermediates in the synthesis of organic compounds. They can also be made from alcohols, which is often more convenient, but the addition of HX to a double bond illustrates a very important reaction of alkenes.

---

Exercise 2.6   Write the names and structures for the product(s) that would form under the conditions shown. Assume that peroxides are absent.

(a) $CH_2{=}CHCH_2CH_3 + HCl \longrightarrow$

(b) $CH_2{=}C{-}CH_3 + HBr \longrightarrow$
$\qquad\quad |$
$\qquad\; CH_3$

(c) $CH_3{-}CH{=}C{-}\langle\text{ring}\rangle + HCl \longrightarrow$ (with $CH_3$ substituent)

(d) $\langle\text{ring with }CH_3\rangle + HBr \longrightarrow$

(e) $\langle\text{ring}\rangle{-}CH{=}CH{-}\langle\text{ring}\rangle + HCl \longrightarrow$

Exercise 2.7   Write the structures and names of alkenes that are best used to prepare each of the following halides. Avoid selecting those yielding isomeric halides that are difficult to separate. If the halide cannot be prepared by the action of HCl or HBr on an alkene (given the limitations of Markovnikov's rule), state this fact.

(a) $CH_3CH_2CHCH_3$
$\qquad\quad\;\; |$
$\qquad\quad\;\; Cl$

(b) $CH_3CH_2CH_2Cl$

(c) $\langle\text{ring}\rangle{-}Br$

(d) $CH_3{-}C{-}CH_2{-}CH_3$
$\qquad\;\; |$
$\qquad\;\; Cl$ (with $CH_3$ above)

(e) $Cl{-}CH_2{-}CH{-}CH_2{-}CH_3$ (with $CH_3$ above)

(f) $CH_3{-}CH{-}CH_2{-}CH_2{-}CH_3$
$\qquad\quad |$
$\qquad\quad Br$

---

[7] When hydrogen bromide is used, it is important that no peroxides, compounds of the type R—O—O—H or R—O—O—R, be present. They catalyze a different mechanism, which we shall not discuss, with the result that in their presence hydrogen bromide (and only this hydrogen halide) adds itself to unsymmetrical alkenes against the Markovnikov rule. Traces of peroxides form when organic chemicals remain exposed to atmospheric oxygen for long periods.

4. **Hydration,** the addition of the elements of water, H—OH. In general:

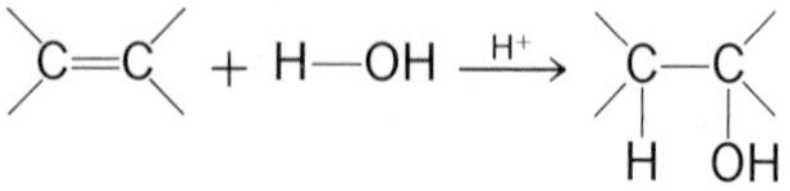

$$\text{C=C} + \text{H—OH} \xrightarrow{H^+} \underset{\underset{\text{H   OH}}{|\quad|}}{\text{C—C}}$$

An alkene            An alcohol

Specific examples:

$$CH_2\text{=}CH_2 + H\text{—OH} \xrightarrow[240°C]{10\% \ H_2SO_4} CH_3\text{—}CH_2\text{—OH}$$

Ethylene                     Ethyl alcohol

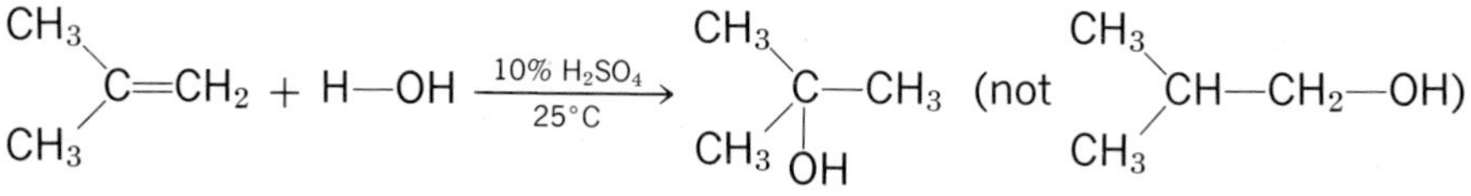

$$\underset{CH_3}{\overset{CH_3}{>}}C\text{=}CH_2 + H\text{—OH} \xrightarrow[25°C]{10\% \ H_2SO_4} \underset{\underset{CH_3 \ OH}{|\quad|}}{C}\text{—}CH_3 \ \ (\text{not} \ \underset{CH_3}{\overset{CH_3}{>}}CH\text{—}CH_2\text{—OH})$$

Isobutylene                t-Butyl alcohol

Markovnikov's rule applies to this reaction in which H—G is now H—OH. The H in H—OH goes to the carbon of the double bond that has the greater number of hydrogens; the —OH goes to the other carbon.

Acid catalysis is the key condition. In fact, the reaction can be completed in two discrete steps if we use concentrated sulfuric acid, another unsymmetrical reagent whose molecules add themselves to double bonds in accordance with Markovnikov's rule:

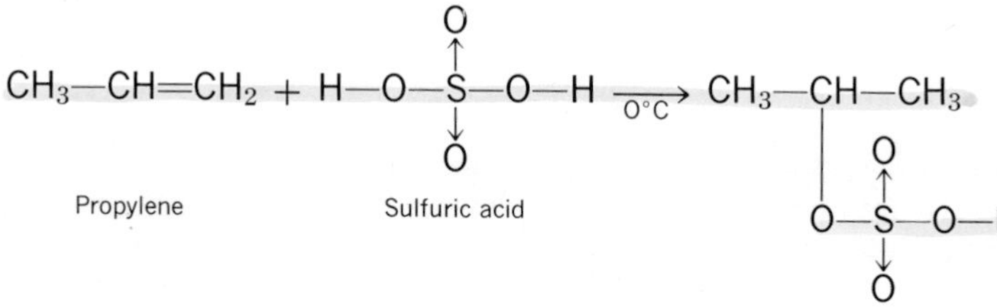

$$CH_3\text{—}CH\text{=}CH_2 + H\text{—O—}\overset{\overset{O}{\uparrow}}{\underset{\downarrow}{S}}\text{—O—H} \xrightarrow{0°C} CH_3\text{—}CH\text{—}CH_3$$

Propylene           Sulfuric acid          O—S—O—H (Isopropyl hydrogen sulfate)

If the product, a member of the class of alkyl hydrogen sulfates, is diluted with water, the sulfate group is replaced by —OH and the product is the alcohol:

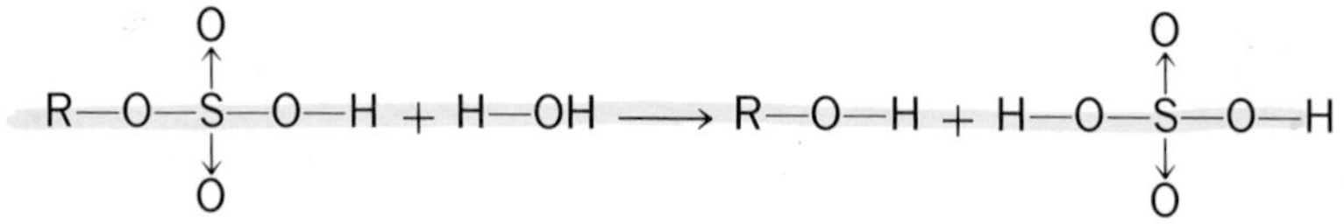

$$R\text{—O—}\overset{\overset{O}{\uparrow}}{\underset{\downarrow}{S}}\text{—O—H} + H\text{—OH} \longrightarrow R\text{—O—H} + H\text{—O—}\overset{\overset{O}{\uparrow}}{\underset{\downarrow}{S}}\text{—O—H}$$

An alkyl hydrogen           An alcohol          Sulfuric acid
sulfate

The addition of water to double bonds is one of the important ways of making alcohols, R—OH. In living organisms this reaction (catalyzed by enzymes) occurs frequently, as we shall see in our study of the metabolism of lipids and the citric acid cycle. The hydration of alkenes is also important commercially. Ethyl alcohol is made from ethylene which is abundantly available from the cracking of petroleum and has many other uses. By 1970 United States production of ethyl alcohol from ethylene is expected to be over 9 million tons per year.

5. **Hydroxylation of Double Bonds. Glycols.** The double bond is vulnerable to attack by oxidizing agents. Oxidation is loss of electrons, reduction gain of electrons. Complete loss and complete gain represent two extremes, and in organic chemistry it is not always obvious what specific molecule or part of it has lost or gained electrons. Between these extremes we could speak of "losing or gaining *control* of some small or large amount of electron density," but this approach is fraught with needless difficulties. We would rather adopt a working definition of oxidation and reduction useful for organic reactions. Whenever a molecule has gained oxygen or lost hydrogen, we say that it has been oxidized. Conversely, if it has lost oxygen or gained hydrogen, it has been reduced. The hydroxylation of a double bond is an oxidation. In general:

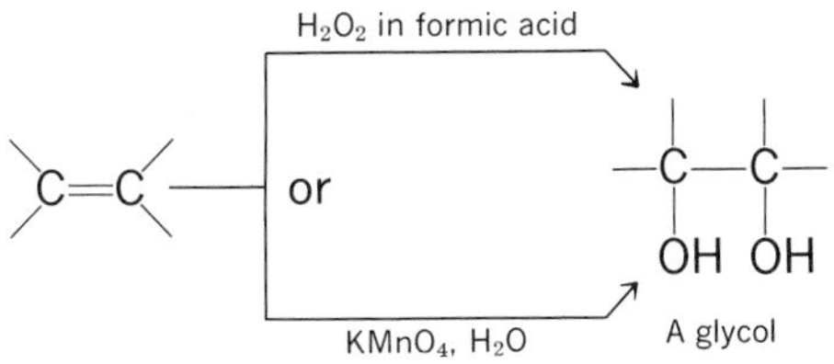


Specific examples:

$$CH_2{=}CH(CH_2)_5CH_3 + H_2O_2 \xrightarrow{\text{formic acid}} \underset{OH}{CH_2}{-}\underset{OH}{CH}(CH_2)_5CH_3$$

1-Octene — Hydrogen peroxide — 1,2-Octanediol (58%)

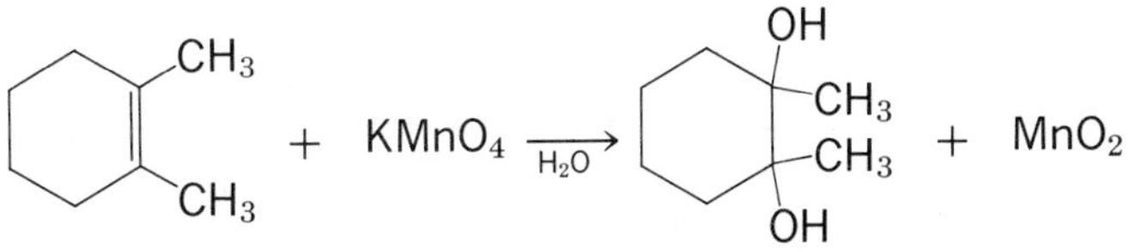


1,2-Dimethylcyclohexene — Potassium permanganate — *cis*-1,2-Dihydroxy-1,2-dimethylcyclohexane (27%) — Manganese dioxide

Compounds whose molecules contain two hydroxyl groups on adjacent carbons are called *glycols*. Ethylene glycol and propylene glycol, both boiling at high temper-

$$\underset{OH}{CH_2}{-}\underset{OH}{CH_2} \qquad CH_3{-}\underset{OH}{CH}{-}\underset{OH}{CH_2}$$

Ethylene glycol (bp 198°C) — Propylene glycol (bp 189°C)

atures and both very soluble in water, are the principal components of commercially available antifreezes. (Neither is made commercially in the ways shown here.)

Potassium permanganate is valuable not so much as a reagent for making glycols from alkenes—the yields are not very high—but rather as a reagent for testing for the presence of an easily oxidized group, such as a carbon-carbon double bond. The inorganic product is manganese dioxide, $MnO_2$, a dark-brown, water-insoluble solid. The permanganate ion in solution is deep purple in color. Thus if an unknown compound with an easily oxidized group is mixed with permanganate solution, its purple color will give way to a brown sludge. Other families that give this test are the aldehydes, certain types of alcohols, acetylenes, and certain types of aromatic compounds.

---

**Exercise 2.8**  Write the structures of the alkenes from which the following glycols can be prepared.

(a) CH₂—CH₂
    |     |
   OH  OH

(b) CH₃—CH—CH₂
        |   |
      OH  OH

(c) [cyclopentane ring with two OH groups]

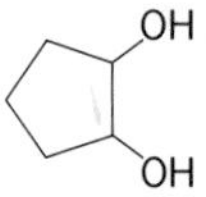

**Exercise 2.9**  An unknown organic compound reacts with aqueous potassium permanganate; the purple color gives way to a brown sludge. Which of the following statements can be made?

(a)  The unknown was an alkene.

(b)  The unknown contained an oxidizable functional group, which might have been a carbon-carbon double bond.

**Exercise 2.10**  How can an unknown be identified by a simple chemical test if it is *known* to be one or the other of each of the following pairs? State what you would do and see. Give *two* tests. Write equations for positive tests.

(a) [benzene ring]  or  [cyclohexane ring]

(b) CH₃—CH=CH—CH₃  or  CH₃—CH₂—CH₂—Br

---

6. **Polymerization.** In general:

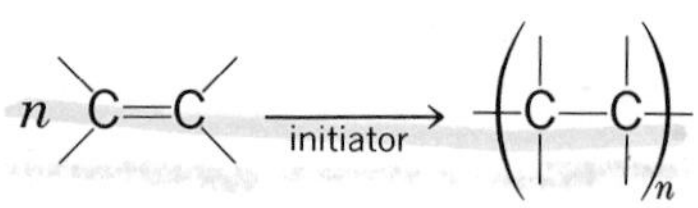

$$n \quad \text{C=C} \xrightarrow{\text{initiator}} \left( \text{C—C} \right)_n$$

Alkene                  Polyalkene

Specific examples:

$$n\text{CH}_2{=}\text{CH}_2 \xrightarrow{\text{initiator}} +(\text{CH}_2{-}\text{CH}_2)_n$$

Ethylene           Polyethylene

($n$ is several hundred—up to several thousand)

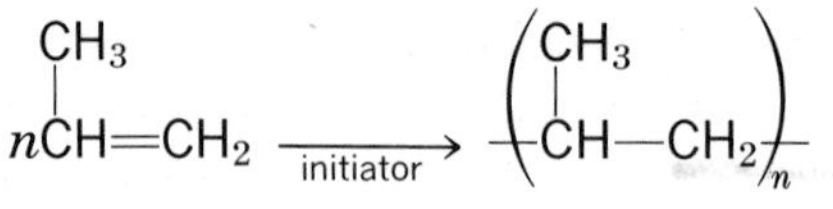

$$n\text{CH}{=}\text{CH}_2 \xrightarrow{\text{initiator}} \left( \begin{array}{c} \text{CH}_3 \\ | \\ \text{CH—CH}_2 \end{array} \right)_n$$

  CH₃

Propylene           Polypropylene

Polymerization of alkenes is accomplished through the use of small amounts of chemicals called *initiators.* For years atmospheric oxygen was the initiator for the polymerization of ethylene, but a high temperature, 100°C, and high pressure, 15,000 lb/in.$^2$, were also necessary. Oxygen, a free radical, acts initially to generate a trace of a free radical of the type R—O·. From then on the reaction proceeds by means of a chain mechanism.

Initiation:        $R—O· + CH_2{=}CH_2 \longrightarrow R—O—CH_2—CH_2·$

Propagation:  (a) $R—O—CH_2—CH_2· + CH_2{=}CH_2 \longrightarrow$
$$R—O—CH_2—CH_2—CH_2—CH_2·$$

(b) $R—O—CH_2CH_2CH_2CH_2· + CH_2{=}CH_2 \longrightarrow$
$$R—O—CH_2CH_2CH_2CH_2CH_2CH_2·$$

(c) product of $b$ + $CH_2{=}CH_2 \longrightarrow$ etc.

Termination:      $R—O—CH_2CH_2(CH_2CH_2)_nCH_2CH_2· + R· \longrightarrow$
$$R—O—CH_2CH_2(CH_2CH_2)_nCH_2CH_2R$$

Polyethylene

The starting alkene is called a monomer, the product a polymer (Greek: *polys,* many; *meros,* part). Polyethylene is largely a straight chain throughout, but there are branches (methyl groups) about every nine carbons because a hydrogen shifts and the odd electron is relocated during the growth of the chain. These branches are like thorns on a stick, inhibiting the close-packing of neighboring chains. As a result the low-density polyethylene (0.940 g/cc or less) that is produced is soft and pliable but unsuitable for spinning into fibers.

In 1963 the Nobel prize in chemistry went to Karl Ziegler of Germany and Giulio Natta of Italy for their work in developing catalysts for the polymerization of ethylene and propylene. Ziegler found that trialkyl derivatives of aluminum, $R_3Al$, plus titanium chloride catalyzed the formation of a higher-density polyethylene at lower temperatures and pressures. This polyethylene softens at temperatures high enough to permit the sterilization of tubes and bottles made of it.

Natta's work produced the symmetrically polymerized polypropylene illustrated in Figure 2.6. The electron clouds of the side chain methyl groups repel each other, causing the chain to coil into a helix. Polymers whose molecules are very long and very symmetrical give promise of being made into serviceable fibers. Natta's

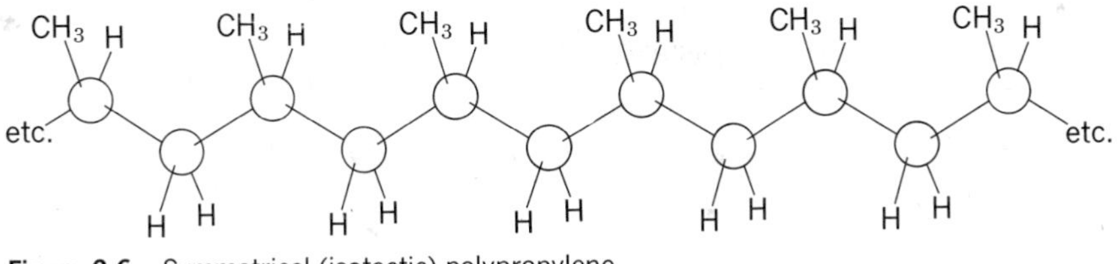

**Figure 2.6**   Symmetrical (isotactic) polypropylene.

symmetrical polypropylene is used more and more to make carpeting for installation around swimming pools, in open entranceways, basements, bathrooms, and even kitchens. One great advantage of both polyethylene and polypropylene is that they are alkanes and have all the chemical unreactivity of this family. They are also thermoplastic polymers; when they are heated, they soften and flow and may be extruded into sheets, molded articles, and tubes. Figures 2.7, 2.8, and 2.9 illustrate some applications of polyethylene and polypropylene. By 1970 the estimated United States production of low- and high-density polyethylene will be 5 billion pounds.

An interesting relative of polyethylene is Teflon, a polymer in which all the hydrogens in polyethylene have been replaced by fluorines. The monomer is tetra-fluorethylene, $F_2C=CF_2$. Teflon is one of the most chemically inert of all organic substances, for the only chemicals known to attack it are molten sodium and potassium. In addition to the well-known Teflon coating for pots and pans, the polymer is used in electrical insulations and antifriction devices.

Another important monomer related to ethylene is vinyl chloride. In the presence of peroxide catalysts it polymerizes to a brittle resin:

$$n CH_2{=}CH \underset{\text{peroxide}}{\xrightarrow{\hspace{1.5cm}}} \;{-}(CH_2{-}CH)_n$$
$$\hspace{1.5cm}|\hspace{4cm}|$$
$$\hspace{1.5cm}Cl\hspace{4cm}Cl$$

Vinyl chloride         Polyvinyl chloride (PVC)

Its brittleness can be overcome and a softer product obtained if during the polymerization we add certain compounds known as *plasticizers;* cresyl phosphate and alkyl phthalates are examples. Polyvinyl chloride is in this way made rubberlike or leatherlike so that it can be widely used in floor tiles, raincoats, tubing, electrical insulation, phonograph records, and protective coatings.

The dichloro derivative of ethylene, vinylidene chloride, is the monomer for saran, used for seat covers and packaging film:

$$\hspace{1.5cm}Cl\hspace{4cm}Cl$$
$$\hspace{1.5cm}|\hspace{4cm}|$$
$$n CH_2{=}C \underset{\text{initiator}}{\xrightarrow{\hspace{1.5cm}}} \;{-}(CH_2{-}C)_n$$
$$\hspace{1.5cm}|\hspace{4cm}|$$
$$\hspace{1.5cm}Cl\hspace{4cm}Cl$$

Vinylidene chloride         Saran

Orlon, a plastic that can be made into fibers for fabrics, is the polymer of a cyano derivative of ethylene called acrylonitrile:

$$n CH_2{=}CH \underset{\text{initiator}}{\xrightarrow{\hspace{1.5cm}}} \;{-}(CH_2{-}CH)_n$$
$$\hspace{1.5cm}|\hspace{4cm}|$$
$$\hspace{1.5cm}CN\hspace{4cm}CN$$

Acrylonitrile         Orlon

**Figure 2.7** Polyethylene has a new use in agriculture, serving as a mulch to protect plants. This photograph was taken at LaCosta, California, in a strawberry field. Because the plants grow through holes in the mulch, weed problems are completely eliminated. The material is being used extensively in Hawaii. (Courtesy of Monsanto Company.)

**Figure 2.8** Low-pressure pipe made of high-density polyalkene resins is chosen for an increasing number of applications. It resists rot and corrosion, and it is relatively easy to install. To connect the sections of this 6-inch pipe for natural gas, a small, motor-powered device operated by two men is all that is needed. To install the same pipe in steel would require a much larger crew and much heavier equipment. (Courtesy of Phillips Petroleum Company.)

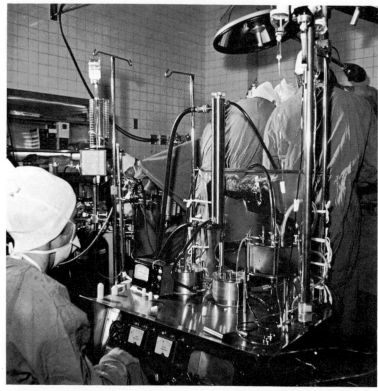

**Figure 2.9** The tubing and containers for blood in this Miniprime heart-lung machine are fashioned from polyvinyl chloride, $-(CH_2CHCl)_n$. The scene is a Los Angeles hospital during open-heart surgery. The machine is manufactured by Travenol Division, Baxter Laboratories. (Courtesy of Monsanto Company.)

## MECHANISM OF ADDITION REACTIONS OF ALKENES

Having studied briefly the principal chemical reactions of the alkene linkage, we turn next to the mechanism for adding reagents such as hydrogen chloride and water to the double bond. We already know that addition reactions happen, that they follow Markovnikov's rule, and that acids or acid catalysts are involved. We must now seek one mechanism to tie these facts together and to explain others still to be introduced.

Let us consider the addition of hydrogen chloride. We already know it as a proton donor, and a proton is an electron-poor reagent. In the terminology of the field, it is said to be an *electrophilic species* (i.e., "electron loving" from *philos*, Greek for loving). The pi-electron cloud at a double bond is "electron-rich"; we should therefore expect some interaction, and we believe it to happen as follows. In step 1 (see Figure 2.10) we imagine a molecule of an alkene to be on collision course with a molecule of hydrogen chloride. If the collision is violent enough and if it is properly

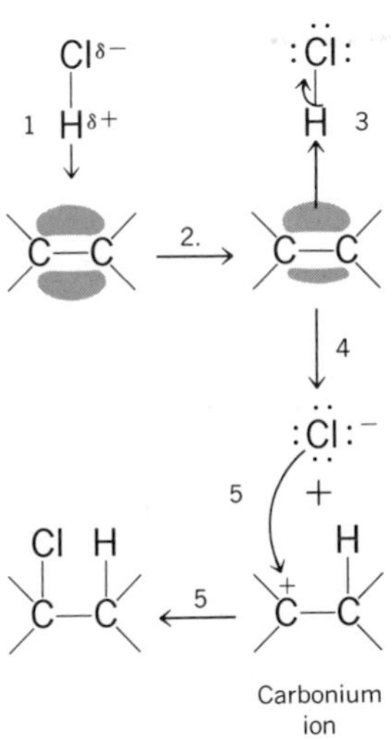

Carbonium
ion

**Figure 2.10**  The addition of hydrogen chloride to an alkene. (The numbers refer to the textual discussion.)

oriented, it will lead (step 2) to a situation in which the pair of pi electrons begins to form a bond to the hydrogen while the chlorine (step 3) begins to disengage from it. The result (step 4) is a pair of ions, one of which has a carbon with a positive charge. Such a species is called a *carbonium ion;* the carbon bearing the positive charge has only six electrons in its valence shell. Carbonium ions have high internal energies and are highly reactive toward electron-rich species. They have only fleeting existences, but they are real and undergo a variety of reactions, *all of them to restore an octet to the carbon.* Hence, in our example, step 5, a chloride ion is very rapidly attached to the site of the positive charge in the carbonium ion to give the final product. To summarize,

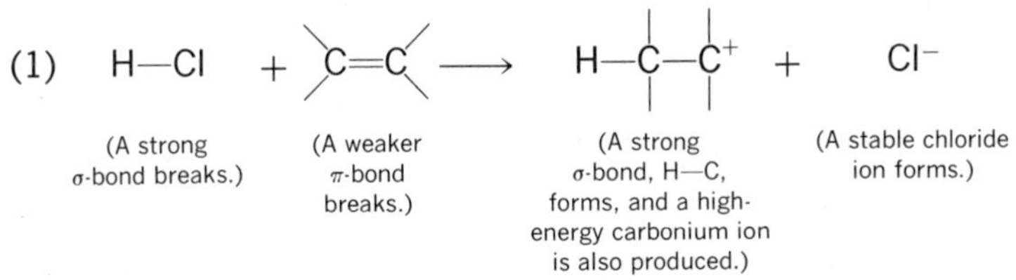

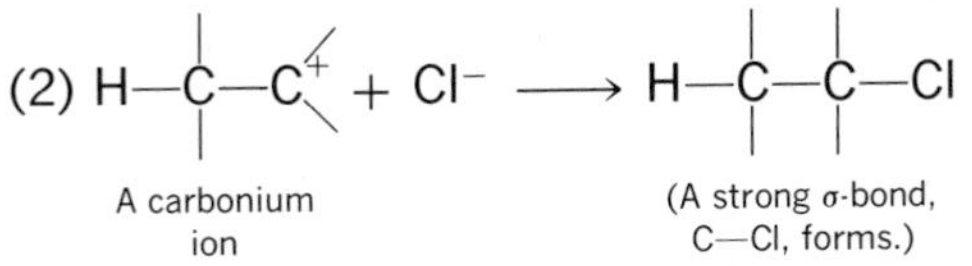

Qualitatively, the energy cost to break a strong and a weak bond (step 1) is more than repaid by the energy lowering that accompanies the formation of two strong bonds.

What has thus far been said qualitatively accounts for two facts: first, the reaction does occur, and second, an acidic, proton-donating reagent rather than a basic, electron-rich species is the type that will attack a double bond. What about the direction or the orientation of the addition by Markovnikov's rule?

Consider the addition of hydrogen chloride to propylene to form isopropyl chloride, not *n*-propyl chloride. We have learned that the first general step in the reaction generates a carbonium ion. Two isomeric carbonium ions could form from propylene:

$$(1) \quad CH_3CH{=}CH_2 + H{-}Cl \longrightarrow CH_3\overset{+}{C}HCH_3 + Cl^-$$

Isopropylcarbonium ion

**1**

$$(2) \quad CH_3CH{=}CH_2 + H{-}Cl \longrightarrow CH_3CH_2CH_2{}^+ + Cl^-$$

*n*-Propylcarbonium ion

**2**

In the final step of the overall addition—combination of the carbonium ion with the chloride ion—only the isopropylcarbonium ion, **1**, can lead to isopropyl chloride:

$$CH_3\overset{+}{C}HCH_3 + Cl^- \longrightarrow CH_3\underset{\underset{Cl}{|}}{C}HCH_3$$

**1**

The result means that **1** rather than **2** is the intermediate. For some reason **1** must form so much more rapidly than **2** that, for all practical purposes, it is the only carbonium ion involved. Perhaps it forms more rapidly because the energy cost of making **1** is far less than that for making **2**. This is, in fact, what a great deal of evidence has led chemists to conclude. In short, a carbonium ion such as **1** is more stable than one such as **2**. We seek now a possible reason for this stability.

Carbonium ions can be classified into three types:

$R{-}CH_2{}^+$      $R{-}\overset{\overset{R'}{|}}{C}H^+$      $R{-}\overset{\overset{R'}{|}}{\underset{\underset{R''}{|}}{C}}{}^+$

Primary or 1°       Secondary or 2°       Tertiary or 3°

The classification relates to the condition of the carbon bearing the positive charge. If only one carbon is attached to it *directly*, it is a primary (1°) carbonium ion.[8] If two carbons are directly attached, it is a secondary (2°) carbonium ion. In a tertiary (3°) carbonium ion three other carbons are directly attached to the carbon bearing the positive charge.

[8] In a context such as this, 1° is not pronounced "one degree" but rather "primary," 2° is secondary," and 3° is "tertiary."

Exercise 2.11   Classify the following carbonium ions as primary, secondary, and tertiary.

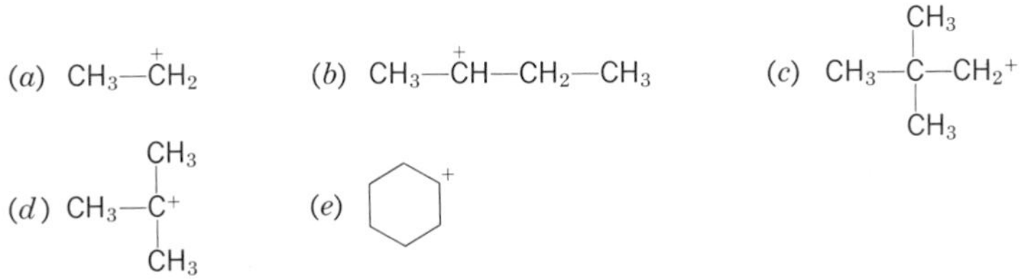

(a) $CH_3-\overset{+}{C}H_2$

(b) $CH_3-\overset{+}{C}H-CH_2-CH_3$

(c) $CH_3-\underset{\underset{CH_3}{|}}{\overset{\overset{CH_3}{|}}{C}}-CH_2^+$

(d) $CH_3-\underset{\underset{CH_3}{|}}{\overset{\overset{CH_3}{|}}{C}}^+$

(e) (cyclohexyl cation ring structure)

In general, *alkyl groups have an electron-releasing inductive effect.* Some evidence for this effect will come later, in Chapter 4, but for our present purposes we make the statement without proof. In a primary carbonium ion only one alkyl group is attached to the positively charged carbon; in the tertiary carbonium ion the number is three. To the extent that alkyl groups induce an electron release toward the positive charge, this charge is partially neutralized. To the extent that it is neutralized, it is stabilized. Three alkyl groups releasing electron density toward a positive site would stabilize it more than two could. And of course two alkyl groups would be better than one. Thus 3° carbonium ions are more stable than 2°, and these are more stable than 1°:

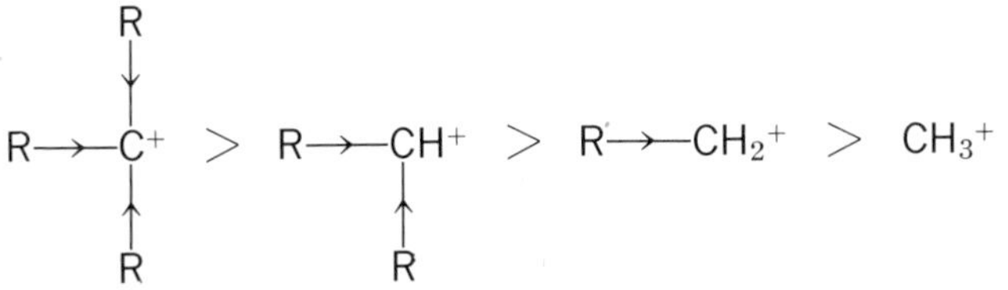

$$R\longrightarrow\underset{\underset{R}{\uparrow}}{\overset{\overset{R}{\downarrow}}{C}}{}^+ \;>\; R\longrightarrow\underset{\underset{R}{\uparrow}}{C}H^+ \;>\; R\longrightarrow CH_2^+ \;>\; CH_3^+$$

Order of stability of some types of carbonium ions

or

*Order of stability of carbonium ions:*   $3° > 2° > 1° > CH_3^+$

Thus propylene reacts with hydrogen chloride to give a secondary carbonium ion (isopropyl) rather than a primary carbonium ion (*n*-propyl) as the intermediate. The more stable ion forms.

Markovnikov's rule would of course be true however we tried to explain it. It is an example of a purely empirical correlation; that is, it depends only on the *observation* of events, however they may be explained. We have seen, however, that this rule can be understood in terms of general theories of physics and chemistry.

The acid-catalyzed addition of water proceeds in much the same way as the addition of hydrogen chloride:

Step 1.   $CH_3CH{=}CH_2 \;+\; \overset{H}{\underset{H}{\overset{+}{O}}}\!\!\overset{H}{\diagup} \;\rightleftharpoons\; CH_3\overset{+}{C}HCH_3 \;+\; :\overset{H}{\underset{H}{O}}\!\!\overset{H}{\diagup}$

Propylene          Catalyst          Isopropylcarbonium ion

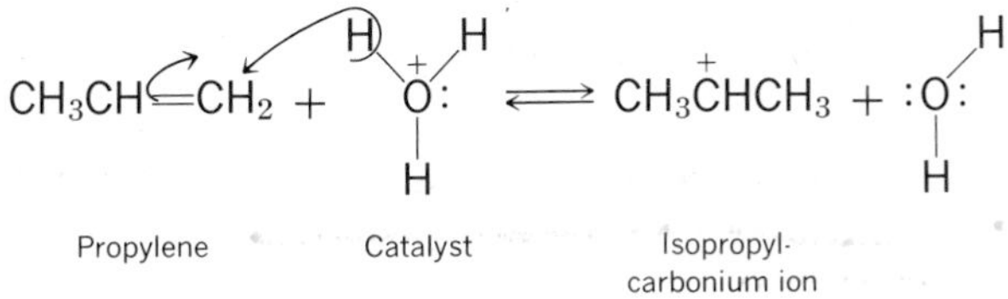

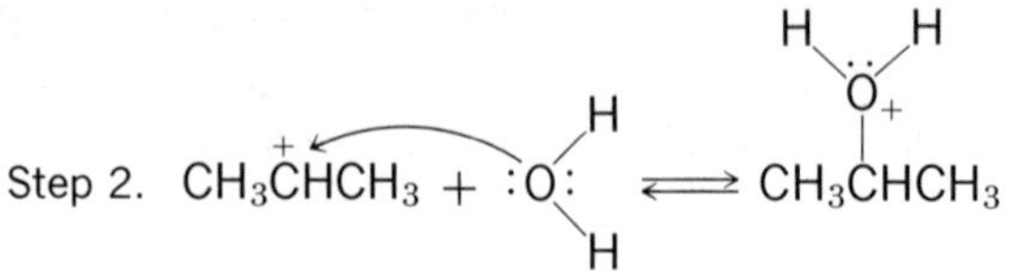

Step 2.  $CH_3\overset{+}{C}HCH_3 + \ddot{:}O\overset{H}{\underset{H}{\diagdown}} \rightleftharpoons CH_3\overset{\overset{H \quad H}{\diagdown \ddot{O} \diagup}}{\underset{+}{C}}HCH_3$

<table>
<tr><td>Isopropyl-<br>carbonium ion</td><td>Water</td><td>Isopropyl alcohol in its<br>protonated form</td></tr>
</table>

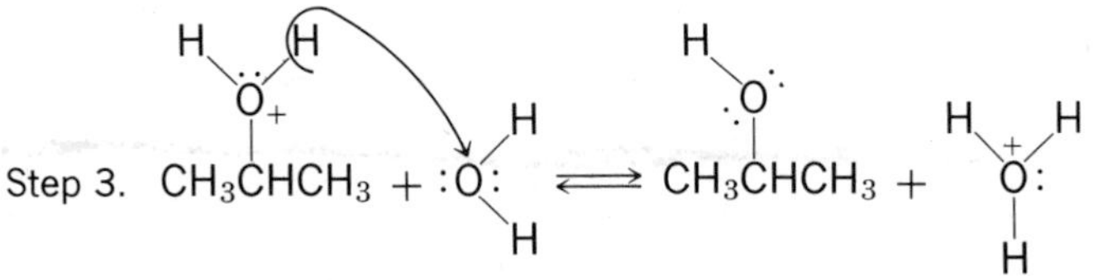

Step 3.  $CH_3\overset{\overset{H \quad H}{\diagdown \ddot{O} \diagup}}{\underset{+}{C}}HCH_3 + \ddot{:}O\overset{H}{\underset{H}{\diagdown}} \rightleftharpoons CH_3\overset{\overset{H}{\diagdown \ddot{O}\ddot{\phantom{.}}}}{C}HCH_3 + \overset{H \quad H}{\underset{H}{\diagdown \overset{+}{O} \diagup}}\ddot{:}$

<table>
<tr><td>Isopropyl<br>alcohol</td><td>Recovered<br>catalyst</td></tr>
</table>

The first step is the generation of a carbonium ion, the more stable of the two possible, which acts like a Lewis acid toward a water molecule (a Lewis base). The alcohol first appears in its protonated form, and then proton transfer gives the final product.

---

**Exercise 2.12**  Write the condensed structures for the two carbonium ions that can conceivably form if a proton becomes attached to each of the following alkenes. Circle the one that is preferred. Where both are reasonable, state that they are. Write the structures of the alkyl chlorides that will form by the addition of hydrogen chloride to each.

(a)  $CH_3-CH_2-CH=CH_2$

(b)  $CH_3-\overset{\overset{\textstyle CH_3}{|}}{C}=CH_2$

(c)  (a cyclohexene ring with a $CH_3$ substituent)

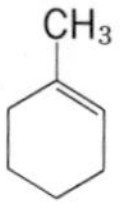

(d)  $CH_3-CH=CH-CH_3$

(e)  $CH_3-CH=CH-CH_2-CH_3$

(f)  Addition of water to 2-pentene (part e) gives a mixture of alcohols. What are they? Why is formation of a mixture to be expected here but not from propylene?

---

With the carbonium ion we have introduced one of the most important intermediates in organic reactions that occur in solution. One of its many properties is that, being a Lewis acid, it is vulnerable to attack by any electron-rich species that happens to be close by. When water is the most abundant of such neighbors, alcohols will form. Another property of the carbonium ion that must be emphasized is its relative instability. With very few exceptions salts cannot be prepared with an intact carbonium ion. Usually in the circumstances of its formation some electron-rich particle can give it the share of an electron-pair needed to complete its octet.

# DIENES

### STRUCTURE AND NOMENCLATURE

Alkenes with two carbon-carbon double bonds are called dienes (i.e., two "ene" functions). If the double bonds are widely enough separated, the diene has the properties of an ordinary alkene. Each double bond behaves independently of the other, and the two are said to be *isolated*. Because the allenic system illustrated in Figure 2.11 is rare, we shall consider it no further.

**Conjugated Dienes.** Dienes in which the two double bonds are separated by only one single bond, as in 1,3-butadiene, $CH_2$=$CH$—$CH$=$CH_2$, are called *conjugated dienes*.[9] In such compounds the alkene properties are modified. Whenever two or more functional groups are adjacent to each other in organic compounds, the properties of the individual groups are changed in important ways.

**Naming the Dienes.** Table 2.3 shows us that dienes are named by the IUPAC system in essentially the same way as the alkenes, except that the ending "-diene" is used. Furthermore, *two* numbers are necessary to specify the locations of the *two* double bonds. The system can obviously be extended to compounds containing three double bonds ("-trienes"), four ("-tetraenes"), etc.

**1,4-Addition to Conjugated Dienes.** When hydrogen bromide is added to a non-conjugated diene such as 1,4-pentadiene, one molecule of hydrogen bromide will add normally:

$$CH_2{=}CH{-}CH_2{-}CH{=}CH_2 + H{-}Br \longrightarrow CH_3{-}\underset{\underset{Br}{|}}{CH}{-}CH_2{-}CH{=}CH_2$$

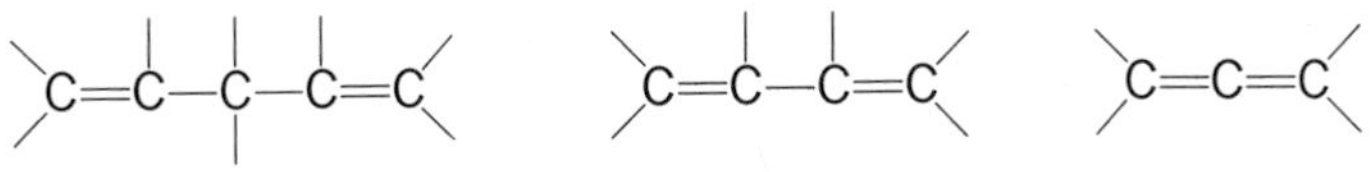

|  (a)  |  (b)  |  (c)  |

**Figure 2.11** (*a*) Nonconjugated or isolated double bonds. (*b*) Conjugated double bonds, meaning that double bonds alternate with single bonds. (*c*) Compounds with adjacent double bonds are allenes.

[9] In general, a conjugated system is any one consisting of alternating double and single bonds.

**Table 2.3  Some Dienes**

| Name | Structure | Bp (°C) | Mp (°C) | Density (in g/cc at 20°C) |
|---|---|---|---|---|
| Allene | $CH_2\!\!=\!\!C\!\!=\!\!CH_2$ | −34.5 | −136 | — |
| 1,3-Butadiene | $CH_2\!\!=\!\!CH\!\!-\!\!CH\!\!=\!\!CH_2$ | −4.4 | −108.9 | — |
| cis-1,3-Pentadiene | 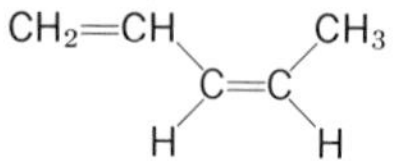 | 44.1 | −140.8 | 0.691 |
| trans-1,3-Pentadiene | 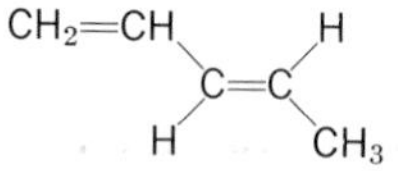 | 42.0 | −87.5 | 0.676 |
| 1,4-Pentadiene | $CH_2\!\!=\!\!CHCH_2CH\!\!=\!\!CH_2$ | 26.0 | −148.3 | 0.661 |
| 2-Methyl-1,3-butadiene ("isoprene") | 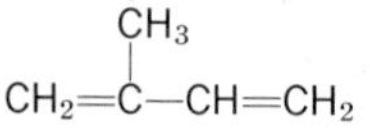 | 34.1 | −146.0 | 0.691 |

Action of one mole of hydrogen bromide on one mole of 1,3-butadiene, a conjugated diene, however, gives the following results:

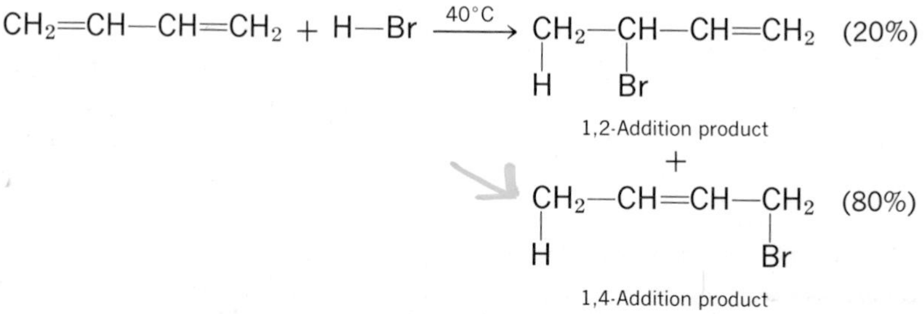

$$CH_2\!\!=\!\!CH\!\!-\!\!CH\!\!=\!\!CH_2 + H\!\!-\!\!Br \xrightarrow{40°C}$$

$$\underset{\substack{|\\H}}{CH_2}\!\!-\!\!\underset{\substack{|\\Br}}{CH}\!\!-\!\!CH\!\!=\!\!CH_2 \quad (20\%)$$
1,2-Addition product

+

$$\underset{\substack{|\\H}}{CH_2}\!\!-\!\!CH\!\!=\!\!CH\!\!-\!\!\underset{\substack{|\\Br}}{CH_2} \quad (80\%)$$
1,4-Addition product

The principal product is not the "expected" 1,2-addition product, the one that would form if the double bonds were isolated. Instead the parts of hydrogen bromide attach at opposite ends of the conjugated system in 1,4-addition. Formation of the 1,2-addition product is not difficult to understand, but to explain the 1,4-product we must refer to the theory of resonance presented in the next chapter. The mechanism of 1,4-addition will be discussed again as an exercise at the end of Chapter 3.

## POLYMERIZATION OF DIENES

Various types of synthetic rubber can be made by the polymerization of dienes. Natural rubber is a polymer of isoprene.

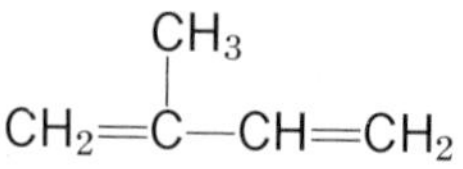

$$\underset{\substack{|\\CH_2=C}}{\overset{CH_3}{\phantom{x}}}\!\!-\!\!CH\!\!=\!\!CH_2$$

Isoprene

To understand its relation to the monomer unit, we may imagine that it forms in the following fashion.

$$
\begin{array}{c}
\underset{\displaystyle CH_2=C-CH=CH_2}{\underset{\displaystyle CH_3}{}} + \underset{\displaystyle CH_2=C-CH=CH_2}{\underset{\displaystyle CH_3}{}} + \underset{\displaystyle CH_2=C-CH=CH_2}{\underset{\displaystyle CH_3}{}} + \underset{\displaystyle CH_2=C-CH=CH_2}{\underset{\displaystyle CH_3}{}} + \text{etc.}
\end{array}
$$

Natural rubber

We note that the main chain emerges cis from each double bond, making this cis-1,4-polyisoprene. The polymer in which the main chain emerges trans from each double bond is gutta-percha. Ziegler's catalysts or lithium metal bring about the all-cis polymerization of isoprene, and thus a synthetic source of natural rubber is available. A cheaper source of a rubber substitute is 1,3-butadiene, which can be polymerized to polybutadiene. Industrial chemists have extensively studied the *copolymerization* of dienes with monoenes (alkenes) to find polymers that have fewer double bonds but are still flexible, elastic, and resistant to solvents and abrasion. One of the most important of these polymers is the copolymer of butadiene and styrene, Buna S rubber. In this substance the polymer molecules are not all identical, but the following illustrates some of their features, without regard to the geometry or the exact proportions of monomers:

$$CH_2=CHCH=CH_2 \qquad\qquad C_6H_5CH=CH_2$$

1,3-Butadiene                                   Styrene

$$-CH_2CH=CHCH_2-\underset{\displaystyle C_6H_5}{CHCH_2}-CH_2CH=CHCH_2-\underset{\displaystyle C_6H_5}{CHCH_2}-$$

Buna S rubber

From butadiene     From styrene     From butadiene     From styrene

This product is also called GRS (Government Rubber Styrene)[10] and "cold rubber."

Another synthetic rubber, *butyl rubber,* is made by polymerizing isobutylene in the presence of a small amount of isoprene. Polyisobutylene would resemble an alkane polymer, with a structure similar to that of polypropylene but containing more methyl groups. By mixing in some isoprene, however, the polymer will have a few double bonds. These sites can enter into reactions that produce cross-linking between chains to make, in effect, an interlacing network of chains and a product that wears well and is resistant to chemicals.

[10] After a government project during World War II.

# ALKYNES

The structural feature characteristic of an alkyne—a carbon-carbon triple bond —does occur among natural products, but it is not important to our study of biochemistry in this text. For that reason we note only its structural features and mention that many of its reactions are of the addition type studied for the alkenes.

## STRUCTURE

The simplest alkyne is commonly called acetylene:

$$H—C\equiv C—H$$

Acetylene
(ethyne)

In it each carbon is bonded to only two other nuclei; in one sense the carbon is divalent. According to one theoretical approach, of the three bonds in a triple bond one is a sigma bond and two are pi bonds (Figure 2.12).

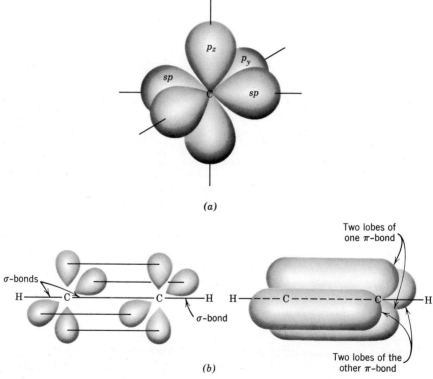

(a)

(b)

**Figure 2.12** The triple bond in acetylene, one theoretical view. (*a*) Orbitals of carbon in its valence state when it is attached to only two other groups, as in an acetylene. (*b*) Two representations of the two $\pi$-bonds in acetylene. The carbon-carbon $\sigma$-bond results from $sp$-$sp$ overlap. The carbon-hydrogen $\sigma$-bonds come from $sp$-$s$ overlap.

# REFERENCES AND ANNOTATED READING LIST

**BOOKS**

J. D. Roberts and M. C. Caserio. *Basic Principles of Organic Chemistry.* W. A. Benjamin, New
York, 1964. Chapter 29, "Polymers," is a survey of the major kinds of polymers,
with emphasis on how their properties are related to their structures.

C. R. Noller. *Chemistry of Organic Compounds,* third edition. W. B. Saunders Company,
Philadelphia, 1965. Chapter 34, "Polyenes, Rubber, and Synthetic Rubbers," provides
not only information about the structures and chemistry of the substances named
but also interesting historical material.


**ARTICLES**

H. L. Fisher. "Rubber." *Scientific American,* November 1956, page 75. A brief history of rubber is followed by a general discussion of polymers that serve as synthetic rubber.

G. Oster. "Polyethylene." *Scientific American,* September 1957, page 139. Polyethylene
plastic is more than an aggregation of linear molecules. Much branching and cross-linking
occur. This article discusses how such structural features arise and how they affect the
properties and enhance the usefulness of the plastic. (The entire September 1957 issue
of *Scientific American* is devoted to "Giant Molecules.")


# PROBLEMS AND EXERCISES

1. Write an illustrated discussion of each of the following terms.

    (*a*)  $sp^2$ hybrid orbital

    (*b*)  pi bond

    (*c*)  cis isomer

    (*d*)  trans isomer

    (*e*)  hydrogenation

    (*f*)  unsymmetrical double bond

    (*g*)  Markovnikov's rule

    (*h*)  glycol

    (*i*)  oxidation (as we shall use the term
    in organic chemistry)

    (*j*)  reduction (as we shall use the term in
    organic chemistry)

    (*k*)  initiator

    (*l*)  polymerization

    (*m*)  electrophilic species

    (*n*)  carbonium ion

    (*o*)  inductive effect

    (*p*)  conjugated system

2. List the families of organic compounds for which we have learned a method of synthesis in
this chapter, and write equations illustrating these reactions.

3. Write IUPAC names for each of the following:

    (*a*)  $CH_3(CH_2)_7CH{=}CH_2$

    (*b*)  $Cl{-}CH{=}CH\overset{\overset{\displaystyle CH_3}{|}}{C}HCH_3$

    (*c*)  $CH_3CH_2CH_2\overset{\overset{\displaystyle CH_2}{\|}}{C}CH_2\overset{\underset{\displaystyle CH_3}{|}}{C}HCHCH_3$

    (*d*)  $CH_3\overset{\overset{\displaystyle CH_3}{|}}{C}CH{=}CH$  with $CH_3$ and $CH_3$ below

    (*e*)  $CH_3CH_2\overset{\overset{\displaystyle CH_2}{\|}}{C}{-}\overset{\underset{\displaystyle CH_2}{\|}}{C}{-}CH_3$

4. Write equations for the reaction of isobutylene with each of the following reagents: (a) cold, concentrated $H_2SO_4$, (b) $H_2$ (Ni, heat, pressure), (c) $H_2O$ ($H^+$ catalysis), (d) H—Cl, (e) H—Br.

5. Write equations for the reaction of 1-methylcyclopentene with each of the reagents of exercise 4.

6. When 1-butene reacts with hydrogen chloride, the product is 2-chlorobutane. The isomer, 1-chlorobutane, does not form. Explain.

7. Ethane is insoluble in concentrated sulfuric acid. Ethylene dissolves readily. Write an equation to show how ethylene is converted into a substance polar enough to dissolve in the highly polar concentrated sulfuric acid.

8. Write out enough of the structures of each of the following substances to show what their molecules are like. Do not use parentheses to condense these structures.

   (a) polyethylene      (b) polypropylene      (c) Teflon
   (d) PVC      (e) polybutadiene      (f) polystyrene

9. Describe a simple chemical test that could be used to distinguish between 2-butene and butane. Describe what you would do and see.

# CHAPTER THREE

*Closed Ring & chiefly from Benzene*

# Aromatic Compounds.
# Theory of Resonance

## AROMATIC AND ALIPHATIC SUBSTANCES

The sixteenth century brought a reorientation of attitudes toward the proper ways for studying and using common substances. The search for the philosopher's stone with which gold might be prepared from baser things was abandoned under the pressure of repeated failures. Almost any ordinary substance was felt to be a proper object of study. The desire to find medicinal agents was certainly one driving motive.

By the seventeenth century a vast amount of information was already available. Furthermore, several techniques—distillation, extraction, pyrolysis, etc.—had been considerably improved. From this period Rudolph Glauber is one figure who stands out. Among his many accomplishments was a study of the action of heat on coal in which he prepared a "pleasant-smelling oil and valuable healing balsams." By treating various plants, Glauber learned the preparation of many aromatic oils. The fragrances of these oils were prized characteristics and made them much sought after.

---

Johann Rudolph Glauber (1604–1668). Born in Karlsdadt, the son of a barber and orphaned while a boy, Glauber was unusually prolific in technical accomplishments. In the many volumes of works he produced are described ways to make sulfuric acid, nitric acid, hydrochloric acid, many salts, including "Glauber's salt" or sodium sulfate decahydrate, and many organic substances, including aromatic oils and alkaloids. Robert Boyle is usually credited with being the father of modern chemistry. To some historians Glauber might justifiably be called a "German Boyle"; he was clearly a pioneer of organic chemistry.

Michael Faraday (1791–1867). The son of a blacksmith, his formal education consisting of but a few years of elementary school, Faraday ranks as a giant in chemistry and physics. Faraday's laws of electrolysis, the introduction of the field concept in magnetism, the conversion of magnetism into electricity—these accomplishments are the chief basis for his reputation. His discovery of benzene, his investigation of many of its chemical reactions, his inventions for liquefying gaseous chemicals, his introduction of the terms *electrolysis, electrolyte, ion, cation, anion, cathode,* and *anode,* and his contributions to the developing ideas concerning equivalent weights ensure him a high place in the history of chemistry.

---

We cannot know with certainty what Glauber's pleasant-smelling oil was. It could have been benzene or a mixture of it and related compounds. His healing balsams were probably phenolic compounds. The efforts of scientists during the next three centuries led to the discovery of a feature common to many of the pleasant-smelling and aromatic oils. In 1825 the Portable Gas Company of London turned over to a British scientist, Michael Faraday, an oily liquid that separated out when they compressed illuminating gas for storage and transportation in cylinders. Faraday purified this oil and analyzed it and, because of his care and accuracy, he is credited with the discovery of benzene. The name, however, was given to the oil a few years later by another chemist.

Several of the aromatic, pleasant-smelling oils obtainable from plant sources were much later found to consist of molecules similar in some ways to those of benzene. With time the term *aromatic compound* was applied to any substance whose molecules contained the benzene structural skeleton, even if the substance had an odor not especially pleasing. In fact, a huge number of aliphatic substances (e.g., esters, terpenes, and essential oils) have pleasant fragrances but do not contain the structural skeleton of benzene.

Although the terms *aliphatic* and *aromatic* originated in relation to natural sources of these materials, they are used by modern chemists to divide organic compounds into two great classes on a structural basis. Aromatic compounds are those that have the benzene structural skeleton or typically show the types of reactions peculiar to this system. Aliphatic compounds are the others we have been studying, the alkanes, alkenes, alkynes, and their cyclic relatives, in fact, any open-chain system or cyclic relative of whatever family.

## STRUCTURE OF BENZENE

Faraday discovered benzene in 1825 before there were successful theories about the ways atoms are joined together in molecules. The pioneers of structural theory in organic chemistry were Kekulé, Couper, Butlerov, and Crum Brown, and their work was still thirty to thirty-five years in the future. They gave us our present symbols, those we call Lewis structures because Lewis identified the straight lines in structural formulas with shared pairs of electrons. Benzene, however, remained a problem in structural theory until about 1931.

August Kekulé (1829–1896). Born in Darmstadt, the son of a grand-ducal Hessian head councillor in the War Office, Kekulé had the advantage of a gymnasium (roughly, high school) and university education. In 1857 he proposed the tetraval- ence of carbon, and the following year he explained the great diversity and number of carbon-containing compounds in terms of carbon's ability to form strong bonds to its own kind while also being able to bond to other atoms. A great lecturer as well as a brilliant scientist, he invented ball and stick models to present his theories of valence and structure.

Archibald Scott Couper (1831–1892). At the same time, and independent of Kekulé, Couper proposed essentially the same theory of carbon chain formation. He had his theory ready for presentation to the French Academy of Science early in 1858, but he was not a member and Academy rules permitted only members to present papers. Couper asked a member Charles Wurtz, to present his, but Profes- sor Wurtz procrastinated. In the meantime, Kekulé published his theory. Couper, understandably upset, apparently berated Wurtz who responded by expelling Couper from his laboratory. Couper then persuaded Jean Baptiste Dumas to present his paper to the French Academy in June of 1858, but Kekulé attacked it, insisting on being credited with earlier conception of a more significant theory. Couper suffered a nervous breakdown and for the rest of his life (thirty-four more years) he was incapable of significant intellectual work. Kekulé's claims were accepted and Couper was ignored by the scientific world of his day. Only much later did scientists realize that he deserved far more credit than he received.

Alexander Mikhailovich Butlerov (1828–1886). Born near Kazan, son of a retired army officer, Butlerov's early specialization in butterflies gave way to an interest in chemistry. By 1857 he was a professor of chemistry at the University of Kazan. Foreign travels took him on visits to both Kekulé and Couper shortly before their publications on organic structures. Butlerov further extended these theories, and he was the first chemist to realize that each compound must have its own unique structure. In fact, he invented the term "chemical structure," and his was the first textbook to use structural ideas throughout. He is also remembered for contributions to our understanding of isomerism and tautomerism. Vladimer Markovnikov was his student.

Alexander Crum Brown (1838–1922). Of Scottish parentage, like Couper, Crum Brown was a student of Bunsen and Kolbe. He was apparently the first to designate every valence in an organic structural formula by a separate line, the line G. N. Lewis decades later identified with a shared pair of electrons.

---

The benzene problem was one of theory making. Structural theory that worked so well for an enormous number of aliphatic compounds simply could not be applied to benzene. Certain known facts about benzene did not correlate well with any simple, Lewis-type structure. Kekulé and Couper worried enough about the benzene problem to devise a theory which, at least for a time, seemed to bring it into line with traditional theory. To understand what they did and why it was not enough, we must survey some of the facts about benzene that show the nature and extent of the benzene problem.

1. The molecular formula of benzene is $C_6H_6$. With six carbons holding only six hydrogens, the molecule should be highly unsaturated. Addition reactions such as those described for the alkenes should be the rule, but instead they are the rare exception.

2. Benzene undergoes substitution reactions which are catalyzed by acids. When monosubstituted benzenes are made (of the general type $C_6H_5$—G; G = —Br, —Cl, —$NO_2$, etc.), no isomers form, which means that all six hydrogens in benzene must be equivalent. The principal aromatic substitution reactions are the following.

Nitration:

$$C_6H_6 + HO-NO_2 \xrightarrow[50-55°C]{H_2SO_4 \text{ (concd)}} C_6H_5-NO_2 + H_2O$$

Nitric acid             Nitrobenzene (85%)
(bp 211°C)

Under these conditions alkenes undergo extensive oxidation and decomposition.

Sulfonation:

$$C_6H_6 + SO_3 \xrightarrow[\text{room temperature}]{H_2SO_4 \text{ (concd)}} C_6H_5-\overset{\overset{\displaystyle O}{\uparrow}}{\underset{\underset{\displaystyle O}{\downarrow}}{S}}-O-H$$

Benzenesulfonic acid (56%)
(mp 525°C)

Sulfuric acid *adds* to double bonds in alkenes to form alkyl hydrogen sulfates.

Halogenation:

$$C_6H_6 + Cl_2 \xrightarrow[FeCl_3]{Fe \text{ or}} C_6H_5-Cl + H-Cl$$

Chlorobenzene (90%)
(bp 132°C)

$$C_6H_6 + Br_2 \xrightarrow{Fe} C_6H_5-Br + H-Br$$

Bromobenzene (59%)
(bp 155°C)

$$C_6H_6 + I_2 \xrightarrow{HNO_3} C_6H_5-I + HI \quad \text{(oxidized as it forms by nitric acid)}$$

Iodobenzene (87%)
(bp 189°C)

Chlorine and bromine add to an alkene with no catalyst required. Iodine is generally unreactive toward a carbon-carbon double bond. Iodination of benzene occurs only if H—I is removed as soon as it forms. That is the function of the nitric acid.

Alkylation, Friedel-Crafts reaction:

$$C_6H_6 + R—Cl \xrightarrow{AlCl_3} C_6H_5—R + H—Cl$$

Alkylbenzenes

(This reaction will be discussed in greater detail later.)

Acylation, Friedel-Crafts acylation:

$$C_6H_6 + R—\overset{\displaystyle O}{\overset{\displaystyle \|}{C}}—Cl \xrightarrow{AlCl_3} C_6H_5—\overset{\displaystyle O}{\overset{\displaystyle \|}{C}}—R + H—Cl$$

Aromatic ketones

(This reaction will be discussed in Chapter 6.)

Benzene is not the only compound reacting in these ways, that is, undergoing substitution reactions for the most part rather than addition reactions. Reactions such as the seven just given are characteristic of aromatic compounds in general. Benzene is unusually stable toward oxidizing agents, which cause extensive changes to alkenes.

3. When a monosubstituted benzene, $C_6H_5$—G, is made to form a disubstituted product, either $C_6H_4G_2$ or $G_6H_4GZ$, three isomeric substances can be isolated.

**The Kekulé Structure of Benzene.** Kekulé proposed structure **1** for benzene as

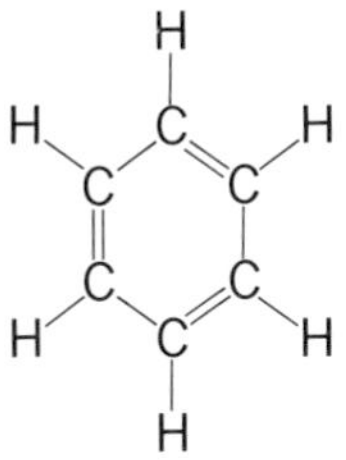

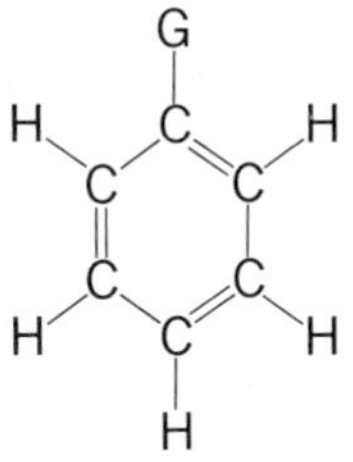

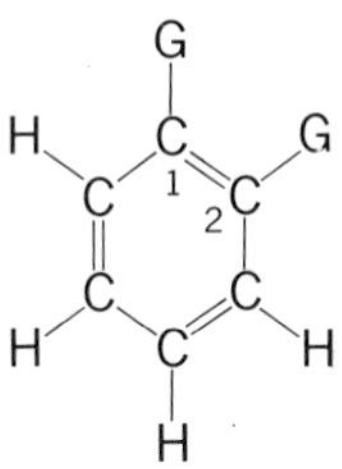

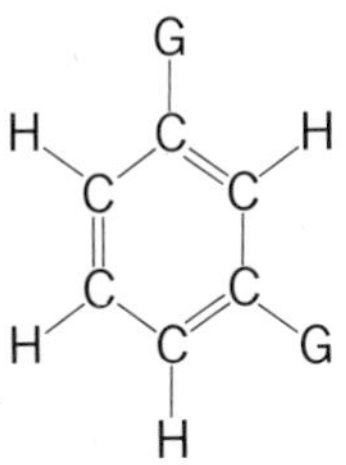

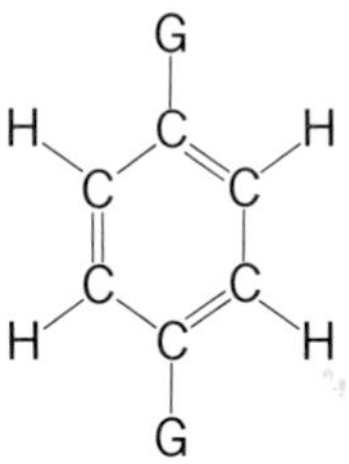

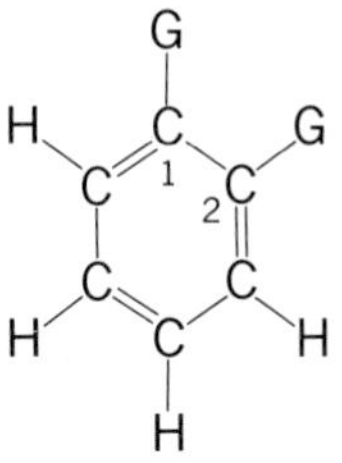

being the most consistent with the evidence. All six hydrogens are equivalent, and the structure of a monosubstitution product would be of type **2**. Possible disubstitution products would be **3, 4,** and **5**. But it was rightly pointed out by Kekulé's critics that if his structure for benzene were correct, the 1,2-disubstituted product **3** should have an isomer **6** which differed from it only in the location of the double bonds. In **3** the carbons holding the two groups are joined by a double bond; in **6** they are joined by a single bond. Kekulé therefore modified his theory by saying that the double bonds rapidly shift back and forth so that **3** and **6** are isomers in such rapid and mobile equilibrium that they can never be separated. Sometime later isomers that readily interconvert at ordinary temperatures but can by very careful work be separated were found in other chemical systems. (Such isomers came to be called *tautomers,* the phenomenon *tautomerism,* and the interconversion a *tautomeric change.* "Tauto-" is another word part meaning "the same"; *meros,* we have learned, means "part.") The fact that tautomers can sometimes be separated lent support to Kekulé's explanation, and it was accepted for decades. But his theory did not solve the problem, for it failed to explain the remarkable nonalkene-like properties of a structure written with three double bonds.

Kekulé's structure for benzene is what we would draw if we were asked to write a structure for 1,3,5-cyclohexatriene. Open-chain trienes are known and are, like dienes, quite reactive toward reagents that add to a double bond. Cyclic dienes are also known, and they undergo addition reactions. But all efforts to make 1,3,5-cyclohexatriene lead instead to benzene. Thus, although Kekulé's formulation of benzene answered many questions and served very usefully for decades, it left some important questions unresolved. Benzene will undergo one important addition reaction, however—hydrogenation to cyclohexane.[1]

**Hydrogenation of Benzene. Heats of Hydrogenation.** Hydrogenation of a double bond is an exothermic process, and it is possible to measure the heat evolved. For carbon-carbon double bonds in most monoalkenes, between 27 and 32 kcal/mole are evolved. Consider, then, the following specific cases. In each the final product is the same, cyclohexane. The heat of hydrogenation of cyclohexene, $-28.6$ kcal/mole, is taken as the value for 1 mole of double bond in a six-membered ring. 1,3-cyclohexadiene with two double bonds should release $2 \times 28.6 = 57.2$ kcal/mole. "1,3,5-Cyclohexatriene," with three ordinary double bonds, should release $3 \times 28.6 = 85.8$ kcal/mole.

Cyclohexene     Cyclohexane     $\Delta H_{obsd} = -28.6$ kcal/mole

1,3-Cyclohexadiene     Cyclohexane     $\Delta H_{obsd} = -55.4$ kcal/mole

$\Delta H_{calcd} = -57.2$ kcal/mole

[1] In the presence of ultraviolet light, benzene will also add chlorine to form "benzene hexachloride" or BHC. This compound is really 1,2,3,4,5,6-hexachlorocyclohexane and consists of several geometric isomers. One of the isomers is a potent insecticide, and for this reason BHC is an important commercial chemical sold under the names gammexane and lindane.

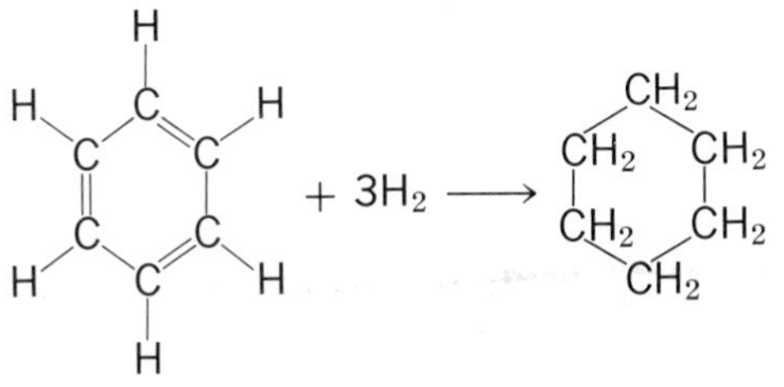

$\Delta H_{obsd} = -49.8$ kcal/mole

$\Delta H_{calcd} = -85.8$ kcal/mole

"1,3,5-Cyclohexatriene"           Cyclohexane
(benzene)

Cyclohexene is rapidly hydrogenated at room temperature over nickel at only 20 lb/in.² of pressure. Benzene requires a pressure several hundred times as great and a much higher temperature. Morever, a great deal less energy is liberated than we might expect on the basis of its being 1,3,5-cyclohexatriene. Since the products in the three examples are identical, the evolution of 36 kcal/mole fewer units of energy for benzene must mean that it has this much less internal energy initially— that is, this much less energy than calculated for a hypothetical 1,3,5-cyclohexatriene. If it has less internal energy than the expected structural system, it must be more stable than that system. Benzene is obviously not well represented by the Kekulé structures.

**Bond Lengths in Benzene.** If benzene molecules had either Kekulé structure with alternating double and single bonds, X-ray diffraction analysis should reveal two different values for carbon-carbon bond distances. Only one is observed, however, 1.39 Å. This result shows that each carbon-carbon bond is equivalent, and that it is intermediate between a single bond (1.54 Å) and a double bond (1.34 Å). Again, the Kekulé structure is inadequate.

## THE THEORY OF RESONANCE

It is clear that a substance such as benzene makes some modification of classical structural theory necessary. Of the approaches available, one is called resonance theory. Organic chemists have found it to be very useful for theorizing about properties of aromatic compounds. According to this theory, whenever two or more structures with the same relative locations of nuclei but different distributions of electrons can be drawn for a molecule, a *condition of resonance* is said to exist.

Two Kekulé structures of benzene, **7** and **8**, differing only in the locations of electrons can be drawn. Two carbons are numbered to show that although **7** and **8** are equivalent, they are not identical. The $C_1—C_2$ bond in **7** is single, in **8** double.

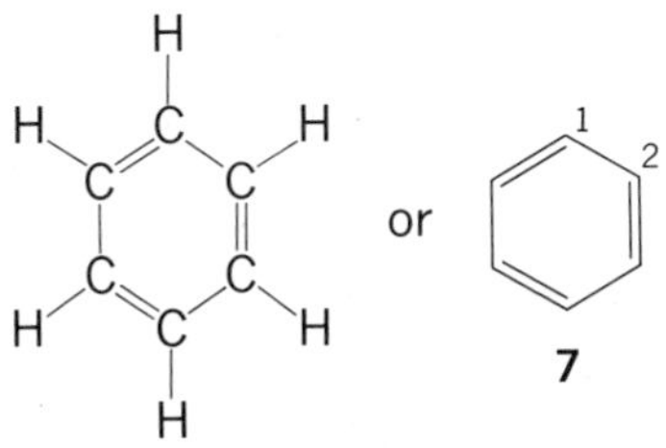

7

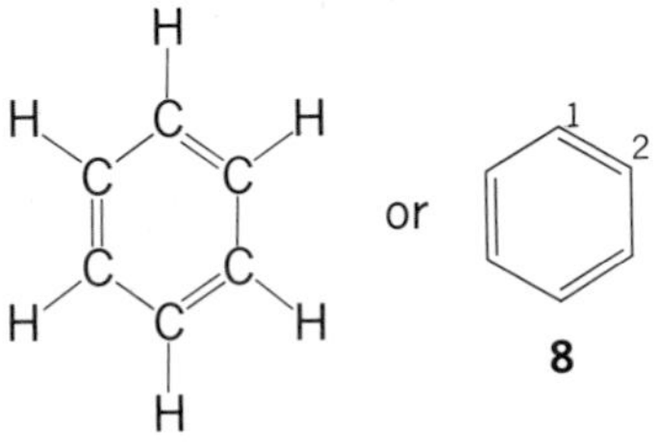

8

Chemists have learned through experience that whenever it is possible to draw two or more such structures differing only in the locations of electrons, neither structure adequately represents the molecule. A good approximation of the real molecular structure, however, can be made by thinking of it as a *hybrid,* a resonance hybrid, of the classical Kekulé or Lewis structures. These, instead of being regarded as real, are said to be *contributing structures.* They "contribute" to the hybrid. That is, they contribute to our mental picture of what the real molecule is probably like. As a good rule of thumb, the more nearly equivalent such contributing structures are, the less likely does any one of them represent the true structural state of affairs and the more seriously must we take the hybrid we mentally put together from the contributors. How the hybrid structure for benzene might be put together and represented on paper is described in Figure 3.1. Note especially the new symbol, a double-headed arrow: ⟷ .We cannot use two arrows pointing in opposite directions because that is our symbol for equilibrium between two or more real substances. The double-headed arrow is our signal that the structures at both ends are contributors to a resonance hybrid. Moreover, we shall usually follow the practice of having the structure on the right develop from the one on the left by the relocation of electrons, indicated in the first structure by curved arrows. If several contributors are drawn, the development will be step by step from left to right.

**Resonance Energy.** When resonance occurs, the substance has less internal energy and is more stable than would be calculated or estimated on the basis of its being any single one of its contributors. The heats of hydrogenation have illustrated this point for benzene, and numerous other examples could be cited. The difference between the predicted and the actual internal energy is called the *resonance energy* of the substance. The substance does *not* have this energy because it does not have the structure used to make the calculation, the best structure that could be written if classical rules of valence were employed. It has instead some nonclassical, hybrid structure.

The more nearly equivalent the contributing structures, the greater this resonance energy of stabilization. The generally accepted value of resonance energy for benzene in which the two contributors are equivalent is 38 kcal/mole—very close to the figure of 36 kcal/mole obtained from heats of hydrogenation. By "equivalent" is meant having equal or nearly equal internal energies—if each contributor had a real existence. We must remember that drawing potential contributing structures is purely a pencil and paper operation. Just because a structure can be drawn does not mean that it is an important contributor. Starting with one Kekulé structure for benzene, we could, on paper, develop the following:

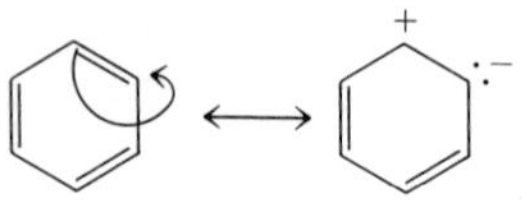

But the second structure is ignored because, if it could exist, it would be much less stable than the first. The separation of a plus and a minus charge is an energy-demanding process. Like charges attract. Thus, although the second structure differs

Step 1. The contributing structures must be drawn before any hybrid can be fashioned. The trick is to be able to write one Kekulé or Lewis structure and then develop the others from it.

$a$          $b$

The curved arrows in the first indicate how pairs of electrons are relocated to write the second.

Step 2. To start the hybrid, examine the contributors and write down everything that is the same in all.

Step 3. Next note the differences. In structure $a$ the $C_1$—$C_2$ bond is single, in $b$ double. Make it a partial double bond in the hybrid by drawing a dotted line (instead of a solid line) for the second of the two lines.

Step 4. Continue this dotted line as in step 3 around each pair of adjacent positions in structures $a$ and $b$; the resulting hybrid for benzene is as follows:

**Figure 3.1**  Developing the hybrid structure of benzene from the contributing structures.

from the first only in the distribution of electrons, it makes no significant contribution to the hybrid, in comparison with the Kekulé structures.

We shall not develop resonance theory any further at this point. The student is not expected to become skilled in writing contributing structures and then selecting those that would contribute most to the hybrid. It is hoped, however, that he will be able to follow arguments based on the use of resonance theory. Before continuing, we shall summarize what we have done and not done in this introduction to the theory of resonance.

The minimum expectation is that the problem making *some* theory necessary be recognized. If a single structure is to mean something, we do not want it to mean everything. If when we draw a carbon-carbon double bond we want it to mean (in part) "addition reactions," we certainly do not want it to mean "no addition reactions" at the same time. But to represent benzene as one or the other of the Kekulé structures confronts us with precisely such a contradiction. Resonance theory is one attempt to solve the contradiction by allowing us to use already familiar symbols, the classical Kekulé or Lewis structures, as a means for making an educated guess at what the actual structure probably is. From the contributors we mentally, or on paper, construct a nonclassical hybrid as a better structure for the substance. We have not, however, developed and illustrated the rules and guidelines by which contributors are drawn and evaluated. We shall formulate them to some extent in our future use of the theory of resonance.

## ORBITAL MODEL OF BENZENE

Where resonance occurs, a stabilizing factor is present, and an alternative theory, the molecular orbital theory, provides insights concerning it. Since the carbons in benzene are individually attached to three other atoms, $sp^2$ hybridization may be assumed for them. A planar system with bond angles of 120° would then be expected, and this is precisely the internal angle of a regular hexagon. With overlapping of unhybridized $p_z$ orbitals (Figure 3.2), benzene acquires a large circular molecular orbital instead of three isolated, localized double bonds. Instead of the pi electrons being localized as implied by a Kekulé structure, they are considerably delocalized throughout a much larger volume of space. It is believed that this delocalization of electrons into a cyclic molecular orbital largely accounts for the "extra" stability of the benzene molecule. In fact, the term *delocalization energy* is sometimes used synonomously with resonance energy.

This orbital theory gives us further insight into the structure of benzene, but only with difficulty can we use this theory to understand the reactions of aromatic systems. Resonance theory, on the other hand, permits us to employ conventional

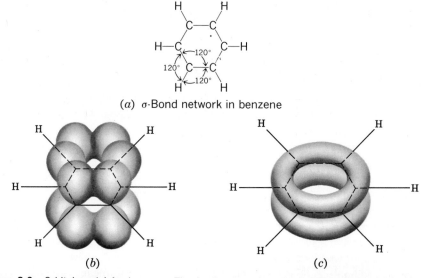

(a) σ-Bond network in benzene

(b)

(c)

**Figure 3.2** Orbital model for benzene. The fundamental framework of the molecule is provided by σ-bonds shown in part a. At each carbon is located an unhybridized $p_z$ orbital which can overlap with $p_z$ orbitals on both neighboring carbons as shown in part b. The result is the double doughnut-shaped molecular orbital shown in part c. This is actually only one of the new molecular orbitals that forms. For our purposes we may treat the shape shown in part c as a molecular shell containing three subshells with two electrons in each.

structures for deducing the hybrid and in the long run is easier to apply. It is common practice, however, to write the structure of benzene and its derivatives, not with three double bonds showing but rather with a dotted or solid circle in the hexagon as indicated by structures **9** and **10**. We follow this practice except when discussing the mechanism of an aromatic substitution reaction.

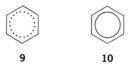

**9**          **10**

## NAMING BENZENE DERIVATIVES

**Monosubstituted Compounds.** For a few compounds the nomenclature has some system. Prefixes naming the substituent are joined to the word "benzene":

Nitrobenzene    Fluorobenzene    Chlorobenzene    Bromobenzene    Iodobenzene

Other derivatives have common names that are always used, even though systematic names are possible:

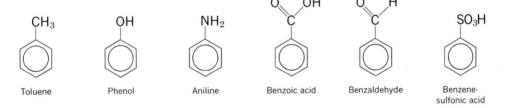

Toluene    Phenol    Aniline    Benzoic acid    Benzaldehyde    Benzene-sulfonic acid

**Disubstituted Compounds.** When two or more groups are attached to the same benzene ring, both their nature and their relative locations must be specified in the name. The prefixes "ortho," "meta," and "para" are used to distinguish, respectively, the 1,2-, the 1,3-, and the 1,4- relations. These prefixes are usually abbreviated *o*-, *m*-, and *p*-; in the following examples both substituent groups are the same.

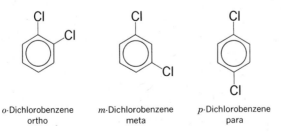

*o*-Dichlorobenzene    *m*-Dichlorobenzene    *p*-Dichlorobenzene
ortho              meta               para

If one of the two groups, alone on the benzene ring, would give it a common name, this name is used and the second group is designated as a substituent. The following examples illustrate this point.

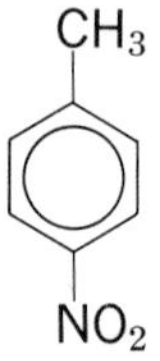

p-Nitrotoluene 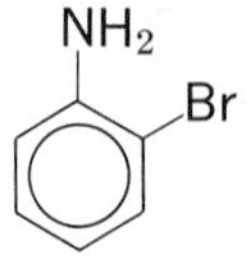     o-Bromoaniline 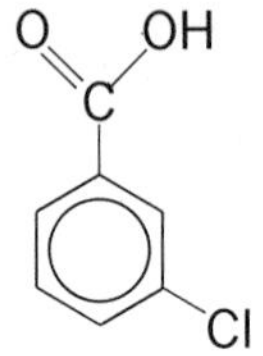     m-Chlorobenzoic acid 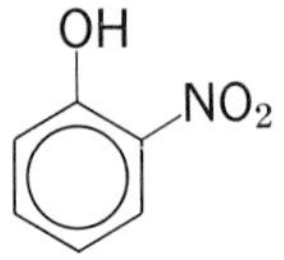     o-Nitrophenol

If neither group would be associated with a common name were it alone on the benzene ring, both groups are named and located in the name. For example,

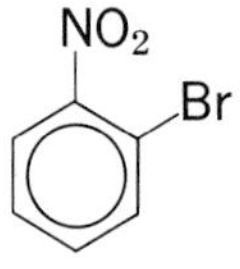

o-Bromonitrobenzene 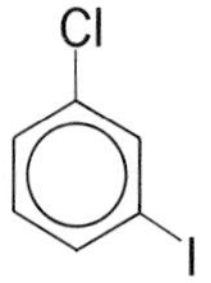     m-Chloroiodobenzene 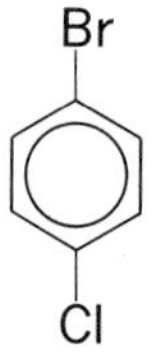     p-Chlorobromobenzene 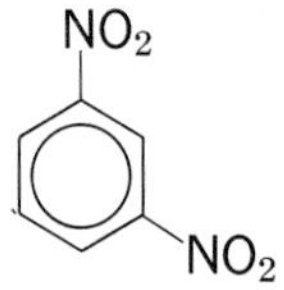     m-Dinitrobenezene

**Polysubstituted Benzenes.** If three or more groups are on a benzene ring, the ring positions must be numbered. When one group is associated with a common name, its position number is 1. For example,

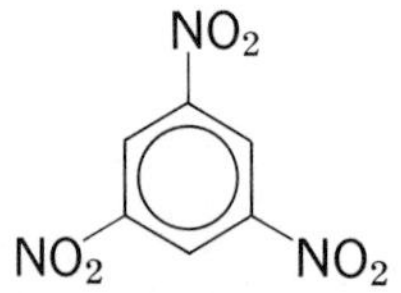

1,3,5-Trinitrobenzene      2,4,6-Trinitrotoluene      2-Bromo-4-nitrophenol
(TNB)                      (TNT)

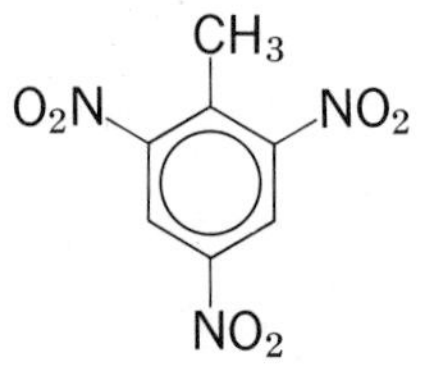

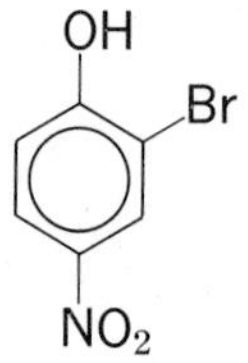

## MECHANISM OF ELECTROPHILIC AROMATIC SUBSTITUTION

The commonly accepted mechanism for nearly all electrophilic aromatic substitutions involves a succession of three steps.

Step 1. Generation of an electrophilic or electron-poor species. The catalyst does its work during this step.

Step 2. Attack by the electrophilic species on the benzene ring. The benzene ring, with its huge pi-electron cloud, is very attractive to the electron-poor electrophilic species. During this step the benzene ring temporarily loses its aromaticity, its circular pi-electron system. A carbonium ion forms which, though resonance-stabilized, is still of relatively high internal energy, like all carbonium ions.

Step 3. The carbonium ion loses a proton from a ring position, and the ring recovers its aromatic character and its pi-electron system.

To illustrate these steps, we shall discuss several mechanisms.

**The Mechanism of the Nitration of Benzene.** When concentrated nitric acid and concentrated sulfuric acid are mixed, freezing point depression studies and other analyses show that the following equilibrium is established. This is step 1 in the nitration of benzene. The nitronium ion which is formed is a powerful electrophilic species.

Step 1.

$$HO-NO_2 + H_2SO_4 \rightleftharpoons NO_2^+ + H_3O^+ + 2HSO_4^-$$

    Nitric acid    Sulfuric acid    Nitronium
                                        ion

Step 2. Attack by the electrophilic nitronium ion on a molecule of benzene.[2]

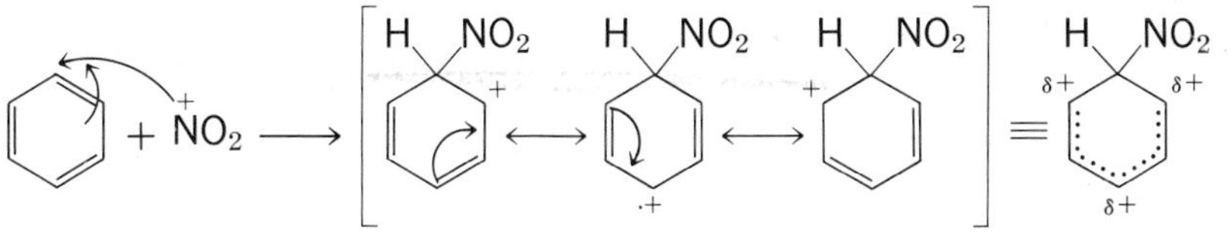

Contributing structures for
resonance-stabilized carbonium ion

Hybrid ion
(The three $\delta+$
total $1+$.)

Step 3. Loss of a proton, recovery of benzene system, and formation of product.

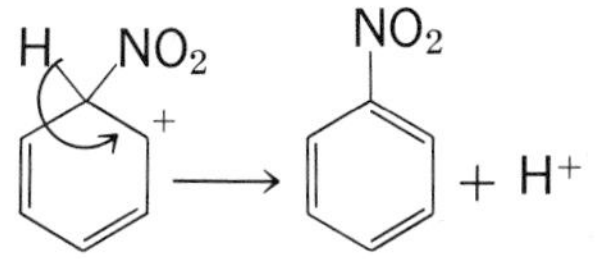

Nitrobenzene

**The Mechanism of the Chlorination of Benzene.** Iron(III) chloride is a Lewis acid and reacts with chlorine to establish the following equilibrium. Because $Cl^+$ is a strongly electrophilic species, this is the first step in the mechanism. (In all likelihood $Cl^+$ and $FeCl_4^-$ exist as a closely associated ion pair.)

Step 1.        $Cl-Cl + FeCl_3 \rightleftharpoons Cl^+ + FeCl_4^-$

Step 2. Attack by $Cl^+$ on the benzene ring.

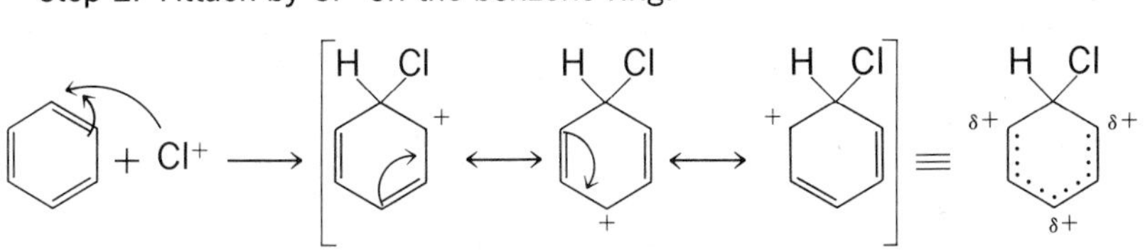

Contributing structures for
resonance-stabilized carbonium ion

Hybrid
ion

[2] In this step the student should be able to follow an argument based on resonance theory, even though he may not be able to develop the contributing structures for the carbonium ion.

Step 3. Loss of a proton from the ring and formation of a resonance-stabilized product.

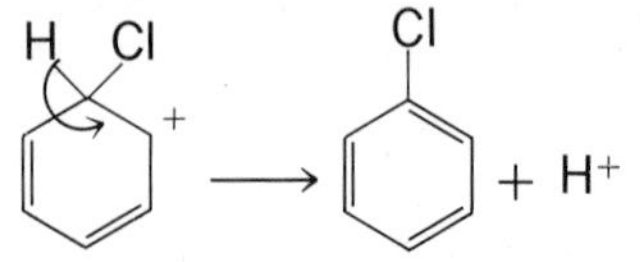

Chlorobenzene

**The Mechanism of the Friedel-Crafts Alkylation of Benzene.**[3] Anhydrous aluminum chloride, a powerful Lewis acid, reacts with alkyl halides in a manner that with some oversimplification may be represented as follows. A carbonium ion, a strongly electrophilic species, is generated.

Step 1.

$$R\text{—}Cl + AlCl_3 \rightleftharpoons R^+AlCl_4^-$$

Step 2. Attack by the carbonium ion, $R^+$, on a molecule of benzene to form a more stable carbonium ion.

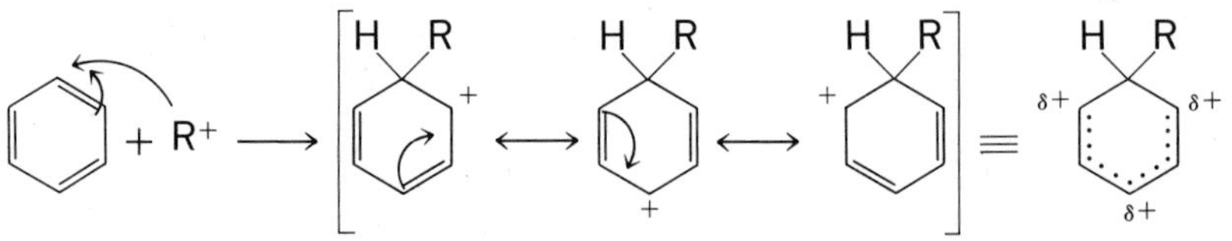

Contributing structures for
resonance-stabilized carbonium ion

Hybrid
ion

Step 3. Loss of a proton from the benzene ring to form an alkylbenzene.

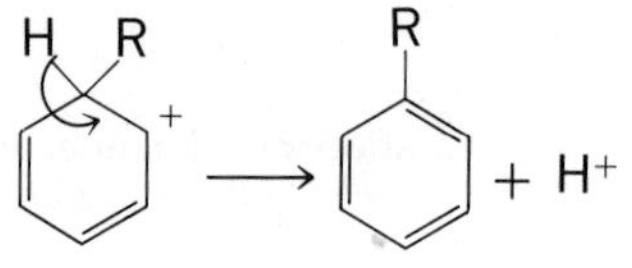

Alkylbenzene

**Carbonium-Ion Rearrangements.** The following examples illustrate the course of some typical Friedel-Crafts alkylations.

$$C_6H_6 + CH_3Cl \xrightarrow{AlCl_3} C_6H_5CH_3 + HCl$$

Toluene

$$C_6H_6 + CH_3CH_2Cl \xrightarrow{AlCl_3} C_6H_5CH_2CH_3 + HCl$$

Ethylbenzene

[3] Charles Friedel (1832–1899), a French chemist, and James Mason Crafts (1839–1917), an American, discovered this reaction in 1877.

$$C_6H_5 + CH_3CH_2CH_2Cl \xrightarrow{\text{AlCl}_3 \ (35\%)}$$

or $CH_3\underset{\underset{Cl}{|}}{CH}CH_3$

$$C_6H_5CH(CH_3)_2 + C_6H_5CH_2CH_2CH_3 + HCl$$

Isopropylbenzene (29%)      *n*-Propylbenzene
                              (19%)

**Reactions of
Monosubstituted
Benzenes**

$$C_6H_6 + \quad \begin{matrix} CH_3CH_2CH_2CH_2Cl \\ \text{or} \\ CH_3CH_2\underset{\underset{Cl}{|}}{CH}CH_3 \end{matrix} \quad \xrightarrow[\text{heat}]{\text{AlCl}_3} \quad C_6H_5\underset{\underset{CH_3}{|}}{CH}CH_2CH_3 + HCl$$

*sec*-Butylbenzene (only monoalkylbenzene
obtained, even from
*n*-butyl chloride) (50%)

Thus, in the Friedel-Crafts alkylation of benzene with *n*-propyl chloride, the principal product is isopropylbenzene; and when *n*-butyl chloride is used, the only monoalkyl product is *sec*-butylbenzene. In both alkylations the carbonium ion we would expect to form in step 1, a primary carbonium ion, cannot be the one that becomes attached to the benzene ring. To account for this discrepancy, the carbonium-ion theory is extended. It is postulated that the initially formed primary carbonium ion rearranges to form a more stable ion—in both instances a secondary carbonium ion—by a hydride-ion shift. Thus, for *n*-butylcarbonium ion,

$$CH_3-CH_2-\overset{\frown}{\underset{}{CH}}-CH_2^+ \longrightarrow CH_3-CH_2-\overset{\cdot\cdot H\cdot\cdot}{\underset{+}{CH}}-CH \longrightarrow CH_3-CH_2-\overset{+}{CH}-\overset{\overset{H}{|}}{CH_2}$$

**11**                              **12**                              *sec*-Butyl-
                                                                        carbonium ion
*n*-Butylcarbonium          Migration of "hydride
ion                          ion" partially complete

Before the hydrogen with its bonding pair of electrons breaks away, it starts to form a new bond at the carbon bearing the initial positive charge (**11** → **12**). A *free* hydride ion, H:⁻, does not actually form, and the "migration" from one side to another proceeds no farther down the chain than from one carbon to the next; hence it is called a 1,2-shift. An alkyl group will migrate if there is no available hydrogen at the carbon adjacent to the positive charge.

Carbonium ions do not rearrange unless a more stable ion can be produced. Thus they may go from 1° to 2° or from 2° to 3°. Since the carbonium ion is one of the most important intermediates in organic chemistry, we must know all the reactions of which it is capable. Those studied thus far are

—Combination with some electron-rich species, as in addition reactions of alkenes or the Friedel-Crafts alkylation.
—Rearrangement to a more stable carbonium ion.

## REACTIONS OF MONOSUBSTITUTED BENZENES

**Orientation of Further Substitution.** Monosubstituted benzenes undergo the same types of substitution reactions as benzene itself. Nitrobenzene can be nitrated to dinitrobenzene. There are three different positions at which the second nitro group could become attached: either of the ortho positions, either of the meta sites, and the para. Statistically, therefore, we might expect the three isomeric dinitro-benzenes, ortho, meta, and para, to form in a ratio of 2:2:1. The experimentally determined ratio, however, is 6:93.5:0.5. The meta isomer clearly predominates.

When toluene is nitrated, however, very little meta isomer forms. The isomeric nitrotoluenes, ortho, meta, and para, form in the ratio 58:38:4. A mixture of o-nitrotoluene and p-nitrotoluene is 96% of the product.

In the two examples given, nitration of nitrobenzene and toluene, the nature of the incoming electrophilic species, nitronium ion, is the same. Only the nature of the group already on the ring is changed. In general, *where* a second substituent becomes attached depends on what the first substituent is, not on the incoming group.

In Table 3.1 are summarized experimental data for the composition of the isomer mixtures that form when various monosubstituted benzenes are nitrated. The substituents already on the ring fall into two groups. The first group consists of the *ortho-para directors*, the second of the *meta directors*.

The data apply chiefly to nitration. Sulfonation, bromination, and other sub-stitutions into a monosubstituted benzene vary in product composition, but for all these reactions the ortho-para directors and the meta directors are the same.

---

Exercise 3.1    Predict the structures of the principal organic products that would form if fur-ther substitution were carried out on the following monosubstituted benzenes. Write the names of the products.

(*a*)  Nitration of toluene.

(*b*)  Chlorination of nitrobenzene.

(*c*)  Sulfonation of phenol.

(*d*)  Bromination of $C_6H_5NH\overset{\overset{\textstyle O}{\|}}{C}\!-\!CH_3$    (acetanilide).

(*e*)  Chlorination of chlorobenzene.

(*f*)  Nitration of $C_6H_5CN$    (benzonitrile).

(*g*)  Friedel-Crafts alkylation of ethylbenzene with methyl iodide.

(*h*)  Sulfonation of nitrobenzene.

(*i*)  Bromination of methyl benzoate ($C_6H_5\overset{\overset{\textstyle O}{\|}}{C}\!-\!O\!-\!CH_3$).

---

**Relative Reactivities in Additional Aromatic Substitutions.** Having first studied *where* the second group enters, we next consider *how rapidly* it goes there, com-pared with how rapidly it would attack benzene itself. Toluene is nitrated ten to twenty times faster than benzene under identical conditions. Nitrobenzene, in con-trast, is much more slowly nitrated than benzene. What is already on the ring therefore affects its reactivity in additional substitutions. Some groups *activate* the

**Table 3.1  Orientation in the Nitration of Monosubstituted Benzenes**

$$C_6H_5G + HNO_3 \xrightarrow{\text{catalyst}} C_6H_4GNO_2 + H_2O$$

Mixture of
o-, m-, and
p-disubstituted
benzenes

| G (Director) | Composition of Mixture of Distributed Benzenes | | | |
| --- | --- | --- | --- | --- |
| | Ortho | Para | Total Ortho plus Para | Meta |
| Ortho-para directors | | | | |
| —OH | 50–55 | 45–50 | 100 | trace |
| —NH—C(=O)—CH$_3$ | 19 | 79 | 98 | 2 |
| —CH$_3$ | 58 | 38 | 96 | 4 |
| —F | 12 | 88 | 100 | trace |
| —Cl | 30 | 70 | 100 | trace |
| —Br | 38 | 62 | 100 | trace |
| —I | 41 | 59 | 100 | trace |
| Meta directors | | | | |
| —$\overset{+}{N}(CH_3)_3$ | 0 | 0 | 0 | 100 |
| —NO$_2$ | 6.4 | 0.3 | 6.7 | 93.3 |
| —C≡N | — | — | 11.5 | 88.5 |
| —SO$_3$H | 21 | 7 | 28 | 72 |
| —C(=O)—OH | 19 | 1 | 20 | 80 |
| —C(=O)—H | — | — | 21 | 79 |

Data from R. T. Morrison and R. N. Boyd, *Organic Chemistry,* second edition, Allyn and Bacon, Boston, 1966, page 344. Used by permission.

ring, and others *deactivate* it. The terms activation and deactivation used in this context imply comparison with benzene.

All the meta directors deactivate the ring in varying degrees. With the exception of one "family," the halogens, all the ortho-para directors activate the ring in varying degrees. The halogens, although ortho-para directors, are deactivators. Table 3.2 summarizes these effects.

---

Exercise 3.2   In preparing disubstituted benzenes, starting with benzene, the order in which the two groups are put on the ring is obviously important. The following questions relate to this problem.

(*a*)  What will the final product be if benzene is first brominated and then that product is nitrated?

(*b*)  What will the final product be if benzene is first nitrated and then that product is brominated?

(*c*)  Predict the principal organic products in each of the following.

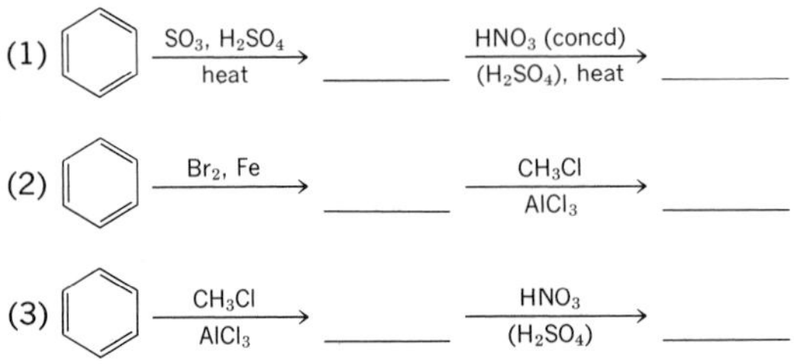

(1)  [benzene] $\xrightarrow[\text{heat}]{\text{SO}_3,\ \text{H}_2\text{SO}_4}$ _____ $\xrightarrow[\text{(H}_2\text{SO}_4),\ \text{heat}]{\text{HNO}_3\ \text{(concd)}}$ _____

(2)  [benzene] $\xrightarrow{\text{Br}_2,\ \text{Fe}}$ _____ $\xrightarrow[\text{AlCl}_3]{\text{CH}_3\text{Cl}}$ _____

(3)  [benzene] $\xrightarrow[\text{AlCl}_3]{\text{CH}_3\text{Cl}}$ _____ $\xrightarrow[\text{(H}_2\text{SO}_4)]{\text{HNO}_3}$ _____

## THEORY OF ORIENTATION AND REACTIVITY IN AROMATIC ELECTROPHILIC SUBSTITUTION

Orientation has to do with *where* a group or atom becomes attached if two or more different sites are available. Reactivity has to do with *how rapidly* the group or atom becomes attached as compared with the rate of reaction with a reference compound. Benzene is the reference for aromatic electrophilic substitutions. Even orientation is a question of relative rate. To say that a nitro group is meta-orienting is to say that an incoming group has a greater number of successful collisions at a position meta to the nitro group than at either of the ortho positions or the para position. The question "How fast?" is in the domain of *chemical kinetics,* a branch of chemistry that attempts to relate reaction rates to properties of reactants and to

**Table 3.2  Aromatic Electrophilic Substitution, Activating or Deactivating Influences of Groups Attached to Benzene**

| Activating groups (all ortho-para directors) | | |
|---|---|---|
| Strong activators | Moderate activators | Weak activators |
| $-NH_2$, $-NHR$, $-NR_2$ | $-OR$ | $-C_6H_5$ |
| $-OH$ | $-NH-\overset{\displaystyle O}{\overset{\displaystyle \|}{C}}-R$ | $-R$   (alkyl) |

Deactivating groups

Meta directors: $-\overset{+}{N}R_3$, $-NO_2$, $-CN$, $-SO_3H$, $-\overset{\displaystyle O}{\overset{\displaystyle \|}{C}}-OH(R)$, $-\overset{\displaystyle O}{\overset{\displaystyle \|}{C}}-H$, $-\overset{\displaystyle O}{\overset{\displaystyle \|}{C}}-R$

Ortho-para directors: $-F$, $-Cl$, $-Br$, $-I$

experimental conditions. To understand the facts of orientation and reactivity in aromatic substitutions or in any other reactions, we must first study some general features of a theory of reaction rates.

**Collision Theory.** Let us take a very general reaction between the two molecules A and B—C:

$$A + B—C \longrightarrow A—B + C$$

We start by mixing together samples of A and B—C, and in any actual experiment (in view of the size of Avogadro's number) hundreds and thousands of billions of molecules of both types will be present.

The molecules are not moving in the same direction but in all directions; there is a random distribution of the directions of motion. The molecules are not moving at the same speeds but at all speeds; there is a random distribution of speeds. Some are moving slowly, some moderately fast, and some with extreme rapidity. Some may be momentarily motionless. Molecules collide at random and at varying angles, or so we assume. Considerable evidence supports our assumption.

Collisions between molecules of A and B—C take place constantly. One experimental factor that affects the frequency of collisions is the concentration of reactants; the greater the concentrations, the more frequent the collisions. Not all collisions will be lined up properly. For A—B to form in our example, A must strike the B end of B—C. This variable cannot be controlled by adjusting the conditions except in some reactions in which catalysts are employed. Because the orientation of collision can vary, only a fraction of all collisions will be successful. Success, of course, means production of A—B and C. Collisions must also occur with sufficient energy if they are to be successful. Some collisions may be mere nudges, some moderate whacks, others violent crashes. Each collision will have a particular *collision energy,* and these values will vary from very low to very high. A small fraction of all collisions at any instant will occur between molecules with very little energy, and a small fraction will be at the other extreme. Figure 3.3 shows graphically how the fraction of collisions having a particular energy varies with that energy.

**Energy of Activation.** For the reaction between A and B—C to take place, the bond in B—C must be broken, but to break a bond requires energy. A new bond

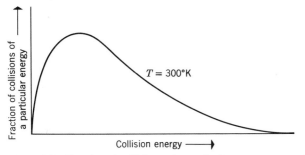

**Figure 3.3** Fraction of collisions of a particular energy versus the collision energy at a given temperature.

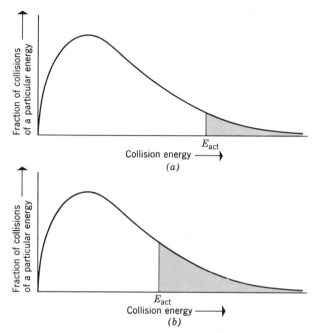

**Figure 3.4** Energy of activation versus fraction of *successful* collisions. The total number of collisions per cubic centimeter per second is represented by all the area under the curve in either part *a* or part *b*. Only the shaded area in each represents the total number of *successful* collisions, those leading to products. The *fraction* of successful collisions is the ratio of the shaded area to the total area. When the energy of activation $E_{act}$ is relatively high, as in part *a*, this fraction is relatively small and the rate of reaction is low. But when $E_{act}$ is low as in part *b*, the fraction of successful collisions is larger and the rate is larger.

from A to B will form, releasing energy, perhaps enough to repay the cost of breaking the bond in B—C. The problem is that the release of energy and the demand for it are not synchronized. It is not enough for A just to tap the B end in B—C to make the change occur. The collision must usually be of some violence. The molecule A must penetrate somewhat into the electron cloud of B—C before the old bond (in B—C) will let go and the new bond (in A—B) will start to form. In other words, we expect that below some definite value of collision energy no chemical change occurs and no reallocation of bonding electrons takes place. At or above this value the collision will be successful. This minimum value of collision energy is called the *energy of activation*. As explained further in the legend to Figure 3.4, if the energy of activation $E_{act}$ is high, collisions must be especially violent for the reaction to be successful. Since only a small fraction of all collisions are this violent, the rate of the reaction will be very slow. In general, a high value of $E_{act}$ means a slow reaction, a low $E_{act}$ means a faster reaction. In order to understand trends in the relative rates of similar reactions, we shall try to discover how changes in structural features affect energies of activation.

The actual value of the energy of activation depends on the particular reaction. Once the reaction has been picked, the energy of activation has also been determined. This variable cannot be changed at will by the experimentalist, although there are some exceptions, the major one being again the phenomenon of catalysis.

**Catalysis.** When a catalyst is used in a reaction, the energy of activation for the reaction is lower and the change takes place much more rapidly under the same conditions. Sometimes the catalyst makes it possible for the reaction to take place by a different mechanism. Sometimes it generates a reactive intermediate as in the aromatic electrophilic substitutions. With the catalyst functioning to generate electrophilic particles (e.g., $NO_2^+$, $Cl^+$, $R^+$) in a first step, the second step can proceed to the carbonium-ion stage with a much lower energy of activation. The bottleneck in the reaction is this second step during which the electrophilic particle attacks the ring. To analyze a rate-determining step such as this one, it is helpful to use "progress of reaction" diagrams.

**Progress of Reaction Diagrams.** To return to the reaction of A with B—C, Figures 3.5 and 3.6 show progress of reaction diagrams for exothermic and endothermic changes. These diagrams resemble graphs, but the units for the coordinates are not well defined. The vertical axis indicates the internal energy of a system of colliding particles before, during, and after a successful collision. The reaction progresses along the horizontal axis as described, in the direction specified by the arrow.

In Figure 3.5 we imagine that a molecule of A is on a head-on collision course with B—C. The curve sweeps upward, signifying that as the collision progresses the kinetic energies of the moving particles are being converted into internal or potential energy. The upward sweep of the curve represents an increase in the total internal energy of the two particles, A and B—C, as kinetic energy is changed to potential

91

Theory of
Orientation
and Reactivity
in Aromatic
Electrophilic
Substitution

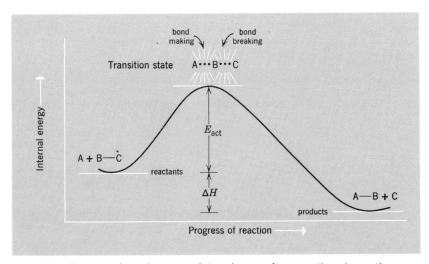

**Figure 3.5** Progress of reaction versus internal energy for an exothermic reaction.

energy. At the summit the total internal energy is sufficient to reallocate bonding electrons and form A—B and C. The bond between A and B begins to form, and the old bond in B—C starts to break. The electron cloud of A has made a sufficient penetration of the electron cloud at the B end of B—C. From this point on, A—B starts to break away and C is released. If these two new particles have kinetic energies higher than those of A and B—C before the collision, the reaction is exothermic. The total internal energy in A + B—C was higher than the total internal energy possessed by A—B + C after the reaction; and the difference, represented by $\Delta H$ in Figure 3.5, appears as energy released. The energy of activation $E_{\text{act}}$ is the difference between the initial energy of the reactants and the energy they possess at the summit.

The unstable aggregate of reactants at the summit is called the transition state; it is the collection of nuclei and electrons with a minimum internal energy that is still high enough to be able to break down to products. Figure 3.6 is a progress of reaction diagram for an endothermic change.

With this background we may now consider how the facts of orientation and reactivity in aromatic electrophilic substitution correlate with general principles. The

H   E

$\delta+$ ⬡ $\delta+$

$\delta+$

13

key intermediate in this reaction is the carbonium ion of type **13**, where E is the initial electrophilic species $E^+$. Any factor that will stabilize this carbonium ion will also stabilize the transition state leading to it, for the two resemble each other. If the

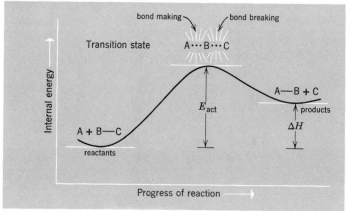

Progress of reaction

**Figure 3.6**   Progress of reaction versus internal energy for an endothermic process.

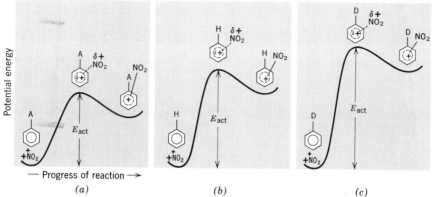

93

Theory of
Orientation
and Reactivity
in Aromatic
Electrophilic
Substitution

— Progress of reaction ⟶

(a)                    (b)                    (c)

**Figure 3.7** Activation and deactivation in aromatic electrophilic substitution: A denotes an activating group, H is hydrogen, and D is a deactivating group. The transition states depicted by structures at the tops of the three potential energy "hills" probably involve partial and developing bonds from the rings to the incoming nitro group. The carbonium ions in the higher-energy "valleys" are slightly more stable than their corresponding transition states. It is reasonable to assume that these transition states *resemble* the carbonium ions to which they convert. Therefore *any factor stabilizing the carbonium ion will also stabilize the transition state.* An activating group (a), by either induction or resonance, helps to stabilize a positive charge; a deactivating group (c) destabilizes a positive charge; substitution into benzene (b) is included for comparison. Consequently, $E_{act}$ is lowest when an activating group is on the ring and the rate of substitution into it is faster than into benzene itself. This figure explains only activation and deactivation. The question of orientation has been avoided by leaving the site of attachment of the nitro group ambiguous.

transition state is made more stable, the energy of activation for reaching it will be lower, and the reaction will be faster. Conversely, any factor that makes ion **13** less stable will raise the energy of activation leading to its transition state. The activating groups make ion **13** more stable than the corresponding ion from unsubstituted benzene; the deactivating groups destabilize **13** compared to the ion from benzene itself. These relations are better seen in the progress of reaction diagrams shown in Figure 3.7. The extent to which the groups already on the ring can influence the stability of **13** depends on the site the incoming group takes. We are dealing here with competing reactions, with attacks at all available ring positions. The product from the most rapid one will predominate. If the energy of activation for ortho (or para) substitution is lower than that for meta substitution, successful collisions will occur more frequently at the ortho (or para) site. To illustrate these principles, we shall study the nitration of toluene.

**The Methyl Group, an Activator and an Ortho-Para Director.** During the nitration of toluene the incoming nitronium ion could become attached at the ortho, the para, or the meta position. In resonance theory the contributing structures for the carbonium ion in each position are as follows.

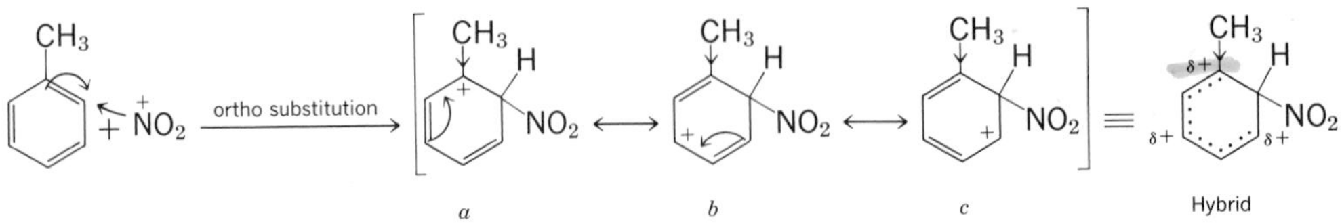

CH₃ ... + NO₂ —ortho substitution→ [ structures a, b, c ] ≡ Hybrid

**14**

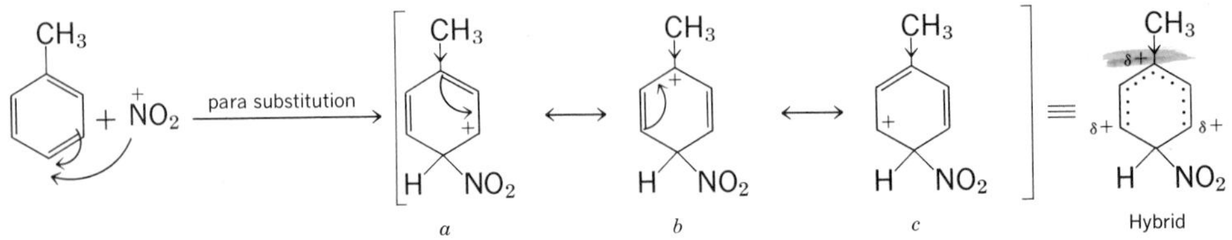

CH₃ ... + NO₂ —para substitution→ [ structures a, b, c ] ≡ Hybrid

**15**

CH₃ ... + NO₂ —meta substitution→ [ structures a, b, c ] ≡ Hybrid

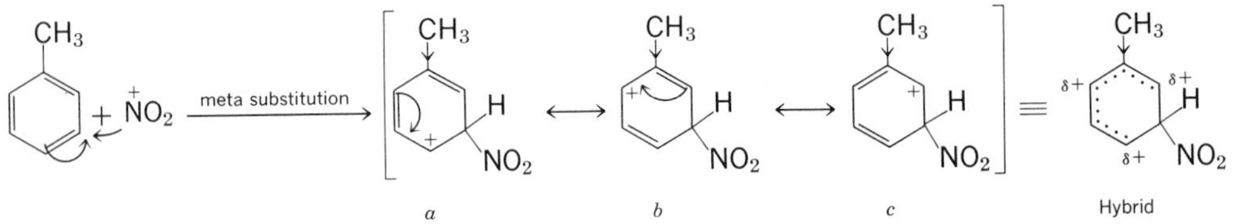

**16**

The methyl group has an electron-releasing inductive effect, indicated by the arrowhead drawn on the line representing its bond to the ring. This inductive effect tends to stabilize any partial positive charge that might develop *at the carbon holding the methyl group. Only for ortho or para substitution does such a partial positive charge appear at this point.* (See structures **14a** and **15b** or the hybrid structures for **14** and **15**.) In meta substitution resonance theory does not predict a partial positive charge where the methyl joins the ring (see **16**). If resonance theory and our use of it have been correct in this prediction, the carbonium ion for meta substitution should be less stable than that for either ortho or para substitution. Since the transition state leading to it resembles the carbonium ion, if the carbonium ion has higher internal energy, the transition state will too. Hence the energy of activation for making it will be higher. The rate of meta substitution must inevitably be slower, much slower in fact, than the rates of either ortho or para substitution. Thus are the ortho-para–directing ability and the activating influence of a methyl group on a benzene ring explained in resonance theory.

**The Nitro Group, a Deactivator and a Meta Director.** Both nitrogen and oxygen are more electronegative than carbon, and the nitro group has a strong, electron-withdrawing inductive effect. With this in mind we shall consider the chlorination of nitrobenzene which produces mostly m-chloronitrobenzene. Again the carbonium ion of step 2 in the reaction is the key intermediate and, depending on whether attack by Cl⁺ is at the ortho, meta, or para position, one of the following structures represents it. In **17** and **18**, according to resonance theory, there is a partial positive charge at the ring carbon holding the nitro group, but its inductive effect is to draw electron density away from this point. Thus attack by Cl⁺ at either the ortho or para position in nitrobenzene tends to put a partial positive charge at a point that already has one. In terms of energy it would be easier to make a successful attack at the meta position because the incoming positive charge need not be distributed to the carbon holding the nitro group, as shown in structure **19**. The new partial positive charges "skip" this point, or so we would predict using resonance theory. Of course, not even **19** is as stable a carbonium ion as would be formed if Cl⁺ attacked

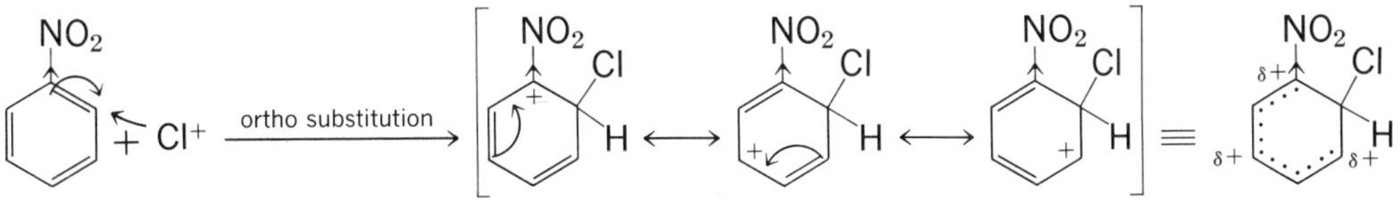

**17**

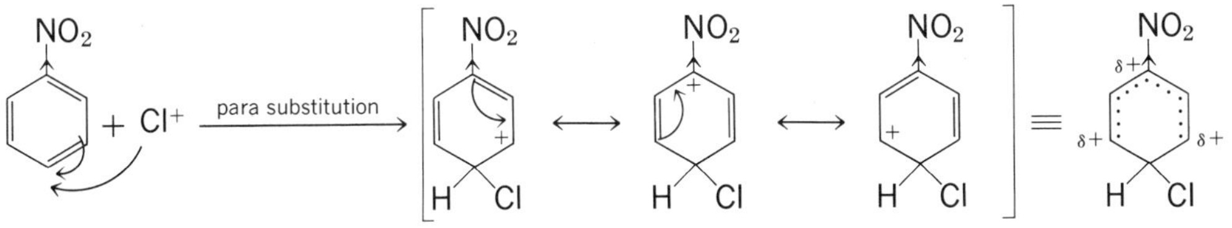

**18**

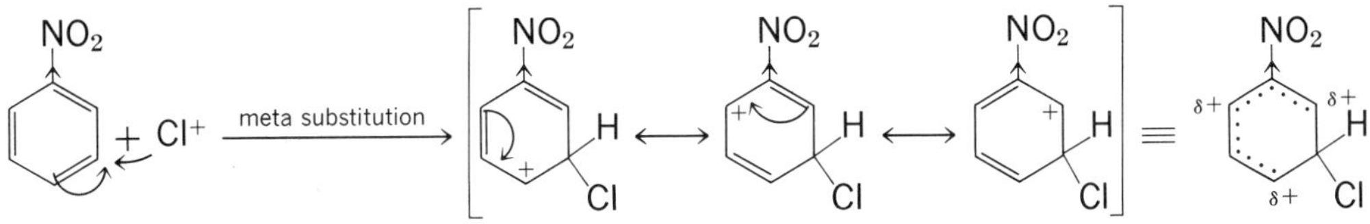

**19**

the benzene ring without the nitro group. Thus resonance theory predicts that the rate of attack by Cl⁺ *anywhere* in nitrobenzene will be slower than its rate of attack on benzene itself, which explains the deactivating influence of the nitro group. Because the meta sites are *relatively* easier to attack, the nitro group can direct incoming groups to them.

**The Phenolic Hydroxyl Group, Resonance versus Induction.** Since oxygen is more electronegative than carbon, we might predict that the phenolic hydroxyl group is electron-withdrawing, somewhat like the nitro group only less strongly so. We would therefore consider it a meta director and a deactivator. Our prediction could not be more incorrect. This group is such a powerful activator (and ortho-para director) that bromination of phenol requires no iron catalyst. In fact, it is virtually impossible to stop bromination at the monobromophenol stage. 2,4,6-Tribromophenol

forms very rapidly if the reaction is carried out at room temperature with water as
the solvent. Only by using a nonpolar solvent such as carbon disulfide at ice bath
temperature can monobromination of phenol be effected. Resonance theory is
successful in explaining this strong activating influence of the phenolic hydroxyl.
The carbonium ion is again the important factor, but with the oxygen attached
directly to the ring another pair of electrons is available for writing contributing structures. Depending on whether Br⁺ (generated in the first step) makes its attack at
the ortho, para, or meta position, one of the following carbonium ions will form. In
ortho or para attack contributors $a$, $c$, and $d$ of **20** and **21** are familiar. Contributor

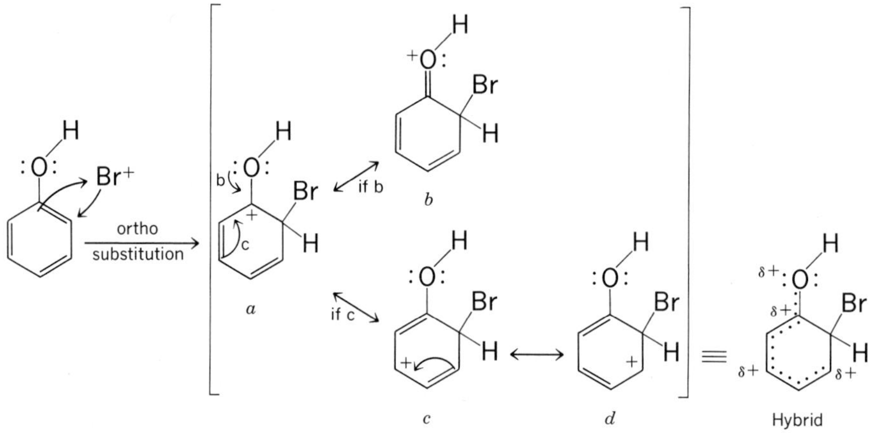

**20**

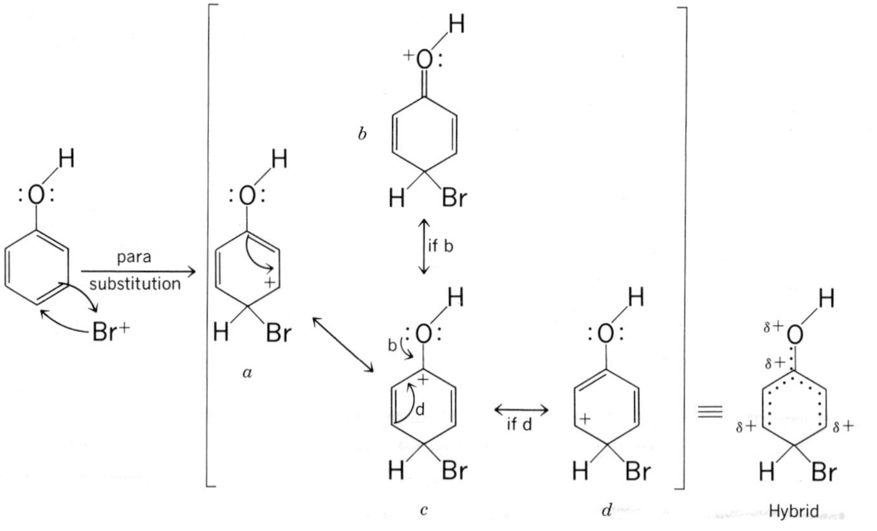

**21**

22

*b* is new and shows the positive charge on oxygen, even while the oxygen still has its octet and all the ring atoms have octets. We have seen a positive charge on oxygen before, in the hydronium ion, $H_3O^+$, and from this ion we learned that if oxygen has an octet it can handle a positive charge. Therefore contributors **20b** and **21b** should be quite important, and we are justified in concluding that the hybrid carbonium ions resulting from ortho and para attacks are more stable than the carbonium ion resulting from the meta attack where the positive charge cannot be delocalized onto oxygen. Moreover, substitution into phenol should for the same reason be easier than substitution into benzene. Thus resonance theory accounts correctly for both the directing and the activating influence of the hydroxyl.

---

**Exercise 3.3**   The amino group is as strong an activator and ortho-para director as the hydroxyl. If aniline, $C_6H_5$—$\ddot{N}H_2$, is added to a mixture of bromine and water, 2,4,6-tribromoaniline forms very rapidly. The explanation for these properties of the amino group, —$\ddot{N}H_2$, when it is attached directly to a benzene ring, is exactly the same as the explanation for the properties of the hydroxyl group. See whether you can write the steps, paying particular attention to the contributing structures for the intermediate carbonium ions for ortho, para, and meta monobrominations of aniline.

---

**The Halogens, Resonance versus Induction.** The halogens are deactivating groups, as might be predicted from their electron-withdrawing inductive effect. (Compare the effect of the nitro group.) This property should also make them meta directors, however, and they are not. With the halogens we are apparently at a borderline. Their inductive effects are not overwhelmed by any resonance effect, as are those of the hydroxyl and amino groups. The relatively strong inductive effect of a halogen on a benzene ring makes pi electrons *anywhere* on the ring less available than those in benzene to some incoming electrophile. But a resonance effect, illustrated for ortho substitution, must be invoked to rationalize the results. In the ortho bromination of chlorobenzene, structures **23a** to *d* contribute to the hybrid. In **23b** the chlorine is shown with a positive charge and an octet, but chlorine is less

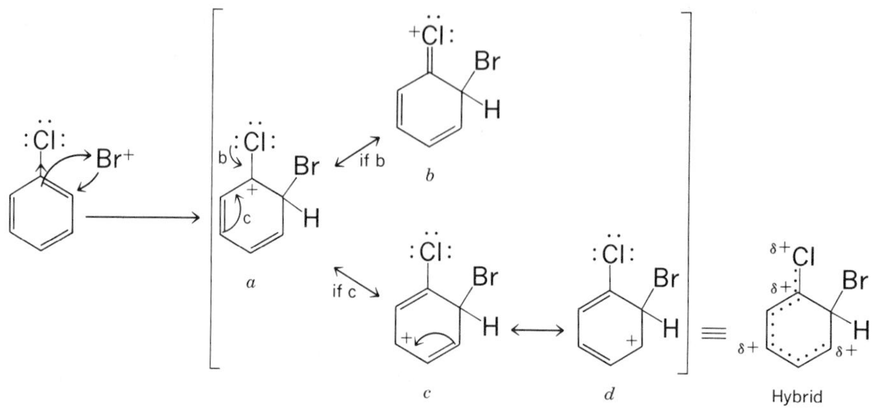

23

able to handle a positive charge than either nitrogen or oxygen. Hence, although **23b** contributes enough to the hybrid to make the carbonium ions for ortho and para substitutions more stable than the carbonium ion for meta substitution, the ring is not more active than that of benzene itself. Such is the argument from resonance theory for the deactivating but ortho-para–directing influence of a halogen atom on a benzene ring.

---

**Exercise 3.4**   Double bonds have shorter distances than single bonds, for example, C=C, 1.34 Å; C—C, 1.54 Å. In *t*-butyl chloride the C—Cl distance is 1.80 Å. In chlorobenzene the C—Cl distance is 1.69 Å. Do these data indicate that the C—Cl bond in chlorobenzene may have some "double-bond character"? Do the data tend to support structure **23b** as a contributor to the carbonium-ion hybrid?

**Exercise 3.5**   The positively charged group, $-\overset{+}{N}R_3$, is the most powerful of all deactivating groups. Explain.

---

## ARENES. ALKYLBENZENES

The names and structures of several alkylbenzenes or "arenes" are given in Table 3.3. Petroleum and coal tar are important sources. Because arenes combine an alkane-like group with a benzene system, we might expect them to exhibit properties of both systems. They do.

### Chemical Properties

1. **Halogenation.** Under conditions that promote the formation of free radicals (sunlight or high temperature), alkylbenzenes undergo halogenation at the side chain, the alkyl group. If an iron (or iron halide) catalyst is used, substitution into the ring occurs. Nitration, sulfonation, and Friedel-Crafts reactions also occur at the ring.

2. **Oxidation.** Benzene is exceptionally stable toward strong oxidizing agents

**Table 3.3    Arenes**

| Name | Structure | Mp (°C) | Bp (°C) | Specific Gravity (at 20°C) |
|------|-----------|---------|---------|----------------------------|
| Benzene |  | 5.5 | 80 | 0.879 |
| Toluene | 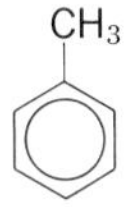 CH₃ | −95 | 111 | 0.866 |
| o-Xylene | 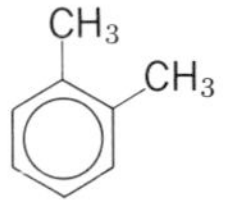 CH₃ CH₃ | −25 | 144 | 0.897 |
| m-Xylene | 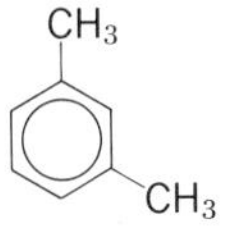 CH₃ CH₃ | −48 | 139 | 0.881 |
| p-Xylene |  CH₃ CH₃ | −13 | 138 | 0.854 |
| Ethylbenzene | 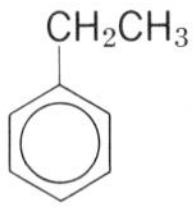 CH₂CH₃ | −95 | 136 | 0.867 |
| Cumene | 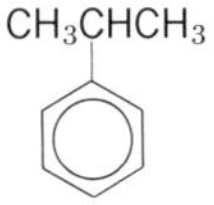 CH₃CHCH₃ | −81 | 152 | 0.862 |
| *Isopropylbenzene* | | | | |
| p-Cymene | 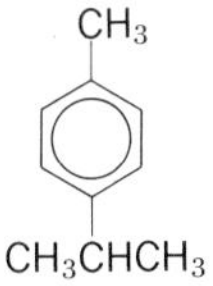 CH₃ CH₃CHCH₃ | −70 | 177 | 0.857 |
| Biphenyl | 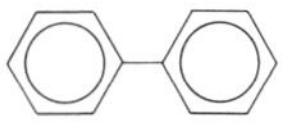 | 70 | 255 | — |

and so are alkanes, but strong oxidizing agents will attack the side chain, however long it is, where it joins the ring. For example,

$n$-Propylbenzene → hot KMnO$_4$ → Benzoic acid $(+ CO_2 + H_2O)$

$o$-Xylene → hot KMnO$_4$ → Phthalic acid $(+ CO_2 + H_2O)$

This process is an important source of aromatic carboxylic acids, and reagents such as potassium permanganate or potassium dichromate may be used for the reaction. In this example the properties of one system are modified by the presence of another. An alkane-like group on a benzene ring is much more susceptible to oxidation than an alkane is.

The benzene ring is not this stable in all situations. If powerful electron-donating groups (for example, amino, hydroxyl) are attached to the ring, it is quite susceptible to oxidation. Atmospheric oxygen slowly attacks anilines and phenols, converting them to deeply colored, complex substances. The first synthetic dye was prepared in 1856 by William Henry Perkin who, in an attempt to convert aniline to quinine, oxidized it with potassium dichromate. From the black product he was able to extract a blue dye. In intermediary metabolism in human beings, phenolic substances obtained from proteins are converted into melanins, the dyes respon-sible for pigmentation of the skin.

### POLYNUCLEAR AROMATIC HYDROCARBONS

Compounds exhibit aromatic properties if their molecules consist of flat, cyclic systems with conjugated multiple bonds when Kekulé structures are written for them. Several hydrocarbons are in this category, and their structures appear to consist of fused or condensed benzene rings. Examples are given in Figure 3.8. Several of these compounds have color, a rare property among hydrocarbons. We shall encounter other examples illustrating that color is associated with an extensive, conjugated pi-electron system.

Naphthalene is a common moth repellant, but its use has declined since the introduction of $p$-dichlorobenzene. Some of these condensed aromatic hydrocarbons occur in the tars of tobacco smoke and in the charred residues on the surfaces of charcoal-broiled steaks. One of them, benzopyrene, is a known carcinogen (cancer inducer).

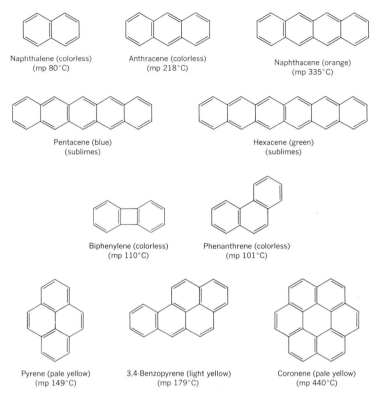

Naphthalene (colorless)
(mp 80°C)

Anthracene (colorless)
(mp 218°C)

Naphthacene (orange)
(mp 335°C)

Pentacene (blue)
(sublimes)

Hexacene (green)
(sublimes)

Biphenylene (colorless)
(mp 110°C)

Phenanthrene (colorless)
(mp 101°C)

Pyrene (pale yellow)
(mp 149°C)

3,4-Benzopyrene (light yellow)
(mp 179°C)

Coronene (pale yellow)
(mp 440°C)

**Figure 3.8**  Some polynuclear aromatic hydrocarbons.

## REFERENCES AND ANNOTATED READING LIST

### BOOKS

O. T. Benfey. *From Vital Force to Structural Formulas.* Houghton Mifflin Company, Boston, 1964. Chapters 8 and 9 discuss the theories of Kekulé and Couper with quotations from their writings. (Paperback.)

O. T. Benfey, editor. *Classics in the Theory of Chemical Combination.* Dover Publications, New York, 1963. Kekulé's famous paper and two by Couper are included, Kekulé's in English translation. (Paperback.)

R. T. Morrison and R. N. Boyd. *Organic Chemistry,* second edition. Allyn and Bacon, Boston, 1966. Chapter 10, "Benzene. Resonance. Aromatic Character" is a good reference among currently available textbooks.

### ARTICLES

L. P. Lessing. "Coal." *Scientific American,* July 1955, page 193. This article describes what aromatic compounds are obtained from coal, an important source, and how they are obtained.

B. E. Schaar. "Aniline Dyes." *Chemistry,* January 1966, page 12. William Henry Perkin was seventeen years old when he accidentally discovered the first aniline dye. Schaar describes what Perkin did and the impact of his work.

## PROBLEMS AND EXERCISES

1. Write structures of each of the following compounds
   (a) m-nitrobenzoic acid      (b) 2,4,6-tribromoaniline      (c) o-chlorophenol
   (d) p-toluenesulfonic acid      (e) m-iodobenzaldehyde
2. Starting with benzene or toluene and any needed inorganic reagents, write reactions for preparing each of the following compounds. (Assume that mixtures of ortho and para isomers can be separated.)
   (a) p-nitrochlorobenzene      (b) benzoic acid      (c) m-nitrobenzoic acid
   (d) p-bromobenzenesulfonic acid      (e) m-bromobenzenesulfonic acid
3. Write equations for the steps in the bromination of benzene, catalyzed by $FeBr_3$.
4. Naphthalene has three important contributing structures. One of them is the following:

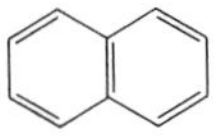

   Write the structures of the other two.
5. In Chapter 2 the 1,4-addition of HBr to 1,3-butadiene was presented. The first step in the mechanism for this reaction is as follows:

$$CH_2{=}CH{-}CH{=}CH_2 + H{-}Br \longrightarrow CH_2{=}CH{-}\underset{+}{C}H{-}CH_3$$
$$\mathbf{24}$$

   (a) Is **24** an adequate structure for the carbonium ion? To answer this question, try to draw a structure identical to **24** except for the location of the pi electrons. *If* you can, and *if* it is enough like **24** to be of roughly similar internal energy, then according to resonance theory your answer to the question will have to be "No." Thus

$$CH_2{=}CH{-}\underset{+}{C}H{-}CH_3 \longleftrightarrow {}^+CH_2{-}CH{=}CH{-}CH_3$$
$$\quad\mathbf{24} \qquad\qquad\qquad\qquad\qquad \mathbf{25}$$

   (b) On the basis of structures **24** and **25,** write a nonclassical structure for the hybrid of the two.
   (c) How does this development explain 1,4-addition of HBr to 1,3-butadiene?
   (d) Does it still allow for some 1,2-addition?
6. Write the structures of the contributing forms for the carbonium ions resulting from ortho, meta, and para attacks by a nitronium ion on benzoic acid. In terms of these structures explain why the carboxyl group, $-\overset{\displaystyle O}{\overset{\|}{C}}-OH$, is a meta director and a deactivator.

# CHAPTER FOUR

# Alcohols

## STRUCTURE

Compounds whose molecules contain a hydroxyl group, —OH, attached to a saturated carbon are called *alcohols*. If this carbon, called the carbinol carbon, has a double (or triple) bond going to another group, the substance is classified in

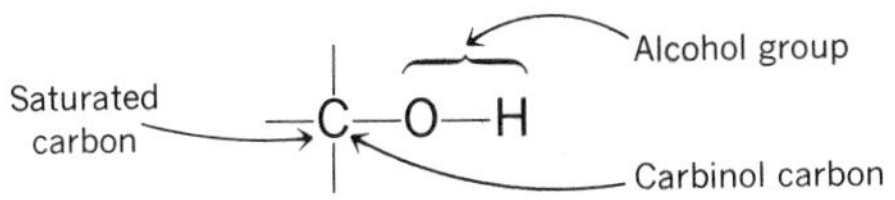

Essential features
of all alcohols

some other way. Several common alcohols are listed in Table 4.1. If the hydroxyl group is attached directly to a benzene ring, the molecule is classified as a *phenol*, which also happens to be the name of a specific compound.

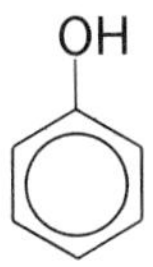

Phenol

If the hydroxyl group is attached to one carbon of a carbon-carbon double bond, the molecule is classified as an *enol*, "ene" to indicate the alk*ene* function, plus "-ol" to designate the hydroxyl group. Most enols are unstable. They rearrange, intramolecularly (within the molecule), to give isomeric molecules containing a *carbonyl* group, $\ce{>C=O}$, a functional group to be studied in later chapters.

**Table 4.1   Some Monohydric Alcohols**

| | Name | Structure | Mp (°C) | Bp (°C) | Density (in g/cc at 20°C) | Solubility (in g/100 g water, 20°C) |
|---|---|---|---|---|---|---|
| $C_1$ | Methyl alcohol | $CH_3OH$ | −98 | 64.5 | 0.791 | soluble |
| $C_2$ | Ethyl alcohol | $CH_3CH_2OH$ | −115 | 78.3 | 0.789 | soluble |
| $C_3$ | n-Propyl alcohol | $CH_3CH_2CH_2OH$ | −127 | 97.2 | 0.803 | soluble |
| | Isopropyl alcohol | $(CH_3)_2CHOH$ | −86 | 82.4 | 0.786 | soluble |
| $C_4$ | n-Butyl alcohol | $CH_3CH_2CH_2CH_2OH$ | −90 | 118 | 0.810 | 8.0 |
| | Isobutyl alcohol | $(CH_3)_2CHCH_2OH$ | −108 | 108 | 0.802 | 10.0 |
| | sec-Butyl alcohol | $CH_3CH_2\underset{\underset{OH}{\mid}}{C}HCH_3$ | −115 | 99.5 | 0.806 | 12.5 |
| | t-Butyl alcohol | $(CH_3)_3COH$ | 25.6 | 82.5 | 0.786 | soluble |
| $C_5$ and higher | n-Amyl alcohol (1-pentanol) | $CH_3(CH_2)_3CH_2OH$ | −79 | 138 | 0.814 | 2.2 |
| | t-Amyl alcohol (2-methyl-2-butanol) | $(CH_3)_2\underset{\underset{OH}{\mid}}{C}CH_2CH_3$ | −8.6 | 102 | 0.806 | 12.2 |
| | Isoamyl alcohol (3-methyl-1-butanol) | $(CH_3)_2CHCH_2CH_2OH$ | | 132 | 0.807 | 2.9 |
| | Neopentyl alcohol (2,2-dimethyl-2-propanol) | $(CH_3)_3CCH_2OH$ | 52 | 113 | | 3.7 |
| | 1-Hexanol | $CH_3(CH_2)_4CH_2OH$ | −52 | 157 | 0.820 | 0.6 |
| | 1-Heptanol | $CH_3(CH_2)_5CH_2OH$ | −34 | 176 | 0.822 | 0.1 |
| | 1-Octanol | $CH_3(CH_2)_6CH_2OH$ | −15 | 195 | 0.826 | 0.04 |
| | 1-Decanol | $CH_3(CH_2)_8CH_2OH$ | 7 | 229 | 0.829 | 0.004 |
| | Cyclopentanol | 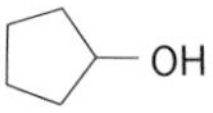 (cyclopentane)—OH | | 140 | 0.949 | slightly soluble |
| | Cyclohexanol | 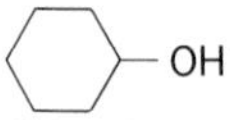 (cyclohexane)—OH | 23 | 161 | 0.962 | 3.6 |
| | Allyl alcohol | $CH_2{=}CHCH_2OH$ | | 97 | 0.852 | soluble |
| | Benzyl alcohol | $C_6H_5CH_2OH$ | −15 | 205 | 1.046 | 4 |

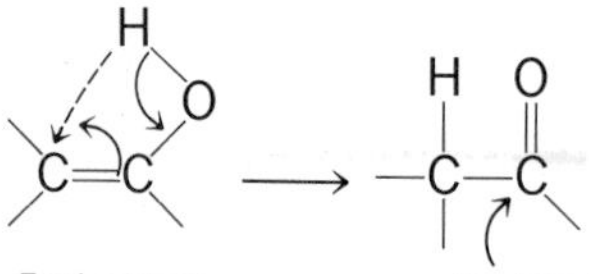

Enol system            Carbonyl group

**Subclasses of Alcohols.** It is frequently useful to classify an alcohol as primary (1°), secondary (2°), or tertiary (3°) according to the condition of the carbinol carbon (cf. p. 103).

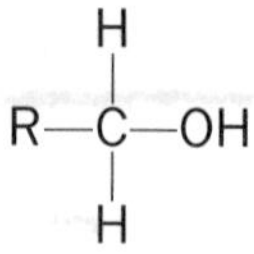

Primary alcohol (1°)          Secondary alcohol (2°)          Tertiary alcohol (3°)

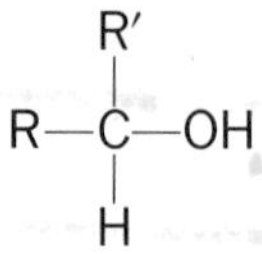

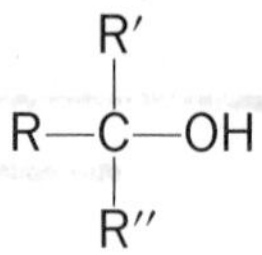

The single and double primes on R (see p. 19) indicate that the alkyl groups may or may not be identical.

The variations of chemical properties from subclass to subclass are not too great, and they often have to do with rates of reactions rather than the kinds of final products that form.

## NOMENCLATURE

Chemists have three systems of nomenclature for this family—common, derived, and IUPAC.

**Common Names.** When the alkyl group to which the hydroxyl is attached has a common name, the corresponding alcohol is designated simply by writing the word "alcohol" after this name.

$$CH_3OH \qquad CH_3\overset{\overset{\displaystyle CH_3}{|}}{\underset{\underset{\displaystyle CH_3}{|}}{C}}OH \qquad \text{⬡}CH_2OH$$

Methyl alcohol        t-Butyl alcohol        Benzyl alcohol

$$CH_3CH_2OH \qquad CH_3\overset{\overset{\displaystyle CH_3}{|}}{C}HOH \qquad CH_3\overset{\overset{\displaystyle CH_3}{|}}{C}HCH_2OH \qquad CH_2\!\!=\!\!CHCH_2OH$$

Ethyl alcohol        Isopropyl alcohol        Isobutyl alcohol        Allyl alcohol

**Derived Names.** Because relatively few alkyl groups have been assigned common names, the "derived" system of nomenclature for alcohols was designed to extend the possibilities for using them. It is not an official system. The word "carbinol" designates the carbinol carbon and its attached hydroxyl group, and the alkyl groups attached to the carbinol carbon are then identified. Carbinol is appended to these names as a suffix, not as a separate word. (Hydrogens attached to the carbinol carbon are "understood.") In the examples given to illustrate this system, the common name, if it is actually the one most frequently used by chemists, has been included in parentheses below the derived name.

$$CH_3OH \qquad CH_3CH_2OH \qquad CH_3\overset{\overset{\displaystyle CH_3}{|}}{C}HOH \qquad CH_3\overset{\overset{\displaystyle CH_3}{|}}{\underset{\underset{\displaystyle CH_3}{|}}{C}}OH$$

Carbinol          Methylcarbinol          Dimethylcarbinol          Trimethylcarbinol
(methyl alcohol)    (ethyl alcohol)        (isopropyl alcohol)        (t-butyl alcohol)

$$CH_3CH_2\overset{\overset{\displaystyle CH_3}{|}}{C}HOH \qquad \text{⬡}\overset{\overset{\displaystyle \text{⬡}}{|}}{\underset{\underset{\displaystyle \text{⬡}}{|}}{C}}\!\!-\!\!OH \qquad CH_3\overset{\overset{\displaystyle CH_3}{|}}{\underset{\underset{\displaystyle CH_3}{|}}{C}}CH_2OH$$

Ethylmethylcarbinol        Triphenylcarbinol        t-Butylcarbinol
(sec-butyl alcohol)                                (neopentyl alcohol)

Names in the derived system, used when alternates are awkward, convey a mental picture of structure in a way that IUPAC names often do not.

**IUPAC Nomenclature Rules for Alcohols**

1. For the "parent" structure, select the longest continuous chain of carbons *that includes the hydroxyl group.* Determine the name of the alkane that corresponds to this carbon chain and replace the terminal "-e" by "-ol."

$$CH_3OH \qquad\qquad CH_3CH_2OH \qquad\qquad \underset{\displaystyle CH_3}{CH_3\overset{|}{C}HCH_2CH_2OH}$$

Methanol          Ethanol          -butanol
(incomplete)

2. When isomerism is possible, designate the location of the hydroxyl group on this chain by numbering the carbons from whichever end will give this location the lower number.

$$CH_3CH_2CH_2OH \qquad \underset{}{CH_3\overset{\overset{\textstyle CH_3}{|}}{C}HOH} \qquad \underset{}{CH_3CH_2\overset{\overset{\textstyle CH_3}{|}}{C}HOH} \qquad \underset{}{CH_3\overset{\overset{\textstyle CH_3}{|}}{C}HCH_2CH_2OH}$$

1-Propanol     2-Propanol     2-Butanol     -1-butanol
(incomplete)

3. Determine the names and the location numbers of any hydrocarbon-like groups (e.g., an alkyl, an aryl) attached to the chain, and assemble these before the parent name developed thus far. Study the examples carefully to determine how commas and hyphens are used to make the final name one word.

$$\underset{}{CH_3\overset{\overset{\textstyle CH_3}{|}}{C}HCH_2CH_2OH} \qquad \underset{}{CH_3CH_2\overset{\overset{\textstyle CH_3}{|}}{C}HCH_2OH} \qquad CH_3CH_2\overset{\overset{\textstyle CH_3}{|}}{\underset{\underset{\textstyle CH_3}{|}}{C}}CH_2OH$$

3-Methyl-1-butanol     2-Methyl-1-butanol     2,2-Dimethyl-1-butanol
(complete)

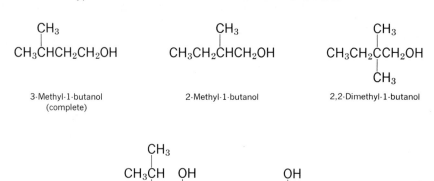

3-Isopropyl-2,3-dimethyl-2-heptanol        3-*t*-Butyl-2-hexanol

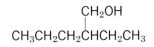

2-Ethyl-1-pentanol
(Note the application of rule 1 here.)

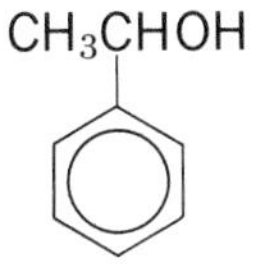

1-Phenylethanol

4. If other atoms or groups are attached to or incorporated into the main chain of the alcohol molecule, first apply rules 1 through 3, and then work other groups into the name.

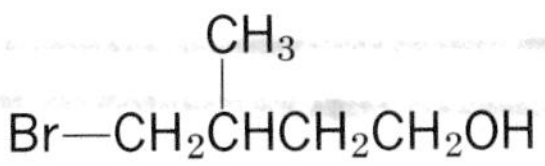

$$CH_3$$
$$Br—CH_2CHCH_2CH_2OH$$

$$Cl$$
$$Cl—CH—CH—CH_2OH$$
$$Cl$$

4-Bromo-3-methyl-1-butanol
(*not* 1-bromo-2-methyl-4-butanol)

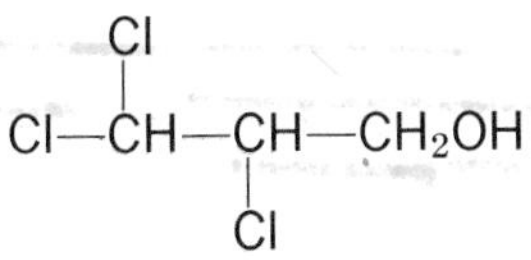

2,3,3-Trichloro-1-propanol

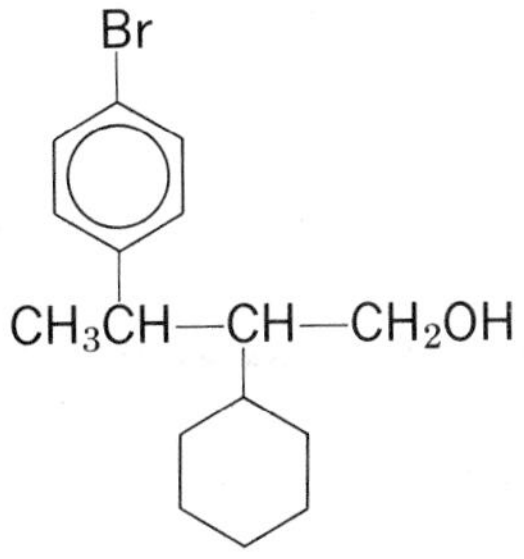

$$Br$$

$$CH_3CH—CH—CH_2OH$$

2-Cyclohexyl-3-(4-bromophenyl)-1-butanol
(Note the use of parentheses to avoid confusing
the numbering of the parent chain and the
numbering of the group.)

**Polyhydric Alcohols.** Compounds containing two or more hydroxyl groups per molecule are quite common, although the hydroxyl groups must be on separate carbon atoms to be stable. *Dihydric alcohols or glycols* contain two hydroxyls per molecule. *Trihydric alcohols* contain three hydroxyl groups per molecule. Glycerol

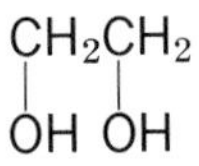

$$CH_2CH_2$$
$$OH\ OH$$

$$CH_3CH—CH_2$$
$$OH\ \ \ OH$$

$$CH_2CH_2CH_2$$
$$OH\ \ \ \ \ OH$$

Ethylene glycol
(1,2-ethanediol,
bp 197°C)

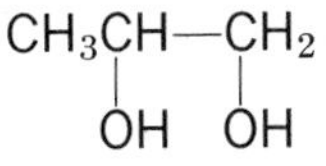

Propylene glycol
(1,2-propanediol,
bp 189°C)

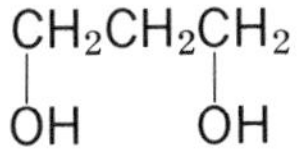

Trimethylene glycol
(1,3-propanediol,
bp 214°C)

is the most common example. The common sugars and polysaccharides consist of polyhydric alcohols, with other groups also present (cf. the structure of glucose).

$$CH_2-CH-CH_2$$
$$\quad|\qquad|\qquad|$$
$$OH\quad OH\quad OH$$

Glycerol
(1,2,3-propanetriol,
bp 290°C)

$$CH_2-CH-CH-CH-CH-\overset{\displaystyle O}{\overset{\displaystyle \|}{C}}-H$$
$$\quad|\qquad|\qquad|\qquad|\qquad|$$
$$OH\quad OH\quad OH\quad OH\quad OH$$

Glucose (open form,
mp 146–150°C)

We know of virtually no compounds that contain both the hydroxyl group and another hydroxyl group, a halo substituent (F, Cl, Br, or I), or an amino group, $-NH_2$, attached to the same carbinol carbon. These systems are apparently too unstable to be isolable, but they are known to exist in solutions under suitable circumstances. Alcohols are therefore largely limited to those in which the carbinol carbon, besides containing the hydroxyl group, holds at the same time only another carbon (or carbons), hydrogens, or both.

$$\overset{\displaystyle OH}{\overset{\displaystyle |}{-C-OH}}\qquad\overset{\displaystyle X}{\overset{\displaystyle |}{-C-OH}}\qquad\overset{\displaystyle NH_2}{\overset{\displaystyle |}{-C-OH}}$$
$$\quad|\qquad\qquad|\qquad\qquad|$$

Unstable systems

## PHYSICAL PROPERTIES. HYDROGEN BONDING. ASSOCIATION

**Boiling Points.** Figure 4.1 shows that the boiling points of the *normal,* saturated alcohols increase regularly as the formula weights increase. The effect of chain branching is indicated in Figure 4.2. In Figure 4.1 the boiling point curve of the homologous series of alkanes starts out much lower than that of the alcohols,

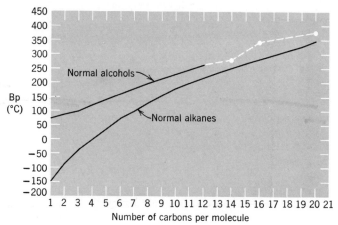

**Figure 4.1** Variation of boiling point with total carbon content in the homologous series of *normal* alkanes and normal primary alcohols. As the alcohols become increasingly alkane-like in structure, their properties also resemble those of the alkanes, as illustrated here for the boiling points.

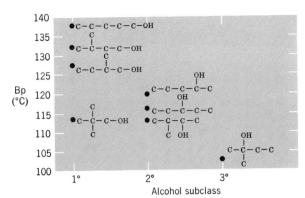

**Figure 4.2** The more compact a molecule, the lower the boiling point of the substance, all other factors being equal. The boiling points of the isomeric five-carbon alcohols illustrate this rule of thumb. The more compact the molecule, the weaker the forces of attraction between molecules, because compactness and symmetry usually mean lower polarity.

but later the curves of the two families draw closer together. The great difference at the lower end of the two series is striking, however. Ethane and methyl alcohol have comparable formula weights (30 versus 32), yet ethane boils 153° below methyl alcohol (−88.6°C versus 64.5°C). Clearly it takes far more energy to separate molecules of the alcohol than to separate those of the alkane. Even more striking is

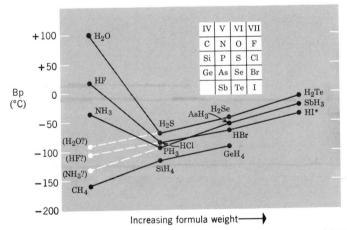

**Figure 4.3** Hydrides—variation of boiling points with formula weights. The sequence from $CH_4$ to $GeH_4$ is the normal trend, that is, boiling points increase with increasing formula weight. The hydrides of group V elements ($NH_3$, $PH_3$, $AsH_3$, and $SbH_3$), the group VII elements (HF, HCl, HBr, and HI), and the group VI elements ($H_2O$, $H_2S$, $H_2Se$, and $H_2Te$) all show abrupt breaks in this trend. The boiling points of $NH_3$, HF, and $H_2O$ are considerably higher than would be predicted on the basis of the rule of thumb that substances consisting of lighter molecules will have lower boiling points. The concept of hydrogen bonding explains these irregularities.

\* The boiling point of HI is taken under pressure to prevent decomposition.

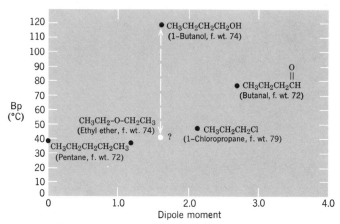

**Figure 4.4**   Boiling point versus dipole moment, formula weights being approximately equal. For a polar molecule the product of the partial charge $\delta$ and the distance separating the partial charge $l$ is the dipole moment $\mu$; $\mu = \delta l$. The larger the magnitude of the partial charge (without regard to algebraic signs) and/or the greater the distance separating them, the higher the dipole moment. Assuming that the dipole moment is a measure of polarity, we might expect the boiling point to increase as the polarity increases. The data plotted here do not support this simple correlation. 1-Butanol has a higher boiling point than would be predicted from polarity alone; only 1-butanol can participate in hydrogen bonding.

the fact that water, of even lower formula weight, boils at a higher temperature than both (100°C). This irregularity is also seen in the trends for the boiling points of simple hydrides (Figure 4.3). Water, hydrogen fluoride, and ammonia are irregular; methane is not.

The data in Figure 4.4 indicate that the high polarity of alcohol molecules as compared to that of alkanes is probably not the whole answer to the abnormally high boiling points of alcohols. Substances of higher polarities that boil at a lower temperature than alcohols of about the same formula weight are known.

**Hydrogen Bonds.** The irregularities just discussed and illustrated in Figures 4.3 and 4.4 occur in molecules in which a hydrogen atom is bonded to a strongly electronegative atom. The most electronegative element of all is fluorine; oxygen is next (cf. the alcohols and water); nitrogen and chlorine nearly tie for third. When fluorine, oxygen, and nitrogen are bonded to hydrogen (of much lower electronegativity), these atoms will surely bear rather large partial negative charges, and the hydrogen will have a correspondingly large partial positive charge. Moreover, a hydrogen substituent is very small, and it apparently can with little difficulty locate itself rather close to a partial negative charge on a neighbor molecule. When the distance between partial (and opposite) charges becomes very small, the force of attraction between the two can become sizable. It is not as great as the attractive force between two opposite, *full* charges (ionic bond), nor is it as much as is involved in the sharing of a pair of electrons between two nuclei (covalent bonding). But the force is large enough, nevertheless, for us to say that a bond exists.

It is called a *hydrogen bond*. It resembles an ionic bond, for it is a force of attraction between separate, oppositely charged particles. The ionic bond exists between *fully* charged and separate ions; the hydrogen bond occurs between *partially* charged sites, usually on separate molecules. The hydrogen bond (Figure 4.5) can be looked upon as a bridge between two highly electronegative atoms (F, O, or N). It is attached to one by a full covalent bond, and it is attracted to the other by a partly ionic bond, between sites of partial and opposite charge.

**Solubility.** Substances in which molecules are attracted to each other by means of hydrogen bonds are said to be *associated*. Such substances often tend to be soluble in each other; ammonia in water and water in methyl alcohol are examples. In order for a nonassociated substance like methane to dissolve in, say, water, hydrogen bonds between water molecules would have to be broken and reduced in

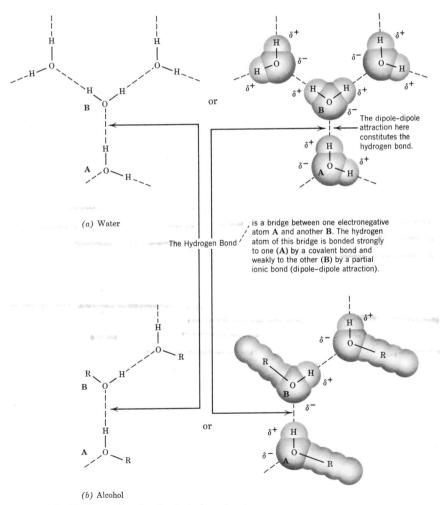

(*a*) Water

The Hydrogen Bond ··· is a bridge between one electronegative atom **A** and another **B**. The hydrogen atom of this bridge is bonded strongly to one (**A**) by a covalent bond and weakly to the other (**B**) by a partial ionic bond (dipole–dipole attraction).

The dipole–dipole attraction here constitutes the hydrogen bond.

(*b*) Alcohol

**Figure 4.5**  Hydrogen bonding in alcohols and water.

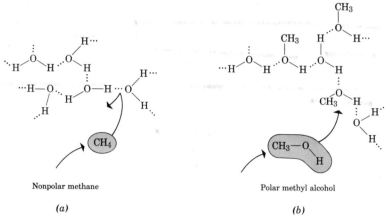

Nonpolar methane         Polar methyl alcohol

(a)            (b)

**Figure 4.6** Understanding the solubility of short-chain alcohols in water. (a) A nonpolar alkane molecule cannot break into the hydrogen-bonded sequence in water. It cannot replace the hydrogen bonds that would have to be broken to let it in. Therefore it is insoluble in water.

(b) A short-chain alcohol molecule, capable of hydrogen bonding, can slip into the sequences in water. It can replace at least some of the hydrogen bonds that must be broken to let it in. In any liquid, of course, the molecules shift from place to place. The fixedness of this figure is not meant to indicate any inflexibility. The drawing may be likened to a high-speed flash photograph that has caught the action of the instant.

number. The energy cost of this would not be repaid by the formation of new hydrogen bonds to the foreign material. Methyl alcohol, in contrast to methane, however, can slip into the hydrogen bond network of water (or vice versa) at much lower energy cost (Figure 4.6). For this reason the low-formula-weight alcohols are soluble in water in all proportions (cf. Table 4.1). As the homologous series of alcohols is ascended, however, the larger and larger hydrocarbon "tails" tend more and more to disturb the hydrogen-bonding opportunities between the solvent and the solute. As the solubility data of Table 4.1 reveal, when straight-chain alcohols have five or six carbons or more, they are virtually insoluble in water.

Since all alcohols are also partly alkane-like it is also possible for them to dissolve in the typical so-called "hydrocarbon solvents"—benzene, ether, carbon tetrachloride, chloroform, gasoline, petroleum ether, and methylene chloride. To generalize, water tends to dissolve water-like molecules, nonpolar solvents to dissolve hydrocarbon-like substances. Even more generally speaking, polar solvents tend to dissolve polar solutes, nonpolar solvents to dissolve nonpolar (or moderately polar) solutes.

Our discussion of solubilities in water has been rather extensive. Since nearly all important reactions in living cells occur between organic molecules in aqueous fluids, we must note very carefully any structural features that help or hinder the solubility of a compound in an aqueous medium. The —OH group is obviously a very important water-solubilizing group.

Exercise 4.1    Compare the boiling points of *n*-propyl alcohol, propylene glycol, and glycerol (cf. p. 107) with the boiling points of alkanes of approximately the same formula weights (cf. p. 15). Include in your list a data the boiling points of the monohydric alcohols that have formula weights approximately the same as those of propylene glycol and glycerol. Explain the trends in boiling points in going from monohydric to dihydric to trihydric alcohols.

## CHEMICAL PROPERTIES OF ALCOHOLS. GENERAL PRINCIPLES

In our study of the chemistry of alkanes we learned that the carbon-carbon single bond and the carbon-hydrogen single bond in an alkane environment are extremely resistant to chemical attack except by certain free radicals (e.g., ·Cl, $O_2$). Ionic reagents such as concentrated sulfuric acid and sodium hydroxide, active metals such as sodium or potassium, and electrically neutral substances such as water do not attack alkanes—at least not at ordinary temperatures and pressures. One way of understanding this phenomenon is to note that the electronegativities of carbon and hydrogen are quite close together (much closer than those of oxygen and hydrogen or of nitrogen and hydrogen). And, of course, the two atoms at the ends of a carbon-carbon single bond do not differ in relative electronegativities. In short, neither carbon-carbon nor carbon-hydrogen single bonds are very polar. Hence molecules or ions of polar or ionic reagents find little in an alkane molecule to which to be electrically attracted. Random collisions occur between the particles, but little chemical interaction takes place. At very high temperatures the situation may change, for under such circumstances the energies of collisions between potential reactants are much more frequently powerful enough to "break and make bonds."

Molecules of alcohols, compared with those of alkanes, have two new bonds, carbon to oxygen and oxygen to hydrogen. Because oxygen has a relatively high electronegativity, compared with those of carbon or hydrogen, both these bonds will be polar. In addition, although alkane molecules have no unshared pairs of

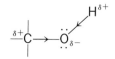

State of polarization of bonds in
an alcohol at the functional group

electrons anywhere, alcohol molecules do have them—on the oxygen. We should expect, therefore, that polar or ionic reagents will be much more successful in interacting with alcohols than with alkanes. Three types of reactions are observed: (1) reactions involving rupture of the O—H bond, (2) reactions involving the unshared electrons on the oxygen, and (3) reactions involving rupture of the C—O bond.

## REACTIONS OF ALCOHOLS INVOLVING THE O—H BOND

Like water, alcohols are weak proton donors. They are weak acids in either the Arrhenius or the Brønsted senses:

$$CH_3CH_2OH \rightleftharpoons CH_3CH_2O^- + H^+ \qquad K_a \approx 10^{-18}$$

Ethanol          Ethoxide ion<br>(anion of ethanol)

The anions of alcohols are known as *alkoxide ions*. They are powerful proton acceptors, as the equilibrium just indicated for ethanol implies.

---

Exercise 4.2   In the liquid state the order of acidity among the subclasses of the alcohols is

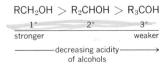

$$RCH_2OH > R_2CHOH > R_3COH$$

1°         2°         3°<br>stronger              weaker<br>——————decreasing acidity——————→<br>of alcohols

What is the *order of basicity* of the corresponding alkoxide ions: $RCH_2O^-$, $R_2CHO^-$, and $R_3CO^-$? *Hint*. What inductive effect of alkyl groups tends to increase the electron density on the oxygen of an alkoxide ion? Or, alternatively, the oxygen of which alkoxide ion, $t$-butoxide, or methoxide is less crowded by alkyl groups and is therefore more open to having its negative charge stabilized by solvation?

---

Like water, alcohols react with the more active metals such as sodium and potassium. In general:

$$H_2O + Na \longrightarrow HO^-Na^+ + \tfrac{1}{2}H_2 \qquad \text{(violent reaction)}$$
$$ROH + Na \longrightarrow RO^-Na^+ + \tfrac{1}{2}H_2 \qquad \text{(moderate reaction)}$$

Specific examples:

$$CH_3OH + Na \longrightarrow CH_3O^-Na^+ + \tfrac{1}{2}H_2 \qquad \text{(moderate reaction)}$$

Methanol          Sodium methoxide

$$(CH_3)_3COH + Na \longrightarrow (CH_3)_3CO^-Na^+ + \tfrac{1}{2}H_2 \qquad \text{(extremely slow reaction)}$$

$t$-Butyl alcohol          Sodium $t$-butoxide

---

Exercise 4.3   What correlation is there between the vigor of this reaction as described in the specific examples and the relative acidities of the various subclasses of alcohols and water?

---

When chemists need a powerful base in a nonaqueous system, they frequently choose the sodium or potassium alkoxides dissolved in the alcohol used to make them.

## REACTIONS OF ALCOHOLS CONFINED TO THE UNSHARED
## PAIRS OF ELECTRONS ON OXYGEN

The reaction of hydrogen bromide or other strong acids with water has its counterpart in the alcohol family. In general, with water:

$$H\!-\!\overset{..}{\underset{H}{O}} + H\!-\!Br \rightleftharpoons H\!-\!\overset{+}{\underset{H}{O}}\!\!\diagup\!\!^H + Br^-$$

Hydronium ion

with alcohols:

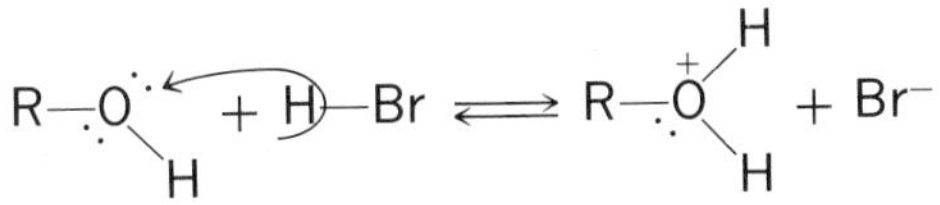

$$R\!-\!\overset{..}{\underset{H}{O}} + H\!-\!Br \rightleftharpoons R\!-\!\overset{+}{\underset{H}{O}}\!\!\diagup\!\!^H + Br^-$$

Alkyloxonium ion

Specific examples:

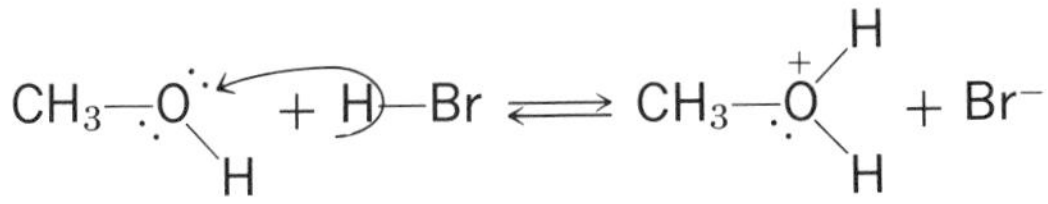

$$CH_3\!-\!\overset{..}{\underset{H}{O}} + H\!-\!Br \rightleftharpoons CH_3\!-\!\overset{+}{\underset{H}{O}}\!\!\diagup\!\!^H + Br^-$$

Methyloxonium bromide
(in solution only)

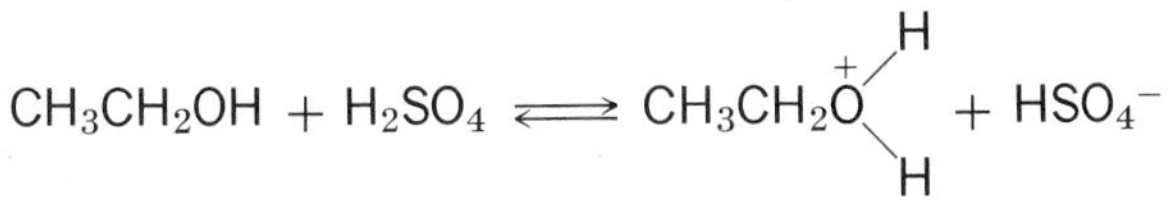

$$CH_3CH_2OH + H_2SO_4 \rightleftharpoons CH_3CH_2\overset{+}{\underset{H}{O}}\!\!\diagup\!\!^H + HSO_4^-$$

Ethyloxonium hydrogen sulfate
(in solution only)

The alkyloxonium salts that form exist only in solution. They are important because the oxygen of the alcohol acquires a positive charge when it picks up a proton. As a result, the pair of electrons this oxygen shares with the carbinol carbon is much more strongly attracted toward the oxygen than before. In fact, the alcohol is on the verge of ionizing as follows:

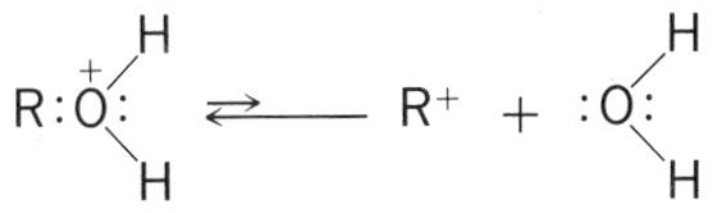

$$R\!:\!\overset{+}{\underset{H}{O}}\!\!\diagup\!\!^H \rightleftharpoons R^+ + :\overset{H}{\underset{H}{O}}\!:$$

| Alkyloxonium ion | Alkyl- | Water |
| (a protonated alcohol) | carbonium ion | |

Whether it will ionize depends on several factors to be discussed, but the possibility leads us to a study of the third type of reactions undergone by alcohols.

## REACTIONS OF ALCOHOLS INVOLVING RUPTURE OF THE C—O BOND

**Nucleophilic Substitution Reactions. General Principles.** As mentioned earlier, the C—O bond is strong enough to resist ionization as follows:

$$R\text{—}OH \xcancel{\longrightarrow} R^+ + OH^-$$

(does not happen)

However, if some acid interacts with the oxygen to put a positive charge on it, $R\text{—}\overset{+}{O}H_2$, the C—O bond is considerably weakened. Whether or not it breaks to form a carbonium ion and water ($R\text{—}\overset{+}{O}H_2 \longrightarrow R^+ + OH_2$) depends on two factors. (1) Can the solvent stabilize the carbonium ion somewhat by solvating it? (2) Are there structural features in the developing carbonium ion that tend to stabilize it?

With respect to the first question, the presence of water or water-like molecules (e.g., low-formula-weight alcohols) will help the formation of $R^+$ because they can solvate this ion. As for the second question, we have already learned that 3° carbonium ions are more stable than 2° which are more stable than 1° carbonium ions.

We might therefore expect that C—O bonds in alcohols will most readily break (1) in the presence of strong acids, (2) in a solvent of good ionizing power, and (3) among those alcohols (3°) capable of yielding the most stable carbonium ions. In a solvent of poor ionizing power and among those alcohols (1°) incapable of forming the more stable carbonium ions, ionization will not occur. The *leaving group*, however, which is water for ROH$_2^+$, can still be "kicked out" by some species

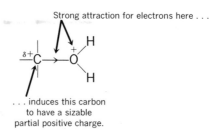

Strong attraction for electrons here . . .

. . . induces this carbon
to have a sizable
partial positive charge.

attracted to the developing partial positive charge on the carbinol carbon. Such a species, to act this way, must be electron-rich. It could be electrically neutral provided it had unshared pairs of electrons (e.g., :NH$_3$, or $R\text{—}\overset{..}{O}\text{—}H$), or it could be negatively charged (e.g., halide ion). Such electron-rich particles capable of bonding to carbon are said to be *nucleophilic reagents* or, simply, *nucleophiles* ("-phile" from *philos,* Greek for loving; "nucleo-," meaning nucleus—in this instance an ambiguous way of talking about a site with only a *partial* positive charge, such as a carbon at the end of a bond polarized so as to place a partial positive charge on it).

Strong nucleophiles are often, but not always, strong Brønsted bases or proton acceptors.[1] This property makes possible another important reaction of alcohols,

---

[1]Basicity in this sense refers to the position of *equilibrium* in an acid-base interaction involving proton donors and acceptors. Strong bases are those capable of coordinating strongly with *protons.* Nucleophilicity refers to the relative *rate* at which the nucleophile can coordinate or bond to *carbon* while (or after) some leaving group departs. Bromide ion (or, better, iodide ion) is a strong nucleophile but a weak base. A more detailed discussion of these factors is not intended, but the instructor may wish to use these statements as a starting point.

elimination of the elements of water to form alkenes. After we have discussed the substitution reactions of alcohols, this elimination reaction will be integrated with general principles covering both types.

What we are about to do, then, is to study in some depth two of the most important types of reactions that organic compounds can undergo, nucleophilic substitution reactions and elimination reactions. Descriptions of these events answer the question "What happens?" The theoretical parts answer the question "How does it happen?" The principles we shall study have wide applicability. Careful and patient attention to them now will be very helpful to our future study. In most courses in organic chemistry these principles are examined during a study of alkyl halides, but this class of organic compounds is of minor significance in biochemistry. We shall study the principles in connection with the alcohols to which they also apply and which are very important to our future work. (Both for convenience and completeness, however, some of the chemistry of alkyl halides is included.)

**Formation of Alkyl Halides by Reaction of Alcohols with Hydrogen Halides.** In general:

$$R—OH \ + \ H—X \longrightarrow R—X + H_2O$$

| Alcohol | Hydrogen | Alkyl |
| (R = alkyl, *not* | halide | halide |
| aryl) | (X = Cl, Br, I) | |

Specific examples:

$$CH_3CH_2CH_2CH_2OH + concd\ HBr \xrightarrow{heat} CH_3CH_2CH_2CH_2Br + H_2O$$

*n*-Butyl alcohol                           *n*-Butyl bromide (95%)

$$CH_3CH_2CH_2CH_2CH_2OH + concd\ HBr \xrightarrow{heat} CH_3CH_2CH_2CH_2CH_2Br + H_2O$$

1-Pentanol                           1-Bromopentane (78%)

$$CH_3-\underset{\underset{\displaystyle CH_3}{|}}{\overset{\overset{\displaystyle CH_3}{|}}{C}}-OH + concd\ HCl \xrightarrow[temperature]{room} CH_3-\underset{\underset{\displaystyle CH_3}{|}}{\overset{\overset{\displaystyle CH_3}{|}}{C}}-Cl + H_2O$$

*t*-Butyl alcohol                         *t*-Butyl chloride (88%)

The presence of a strong acid is essential for these reactions. Salts such as sodium bromide or potassium chloride, also suppliers of halide ions, do not react with alcohols.

The order of reactivity of the subclasses of alcohols is

$$3° > 2° > 1° > CH_3OH$$

which is the order of stability of the corresponding carbonium ions. These two facts, the need for acid and the order of reactivity, are understood in terms of the most probable mechanism for the reaction. Tertiary alcohols normally react by one mechanism, primary alcohols by another. For tertiary alcohols the following steps are believed to occur.

Step 1. The alcohol is protonated by the acid.

$$R\overset{\cdot\cdot}{\underset{\cdot\cdot}{O}}H + H\!\!-\!\!X \rightleftharpoons R\!\!-\!\!\overset{\overset{\displaystyle H}{|}}{\underset{\displaystyle H}{\overset{+}{O}}}\!\!: \;+\; X^- \qquad \text{(rapid formation of equilibrium)}$$

Step 2. The C—O bond is now much weaker and breaks, forming a carbonium ion.

$$R\!\!-\!\!\overset{\overset{\displaystyle H}{\diagup}}{\underset{\displaystyle H}{\overset{+}{O}}}\!\!: \;\rightleftharpoons\; R^+ + H_2O \qquad \text{(slow rate of ionization)}$$

Carbonium ion

Step 3. The carbonium ion combines with a halide ion.

$$R^+ + X^- \rightleftharpoons R\!\!-\!\!X \qquad \text{(rapid)}$$

The slow step, step 2, is the bottleneck in the reaction, the *rate-determinining step*. The final product cannot form any faster than carbonium ions can be generated. It does not matter how concentrated the solution is in halide ions; they do not enter the reaction until step 2 provides carbonium ions. Since this slow step involves something happening to only one molecule, the entire reaction is called a *unimolecular* process. Just how the C—O bond breaks may be visualized as follows. This bond, as any bond, can be likened to a tiny spring.[2] The groups joined by the bond vibrate—stretch in and out and back and forth—at the ends of the spring. But any spring has its limit. If sufficient energy is provided, it may become stretched beyond this limit and break. In any large collection of alkyloxonium ions, $R\!\!-\!\!OH_2^+$, a certain fraction of them will at any moment receive enough energy of activation from collisions and absorption to break the C—O spring, separating into the fragments $R^+$ and $H_2O$. Since the organic ion, $R^+$, is not preformed in the alcohol, R—OH, and since no carbonion ion, not even a 3° one, is very stable, it takes energy and therefore time for its formation. A progress of reaction diagram, Figure 4.7, explains this further.

The order of reactivity of alcohols is understandable in terms of this mechanism. The most reactive alcohols are those that can form the most stable carbonium ions. The mechanism has a useful shorthand symbol, $S_N1$, with $S$ standing for substitution, $N$ for nucleophilic, and 1 for unimolecular. This symbol could also be read "carbonium-ion mechanism" or "order of reactivity will be the same as order of stability of carbonium ions" or "this reaction pathway will be favored when the reaction occurs in solvents of good ionizing power." The one symbol $S_N1$ carries all this meaning.

Primary alcohols would yield primary carbonium ions if they reacted by an $S_N1$ mechanism. Probably because these are the least stable of carbonium ions,

[2]This is a model for a bond that is useful in other situations too, for example, in understanding certain aspects of infrared spectroscopy.

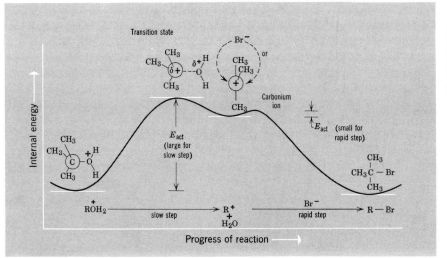

**Figure 4.7** The $S_N1$ mechanism. The rate-determining step (slow step) is the ionization of the bond between carbon and the group leaving, which is here a water molecule. The $E_{act}$ for this is relatively large, which means that the rate is slow in relation to other steps. Once the carbonium ion forms, it rapidly combines with any nucleophile nearby—bromide ion in the example.

primary alcohols normally react by another mechanism, one not involving a carbonium ion. Step 1 in this new mechanism is the same as before.

Step 1.

$$R-CH_2-OH + H-X \rightleftharpoons R-CH_2-\overset{+}{\underset{H}{O}}{}^{H} + X^- \quad \text{(rapid)}$$

Step 2. The nucleophile, $X^-$, attacks the carbon holding the oxygen.

$$X^- + CH_2-\overset{+}{\underset{H}{O}}{}^{H}_R \longrightarrow X-\overset{R}{\underset{}{CH_2}} + H_2O \quad \text{(slow)}$$

Alkyl halide

This is the rate-determining step, and because two particles participate directly in it, $X^-$ and $RCH_2\overset{+}{O}H_2$, the overall process is called *bimolecular*. Experimentally, the rate at which alkyl halide forms will be doubled by doubling the concentration of either $X^-$ or $RCH_2\overset{+}{O}H_2$. Or the rate can be cut in half by cutting either concentration in half. The shorthand symbol for this mechanism is $S_N2$, with $S$ for substitution, $N$ for nucleophilic, and 2 for bimolecular. Also implied by this symbol is "no carbonium ion," as well as other aspects to be discussed.

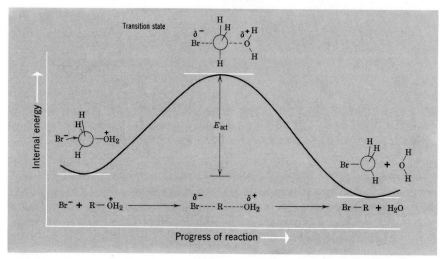

**Figure 4.8** The $S_N2$ mechanism. The rate-determining step (slow step) produces a transition state in which carbon is temporarily holding five groups. The bulkier these are, the higher $E_{act}$ is and the slower the rate will be. Note that as the reaction proceeds the bonds from the central carbon are turned inside out like an umbrella in a gale (inversion of configuration).

The $S_N2$ mechanism is discussed further in Figure 4.8. The transition state for the reaction is an unstable species in which five groups are for a very brief time partially or fully attached to one carbon. There is considerable crowding in this

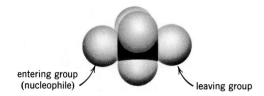

entering group
(nucleophile)                                    leaving group

The larger these three groups around the
middle, the more energetic a collision
between the nucleophile and the
carbon must be to be successful.

particle, which helps us understand how it is that 1° alcohols have a much greater tendency than 3° alcohols to react by an $S_N2$ process. As Figure 4.9 shows, the carbon that must be attacked by the nucleophile is much more open to this attack in a 1° alcohol than in a 3° alcohol. Furthermore, one factor favoring attack at this carbon is the partial positive charge there. In a 3° alcohol this partial charge is reduced by the inductive effect of the alkyl groups attached to the 3° carbon, but in the 1° alcohol no such reduction occurs. Therefore the 1° alcohols have a greater tendency than the 3° alcohols to react by an $S_N2$ mechanism for three reasons: (1) the 1° carbonium ion is less stable than the 3°; (2) the partial positive charge on the carbon of the C—O system is greater in the 1° alcohol than in the 3°; (3) this carbon in the 1° alcohol is more open to attack than that in the 3°. The last factor is called a *steric factor*—it has to do with spatial requirements for certain reaction pathways.

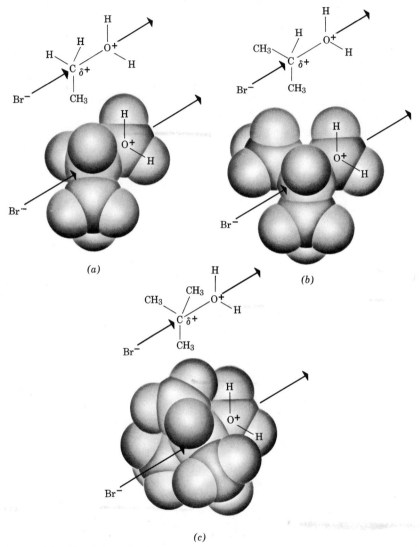

(a)

(b)

(c)

**Figure 4.9** The $S_N2$ mechanism, steric factors. Attack by a bromide ion on the protonated form of a 1° alcohol, part *a*, encounters relatively little hindrance from groups already attached to the carbinol carbon. For a 2° alcohol, part *b*, there are two alkyl groups on the carbinol carbon, and they cause some hindrance. With the 3° alcohol, part *c*, this steric hindrance is at its maximum; such alcohols usually react by an $S_N1$ process.

Since ionization is not a step in the $S_N2$ mechanism, the solvent need not have good solvating power.

The symbol $S_N2$ therefore carries all the following meanings:

—Substitution, nucleophilic, bimolecular.
—Order of reactivity: $CH_3OH > 1° > 2° > 3°$ (just the reverse of $S_N1$).
—Carbonium ion not involved, but rather a transition state with the attacked carbon temporarily pentavalent.
—The configuration of the groups at the attacked carbon being turned inside out or inverted (cf. Figure 4.8).

Since 2° alcohols are capable of forming carbonium ions of intermediate stability, and since they are not as "crowded" at the carbinol carbon as in a 3° alcohol, some molecules may react by an $S_N1$ process and others in the same mixture by an $S_N2$ process.

**Conversion of Alcohols into Alkenes. Elimination Reactions.** In general:

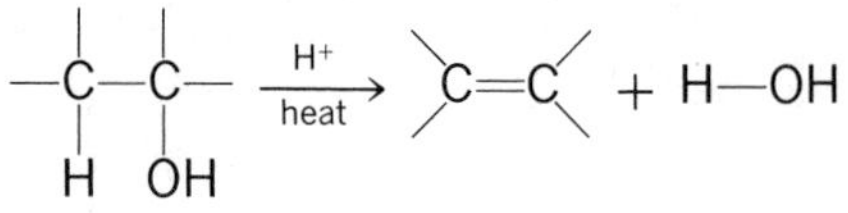

$$-\overset{|}{\underset{\underset{H}{|}}{C}}-\overset{|}{\underset{\underset{OH}{|}}{C}}- \xrightarrow[\text{heat}]{H^+} \quad {\diagup}C{=}C{\diagdown} + H{-}OH$$

Alcohol                 Alkene

Specific examples:

$$CH_3CH_2{-}OH \xrightarrow[\text{170–180°C}]{\text{concd } H_2SO_4} CH_2{=}CH_2 + H_2O$$

Ethyl alcohol                             Ethylene

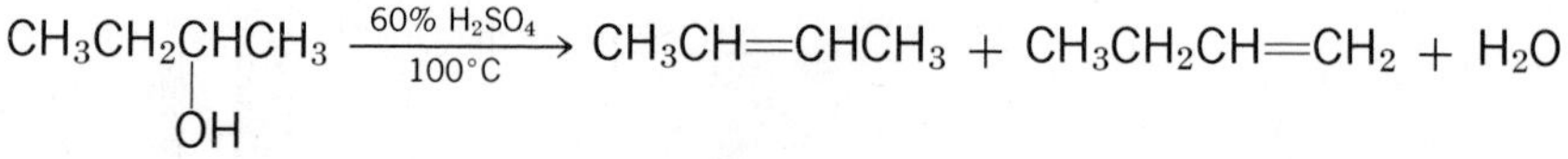

$$CH_3CH_2\underset{\underset{OH}{|}}{C}HCH_3 \xrightarrow[\text{100°C}]{60\% \ H_2SO_4} CH_3CH{=}CHCH_3 + CH_3CH_2CH{=}CH_2 + H_2O$$

*sec*-Butyl alcohol                  2-Butene           1-Butene
                         (principal product)

$$CH_3{-}\overset{\overset{CH_3}{|}}{\underset{\underset{CH_3}{|}}{C}}{-}OH \xrightarrow[\text{80–90°C}]{20\% \ H_2SO_4} CH_2{=}\overset{\overset{CH_3}{\diagup}}{C}{\diagdown_{CH_3}} + H_2O$$

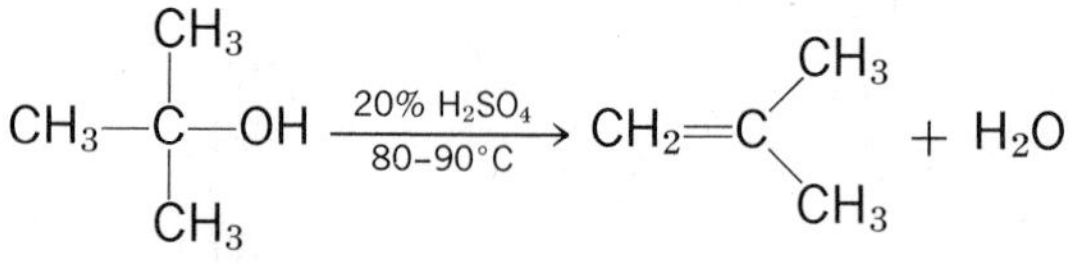

*t*-Butyl alcohol                 Isobutylene

In the general example we see that the elements of water, H and OH, are taken from *adjacent* carbons. In the specific examples it is much easier to form the alkene from *t*-butyl alcohol than from ethyl alcohol. The examples illustrate something that is generally true about this reaction: the order of ease of dehydration of alcohols to form alkenes is

$$3° > 2° > 1°$$

Exercise 4.4 Methyl alcohol is not included in this order. Why?

123

Reactions of
Alcohols
Involving Rupture
of the C—O Bond

Two more facts are illustrated by the examples. An acid catalyst is needed. Furthermore, whenever it is possible for more than one alkene to form, the one more highly branched will be the major product (cf. dehydration of *sec*-butyl alcohol). By "more highly branched" is meant one in which the greater number of bonds to other carbons go out from the C=C system. 2-Butene has two branches by this definition, 1-butene only one.

(We shall mention one additional fact. During the dehydration of some alcohols, the intermediate carbonium ions rearrange, forming alkenes with structures that could not have been predicted from the general example given at the beginning of this section. We encountered carbonium-ion rearrangements in Chapter 3. Nothing further will be said about them.)

The facts of the dehydration reaction are correlated by the mechanisms, two of which are recognized as important. One is very similar to the $S_N1$ mechanism and is designated $E_1$, the $E$ for elimination, the 1 for unimolecular. The other resembles the $S_N2$ and is called $E_2$ or elimination, bimolecular.

**The $E_1$ Mechanism.** The first two steps are identical with the first two steps in an $S_N1$ process, and they may be illustrated by the dehydration of *t*-butyl alcohol.

Step 1. Protonation of the alcohol, a rapid acid-base equilibrium. Protona-

$$(CH_3)_3COH + H_2SO_4 \rightleftharpoons (CH_3)_3C\overset{+}{-}O\overset{H}{\underset{H}{\diagdown}} + HSO_4^-$$

*t*-Butyl alcohol

tion of the oxygen places on it a positive charge, which attracts electrons to it from the C—O bond. A molecule of water is ready to leave.

Step 2. Ionization at the C—O bond, carbonium-ion formation, the rate-determining step. Since primary carbonium ions are the least stable types, it is under-

$$(CH_3)_3C\overset{+}{-}O\overset{H}{\underset{H}{\diagdown}} \longrightarrow \left[(CH_3)_3\overset{\delta+}{C}\cdots\overset{\delta+}{O}\overset{H}{\underset{H}{\diagdown}}\right] \longrightarrow (CH_3)_3C^+ + H_2O$$

Transition state      *t*-Butylcarbonium
ion

standable why 1° alcohols dehydrate much more slowly than 3° alcohols. Having formed, the carbonium ion can now be attacked by some nucleophile, but this would lead to a by-product, not to an alkene. It is alkene formation that we seek to explain here.

Step 3. Removal of a proton from the carbonium ion. Monohydrogen sulfate ion ($HSO_4^-$) is shown acting as the proton acceptor, although a water molecule or

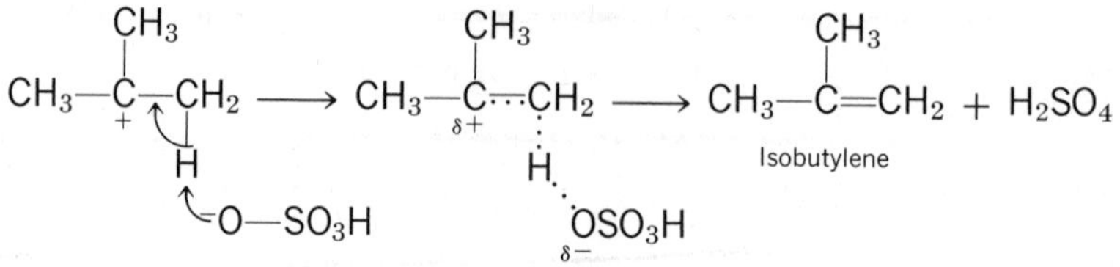

$$CH_3-\underset{\underset{\displaystyle \overset{\displaystyle H}{\underset{\displaystyle O-SO_3H}{|}}}{+}}{\overset{\displaystyle \overset{\displaystyle CH_3}{|}}{C}}-CH_2 \longrightarrow CH_3-\underset{\overset{\displaystyle H}{\underset{\displaystyle OSO_3H}{\delta+}}}{\overset{\displaystyle \overset{\displaystyle CH_3}{|}}{C}}-CH_2 \longrightarrow CH_3-\overset{\displaystyle \overset{\displaystyle CH_3}{|}}{C}=CH_2 + H_2SO_4$$

Isobutylene

Second transition state of
the $E_1$ mechanism

another alcohol molecule could serve the same function. As the proton is pulled off, the pair of electrons holding it to the carbon pivots toward the positively charged carbon and the pi bond forms.

### The $E_2$ Mechanism. Illustrated with Ethyl Alcohol

Step 1. Rapidly established acid-base equilibrium, protonation of the alcohol. This step is identical with the first step in either the $E_1$ or the $S_N1$ reactions. If the

$$CH_3CH_2OH + H_2SO_4 \rightleftharpoons CH_3CH_2-\overset{\displaystyle \overset{+}{\underset{\displaystyle H}{O}}\nearrow^H}{} + HSO_4^-$$

temperature of the medium is unusually high, enough energy may be available for the reaction of a $1°$ alcohol to continue by means of a carbonium-ion mechanism, but at lower temperatures (below the boiling point of the mixture) the next step is bimolecular.

Step 2. A proton acceptor strikes the hydrogen that is to be pulled off *while* the molecule of water leaves and the pi bond forms. It is said to be a *concerted* process.

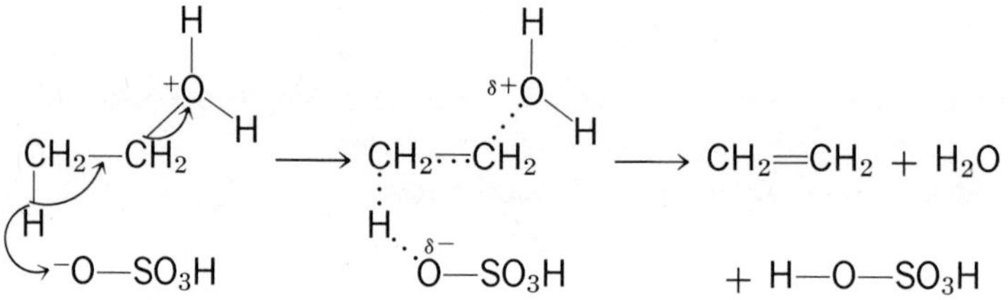

$$CH_2-CH_2 \longrightarrow CH_2\cdots CH_2 \longrightarrow CH_2=CH_2 + H_2O$$

Bisulfate ion      Transition state of
the $E_2$ mechanism      Sulfuric acid catalyst
is recovered

$$+ H-O-SO_3H$$

These mechanisms explain the necessity for an acid catalyst to protonate the alcohol, weaken the C—O bond, and create a stable leaving group, water. They also explain the order of ease of dehydration of alcohols. We have yet to explain the formation of the most highly branched alkene.

**Relative Stabilities of Alkenes.** The following reactions, including the heats of hydrogenation, illustrate a general trend. The more highly branched the alkene, the more stable it is. Since the same product forms in each case, if the 2-butene isomers give off less energy than 1-butene, they must initially have had less energy. Hence

$$CH_3CH_2CH=CH_2 + H_2 \xrightarrow{\text{catalyst, etc.}} CH_3CH_2CH_2CH_3 \quad \Delta H = -30.3 \text{ kcal/mole}$$

1-Butene                             Butane

$$CH_3CH\!\!=\!\!CHCH_3 + H_2 \longrightarrow CH_3CH_2CH_2CH_3$$

2-Butene
(cis or trans)

Butane

$\Delta H = -27.6$ kcal/mole
(from the trans alkene)

$\Delta H = -28.6$ kcal/mole
(from the cis alkene)

they must be more stable. They are also the more highly branched alkenes, which does not explain *why* they are more stable but simply provides some evidence that they are.

In both the $E_1$ and the $E_2$ mechanisms a transition state with a partial double bond emerges. We may assume that whatever stabilizes a full double bond will also stabilize a partial one, which means that the transition state leading to the more highly branched alkene is of lower internal energy and more stable than the one leading to the less highly branched isomer. Consequently, the energy of activation needed for the more stable alkene is less and the rate of its formation is faster than that for the less stable alkene. In the dehydration of *sec*-butyl alcohol, shown in the specific examples (p. 122), 2-butene, is the major product because its rate of formation is faster than that for 1-butene, And this rate is faster because the energy of activation leading to it is less. We have just explained why this energy of activation is lower. Whenever there are competing reactions—and they are the rule rather than the exception in organic chemistry—the product forming at the faster rate will obviously be the major product. Figure 4.10 illustrates the dehydration of *sec*-butyl alcohol by an $E_2$ mechanism.

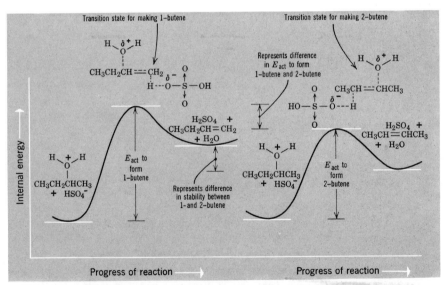

**Figure 4.10** The $E_2$ mechanism, competing reactions. In the acid-catalyzed dehydration of 2-butanol, 2-butene rather than 1-butene is the major product. 2-Butene is known to be slightly more stable than 1-butene, and whatever provides this extra stability probably stabilizes the transition state leading to it. As shown here, the $E_{act}$ for forming 2-butene is therefore probably less than that for forming 1-butene; and this automatically means that 2-butene forms at a faster rate than its isomer.

Competing reactions are usually the rule in the chemical systems of living cells. Which of several possible reactions takes place is determined not only by the availability of the chemicals and by how relatively favorable the internal energy changes may be, but also by the presence of a catalyst (enzyme) that somehow lowers the energy of activation for one of the possible reactions and not for any of the others.

## OXIDATION OF ALCOHOLS

Primary and secondary alcohols may be oxidized by a variety of reagents to compounds whose molecules contain a carbonyl group, one of the most widely

$$:O:$$
$$\overset{\|}{\underset{/\,\,\backslash}{C}}$$

Carbonyl
group

occurring functional groups in organic chemistry. A variety of families whose names and properties are studied in later chapters are characterized by it. We shall concentrate here only on the fact that alcohols are a potential source of the following types:

$$\overset{O}{\underset{\|}{R-C-H}} \qquad \overset{O}{\underset{\|}{R-C-O-H}} \qquad \overset{O}{\underset{\|}{R-C-R'}}$$

Aldehydes          Carboxylic            Ketones
                     acids

When an alcohol is oxidized, the "elements" of hydrogen, $(H:^- + H^+)$,[3] are lost:

$$\overset{\backslash}{\underset{H}{C}}-O \underset{H}{} \longrightarrow \overset{\backslash}{C}=O + (H:^- + H^+)$$

1° or 2° alcohol        Aldehyde         "Elements" of
                      or ketone          hydrogen

It is only a matter of convenience to write the elements of hydrogen as shown. In most oxidations the oxidizing agent converts them to water, but in one procedure hydrogen gas, $H_2$, is produced. In higher animals the elements of hydrogen are passed along a series of enzymes until they eventually combine with oxygen taken in at the lungs and form water. It is interesting that the word "aldehyde" is a contraction of "*alcohol-dehydr*ogenation product."

**Oxidation of 1° Alcohols. Synthesis of Aldehydes and Carboxylic Acids.** Primary alcohols are oxidized first to aldehydes, but aldehydes are even more readily oxidized than alcohols, with the product of this second oxidation a carboxylic acid.

---

[3] The expression "elements of hydrogen," a contraction of "elements of the element hydrogen, $H_2$," refers to particles which, if somehow formed and mixed, would combine to give hydrogen gas.

As soon as some aldehyde forms, it competes for unchanged oxidizing agent with unchanged 1° alcohol, and it wins, provided it is not removed from the solution. Primary alcohols are therefore an excellent source of carboxylic acids. The oxidation can be stopped at the aldehyde stage by carefully choosing the oxidizing agent or by removing the aldehyde as soon as it forms. Since the aldehydes of lower formula weight usually have much lower boiling points than the parent alcohols, they can be distilled continuously from the mixture. In general (the equations are not balanced):

$$RCH_2OH \xrightarrow[200–300°C]{Cu} R-\overset{\overset{O}{\|}}{C}-H + H_2$$

1° alcohol          Aldehyde

$$RCH_2OH + KMnO_4 \longrightarrow R-\overset{\overset{O}{\|}}{C}-H \xrightarrow{KMnO_4} R-\overset{\overset{O}{\|}}{C}-O^-K^+ + MnO_2 + KOH$$

Potassium              Salt of carb-     Manganese
permanganate           oxylic acid       dioxide
                                          (brown sludge)

$$\Big\downarrow H_3O^+$$

$$R-\overset{\overset{O}{\|}}{C}-O-H + H_2O + K^+$$

Carboxylic acid

$$RCH_2OH + Na_2Cr_2O_7 \xrightarrow{H^+} R-\overset{\overset{O}{\|}}{C}-H \xrightarrow{Cr_2O_7{}^{2-}} R-\overset{\overset{O}{\|}}{C}-OH + Cr^{3+}$$

Sodium                             Chromium
dichromate                         ion
(bright orange)                    (bright green)

Specific examples (equations are not balanced, sufficient oxidizing agent is assumed available):

Aldehydes from 1° alcohols:

$$CH_3CH_2OH \xrightarrow{(O)} CH_3\overset{\overset{O}{\|}}{C}H \qquad (O) = Cr_2O_7{}^{2-}, H^+;\ 74\% \text{ yield}$$

Ethyl alcohol          Acetaldehyde
(bp 78°C)              (bp 20°C)

$$CH_3CH_2CH_2OH \xrightarrow{(O)} CH_3CH_2\overset{\overset{O}{\|}}{C}H \qquad \begin{array}{l}(O) = Cu, \text{ high temperature; } 67\% \text{ yield} \\ = Cr_2O_7{}^{2-}, H^+;\ 49\% \text{ yield}\end{array}$$

n-Propyl alcohol       Propionaldehyde
(bp 97°C)             (bp 55°C)

$$CH_3CH_2CH_2CH_2OH \xrightarrow{(O)} CH_3CH_2CH_2\overset{\displaystyle O}{\overset{\displaystyle \|}{C}}H$$

(O) = Cu, heat; 62% yield
    = $Cr_2O_7^{2-}$, $H^+$; 72% yield

*n*-Butyl alcohol
(bp 118°C)

*n*-Butyraldehyde
(bp 82°C)

$$CH_3(CH_2)_7CH_2OH \xrightarrow{(O)} CH_3(CH_2)_7\overset{\displaystyle O}{\overset{\displaystyle \|}{C}}H$$

(O) = Cu, heat; 90% yield

1-Nonanol
(bp 212°C)

Nonanal
(bp 185°C)

Carboxylic acids from 1° alcohols:

$$CH_3CH_2CH_2OH \xrightarrow{Cr_2O_7^{2-},\ H^+} CH_3CH_2\overset{\displaystyle O}{\overset{\displaystyle \|}{C}}OH$$    (65% yield)

*n*-Propyl alcohol

Propionic acid

$$\underset{\text{2-Ethyl-1-hexanol}}{CH_3CH_2CH_2CH_2\overset{\displaystyle CH_3CH_2}{\overset{\displaystyle |}{C}}HCH_2OH} \xrightarrow[\text{followed by } H^+]{MnO_4^-,\ OH^-} \underset{\text{2-Ethylhexanoic acid}}{CH_3CH_2CH_2CH_2\overset{\displaystyle CH_3CH_2}{\overset{\displaystyle |}{C}}H-\overset{\displaystyle O}{\overset{\displaystyle \|}{C}}OH}$$    (74% yield)

$$\underset{\underset{\text{6-Methyl-1-octanol}}{\displaystyle |}}{CH_3CH_2\overset{}{C}H(CH_2)_4CH_2OH} \xrightarrow[\text{followed by } H^+]{MnO_4^-,\ OH^-} \underset{\underset{\text{6-Methyloctanoic acid}}{\displaystyle |}}{CH_3CH_2\overset{}{C}H(CH_2)_4\overset{\displaystyle O}{\overset{\displaystyle \|}{C}}OH}$$    (66% yield)
$$\quad CH_3 \qquad\qquad\qquad\qquad\qquad CH_3$$

**Ketones from 2° Alcohols.** The ketone system is characterized by a carbonyl

$$C-\overset{\displaystyle O}{\overset{\displaystyle \|}{C}}-C$$

Ketone system

group flanked on both sides by carbons, and it is resistant to further oxidation except under vigorous conditions.

$$\underset{\text{2-Butanol}}{CH_3\overset{\displaystyle OH}{\overset{\displaystyle |}{C}}HCH_2CH_3} \xrightarrow[\text{warm}]{Cr_2O_7^{2-},\ H^+} \underset{\text{2-Butanone}}{CH_3\overset{\displaystyle O}{\overset{\displaystyle \|}{C}}CH_2CH_3}$$    (74% yield)

Cyclohexanol $\xrightarrow[\text{warm}]{Cr_2O_7^{2-},\ H^+}$ Cyclohexanone    (85% yield)

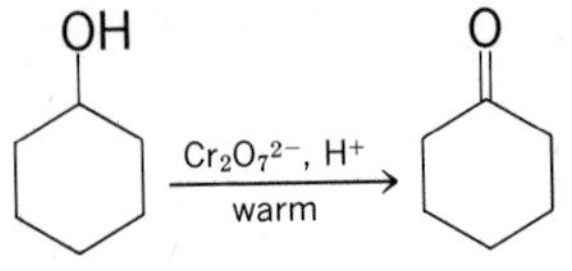

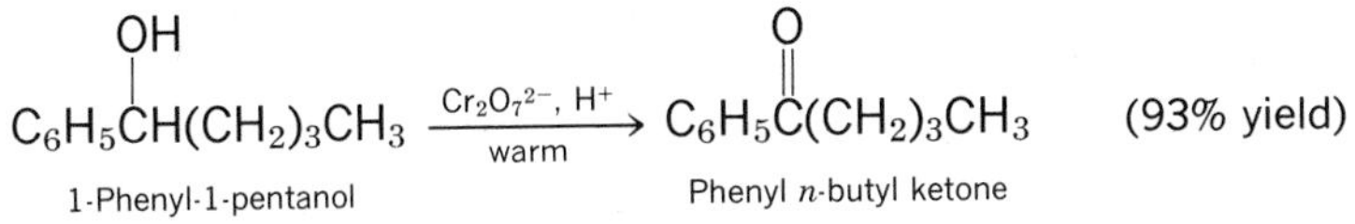

$$\underset{\text{1-Phenyl-1-pentanol}}{C_6H_5\overset{\displaystyle OH}{\underset{\displaystyle |}{C}}H(CH_2)_3CH_3} \xrightarrow[\text{warm}]{Cr_2O_7{}^{2-},\ H^+} \underset{\text{Phenyl } n\text{-butyl ketone}}{C_6H_5\overset{\displaystyle O}{\overset{\displaystyle \|}{C}}(CH_2)_3CH_3} \qquad (93\%\ \text{yield})$$

**Examples from molecular biology.** During the degradation of a long-chain carboxylic acid (delivered to the system by fats and oils), the following reaction takes place.

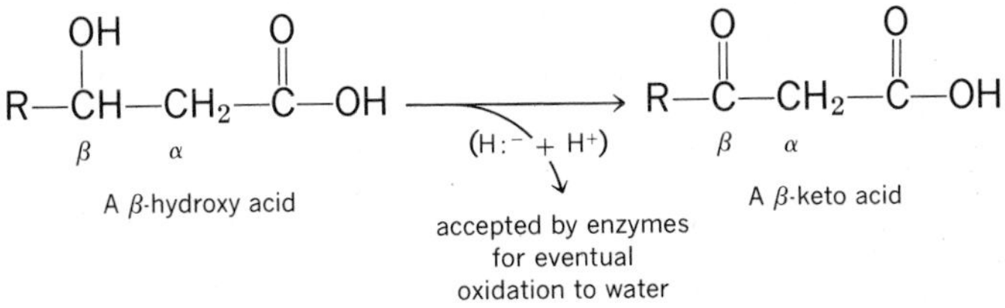

$$\underset{\text{A } \beta\text{-hydroxy acid}}{R-\overset{\displaystyle OH}{\underset{\displaystyle |}{C}}H-\underset{\alpha}{\underset{\beta}{CH_2}}-\overset{\displaystyle O}{\overset{\displaystyle \|}{C}}-OH} \xrightarrow[\substack{(H:^- + H^+) \\ \downarrow \\ \text{accepted by enzymes} \\ \text{for eventual} \\ \text{oxidation to water}}]{} \underset{\text{A } \beta\text{-keto acid}}{R-\overset{\displaystyle O}{\overset{\displaystyle \|}{C}}-\underset{\alpha}{\underset{\beta}{CH_2}}-\overset{\displaystyle O}{\overset{\displaystyle \|}{C}}-OH}$$

Two other instances of the oxidation of a 2° alcohol occur during the citric acid cycle, a series of reactions by which acetic acid units ($CH_3\overset{\displaystyle O}{\overset{\displaystyle \|}{C}}-OH$) are degraded to carbon dioxide and water. The details are discussed more fully in Chapter 13, but two of its steps (shown here for purposes of illustration—they need not be memorized) are as follows.

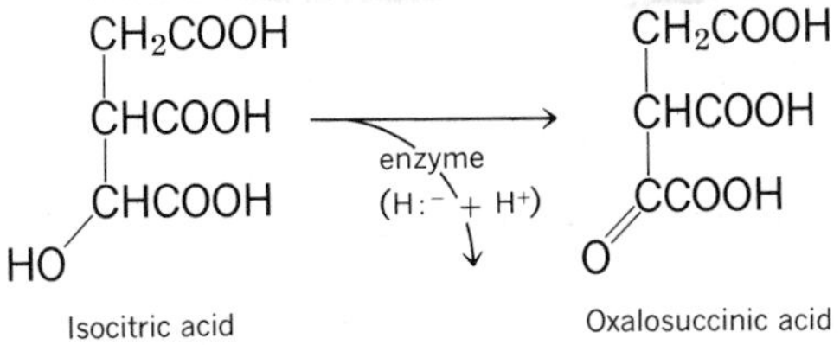

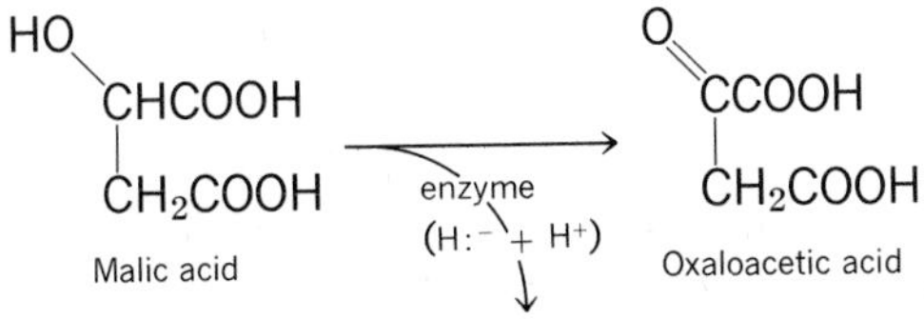

$$\underset{\text{Isocitric acid}}{\begin{array}{c} CH_2COOH \\ | \\ CHCOOH \\ | \\ CHCOOH \\ \overset{|}{HO} \end{array}} \xrightarrow[\substack{\text{enzyme} \\ (H:^- + H^+) \\ \downarrow}]{} \underset{\text{Oxalosuccinic acid}}{\begin{array}{c} CH_2COOH \\ | \\ CHCOOH \\ | \\ CCOOH \\ \overset{|}{O} \end{array}}$$

$$\underset{\text{Malic acid}}{\begin{array}{c} \overset{HO}{\diagdown} \\ CHCOOH \\ | \\ CH_2COOH \end{array}} \xrightarrow[\substack{\text{enzyme} \\ (H:^- + H^+) \\ \downarrow}]{} \underset{\text{Oxaloacetic acid}}{\begin{array}{c} \overset{O}{\diagdown} \\ CCOOH \\ | \\ CH_2COOH \end{array}}$$

*Tertiary alcohols are not oxidized if the medium is alkaline.* Under acidic conditions tertiary alcohols easily dehydrate to alkenes, and these are readily attacked by oxidizing agents.

## SYNTHESES OF ALCOHOLS

The alcohol function may be introduced into a molecule in the following ways. This list of methods is by no means complete.

1. **Reduction of a carbonyl group.** Reduction of an aldehyde produces a 1° alcohol.[4]

[4] Just as (O) was a symbol for an oxidizing agent, (2H) stands for a reducing agent that will accomplish the reaction.

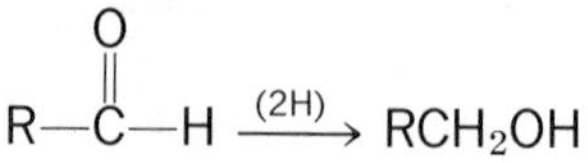

$$R-\overset{\overset{\displaystyle O}{\|}}{C}-H \xrightarrow{\text{(2H)}} RCH_2OH$$

1° alcohol

Reduction of a ketone produces a 2° alcohol.

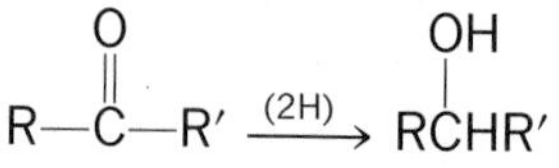

$$R-\overset{\overset{\displaystyle O}{\|}}{C}-R' \xrightarrow{\text{(2H)}} R\overset{\overset{\displaystyle OH}{|}}{C}HR'$$

2° alcohol

Hydrogen over a metal catalyst such as powdered nickel or platinum (catalytic reduction), lithium aluminum hydride (LiAlH$_4$), and sodium borohydride (NaBH$_4$) are a few of the reducing agents available. The reduction of a keto group to a 2° alcohol has its counterpart in molecular biology:

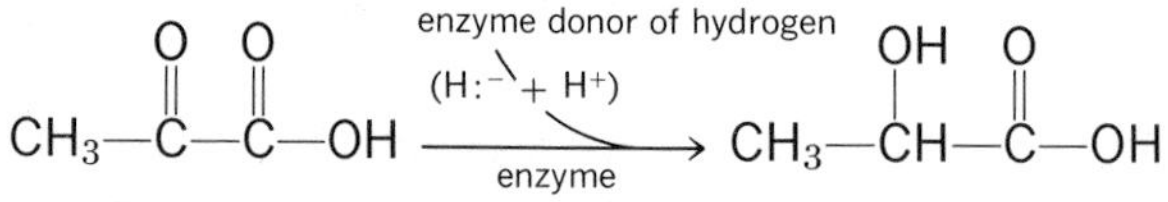

$$CH_3-\overset{\overset{\displaystyle O}{\|}}{C}-\overset{\overset{\displaystyle O}{\|}}{C}-OH \xrightarrow[\text{enzyme}]{\substack{\text{enzyme donor of hydrogen} \\ (H:^- + H^+)}} CH_3-\overset{\overset{\displaystyle OH}{|}}{C}H-\overset{\overset{\displaystyle O}{\|}}{C}-OH$$

Pyruvic acid          Lactic acid

Pyruvic acid is an intermediate in the metabolism of glucose. The reaction is reversible. The source of hydrogen is another organic compound, an enzyme, which is itself dehydrogenated in the reaction.

2. **The addition of water to a carbon-carbon double bond.** This method was discussed in Chapter 2. The reaction occurs in molecular biology, for example, during the metabolism of a carboxylic acid (fatty acid) when the elements of water add to the double bond of an unsaturated acid:

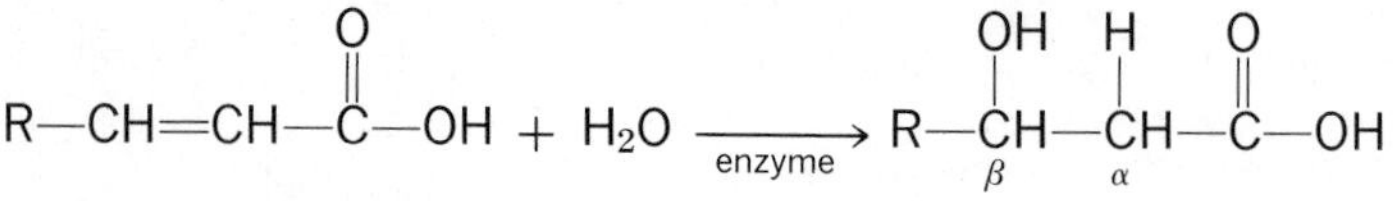

$$R-CH{=}CH-\overset{\overset{\displaystyle O}{\|}}{C}-OH + H_2O \xrightarrow{\text{enzyme}} R-\underset{\beta}{\overset{\overset{\displaystyle OH}{|}}{C}}H-\underset{\alpha}{\overset{\overset{\displaystyle H}{|}}{C}}H-\overset{\overset{\displaystyle O}{\|}}{C}-OH$$

Unsaturated fatty acid        A β-hydroxy acid

It also occurs at two places in the citric acid cycle, in the conversions of aconitic acid to isocitric acid and fumaric acid to malic acid; they are shown here only for purposes of illustrating how *in vitro* reactions have counterparts *in vivo*.[5]

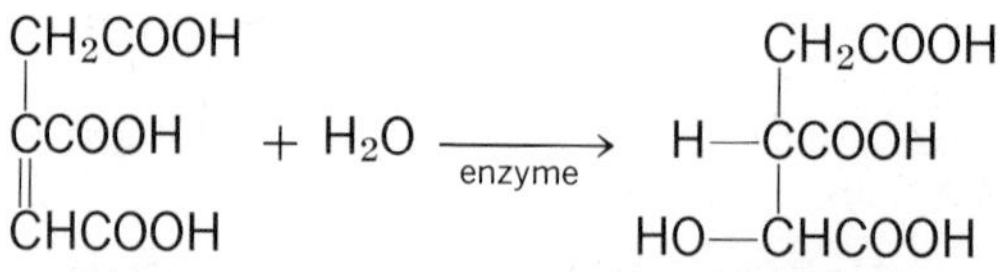

$$\begin{array}{l} CH_2COOH \\ | \\ CCOOH \quad + \; H_2O \xrightarrow{\text{enzyme}} \\ \| \\ CHCOOH \end{array} \qquad \begin{array}{l} CH_2COOH \\ | \\ H-CCOOH \\ | \\ HO-CHCOOH \end{array}$$

Aconitic acid              Isocitric acid

---

[5] *In vitro* means "in a glass vessel." The expression applies to reactions and processes performed with laboratory glassware. *In vivo* refers to processes in living systems.

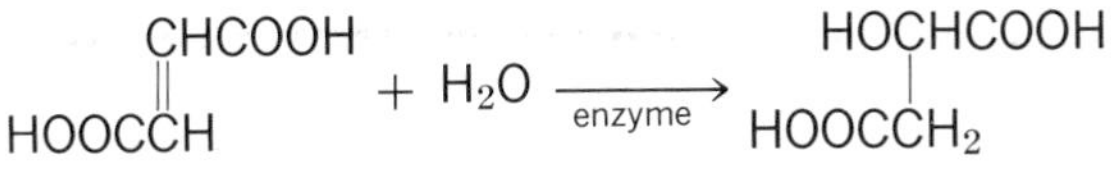

$$\underset{\text{Fumaric acid}}{\overset{\displaystyle \text{CHCOOH}}{\underset{\displaystyle \text{HOOCCH}}{\|}}} + H_2O \xrightarrow{\text{enzyme}} \underset{\text{Malic acid}}{\overset{\displaystyle \text{HOCHCOOH}}{\underset{\displaystyle \text{HOOCCH}_2}{|}}}$$

3. **Hydrolysis of alkyl halides.** This reaction,

$$R\!-\!X + H_2O \longrightarrow R\!-\!OH + HX$$

will be studied in Chapter 5. Phenols generally cannot be made this way.

4. **The Grignard synthesis of alcohols.** The Grignard reagent, discovered about 1900 by Victor Grignard, is one of the most versatile in all of organic chemistry. Its

---

Victor Grignard (1871–1935). French chemist and co-winner (with Paul Sabatier) of the 1912 Nobel prize in chemistry for his discovery of the reagent named after him and for his extensive work demonstrating that several types of organic compounds can be made from it.

---

structure is commonly written as RMgX, but the reagent is actually more complicated. It is prepared by the action of magnesium metal on an alkyl (or aryl) halide, usually in anhydrous diethyl ether:

$$\underset{\substack{\text{Alkyl halide} \\ (X = \text{Cl, Br, or I})}}{R\!-\!X + Mg} \xrightarrow{\text{CH}_3\text{CH}_2\text{OCH}_2\text{CH}_3} \underset{\text{Grignard reagent}}{RMgX}$$

The most general *in vitro* synthesis of alcohols is the action of a Grignard reagent on a carbonyl group. The synthesis does not occur *in vivo*.

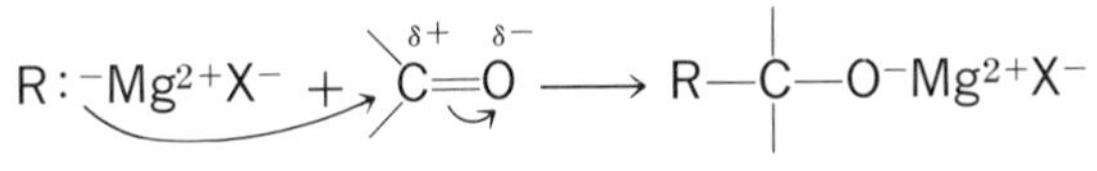

$$\underset{\text{Grignard reagent}}{R:^-Mg^{2+}X^-} + \underset{\substack{\text{Carbonyl group} \\ \text{of aldehyde or} \\ \text{ketone}}}{\overset{\delta+ \quad \delta-}{C\!=\!O}} \longrightarrow \underset{\text{Salt of an alcohol}}{R\!-\!\overset{|}{\underset{|}{C}}\!-\!O^-Mg^{2+}X^-}$$

Then the mixture is acidified:

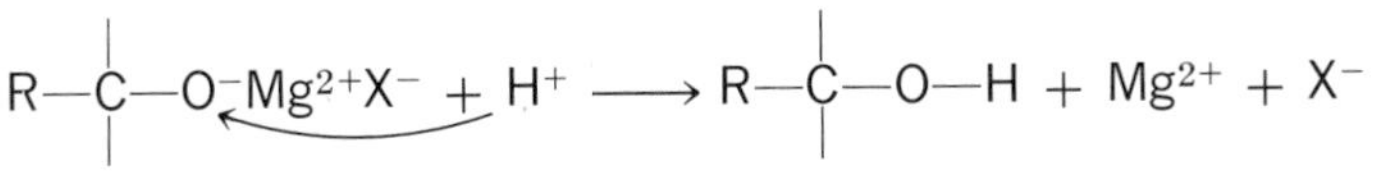

$$R\!-\!\overset{|}{\underset{|}{C}}\!-\!O^-Mg^{2+}X^- + H^+ \longrightarrow \underset{\text{Alcohol}}{R\!-\!\overset{|}{\underset{|}{C}}\!-\!O\!-\!H} + Mg^{2+} + X^-$$

In this reaction the Grignard reagent may be regarded as a source of *carbanions,* species in which a carbon has an unshared pair of electrons and a negative charge, as illustrated for the methyl carbanion:

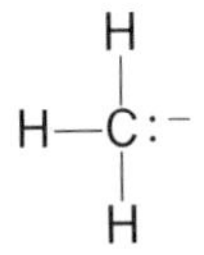

$$H\!-\!\overset{\displaystyle H}{\underset{\displaystyle H}{\overset{|}{\underset{|}{C}}}}\!:^-$$

Methyl carbanion

Primary alcohols are formed by combining Grignard reagents with formaldehyde. In general:

$$RMgX + H\overset{\overset{\displaystyle O}{\|}}{-C}-H \xrightarrow{\text{ether}} RCH_2O^-Mg^{2+}X^- \xrightarrow{H^+} RCH_2OH$$

Grignard    Formaldehyde                    Salt of alcohol           1° alcohol
reagent                                 (not isolated)

Specific examples:

$$CH_3CH_2CH_2CH_2MgBr + H\overset{\overset{\displaystyle O}{\|}}{-C}-H \xrightarrow[\text{by } H^+]{\text{followed}} CH_3CH_2CH_2CH_2CH_2OH$$

n-Butylmagnesium bromide                         1-Pentanol (68%)

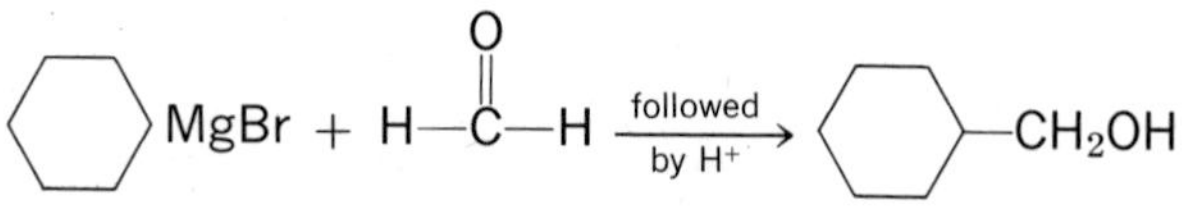

Cyclohexylmagnesium bromide                 Cyclohexylcarbinol (69%)

Grignard reagents combine with other aldehydes to form secondary alcohols. In general:

$$RMgX + R'\overset{\overset{\displaystyle O}{\|}}{-C}-H \xrightarrow{\text{ether}} R'\overset{\overset{\displaystyle O^-Mg^{2+}X^-}{|}}{-CH}-R \xrightarrow{H^+} R'\overset{\overset{\displaystyle OH}{|}}{-CH}-R$$

Aldehyde                   Salt of alcohol           2° alcohol
                            (not isolated)

Specific examples:

$$(CH_3)_2CHCH_2CH_2MgBr + CH_3\overset{\overset{\displaystyle O}{\|}}{C}H \xrightarrow[H^+]{\text{followed by}} (CH_3)_2CHCH_2CH_2\overset{\overset{\displaystyle OH}{|}}{C}HCH_3$$

Isoamylmagnesium bromide     Acetaldehyde             5-Methyl-2-hexanol (65%)

$$C_6H_5MgBr + C_6H_5\overset{\overset{\displaystyle O}{\|}}{C}H \xrightarrow[H^+]{\text{followed by}} C_6H_5\overset{\overset{\displaystyle OH}{|}}{C}HC_6H_5$$

Phenylmagnesium bromide   Benzaldehyde             Diphenylcarbinol (70%)
                                    ("benzhydrol")

Tertiary alcohols are formed by combining Grignard reagents with ketones. In general:

$$RMgX + R'\overset{\overset{\displaystyle O}{\|}}{-C}-R'' \xrightarrow{\text{ether}} R'\overset{\overset{\displaystyle O^-Mg^{2+}Br^-}{|}}{\underset{\underset{\displaystyle R}{|}}{-C}}-R'' \xrightarrow{H^+} R'\overset{\overset{\displaystyle OH}{|}}{\underset{\underset{\displaystyle R}{|}}{-C}}-R''$$

Specific examples:

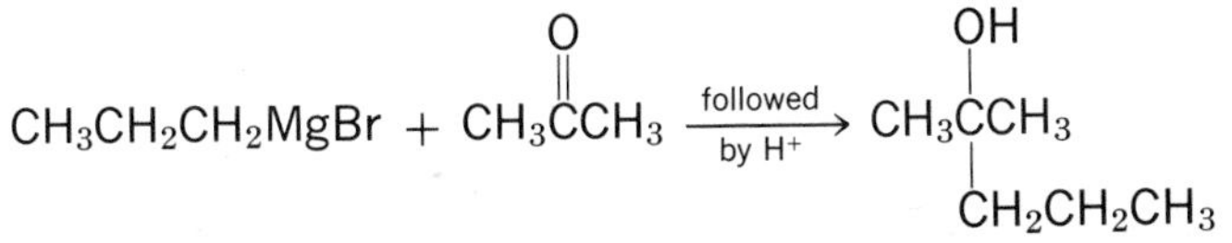

$$CH_3CH_2CH_2MgBr + CH_3\overset{O}{\underset{\|}{C}}CH_3 \xrightarrow[\text{by } H^+]{\text{followed}} CH_3\underset{\underset{CH_2CH_2CH_3}{|}}{\overset{\overset{OH}{|}}{C}}CH_3$$

*n*-Propylmagnesium bromide          Acetone          2-Methyl-2-pentanol (50%)

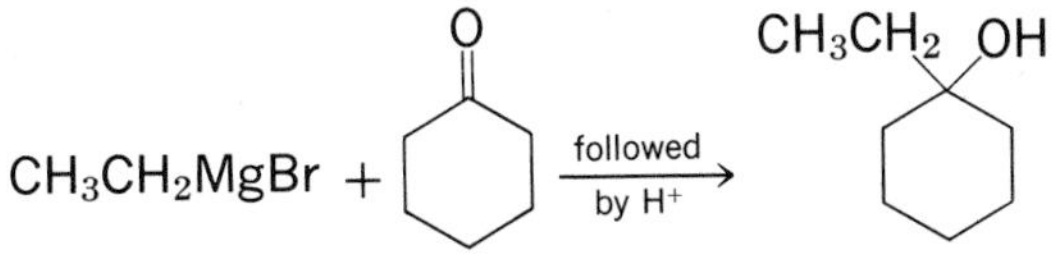

$$CH_3CH_2MgBr + \quad \overset{O}{\underset{}{\bigcirc}} \quad \xrightarrow[\text{by } H^+]{\text{followed}}$$

Ethylmagnesium bromide    Cyclohexanone    1-Ethylcyclohexanol (62%)

---

**Exercise 4.5**  Grignard reagents are formed by the action of magnesium metal turnings on an alkyl halide in dry ether. Write the structures of an alkyl halide and an aldehyde or ketone that together can be used to prepare each of the following alcohols by means of the Grignard reaction.

(a)  $CH_3\overset{\overset{OH}{|}}{C}HCH_3$

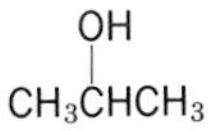

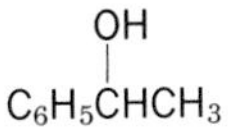

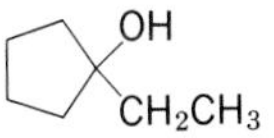

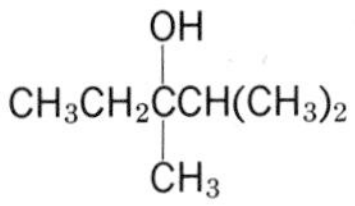

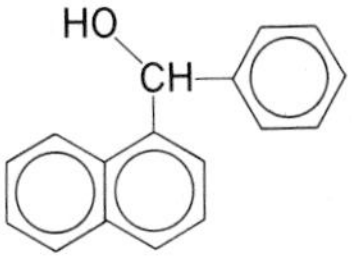

(b)  $C_6H_5\overset{\overset{OH}{|}}{C}HCH_3$

(c)  [cyclopentane ring with] OH, CH$_2$CH$_3$

(d)  $CH_3CH_2\overset{\overset{OH}{|}}{\underset{\underset{CH_3}{|}}{C}}CH(CH_3)_2$

(e)  [naphthalene ring with] HO, CH—[benzene ring]

(f)  [cyclohexane ring with OH attached to another ring]

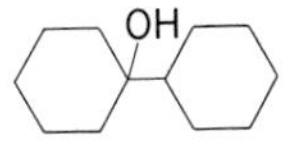

(g)  $(CH_3)_2CHCH_2CH_2OH$

(h)  $CH_3$—[benzene ring]—$\overset{\overset{OH}{|}}{C}H$—$\overset{\overset{CH_3}{|}}{C}HCH_2CH_3$

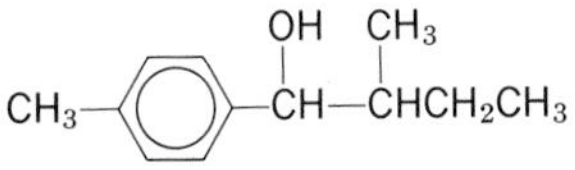

**Exercise 4.6**  Find two more ways in which the compound in part *d*, can be made.

**Exercise 4.7**  Recalling that aldehydes and ketones may be made by the oxidation of alcohols and that alkyl halides may also be made from alcohols, outline reaction sequences for the preparation of each of the following compounds, starting with alcohols containing four carbons or fewer and any needed inorganic reagents. The key for handling problems of multiple-step synthesis, is to work backward from the given product to the specified starting materials. For example, for the compound of part *a* first note the family and subfamily (if any): 1° alcohol. Next, consider all the ways in which 1° alcohols can be prepared in one step, whatever the starting substances. The following one-step synthesis of alcohols should come to mind.

—Addition of water to a double bond. (This synthesis will not work here because of Markovnikov's rule.)

—Hydrolysis of an alkyl halide. (This would require $CH_3CH_2CH_2CH_2CH_2X$, which has five carbons.)

—Reduction of an aldehyde. (So far, the only way we have learned to make them is by oxidation of a 1° alcohol; but making one of these is our present problem.)

—Action of a Grignard reagent on formaldehyde. (This would require $CH_3CH_2CH_2CH_2MgX$ and $CH_2=O$. Each contains four or fewer carbons.)

From this list select the one method most workable from the standpoint of initial specifications about starting materials. In our example the Grignard synthesis is the best. Writing more or less backward, from right to left, we record the following as the answer.

$$CH_3CH_2CH_2CH_2MgBr + H\overset{O}{\overset{\|}{C}}H \longrightarrow CH_3CH_2CH_2CH_2CH_2O^-Mg^{2+}Br^- \xrightarrow{H^+}$$

$$\uparrow \text{Mg, ether} \qquad \uparrow \text{Cu, heat} \qquad\qquad CH_3CH_2CH_2CH_2CH_2OH$$

$$CH_3CH_2CH_2CH_2Br \qquad CH_3OH$$

$$\uparrow \text{HBr, heat}$$

$$CH_3CH_2CH_2CH_2OH$$

Some of the following compounds are formed in several steps, some in a single step.

(a) $CH_3CH_2CH_2CH_2CH_2OH$
(worked above)

(b) $CH_3\overset{OH}{\overset{|}{C}}HCH_2CH_2CH_3$

(c) $CH_3CH_2\overset{O}{\overset{\|}{C}}CH_3$

(d) $CH_3CH_2\overset{O}{\overset{\|}{C}}CH_2CH_3$

(e) $CH_3\overset{CH_3}{\overset{|}{C}}=CHCH_3$

(f) $CH_3CH_2CH\!-\!CHCH_3$ with $\overset{|}{OH}$ $\overset{|}{OH}$

(g) $CH_3\overset{CH_3}{\overset{|}{C}}H\!-\!\overset{CH_3}{\overset{|}{C}}HCH_3$

(h) $CH_3CH_2CH_2COOH$

(i) $CH_3\overset{CH_3}{\overset{|}{C}}CH_2CH_3$ with $\overset{}{\underset{OH}{|}}$

(j) $CH_3\overset{}{\overset{|}{C}}HCH_2CH_2Br$ with $\overset{}{\underset{CH_3}{|}}$

---

## IMPORTANT INDIVIDUAL ALCOHOLS

**Methyl alcohol,** $CH_3\!-\!OH$ (wood alcohol, methanol). When selected dried hardwoods are heated in an oven or in a vertical retort, the subsequent exothermic process produces gases and a liquid condensate, leaving behind a residue of charcoal. The liquid, partly tarry and partly aqueous, contains in its aqueous portion, called pyroligneous acid, a small amount of methyl alcohol (Greek *methu*, wine; *hule*, wood or material). This process was formerly the principal means of producing methyl alcohol, but in 1923 chemists of the Badische Company in Germany discovered a synthetic route. Carbon monoxide and hydrogen, under a pressure of 200 to 300 atmospheres, at a temperature of 300 to 400°C, and in the presence of a mixed metal oxide catalyst, combine directly to form methyl alcohol:

$$2H_2 + CO \xrightarrow[\substack{300-400°C \\ ZnO-Cr_2O_3}]{200-300\ \text{atm}} CH_3OH$$

Methyl alcohol made this way is much cheaper than that available from the heating of wood, but archaic laws keep the latter method alive; federally approved formulas for denatured alcohol specify methyl alcohol prepared from wood. The annual United States production of synthetic methyl alcohol is about 3.9 billion pounds, or about 590 million gallons. Both the carbon monoxide and hydrogen needed for its synthesis can be produced from the partial oxidation of methane, obtained from petroleum reserves:

$$2CH_4 + O_2 \longrightarrow 2CO + 4H_2$$

Methyl alcohol is a dangerous poison, causing blindness or death. About half the annual production goes for making formaldehyde, an important raw material for plastics. Methyl alcohol is also used as a temporary antifreeze, as a component of jet fuel, and as a laboratory and commercial solvent.

**Ethyl alcohol,** $CH_3CH_2$—OH, grain alcohol; ethanol. Ethyl alcohol is made by the fermentation of sugars (discussed in Chapters 13 and 14) or by the hydration of ethylene. The source of ethylene for the second process is the cracking of ethane and other components of petroleum. Ethylene is absorbed in concentrated sulfuric acid at about 100°C, and a mixture of ethyl hydrogen sulfate and diethyl sulfate forms. When this mixture is diluted with water, the sulfates are hydrolyzed and ethyl alcohol is formed. In the year 1967, 700 million gallons of ethyl alcohol were produced in the United States, and of this 58% was from ethylene and only 15% from sources of sugar (grains, fruits, molasses). Another method, about 18% of the production, was the direct, vapor phase hydration of ethylene.

Industrially, ethyl alcohol is used as a solvent and in the compounding of pharmaceuticals, perfumes, lotions, tonics, and rubbing compounds. For these purposes it is adulterated (*denatured*) by poisons that are difficult to remove and that make it thoroughly undesirable for drinking purposes. Nearly all governments derive considerable revenue by taxing potable alcohol. Tax-free alcohol in the United States costs about 70 cents a gallon; the tax on a gallon of potable alcohol is slightly over 20 dollars. In days of old whiskey was tested by pouring it on gunpowder and seeing whether the gunpowder would ignite after the alcohol burned away. If it did, the tester had *proof* that the whiskey did not contain too much water. This was the origin of the term used in connection with alcohol-water solutions. Pure alcohol is 200 proof. Alcohol containing 5% water (190 proof) cannot be purified by simple distillation and exists as a constantly boiling mixture or *azeotrope.* When taken internally alcohol gives the illusion that it is a stimulant through its first effect, which is to depress activity in the uppermost level of the brain, the center of judgment, inhibition, and restraint.

**Isopropyl alcohol,** $CH_3CHCH_3$, 2-propanol. This alcohol, twice as toxic as ethyl
$\phantom{CH_3CH}\overset{|}{O}H$
alcohol, is a common substitute for it as a rubbing compound.

**Ethylene glycol,** $HOCH_2CH_2OH$; **propylene glycol,** $CH_3CHCH_2OH$. Both liquids
$\phantom{CH_3CHCH_2}\overset{|}{O}H$
have high boiling points, are soluble in water in all proportions, and are used chiefly as permanent antifreezes and coolants in refrigerator systems. The 1966 production

of ethylene glycol in the United States was 2.1 billion pounds, most of it made by the hydrolysis of ethylene oxide:

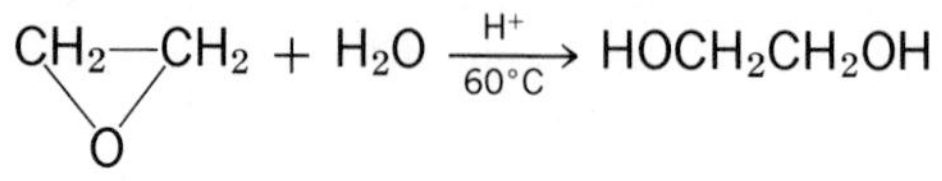

$$CH_2\!-\!CH_2 + H_2O \xrightarrow[60°C]{H^+} HOCH_2CH_2OH$$
$$\diagdown O \diagup$$

Ethylene oxide          Ethylene glycol
(made from ethylene)

Production of propylene glycol, largely from the hydrolysis of propylene oxide, was about 260 million pounds in the United States in 1966.

**Glycerol,** $CH_2\!-\!CH\!-\!CH_2$, glycerin. This colorless, syrupy liquid with a sweet
        $OH$   $OH$   $OH$

taste is freely soluble in water and insoluble in nonpolar solvents. A by-product in the manufacture of soap, over half the annual United States production of about 300 million pounds is made from one of two synthetic processes which start from propylene:

$$CH_2\!=\!CHCH_3 + Cl_2 \xrightarrow[500\text{--}600°C]{} CH_2\!=\!CHCH_2Cl \xrightarrow[NaOH]{Na_2CO_3} CH_2\!=\!CHCH_2OH$$

Propylene                    Allyl chloride             Allyl alcohol

Then

$$CH_2\!=\!CHCH_2OH + Cl_2 + H_2O \longrightarrow \underset{\overset{|}{Cl}\ \overset{|}{OH}\ \overset{|}{OH}}{CH_2CH\!-\!CH_2} \xrightarrow[NaOH]{Na_2CO_3} \underset{\overset{|}{OH}\ \overset{|}{OH}\ \overset{|}{OH}}{CH_2\!-\!CH\!-\!CH_2}$$

Glycerol

Or:

$$CH_2\!=\!CHCH_2OH + H_2O_2 \xrightarrow[65°C]{WO_3} \underset{\overset{|}{OH}\ \overset{|}{OH}\ \overset{|}{OH}}{CH_2\!-\!CH\!-\!CH_2}$$

Allyl alcohol                      Glycerol

Glycerol is used in the industrial synthesis of glyptal resins, as a humectant for tobacco, as a softening agent for cellophane, and in the compounding of cosmetics and drugs. It is also a raw material for the manufacture of nitroglycerin, a powerful explosive, and it is an important intermediate in the metabolism of lipids.

## REFERENCES AND ANNOTATED READING LIST

C. R. Noller. *Chemistry of Organic Compounds,* third edition. W. B. Saunders Company, Philadelphia, 1965. This textbook for a first-year course in organic chemistry is a rich source of information, not only about the subject in general but also about the history, manufacturing processes, and uses of individual compounds.

R. T. Morrison and R. N. Boyd, *Organic Chemistry,* second edition. Allyn and Bacon, Boston, 1966. This textbook contains two chapters on alcohols.

## PROBLEMS AND EXERCISES

1. Write condensed structural formulas for each of the following compounds.
   (*a*) allyl alcohol
   (*b*) benzyl alcohol
   (*c*) isopropyl alcohol
   (*d*) *sec*-butyl alcohol
   (*e*) *n*-octyl alcohol
   (*f*) *p*-nitrophenol
   (*g*) glycerol
   (*h*) propylene glycol
   (*i*) *t*-butyl alcohol
   (*j*) *n*-propyl alcohol
   (*k*) ethylene glycol
   (*l*) formaldehyde

2. Write names according to the IUPAC rules for the following compounds.

(*a*) $CH_3CH_2CH_2CH_2CH_2CH_2OH$

(*b*) $CH_3CHCH_2OH$ with $CH_3$ substituent on the middle carbon

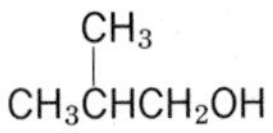

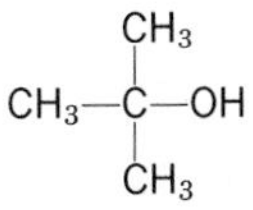

(*c*) $CH_3-\overset{\overset{\displaystyle CH_3}{|}}{\underset{\underset{\displaystyle CH_3}{|}}{C}}-OH$

(*d*) $CH_3CH_2CCH_2CHCH_2CHCHCH_3$ with $CH_3$, $OH$, $CH_2CH_2CH_3$ on the chain, $CH$ below with $CH_3$ and $CH_3$, and $CH_3$

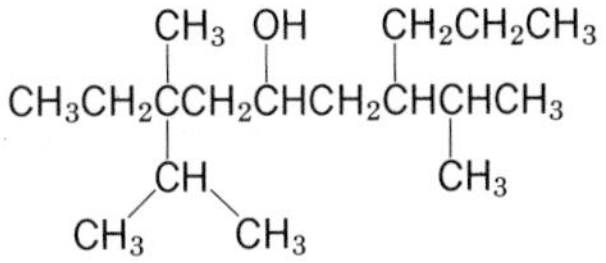

(*e*) $CH_3CH-OH$ with $CH_3$ substituent

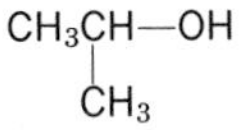

3. Arrange the following in order of their increasing boiling points and explain the reasoning behind your choices.

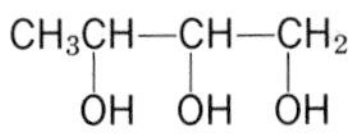

$CH_3(CH_2)_5CH_3$      $CH_3CH-CH-CH_2$ with $OH$ $OH$ $OH$      $CH_3CH_2CH_2CH_2CH_2CH_2OH$

A      B      C

$CH_3CH_2CH_2CH-CH_2$ with $OH$ $OH$

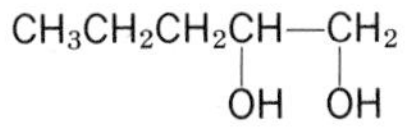

D

4. What do the following terms mean? Illustrate them with structures or reactions as the case may be.
   (*a*) hydrogen bond
   (*b*) dipole-dipole attraction
   (*c*) alkoxide ion
   (*d*) carbinol carbon
   (*e*) $S_N1$
   (*f*) alkyloxonium ion
   (*g*) methylcarbonium ion
   (*h*) elements of hydrogen
   (*i*) $S_N2$
   (*j*) nucleophile
   (*k*) rate-determining step
   (*l*) $E_1$
   (*m*) $E_2$
   (*n*) steric factor
   (*o*) Grignard reagent

5. Write an equation (not necessarily balanced) for the reaction of *n*-propyl alcohol with each of the following reagents.
   (*a*) potassium
   (*b*) sodium dichromate, $H^+$
   (*c*) copper (high temperature)
   (*d*) concentrated HBr
   (*e*) concentrated $H_2SO_4$, heat

6. Assume the availability of any needed alkyl halides and any needed aldehydes or ketones of four carbons or fewer. Outline the steps in the Grignard syntheses of 3-methyl-4-heptanol, 5-methyl-3-hexanol, and 2-methyl-2-pentanol.

7. Outline steps for the conversions indicated. More than one step will be needed for each.

(a) synthesis of propane from isopropyl alcohol

(b) synthesis of $CH_3\overset{O}{\overset{\|}{C}}CH_2CH_3$ from ethylene as the sole source of carbons

8. Write the structure of an alcohol from which each of the following compounds could be made; specify a reagent for carrying out the synthesis.

(a) 3-pentanone, $CH_3CH_2\overset{O}{\overset{\|}{C}}CH_2CH_3$    (b) $C_6H_5-\overset{O}{\overset{\|}{C}}-H$    (c) cyclohexene

(d) $(CH_3)_3C-Cl$    (e) isobutylene

9. Write all the steps in the mechanism of the dehydration of 1-propanol assuming that the acid-catalyzed process is $E_2$. Be sure to write representations of the transition state(s) and then construct a progress of reaction diagram.

10. Write all the steps in the mechanism of the conversion of benzyl alcohol to benzyl bromide by the action of concentrated hydrobromic acid. Assume that the rate of this reaction is not affected by the concentration of bromide ion.

11. Write all the steps in the mechanism of the dehydration of 2-methyl-2-butanol to 2-methyl-2-butene, assuming that the process is $E_1$. Draw a progress of reaction diagram.

12. Follow the instructions for problem 11, except take the reaction to the formation of the less stable alkene, 2-methyl-1-butene. Draw a progress of reaction diagram *on the same scale* used for problem 11.

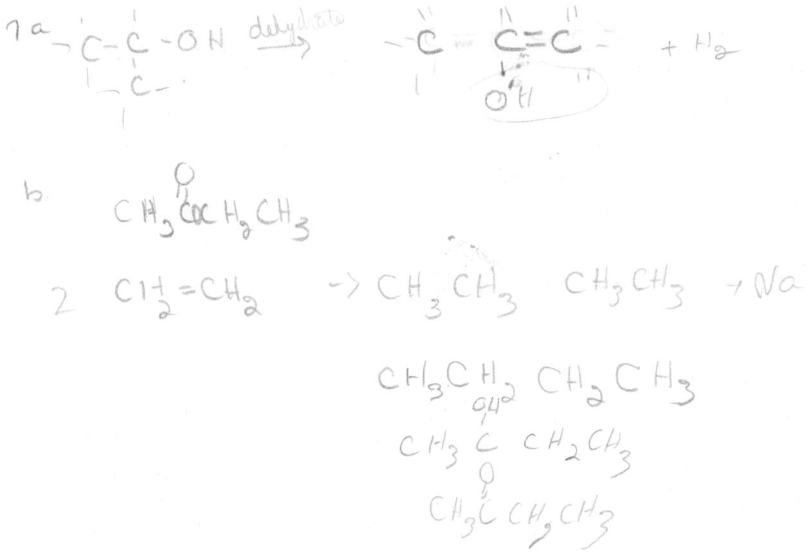

# CHAPTER FIVE

# Optical Isomerism.
# Organohalogen Compounds.
# Ethers. Amines. Mercaptans

Even though two (or more) compounds have the same general structure, they may be different in a very subtle way, a way even more subtle than cis-trans isomerism, which is merely a geometric difference. This new isomerism, optical isomerism, occurs widely in nature, and to interpret some of nature's workings we must understand it. We shall find examples of optical isomerism and its implications among the organic compounds and reactions yet to be studied.

Several families of organic compounds have reactions that are of particular importance in biochemistry but need not be studied in great depth. The alkyl halides and other organohalogen compounds, although seldom encountered in the body, nevertheless provide a basis for illustrating some important principles. Another family, the ethers, is also rarely represented in the molecules of living things, but when it occurs we need to know what to expect of its functional group. Mercaptans are sulfur analogs of alcohols, and the —SH group is important in protein chemistry. Finally, the amines, the organic derivatives of ammonia, are essential to a study of protein chemistry. Because only a few of their reactions have a bearing on the molecular basis of life, our study of them will be brief. Rather than creating separate, very short chapters for each of these topics, we have gathered them together in this one.

## OPTICAL ISOMERISM

### TYPES OF ISOMERISM

We have already encountered several ways in which substances can have the same molecular formula and still be different. The more obvious examples were *chain isomers, position isomers,* and *functional isomers.*

**140**

Optical Isomerism.
Organohalogen
Compounds.
Ethers.
Amines.
Mercaptans

Chain isomers: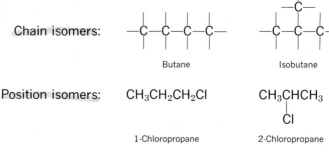

Butane      Isobutane

Position isomers:    $CH_3CH_2CH_2Cl$     $CH_3CHCH_3$
                                                 $Cl$

1-Chloropropane      2-Chloropropane

Functional group isomers:    $CH_3CH_2OH$      $CH_3$—O—$CH_3$

                                        Ethyl alcohol          Methyl ether

Less obvious was stereoisomerism, and we have studied only one of the two types, cis-trans or geometrical. For two compounds to be related as stereoisomers, they must have not only the same molecular formula but also the same nucleus-to-nucleus sequence. If they are then different because one molecular part does not rotate freely about a bond, this particular stereoisomerism is cis-trans. It commonly occurs among the alkenes and with cyclic compounds.

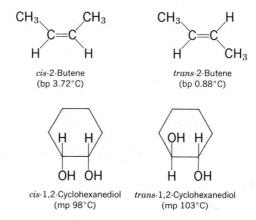

*cis*-2-Butene          *trans*-2-Butene
(bp 3.72°C)          (bp 0.88°C)

*cis*-1,2-Cyclohexanediol    *trans*-1,2-Cyclohexanediol
(mp 98°C)            (mp 103°C)

The second kind of stereoisomerism has to do with the lack of symmetry in the molecule.

## "ONE SUBSTANCE, ONE STRUCTURE"

Chemists have long worked with the principle "one substance, one structure." If two substances are identical in their physical and chemical properties, they must be identical at the level of their individual molecules. If two substances differ in even one way, their individual molecules must also differ in some way.

Consider now a substance known as asparagine (ăs-păr'-à-jĭn), a white solid with a bitter taste, first isolated in 1806 from the juice of asparagus. Its molecular formula is $C_4H_8N_2O_3$, and its structure is now known to be **1**. Eighty years later, in

$$NH_2-\overset{\overset{\displaystyle O}{\|}}{C}-CH_2-\underset{\underset{\displaystyle NH_2}{|}}{CH}-\overset{\overset{\displaystyle O}{\|}}{C}-OH$$

**1**

Asparagine

1886, a chemist isolated from sprouting vetch[1] a substance of the same molecular formula and structure, but it had a sweet taste. To distinguish these two substances from each other, we call the one isolated from asparagus L-asparagine and the one from vetch sprouts D-asparagine. The small capitals D and L, although arbitrary here, acquire significance later.

These two samples of asparagine have the same solubilities in solvents. Toward ordinary chemicals they have identical reactions with identical rates. Their solutions have identical spectra (which are comparable to identical fingerprints). From all the data the same structure has to be assigned the molecules in both samples, namely structure **1**. But they have this rather dramatic difference in taste. (Ordinarily chemists do not taste chemicals, but the fact remains that this difference in taste was discovered decades ago.)

Taste is a chemical sense. Chemical reactions between the substance being tasted and chemicals in the taste buds result in patterns of signals sent to the brain. We label these signals by words such as sweet, sour, bitter, salty, etc. The two samples of asparagine therefore show one difference in chemical property. (One physical property, to be described later, is also different.) Under the one-substance, one-structure doctrine, the molecules of D- and L-asparagine must be structurally different in some way, even though they both "answer" to structure **1**. Although all the forms of isomerism thus far discussed are ruled out by the physical data alone, another type must be possible.

Nature is rich in substances such as asparagine. Monosodium glutamate is a common, widely used flavor-enhancing agent. Yet two forms, identical in the same

$$HO-\overset{\overset{\displaystyle O}{\|}}{C}-CH_2CH_2\underset{\underset{\displaystyle NH_2}{|}}{CH}\overset{\overset{\displaystyle O}{\|}}{C}-O^-Na^+$$

Monosodium glutamate

$$HO-\overset{\overset{\displaystyle O}{\|}}{C}-\underset{\underset{\displaystyle OH}{|}}{CH}-\underset{\underset{\displaystyle OH}{|}}{CH}-\overset{\overset{\displaystyle O}{\|}}{C}-OH$$

Tartaric acid

ways that the two forms of asparagine are identical, are known. One is a flavor enhancer, the other not. Tartaric acid is another example. One form is fermented through the help of the enzymes in the organism *Penicillium glaucum,* another not. Chloromycetin, an important antibiotic of very complicated structure, also exists in two forms. One of the forms is effective as an antibiotic, the other not. These examples are only a few of the great many in molecular biology. Organic chemistry has still more. To understand how it might be possible for these differences to exist,

---

[1] Vetch is a member of a genus of herbs, some of which are useful as fodder.

we must understand the most fundamental way of deciding whether two molecular structures (in spatial models) represent identical compounds.

142

Optical Isomerism.
Organohalogen
Compounds.
Ethers.
Amines.
Mercaptans

## MOLECULAR SYMMETRY AND ASYMMETRY

Two partial ball and stick models of a molecule of asparagine are shown in Figure 5.1*a*. Examine each carefully to assure yourself that each is a representation of structure **1**. For convenience, each model in part *a* of the figure has been simplified as shown in part *b*. In each the same four groups are attached to a carbon, yet the two structures are not identical. For two structures to be identical it must be possible, mentally, to complete an operation known as superimposition, illustrated in Figure 5.1*c*. This is the fundamental criterion of identity for two molecular structures. Each must have a conformation by which the two become superimposable.

In asparagine the two nonidentical structures that cannot be superimposed are related in a significant way—as an object to its mirror image. Consider now what is characteristic of an object whose mirror image is superimposable. Imagine a perfect cube before a mirror, and imagine that the image in the mirror can be taken out and a model made of it. The two cubes would be superimposable and therefore essentially identical. The same "thought experiment" could be performed with a sphere, a pyramid, or any regular solid. By "regular" of course is meant symmetrical. Although we may not think about it this way, when we use the word symmetrical to designate an object, we are saying in effect that the object and its mirror image are superimposable. The two forms of asparagine, have the same molecular formula, the same skeletons, the same functional groups located at the same places on the chain. Yet they are not in a cis-trans relation; simple, free rotation cannot convert one into a form superimposable with the other. The two asparagines differ structurally only in the *relative* directions by which the four groups attached to the carbon shown in Figure 5.1 project into space. Chemists use the word *configuration* to denote a relative arrangement of nuclei in three-dimensional space. The two forms of asparagine could therefore be called configurational isomers, but instead, for historical reasons which will soon be apparent, the term *optical isomers* is used almost exclusively, There are several types of optical isomers, and the two asparagines illustrate one.

**Enantiomers.** Isomers which, like the two asparagines, are related as object and mirror image that cannot be superimposed are called *enantiomers* (Greek: *enantios,* opposite; *meros,* part). To deal as swiftly as possible with one source of misunderstanding, *any* object will have a mirror image. Only when the object and mirror image cannot be superimposed are they said to be enantiomers. With this definition of enantiomers, optical isomers in general can be better defined. Optical isomers are members of a set of stereoisomers that includes at least two related as enantiomers.

Physical properties are identical for two enantiomers. This *must* be so. Within two enantiomers intranuclear distances and bond angles are identical, and polarities must therefore be identical. Melting points, boiling points, densities, solubilities, spectra, and of course molecular weights have to be the same for two enantiomers.

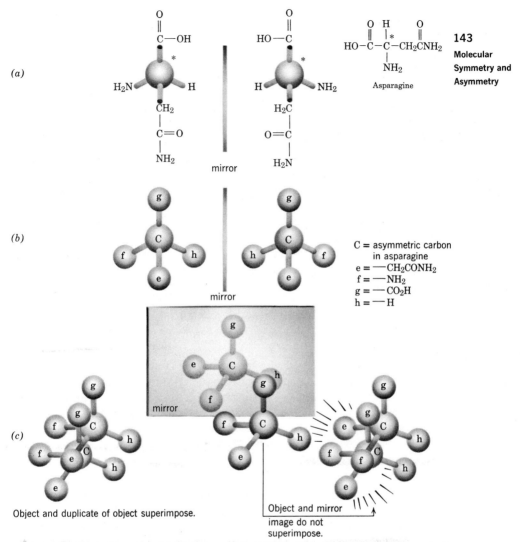

(a)

(b)

C = asymmetric carbon
   in asparagine
e = ─CH₂CONH₂
f = ─NH₂
g = ─CO₂H
h = ─H

(c)

Object and duplicate of object superimpose.

Object and mirror
image do not
superimpose.

**Figure 5.1** The enantiomers of asparagine are shown in part $a$. When the groups attached to an asymmetric carbon do not themselves contribute further to asymmetry in the molecule, they may be more simply represented as in part $b$. Checking superimposability requires the imaginary operation begun in part $c$ but prevented by the impenetrability of wood.

In thinking about intranuclear distances and angles in enantiomers, let us examine our two hands, disregarding differences in fingerprints and wrinkles. The left hand and the right hand can serve as enantiomers because they are related as object and mirror image that cannot be superimposed. If we try to superimpose them with both palms in the same position, facing toward us, the fingers and thumb in one hand do not line up with the corresponding fingers and thumb in the other hand. If

**144**

Optical Isomerism.
Organohalogen
Compounds.
Ethers.
Amines.
Mercaptans

we try to superimpose them with the fingers and thumbs lined up, the palm of the first hand, when we imagine it going "through" the other, comes out on the back side of the second. Yet in each hand each finger has the same position in relation to the others. The only difference is that with the two hands held palms facing us, we "read" clockwise from thumb to little finger in the left hand and counterclockwise in the right hand.

Chemical properties of enantiomers are also identical, provided other reagents are symmetrical in the sense used here. If the reaction is with an unsymmetrical reagent, we observe striking differences in properties. A widely used illustration involves a pair of gloves. Because two hands are related as enantiomers, so too are a pair of ordinary gloves:

$$\text{Right hand} + \text{right glove} \longrightarrow \text{gloved right hand}$$
$$\text{Right hand} + \text{left glove} \longrightarrow \text{no fit}$$

Taking the gloves as the unsymmetrical "reagents," we see that the ease and comfort with which each glove can "react" with each hand is certainly different. By contrast, if we use a pair of, say, ski poles which are superimposable and symmetrical as the reagent system, no differences can be noted in the way these "react" with either hand.

**Deciding Whether a Molecule Is Symmetrical.** Any structure that is identical with its mirror image is symmetrical in some way. Three ways are recognized, but for our purposes we need only one, the existence of a *plane of symmetry*.[2] When optical isomerism is a possibility, we search the structures in question for a plane of symmetry. If we can discover that it is either present or absent, our work is greatly simplified, for we can dispense with a tedious task, making or drawing a three-dimensional model, reflecting it in an imaginary mirror, constructing a model of the mirror image, and finally checking the superimposability of the object with the mirror image. If we find a plane of symmetry, we are guaranteed that the object and mirror images are identical (superimposable). On the other hand, if we cannot discover this (or any) symmetry in a structure, in at least one conformation of it, we are guaranteed that the object and mirror images are not superimposable. It is apparent that ways to discover the symmetry element will save considerable time in trying to decide whether the system supports optical isomerism.

*A plane of symmetry is an imaginary plane that divides an object into two halves that are mirror images of each other.* In Figure 5.2 some simple molecules that have planes of symmetry are shown. Some of them have several planes, but we require only one to guarantee that object and mirror image are superimposable and therefore identical. (Figure 5.2c has only one plane of symmetry.) Only when there is no element of symmetry do the interesting possibilities seen in asparagine, glutamic acid, tartaric acid, and chloromycetin materialize.

**The Asymmetric Carbon—A Shortcut.** Chemists have noted that in nearly all examples of optical isomerism the molecules possess at least one carbon to which are attached four *different* atoms or groups. In asparagine (Figure 5.1) one and

---

[2] For the record, the other two elements of symmetry are a center of symmetry and a fourfold alternating (or mirror) axis of symmetry.

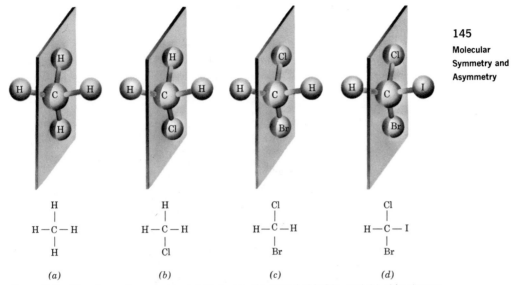

|   |   |   |   |
|---|---|---|---|
| H | H | Cl | Cl |
| \| | \| | \| | \| |
| H—C—H | H—C—H | H—C—H | H—C—I |
| \| | \| | \| | \| |
| H | Cl | Br | Br |
| *(a)* | *(b)* | *(c)* | *(d)* |

**Figure 5.2** The plane of symmetry. (*a*) Methane, (*b*) methyl chloride, and (*c*) chlorobromo-methane each possess at least one plane of symmetry. This plane, serving as an imaginary mirror, splits each molecule into two halves, one being the mirror image of the other. No plane of symmetry exists for molecule *d*, which is asymmetric. If a mirror image were made of it, the object and it would not be superimposable.

only one carbon has attached to it four *different* substituents: —H, —NH$_2$, —COOH, and NH$_2$COCH$_2$—. Such a carbon is defined as an *asymmetric* carbon. Its presence in a structure nearly always guarantees that there will be no element of symmetry in any conformation of the molecule. Therefore molecules with an asymmetric carbon will nearly always be asymmetric themselves and will belong to a set of optical isomers.

The qualification "nearly always" used repeatedly in these statements is an important one. Many substances with no asymmetric carbons have asymmetric molecules (Figure 5.3), and a few substances with molecules containing asymmetric carbons are yet symmetrical in some way (Figure 5.4). Some of these substances will be studied in more detail later. In general, if a molecule has *only one* asymmetric carbon, it will be asymmetric as a whole and it will belong to a set of optical isomers. We need not qualify this statement with "nearly always."

When it works, the concept of an asymmetric carbon is extremely useful, for we can usually consult an ordinary structural formula in trying to discover whether optical isomerism is a possibility. Thus we have a shortcut to replace the more difficult effort of seeking an element of symmetry, a method which in turn is a great improvement over making or drawing two or more structures. It must be remembered that the shortcut is fallible (cf. Figures 5.3 and 5.4), but with a little experience the exceptions are easily recognized. Moreover, they do not occur frequently.

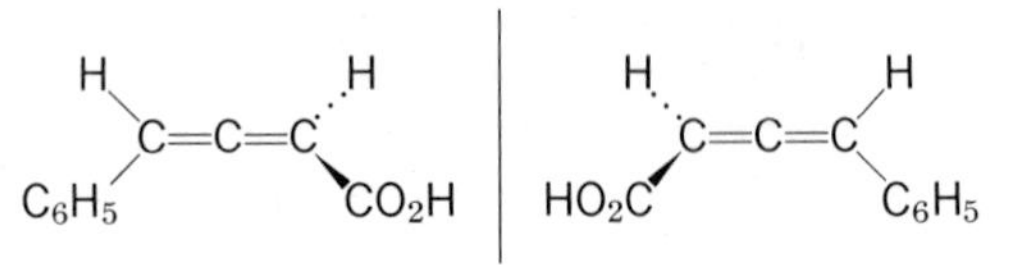

146

Optical Isomerism.
Organohalogen
Compounds.
Ethers.
Amines.
Mercaptans

Enantiomeric allenes

··· Bond is in a plane perpendicular to and behind page.

► Bond is in a plane perpendicular to and in front of page.

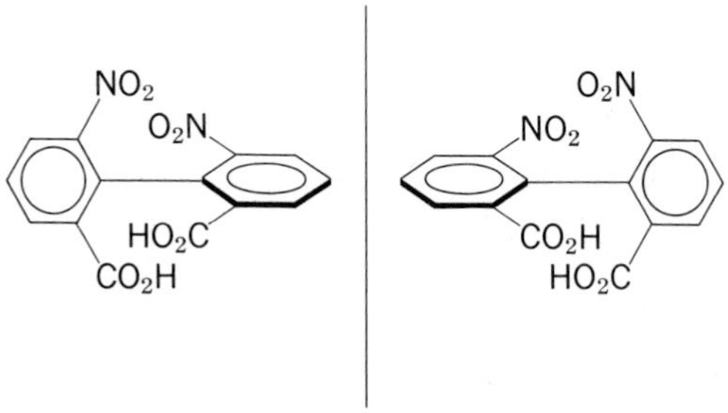

Enantiomeric biphenyls

**Figure 5.3** Asymmetric molecules having no asymmetric carbon atoms. In the allenes the rigid geometry about the ends of the double bonds makes asymmetry possible with the appropriate end groups. In the biphenyls the bulkiness of the groups located at the ortho positions prevents free rotation about the single bond joining the two benzene rings. With the appropriate choice and distribution of these ortho substituents, the molecules can be asymmetric.

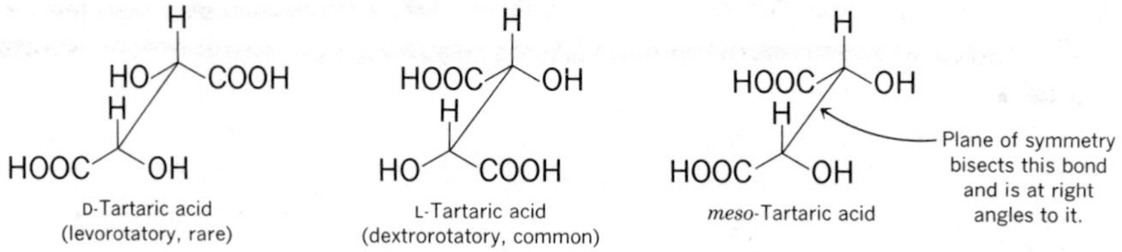

D-Tartaric acid
(levorotatory, rare)

L-Tartaric acid
(dextrorotatory, common)

meso-Tartaric acid

Plane of symmetry bisects this bond and is at right angles to it.

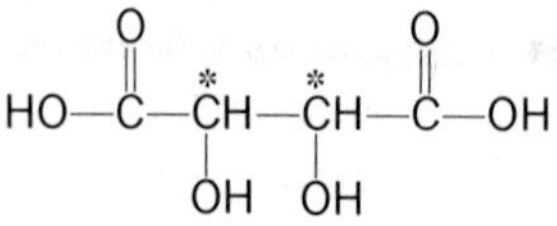

Tartaric acid

**Figure 5.4** The presence of an asymmetric carbon does not always guarantee that the molecule will be asymmetric. For tartaric acid there are two enantiomeric forms, D- and L-tartaric acid, related as object and mirror image that cannot be superimposed. Another form of tartaric acid, the meso form, has a plane of symmetry (Greek *meso*, in the middle). Meso forms have molecules possessing two (or more) asymmetric carbons, to each of which are attached the same set of four different groups.

Exercise 5.1    Examine the given structures. Place an asterisk above each carbon to which four different groups are attached. If ball and stick sets or their equivalent are available to you, make molecular models of each structure and try to discover planes of symmetry. Make molecular models of the mirror images and check to see whether the objects and mirror images are superimposable.

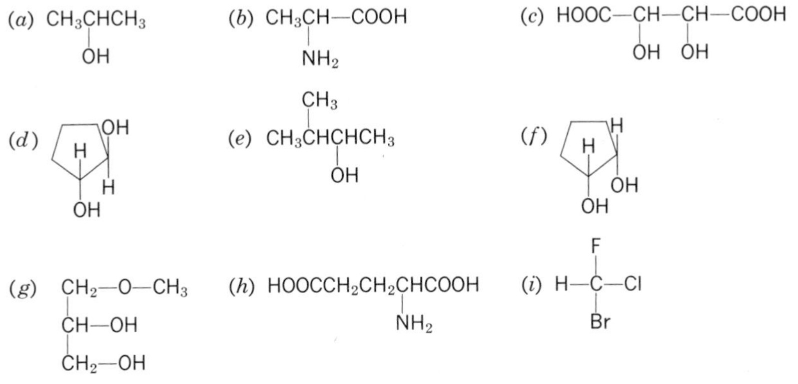

(a) $CH_3CHCH_3$
    |
    $OH$

(b) $CH_3CH-COOH$
    |
    $NH_2$

(c) $HOOC-CH-CH-COOH$
    |      |
    $OH$  $OH$

(d)

(e) $CH_3CHCHCH_3$  with $CH_3$ and $OH$ substituents
    |
    $CH_3$ ... $OH$

(f)

(g) $CH_2-O-CH_3$
    |
    $CH-OH$
    |
    $CH_2-OH$

(h) $HOOCCH_2CH_2CHCOOH$
    |
    $NH_2$

(i) $H-\underset{Br}{\overset{F}{C}}-Cl$

## OPTICAL ACTIVITY

Taste happens to be one way in which the asparagine enantiomers differ, but for general use the taste test is unreliable and physically dangerous. We must be able to detect enantiomers in some other way. Fortunately, one physical measurement, the effect of an enantiomer on plane-polarized light, is available.

Light is an electromagnetic radiation involving oscillations in the strengths of electric and magnetic fields set up by the light source. In ordinary light these oscillations or vibrations occur equally in all directions about the line that may be used to define the path of the light ray.

Certain materials, for example, the Polaroid film used in some sunglasses, affect ordinary light in a special manner. They interact with the oscillating electrical field in such a way that the emerging light vibrates in one plane. Such light is called *plane-polarized light* (Figure 5.5). If we look at some lighted object through a piece of Polaroid film and then place in front of the first film (or behind it) a second film, we can by rotating the two films in relation to each other find one orientation in which the lighted object has maximum brightness and one in which it can no longer be seen. To go from the first orientation to the second, one film is simply rotated 90° with respect to the other. It is as though, to use a rather unsatisfactory analogy, the film acts as a lattice fence forcing any emerging light to vibrate in the plane of the long axis of the slats. If, when this light reaches the second film or lattice fence, the second slats are related perpendicular to those of the first fence, the light cannot go through. At intermediate angles fractional amounts of light can go through the second lens or fence. Even though this analogy is not completely valid, we can take solace in the fact that a complete understand-

**148**

Optical Isomerism.
Organohalogen
Compounds.
Ethers.
Amines.
Mercaptans

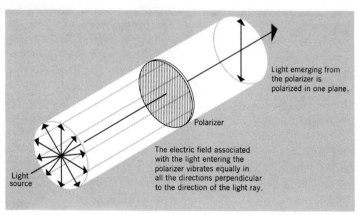

**Figure 5.5**  Light becomes polarized into vibrations in one plane when it passes through certain materials generally called polarizers. Polaroid film is an example.

ing of polarized light is unnecessary to our comprehension of the structural and chemical aspects of optical isomerism. Very few chemists are deeply versed in the mathematicophysical theories of plane-polarized light.

**Optical Rotation.** If a sample of an organic substance in which the molecules of one enantiomer predominate, like D-asparagine, is dissolved in an ordinary solvent, like water, and if this solution is placed in the path of plane-polarized light, the plane of polarization will be rotated. Any substance that will do this to plane-polarized light—rotate the plane—is said to be *optically active.*

**The Polarimeter.** The instrument used to detect optical activity and to measure the number of degrees of rotation of plane-polarized light is called a polarimeter. Figure 5.6 illustrates its principal working parts, which consist of a polarizer, a tube for holding solutions in the light path, an analyzer (actually, just another polarizing device), and a scale for measuring degrees of rotation. As shown in Figure 5.6a, when polarizer and analyzer are "parallel" and the tube contains no optically active material, the polarized light goes through to the observer. If, now, a solution containing an optically active material is placed in the light path, the plane-polarized light encounters asymmetric molecules, and its plane of oscillation is rotated. The polarized light that leaves the solution is no longer "parallel" with the analyzer (cf. Figures 5.6a and c), and the intensity of the observed light is reduced. To restore the original intensity, the analyzer is rotated to the right or to the left until it is again parallel with the light emerging from the tube. As the operator looks toward the light source, he will, if he rotates the analyzer to the right, record the degrees as positive, and the optically active substance will be called *dextrorotatory.* If the rotation is counterclockwise, the degrees will be recorded as negative, and the substance will be called *levorotatory.* (The direction requiring the fewer degrees to restore the original brightness is taken.) In the example of figure 5.6c, the record will read $\alpha = -40°$, where $\alpha$ stands for the *optical rotation,* the observed number of degrees of rotation.

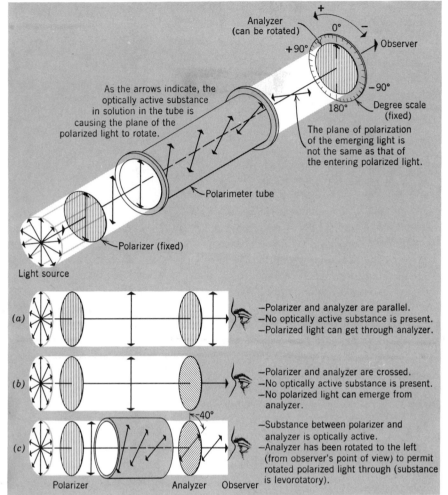

**Figure 5.6** Principal working parts of a polarimeter and the measurement of optical rotation.

Since the optical rotation varies with both the temperature of the solution and the frequency of the light, the record actually notes both data, for example, $\alpha_D^{20} = -40°$. The subscript D stands for the D-line of the spectrum of sodium, which is one particular frequency that electromagnetic radiation can have, and 20 means that the solution has a temperature of 20°C.

**Specific Rotation.** The extent of the optical rotation also depends on the population of asymmetric molecules that the light beam encounters. If it travels through a longer tube, the plane of oscillation of the light will be rotated more. Or if the tube length is held constant but the concentration of the solution is increased, the degrees of rotation also increase. To incorporate such effects into the record, the observer converts the observed value of $\alpha$ into a ratio expressing the number of degrees per unit concentration per unit of path length. The *specific rotation* is one way of expressing this ratio. Its symbol is $[\alpha]_\lambda^{t°}$, or with temperature and frequency specified $[\alpha]_D^{20}$. By definition,

$$[\alpha]_\lambda^{t°} = \frac{100\alpha}{cl}$$

where $c$ = concentration in grams per 100 cc of solution
$l$ = length of the light path in the solution, measured in decimeters (1 decimeter = 10 cm)
$\alpha$ = observed rotation in degrees (plus or minus)
$\lambda$ = wavelength of light
$t°$ = temperature of the solution in degrees centigrade

The specific rotation of a compound is an important physical constant, comparable to its melting point, boiling point, etc. It is one more physical characteristic that a chemist can use to identify a substance. Moreover, by measuring the actual rotation $\alpha$ of a solution of a substance of known specific rotation in a tube of fixed length, we can calculate the concentration of the solution. That is, with $[\alpha]$, $\alpha$, and $l$ known, the equation readily permits calculation for $c$.

Table 5.1 lists specific rotations for several materials, Table 5.2 physical constants for some sets of optical isomers. *Pairs of enantiomers differ physically only in the sign or direction in which each rotates the plane of plane-polarized light.* All other physical constants are the same, including even the number of degrees of rotation.

**Table 5.1   Specific Rotations of Various Substances**

| Physical State | Substance | Specific Rotation (sodium D light at 20°C in solvent specified) |
|---|---|---|
| Solutions* | asparagine | +5.41 (water) |
| | albumin (a protein) | −25 to −38 (water) |
| | cholesterol | −31.61 (chloroform) |
| | glucose | +52.5 (aged solution in water) |
| | sucrose (table sugar) | +66.4 (water) |
| | 3-methyl-2-butanol ("active amyl alcohol") | +5.34 (ethyl alcohol) |
| | quinine sulfate | −214 (water, at 17°C) |
| Pure liquids* | turpentine | −37 |
| | cedar oil | −30 to −40 |
| | citron oil | +62 (at 15°C) |
| | nicotine | −162 |
| Pure solids† | quartz | +21.7 |
| | cinnabar (HgS) | +32.5 |
| | sodium chlorate | +3.13 |

* Specific rotation in degrees per decimeter.
† Specific rotation in degrees per millimeter of path length in the crystal.

150

Optical Isomerism.
Organohalogen
Compounds.
Ethers.
Amines.
Mercaptans

**Table 5.2  Physical Constants of Optical Isomers**

| Set | Members of Set | Mp (°C) | $[\alpha]_D^{20}$ (degrees) | Miscellaneous |
|---|---|---|---|---|
| Mandelic acids $C_6H_5CHCO_2H$ $\overset{|}{OH}$ | (+) mandelic acid | 132.8 | +155.5 | Solubility: 8.54 g/100 cc water (at 20°C) |
| | (−) mandelic acid | 132.8 | −155.4 | Solubility: 8.64 g/100 cc water (at 20°C) |
| Asparagine | (+) asparagine | 234.5 (decomposes) | +5.41 | $d_4^{15} = 1.534$ g/cc |
| | (−) asparagine | 235 (decomposes) | −5.41 | $d_4^{15} = 1.54$ g/cc |
| Tartaric acids | (+) tartaric acid | 170 | +11.98 | $d_4^{15} = 1.760$ g/cc |
| | (−) tartaric acid | 170 | −11.98 | $d_4^{15} = 1.760$ g/cc |
| | meso-tartaric acid | 140 | 0 | $d_4^{15} = 1.666$ g/cc |
| | racemic tartaric acid* | 205 | 0 | $d_4^{15} = 1.687$ g/cc |

* This is a 50:50 "mixture" of the plus and minus forms of tartaric acid, which in this case happens to form what is known as a racemic compound.

## OTHER KINDS OF OPTICAL ISOMERS

**Meso Compounds.** In some sets of optical isomers, for example, the tartaric acids of Figure 5.4, one member is found to be optically inactive, even though it possesses asymmetric carbons. We are reminded that the fundamental criterion for asymmetry in a molecule is not the asymmetric carbon but rather the impossibility of superimposing object and mirror images. The isomer that belongs to a set of optically active substances but is optically inactive itself is called the meso isomer. The mirror image of meso-tartaric acid would be superimposable on the object. The acid also possesses within the molecule a plane of symmetry. Behind this plane is a reflection of what is in front. Hence, as one-half of the molecule tends to rotate polarized light, say, to the left, the other half cancels this with an equal rightward rotation. The net effect is optical inactivity.

**Diastereomers.** Optical isomers that are not related as object to mirror image are called diastereomers. Figure 5.7 illustrates examples, and we shall encounter them again in carbohydrate chemistry. For two molecules to qualify as diastereomers, they must have identical molecular formulas, they must have identical nucleus-to-nucleus sequences (without regard to geometry), they must be members of a set of optical isomers, but they may not be related as object to mirror image. Generally, they will have different physical properties and sometimes different chemical properties.

**Racemic Mixtures.** A substance that is composed of a 50:50 mixture of enantiomers is called a racemic mixture. One enantiomer cancels the effects of the other on polarized light, and a racemic mixture is optically inactive, emphazing the point that the phenomenon of optical activity relates to a measurement we can make on a substance. The explanation of the phenomenon, in terms of asymmetric molecules, relates to our theory about it.

152

Optical Isomerism.
Organohalogen
Compounds.
Ethers.
Amines.
Mercaptans

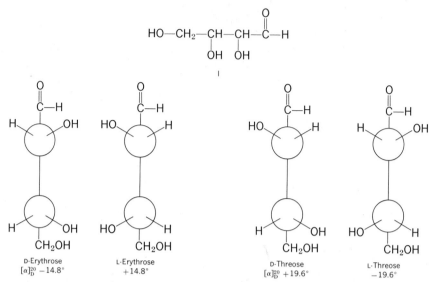

**Figure 5.7** Optical isomers that illustrate pairs of enantiomers and diastereomers. The pairs of enantiomers of I are D- and L-erythrose and D- and L-threose. Any member of the first pair is a diastereomer of any member of the second. Compounds related as diastereomers are optical isomers that are not related as object and mirror image. As the values of specific rotation indicate, diastereomers may have different physical properties. Both erythrose and threose are carbohydrates.

A racemic mixture must not be confused with a meso compound. In a meso compound optical inactivity is the result of an internal plane of symmetry. In a racemic mixture optical inactivity is the result of both enantiomers' being present in a 1:1 ratio. Racemic comes from racemic acid, the name once given to a form of tartaric acid that Louis Pasteur demonstrated consisted of a 50:50 mixture of enantiomers.

## CONFIGURATIONAL CHANGES IN CHEMICAL REACTIONS

$n$-Butane has no asymmetric carbon, and it has a plane of symmetry. A molecule of bromine, $Br_2$, is also symmetrical. Yet if bromine can be made to react with $n$-butane at the second carbon, this carbon becomes asymmetric in the product.

$$CH_3CH_2CH_2CH_3 + Br_2 \longrightarrow CH_3CH_2\underset{\underset{Br}{|}}{C}HCH_3 + HBr$$

$n$-Butane

$sec$-Butyl bromide

The question is whether the substance actually isolated as a product of this reaction and identified as $sec$-butyl bromide is optically active. The answer is *no*. We must distinguish between a *substance* that is optically active and one that has asymmetric molecules. Optical activity is something we can measure with a polarimeter, completely without regard for an interpretation of the observation. The reaction produces an asymmetric carbon but in such a way that the product is a 50:50 mixture of enantiomers, a racemic or optically inactive mixture. We can understand the reason for this outcome by examining Figure 5.8. In *n*-butane, at the site where bromine will eventually become affixed, are two equivalent hydrogens. In a substitution occurring to any one molecule there is an equal chance that each hydrogen will be the one replaced. In a collection approaching Avogadro's number in size, the statistics are such that production of a racemic mixture is a certainty.

It is, in general, true that optically active *substances* cannot be synthesized from optically inactive reagents. Asymmetric centers can easily be made, but the final substance will be either a racemic mixture or a meso compound.

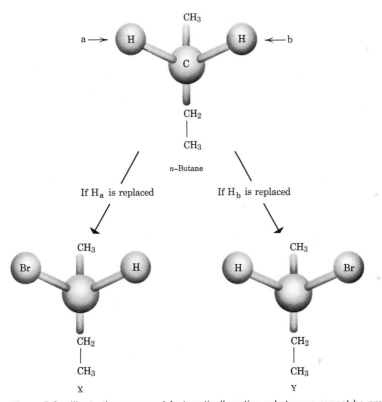

**Figure 5.8**  Illustrating a general fact: optically active substances cannot be prepared by the action of an optically inactive reagent on an optically inactive compound. In *n*-butane the two hydrogens that are candidates for replacement, if *sec*-butyl bromide is to form, have the same chance to be replaced. Therefore a 50:50 mixture of the two enantiomers of *sec*-butyl bromide, X and Y, is produced. Although the enantiomers would individually be optically active, the racemic mixture of them that actually forms is not.

# ORGANOHALOGEN COMPOUNDS

TYPES OF ORGANOHALOGEN COMPOUNDS. STRUCTURE. NOMENCLATURE

Optical Isomerism.
Organohalogen
Compounds.
Ethers.
Amines.
Mercaptans

The monohalogen derivatives of hydrocarbons of widest use in organic chemistry are the monochloro, monobromo, and monoiodo derivatives of alkanes and aromatic hydrocarbons. Several examples are given in Table 5.3. The fluoro derivatives are virtually in a class by themselves, partly because they are more difficult to make. They are, however, of great industrial and commercial importance, and some of the more widely used examples are listed in Table 5.4. Although alkyl halides are not found in the body, they do illustrate principles important in living systems.

**Table 5.3    Some Classes of Organohalogen Compounds**

| Halides | Example | Structure | Bp (°C) |
|---|---|---|---|
| Alkyl halides | methyl bromide | $CH_3-Br$ | 4 |
| | isopropyl chloride | $(CH_3)_2CH-Cl$ | 35 |
| | $t$-butyl chloride | $(CH_3)_3C-Cl$ | 51 |
| | cyclohexyl chloride | 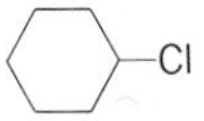 [cyclohexyl–Cl structure] | 164 |
| | allyl chloride | $CH_2=CHCH_2-Cl$ | 45 |
| | $\beta$-phenylethyl bromide | $C_6H_5CH_2CH_2-Br$ | 217 |
| | benzyl chloride | $C_6H_5CH_2-Cl$ | 179 |
| Vinyl halides | | | |
| | vinyl bromide | $CH_2=CH-Br$ | -14 |
| | 2-chloropropene | $CH_2=\underset{\underset{Cl}{\mid}}{C}-CH_3$ | 23 |
| Aryl halides | | | |
| | bromobenzene | 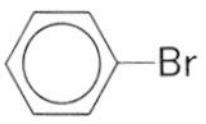 [benzene ring–Br structure] | 155 |
| | $o$-chlorotoluene | 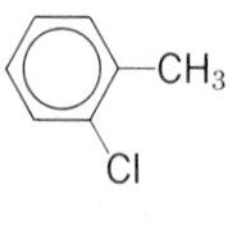 [benzene ring–$CH_3$, Cl structure] | 159 |
| Polyhalogen compounds | | | |
| | methylene chloride | $CH_2Cl_2$ | 40 |
| | chloroform | $CHCl_3$ | 61 |
| | carbon tetrachloride | $CCl_4$ | 77 |
| | iodoform | $CHI_3$ | 119 (mp) |
| | trichloroethylene (Triclene) | $Cl-CH=CCl_2$ | 87 |

**Table 5.4  Important Organofluorine Compounds**

| Name | Structure | Uses | |
|------|-----------|------|---|
| Freon 11 | $CCl_3F$ (bp 24°C) | propellant; refrigerant | **155**<br><br>**Types of Organohalogen Compounds.**<br>**Structure.**<br>**Nomenclature** |
| Freon 12 | $CCl_2F_2$ (bp −30°C) | first fluorocarbon aerosol propellant; refrigerant | |
| Freon 114 | $CClF_2CClF_2$ (bp 4°C) | refrigerant (very stable, odorless); propellant for shaving creams, colognes | |
| Bromotrifluoromethane | $CBrF_3$ (bp −58°C) | fire extinguishers (low toxicity) | |
| Perfluorocyclobutane | $\begin{array}{c} F_2C-CF_2 \\ \mid \quad \mid \\ F_2C-CF_2 \end{array}$ (bp −4°C) | propellant for aerosol cans of food products (e.g., whipped cream) | |
| Teflon | $-(CF_2-CF_2)_{\overline{n}}$ | cookware; bearings | |
| Tedlar (polyvinyl fluoride) | $-(CH_2-CHF)_{\overline{n}}$ | weatherproofing material | |
| Scotchgard | Organofluorine substance (structure is not publicly known) | grease and oil repellent for fabrics and paper (see photograph) | |

This closeup shows how oil and water bead on a fabric treated with Scotchgard fluorocarbon repeller, which does not close the pores of the fabric. (Courtesy of the 3M Company.)

## SYNTHESIS

156

Optical Isomerism.
Organohalogen
Compounds.
Ethers.
Amines.
Mercaptans

In Chapter 4 we learned that alkyl halides may be made by the action of concentrated H—X on an alcohol. Earlier we learned that H—X can also be made to add to an alkene linkage. Other reagents that will convert alcohols to halides include thionyl chloride ($SOCl_2$), phosphorus tribromide ($PBr_3$), and a mixture of potassium iodide and phosphoric acid:

$$ROH + SOCl_2 \longrightarrow R—Cl + SO_2 + HCl$$
$$3ROH + PBr_3 \longrightarrow 3R—Br + H_3PO_3$$
$$ROH + KI + H_3PO_4 \longrightarrow R—I + KH_2PO_4 + H_2O$$

## NUCLEOPHILIC SUBSTITUTION REACTIONS OF ALKYL HALIDES

Several types of nucleophilic substitution reactions are tabulated in Table 5.5. For the most part the nucleophiles are also bases. Therefore an important side reaction is the elimination of the elements of H—X and the formation of an alkene (see the next section). Tertiary halides are particularly susceptible to this reaction, making it very difficult to use them successfully in a nucleophilic substitution reaction.

**Stereochemistry of Nucleophilic Substitutions.** If the carbon holding the halogen in an alkyl halide is asymmetric, what will be the fate of this center of asymmetry when it experiences nucleophilic substitution? If the mechanism is $S_N1$, the product will be a racemic mixture, or very nearly one, because the formation of a carbonium ion at a carbon that is initially asymmetric destroys that asymmetry. The carbon in the ion has only three groups attached to it. The nucleophile can approach the positive charge from either side of the ion, and the product will be a racemic mixture (Figure 5.9).

**Table 5.5  Some Nucleophilic Substitution Reactions of Alkyl Halides***

| Nucleophilic Agent | R—X + Y: ⟶ R—Y + X: |
|---|---|
| Negatively charged nucleophiles | |
| Hydroxide ion, $OH^-$ | $R—X + HO^- \longrightarrow R—OH + X^-$ <br> Alcohols |
| Alkoxide ion, $RO^-$ | $+ RO^- \longrightarrow R—O—R + X^-$ <br> Ethers |
| Cyanide ion, $^-CN$ | $+ {}^-CN \longrightarrow R—CN + X^-$ <br> Nitriles |
| Amide ion, $^-NH_2$ | $+ {}^-NH_2 \longrightarrow R—NH_2 + X^-$ <br> Amines |
| Neutral nucleophiles | |
| Ammonia, $:NH_3$ | $+ :NH_3 \longrightarrow R—\ddot{N}H_2 + (HX)$ <br> Amines |
| Water, $:\ddot{O}H_2$ | $+ H_2\ddot{O}: \longrightarrow R—\ddot{O}H + (HX)$ <br> Alcohols |

\* The alkyl halide may be 1° or 2°, alkyl or cycloalkyl, allylic or benzylic. It may not be vinyl or aryl, which are usually inert toward nucleophilic substitution.

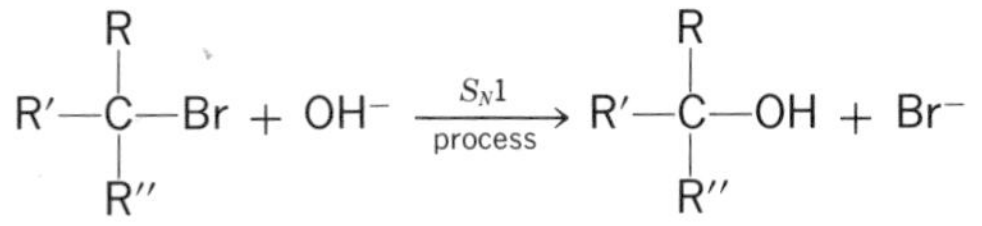

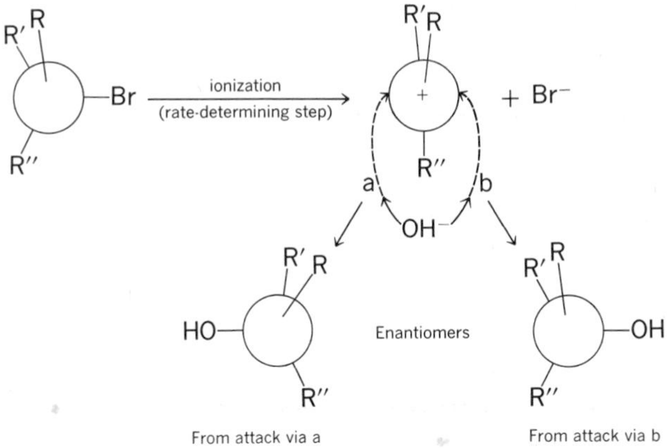

**Figure 5.9** Stereochemistry in a $S_N1$ reaction. If an optically active bromide is converted to the corresponding alcohol under conditions that assure an $S_N1$ mechanism, asymmetry will be lost at the carbonium-ion stage. Approach to the positively charged carbon by the nucleophile (OH$^-$ in our example) is equally likely from either side. The enantiomers of the product will form in a ratio of 1:1, and the product will be optically inactive. If the departing bromide ion stays for a while close by one side of the positively charged carbon, the nucleophile may more frequently attack the other side, and the product will have some optical activity. Only when the opportunities for approach to either side are the same will a fully racemic mixture form.

If the mechanism is $S_N2$, the configuration of the asymmetric carbon will be inverted. The product will be optically active, but its configuration will not be the same as in the starting material (Figure 5.10).

In the last chapter we learned relative orders of reactivity of alcohols in $S_N1$ and $S_N2$ mechanisms. These same orders apply to the nucleophilic substitution reactions of alkyl halides and for the same reasons. Aryl and vinyl halides are exceptionally inert toward nucleophilic substitution.

Transition state

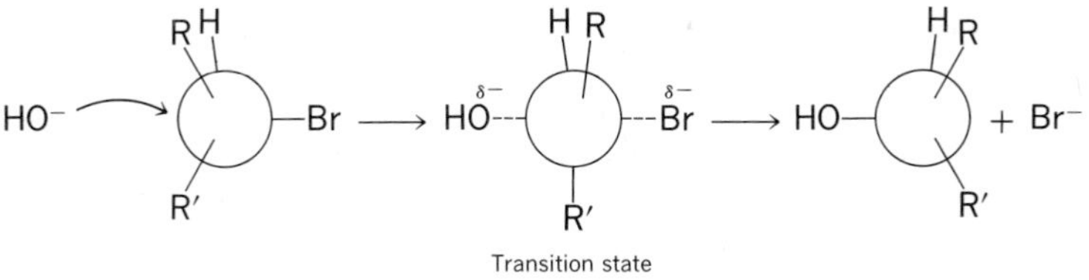

**Figure 5.10** Stereochemistry in an $S_N2$ reaction. If an optically active bromide is converted to the corresponding alcohol under conditions that assure an $S_N2$ mechanism, asymmetry will be preserved, but the configuration of the asymmetric carbon will be inverted. The product will be optically active.

**158**
Optical Isomerism.
Organohalogen
Compounds.
Ethers.
Amines.
Mercaptans

A carbon-carbon double bond can be introduced into a molecule by splitting the elements of H—X out of an alkyl halide:

$$-\overset{\displaystyle |}{\underset{\displaystyle H}{C}}-\overset{\displaystyle |}{\underset{\displaystyle X}{C}}- + \text{base} \longrightarrow \,\,\, \overset{\displaystyle }{C}{=}\overset{\displaystyle }{C} + (HX)$$

$$\xrightarrow{\text{base}} \begin{array}{l} \text{salt between} \\ \text{acid and base} \end{array}$$

The elements lost, H—X, must be on adjacent carbons, as shown. A common reagent for accomplishing this splitting is a solution of potassium hydroxide in alcohol. Since hydroxide ion is also a nucleophile, some molecules of halide can lose H—X to form an alkene, while others, undergoing different kinds of collisions, can take part in a substitution reaction to form R—OH. By suitable selection of conditions, however, chemists have been able to make one or the other of these two processes, either elimination or substitution, dominate the course of the reaction. In general:

> —Elimination is favored by high concentration of hydroxide ion in *alcohol* solvent at relatively high temperatures.
> —Substitution is favored by low concentration of hydroxide ion in *water* at lower temperatures.

In more advanced courses these observations are explained in terms of general principles. The student is not expected to memorize these facts. They are mentioned simply to show that variations in experimental conditions (concentrations, solvents, temperatures for example) can greatly affect the course of a reaction.

The experimental work of many chemists is devoted to studies of these factors, not only what works (practical) but how it works (theoretical). If a particular synthesis is to be put on stream in an around-the-clock industrial operation employing many people and producing organic chemicals in tank car quantities for sale in a competitive business, an experimental factor that increases the yield by even a few percent can make a great difference in the profit-loss columns.

## ETHERS

### STRUCTURE AND NOMENCLATURE

The ethers are substances whose molecules are of the general formulas R—O—R, R—O—Ar, and Ar—O—Ar, with R any alkyl group and Ar any aromatic group. They are generally designated by first naming the two groups attached

$CH_3{-}O{-}CH_3$                $CH_3{-}O{-}CH_2CH_3$                $CH_3CH_2{-}O{-}CH_2CH_3$

Dimethyl ether[3]                Ethyl methyl ether                Diethyl ether
(anesthetic ether)

[3] The names of symmertical ethers frequently do not contain the prefix "di-"; thus dimethyl ether can be named methyl ether. Diethyl ether is often called simply ethyl ether and sometimes, as here, it can be called just ether.

$$CH_2{=}CH{-}O{-}CH{=}CH_2$$

Divinyl ether
(vinethene, another
anesthetic)

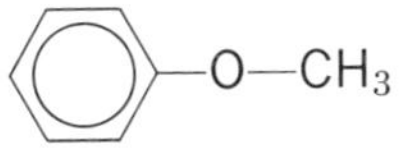

Methyl phenyl ether
(anisole)

to oxygen and then adding the separate word *ether,* as illustrated by the following examples. When the two groups attached to oxygen are identical, the ether is said to be *symmetrical;* otherwise it is said to be *unsymmetrical.* The IUPAC has rules for naming more complicated systems, but we shall not study them.

## SYNTHESIS OF ETHERS

Ethers can be made in two important ways—through the action of the salt of an alcohol on an alkyl halide, or through the action of heat and an acid catalyst on an alcohol. The first is important in organic synthesis in research; the second is important commercially for making certain ethers.

**The Williamson Ether Synthesis.** In general:

First:    $R{-}O{-}H + Na \longrightarrow R{-}O^-Na^+ + \frac{1}{2}H_2$

A sodium alkoxide

Then:    $R{-}O^- + R'{-}X \longrightarrow R{-}O{-}R' + X^-$

$$CH_3{-}I + {}^-O{-}CH_2CH_2CH_2CH_3 \longrightarrow CH_3{-}O{-}CH_2CH_2CH_2CH_3 + I^-$$

Methyl
iodide

*n*-Butoxide ion

Methyl *n*-butyl ether (71%)

$$CH_3CH{-}I + {}^-O{-}CH_2CH_2CH_2CH_3 \longrightarrow$$

Ethyl
iodide

$$CH_3CH_2{-}O{-}CH_2CH_2CH_2CH_3 + I^-$$

Ethyl *n*-butyl ether (71%)

$$CH_3CH_2{-}I + {}^-O{-}CH_2CH_2CH_2CH_2CH_3 \longrightarrow$$

Ethyl
iodide

*n*-Pentoxide ion

$$CH_3CH_2{-}O{-}CH_2CH_2CH_2CH_2CH_3 + I^-$$

Ethyl *n*-pentyl ether (47%)

Because alkoxide ions are powerful bases as well as nucleophiles, 3° halides will undergo dehydrohalogenation, not ether formation. The Williamson synthesis is very successful when the halide is primary and then the mechanism is usually $S_N2$.

**Ethers from Dehydration of Alcohols.** In general:

$$2ROH \xrightarrow{\text{H}^+} R{-}O{-}R + H_2O$$

or

$$R{-}O{-}H + H{-}O{-}R \xrightarrow{\text{H}^+} R{-}O{-}R + H_2O$$

**160**

Optical Isomerism.
Organohalogen
Compounds.
Ethers.
Amines.
Mercaptans

Specific example:

$$2CH_3CH_2OH \xrightarrow[H^+]{140°C} CH_3CH_2-O-CH_2CH_3 + H_2O$$

Ethyl alcohol                Diethyl ether

Mechanism. Type $S_N2$.

Step 1.     $CH_3CH_2-OH + H_2SO_4 \rightleftharpoons CH_3CH_2-\overset{+}{O}H_2 + HSO_4^-$     (rapid)

         Ethyl alcohol                 Ethyloxonium ion

Step 2.

Unprotonated ethyl alcohol, the nucleophile      Ethyloxonium ion

Transition state            Protonated form of diethyl ether, diethyloxonium ion

Step 3.

Diethyloxonium ion                        Diethyl ether

Step 1 is simply an acid-base equilibrium involving ethanol as the stronger proton acceptor and sulfuric acid as the stronger proton donor. Acid-base equilibria of this type are rapidly established. Step 2 is the rate-determining or slow step—the attack by the nucleophile ethanol on the carbinol carbon of the ethyloxonium ion. Because its oxygen is protonated and positively charged, this carbinol carbon bears a significant partial positive charge. The product of this step is the diethyloxonium ion. Step 3 is, again, a rapidly established acid-base equilibrium, wherein bisulfate ion is shown as the proton acceptor. Actually, this is only one of the proton acceptors present. Molecules of water or ethanol could serve here too.

Exercise 5.2   When two primary alcohols, R—O—H and R′—O—H, are subjected to conditions favoring the formation of ether, three are usually isolable: R—O—R, R—O—R′, and
R′—O—R′. By writing step 1 for each alcohol, by combining each oxonium ion obtained
from this reaction with each alcohol as the nucleophile in step 2, show how these three
ethers can be expected to form in step 3. What are the relative proportions of each if a
solely statistical distribution is obtained—1:1:1 or 2:1:2 or 1:2:1 (in the order ROR,
ROR′, R′OR′)?

The exercise reveals the impracticality of preparing mixed ethers by dehydration between different primary alcohols. Certain other combinations, however, work
quite well and may be used to illustrate general principles. A mixture of $t$-butyl alcohol
and ethyl alcohol, when warmed in the presence of 15% sulfuric acid, yields the
mixed ether, $t$-butyl ethyl ether, as virtually the only product (95%).

Exercise 5.3   What other ethers might also have formed, if statistical factors govern the course
of the reaction? Write their structures.

The mechanism for this reaction is probably the $S_N1$ type.

Step 1. $(CH_3)_3COH + H_3O^+ \rightleftharpoons (CH_3)_3C\overset{+}{O}H_2 + H_2O$

Step 2. $(CH_3)_3C\overset{+}{O}H_2 \longrightarrow (CH_3)_3C^+ + H_2O$     (probably the
slowest step)

Step 3. $(CH_3)_3C^+ + CH_3CH_2OH \longrightarrow (CH_3)_3C\overset{+}{\underset{H}{-}}O—CH_2CH_3$

Step 4. $(CH_3)_3C\overset{+}{\underset{H}{-}}O—CH_2CH_3 + H_2O \rightleftharpoons (CH_3)_3C—O—CH_2CH_3 + H_3O^+$

Thus the relatively easy conversion of a tertiary alcohol into its carbonium ion (steps
1 and 2) and the rapid nucleophilic attack on it by ethyl alcohol, whose oxygen is
certainly much less crowded by neighboring groups than the oxygen in $t$-butyl alcohol,
indicate why this reaction takes place. To state the general principal, reactions funnel
through paths of minimum energy and maximum probability.
    The formation of ethers by acid-catalyzed dehydration of alcohols depends on
careful control of conditions. If the temperature is allowed to climb too high,
an alternative reaction, the *internal* dehydration of an alcohol to produce an alkene,
becomes significant. In fact, the internal dehydration of alcohols is one of the
important ways of introducing a carbon-carbon double bond.

**Table 5.6  Some Ethers**

162
Optical Isomerism.
Organohalogen
Compounds.
Ethers.
Amines.
Mercaptans

| Name | Structure | Mp (°C) | Bp (°C) | Solubility in Water |
|---|---|---|---|---|
| Dimethyl ether | $CH_3$—O—$CH_3$ | −138 | −23 | 37 volumes gas dissolve in 1 volume $H_2O$ at 18°C |
| Methyl ethyl ether | $CH_3$—O—$CH_2CH_3$ | −116 | 11 | — |
| Diethyl ether | $CH_3CH_2$—O—$CH_2CH_3$ | −116 | 34.5 | 8 g/100 cc (16°C) |
| Di-$n$-propyl ether | $CH_3CH_2CH_2$—O—$CH_2CH_2CH_3$ | −122 | 91 | slightly soluble |
| Methyl phenyl ether | $CH_3$—O—$C_6H_5$ | −38 | 155 | insoluble |
| Diphenyl ether | $C_6H_5$—O—$C_6H_5$ | 28 | 259 | insoluble |

We shall later encounter ether formation in a special reaction of an aldehyde with an alcohol; our study of the mechanism whereby ethers may form from alcohols will be important at that time.

## PHYSICAL PROPERTIES OF ETHERS

Several representative ethers are listed in Table 5.6 together with certain of their physical constants. Ether molecules are slightly polar. Since they possess no O—H groups, they are incapable of the hydrogen bonding between two molecules that alcohols engage in. Hence the boiling points of ethers and alkanes of comparable formula weights are much the same. Diethyl ether (formula weight 74) boils at 34.6°C; $n$-pentane (formula weight 72) boils at 36°C; $n$-butyl alcohol (formula weight 74) boils much higher, at 117°C.

Since ether molecules do contain an oxygen, they can be involved in hydrogen bonding with water molecules. Even though an ether molecule cannot "donate" a hydrogen bridge, it can, of course, "accept" one (cf. Figure 5.11). The solubilities in water of ethers and alcohols of comparable formula weight are therefore found to be similar. For example, both $n$-butyl alcohol and diethyl ether dissolve to the extent of about 8 g in 100 cc of water.

Neither boiling points nor the solubilities are to be memorized, for they are cited only to help indicate how structural factors are related to these physical properties.

## CHEMICAL PROPERTIES OF ETHERS

Although ethers undergo a few reactions, none is very important to our later study of biochemistry. We simply note the fact that the ether linkage is chemically quite stable. At room temperature strong aqueous bases do not react with ethers. Concentrated, strong acids, for example, hydrobromic and hydriodic acid, will cleave ethers:

$$R—O—R + 2H—Br \xrightarrow{heat} 2R—Br + H_2O$$

Such conditions are obviously not found in living organisms. Only a special kind of ether (an acetal) is easily cleaved, a fact essential to understanding the digestion of sugars.

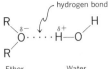

Ether
molecule

Water
molecule

**Figure 5.11** The oxygen of an ether molecule can serve as a hydrogen bond acceptor, which makes ethers somewhat more soluble in water than alkanes.

## HAZARDS

Most liquid, aliphatic ethers (R—O—R), when exposed to air, react slowly to form unstable peroxides which even in small concentrations can be quite dangerous. They can explode violently when exposed to air, setting fire to the ether. The presence of peroxides can be checked easily by shaking the ether with an aqueous solution of iron(II) ammonium sulfate plus potassium thiocyanate. If peroxides are present, a red color is produced.

Anesthetic ether, since it has such a low boiling point, readily volatilizes in an open container. This mixture of air and ether vapor is highly explosive. When diethyl ether is used as an anesthetic, precautions must be taken to ensure that no stray spark will set off the reaction. Because ether vapors are denser than air and tend to sink to the floor, workers using this substance should never wear shoes with cleats. Open flames are obviously to be kept at a considerable distance.

## IMPORTANT INDIVIDUAL ETHERS

**Diethyl Ether ("Ether").** Diethyl ether, a colorless, volatile liquid with a pungent, somewhat irritating odor, is a depressant for the central nervous system and is at the same time somewhat of a stimulant for the sympathetic system. It exerts an effect on nearly all tissues of the body, but it is still one of the safest anesthetics for human use.

**Divinyl Ether ("Vinethene").** This ether, another anesthetic, is more rapid in its action than diethyl ether. It too forms an explosive mixture with air.

# MERCAPTANS AND RELATED COMPOUNDS

## STRUCTURE AND NOMENCLATURE. OCCURRENCE

Mercaptans, sulfur analogs of alcohols, are sometimes called thio alcohols.[4] Table 5.7 lists a few representative examples together with names and certain physical properties. The —SH group that characterizes mercaptans is found among many of the protein molecules in the chemical inventories of living cells. The low-formula-weight mercaptans, which are therefore the more volatile, possess some of the most disagreeable odors of all organic materials. n-Butyl mercaptan is the principle constituent of the well-known fluid that is the defensive weapon of the skunk.

---

[4] Mercaptan is derived from the Latin term *mercurium captans,* meaning "seizing mercury." Mercuric ions react and form a precipitate (a mercuric salt) with mercaptans.

Table 5.7 Some Mercaptans

164
Optical Isomerism.
Organohalogen
Compounds.
Ethers.
Amines.
Mercaptans

| Name | Structure | Mp (°C) | Bp (°C) | Solubility in Water |
|---|---|---|---|---|
| Methyl mercaptan | $CH_3SH$ | −123 | 6 | soluble |
| Ethyl mercaptan | $CH_3CH_2SH$ | −144 | 37 | slightly soluble |
| n-Propyl mercaptan | $CH_3CH_2CH_2SH$ | −113 | 67 | slightly soluble |
| Isopropyl mercaptan | $(CH_3)_2CHSH$ | −131 | 58 | slightly soluble |
| n-Butyl mercaptan | $CH_3CH_2CH_2CH_2SH$ | −116 | 98 | slightly soluble |
| Cysteine (an amino acid) | $\underset{\underset{NH_2}{\vert}}{HSCH_2CHCO_2H}$ | 258 (decomposes) | — | moderately soluble |

## OXIDATION OF MERCAPTANS TO DISULFIDES

Mercaptans are reducing agents easily oxidized by hydrogen peroxide or sodium hypochlorite to disulfides:

$$2R{-}S{-}H + H_2O_2 \longrightarrow R{-}S{-}S{-}R + H_2O$$

Disulfides are easily reduced to mercaptans:

$$R{-}S{-}S{-}R + (2H) \longrightarrow 2R{-}S{-}H$$

These two reactions, interconversions of mercaptan and disulfide groups, are quite important in protein chemistry.

# AMINES

## STRUCTURE AND NOMENCLATURE

Amines are alkyl or aryl derivatives of ammonia. One, two, or all three of the hydrogens on a molecule of ammonia may be replaced by such groups. For example,

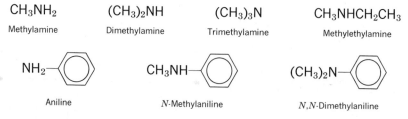

| $CH_3NH_2$ | $(CH_3)_2NH$ | $(CH_3)_3N$ | $CH_3NHCH_2CH_3$ |
|---|---|---|---|
| Methylamine | Dimethylamine | Trimethylamine | Methylethylamine |

Aniline      N-Methylaniline      N,N-Dimethylaniline

*Aliphatic amines* are those in which the carbon(s) attached directly to the nitrogen has (or have) only single bonds to other groups. Thus the compound $R{-}\overset{\overset{O}{\parallel}}{C}{-}NH_2$ is not an amine, but the compound $R{-}\overset{\overset{O}{\parallel}}{C}{-}CH_2{-}NH_2$ has an amine function. If even one of the groups attached directly to the nitrogen is aromatic, the amine is classified as an *aromatic amine.* This subclassification is useful because aromatic amines have benzene rings highly reactive toward aromatic substitution reactions and because they are much weaker bases than aliphatic amines. In other words, the differences between the two are enough to make two subclasses convenient. They also share several types of reactions.

Another way of subclassifying amines is by the designations 1° (primary), 2° (secondary), and 3° (tertiary) amines. The numerals denote the number of groups attached to the *nitrogen:*

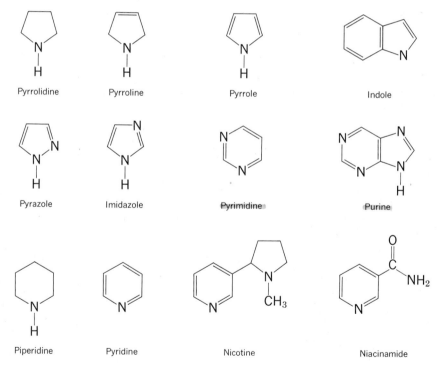

Because of structural similarity to amines, tetraalkyl derivatives of the ammonium ion, $NH_4^+$, called *quaternary ammonium ions,* $R_4N^+$, are also included here.

**Heterocyclic Amines.** Many compounds of biochemical importance consist, at least in part, of ring systems in which one of the ring atoms is a nitrogen. Some of the important heterocyclic amines are

The pyrimidine and purine systems are particularly important in biochemical genetics, the chemistry of heredity, but little more need be said about these heterocyclic amines at this point. It can be noted, however, that the saturated

166
Optical Isomerism.
Organohalogen
Compounds.
Ethers.
Amines.
Mercaptans

systems, pyrrolidine and piperidine, behave like ordinary 2° amines. Pyridine is a weakly basic 3° amine with a benzene-like ring. In fact, it has many aromatic properties.

**Nomenclature.** In the *common system* of nomenclature, aliphatic amines are named by designating the alkyl groups attached to the nitrogen and following this series by the word part "-amine."

$$(CH_3)_2CHCH_2—NH_2$$

Isobutylamine

$$(CH_3)_2CH—NH—CH_2CH_3$$

Ethylisopropylamine

$$CH_3$$
$$|$$
$$CH_3CH_2CH_2—N—CH_2CH_2CH_2CH_3$$

Methyl-*n*-propyl-*n*-butylamine

Where this system breaks down, the IUPAC system can be employed. In this system the $—NH_2$ group is named the *amino* group, or *N*-alkyl or *N,N*-dialkylamino, and the specific groups are of course named.

$$CH_3—CH—CH_2—OH$$
$$|$$
$$NH_2$$

2-Amino-1-propanol

$$CH_3CH_2NH—CH_2CH_2CH_2CH_2CH_2CH_2CH_2CH_3$$

1-(*N*-Ethylamino)octane (That is, at the 1-position in octane an *N*-ethylamino group is attached.)

The capital *N* designation means that the group immediately following it is attached to nitrogen. (See also the earlier examples of *N*-methylaniline and *N,N*-dimethylaniline.)

The simplest aromatic amine is always called *aniline*.

---

✓ Exercise 5.4  Write structures for the following compounds.

(*a*) triethylamine
(*b*) dimethylethylamine
(*c*) *t*-butyl-*sec*-butyl-*n*-butylamine
(*d*) *p*-nitroaniline
(*e*) *p*-aminophenol
(*f*) 1,3-diamino-2-pentanol
? (*g*) cyclohexylamine
(*h*) diphenylamine
(*i*) 2-(*N*-methyl-*N*-ethylamino)butane
(*j*) benzylmethylamine

✓ Exercise 5.5  Classify the amines of the preceding exercise according to (*a*) aromatic versus aliphatic subclasses, (*b*) 1°, 2°, or 3° subclasses.

✓ Exercise 5.6  Write names for each of the following according to the common system whenever possible, otherwise according to the IUPAC system.

*ethyl-di-n-butylamine*

(*a*) $CH_3CH_2N(CH_2CH_2CH_2CH_3)_2$

(*b*) $NH_2CH_2CH_2NH_2$

(*c*) [benzene ring]—N⟨$CH_2CH_3$ / $CH_2CH_3$⟩

$$NH_2$$
(*d*) [cyclopentane ring with NH₂]

*cyclopentylamine*

(*e*) $(CH_3)_4N^+Br^-$

*tetramethylammonium bromide*

## PHYSICAL PROPERTIES

In Table 5.8 several representative amines are listed together with important
physical properties. Molecules of amines are moderately polar. Hydrogens on nitrogen in amines can form hydrogen bonds to nitrogens on neighboring amines or to oxygens on neighboring molecules of water if that is present. As a result, when compounds of similar formula weights are compared, amines have higher boiling points than alkanes but lower boiling points than alcohols.

|  | F.Wt. | Bp (°C) |
|---|---|---|
| $CH_3CH_3$ | 30 | $-89$ |
| $CH_3NH_2$ | 31 | $-6$ |
| $CH_3OH$ | 32 | 65 |

Like the —OH group in alcohols, the —$NH_2$ group in amines helps make them soluble in water (Figure 5.12). Amines are also soluble in less polar solvents.

The lower-formula-weight amines (e.g., methylamine, ethylamine) have odors very much like that of ammonia. At higher formula weights the odors become "fishy." Aromatic amines have moderately pleasant, pungent odors, but they are at the same time very toxic. They are absorbed directly through the skin with sometimes fatal results. Aromatic amines are also easily oxidized by atmospheric oxygen. Dark-colored substances form during this process, and although highly purified aromatic amines are usually colorless, after standing for a time on the shelf they are discolored or sometimes black.

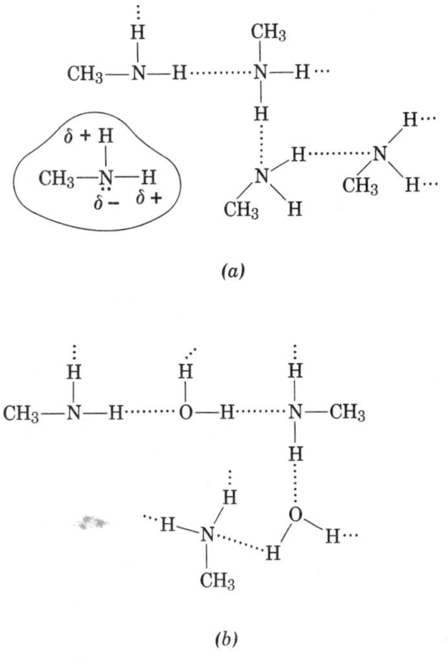

(a)

(b)

**Figure 5.12** Amines as hydrogen bond donors and acceptors. (a) Hydrogen bonds (dotted lines) can exist between molecules of an amine. (b) Low-formula-weight amines are soluble in water because they can slip into the hydrogen-bonding network of water.

**Table 5.8 Amines**

| Name | Structure | Mp (°C) | Bp (°C) | Solubility in $H_2O$ | $K_b$ (at 25°C) |
|---|---|---|---|---|---|
| Ammonia | $NH_3$ | −78 | −33 | very soluble | $1.8 \times 10^{-5}$ |
| Methylamine | $CH_3NH_2$ | −94 | −6 | very soluble | $4.4 \times 10^{-4}$ |
| Dimethylamine | $(CH_3)_2NH$ | −96 | 7 | very soluble | $5.2 \times 10^{-4}$ |
| Trimethylamine | $(CH_3)_3N$ | −117 | 3.5 | very soluble | $0.5 \times 10^{-4}$ |
| Ethylamine | $CH_3CH_2NH_2$ | −84 | 17 | very soluble | $5.6 \times 10^{-4}$ |
| Diethylamine | $(CH_3CH_2)_2NH$ | −48 | 56 | very soluble | $9.6 \times 10^{-4}$ |
| Triethylamine | $(CH_3CH_2)_3N$ | −115 | 90 | 14 g/100 cc | $5.7 \times 10^{-4}$ |
| n-Propylamine | $CH_3CH_2CH_2NH_2$ | −83 | 49 | very soluble | $4.7 \times 10^{-4}$ |
| Di-n-propylamine | $(C_3H_7)_2NH$ | −40 | 110 | very soluble | $9.5 \times 10^{-4}$ |
| Tri-n-propylamine | $(C_3H_7)_3N$ | −90 | 156 | slightly soluble | $4.4 \times 10^{-4}$ |
| Isopropylamine | $(CH_3)_2CHNH_2$ | −101 | 33 | very soluble | $5.3 \times 10^{-4}$ |
| n-Butylamine | $CH_3CH_2CH_2CH_2NH_2$ | −51 | 78 | very soluble | $4.1 \times 10^{-4}$ |
| Isobutylamine | $(CH_3)_2CHCH_2NH_2$ | −86 | 68 | very soluble | $3.1 \times 10^{-4}$ |
| sec-Butylamine | $CH_3CH_2CH(CH_3)NH_2$ | <−72 | 63 | very soluble | $3.6 \times 10^{-4}$ |
| t-Butylamine | $(CH_3)_3CNH_2$ | −68 | 45 | very soluble | $2.8 \times 10^{-4}$ |
| Ethylenediamine | $NH_2CH_2CH_2NH_2$ | 9 | 117 | very soluble | $8.5 \times 10^{-4}$ |
| Tetramethylammonium hydroxide | $(CH_3)_4N^+OH^-$ | 130–135 (decomposes) | — | very soluble | very strong base |
| **Aromatic Amines** | | | | | |
| Aniline | 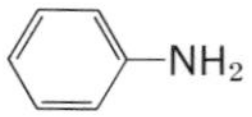 $\phantom{}$—$NH_2$ (phenyl) | −6 | 184 | 3.7 g/100 cc | $3.8 \times 10^{-10}$ |
| N-Methylaniline | 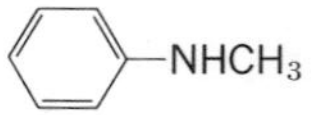 $\phantom{}$—$NHCH_3$ (phenyl) | −57 | 196 | slightly soluble | $5 \times 10^{-10}$ |
| N,N-Dimethylaniline | 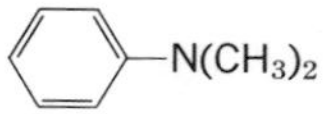 $\phantom{}$—$N(CH_3)_2$ (phenyl) | 3 | 194 | slightly soluble | $11.5 \times 10^{-10}$ |
| Sulfanilamide | 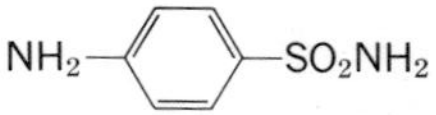 $NH_2$—(phenyl)—$SO_2NH_2$ | 163 | — | 0.4 g/100 cc | |

# BASICITY OF AMINES

Ammonia is a stronger base than water, for it holds a proton better than a water molecule can.

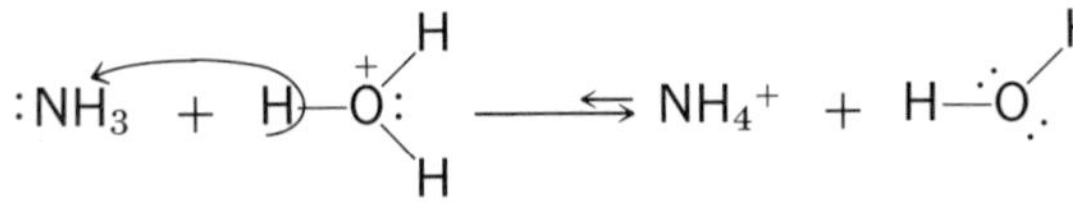

$$:NH_3 \ + \ H\!-\!\overset{+}{\underset{H}{O}}\!: \ \rightleftharpoons \ NH_4^+ \ + \ H\!-\!\overset{..}{O}$$

| Ammonia | Hydronium | Ammonium ion | Water |
|---|---|---|---|
| (stronger base) | ion | (weaker acid) | (weaker base) |
| | (stronger acid) | | |

The ammonium ion, as it occurs in ammonium salts, will give up its extra proton to hydroxide ion. Hence ammonia is not as strong a base as the hydroxide ion.

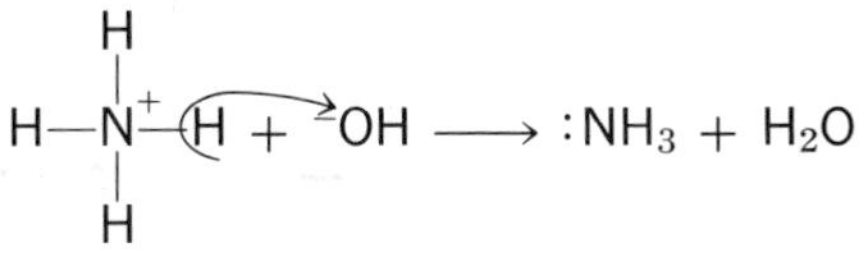

$$H\!-\!\overset{H}{\underset{H}{\overset{|}{N^+}}}\!-\!H \ + \ {}^-OH \ \longrightarrow \ :NH_3 \ + \ H_2O$$

| Stronger | Stronger | Weaker | Weaker |
|---|---|---|---|
| acid | base | base | acid |

Much of the chemistry of amines can be understood by comparing them to ammonia and to ammonium ions. Amines are simply close relatives of ammonia.

$$R\!-\!\overset{..}{N}H_2 \ + \ H_3O^+ \ \rightleftharpoons \ R\!-\!NH_3^+ \ + \ H_2O$$

1° amine
Stronger base    Stronger acid      Weaker acid    Weaker base

$$R_2\overset{..}{N}H \ + \ H_3O^+ \ \rightleftharpoons \ R_2NH_2^+ \ + \ H_2O$$

2° amine

$$R_3N: \ + \ H_3O^+ \ \rightleftharpoons \ R_3NH^+ \ + \ H_2O$$

3° amine

$$R_4N^+ \ + \ H_3O^+ \ \overset{/\!/\!/}{\longrightarrow} \ \text{no reaction}$$

4° ammonium
ion

Since the 4° ammonium ion has no unshared pair of electrons, and since like the hydronium ion it is positively charged, no reaction can occur with acid. *The basicity of amines and of ammonia is correlated with the unshared pair of electrons on nitrogen.* Any structural factor in an amine that makes this pair more "available" to a proton than it is in ammonia will make the amine a better proton acceptor, that is, a better or stronger base than ammonia. Any structural factor in an amine that makes this unshared pair less available than it is in ammonia will make the amine a weaker proton acceptor or a weaker base.

**170**

**Optical Isomerism.
Organohalogen
Compounds.
Ethers.
Amines.
Mercaptans**

$$G \longrightarrow \overset{..}{N}H_2 \qquad\qquad G \longleftarrow \overset{..}{N}H_2$$

If G is electron-donating,
the amine is a stronger
base than ammonia.

If G is electron-withdrawing,
the amine is a weaker base
than ammonia.

Relative basicities of amines can be seen by comparing the basicity constants for their reaction with water.

$$\overset{..}{R}NH_2 + H_2O \rightleftharpoons RNH_3^+ + OH^-$$

$$K_b = \frac{[RNH_3^+][OH^-]}{[RNH_2]}$$

*The higher the $K_b$, the stronger the basicity of the amine.* The $K_b$ values for several amines are listed in Table 5.8, and it is apparent that the aliphatic amines for the most part are *stronger* bases than ammonia.

**Basicity and Structure.** The difference, structurally, between ammonia and an aliphatic amine such as methylamine is, of course, the presence of the methyl group. We have learned that alkyl groups have an electron-releasing effect. Being directed toward nitrogen in methylamine, the unshared pair of electrons on nitrogen is made more available for sharing with a proton and accepting it. Once the proton is accepted, the same electron-releasing effect of the methyl group pushes electrons toward a positively charged site, stabilizing it.

$$CH_3 \rightarrow \overset{\overset{\displaystyle H}{|}}{\underset{\displaystyle H}{N}}: \; + H^+ \rightleftharpoons CH_3 \rightarrow \overset{\overset{\displaystyle H}{|}}{\underset{\displaystyle H}{\overset{+}{N}}} - H$$

Methyl releases
electron density,
making the unshared pair
of electrons more
available. Hence
methylamine is a stronger
base than ammonia.

Methyl releases
electron density, tending
to stabilize the
positive charge. Hence
methylamine is a stronger
base than ammonia.

According to the $K_b$ values in Table 5.8, aromatic amines are weaker bases than ammonia or aliphatic amines by factors of about one million ($10^{-4}$ versus $10^{-10}$). Apparently the unshared pair on nitrogen in an aniline is much less available for accepting a proton than that in an aliphatic amine. The resonance effect helps explain this phenomenon and is at the same time consistent with several facts known about aromatic amines. The principal contributing resonance forms for aniline are structures **A** through **E**. Of these **B**, **C**, and **D** involve both a

A          B          C          D          E

separation of unlike charges (energy-demanding) and a disruption of the alternating double-bond—single-bond network of the benzene ring (also energy-demanding). We therefore conclude that these three cannot contribute significantly to the true state of the aniline molecule. Forms **A** and **E** must make the principal contributions. However, *to the extent that* **B, C,** *and* **D** *make any contribution,* the unshared pair of electrons on the nitrogen is *at least somewhat delocalized into the ring.* The pair is to that extent less available for accepting a proton, and aniline is to that extent a weaker base than ammonia or aliphatic amines. This theory at the same time accounts for the high reactivity of the benzene ring in aniline (compared with that of benzene itself) toward electron-seeking reagents ($NO_2^+$ in nitration, $Br^+$ in bromination, etc.) Such attacking species find a higher electron density in the ring in aniline than they find in benzene, which cannot benefit from an electron "feeding" from an attached group. Moreover, the theory also accounts for the ortho-para–directing ability of the amino group. Resonance contributors B, C, and D enable us to predict that the increased electron density of the ring will be greatest at the ortho and para positions.

## AMINE SALTS

Table 5.9 contains a list of a few representative salts of amines. These are salts in the true sense, that is, they are typical ionic compounds. All are solids, all have relatively high melting points, especially compared with the melting points of their corresponding amines, and all are like ammonium salts generally soluble in water and insoluble in nonpolar compounds such as carbon tetrachloride.

In amine salts the positive ion can be called a *protonated amine;* such positively charged systems are of considerable importance in protein chemistry as proton donors. Hence they can act in cells to neutralize bases and help control the pH of body fluids. They also help make the substance soluble in water.

## CHEMICAL PROPERTIES OF AMINES

Several of the reactions of amines with organic compounds will be studied in later sections on these materials. Anticipating these future discussions, we call attention to the most important way of regarding an amine molecule. The unshared pair of electrons makes it not only a basic substance (that is, a proton-seeking molecule) but also a nucleophilic substance. Whenever an organic molecule contains a partial positive charge on carbon, it must be regarded as a likely site for attack by an amine.

**Table 5.9    Amine Salts**

| Name | Structure | Mp (°C) |
|---|---|---|
| Methylammonium chloride | $CH_3NH_3^+Cl^-$ | 232 |
| Dimethylammonium chloride | $(CH_3)_2NH_2^+Cl^-$ | 171 |
| Dimethylammonium bromide | $(CH_3)_2NH_2^+Br^-$ | 134 |
| Dimethylammonium iodide | $(CH_3)_2NH_2^+I^-$ | 155 |
| Tetramethylammonium hydroxide (base as strong as KOH) | $(CH_3)_4N^+OH^-$ | 130–135 (decomposes) |

Optical Isomerism.
Organohalogen
Compounds.
Ethers.
Amines.
Mercaptans

**Reaction of Amines with Alkyl Halides.** This reaction is an example of an amine attacking an organic molecule with a partial positive charge on carbon. In general:

$$R\!-\!\overset{..}{N}H_2 + R'\!-\!X \longrightarrow R\!-\!\overset{+}{N}H_2\!-\!R' + X^-$$

1° amine

$$\xrightarrow{\;\;OH^-\;\;} R\!-\!\overset{..}{N}H\!-\!R' + H_2O$$

2° amine

Ordinarily, sodium hydroxide is added after the initial reaction is over to liberate the amine.

$$R\!-\!\underset{\underset{R'}{|}}{\overset{..}{N}}\!-\!H + R''\!-\!X \xrightarrow[\text{hydroxide ion}]{\text{followed by}} R\!-\!\underset{\underset{R'}{|}}{\overset{..}{N}}\!-\!R''$$

2° amine              3° amine

$$R\!-\!\underset{\underset{R'}{|}}{\overset{..}{N}}\!-\!R'' + R^*\!-\!X \longrightarrow R\!-\!\underset{\underset{R'}{|}}{\overset{\overset{\displaystyle R^*}{|}}{N}}{}^{+}\!-\!R''\;\; X^-$$

3° amine              4° ammonium salt

When a 1° amine is allowed to react with an alkyl halide, a mixture consisting principally of 2° and 3° amines plus smaller amounts of 4° ammonium salt and unchanged 1° amine results. The relative amounts of these products can be controlled by adjusting the proportions of reactants. A large excess of 1° amine over alkyl halide, for example, will favor the 2° amine with only traces of higher alkylated amines. (Why?)

Specific examples:

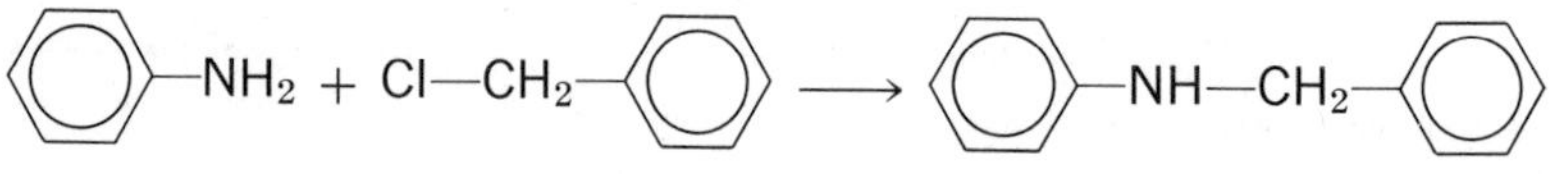

Aniline         Benzyl chloride            Phenylbenzylamine
4 parts : 1 part                     (96%)

$$(CH_3CH_2)_2NH + Br\!-\!CH(CH_3)_2 \longrightarrow (CH_3CH_2)_2NCH(CH_3)_2$$

Diethylamine       Isopropyl bromide       Diethylisopropylamine
1.3 parts : 1 part                (60%)

$$(CH_3CH_2)_2NH + ClCH_2CH_2OH \longrightarrow (CH_3CH_2)_2NCH_2CH_2OH$$

Diethylamine         Ethylene       β-Diethylaminoethyl alcohol (70%)
chlorohydrin

Mechanism: In the alkyl halide the bond is polarized as shown. The carbon holding the halogen has a partial positive charge, a likely site for an electron-rich reagent (the amine) to attack. The attack is therefore usually of the $S_N2$ type:

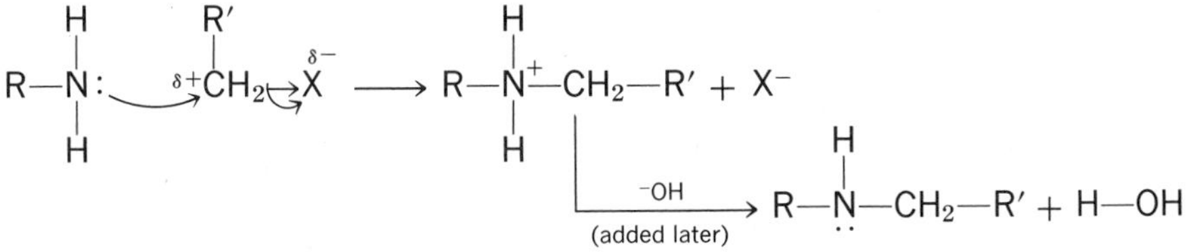

$$R—\overset{\overset{\displaystyle H}{|}}{\underset{\underset{\displaystyle H}{|}}{N}}:\overset{\delta-}{\underset{\delta+}{CH_2}}X \longrightarrow R—\overset{\overset{\displaystyle H}{|}}{\underset{\underset{\displaystyle H}{|}}{N^+}}—CH_2—R' + X^-$$

$$\xrightarrow[\text{(added later)}]{-OH} R—\overset{\overset{\displaystyle H}{|}}{\underset{\underset{\displaystyle \cdot\cdot}{N}}{}}—CH_2—R' + H—OH$$

**Reactions of Amines with Nitrous Acid.** The subclasses of amines behave differently toward nitrous acid. (This substance is usually prepared by mixing sodium nitrite, $NaNO_2$, and hydrochloric acid.)

**1° Amines**

Aliphatic: $R—\ddot{N}H_2 \xrightarrow[HX]{HONO} [R—\overset{+}{N}\equiv\overset{\cdot\cdot}{N}]X^- \longrightarrow R^+ + :N\equiv N: + X^-$

Alkyl diazonium ion     Carbonium    Nitrogen gas
(unstable even         ion
at 0°C)

$\Big\downarrow H_2O$

$\longrightarrow R—OH + $ by-products
       Alcohol

Aromatic: $Ar—\ddot{N}H_2 \xrightarrow{HONO} Ar—\overset{+}{N}\equiv\overset{\cdot\cdot}{N} \xrightarrow[\text{warm the}]{H_2O} Ar—OH + N_2 + H^+$

Aryl diazonium ion   solution      Phenol
(stable at 0°C)

**2° Amines.** Both aliphatic and aromatic:

$$(R)Ar—\underset{\cdot\cdot}{N}H \xrightarrow{HONO} (R)Ar—\overset{\overset{\displaystyle R}{|}}{N}—\ddot{N}{=}O + H_2O \qquad \text{(no evolution of } N_2\text{)}$$

*N*-Nitrosamine
(neutral compounds, insoluble
in acid; usually yellow or
orange-colored oils or solids)

**3° Amines.** Aliphatic amines. These dissolve in the acidic medium containing the nitrous acid. No nitrogen evolves. Products of varying complexity may or may not be generated.

Aromatic amines:

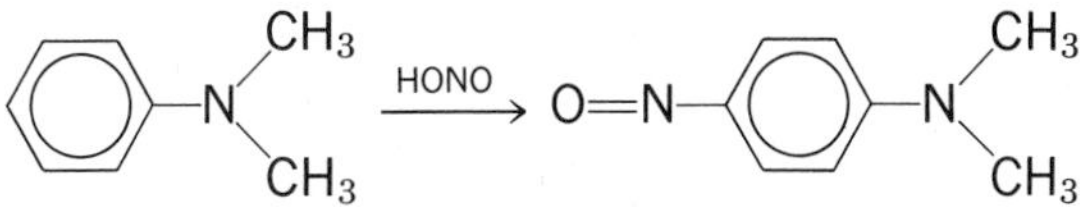

$$\underset{}{\bigcirc}—N\overset{\diagup CH_3}{\diagdown CH_3} \xrightarrow{HONO} O{=}N—\underset{}{\bigcirc}—N\overset{\diagup CH_3}{\diagdown CH_3}$$

*p*-Nitroso-*N,N*-dimethylaniline

In this reaction the benzene ring is so activated that ring substitution occurs. No nitrogen is evolved.

Optical Isomerism.
Organohalogen
Compounds.
Ethers.
Amines.
Mercaptans

The variety of responses of the subclasses of amines to nitrous acid makes this reagent useful in placing an "unknown" amine in its appropriate category.

—If reaction occurs at 0° and nitrogen evolves, the unknown is a 1° aliphatic amine.

—If reaction occurs at 0° but nitrogen does not evolve until the solution is warmed, a 1° aromatic amine is indicated.

—If reaction occurs without evolution of nitrogen and a yellow or orange oil (or solid) separates, it is a 2° aliphatic or aromatic amine.

—If the amine dissolves without evolution of nitrogen and without separation of a yellow or orange oil (or solid), the product is a 3° aliphatic or aromatic amine.

Even though several types of reactions with nitrous acid are possible, depending on the type of amine, the mechanism for most of them is initially the same. Freshly generated nitrous acid reacts with itself to form dinitrogen trioxide:

$$O=N-O-(H + H-O-)N=O \rightleftharpoons O=N-O-N=O + H-OH$$

Nitrous acid      Nitrous acid      Dinitrogen trioxide

This species appears to be the active one. The nitrogens bear partial positive charges, making them attractive to an amine molecule with unshared pairs of electrons:

$$-N: + \overset{\delta-}{O}\overset{\delta+}{=}N-O-\overset{\delta+}{N}\overset{\delta-}{=}O \rightleftharpoons {}^+N-N-O-N=O \rightleftharpoons {}^+N-N=O + NO_2^-$$

Amine
(any class)      N-Nitrosamine salt      Nitrite ion

If the original amine is 1°, the intermediate N-nitroso derivative breaks down as follows:

$$R-\overset{+}{N}-N=O \xrightarrow{-H^+} R-N-N=O \longrightarrow R-N=N \xrightarrow{H^+} R-N=N$$

An alkyl diazotic acid

$$ROH \xleftarrow[-H^+]{H_2O}$$
$$R-X \xleftarrow{X^-}$$
$$\text{Olefin} \xleftarrow{-H^+}$$
(mixture of products)

$$R^+ + N_2 \longleftarrow R-\overset{+}{N}\equiv N: + O$$

Carbonium ion      Diazonium ion

If the original amine is 2°, the following happens:

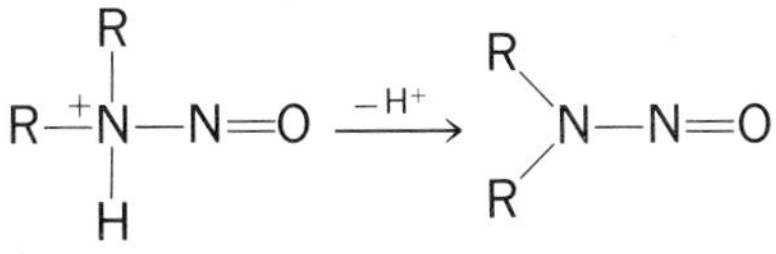

$$\underset{\underset{H}{|}}{\overset{\overset{R}{|}}{R-\overset{+}{N}}}-N=O \xrightarrow{-H^+} \underset{R}{\overset{R}{\diagdown}}N-N=O$$

N-Nitrosamine

If the original amine is 3°, the events that occur are more complicated and will not be discussed.

**The Van Slyke Determination of Amino Groups.** Proteins usually have several free amino groups ($-NH_2$) per molecule. Their relative number can be estimated because they react quantitatively with nitrous acid to yield one mole of nitrogen gas per mole of $-NH_2$ group. The nitrogen can be trapped in a gas buret and its volume measured and then converted by simple gas law calculations into moles of nitrogen. The determination is named after its developer, Donald Van Slyke.

---

✓ Exercise 5.7   Write the structures of the principal products that can be expected to form in the following reactions. For the purposes of this exercise assume that 1° aliphatic amines react with nitrous acid to produce an alcohol plus nitrogen gas: $R-NH_2 \xrightarrow{HNO_2} R-OH + N_2$.

(a) $(CH_3)_2NH \xrightarrow{NaNO_2,\ HCl}$

(b) $CH_3-\!\!\bigcirc\!\!-NH_2 \xrightarrow[0°]{NaNO_2,\ HCl}$

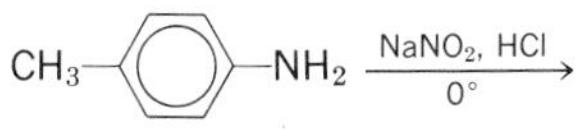

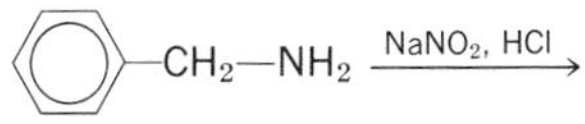

(c) $\bigcirc\!\!-CH_2-NH_2 \xrightarrow{NaNO_2,\ HCl}$

(d) $\bigcirc\!\!-NHCH_3 \xrightarrow{NaNO_2,\ HCl}$

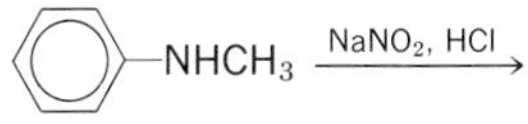

(e) $\bigcirc\!\!-N(CH_2CH_3)_2 \xrightarrow{NaNO_2,\ HCl}$

---

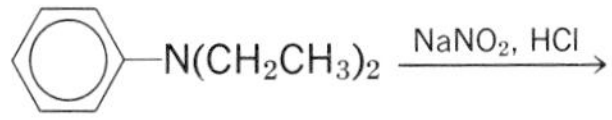

## REACTIONS OF AROMATIC DIAZONIUM SALTS

Aromatic diazonium salts can be made to undergo a variety of reactions leading to other aromatic compounds. Through these diazonium compounds entry into aromatic systems that are otherwise inaccessible can be initiated. Direct substitution into a benzene ring is for the most part possible only through nitration, bromination, chlorination, sulfonation, alkylation, and acylation. Although this list is impressive, it does not include direct introduction of $-NH_2$, $-OH$, $-F$, and $-CN$. The aromatic amino group is easily accessible through the corresponding nitro compound (p. 177).

$$Ar-NO_2 \xrightarrow{reduce} Ar-NH_2 \xrightarrow{HONO} Ar-N_2^+$$

From the amine the diazonium salt is easily generated. Depending on the selection of the next reagent to be added, the following transformations occur.

**Reactions in Which Nitrogen Is Lost.**

Optical Isomerism.
Organohalogen
Compounds.
Ethers.
Amines.
Mercaptans

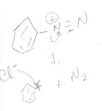

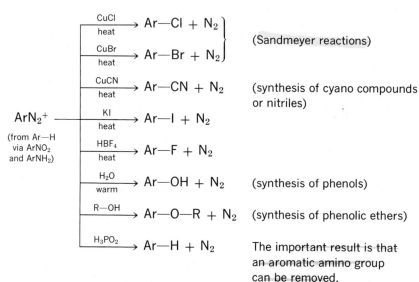

| | |
|---|---|
| $\xrightarrow[\text{heat}]{\text{CuCl}}$ Ar—Cl + $N_2$ | (Sandmeyer reactions) |
| $\xrightarrow[\text{heat}]{\text{CuBr}}$ Ar—Br + $N_2$ | |
| $\xrightarrow[\text{heat}]{\text{CuCN}}$ Ar—CN + $N_2$ | (synthesis of cyano compounds or nitriles) |
| $\xrightarrow[\text{heat}]{\text{KI}}$ Ar—I + $N_2$ | |
| $\xrightarrow[\text{heat}]{\text{HBF}_4}$ Ar—F + $N_2$ | |
| $\xrightarrow[\text{warm}]{\text{H}_2\text{O}}$ Ar—OH + $N_2$ | (synthesis of phenols) |
| $\xrightarrow{\text{R—OH}}$ Ar—O—R + $N_2$ | (synthesis of phenolic ethers) |
| $\xrightarrow{\text{H}_3\text{PO}_2}$ Ar—H + $N_2$ | The important result is that an aromatic amino group can be removed. |

ArN$_2^+$ (from Ar—H via ArNO$_2$ and ArNH$_2$)

**Reactions in Which Nitrogen Is Retained. Coupling.** With proper control of experimental conditions, aryl diazonium ions will react with phenols or aromatic amines to give compounds of the general type Ar—N=N—Ar′, called *azo compounds*. The following are examples.

Benzene-diazonium ion + Phenol → *p*-Hydroxyazobenzene (orange) + H$^+$

Diazotized sulfanilic acid + *N,N*-Dimethyl-aniline →

Methyl orange
(an acid-base indicator: above pH 4, yellow; below pH 3, red)

The reaction is little more than an aromatic substitution into the highly activated benzene ring of the phenol or the aromatic amine. The diazonium ion acts as the electron-seeking (electrophilic) reagent.

These aromatic azo compounds are all highly colored, and many are used as dyes. Close to half of all commercial dyes are azo compounds. In biochemistry the coupling reaction is the basis of the van der Bergh test, a method of identifying tissues that are rich in enzymes capable of catalyzing the hydrolysis of esters. An ester of phenol is added to the tissue, together with a diazonium salt. The ester will be hydrolyzed by action of the enzyme, if it is present, to liberate phenol, which as soon as it forms couples with the diazonium ion to give a highly colored product:

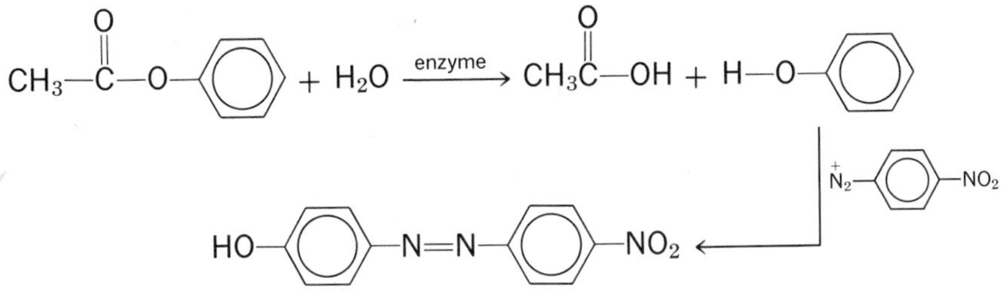

Colored azo
compound

If the tissue does not contain the enzyme capable of hydrolyzing the ester (which cannot couple with the diazonium ion), no phenol is produced and no colored spot develops.

## SYNTHESIS OF AMINES

A few representative syntheses are noted here, although a fuller discussion of each method is given elsewhere.

1. Alkylation of ammonia (see p. 156) or amines (see p. 172):

$$R-X + NH_3 \text{ (excess)} \longrightarrow R-NH_2 \quad \text{(as a salt which is allowed to react with a strong base to liberate the amine)}$$

$$R-X + R'-NH_2 \longrightarrow R-NH-R'$$

2. Reduction of nitro compounds:

$$Ar-NO_2 \xrightarrow{\text{reducing agent}} Ar-NH_2$$

Reducing agents that have been used include iron or tin with hydrochloric acid and catalytic hydrogenation.

3. Reductive amination (discussed on p. 207).

Optical Isomerism.
Organohalogen
Compounds.
Ethers.
Amines.
Mercaptans

**Exercise 5.8** Outline by means of reaction sequences the steps that will accomplish the following overall transformations.

(a) $CH_3$—⬡—$NH_2$  to  $CH_3$—⬡—I

(b) $NO_2$—⬡($NH_2$)  to  $NO_2$—⬡(Br)

(c) $NO_2$—⬡($NH_2$)  to  Cl—⬡(Br)

(d) $CH_3$—⬡—$NO_2$  to  $CH_3$—⬡—OH

(e) $CH_3$—⬡—$NO_2$  to  $CH_3$—⬡(Br)—OH

(f) $CH_3$—⬡—$NO_2$  to  $CH_3$—⬡(Br)(Br)

(g) $CH_3NH_2$  to  $(CH_3)_2NH$

(h) ⬡—$NHCH_3$  to  ⬡—$\overset{+}{N}(CH_3)_3I^-$

(i) $CH_3CH_2NH_2$  to  $CH_3CH_2OH$

(j) ⬡—$NH_2$  to  ⬡—N=N—⬡—OH

## IMPORTANT INDIVIDUAL AMINES

**Dimethylamine.** $(CH_3)_2NH$. One of the constituents that gives herring brine its distinctive odor, dimethylamine is synthesized and used industrially in the manufacture of fungicides and accelerators for the vulcanization of rubber. It is also used to make dimethylformamide (DMF), $(CH_3)_2N—\overset{\overset{\textstyle O}{\|}}{C}—H$, a powerful solvent capable of dissolving both water and many types of organic substances.

Several aliphatic amines (e.g., ethyl-, isobutyl-, isopentyl-, and $\beta$-phenylethylamine) are synthesized in animal cells. They act to stimulate the central nervous system.

**Ethylenediamine.** $NH_2CH_2CH_2NH_2$. The action of ammonia on 1,2-dichloro-ethane (from ethylene and chlorine) produces this amine. If it is allowed to react with the sodium salt of chloroacetic acid, one of the most versatile complexing (or chelating) agents known forms:

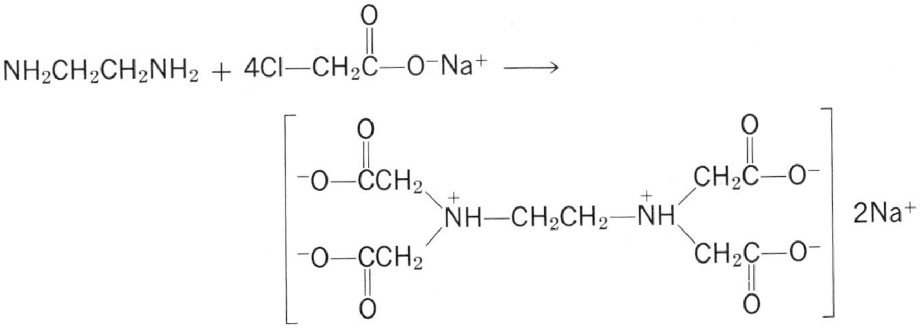

$$NH_2CH_2CH_2NH_2 + 4Cl-CH_2\overset{\displaystyle O}{\overset{\|}{C}}-O^-Na^+ \longrightarrow$$

$$\left[\begin{array}{c} {}^-O-\overset{O}{\overset{\|}{C}}CH_2 \\ {}^-O-\overset{O}{\overset{\|}{C}}CH_2 \end{array}\overset{+}{N}H-CH_2CH_2-\overset{+}{N}H\begin{array}{c} CH_2\overset{O}{\overset{\|}{C}}-O^- \\ CH_2\overset{\|}{\underset{O}{C}}-O^- \end{array}\right] 2Na^+$$

Ethylenediaminetetraacetic acid
(EDTA, as its sodium salt)

**Choline Chloride.** $(CH_3)_3\overset{+}{N}CH_2CH_2OHCl^-$. The positive ion *choline* is known to be necessary for growth and for the pathways of fat, carbohydrate, and protein metabolism. A derivative of choline, called *acetylcholine,* is one of the chemical transmitters of nerve impulses between cells of motor nerves (additional discussion on p. 391).

$$(CH_3)_3\overset{+}{N}CH_2CH_2O-\overset{\displaystyle O}{\overset{\|}{C}}CH_3$$

Acetylcholine

**Invert Soaps.** $CH_3(CH_2)_n\overset{+}{N}(CH_3)_3Cl^-$. Quaternary ammonium salts in which one of the alkyl groups is long (e.g., $n = 15$) have detergent and germicidal properties. In ordinary soaps the detergent action is associated with a negatively charged organic ion. In an invert soap a positive ion has the detergent property. (Detergent action is discussed on page 306.)

**Aniline.** $C_6H_5NH_2$. The annual production of aniline in the United States is well over 100,000 tons. Virtually all of it is used in the manufacture of aniline dyes, pharmaceuticals, and chemicals for the plastics industry. Two methods of synthesizing aniline are used industrially. In one benzene is first converted to nitrobenzene, which is in turn reduced by the action of iron and hydrochloric acid.

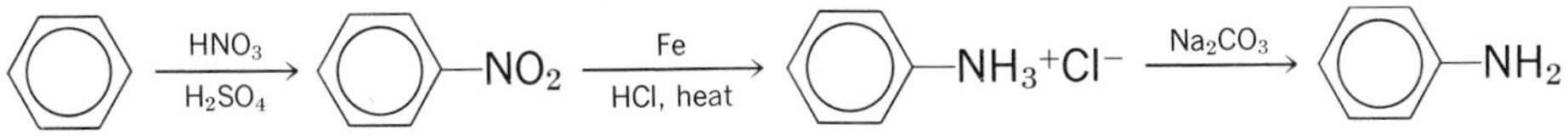

Benzene          Nitrobenzene          Anilinium chloride          Aniline

The other method begins with benzene which is chlorinated to produce chlorobenzene. Then ammonia in the presence of copper(II) oxide, high pressure, and an elevated temperature converts the chlorobenzene into aniline:

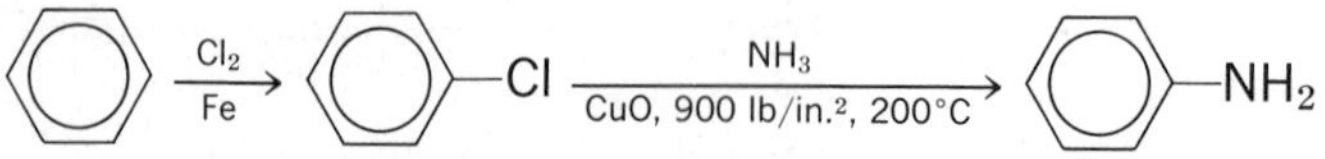

Optical Isomerism.
Organohalogen
Compounds.
Ethers.
Amines.
Mercaptans

$$\bigcirc \xrightarrow[\text{Fe}]{\text{Cl}_2} \bigcirc\text{-Cl} \xrightarrow[\text{CuO, 900 lb/in.}^2,\ 200°\text{C}]{\text{NH}_3} \bigcirc\text{-NH}_2$$

**Alkaloids.** The list of amino compounds that are important in biochemistry is very extensive, and several will be encountered from time to time in later sections. One group, the alkaloids, engages in considerable physiological activity and is largely produced by plants. Several are important drugs.

**Opium alkaloids.** Opium is isolated from the milky fluid that can be expressed from unripe Oriental poppy seeds (*Papaver somniferum*). It is a mixture of several alkaloids, among which codeine, papaverine, and morphine are the most important. A dose of 20 mg of morphine (Greek *Morpheus,* god of dreams) slows respiration, constricts pupils, and produces a deep, dreamless sleep. The drug is used principally to relieve severe pain.

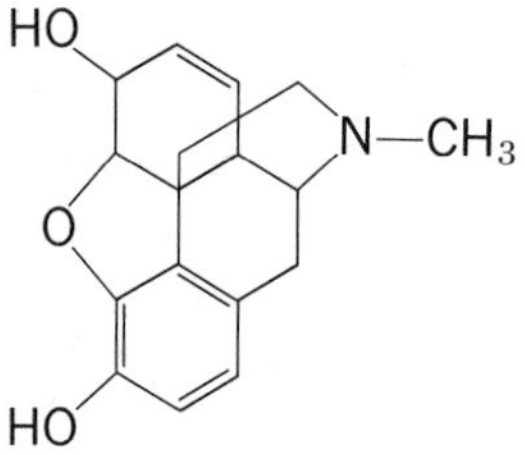

Morphine

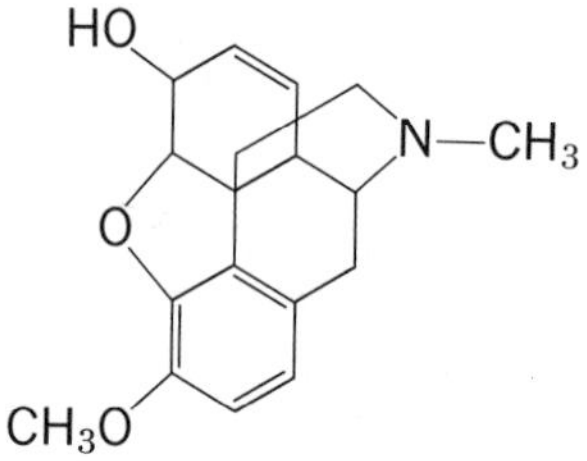

Codeine

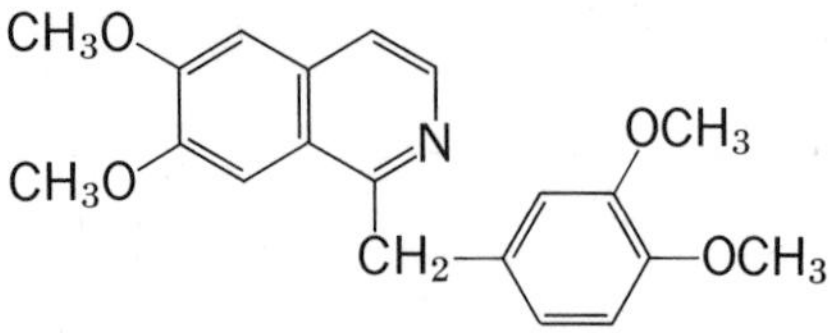

Papaverine

Codeine, which differs from morphine by only one methyl group, is used principally to relieve coughing; many cough medicines contain it in small amounts. In marked contrast to both morphine and codeine, papaverine has never been known to cause addiction and has but a slight sedative effect. It is useful as an antispasmodic and as a vasodilator.

Heroin, which does not occur in nature, is made by acetylating the two hydroxyl groups on morphine.

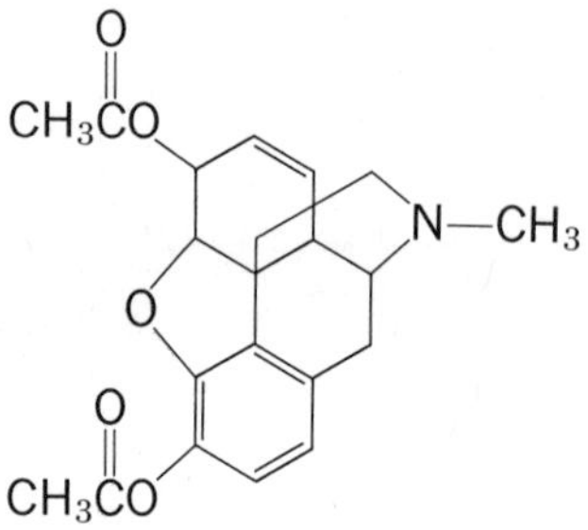

Heroin

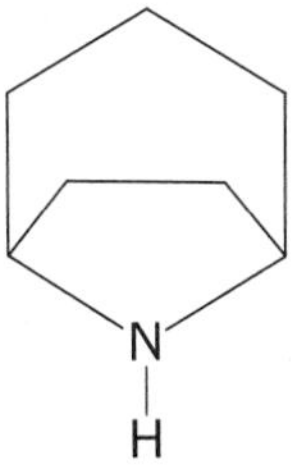

Tropane ring system

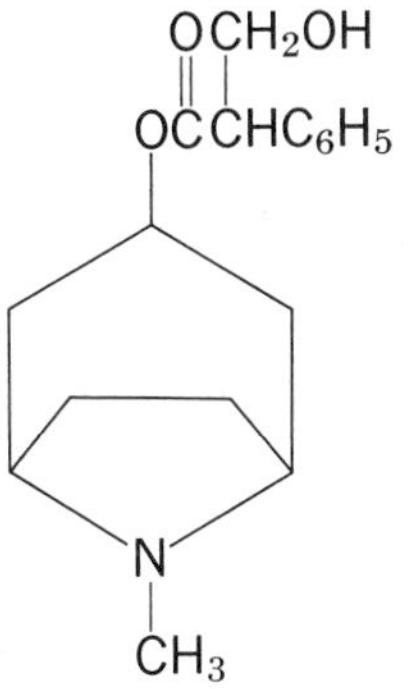

Atropine

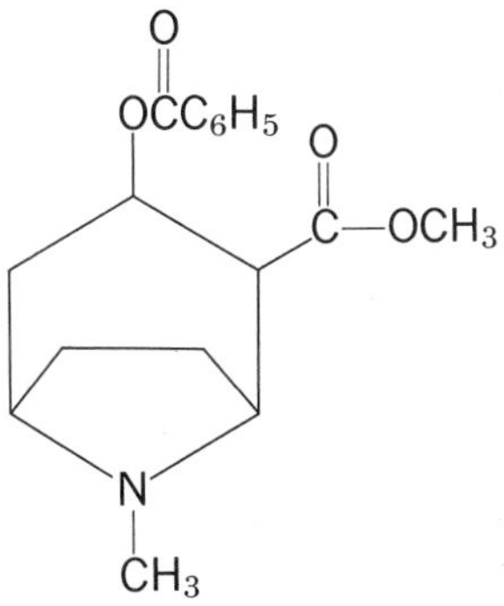

Cocaine

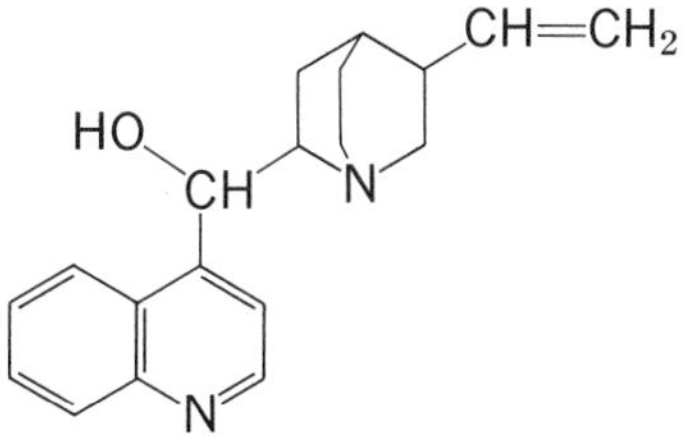

Quinine

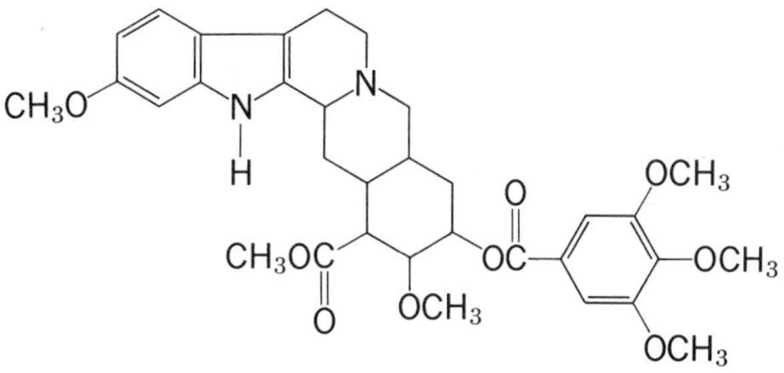

Reserpine (Serpasil)

**Tropane alkaloids.** Atropine and cocaine are the two most important members of this family of drugs, all of which possess the characteristic tropane ring system. Atropine, obtained from such exotic plants as the deadly nightshade (*Datura stramonium*) and belladonna (*Atropa belladona*), is an antidote for the cholinergic nerve stimulation produced, for example, by nerve gases (see p. 392). Cocaine, extracted from the leaves of a Peruvian bush, *Erythroxylon coca*, was the first local anesthetic used in medicine and is the only one still supplied from natural sources.

**Cinchona bark alkaloids.** High on the eastern slopes of the Andes mountains in South America grows a tree, *Cinchona officinalis*, the bark of which yields an alkaloid, quinine. Quinine was found to be specific for one of the most widespread diseases in the history of the world, malaria.

**182**

Optical Isomerism.
Organohalogen
Compounds.
Ethers.
Amines.
Mercaptans

**Rauwolfia alkaloids.** The roots of a shrub found in the hot, humid regions of India, *Rauwolfia serpentina* (Indian snake root), yield a drug that has become an important tranquilizer and sedative, reserpine or Serpasil. The crude root extract has long been used for dysentery, snake bite, and fever.

## REFERENCES AND ANNOTATED READING LIST

### BOOKS

R. T. Morrison and R. N. Boyd. *Organic Chemistry*, second edition. Allyn and Bacon, Boston, 1966. The following chapters in this standard first-year text for organic chemistry make a helpful supplement going beyond the material we have covered but not in a way that is difficult to handle: Chapter 3, "Stereochemistry I"; Chapter 7, "Stereochemistry II"; Chapter 14, "Alkyl Halides. Nucleophilic Aliphatic Substitution."

F. G. Bordwell. *Organic Chemistry*. The Macmillan Company, New York, 1963. Chapter 17, "Stereoisomerism," includes some interesting information about the discovery of optical activity.

### ARTICLE

M. Gates. "Analgesic Drugs." *Scientific American,* November 1966, page 131. Morphine is the most widely used drug for killing intense pain. This article gives its history and tells the story of the search for other pain-killers that do not have morphine's bad qualities.

In addition to these sources, any standard textbook on organic chemistry will have one or more chapters devoted to optical isomerism.

## PROBLEMS AND EXERCISES

1. Write definitions for each of the following terms and when appropriate provide suitable illustrations.

   (*a*)  plane of symmetry       (*b*)  optical isomers
   (*c*)  enantiomers       (*d*)  superimposability
   (*e*)  asymmetric carbon       (*f*)  optical activity
   (*g*)  polarimeter       (*h*)  dextrorotatory
   (*i*)  levorotatory       (*j*)  optical rotation (and symbol)
   (*k*)  specific rotation       (*l*)  meso compound
   (*m*)  diastereomers       (*n*)  racemic mixture
   (*o*)  inversion of configuration       (*p*)  Williamson reaction
   (*q*)  2° amine       (*r*)  heterocyclic amine
   (*s*)  basicity constant       (*t*)  diazotization
   (*u*)  Sandmeyer reaction       (*v*)  alkaloid

2. Examine the following structures and predict whether optical isomerism is a possibility.

$$(a) \quad CH_3\overset{\displaystyle CH_3}{\underset{\displaystyle OH}{C}}CH_2CH_3 \qquad (b) \quad CH_3\overset{\displaystyle CH_3}{\underset{\displaystyle OH}{CH}}CHCH_3 \qquad (c) \quad HO\overset{\displaystyle O}{\overset{\|}{C}}\underset{\displaystyle Br}{CH}CH_3$$

(*d*) $HO\overset{O}{\overset{\|}{C}}CH_2\underset{Br}{\underset{|}{C}}H_2$ (*e*) *cis*-1,2-cyclopentanediol (*f*) *trans*-1,2-cyclopentanediol

3. Explain why enantiomers *should* have identical physical properties (except for the sign of rotation of plane-polarized light).
4. Explain why diastereomers *should not* be expected to have identical physical properties (except coincidentally).
5. A solution of sucrose in water at 25°C in a tube 10 cm long gives an observed rotation of +2.0°. The specific rotation of sucrose in this solvent at this temperature is +66.4°. What is the concentration of the sucrose solution in grams per 100 cc?
6. Complete the following equations by supplying the structures of the principal organic products. Do not balance the equations. If no reaction is to be expected, write "No reaction."

   (*a*) $CH_3CH_2OH + SOCl_2 \longrightarrow$

   (*b*) $CH_3Br + NaCN \longrightarrow$

   (*c*) $CH_3CH_2SH + H_2O_2 \longrightarrow$

   (*d*) $CH_3CH_2OCH_2CH_3 + Na \longrightarrow$

   (*e*) $CH_3CH_2CH_2OH + PBr_3 \longrightarrow$

   (*f*) $(CH_3)_2CHBr + NH_3\ (\text{excess}) \xrightarrow[\text{[followed by NaOH}(aq)\text{]}]{}$

   (*g*) $C_6H_5NH_2 + NaNO_2 + HCl \xrightarrow[0°C,\ H_2O]{}$

   (*h*) $(CH_3)_3CCl + Na^{+-}OCH_3 \xrightarrow[HOCH_3]{}$

   (*i*) $CH_3CH_2CH_2CH_2Cl + Na^{+-}OCH_3 \xrightarrow[HOCH_3]{}$

   (*j*) $(CH_3)_2CHCH_2{-}S{-}S{-}CH_2CH(CH_3)_2 + (2H) \longrightarrow$

   (*k*) $CH_3NH_2 + HCl(aq) \xrightarrow[\text{room temperature}]{}$

   (*l*) $CH_3CH_2NH_3^+Cl^- + NaOH(aq) \longrightarrow$

   (*m*) $CH_3NH_2 + C_6H_5CH_2Cl \xrightarrow[\text{[followed by NaOH}(aq)\text{]}]{}$

7. The number of one-step organic reactions is growing rather rapidly during the course of our study. If they are to be of use to us in the biochemistry section, some way of organizing them is needed. Many students have found it helpful to make cards. For example, on one side of a 5 × 8-in. card the following question might appear: "What are all the ways we have learned for introducing the hydroxyl group into an organic molecule in one step from anything?" The reverse side of the card might look like this:

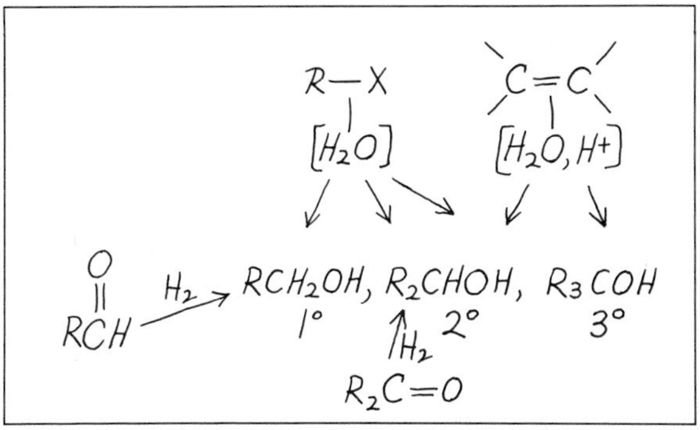

184

Optical Isomerism.
Organohalogen
Compounds.
Ethers.
Amines.
Mercaptans

Prepare similar cards for the following functional groups: carbon-carbon double bond, aldehyde, ketone, carboxylic acid, aliphatic amine, ether, and alkyl halide. Keep these cards current as additional methods are discussed in later chapters.

8. Students have found collecting the one-step reactions of each functional group useful. Thus on one side of a 5 × 8-in. card might appear the question "What reactions that alcohols undergo have we studied?" On the reverse side might appear something like this:

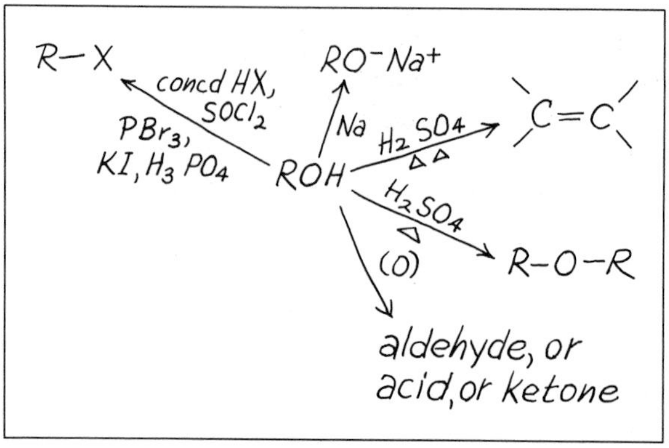

Prepare similar cards for the following functional groups: carbon-carbon double bond, aldehyde, ketone, carboxylic acid, aliphatic amines (all subclasses), aromatic amines, alcohols (all subclasses), alkyl halides, alkanes, and benzene. Keep these cards current as new reactions are described in later chapters.

9. Still another useful way of studying organic reactions is to look at them by reagents. Thus on one side of a 5 × 8-in. card might appear the question "What are the functional groups that react with water with ($a$) no added catalyst, ($b$) acid catalyst, ($c$) base catalyst?" Go back through all the functional groups we have studied and name those that answer the question. Write a typical reaction to illustrate each.

Prepare similar cards for the following reagents: water; water plus acid catalyst; water plus base catalyst; aqueous acid [e.g., HCl($aq$) as in the reaction with an amine]; aqueous base (e.g., NaOH, room temperature); oxidizing agents; reducing agents; sodium; ammonia; concentrated sulfuric acid (and heat); concentrated HX (and heat); hydrogen; nitrous acid; bromine; and peroxides. Keep these cards current, making new ones as new reagents are described, throughout the rest of our study of organic chemistry.

10. Write structures for each of the following compounds.

($a$)  diisopropyl ether  
($b$)  $n$-butyl mercaptan  
($c$)  $t$-butyl bromide  
($d$)  $p$-nitroaniline  
($e$)  methylethyl-$n$-propylamine  
($f$)  ethyl phenyl ether  
($g$)  tetramethylammonium iodide  
($h$)  $N$-nitroso-$N$-methylaniline  
($i$)  methyl isobutyl ether  
($j$)  an optically active ether of formula $C_5H_{12}O$  
($k$)  ethylammonium chloride

# CHAPTER SIX

# Aldehydes and Ketones

## THE CARBONYL GROUP

The carbonyl group consists of a carbon-to-oxygen double bond. The oxygen atom of this group, being more electronegative than carbon, carries a partial negative charge, which leaves the carbonyl carbon with a partial positive charge. These two simple considerations will be of great help to us in understanding several of the reactions of this group.

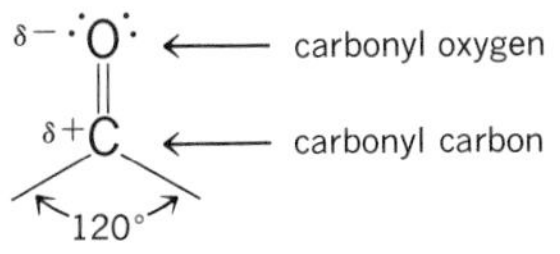

Carbonyl group

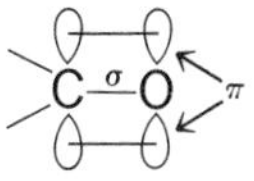

The two carbon-oxygen bonds in the carbonyl group consist of one σ-bond and one π-bond.

## FAMILIES OF ORGANIC COMPOUNDS WITH CARBONYL GROUPS

Several important families of organic compounds consist of molecules containing the carbonyl group, each family being characterized by the nature of the two groups attached to the carbonyl carbon. The essential characteristics of these families, outlined in Table 6.1 and discussed in the following paragraphs,[1] should be memorized.

**Aldehydes.** (H)R—C̈—H. To be classified as an aldehyde, the molecules of the substance must have a hydrogen attached to the carbonyl carbon. The other group

---

[1] In the examples given, R— is not limited just to alkyl groups. It may also be aryl, or any other group in which the bond to the carbonyl group is from a *carbon*, saturated or otherwise.

must involve a bond from the carbonyl carbon to another *carbon* (or a hydrogen in

formaldehyde, $H-\overset{\overset{\displaystyle O}{\|}}{C}-H$). The *aldehyde group*, then, is $-\overset{\overset{\displaystyle O}{\|}}{C}-H$. (Specific examples are listed in Table 6.3, page 197.)

**Ketones.** $R-\overset{\overset{\displaystyle O}{\|}}{C}-R'$. The molecules in ketones must have a carbonyl group flanked on both sides of the carbonyl carbon by bonds to other *carbons*. (Specific examples are listed in Table 6.4, page 198.) When it occurs in ketones, the carbonyl group is frequently called the *keto group*.

**Carboxylic Acids.** $(H)R-\overset{\overset{\displaystyle O}{\|}}{C}-OH$. The molecules in substances classified as carboxylic acids must have a hydroxyl group attached to the carbonyl carbon. The

Table 6.1   The Carbonyl Group in Families of Organic Compounds

| Family Name | Generic Family Structure* |
|---|---|
| Aldehydes | $(H)R-\overset{\overset{\displaystyle O}{\|}}{C}-H$ |
| Ketones | $R-\overset{\overset{\displaystyle O}{\|}}{C}-R'$ |
| Carboxylic acids | $(H)R-\overset{\overset{\displaystyle O}{\|}}{C}-OH$ |
| Derivatives of carboxylic acids | |
| Acid chlorides | $R-\overset{\overset{\displaystyle O}{\|}}{C}-Cl$ |
| Anhydrides | $R-\overset{\overset{\displaystyle O}{\|}}{C}-O-\overset{\overset{\displaystyle O}{\|}}{C}-R'$ |
| Esters | $(H)R-\overset{\overset{\displaystyle O}{\|}}{C}-O-R'$ |
| Amides | $(H)R-\overset{\overset{\displaystyle O}{\|}}{C}-NH_2$   (simple amides) |
| | $(H)R-\overset{\overset{\displaystyle O}{\|}}{C}-NH-R'$   (*N*-alkyl amides) |
| | $(H)R-\overset{\overset{\displaystyle O}{\|}}{C}-\underset{\underset{\displaystyle R''}{\|}}{N}-R'$   (*N,N*-dialkyl amides) |

* When the symbol for hydrogen is placed in parentheses before (or after) R [e.g., (H)R— or —R(H)], the specific R— group may be either an alkyl group or hydrogen. In these compounds, R— may also be aryl. When R— is primed (e.g., R'— or R''—), the two or more R— groups in a given symbol may be different.

other bond from the carbonyl carbon must be to another *carbon* (or to hydrogen in

formic acid, $H—\overset{O}{\overset{\|}{C}}—OH$). The group, $—\overset{O}{\overset{\|}{C}}—OH$, is called the *carboxylic acid group* or *carboxyl group*. It is frequently condensed to —COOH or to —CO$_2$H, but when this is done the reader must remember the presence of the carbon-oxygen double bond and the hydroxyl group. Although they possess hydroxyl groups, carboxylic acids are definitely not alcohols. The carbonyl group so changes the properties of the —OH attached to it (and vice versa) that it is more convenient to designate a new class. (Specific examples of carboxylic acids are listed in Table 7.1, page 225.)

**Carboxylic Acid Chlorides. Acid Chlorides.** $R—\overset{O}{\overset{\|}{C}}—Cl$. These molecules have carbonyl-to-chlorine bonds, and the group, $—\overset{O}{\overset{\|}{C}}—Cl$, is called the *acid chloride function*. (Specific examples are given in Table 7.5, page 243.)

**Carboxylic Acid Anhydrides. Anhydrides.** $R—\overset{O}{\overset{\|}{C}}—O—\overset{O}{\overset{\|}{C}}—R$. In these molecules an oxygen atom is flanked on both sides by carbonyl groups. They are called anhydrides ("not hydrated") because they can be considered to result from the loss of water between two molecules of a carboxylic acid:

$$R—\overset{O}{\overset{\|}{C}}\underset{\text{-----------}}{(O—H + H)}O—\overset{O}{\overset{\|}{C}}—R \longrightarrow R—\overset{O}{\overset{\|}{C}}—O—\overset{O}{\overset{\|}{C}}—R + H—OH$$

In both acid chlorides and anhydrides the R— groups may not be replaced by hydrogens. Such substances, for example, $H—\overset{O}{\overset{\|}{C}}—Cl$, are not known, for they are too unstable to exist. (A few examples of anhydrides are given in Table 7.5, page 243.)

**Carboxylic Acid Esters. Esters.** $(H)R—\overset{O}{\overset{\|}{C}}—O—R'$ or (H)RCO$_2$R'. The *ester group* has the features shown, and the bond drawn with a heavier line is called the *ester linkage*. To be an ester the molecule must have a carbonyl-oxygen-carbon network. (Several are listed in Tables 7.6 and 7.7, pages 247 and 248.)

Ester group

**Carboxylic Acid Amides. Amides.** $(H)R—\overset{O}{\overset{\|}{C}}—\overset{R''(H)}{\overset{|}{N}}—R'(H)$. So-called "simple amides" are ammonia derivatives of carboxylic acids, for they may be regarded as having formed by the splitting out of water between an acid and ammonia:

$$R—\overset{O}{\overset{\|}{C}}\underset{\overset{\frown}{(-OH}} + \underset{}{(H)}-\overset{H}{\overset{\|}{N}}-H \xrightarrow[-H_2O]{} R—\overset{O}{\overset{\|}{C}}—NH_2 \quad (or\ RCONH_2)$$

Simple amides

Other amides (the so-called $N$-alkyl or $N,N$-dialkyl amides, where $N$ denotes attachment to the nitrogen of a simple amide) are also well-known types of compounds:

$$R—\overset{O}{\overset{\|}{C}}—\overset{H}{\overset{\|}{N}}—R' \quad or \quad R—\overset{O}{\overset{\|}{C}}—NHR' \quad (or\ RCONHR')$$

$$R—\overset{O}{\overset{\|}{C}}—\overset{R'}{\overset{\|}{N}}—R'' \quad or \quad R—\overset{O}{\overset{\|}{C}}—NR'R'' \quad (or\ RCONR'R'')$$

To be classified as an amide, the molecule must have a carbonyl-to-nitrogen bond, called the *amide linkage*. In proteins the name peptide bond is synonymous with amide linkage. (Several examples of amides are given in Table 7.8, page 262.)

$$—\overset{O}{\overset{\|}{C}}—N\diagdown \quad \text{— amide linkage}$$

The amide group

---

✗ Exercise 6.1    The ability to recognize quickly the family to which a structure belongs is important if we are to have any degree of success in correlating structures and properties. Examine each of the following structural formulas and place each in the correct family. In some more than one functional group is present. List such examples in all classes to which they belong.

*Examples.*

$$CH_3\overset{O}{\overset{\|}{C}}OH \qquad \text{Carboxylic acid}$$

$$C_6H_5\overset{O}{\overset{\|}{C}}H \qquad \text{Aldehyde}$$

$$CH_3\overset{O}{\underset{\overset{|}{OH}}{\overset{\|}{C}H}}COH \qquad \begin{array}{l}\text{Carboxylic acid}\\ \text{and alcohol}\end{array}$$

$$CH_3\overset{O}{\overset{\|}{C}}CH_2\overset{O}{\overset{\|}{C}}CH_3 \qquad \text{Two keto groups}$$

$$CH_3\overset{O}{\overset{\|}{C}}O\overset{O}{\overset{\|}{C}}CH_3 \qquad \text{Anhydride}$$

$$H\overset{O}{\overset{\|}{C}}OH \qquad \begin{array}{l}\text{Carboxylic acid (This acid is}\\ \text{not in the aldehyde class,}\\ \text{but it does have some}\\ \text{aldehyde properties.)}\end{array}$$

$$(a)\ \ HO\overset{O}{\overset{\|}{C}}CH_3 \qquad\qquad (b)\ \ H\overset{O}{\overset{\|}{C}}C_6H_5 \qquad\qquad (c)\ \ CH_3\overset{O}{\overset{\|}{C}}Cl$$

$$(d)\ \ H\overset{O}{\overset{\|}{C}}NH_2 \qquad\qquad (e)\ \ CH_3O\overset{O}{\overset{\|}{C}}CH_3 \qquad\qquad (f)\ \ CH_3\overset{O}{\overset{\|}{C}}CH_3$$

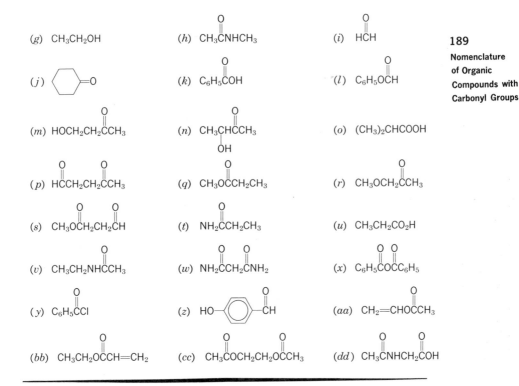

(g) CH₃CH₂OH

(h) CH₃CNHCH₃

(i) HCH

(j) (cyclohexanone) =O

(k) C₆H₅COH

(l) C₆H₅OCH

(m) HOCH₂CH₂CCH₃

(n) CH₃CHCCH₃ with OH

(o) (CH₃)₂CHCOOH

(p) HCCH₂CH₂CCH₃

(q) CH₃OCCH₂CH₃

(r) CH₃OCH₂CCH₃

(s) CH₃OCH₂CH₂CH

(t) NH₂CCH₂CH₃

(u) CH₃CH₂CO₂H

(v) CH₃CH₂NHCCH₃

(w) NH₂CCH₂CNH₂

(x) C₆H₅COCC₆H₅

(y) C₆H₅CCl

(z) HO—⟨benzene ring⟩—CH

(aa) CH₂=CHOCCH₃

(bb) CH₃CH₂OCCH=CH₂

(cc) CH₃COCH₂CH₂OCCH₃

(dd) CH₃CNHCH₂COH

## NOMENCLATURE OF ORGANIC COMPOUNDS WITH CARBONYL GROUPS

Although discussions of families with carbonyl groups will be spread out over several chapters, it will be convenient for our work to assemble in one place the rules for naming them. Our major concern is knowing how to write a structure from a name.

All families have both a common and an IUPAC system of nomenclature. We shall concentrate mainly on the common system, for the IUPAC names are very seldom applied to the simpler members of the families.

Common names for aldehydes and for the carboxylic acids and their derivatives share common prefix portions. These are summarized in Table 6.2.

**Common Names.** Considering first the carboxylic acids and their derivatives, we define the carbonyl portion or acid portion of the compound as that part in which

Cl—C—CH₂CH₃
carbonyl portion

CH₃C—O—CH₂CH₃
carbonyl portion

CH₃CH₂OCCH₂CH₃
carbonyl portion

the carbon skeleton has the carbonyl group. It is for this part of the molecule that the prefix portions given in Table 6.2 are characteristic. In learning these word

**Table 6.2 Common Names for Aldehydes, Acids, and Acid Derivatives**

| Class | Characteristic Suffix for the Name | Characteristic Prefix in the Carbonyl Portion of the Name | | | | |
|---|---|---|---|---|---|---|
| | | $C_1$ form- | $C_2$ acet- | $C_3$ propion- | $C_4$ $n$-butyr-* | $C_4$ isobutyr- |
| Aldehydes | -aldehyde | $\overset{O}{HCH}$ formaldehyde | $\overset{O}{CH_3CH}$ acetaldehyde | $\overset{O}{CH_3CH_2CH}$ propionaldehyde | $\overset{O}{CH_3CH_2CH_2CH}$ $n$-butyraldehyde | $\overset{O}{(CH_3)_2CHCH}$ isobutyraldehyde |
| Carboxylic acids | -ic acid | $HCO_2H$ formic acid | $CH_3CO_2H$ acetic acid | $CH_3CH_2CO_2H$ propionic acid | $CH_3CH_2CH_2CO_2H$ $n$-butyric acid | $(CH_3)_2CHCO_2H$ isobutyric acid |
| Carboxylic acid salts | -ate (with the name of the positive ion written first as a separate name) | $HCO_2{}^-Na^+$ sodium formate | $CH_3CO_2{}^-Na^+$ sodium acetate | $CH_3CH_2CO_2{}^-K^+$ potassium propionate | $CH_3CH_2CH_2CO\text{-}NH_4{}^+$ ammonium $n$-butyrate | $(CH_3)_2CHCO_2{}^-Na^+$ sodium isobutyrate |
| Acid chlorides | -yl chloride | (unstable) | $\overset{O}{CH_3CCl}$ acetyl chloride | $\overset{O}{CH_3CH_2CCl}$ propionyl chloride | $\overset{O}{CH_3CH_2CH_2CCl}$ $n$-butyryl chloride | $\overset{O}{(CH_3)_2CHCCl}$ isobutyryl chloride |
| Anhydrides | -ic anhydride | (unstable) | $\overset{O\quad O}{CH_3COCCH_3}$ acetic anhydride | $\overset{O\quad\quad O}{CH_3CH_2COCCH_2CH_3}$ propionic anhydride | $\overset{O}{(CH_3CH_2CH_2C)_2O}$ $n$-butyric anhydride | $\overset{O}{(CH_3)_2CHC\text{)}_2O}$ isobutyric anhydride |
| Esters | -ate (with the name of the alkyl group on oxygen written first as a separate word) | $\overset{O}{HCOCH_2CH_3}$ ethyl formate | $CH_3COCH(CH_3)_2$ isopropyl acetate | $CH_3CH_2COCH_2CH_3$ ethyl propionate | $CH_3CH_2CH_2COCH_3$ methyl $n$-butyrate | $(CH_3)_2CHCOCH(CH_3)_2$ isopropyl isobutyrate |
| Simple amides | -amide | $\overset{O}{HCNH_2}$ formamide | $\overset{O}{CH_3CNH_2}$ acetamide | $CH_3CH_2CNH_2$ propionamide | $CH_3CH_2CH_2CNH_2$ $n$-butyramide | $(CH_3)_2CHCNH_2$ isobutyramide |

* It is also very common to omit $n$- from $n$-butyr- in all these names. Thus "butyr-" without the "$n$-" means the normal isomer.

parts, we must be certain to associate the *total* carbon content *of the acid portion* with the prefixes. In counting carbon content, the carbonyl carbon is included.

Exercise 6.2    In the structures given, circle the carbonyl portions. Write the prefix associated with it.

*Examples.*

acet-

$CH_3CH_2CH_2COCH_2CH_2CH_2CH_3$

*n*-butyr-

Note that the oxygen that has only single
bonds and the R— group attached to it
are not included.

(a) HCOCH₃

(b) CH₃CH₂CO⁻Na⁺

(c) CH₃CH₂OCH

(d) CH₃CH₂OCCH₃

(e) CH₃CH₂COCH₃

(f) (CH₃)₂CHCOCH(CH₃)₂

In esters the name of the group attached by a *single* bond to oxygen is determined first and designated. Then the carbonyl portion must be recognized and named by attaching the suffix "-ate" to the appropriate prefix.

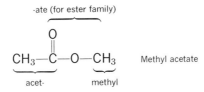

-ate (for ester family)

$CH_3$—$C$—$O$—$CH_3$    Methyl acetate

acet-            methyl

Common names for aldehydes are based on common names for their corresponding carboxylic acids. To the prefix for this acid we attach the suffix "-aldehyde."

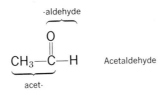

-aldehyde

$CH_3$—$C$—$H$    Acetaldehyde

acet-

Exercise 6.3    Complete the following table according to the example given.

| Structure | Family | Common Name |
|---|---|---|
| CH₃COCH₃ | ester | methyl acetate |

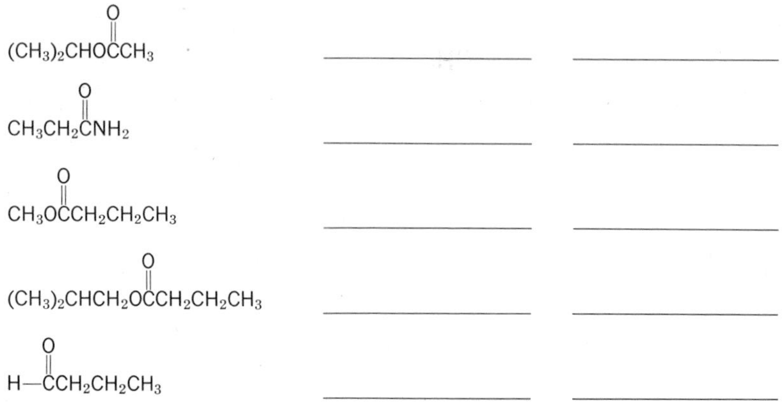

$(CH_3)_2CHOCCH_3$ with O double-bonded above the second C

_____  _____

$CH_3CH_2CNH_2$ with O double-bonded above the C

_____  _____

$CH_3OCCH_2CH_2CH_3$ with O double-bonded above the C

_____  _____

$(CH_3)_2CHCH_2OCCH_2CH_2CH_3$ with O double-bonded above the C

_____  _____

$H-CCH_2CH_2CH_3$ with O double-bonded above the C

_____  _____

Common names for slightly more complicated carbonyl compounds are derived by designating with Greek letters the positions in the carbon skeleton as it extends away from the carbonyl group:

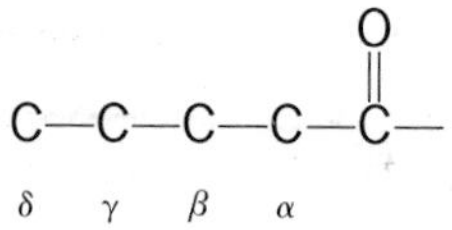

$$C-C-C-C-C-$$
$$\delta \quad \gamma \quad \beta \quad \alpha$$

The $\alpha$-position is always the carbon attached directly to the carbonyl. The carbonyl carbon is *not* lettered. (The IUPAC system, discussed later, will be different at this point.) The following examples illustrate how the common system can be extended by using these designations.

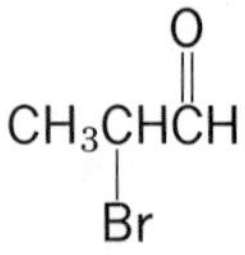

$CH_3CHCH$ with O double-bonded above terminal C, Br below middle C

$\alpha$-Bromopropionaldehyde

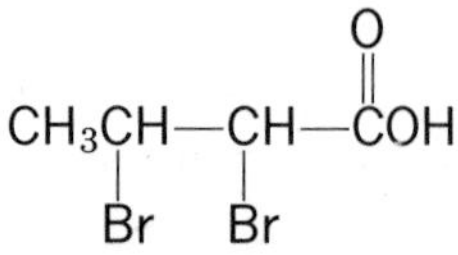

$CH_3CH-CH-COH$ with O double-bonded above C, Br, Br below

$\alpha,\beta$-Dibromobutyric acid

$CH_3CH_2OCCH_2CH_2OH$ with O double-bonded above C

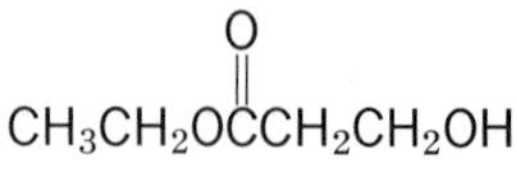

Ethyl $\beta$-hydroxypropionate

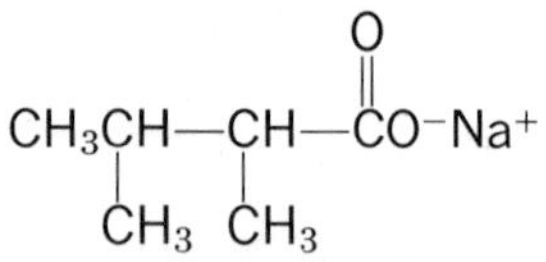

$CH_3CH-CH-CO^-Na^+$ with O double-bonded above C, $CH_3$ $CH_3$ below

Sodium $\alpha,\beta$-dimethylbutyrate

Common names for ketones are devised by giving the name ketone to the carbonyl group and then by designating the groups attached to it. The following examples illustrate the method. The simplest ketone possible, dimethyl ketone, is, however, always called by its trivial name, acetone.

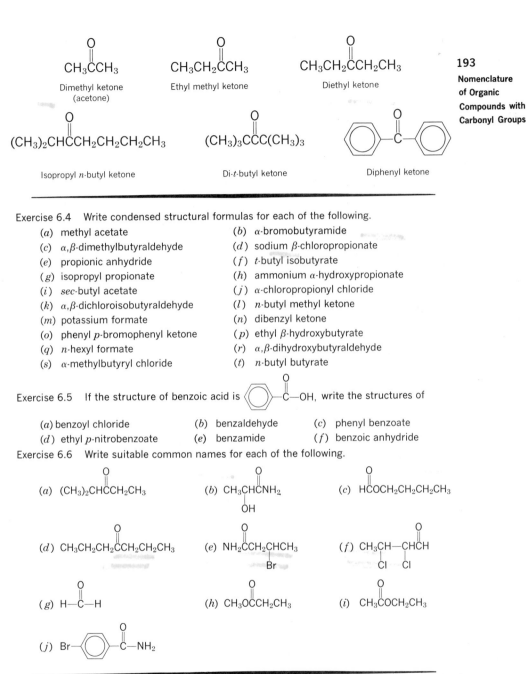

$CH_3\overset{O}{\overset{\|}{C}}CH_3$

Dimethyl ketone
(acetone)

$CH_3CH_2\overset{O}{\overset{\|}{C}}CH_3$

Ethyl methyl ketone

$CH_3CH_2\overset{O}{\overset{\|}{C}}CH_2CH_3$

Diethyl ketone

$(CH_3)_2CH\overset{O}{\overset{\|}{C}}CH_2CH_2CH_3$

Isopropyl n-butyl ketone

$(CH_3)_3C\overset{O}{\overset{\|}{C}}C(CH_3)_3$

Di-t-butyl ketone

Diphenyl ketone

Exercise 6.4    Write condensed structural formulas for each of the following.

(a)  methyl acetate
(b)  α-bromobutyramide
(c)  α,β-dimethylbutyraldehyde
(d)  sodium β-chloropropionate
(e)  propionic anhydride
(f)  t-butyl isobutyrate
(g)  isopropyl propionate
(h)  ammonium α-hydroxypropionate
(i)  sec-butyl acetate
(j)  α-chloropropionyl chloride
(k)  α,β-dichloroisobutyraldehyde
(l)  n-butyl methyl ketone
(m)  potassium formate
(n)  dibenzyl ketone
(o)  phenyl p-bromophenyl ketone
(p)  ethyl β-hydroxybutyrate
(q)  n-hexyl formate
(r)  α,β-dihydroxybutyraldehyde
(s)  α-methylbutyryl chloride
(t)  n-butyl butyrate

Exercise 6.5    If the structure of benzoic acid is   , write the structures of

(a) benzoyl chloride
(b)  benzaldehyde
(c)  phenyl benzoate
(d) ethyl p-nitrobenzoate
(e)  benzamide
(f)  benzoic anhydride

Exercise 6.6    Write suitable common names for each of the following.

(a)  $(CH_3)_2CH\overset{O}{\overset{\|}{C}}CH_2CH_3$

(b)  $CH_3\underset{OH}{CH}\overset{O}{\overset{\|}{C}}NH_2$

(c)  $H\overset{O}{\overset{\|}{C}}OCH_2CH_2CH_2CH_3$

(d)  $CH_3CH_2CH_2\overset{O}{\overset{\|}{C}}CH_2CH_2CH_3$

(e)  $NH_2\overset{O}{\overset{\|}{C}}CH_2\underset{Br}{CH}CH_3$

(f)  $CH_3\underset{Cl}{CH}—\underset{Cl}{CH}\overset{O}{\overset{\|}{C}}H$

(g)  $H—\overset{O}{\overset{\|}{C}}—H$

(h)  $CH_3O\overset{O}{\overset{\|}{C}}CH_2CH_3$

(i)  $CH_3\overset{O}{\overset{\|}{C}}OCH_2CH_3$

(j)  Br—   —$\overset{O}{\overset{\|}{C}}$—$NH_2$

**IUPAC Nomenclature.** For aldehydes, ketones, and carboxylic acids, the longest continuous sequence of carbons, including the carbonyl carbon, is taken as the base. This chain is numbered from the end that will give the carbonyl carbon the lower number, regardless of the other groups that may be present.

For aliphatic or cycloaliphatic systems, the name of the saturated hydrocarbon corresponding to the selected longest chain or largest ring is determined. This name is then altered in a way characteristic for the kind of carbonyl compound involved.

*For aldehydes, the "-e" at the end of the alkane name is changed to "-al."* The "-al" function requires no number to designate its location because it can occur only at the end of a chain and the 1-position for the aldehyde carbon is understood.

$$\underset{\substack{\text{Methanal}\\\text{(formaldehyde)}}}{H-\overset{\displaystyle O}{\overset{\|}{C}}-H} \qquad \underset{\substack{\text{Ethanal}\\\text{(acetaldehyde)}}}{CH_3\overset{\displaystyle O}{\overset{\|}{C}}-H} \qquad \underset{\substack{\text{Propanal}\\\text{(propionaldehyde)}}}{CH_3CH_2\overset{\displaystyle O}{\overset{\|}{C}}-H}$$

*For ketones, the "-e" ending of the alkane corresponding to the longest chain in the ketone is changed to "-one."* Whenever it is possible for the keto group to be located in different positions, the number of its carbon must be designated.

$$\underset{\substack{\text{Propanone}\\\text{(acetone)}}}{CH_3-\overset{\displaystyle O}{\overset{\|}{C}}-CH_3} \qquad \underset{\substack{\text{Butanone}\\\text{(ethyl methyl}\\\text{ketone)}}}{CH_3-\overset{\displaystyle O}{\overset{\|}{C}}-CH_2CH_3} \qquad \underset{\substack{\text{2-Pentanone}\\\text{(methyl }n\text{-propyl}\\\text{ketone)}}}{CH_3-\overset{\displaystyle O}{\overset{\|}{C}}-CH_2CH_2CH_3}$$

$$\underset{\substack{\text{3-Pentanone}\\\text{(diethyl ketone)}}}{CH_3CH_2-\overset{\displaystyle O}{\overset{\|}{C}}-CH_2CH_3}$$

*For carboxylic acids, the "-e" ending of the alkane corresponding to the longest chain in the acid is changed to "-oic acid".* As with the aldehydes, the location of the carbonyl carbon is not specified because it can occur only at the end of

$$\underset{\substack{\text{Methanoic acid}\\\text{(formic acid)}}}{H-\overset{\displaystyle O}{\overset{\|}{C}}-OH} \qquad \underset{\substack{\text{Ethanoic acid}\\\text{(acetic acid)}}}{CH_3-\overset{\displaystyle O}{\overset{\|}{C}}-OH} \qquad \underset{\substack{\text{Butanoic acid}\\\text{(butyric acid)}}}{CH_3CH_2CH_2-\overset{\displaystyle O}{\overset{\|}{C}}-OH}$$

$$\underset{\substack{\text{2-Methylpropanoic acid}\\\text{(isobutyric acid)}}}{\overset{\displaystyle CH_3}{\underset{3}{CH_3}-\underset{}{\overset{|}{C}H}-\overset{\displaystyle \overset{O}{\|}}{\underset{}{C}}-OH}}$$

the chain and its 1-position is understood. Substituents on the basic chain are named and located in the usual way, as illustrated by 2-methylpropanoic acid.

The IUPAC names for acid derivatives (salts, acid chlorides, anhydrides, esters, and amides) are based on the name of the acid itself. To name any of these derivatives, first write the name of the parent acid. Then drop the "-ic acid" suffix portion and replace it by the suffix specified for the derivative in the second column of Table 6.2. These word parts are the same in both the common and the IUPAC system. For example,

$$CH_3\overset{O}{\overset{\|}{C}}O^-Na^+$$

Sodium ethanoate
(sodium acetate)

$$CH_3O\overset{O}{\overset{\|}{C}}CH_2CH_3$$

Methyl propanoate
(methyl propionate)

$$CH_3CH_2CH_2\overset{O}{\overset{\|}{C}}Cl$$

Butanoyl chloride
(butyryl chloride)

$$CH_3\overset{O}{\overset{\|}{C}}O\overset{O}{\overset{\|}{C}}CH_3$$

Ethanoic anhydride
(acetic anhydride)

$$H\overset{O}{\overset{\|}{C}}NH_2$$

Methanamide
(formamide)[2]

$$CH_3CH_2CH_2\overset{O}{\overset{\|}{C}}OCH_2CH_2CH_3$$

Propyl butanoate
(n-propyl butyrate)

$$ClCH_2\overset{O}{\overset{\|}{C}}OCH_3$$

Methyl 2-chloroethanoate
(methyl α-chloroacetate)

$$Na^+{}^-O\overset{O}{\overset{\|}{C}}CH_2CH-CHCH_2CH_3$$
$$\quad\quad\quad\quad\quad\quad | \quad\quad |$$
$$\quad\quad\quad\quad\quad\quad Br \quad Br$$

Sodium 3,4-dibromohexanoate
(sodium β,γ-dibromocaproate)

---

**Exercise 6.7** Write condensed structural formulas for each of the following names.

(a) ammonium pentanoate
(b) isopropyl propanoate
(c) butanoic anhydride
(d) 3,4-dimethylhexanal
(e) 2,4-dibromo-3-pentanone
(f) decanamide
(g) methyl 2,3-dimethylhexanoate
(h) butyl methanoate
(i) 2-hydroxypropanal
(j) 1-phenyl-1-propanone
(k) butanoyl chloride
(l) sodium 2-methylpropanoate
(m) 2,3-dibromopentanal
(n) t-butyl pentanoate
(o) cyclohexanone
(p) 2,5-dichlorocyclopentanone
(q) N,2-dimethylpropanamide

**Exercise 6.8** Write IUPAC and common names for each of the following.

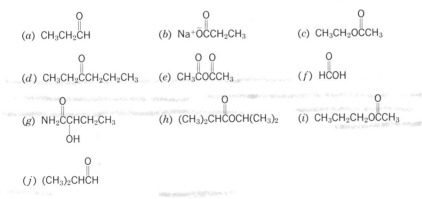

(a) $CH_3CH_2\overset{O}{\overset{\|}{C}}H$

(b) $Na^+{}^-O\overset{O}{\overset{\|}{C}}CH_2CH_3$

(c) $CH_3CH_2O\overset{O}{\overset{\|}{C}}CH_3$

(d) $CH_3CH_2\overset{O}{\overset{\|}{C}}CH_2CH_2CH_3$

(e) $CH_3\overset{O}{\overset{\|}{C}}\overset{O}{\overset{\|}{C}}CH_3$

(f) $H\overset{O}{\overset{\|}{C}}OH$

(g) $NH_2\overset{O}{\overset{\|}{C}}CHCH_2CH_3$
$\quad\quad\quad\quad | $
$\quad\quad\quad\quad OH$

(h) $(CH_3)_2CH\overset{O}{\overset{\|}{C}}OCH(CH_3)_2$

(i) $CH_3CH_2CH_2O\overset{O}{\overset{\|}{C}}CH_3$

(j) $(CH_3)_2CH\overset{O}{\overset{\|}{C}}H$

[2] Note that the "o" is also dropped. The name is not methanoamide.

Exercise 6.9    If the common name for $CH_3CH_2CH_2CH_2\overset{\displaystyle O}{\overset{\|}{C}}OH$ is valeric acid, write common and IUPAC names for the following.

(a) $CH_3CH_2CH_2CH_2\overset{\displaystyle O}{\overset{\|}{C}}H$

(b) $CH_3O\overset{\displaystyle O}{\overset{\|}{C}}CH_2CH_2CH_2CH_3$

(c) $Cl\overset{\displaystyle O}{\overset{\|}{C}}CH_2CH_2CH_2CH_3$

(d) $CH_3CH_2CH_2CH_2\overset{\displaystyle O}{\overset{\|}{C}}NH_2$

(e) $CH_3CH_2CH_2CH_2\overset{\displaystyle O}{\overset{\|}{C}}O^-Na^+$

---

## PROPERTIES OF ALDEHYDES AND KETONES

### PHYSICAL PROPERTIES

Aldehydes and ketones are moderately polar compounds. An approximate idea of how their polarity relates to that of other families may be obtained from the following boiling point data for compounds of closely similar formula weights, given in parentheses.

| | | | |
|---|---|---|---|
| Propane | $CH_3CH_2CH_3$ | (44) | bp $-45°C$ |
| Dimethyl ether | $CH_3OCH_3$ | (46) | bp $-25$ |
| Methyl chloride | $CH_3Cl$ | (50) | bp $-24$ |
| Ethylamine | $CH_3CH_2NH_2$ | (45) | bp $17$ |
| Acetaldehyde | $CH_3CH=O$ | (44) | bp $21$ |
| Ethyl alcohol | $CH_3CH_2OH$ | (46) | bp $78.5$ |

Low-formula-weight aldehydes and ketones are soluble in water, but by the time the molecule has five carbons, this solubility has become quite low. The carbonyl group cannot act as a hydrogen bond donor, but the carbonyl oxygen, of course, is a hydrogen bond acceptor. Trends in typical properties are indicated by the data in Tables 6.3 and 6.4.

### OXIDATION

At room temperature molecular oxygen (from air) is not considered a vigorous oxidizing agent. Yet the aldehyde group is sensitive enough to oxidation that aldehydes cannot be stored exposed to air or they will slowly convert to their corresponding carboxylic acids. In sharp contrast, ketones are ordinarily very stable to further oxidation. In fact, this is the principal difference between aldehydes and ketones. Otherwise, the two families exhibit very similar chemical properties.

The ease with which the aldehyde group is oxidized makes it possible to detect its presence with reagents that leave other groups unaffected.

**Tollens' Test. Silvering Mirrors.** Tollens' reagent is a solution of the diammonia complex of the silver ion in dilute ammonium hydroxide. (The silver ion is precipitated

## Table 6.3 Aldehydes

| Common Name | Structure | Mp (°C) | Bp (°C) | Solubility in Water | Oxidation |
|---|---|---|---|---|---|
| Formaldehyde | $CH_2\!=\!O$ | −92 | −21 | very soluble | |
| Acetaldehyde | $CH_3CH\!=\!O$ | −125 | 21 | very soluble | |
| Propionaldehyde | $CH_3CH_2CH\!=\!O$ | −81 | 49 | 16 g/100 cc (25°) | |
| n-Butyraldehyde | $CH_3CH_2CH_2CH\!=\!O$ | −99 | 76 | 4 g/100 cc | |
| Isobutyraldehyde | $(CH_3)_2CHCH\!=\!O$ | −66 | 65 | 9 g/100 cc . | |
| Valeraldehyde | $CH_3CH_2CH_2CH_2CH\!=\!O$ | −92 | 102 | slightly soluble | |
| Caproaldehyde | $CH_3CH_2CH_2CH_2CH_2CH\!=\!O$ | −56 | 128 | very slightly soluble | |
| Acrolein | $CH_2\!=\!CHCH\!=\!O$ | −87 | 53 | 40 g/100 cc | |
| Crotonaldehyde | $CH_3CH\!=\!CHCH\!=\!O$ | −77 | 104 | moderately soluble | |
| Benzaldehyde | 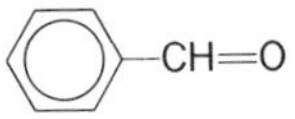 ⬡—CH=O | −56 | 178 | 0.3 g/100 cc | |
| Salicylaldehyde | 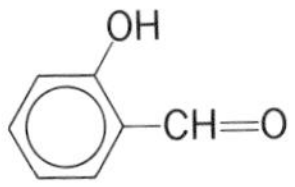 ⬡(OH)—CH=O | −10 | 197 | slightly soluble | |
| Vanillin | 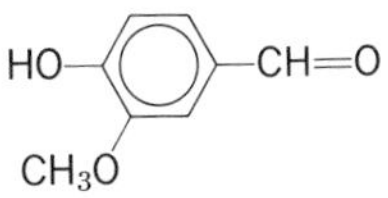 HO—⬡—CH=O ($CH_3O$) | 81 | 285 | 1 g/100 cc | |
| Cinnamaldehyde (trans) | 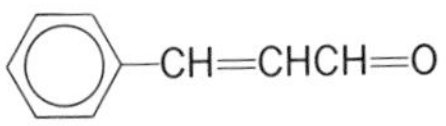 ⬡—CH=CHCH=O | −8 | 253 | insoluble | |
| Furfural | 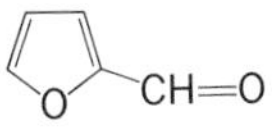 (furan ring)—CH=O | −31 | 162 | 9 g/100 cc | |
| Glucose* (dextrose) | $CH_2\!-\!CH\!-\!CH\!-\!CH\!-\!CH\!-\!CH\!=\!O$ / OH OH OH OH OH | 146 (decomposes) | — | very soluble | |

* Only the general structural features of the open-chain form of glucose are shown. Additional details are discussed in Chapter 8.

as silver oxide in the presence of hydroxide ion unless it is complexed with ammonia molecules.) In general:

$$RCH\!=\!O + 2Ag(NH_3)_2^+ + 3OH^- \longrightarrow RCO_2^- + 2Ag\!\downarrow + 2H_2O + 4NH_3$$

Aldehyde          Carboxylate ion

Specific example:

$$CH_3CH\!=\!O + 2Ag(NH_3)_2^+ + 3OH^- \longrightarrow CH_3CO_2^- + 2Ag\!\downarrow + 2H_2O + 4NH_3$$

Acetaldehyde          Acetate ion    Silver

Any good test for a functional group must provide sound, positive evidence that reaction occurs with that functional group. The positive evidence in the Tollens

**Table 6.4   Ketones**

| Name | Structure | Mp (°C) | Bp (°C) | Solubility in Water |
|------|-----------|---------|---------|---------------------|
| Acetone | $CH_3\overset{O}{\overset{\|}{C}}CH_3$ | −95 | 56 | very soluble |
| Methyl ethyl ketone | $CH_3\overset{O}{\overset{\|}{C}}CH_2CH_3$ | −87 | 80 | 33 g/100 cc (25°) |
| 2-Pentanone | $CH_3\overset{O}{\overset{\|}{C}}CH_2CH_2CH_3$ | −84 | 102 | 6 g/100 cc |
| 3-Pentanone | $CH_3CH_2\overset{O}{\overset{\|}{C}}CH_2CH_3$ | −40 | 102 | 5 g/100 cc |
| 2-Hexanone | $CH_3\overset{O}{\overset{\|}{C}}CH_2CH_2CH_2CH_3$ | −57 | 128 | 1.6 g/100 cc |
| 3-Hexanone | $CH_3CH_2\overset{O}{\overset{\|}{C}}CH_2CH_2CH_3$ | — | 124 | 1.5 g/100 cc |
| Cyclopentanone | 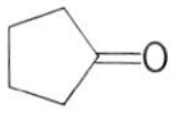 | −53 | 129 | slightly soluble |
| Cyclohexanone | 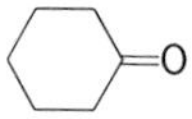 |  | 156 | slightly soluble |
| Muscone* | 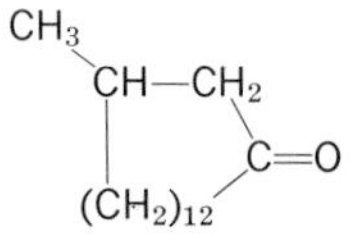 |  | 328 | insoluble |
| Civetone† | 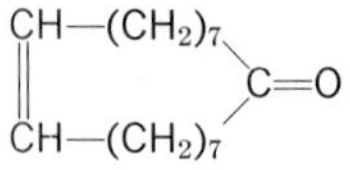 | 33 | 342 | insoluble |
| Camphor | 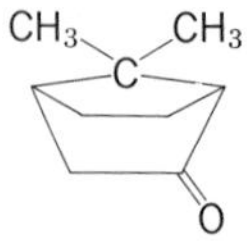 | 176 | 209 | insoluble |
| Acetophenone | $C_6H_5-\overset{O}{\overset{\|}{C}}CH_3$ | −23 | 137 | insoluble |
| Fructose (levulose) | $CH_2-CH-CH-CH-\overset{O}{\overset{\|}{C}}CH_2$ <br> $\quad\,\, \|\quad\,\, \|\quad\,\, \|\quad\,\, \|\qquad\quad \|$ <br> $\quad\, OH\ \ OH\ \ OH\ \ OH\qquad OH$ | 48 |  | soluble |

* Active principle of musk, a substance of powerful odor used in trace amounts as a perfume base and obtained from an abdominal gland of the Himalayan male musk deer. The male uses musk to attract the female deer.

† Also a rare perfume base, civetone is the active fragrant principle of civet, a substance found in the African civet cat.

test is the appearance of metallic silver in a previously clear, colorless solution. If the inner wall of the test vessel is clean and grease-free, silver deposits to form a beautiful mirror, a reaction that serves as the basis for silvering mirrors. If the glass surface is not clean, the silver separates as a gray, finely divided, powdery precipitate. Tollens' reagent does not keep well and is prepared just before use.

**Benedict's Test and Fehling's Test.** Benedict's solution and Fehling's solution are both alkaline and contain the copper(II) ion, $Cu^{2+}$, stabilized by a complexing agent. This agent is the citrate ion in Benedict's solution and the tartrate ion in Fehling's solution. Without these complexing agents, copper(II) ions would be precipitated as copper(II) hydroxide by the alkali present. Benedict's solution is more frequently used because it stores well. Fehling's solution must be prepared just before use.

With either solution the $Cu^{2+}$ ion is the oxidizing agent. In the presence of certain easily oxidized groups it is reduced to the copper(I) state, but copper(I) ions cannot be solubilized by either citrate or tartrate ions and a precipitate of copper(I) oxide, $Cu_2O$, forms. What we see in a positive test, then, is a change from the brilliant blue color of the test solution to the bright orange-red color of precipitated copper(I) oxide. The principal systems that give this result are $\alpha$-hydroxy aldehydes, $\alpha$-keto aldehydes, and $\alpha$-hydroxy ketones. $\alpha$-Hydroxy aldehydes and $\alpha$-hydroxy ketones

| $\alpha$-Hydroxy aldehyde | $\alpha$-Keto aldehyde | $\alpha$-Hydroxy ketone |
|---|---|---|

are common to the sugar families (cf. the last compounds listed in Tables 6.3 and 6.4, glucose and fructose). Benedict's test, or Fehling's test, is a common method for detecting the presence of glucose in urine. Normally, urine does not contain glucose, but in certain conditions, for example, diabetes, it does, A positive test for glucose varies from a bright green color (0.25% glucose), to yellow-orange (1% glucose), to brick red (over 2% glucose).

Some ordinary aliphatic aldehydes react with these reagents, but the changes are complex. A dark, gummy precipitate may form, which apparently is not copper(I) oxide, while the color of the test solution remains blue or becomes any hue from yellow to green to brown. Aromatic aldehydes, in general, do not react.[3]

**Haloform Reaction. Iodoform Test.** Methyl ketones react with the halogens in the presence of sodium hydroxide and form trihalomethane (e.g., chloroform, bromoform, or iodoform) plus the salt of that carboxylic acid resulting from loss of the methyl carbon:

$$R-\overset{O}{\overset{\|}{C}}-CH_3 + 3X_2 + 4OH^- \longrightarrow R-\overset{O}{\overset{\|}{C}}-O^- + CHX_3 + 3H_2O + 3X^-$$

(X = Cl, Br, or I)    Haloform

[3] R. Daniels, C. C. Rush, and L. Bauer, *Journal of Chemical Education,* Vol. 37, 1960, page 205.

When iodine is the halogen, the product is iodoform, a yellow solid with a characteristic odor and color. Its formation in the iodoform test is the basis for the detection of methyl ketones. In a positive test the deep brown color of molecular iodine disappears and the yellow iodoform is evident.

The test is also positive for methylcarbinols. Thus the systems giving the test are

$$\underset{\substack{\text{Methylcarbinols} \\ \text{(or ethanol, if R = H)}}}{\overset{\displaystyle \overset{\text{OH}}{|}}{(H)R-CH-CH_3}} \qquad \underset{\substack{\text{Methyl ketones} \\ \text{(or acetaldehyde, if R = H)}}}{\overset{\displaystyle \overset{\text{O}}{\|}}{(H)R-C-CH_3}}$$

Methylcarbinols are believed to be oxidized by the reagent to their corresponding methyl ketones, which are then further attacked. The reaction with methyl ketones apparently involves halogenation of the methyl group followed by cleavage of the molecule:

$$\overset{\displaystyle \overset{\text{O}}{\|}}{R-C-CH_3} \xrightarrow{\text{I}_2,\ \text{NaOH},\ \text{H}_2\text{O}} \overset{\displaystyle \overset{\text{O}}{\|}}{R-C-Cl_3}$$

$$\overset{\displaystyle \overset{\text{O}}{\|}}{R-C-Cl_3} + OH^- \longrightarrow (R-\overset{\displaystyle \overset{\text{O}}{\|}}{C}-O-H) + :Cl_3^-$$

$$\downarrow$$

$$\underset{\substack{\text{Carboxylate ion} \qquad \text{Iodoform}}}{\overset{\displaystyle \overset{\text{O}}{\|}}{R-C-O^-} + HCl_3}$$

**Strong Oxidizing Agents.** Aldehydes are of course readily oxidized to carboxylic acids by permanganate ion, dichromate ion, and other oxidizing agents. Ketones also eventually yield to such reagents, but except in isolated instances the reaction is not particularly useful. Ketones tend to fragment somewhat randomly on both sides of the carbonyl carbon:

$$\overset{\displaystyle \overset{\text{O}}{\|}}{R-C-R'} \xrightarrow{\text{strong oxidizing agent, heat}}$$

$$RCO_2H + R'CO_2H \qquad \text{(as well as acids of lower formula weights)}$$

## REDUCTION

Several methods are available for reducing aldehydes and ketones to their corresponding primary and secondary alcohols.

**Catalytic Hydrogenation.** In general:

$$\underset{\text{Aldehyde}}{\overset{\displaystyle \overset{\text{O}}{\|}}{R-C-H}} + H_2 \xrightarrow[\substack{\text{pressure,} \\ \text{heat}}]{\text{Ni}} \underset{\text{1° alcohol}}{R-CH_2-OH}$$

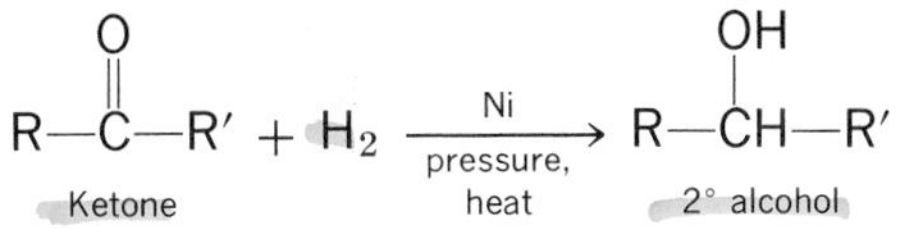

$$R-\overset{\displaystyle O}{\overset{\|}{C}}-R' + H_2 \xrightarrow[\substack{\text{pressure,} \\ \text{heat}}]{\text{Ni}} R-\overset{\displaystyle OH}{\overset{|}{C}H}-R'$$

Ketone             2° alcohol

Specific examples:

$$CH_3CH_2CH_2CH{=}O + H_2 \longrightarrow CH_3CH_2CH_2CH_2OH$$

n-Butyraldehyde          n-Butyl alcohol (85%)

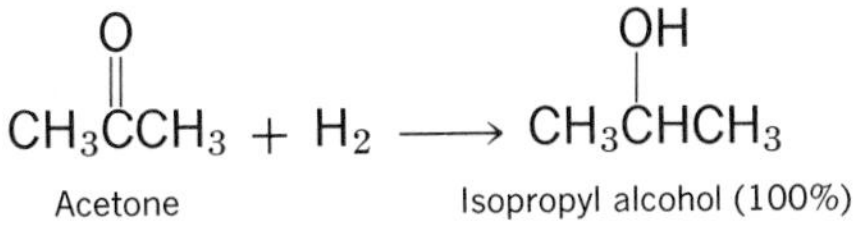

$$CH_3\overset{\displaystyle O}{\overset{\|}{C}}CH_3 + H_2 \longrightarrow CH_3\overset{\displaystyle OH}{\overset{|}{C}H}CH_3$$

Acetone          Isopropyl alcohol (100%)

**Other Methods.** Metallic hydrides such as lithium aluminum hydride ($LiAlH_4$) and sodium borohydride ($NaBH_4$) are excellent but expensive reducing agents for aldehydes and ketones.

## ADDITIONS OF HYDROXYLIC COMPOUNDS TO ALDEHYDES AND KETONES

The analogy between the alkene double bond (C=C) and the carbonyl double bond (C=O) does not extend too far. Although both will add hydrogen in the presence of a suitable catalyst, the carbonyl group does not add halogens ($Cl_2$, etc) or water to form stable addition compounds. Yet the interaction of the carbonyl group with water, and especially with alcohols, is of primary importance to our future study of carbohydrates.

**Hydrates.** Most aldehydes react in aqueous solutions with water to establish an equilibrium mixture with the hydrated form:

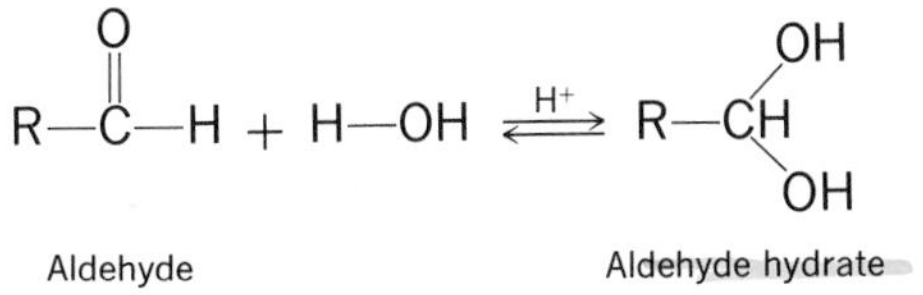

$$R-\overset{\displaystyle O}{\overset{\|}{C}}-H + H-OH \overset{H^+}{\rightleftharpoons} R-\overset{\displaystyle OH}{\underset{\displaystyle OH}{CH}}$$

Aldehyde          Aldehyde hydrate

Formaldehyde is especially prone to form its hydrate. Of course, of all simple aldehydes and ketones, formaldehyde has a carbonyl carbon that is least hindered or crowded by neighboring groups. Because the carbonyl carbon has no alkyl groups with their electron-releasing effects, it bears in formaldehyde the largest partial positive charge of all simple aldehydes and ketones. Water is probably added as follows:

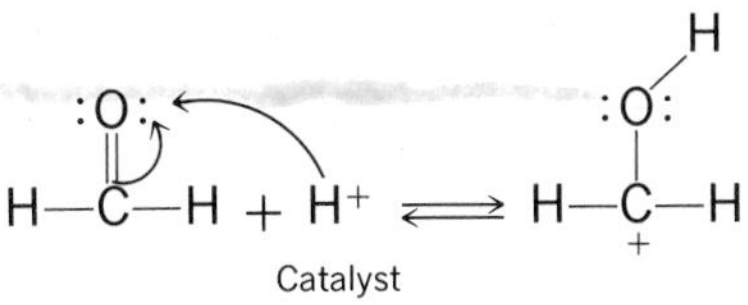

$$H-\overset{\displaystyle :\overset{..}{O}:}{\overset{\|}{C}}-H + H^+ \rightleftharpoons H-\overset{\displaystyle :\overset{..}{O}:{-}H}{\underset{\displaystyle +}{C}}-H$$

Catalyst

$$H-\overset{\overset{\displaystyle H}{\displaystyle |}}{\underset{+}{\overset{\ddot{O}\cdot}{\overset{|}{CH}}}} + H \quad \overset{\ddot{O}\cdot}{\underset{H}{\overset{|}{\quad}}} \quad H \rightleftharpoons H-\overset{\overset{\displaystyle O-H}{\displaystyle |}}{\underset{\underset{H}{\overset{|}{\overset{+}{\ddot{O}}-H}}}{\overset{|}{CH}}} \rightleftharpoons H-\overset{\overset{\displaystyle OH}{\displaystyle |}}{\underset{OH}{\overset{|}{CH}}} + H^+$$

Proton loss        Recovery of
catalyst

**Hemiacetals and Acetals.** In a similar manner, solutions of most aldehydes in alcohols consist of equilibrium mixtures:

$$R-\overset{\overset{\displaystyle O}{\displaystyle \|}}{C}-H + H-O-R' \overset{H^+}{\rightleftharpoons} R-\overset{\overset{\displaystyle OH}{\displaystyle |}}{\underset{OR'}{\overset{|}{CH}}}$$

Aldehyde      Alcohol

original carbonyl
carbon

A hemiacetal

From the alcohol the hydrogen goes to the carbonyl oxygen; the alcohol oxygen goes to the carbonyl carbon. Most hemiacetals cannot be isolated, for efforts to do so cause them to convert back to the original aldehyde and alcohol. In fact, the hemiacetal group is often called a *potential aldehyde group*. The hemiacetals that can be isolated are, however, very important, for they occur among carbohydrate molecules.

The hemiacetal system is defined by

$$\overset{\diagup}{\underset{\diagup}{C}}\overset{\overset{\displaystyle O-H}{\displaystyle }}{\underset{O-\overset{|}{\underset{|}{C}}-}{\big\backslash}}$$

*Alc. & ether*

One carbon, in boldface, carries both an OH and an OR group. This carbon was the original carbonyl carbon. The hemiacetal is both an alcohol and an ether. Since these two groups occur so close together, however, each modifies the properties of the other. For example, ether linkages are difficult to break, yet in hemiacetals they are seldom strong enough to hold the system together. These compounds usually exist only as part of an equilibrium in solution. The properties of the —OH group are similarly changed by the neighboring ether linkage. Ordinary alcohols interact and form ordinary ethers only under rather extreme conditions (cf. p. 160). As we shall see later, however, the —OH group in a hemiacetal is easily converted into a second ether linkage.

The key to this behavior, at least under certain conditions, is the relative stability of the carbonium ion formed as follows:

203

Additions of
Hydroxylic
Compounds to
Aldehydes and
Ketones

$$R-\overset{\overset{\displaystyle \overset{\frown}{O}H}{|}}{\underset{\underset{\displaystyle \overset{..}{O}R'}{|}}{CH}} \xrightarrow[\text{(H}_2\text{O leaves)}]{H^+} \left[ \begin{array}{c} R-\overset{+}{C}H\overset{\frown}{\underset{..}{O}}-R' \\ \updownarrow \\ R-CH=\overset{+}{\underset{..}{O}}-R \end{array} \right] + H_2O$$

The carbonium ion
is a resonance-stabilized
hybrid of these two contributors.

Hybrid: $R-\overset{\delta+}{C}\!=\!\!=\!\overset{\delta+}{O}-R'$

This carbonium ion, in which the positive charge is delocalized between carbon and oxygen, is much more stable than ordinary carbonium ions for which delocalization by resonance cannot occur. Alteratively,

$$R-\overset{\overset{\displaystyle \overset{..}{O}H}{|}}{\underset{\underset{\displaystyle \overset{..}{O}R'}{|}}{CH}} \xrightarrow[\text{(R'OH leaves)}]{H^+} \left[ \begin{array}{c} R-\overset{+}{C}H\overset{\frown}{\underset{..}{O}}H \\ \updownarrow \\ R-CH=\overset{+}{\underset{..}{O}}-H \end{array} \right] + H-O-R'$$

In this example the ether linkage is broken; the driving force is again the relative stability of the carbonium ion.

Hemiacetals can be converted into acetals by the action of additional alcohol in the presence of an acid catalyst such as dry hydrogen chloride:

$$R-\overset{\overset{\displaystyle O-H}{|}}{\underset{\underset{\displaystyle O-R'}{|}}{CH}} + H-O-R' \xrightarrow[\substack{\text{(via resonance-stabilized} \\ \text{carbonium ion)}}]{H^+} R-\overset{\overset{\displaystyle O-R'}{|}}{\underset{\underset{\displaystyle O-R'}{|}}{CH}} + H_2O$$

Hemiacetal                  Alcohol                                                                    Acetal

The acetal system is defined as

$$\overset{}{\underset{}{\text{C}}} \begin{array}{c} O-\overset{|}{C}- \\ \diagdown \\ \diagup \\ O-\overset{|}{C}- \end{array}$$

The carbon shown in boldface was the original carbonyl carbon. Wherever two ether linkages come to the *same* carbon, the acetal system exists. It occurs widely in carbohydrates and is the key to understanding such substances as sucrose, lactose, maltose, cellulose, starch, and glycogen.

Acetals are stable in neutral and alkaline media. They can be isolated and stored, in great contrast to the hemiacetals. The one reaction of importance to our future studies is their ready *acid-catalyzed* hydrolysis—first to the hemiacetal stage, which of course is immediately in equilibrium with the original alcohol and carbonyl compound. In general:

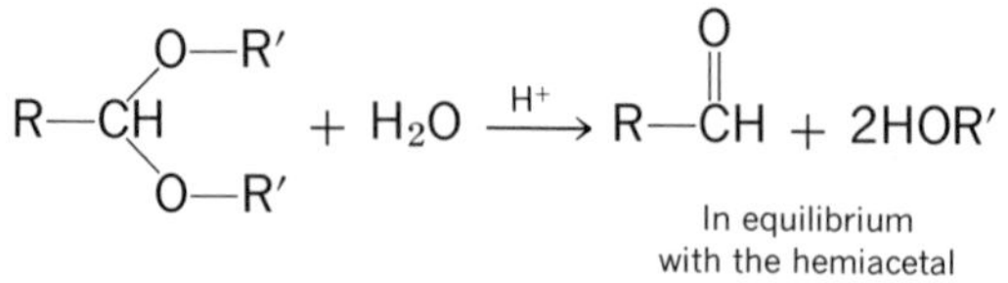

$$R-\underset{\underset{O-R'}{|}}{\overset{\overset{O-R'}{|}}{CH}} \quad + H_2O \xrightarrow{H^+} R-\overset{\overset{O}{\|}}{CH} + 2HOR'$$

In equilibrium
with the hemiacetal

Specific example:

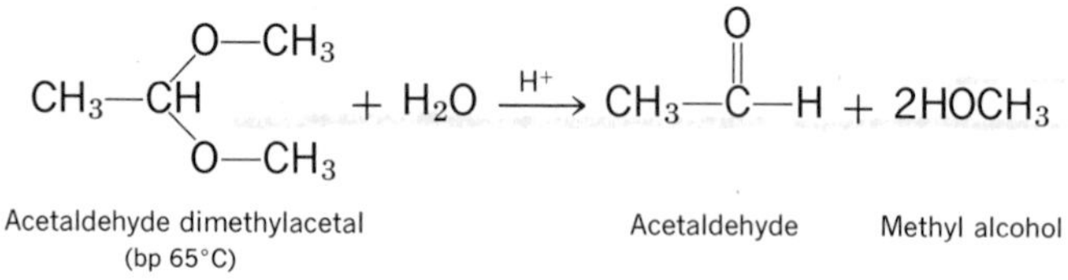

$$CH_3-\underset{\underset{O-CH_3}{|}}{\overset{\overset{O-CH_3}{|}}{CH}} \quad + H_2O \xrightarrow{H^+} CH_3-\overset{\overset{O}{\|}}{C}-H + 2HOCH_3$$

Acetaldehyde dimethylacetal           Acetaldehyde     Methyl alcohol
(bp 65°C)

---

**Exercise 6.10**   Write the structure of the hemiacetal that will exist in equilibrium with each of the following pairs of compounds.

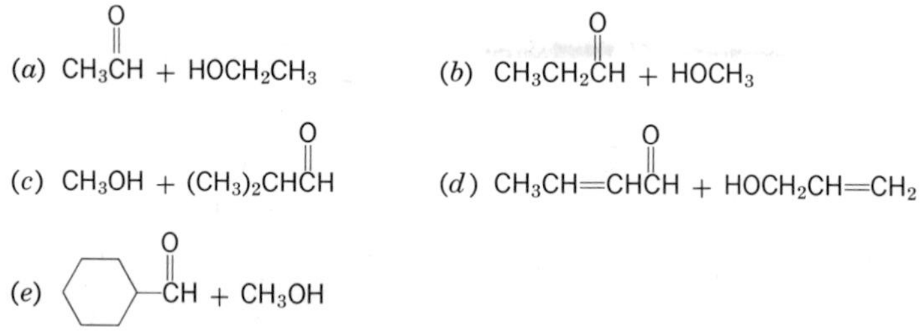

(a) $CH_3\overset{\overset{O}{\|}}{CH} + HOCH_2CH_3$          (b) $CH_3CH_2\overset{\overset{O}{\|}}{CH} + HOCH_3$

(c) $CH_3OH + (CH_3)_2CH\overset{\overset{O}{\|}}{CH}$       (d) $CH_3CH=CH\overset{\overset{O}{\|}}{CH} + HOCH_2CH=CH_2$

(e)  [cyclohexyl]$-\overset{\overset{O}{\|}}{CH} + CH_3OH$

**Exercise 6.11**   Write the structure of the acetal that will form if the aldehyde in each of the parts of exercise 6.10 combines with two molecules of the alcohol that is shown with it.

**Exercise 6.12**   Write the structures of the original aldehyde and the original alcohol that will form if the following acetals are hydrolyzed. One example is not an acetal. Which one is it?

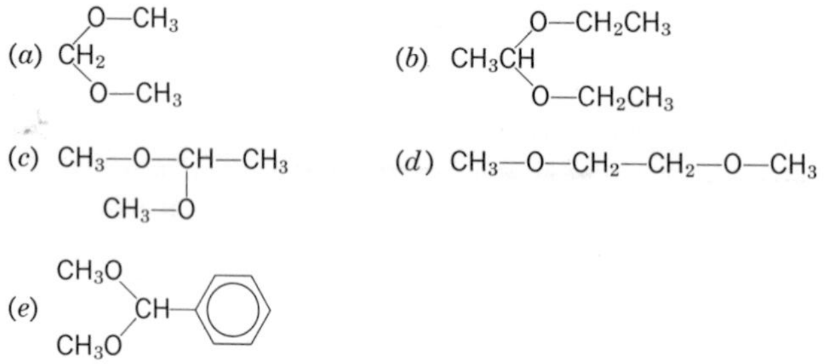

(a) $CH_2\underset{\underset{O-CH_3}{}}{\overset{\overset{O-CH_3}{}}{\big\langle}}$       (b) $CH_3CH\underset{\underset{O-CH_2CH_3}{}}{\overset{\overset{O-CH_2CH_3}{}}{\big\langle}}$

(c) $CH_3-O-\underset{\underset{CH_3-O}{|}}{CH}-CH_3$       (d) $CH_3-O-CH_2-CH_2-O-CH_3$

(e) $\underset{CH_3O}{\overset{CH_3O}{\big\langle}}CH-$[phenyl]

Exercise 6.13    The following compound is one of the forms in which glucose can exist. The carbons are numbered. Which is the hemiacetal carbon (the one that was "originally" a carbonyl carbon)?

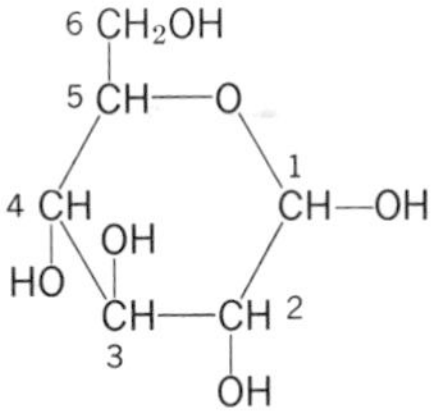

205

Additions of
Hydroxylic
Compounds to
Aldehydes and
Ketones

If this hemiacetal were to undergo a regeneration of the original carbonyl group and the original —OH group, what would the structure of the resulting substance be? In this regeneration the new alcohol portion is elsewhere on the same new aldehyde molecule.

Exercise 6.14    Suggest a mechanism for the acid-catalyzed conversion of a hemiacetal to an acetal.

---

**Condensation Reactions Between Aldehydes or Ketones and Derivatives of Ammonia.** Aldehydes and ketones will undergo a wide variety of reactions of the following general type:

$$\text{C=O} + H_2N\text{—G} \longrightarrow \text{C=N—G} + H_2O$$

The group, —G, attached to nitrogen may vary as illustrated in the examples that follow. The reaction is usually catalyzed by acids, but the exact pH that works best varies with the nature of the group. Shown below is a probable mechanism.

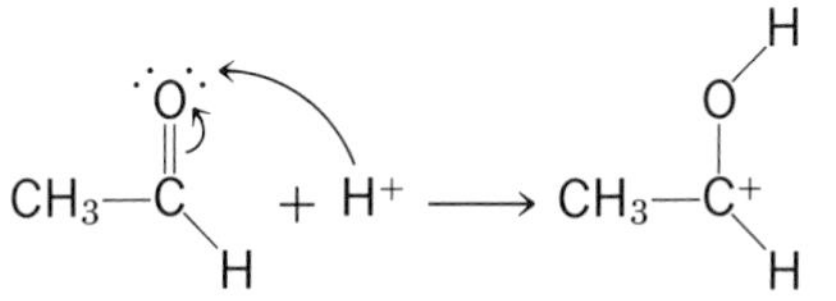

This reaction makes the carbonyl carbon more attractive to an electron-rich species.

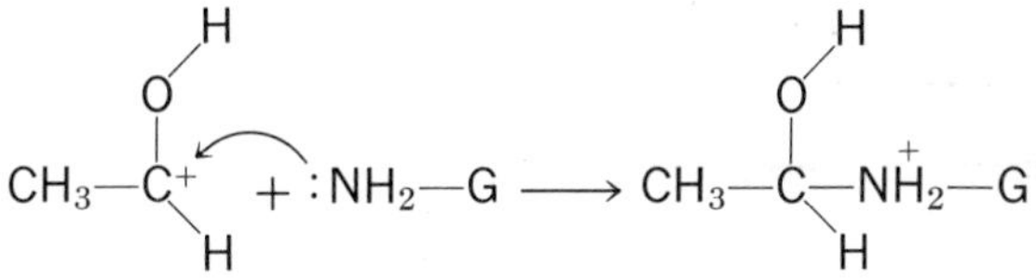

The nucleophilic amino group attacks.

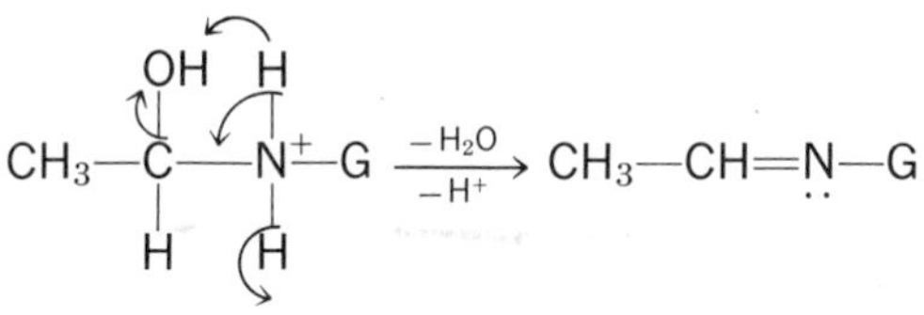

Both water and the proton are lost.

**Examples.** All reactions are general for both aldehydes and ketones.

1. Formation of *substituted imines.* G = alkyl.

$$R—CH{=}O + NH_2—R' \longrightarrow R—CH{=}N—R + H_2O$$

An aldehyde      A 1° amine      Imines<br>(only rarely<br>stable)

2. Formation of *phenylhydrazones.* G = —$NHC_6H_5$.

$$R—C{=}O + NH_2NHC_6H_5 \longrightarrow R—C{=}NNHC_6H_5 + H_2O$$

*ketone* R'    C-N-N

$$C = N - N$$

Phenylhydrazine      A phenylhydrazone

3. Formation of *2,4-dinitrophenylhydrazones.* G = —NH—C$_6$H$_3$(NO$_2$)$_2$

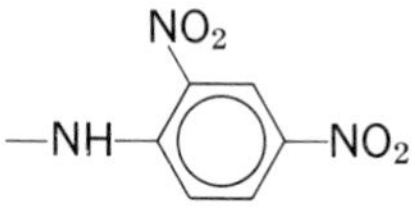

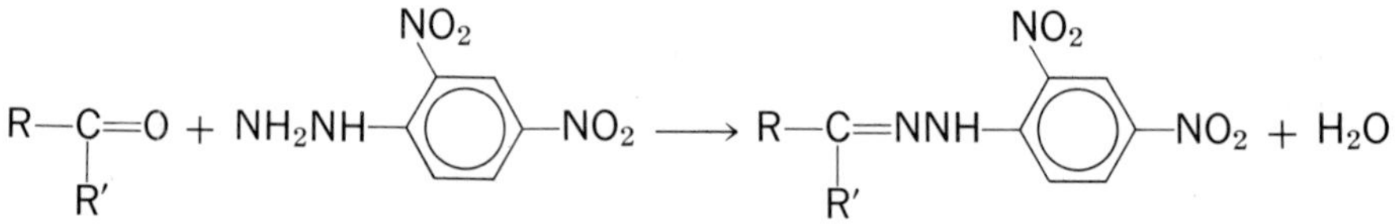

$$R—C{=}O + NH_2NH—C_6H_3(NO_2)_2 \longrightarrow R—C{=}NNH—C_6H_3(NO_2)_2 + H_2O$$

with R' substituent

2,4-Dinitro-<br>phenylhydrazine      A 2,4-dinitrophenylhydrazone

4. Formation of *semicarbazones.* G = —NH—C(=O)—$NH_2$.

$$R—CH{=}O + NH_2NHCNH_2 \longrightarrow RCH{=}NNHCNH_2 + H_2O$$

Semicarbazide      A semicarbazone

5. Formation of *oximes.* G = —OH.

$$RCH{=}O + NH_2OH \longrightarrow RCH{=}NOH + H_2O$$

Hydroxylamine      An oxime

6. Formation of *unsubstituted imines.* G = —H.

$$C{=}O + NH_3 \longrightarrow C{=}NH$$

Unsubstituted imine<br>(only rarely stable)

This reaction is listed after the others because unsubstituted imines are seldom stable, although they apparently appear as intermediates in some reactions in living cells. For example,

$$\underset{\text{An } \alpha\text{-keto acid}}{R-\overset{\overset{\textstyle O}{\|}}{C}-CO_2H} \underset{+H_2O}{\overset{+NH_3}{\rightleftharpoons}} \underset{\text{An } \alpha\text{-imino acid}}{R-\overset{\overset{\textstyle NH}{\|}}{C}-CO_2H} \underset{(O)}{\overset{+2(H)}{\rightleftharpoons}} \underset{\text{An } \alpha\text{-amino acid}}{R-\overset{\overset{\textstyle NH_2}{|}}{CH}-CO_2H}$$

The sequence from left to right illustrates one way in which cells can synthesize amino acids (the building blocks for proteins) from organic sources that do not contain nitrogen. The reverse reactions illustrate one way in which cells degrade amino acids to other compounds.

The overall process of converting a carbonyl group to an amine is called *reductive amination*, a good means of synthesizing amines *in vitro*.

7. *Reductive amination.*

*amine*
*Single bond C—N*

$$R-\overset{\overset{\textstyle O}{\|}}{C}-R'(H) + H-N\overset{\textstyle R(H)}{\underset{\textstyle R(H)}{\big\langle}} + H_2 \xrightarrow[\substack{\text{pressure,} \\ \text{heat}}]{\text{Ni}} R-\overset{\overset{\textstyle (H)R \quad R(H)}{\overset{\textstyle \diagdown \diagup}{N}}}{\underset{}{CH}}-R'(H)$$

$$\underset{\text{Benzaldehyde}}{C_6H_5\overset{\overset{\textstyle O}{\|}}{C}-H} + NH_3 + H_2 \xrightarrow[\substack{\text{pressure,} \\ \text{heat}}]{\text{Ni}} \underset{\text{Benzylamine (89\%)}}{C_6H_5CH_2NH_2}$$

$$\underset{\text{2-Heptanone}}{CH_3(CH_2)_4\overset{\overset{\textstyle O}{\|}}{C}CH_3} + NH_3 + H_2 \xrightarrow[\substack{\text{pressure,} \\ \text{heat}}]{\text{Ni}} \underset{\text{2-Aminoheptane (80\%)}}{CH_3(CH_2)_4\underset{\underset{\textstyle NH_2}{|}}{C}HCH_3}$$

$$\underset{\text{Acetone}}{CH_3\overset{\overset{\textstyle O}{\|}}{C}CH_3} + NH_2CH_3 + H_2 \xrightarrow[\substack{\text{pressure,} \\ \text{heat}}]{\text{Ni}} \underset{\text{Methylisopropylamine (65\%)}}{CH_3\underset{\underset{\textstyle NHCH_3}{|}}{C}HCH_3}$$

$$\underset{\text{Acetaldehyde}}{CH_3\overset{\overset{\textstyle O}{\|}}{C}H} + HN(CH_2CH_3)_2 + H_2 \xrightarrow[\substack{\text{pressure,} \\ \text{heat}}]{\text{Ni}} \underset{\text{Diethylmethylamine (92\%)}}{CH_3CH_2N(CH_2CH_3)_2}$$

**Exercise 6.15** Write the structure of the amine that would form by reductive amination involving the following pairs of compounds: (*a*) propionaldehyde and ethylamine, (*b*) benzaldehyde and methylamine, (*c*) 2-pentanone and dimethylamine, (*d*) acetone and ammonia, and (*e*) methyl phenyl ketone and dimethylamine.

**Exercise 6.16** Write the structures of the starting materials that can be used to prepare each of the following amines by reductive amination: (*a*) *sec*-butylamine, (*b*) cyclohexylamine, (*c*) triethylamine, (*d*) 3-(*N*-methylamino)hexane, and (*e*) cyclopentylmethylamine.

**The Use of Solid Derivatives to Identify Aldehydes and Ketones.** When a chemist is faced with the task of identifying an aldehyde or a ketone, the problem is complicated by the fact that a great number of these compounds are liquids. Experimentally, solids are usually easier to purify than liquids. Therefore the carbonyl compound is often converted into a solid derivative whose melting point identifies the original material. The oximes, semicarbazones, phenylhydrazones, and especially the 2,4-dinitrophenylhydrazones of nearly all aldehydes and ketones are crystalline solids with well-defined melting points, their most serviceable property. These carbonyl derivatives are used principally for identification purposes.

Suppose, for example, that a chemist has a compound he believes is an aldehyde (it gives a positive Tollens test). It boils at 199°C. Extensive tables list known aldehydes according to their increasing boiling points. Table 6.5 is a portion of one such table. It will be noted that four or five aldehydes boil reasonably close to 199°C. The boiling point alone is not characteristic enough to narrow the choice down to one most likely candidate for the unknown. The chemist will therefore take a small portion of the unknown aldehyde and convert it into a solid derivative, perhaps the semicarbazone. Suppose that this derivative of his unknown melts at 209°C. The choice is considerably narrowed, but still two candidates remain, o-tolualdehyde and possibly m-tolualdehyde. The chemist might next prepare the 2,4-dinitrophenylhydrazone. If it melts at 195°C, his unknown and a known compound have three points of coincidence. Within the limits of experimental error both have the same boiling point and the same melting points for the two solid derivatives. He will be quite safe in declaring the unknown to be o-tolualdehyde. The situation is roughly analogous to our custom of giving three names to children. There are many with the forename John, there are fewer with the forename and surname John Thornton, and there is very likely only one with the three names John Wilberforce Thornton. A chemist has many other approaches, however, for he can check for coincidence between his unknown and data for a known compound on many points—boiling point, density, melting point, four kinds of spectra, refractive index, solid derivatives, and any number of chemical properties.

**Table 6.5  Derivatives of Some Aldehydes***

| Aldehyde | Bp (°C) | Melting Points of Derivatives (°C) | | |
|---|---|---|---|---|
| | | Oxime | Semicarbazone | 2,4-Dinitrophenyl-hydrazone |
| Phenylacetaldehyde | 194 | 103 | 156 | 121 |
| Salicylaldehyde | 196 | 57 | 231 | 252 (decomposes) |
| m-Tolualdehyde | 199 | 60 | 213 | 211 |
| o-Tolualdehyde | 200 | 49 | 208 | 195 |
| p-Tolualdehyde | 204 | 79 | 221 | 239 |
| Citronellal | 206 | liquid | 82 | 77 |

* Data from R. L. Shriner, R. C. Fuson, and D. Y. Curtin, *The Systematic Identification of Organic Compounds*, fifth edition, John Wiley and Sons, New York, 1964, page 320.

REACTIONS AT THE α-POSITION. THE ALDOL CONDENSATION
AND RELATED REACTIONS
209

Reactions at the
α-Position.
The Aldol
Condensation and
Related Reactions

When heated in the presence of 10% sodium hydroxide, acetaldehyde reacts to form β-hydroxybutyraldehyde, a reaction called the *aldol condensation*.

$$2CH_3CH=O \xrightarrow[\text{heat}]{\text{NaOH}(aq)} \underset{\underset{\text{OH}}{|}}{CH_3CHCH_2CH=O}$$

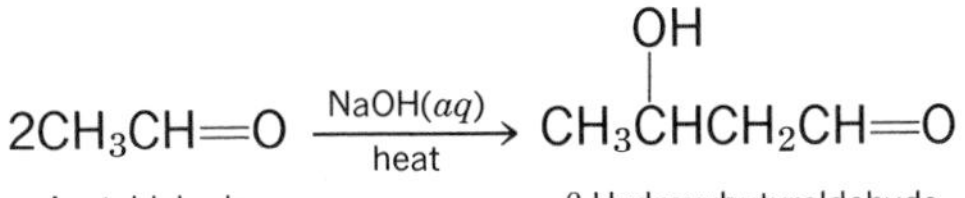

Acetaldehyde      β-Hydroxybutyraldehyde
(aldol)

In general:

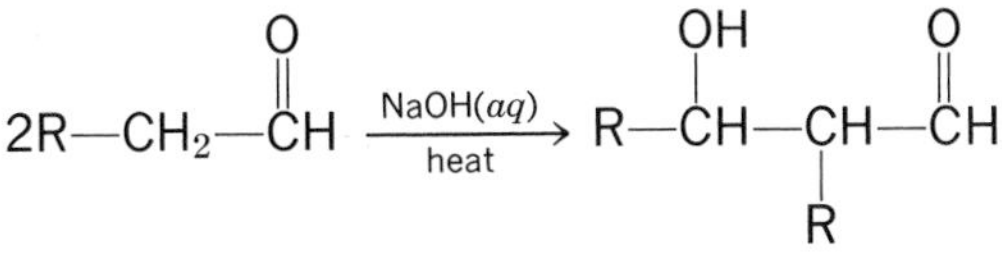

$$2R-CH_2-\overset{\overset{\text{O}}{||}}{CH} \xrightarrow[\text{heat}]{\text{NaOH}(aq)} R-\underset{\underset{\text{R}}{|}}{\overset{\overset{\text{OH}}{|}}{CH}}-CH-\overset{\overset{\text{O}}{||}}{CH}$$

This reaction is illustrative of a large number that have the following feature in common:

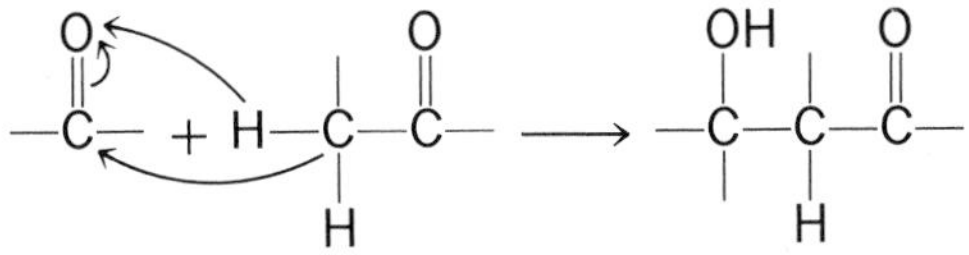

Carbonyl          Carbonyl              A β-hydroxy
compound       compound with      carbonyl compound
                two α-hydrogens

In words, one aldehyde adds across the carbonyl group of another. Most other examples of this type of reaction continue beyond the β-hydroxy carbonyl stage. Dehydration occurs and an unsaturated carbonyl compound forms:

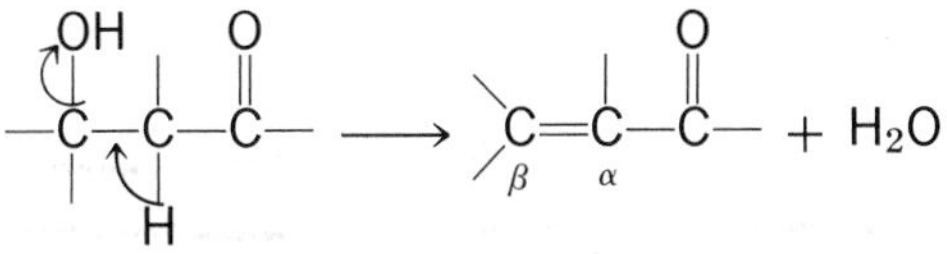

A β-hydroxy       An α,β-unsaturated carbonyl
carbonyl compound          compound

Before examining specific examples, including some from metabolic reactions in living cells, the mechanism for this general reaction will be described.

**General Mechanism for Aldol-Type Condensations**

Step 1. Acid-base equilibration between the base and a molecule of carbonyl compound. The base takes a proton from the α-position and only from the α-position.

In general:

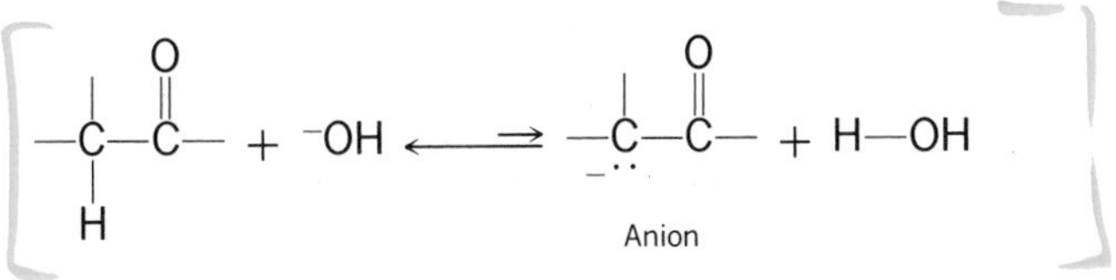

Example:

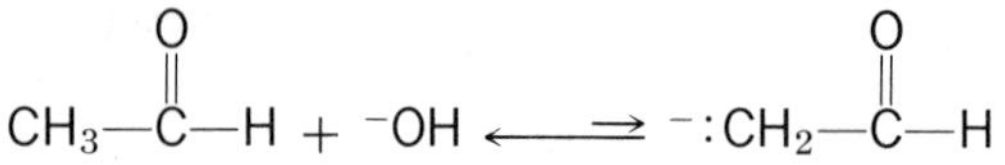

$$CH_3-\overset{\overset{\displaystyle O}{\|}}{C}-H + {}^-OH \; \rightleftharpoons \; {}^-\!:CH_2-\overset{\overset{\displaystyle O}{\|}}{C}-H$$

Acetaldehyde                  An anion

Step 2. Attack by the anion[4] formed in step 1 at the carbonyl carbon of another carbonyl-containing molecule.

In general:

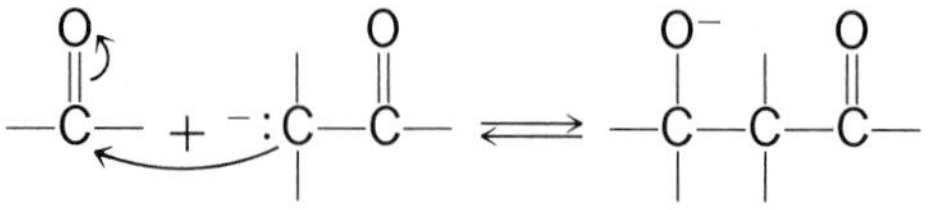

$$-\overset{\overset{\displaystyle O}{\|}}{C}- + {}^-\!:\overset{\overset{\displaystyle O}{\|}}{C}-\overset{}{C}- \; \rightleftharpoons \; -\overset{\overset{\displaystyle O^-}{|}}{C}-\overset{}{C}-\overset{\overset{\displaystyle O}{\|}}{C}-$$

A new, more stable anion

Example:

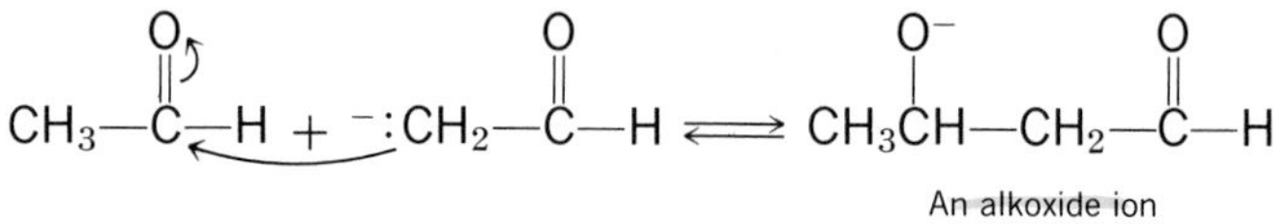

$$CH_3-\overset{\overset{\displaystyle O}{\|}}{C}-H + {}^-\!:CH_2-\overset{\overset{\displaystyle O}{\|}}{C}-H \; \rightleftharpoons \; CH_3\overset{\overset{\displaystyle O^-}{|}}{C}H-CH_2-\overset{\overset{\displaystyle O}{\|}}{C}-H$$

An alkoxide ion

Step 3. Acid-base equilibration between the alkoxide ion formed in step 2 and a proton donor (such as water).

In general:

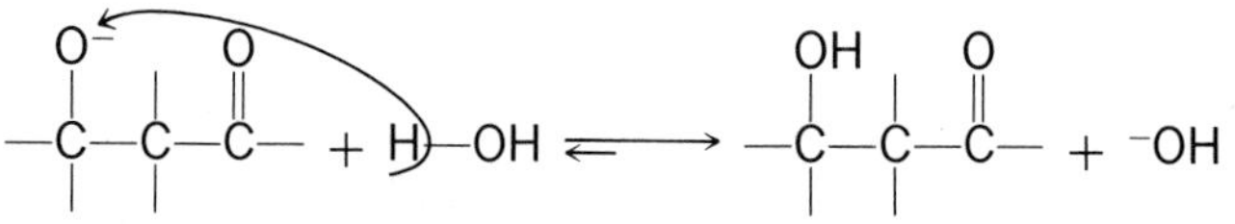

$$-\overset{\overset{\displaystyle O^-}{|}}{C}-\overset{}{C}-\overset{\overset{\displaystyle O}{\|}}{C}- + H-OH \; \rightleftharpoons \; -\overset{\overset{\displaystyle OH}{|}}{C}-\overset{}{C}-\overset{\overset{\displaystyle O}{\|}}{C}- + {}^-OH$$

Example:

$$CH_3-\overset{\overset{\displaystyle O^-}{|}}{C}H-CH_2-\overset{\overset{\displaystyle O}{\|}}{C}-H + H-OH \; \rightleftharpoons$$

$$CH_3-\overset{\overset{\displaystyle OH}{|}}{C}H-CH_2-\overset{\overset{\displaystyle O}{\|}}{C}-H + {}^-OH$$

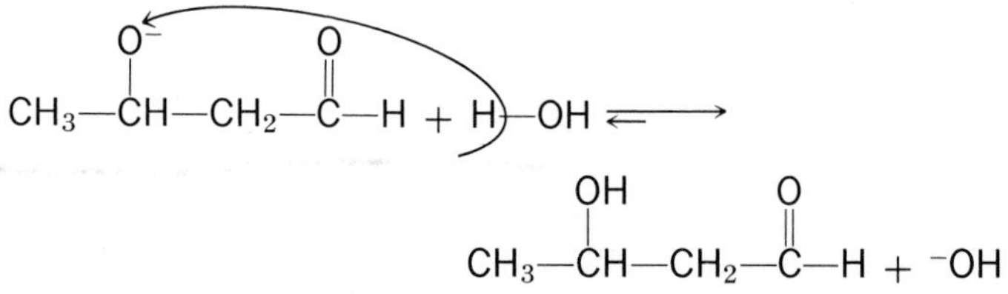

β-Hydroxybutyraldehyde
(aldol)

*c has ⁻charge*

In step 1 a carbanion forms. Thus far we have learned that carbanions are extremely difficult to form and that a carbon holding a hydrogen is simply not a proton donor if a base as relatively weak as the hydroxide ion comes seeking a proton. The situation is changed, however, when the carbon holding the hydrogen is *alpha* to a carbonyl group. In this situation the negative charge can be delocalized, and the carbanion formed thereby is much more stable than ordinary ones.

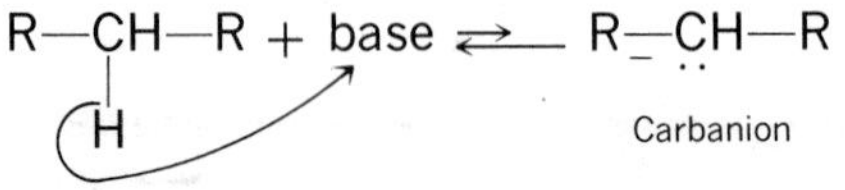

$$R-\overset{\overset{\displaystyle }{|}}{\underset{\underset{\displaystyle H}{|}}{C}}H-R + base \; \rightleftharpoons \; R-\overset{..}{C}H-R$$

                                      Carbanion

—Charge cannot be delocalized
—Carbanion difficult to form

[4] The term *anion* denotes any negatively charged ion.

211

Reactions at the
α-Position.
The Aldol
Condensation and
Related Reactions

$$R-\underset{\underset{H}{|}}{\overset{\overset{O}{\|}}{C}}-\underset{}{C}-R + base \rightleftharpoons$$

$$\left[ \begin{array}{c} \overset{:O:}{\overset{\|}{R-CH-C-R}} \\ \downarrow \\ \overset{:\ddot{O}:^-}{R-CH=C-R} \end{array} \right]$$

The carbanion is
a hybrid of
these two
contributors.

—Charge is delocalized and
shared between the α-carbon
and the oxygen

—Anion easier to form

—The hybrid is $R-\overset{\delta-}{CH}\cdots\overset{\overset{\delta-}{O}}{\underset{R}{C}}$

In an aldol condensation the carbonyl compound must have an α-hydrogen. The base cannot take a proton from a β-position or one farther down the chain.

The anion formed in step 1 is still, however, a species actively attracted to positively charged sites. The carbonyl carbon of another molecule is such a site, and it is attacked by the anion. An alkoxide ion (with a carbonyl group elsewhere in the same molecule) is the product. This, of course, is a powerful proton acceptor; in step 3 it takes a proton and the final product, the aldol, forms. Aldol is a specific name for β-hydroxybutyraldehyde and a general name for β-hydroxy aldehydes.

Step 4. The aldol suffers loss of the elements of water.

In general:

$$-\underset{\underset{H}{|}}{\overset{\overset{OH}{|}}{C}}-\underset{}{C}-\overset{\overset{O}{\|}}{C}- \longrightarrow -\overset{}{C}=\overset{}{C}-\overset{\overset{O}{\|}}{C}- + H_2O$$

Example:

$$CH_3-\underset{\overset{|}{OH}}{CH}-CH_2-\overset{\overset{O}{\|}}{C}-H \longrightarrow CH_3-CH=CH-\overset{\overset{O}{\|}}{C}-H + H_2O$$

Crotonaldehyde

This last step is essentially irreversible, but dehydration is not spontaneous with the aldols formed by condensation between aldehyde molecules. For these an acid catalyst is usually necessary. In many variations of the aldol condensation dehydration is spontaneous, however, and the final isolable products are unsaturated carbonyl compounds.

**Crossed Aldol Condensations.** If a mixture of two different aldehydes is heated in the presence of aqueous sodium hydroxide, four products form in a "crossed" aldol condensation. "Crossing over" inevitably occurs in the random interactions. For example, with a mixture of acetaldehyde and propionaldehyde in the presence of a base, two anions are possible:

$$CH_3-\overset{\overset{O}{\|}}{C}-H + {}^-OH \rightleftharpoons {}^-:CH_2-\overset{\overset{O}{\|}}{C}-H$$

Acetaldehyde

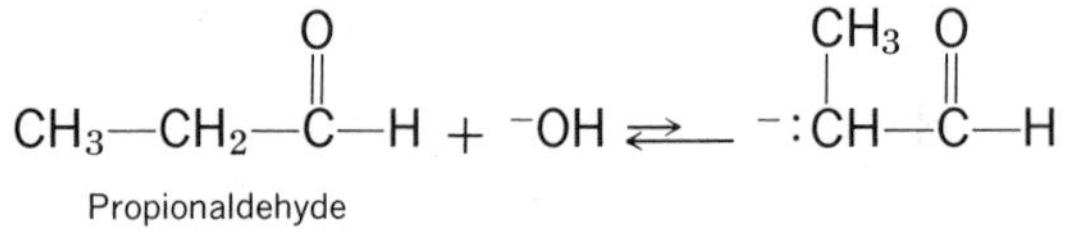

$$\underset{\text{Propionaldehyde}}{CH_3-CH_2-\overset{\displaystyle O}{\overset{\|}{C}}-H} + {}^-OH \rightleftharpoons {}^-:\underset{}{\overset{\displaystyle CH_3}{\overset{|}{CH}}}-\overset{\displaystyle O}{\overset{\|}{C}}-H$$   Note that the event occurs at the α-carbon.

These two anions form in the presence of both of the unchanged aldehydes. A carbonyl carbon is a carbonyl carbon, and the anions will attack any accidentally near it:

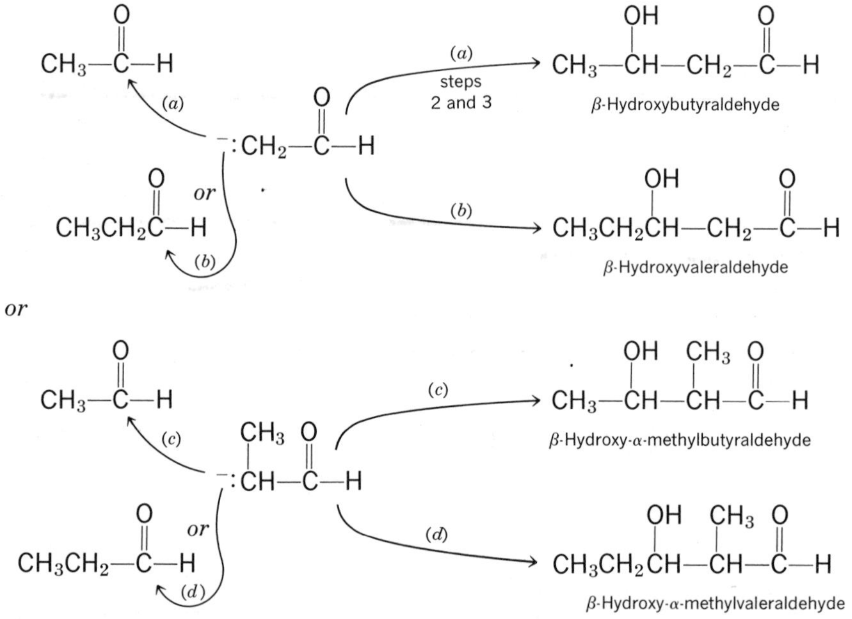

Thus, whenever *both* aldehydes have α-hydrogens, the crossed aldol gives a mixture of four aldols. They are difficult to separate, however, and the method is seldom useful in synthesis. The problem is somewhat simplified if only one of the aldehydes has an α-hydrogen. Then only one anion can form in the first step. By carefully regulating the experimental conditions, the statistical chances of this anion attacking the carbonyl of the *other* compound can be made great enough to render such a crossed aldol condensation useful.

**Examples of crossed aldol condensations.** In these examples we assume that the fourth step, loss of water, takes place, and usually it does. The net effect in each is

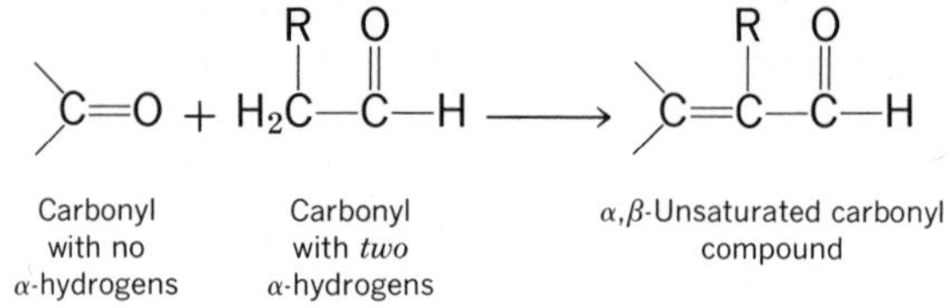

**213**

Reactions at the
α-Position.
The Aldol
Condensation and
Related Reactions

$$\underset{H}{\overset{\displaystyle H}{C_6H_5-\overset{|}{C}=O}} + CH_3-\overset{\displaystyle O}{\overset{\|}{C}}-H \xrightarrow[20°C]{OH^-} C_6H_5-CH=CH-\overset{\displaystyle O}{\overset{\|}{C}}-H$$

$$+ CH_3-\overset{\displaystyle O}{\overset{\|}{C}}-CH_3 \xrightarrow[100°C]{OH^-} C_6H_5-CH=CH-\overset{\displaystyle O}{\overset{\|}{C}}-CH_3$$

$$+ CH_3-\overset{\displaystyle O}{\overset{\|}{C}}-C_6H_5 \xrightarrow[20°C]{OH^-} C_6H_5-CH=CH-\overset{\displaystyle O}{\overset{\|}{C}}-C_6H_5$$

$$+ \underset{C_6H_5}{CH_2}-\overset{\displaystyle O}{\overset{\|}{C}}-H \xrightarrow[20°C]{OH^-} C_6H_5-CH=\underset{C_6H_5}{C}-\overset{\displaystyle O}{\overset{\|}{C}}-H$$

Thus far our discussion of the aldol condensation has concerned *in vitro* reactions. Reactions in cells are under the control of highly specialized catalysts, enzymes, which can accomplish changes that are not possible in ordinary catalysis. One crossed aldol condensation that takes place in plant and animal cells occurs as a step in the synthesis of glucose units, the building blocks for starches, glycogen, and cellulose.

A is dihydroxyacetone phosphate
B is glyceraldehyde 3-phosphate

Fructose 1,6-diphosphate

The enzyme in this reaction is called aldolase to denote that it catalyzes an aldol-type condensation. *The aldol condensation is reversible* (step 3 backward to the start of step 1, in our mechanism), and the reverse aldol condensation is a key step in the degradation of a glucose unit which is eventually converted into carbon dioxide, water, and chemical energy plus some heat.

Another reaction that resembles the aldol condensation and occurs during the breakdown of glucose units is the first step in a series called the citric acid cycle, which we shall study in much greater detail later. These two examples from the field of biochemistry need not be memorized. They are shown simply because they illustrate that our understanding of biochemical events in cells can be greatly increased if we have previously seen similar examples as they occur in less complicated systems. In this first step in the citric acid cycle a H—C system that is alpha to a carbonyl group (in acetic acid) adds across the carbonyl unit in another molecule (the keto group in $\alpha$-ketosuccinic acid).

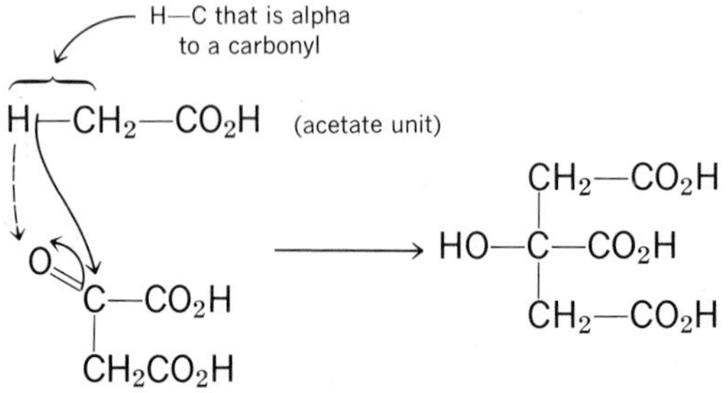

H—C that is alpha to a carbonyl

$$H\text{—}CH_2\text{—}CO_2H \quad \text{(acetate unit)}$$

$$\underset{\underset{\text{$\alpha$-Ketosuccinic acid}}{CH_2CO_2H}}{\overset{O}{\underset{\|}{C}}\text{—}CO_2H} \longrightarrow \underset{\underset{\text{Citric acid}}{CH_2\text{—}CO_2H}}{HO\text{—}\overset{CH_2\text{—}CO_2H}{\underset{}{C}}\text{—}CO_2H}$$

Exercise 6.17 Write the structure of the product of the aldol condensation involving each of the following aldehydes: (*a*) propionaldehyde, (*b*) butyraldehyde, and (*c*) phenylacetaldehyde ($C_6H_5CH_2CH{=}O$).

Exercise 6.18 Write the structure of the product of the crossed aldol condensation between the following pairs of compounds: (*a*) benzaldehyde and propionaldehyde, (*b*) benzaldehyde and butyraldehyde, and (*c*) benzaldehyde and $CH_3CH_2\overset{O}{\overset{\|}{C}}C_6H_5$ (propiophenone).

## SYNTHESIS OF ALDEHYDES AND KETONES

1. By oxidation of 1° and 2° alcohols (see p. 126). The 1° alcohols, under conditions carefully regulated to inhibit oxidation of the newly formed product, yield aldehydes:

$$RCH_2\text{—}O\text{—}H \xrightarrow[\text{heat}]{Cu} RCH{=}O + H_2$$

The 2° alcohols can be oxidized to ketones:

$$R_2CH\text{—}O\text{—}H \xrightarrow[\substack{\text{or } CrO_3, \\ \text{or other strong oxidizing agents}}]{Cu, \text{ heat},} R_2C{=}O \quad (+H_2 \text{ or } H_2O)$$

As discussed and illustrated in Chapter 4 (p. 129), the conversion of alcohols into aldehydes or ketones is a fairly common reaction in cells.

2. Friedel-Crafts acylation, a synthesis of aromatic ketones. In general:

$$Ar\text{—}H + Cl\text{—}\overset{O}{\overset{\|}{C}}\text{—}R \xrightarrow{AlCl_3} Ar\text{—}\overset{O}{\overset{\|}{C}}\text{—}R + HCl$$

Aromatic hydrocarbon     Acid chloride     Aromatic ketone

Specific examples:

$$C_6H_6 + Cl\overset{O}{\underset{\|}{-}}C-CH_2C_6H_5 \xrightarrow{AlCl_3} C_6H_5\overset{O}{\underset{\|}{-}}C-CH_2C_6H_5$$

Benzyl phenyl ketone (83%)

$$+ Cl\overset{O}{\underset{\|}{-}}C-CH_2CH_2CH_3 \longrightarrow C_6H_5\overset{O}{\underset{\|}{-}}C-CH_2CH_2CH_3$$

Phenyl *n*-propyl ketone (65%)

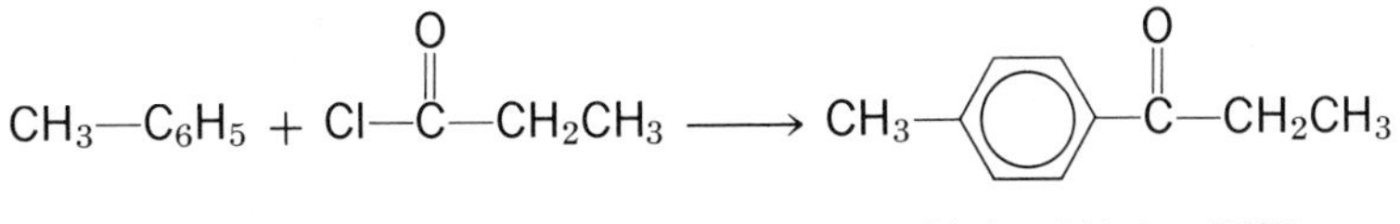

$$CH_3-C_6H_5 + Cl\overset{O}{\underset{\|}{-}}C-CH_2CH_3 \longrightarrow CH_3-\boxed{\bigcirc}\overset{O}{\underset{\|}{-}}C-CH_2CH_3$$

Ethyl *p*-tolyl ketone (86%)

3. The Rosenmund reduction of acid chlorides, a synthesis of aldehydes. In general:

$$R\overset{O}{\underset{\|}{-}}C-Cl + H_2 \xrightarrow{catalyst} R\overset{O}{\underset{\|}{-}}C-H + HCl$$

Specific examples:

$$C_6H_5\overset{O}{\underset{\|}{-}}C-Cl + H_2 \longrightarrow C_6H_5\overset{O}{\underset{\|}{-}}C-H$$

Benzaldehyde (96%)

$$(CH_3)_2CHCH_2\overset{O}{\underset{\|}{-}}C-Cl + H_2 \longrightarrow (CH_3)_2CHCH_2\overset{O}{\underset{\|}{-}}CH$$

Isovaleraldehyde (100%)

Aldehydes and ketones can be prepared in dozens of other ways. In fact, one reference[5] discusses thirty-seven ways for making aldehydes and thirty-seven ways for making ketones. The few methods just presented illustrate some of those most commonly used.

---

Exercise 6.19   Write the structures of the principal organic products that would form in each of the following reactions.

(*a*) $(CH_3)_2CHOH \xrightarrow[H^+]{CrO_3}$

(*b*) $C_6H_6 + CH_3\overset{O}{\underset{\|}{C}}Cl \xrightarrow{AlCl_3}$

[5] R. B. Wagner and H. D. Zook, *Synthetic Organic Chemistry*, John Wiley and Sons, New York, 1953.

(c) $CH_3\overset{O}{\overset{\|}{C}}Cl + H_2 \xrightarrow{\text{catalyst}}$

(d) $CH_3CH_2CH_2CH_2OH \xrightarrow[\text{heat}]{\text{Cu}}$

(e) $CH_3\text{—}\langle\bigcirc\rangle\text{—}CH_3 + Cl\text{—}\overset{O}{\overset{\|}{C}}\text{—}C_6H_5 \xrightarrow{\text{AlCl}_3}$

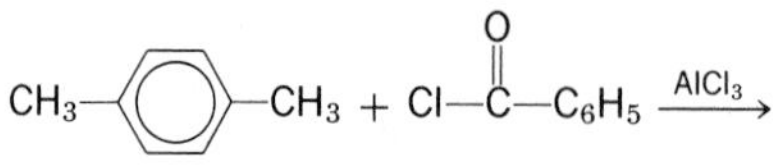

(f) $C_6H_5\text{—}\underset{\underset{OH}{|}}{CH}\text{—}C_6H_5 \xrightarrow[\text{H}^+]{\text{CrO}_3}$

(g) $CH_3\text{—}\overset{O}{\overset{\|}{C}}\text{—}C_6H_5 \xrightarrow[\text{NaOH, heat}]{\text{I}_2}$

(h) $C_6H_5\text{—}\overset{O}{\overset{\|}{C}}\text{—}H + NH_2OH \xrightarrow{\text{H}^+}$

(i) $C_6H_5\text{—}\overset{O}{\overset{\|}{C}}\text{—}H + CH_3\text{—}\overset{O}{\overset{\|}{C}}\text{—}H \xrightarrow{\text{OH}^-}$

(j) $CH_3\overset{O}{\overset{\|}{C}}CH_3 + NH_2NH\overset{O}{\overset{\|}{C}}NH_2 \xrightarrow{\text{H}^+}$

---

## IMPORTANT ALDEHYDES AND KETONES

**Formaldehyde.** $H\text{—}\overset{O}{\overset{\|}{C}}\text{—}H$. At room temperature formaldehyde is a gas with a very irritating and distinctive odor. In one industrial method it is prepared by passing a mixture of methyl alcohol and air over a silver catalyst at a temperature of over 600°C. The newly formed formaldehyde together with unchanged methanol is absorbed in water until the solution has a concentration of about 37% formaldehyde. In this form it is marketed as Formalin. Little free formaldehyde exists in Formalin. The hydrated form $CH_2(OH)_2$ (cf. p. 201), and a polymeric form, $HO(CH_2O)_nH$, with $n$ having an average value of 3, are the major organic constituents.

If Formalin is concentrated still further, at diminished pressure, the size of $n$ in the polymer, $HO(CH_2O)_nH$, grows to an average value of 30, and it precipitates. This longer-chained, solid polymer is called paraformaldehyde and is another form in which formaldehyde is shipped and stored. At temperatures above 137°C the polymer decomposes to gaseous formaldehyde.

Another solid form of formaldehyde, the trimer 1,3,5-trioxane, is produced if Formalin is concentrated and distilled in the pressure of a small amount of sulfuric acid catalyst.

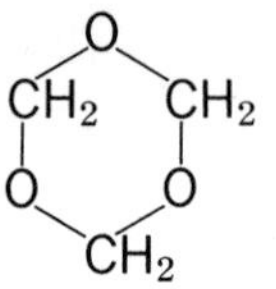

1,3,5-Trioxane

It is practically impossible to inhibit the polymerization of formaldehyde during shipping and storage, but the relative ease of depolymerization makes formaldehyde readily available from these polymers. Formaldehyde, as its Formalin solution, is used as a disinfectant and as a preservative for biological samples and specimens. The largest amounts of Formalin, however, are consumed in the manufacture of various resins. Bakelite, a phenol-formaldehyde plastic developed in a commercially useful way by Baekeland (see Figure 6.1), and melamine resins, made from melamine and formaldehyde, are the most familiar. Exceptional resistance to heat and water makes melamine resin a favored material for plastic dinnerware (see

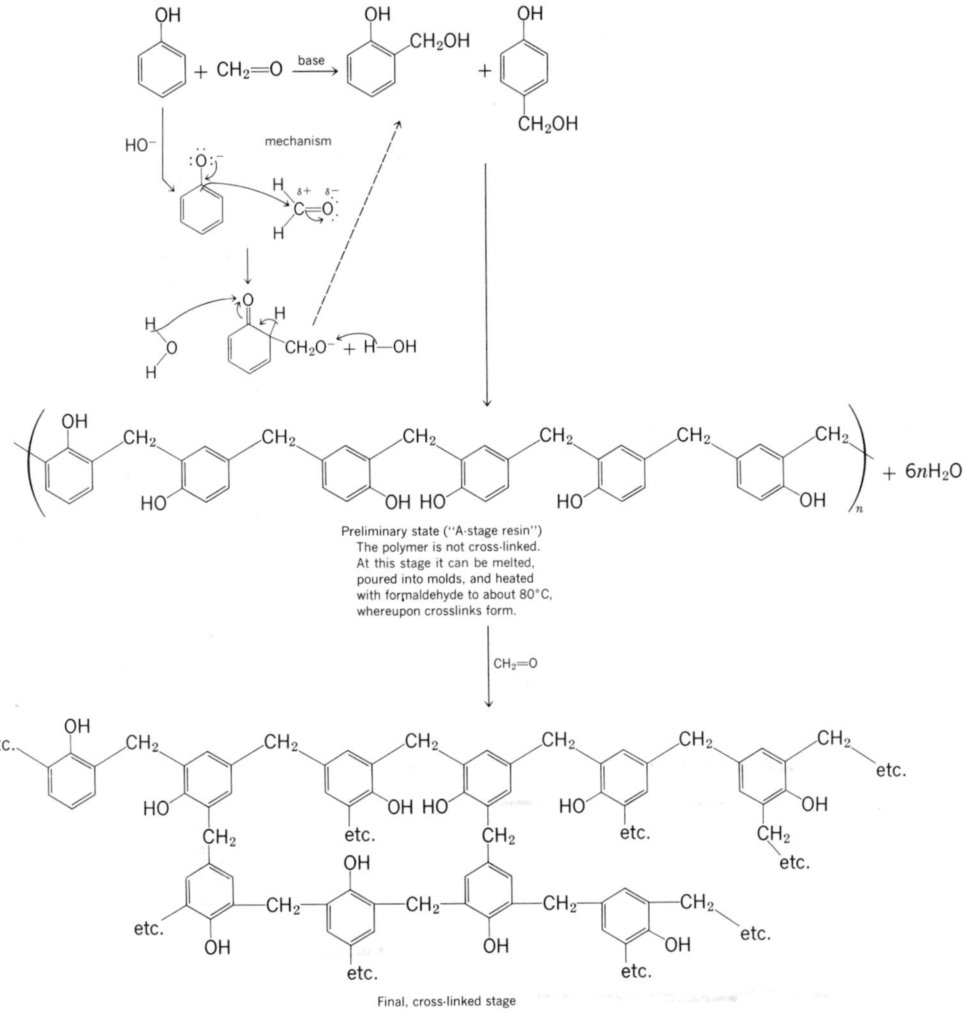

Preliminary state ("A-stage resin")
The polymer is not cross-linked.
At this stage it can be melted,
poured into molds, and heated
with formaldehyde to about 80°C,
whereupon crosslinks form.

Final, cross-linked stage

**Figure 6.1**  Bakelite, a phenol-formaldehyde resin. Only the principal structural features are indicated here.

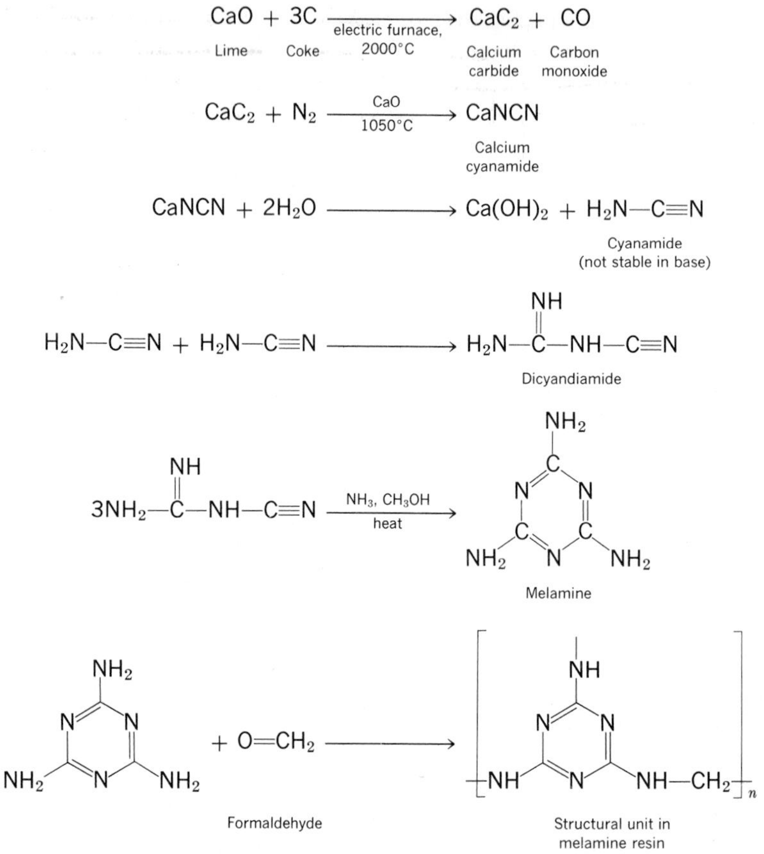

$$CaO + 3C \xrightarrow[\text{2000°C}]{\text{electric furnace,}} CaC_2 + CO$$

Lime   Coke             Calcium  Carbon
carbide  monoxide

$$CaC_2 + N_2 \xrightarrow[\text{1050°C}]{CaO} CaNCN$$

Calcium
cyanamide

$$CaNCN + 2H_2O \longrightarrow Ca(OH)_2 + H_2N-C\equiv N$$

Cyanamide
(not stable in base)

$$H_2N-C\equiv N + H_2N-C\equiv N \longrightarrow \overset{\overset{\displaystyle NH}{\|}}{H_2N-C}-NH-C\equiv N$$

Dicyandiamide

$$3\,\overset{\overset{\displaystyle NH}{\|}}{NH_2-C}-NH-C\equiv N \xrightarrow[\text{heat}]{NH_3,\ CH_3OH}$$

Melamine

$$+\ O{=}CH_2 \longrightarrow$$

Formaldehyde                     Structural unit in
melamine resin

**Figure 6.2** Melamine resin. Its manufacture illustrates one of several industrial processes that start with abundant, relatively inexpensive natural resources: lime from limestone, coke from coal and wood, nitrogen from air, energy from hydroelectric sources, and other materials.

Figure 6.2). Formaldehyde is also used in the manufacture of high explosives such as cyclonite and pentaerythritol nitrate (PETN).

**Acetaldehyde.** $CH_3\overset{\overset{\displaystyle O}{\|}}{C}-H$. Acetaldehyde is manufactured by a variety of processes, by the hydration of acetylene,

$$H-C\equiv C-H + H-OH \xrightarrow[Hg^{2+}]{H^+} \left[ H-\overset{\overset{\displaystyle H}{|}}{C}{=}\overset{\overset{\displaystyle OH}{|}}{C}-H \right] \longrightarrow CH_3-\overset{\overset{\displaystyle O}{\|}}{C}-H$$

An enol

by the air oxidation of ethanol,

$$CH_3-CH_2-OH \xrightarrow[\text{Ag, heat}]{O_2} CH_3-\overset{\overset{\displaystyle O}{\|}}{C}-H$$

and by the controlled air oxidation of propane-butane fractions of natural gas. It is an important raw material for the manufacture of acetic acid, acetic anhydride, and ethyl acetate, and of a solid trimer, *paraldehyde*.

Paraldehyde

Paraldehyde is a convenient form for shipping and storing acetaldehyde (boiling point 20°C), to which it can readily be depolymerized when heated in the presence of a trace of acid. Although sometimes prescribed as a hypnotic or soporific (sleep inducer), one with a commendably low toxicity, the burning taste and the disagreeable odor of paraldehyde as it is eliminated through the lungs into the patient's breath make it somewhat objectionable.

A tetramer of acetaldehyde, commercially known as *metaldehyde,* is sold as a pesticide to control garden slugs and snails.

**Vanillin.** A tropical climbing orchid, *Vanilla planifolia,* produces beans from which vanillin is extracted. In contrast with most aldehydes, vanillin is quite resistant to air oxidation. Because of its importance as a flavoring and odor-masking material, methods for producing vanillin synthetically have been developed. Large amounts are also made by processing lignin, a polymeric material that forms an important part of wood (cellulose is another constituent). Both the natural and the synthetic vanillin are marketed as a solution in alcohol, the familiar vanilla extract. There is no difference between "pure vanilla extract," that containing the vanillin extracted from vanillin beans, and "artificial vanilla flavor" or "imitation vanilla flavor." The concentration of vanillin in the extract may vary of course, but vanillin is vanillin whatever its source.

**Acetone.** $CH_3CCH_3$ One of the most important organic solvents, acetone not only dissolves a wide variety of organic substances but is also miscible with water in all proportions. Industrially, acetone is used as a solvent in processing cellulose acetate (for fiber), cellulose nitrate (an explosive), and acetylene (which would present great hazards shipped in the undiluted form). Acetone is made in many ways, including the air oxidation of isopropyl alcohol at elevated temperatures.

Acetone, a normal product of fat metabolism, is further oxidized by the body to carbon dioxide and water. In diabetes mellitus one malfunction is an abnormally rapid fat metabolism, with acetone accumulating to the point that its odor can be detected on the breath ("acetone breath"). Under these circumstances the kidneys will remove it and place it in urine. Acetone is easily detected in urine with the sodium nitroprusside test.

**Glucose and Fructose.** Glucose in one form is a pentahydroxy aldehyde, fructose a pentahydroxy ketone. Of obvious importance in nutrition and metabolism, they will be discussed more fully in Chapter 8.

## REFERENCES

All the standard textbooks in organic chemistry have one or more chapters on the chemistry of aldehydes and ketones.

## PROBLEMS AND EXERCISES

1. Write the structures for each of the following.
   (a) 2,3-dimethylhexanal    (b) acetic anhydride    (c) chloroacetic acid
   (d) ethyl propionate    (e) di-n-butyl ketone    (f) sodium isobutyrate
   (g) α-methylbutyryl chloride    (h) diisopropyl ketone    (i) α,β-dibromopropionaldehyde
   (j) p-bromobenzaldehyde

2. Write common names for each of the following.

   (a) $CH_3CH_2\overset{\displaystyle O}{\overset{\|}{C}}H$

   (b) $CH_3CH_2CH_2\overset{\displaystyle O}{\overset{\|}{C}}C_6H_5$

   (c) $BrCH_2CH_2CH_2\overset{\displaystyle O}{\overset{\|}{C}}H$

   (d) $Na^+{}^-O\overset{\displaystyle O}{\overset{\|}{-C}}H$

   (e) $(CH_3)_2CH\overset{\displaystyle O}{\overset{\|}{C}}OCH_2CH_3$

   (f) $CH_3CH_2\overset{\displaystyle O}{\overset{\|}{C}}OCH(CH_3)_2$

   (g) $CH_3\overset{\displaystyle OH}{\overset{\|}{C}}HCH_2\overset{\displaystyle O}{\overset{\|}{C}}OH$

   (h) [structure: benzene ring with NO$_2$ substituent and $-\overset{\displaystyle O}{\overset{\|}{C}}H$ substituent]

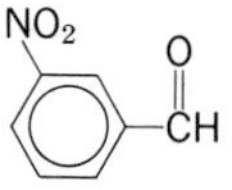

   (i) $CH_3CH_2O\overset{\displaystyle O}{\overset{\|}{C}}H$

   (j) $CH_3CH_2OCH_2CH_3$

3. Write the structure(s) of the principal organic product(s) that might reasonably be expected to form if *acetaldehyde* were subjected to each of the following reagents and conditions.
   (a) $H_2$, Ni, pressure, heat    (b) excess $CH_3OH$, dry HCl
   (c) $CH_3OH$ (as a solvent)    (d) $NH_2OH$

   (e) 10% NaOH, heat    (f) $NH_2NH\overset{\displaystyle O}{\overset{\|}{C}}NH_2$
   (g) Tollens' reagent    (h) $CrO_3$, $H^+$
   (i) excess $I_2$, NaOH    (j) $NH_2NHC_6H_5$

4. Repeat question 3 using *acetone* instead of acetaldehyde.

✗ 5. Which of the following would be expected to give a positive Benedict or Fehling test?

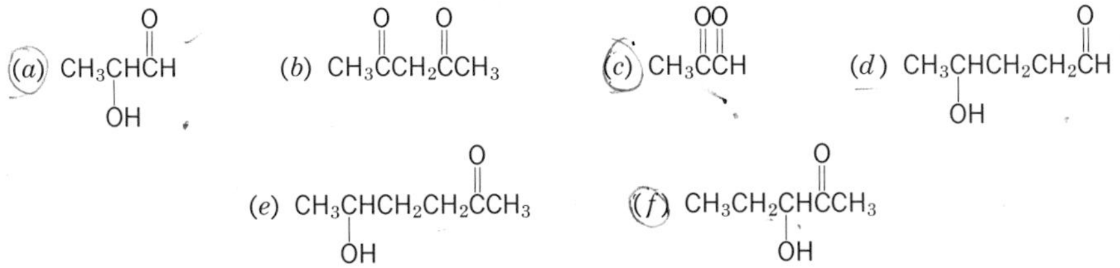

(a) $\underset{\underset{\text{OH}}{|}}{CH_3CHCH}=O$

(b) $CH_3\overset{O}{\overset{||}{C}}CH_2\overset{O}{\overset{||}{C}}CH_3$

(c) $CH_3\overset{OO}{\overset{||}{C}}CH$

(d) $\underset{\underset{\text{OH}}{|}}{CH_3CHCH_2CH_2}\overset{O}{\overset{||}{C}}H$

(e) $\underset{\underset{\text{OH}}{|}}{CH_3CHCH_2CH_2}\overset{O}{\overset{||}{C}}CH_3$

(f) $CH_3CH_2\underset{\underset{\text{OH}}{|}}{CH}\overset{O}{\overset{||}{C}}CH_3$

6. Shown below is the structure of a hypothetical molecule.

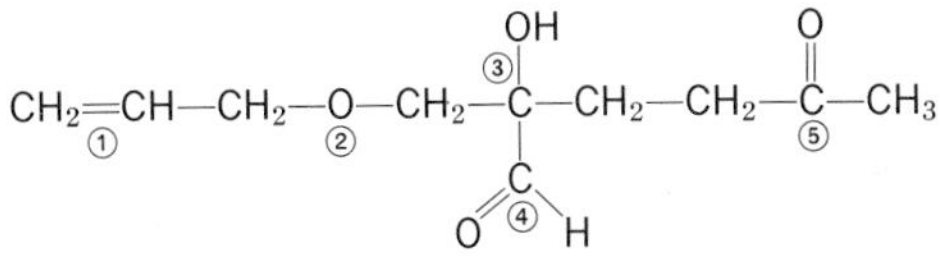

$$CH_2\!\!=\!\!\underset{①}{CH}\!-\!CH_2\!-\!\underset{②}{O}\!-\!CH_2\!-\!\underset{\substack{|③\\\underset{O}{\overset{C}{\diagdown}}{}^{④}H}}{\overset{OH}{C}}\!-\!CH_2\!-\!CH_2\!-\!\underset{⑤}{\overset{O}{\overset{||}{C}}}\!-\!CH_3$$

Predict the chemical reactions it would probably undergo if it were subjected to the reagents and conditions listed. To simplify the writing of the equations illustrating these reactions, isolate the portion of the molecule that would be involved, replace the other non-involved portions by symbols such as G, G′, G″, etc., and then write the equation (see example). You are expected to make reasonable predictions based only on the reactions we have studied thus far. If no reaction is to be predicted, write "None."

*Example.* Reagent: excess $H_2$, Ni, heat, pressure. We approach this problem in the following steps.

(a) First identify *by name* the functional groups in the molecule. (By doing this you take advantage of the way you mentally store information about the reactions of functional groups. Thus the *name* for —CH=O is aldehyde or aldehyde group. In your mental "file" on this group there should be such statements as "Aldehydes can be reduced to 1° alcohols" or "Aldehydes can be easily oxidized to acids," etc.

In the structure given, you should recognize ① alkene, ② ether, ③ 3° alcohol, ④ aldehyde, ③ and ④ α-hydroxy aldehyde, ⑤ ketone.
(b) To continue the example, ask yourself how each group would respond to the reagent given.
   ① Alkenes will add hydrogen catalytically under heat and pressure.
   ② Ethers do not react with hydrogen.
   ③ There is no reaction of 3° alcohols with hydrogen.
   ④ Aldehydes can be reduced to 1° alcohols.
   ⑤ Ketones can be reduced to 2° alcohols.
(c) Third, write the equations. For example,

$$CH_2\!\!=\!\!CH\!-\!G + H_2 \xrightarrow[\substack{\text{heat,}\\\text{pressure}}]{\text{Ni}} CH_3\!-\!CH_2\!-\!G$$

where G in this reaction is $-CH_2OCH_2\underset{\underset{\text{OH}}{|}}{C}\overset{\overset{\displaystyle CH=O}{|}}{{}}CH_2CH_2\overset{O}{\overset{||}{C}}CH_3$.

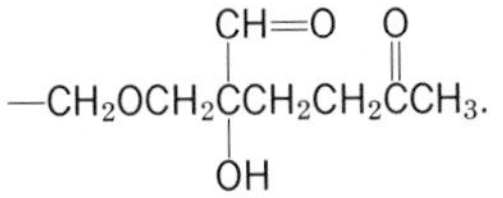

$$G'-\overset{\overset{\text{O}}{\|}}{C}H + H_2 \xrightarrow[\substack{\text{heat,}\\\text{pressure}}]{\text{Ni}} G'-CH_2OH \qquad \text{(What is G' in this reaction?)}$$

$$G''-\overset{\overset{\text{O}}{\|}}{C}CH_3 + H_2 \xrightarrow[\substack{\text{heat,}\\\text{pressure}}]{\text{Ni}} G''-\overset{\overset{\text{OH}}{|}}{C}HCH_3$$

The reagents and conditions for this exercise are

(1) $Br_2$ in $CCl_4$ solution, room temperature.

(2) Excess $CH_3OH$ in the presence of dry HCl.

(3) Tollens' reagent.

(4) Benedict's reagent (do not attempt to write an equation).

(5) Phenylhydrazine

(6) Dilute sodium hydroxide at room temperature.

7. Repeat the directions given in exercise 6 for the following hypothetical structure and the reagents given.

$$CH_3-O-\overset{\overset{\text{O}-CH_3}{|}}{C}H-CH_2-O-CH_2CH_2\overset{\overset{\text{O}}{\|}}{C}H$$

(a) Tollens' reagent (which is a basic solution).

(b) Excess water, $H^+$ catalyst.

(c) $MnO_4^-$, $OH^-$.

8. One of our goals is to be able to understand chemical events occurring among molecules of biochemical interest. Our chief purpose in studying organic chemistry has been to learn several of these reactions as they occur in simple systems. If we are to apply what we have learned, we must be able to recognize the simple organic reactions even when they take place with complicated structures.

The following reactions have been established for one or more biological systems. (Enzymes and other reagents have been omitted. Only the overall results are shown.) Classify each according to the following types.

A. Oxidation of a 2° alcohol to a ketone.

B. Oxidation of a mercaptan to a disulfide.

C. Oxidation of an aldehyde to a carboxylic acid.

D. Oxidation of a 1° alcohol to an aldehyde.

E. Dehydration of an alcohol.

F. Reduction of a ketone to a 2° alcohol.

G. Addition of water to an alkene linkage.

H. Reduction of an alkene linkage.

I. Reverse aldol condensation.

J. Hemiacetal formation.

(a) 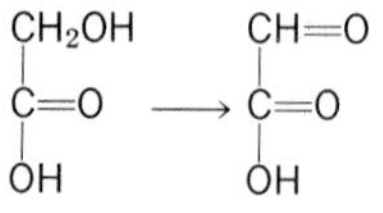
$$\begin{array}{ccc}
CH_2OH & & CH=O \\
| & & | \\
C=O & \longrightarrow & C=O \\
| & & | \\
OH & & OH \\
\text{Glycolic acid} & & \text{Glyoxylic acid}
\end{array}$$

(b) 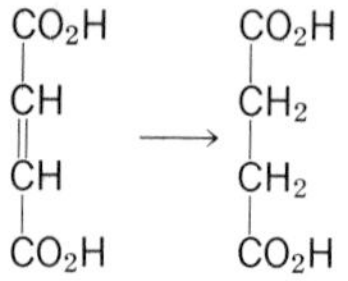
$$\begin{array}{ccc}
CO_2H & & CO_2H \\
| & & | \\
CH & & CH_2 \\
\| & \longrightarrow & | \\
CH & & CH_2 \\
| & & | \\
CO_2H & & CO_2H \\
\text{Fumaric acid} & & \text{Succinic acid}
\end{array}$$

(c)

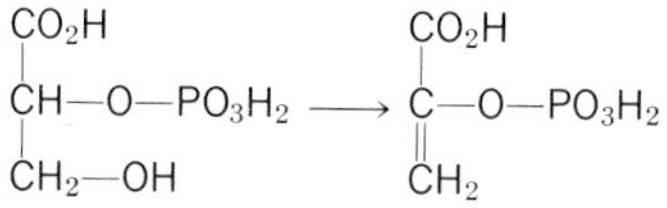

E

$$CH-O-PO_3H_2 \longrightarrow C-O-PO_3H_2$$

with $CO_2H$ above each, $CH_2-OH$ below left and $CH_2$ below right

Phosphoglyceric acid    Phosphoenolpyruvic acid

(d)

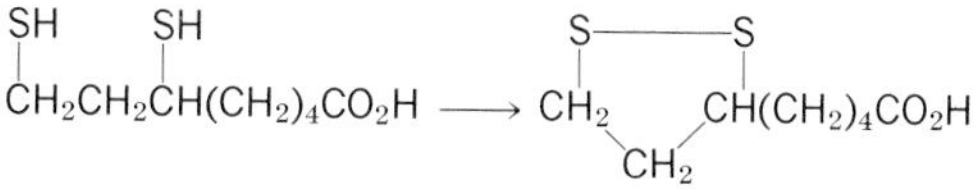

B

$$SH \quad SH$$
$$CH_2CH_2CH(CH_2)_4CO_2H \longrightarrow$$

$$S-S$$
$$CH_2 \quad CH(CH_2)_4CO_2H$$
$$CH_2$$

Dihydrolipoic acid          Lipoic acid

(e)

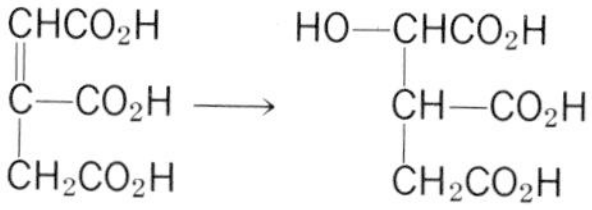

G

$$CHCO_2H \qquad HO-CHCO_2H$$
$$C-CO_2H \longrightarrow CH-CO_2H$$
$$CH_2CO_2H \qquad CH_2CO_2H$$

Aconitic acid          Isocitric acid

(f)

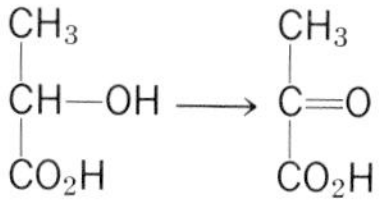

A

$$CH_3 \qquad CH_3$$
$$CH-OH \longrightarrow C=O$$
$$CO_2H \qquad CO_2H$$

Lactic acid    Pyruvic acid

(g)

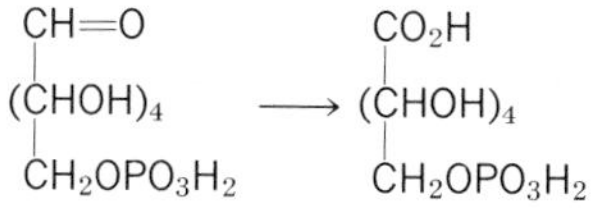

$$CH=O \qquad CO_2H$$
$$(CHOH)_4 \longrightarrow (CHOH)_4$$
$$CH_2OPO_3H_2 \qquad CH_2OPO_3H_2$$

Glucose 6-phosphate   6-Phosphogluconic acid

(h)

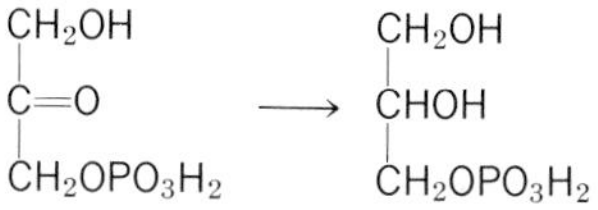

$$CH_2OH \qquad CH_2OH$$
$$C=O \longrightarrow CHOH$$
$$CH_2OPO_3H_2 \qquad CH_2OPO_3H_2$$

Dihydroxyacetone phosphate  α-Glycerophosphate

(i)

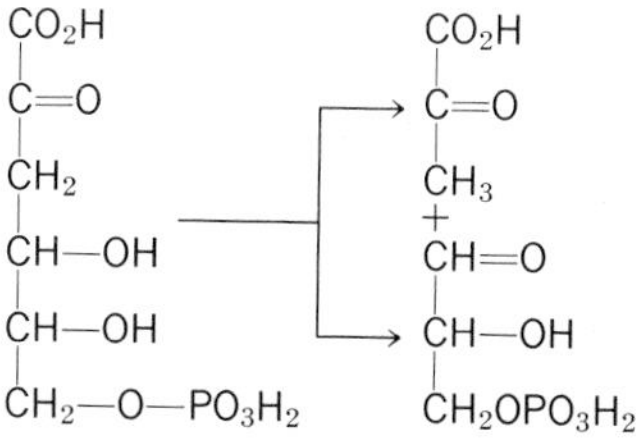

$$CO_2H$$
$$C=O$$
$$CH_2$$
$$CH-OH$$
$$CH-OH$$
$$CH_2-O-PO_3H_2$$

2-Keto-3-deoxy-6-phosphogluconic acid

$$CO_2H$$
$$C=O \qquad \text{Pyruvic acid}$$

$$CH_3$$
$$+$$
$$CH=O$$

$$CH-OH \qquad \text{Glyceraldehyde 3-phosphate}$$
$$CH_2OPO_3H_2$$

(j)

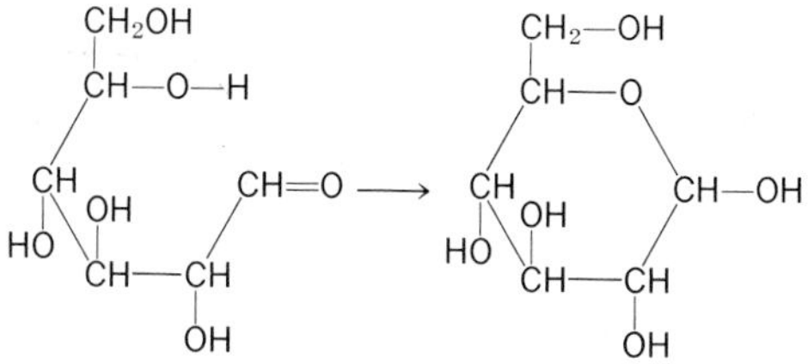

$$CH_2OH$$
$$CH-O-H$$
$$CH \quad CH=O \longrightarrow$$
$$HO \quad OH$$
$$CH-CH$$
$$OH$$

Glucose (open form)

$$CH_2-OH$$
$$CH-O$$
$$CH \quad CH-OH$$
$$HO \quad OH$$
$$CH-CH$$
$$OH$$

Glucose (closed form)

# CHAPTER SEVEN

# Carboxylic Acids
# and Related Compounds

The carboxylic acid group occurs widely among substances of biochemical and industrial importance. Three closely related types of substances, salts of carboxylic

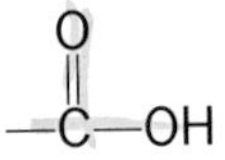

$$
\begin{array}{c}
O \\
\| \\
-C-OH
\end{array}
$$

Carboxyl group
(*carbonyl* + hydro*xyl*)

acids, esters, and amides are also important. The chemistry of esters is basic to an understanding of simple lipids, and the chemistry of amides provides many of our insights into protein chemistry.

The nomenclature of these substances was described in Chapter 6. The several tables in this chapter provide additional illustrations.

## CARBOXYLIC ACIDS

Several representative carboxylic acids are listed in Table 7.1. They are often called fatty acids because many are obtainable from fats and oils. Formic acid is a sharp-smelling, irritating liquid responsible for the sting of certain ants and the nettle. Acetic acid is the constituent of vinegar. The odor of rancid butter is produced by the presence of butyric acid. What is "strong" about valeric acid, from the Latin *valerum,* meaning "to be strong," is its odor. The same is true of caproic, caprylic, and capric acids, from the Latin *caper,* meaning "goat." Smelly goats and locker rooms telegraph their condition by molecules of these oily acids and their corresponding aldehydes. The vapor pressures of the acids with twelve carbons and more are low enough at room temperature that odors are slight.

**Table 7.1  Carboxylic Acids**

| $n$ | Structure | Name | Origin of Name | Mp (°C) | Bp (°C) | Solubility (in g/100 g water at 20°C) | $K_a$ (at 25°C) |
|---|---|---|---|---|---|---|---|
| Straight-chain saturated acids, $C_nH_{2n}O_2$ | | | | | | | |
| 1 | $HCO_2H$ | Formic acid (methanoic) | L.*formica*, ant | 8 | 101 | ∞ | $1.77 \times 10^{-4}$ (20°C) |
| 2 | $CH_3CO_2H$ | Acetic acid (ethanoic) | L.*acetum*, vinegar | 17 | 118 | ∞ | $1.76 \times 10^{-5}$ |
| 3 | $CH_3CH_2CO_2H$ | Propionic acid (propanoic) | Gr.*proto*, first *pion*, fat | −21 | 141 | ∞ | $1.34 \times 10^{-5}$ |
| 4 | $CH_3CH_2CH_2CO_2H$ | Butyric acid (butanoic) | L.*butyrum*, butter | −6 | 164 | ∞ | $1.54 \times 10^{-5}$ |
| 5 | $CH_3(CH_2)_3CO_2H$ | Valeric acid (pentanoic) | L.*valere*, to be strong (valerian root) | −35 | 186 | 4.97 | $1.52 \times 10^{-5}$ |
| 6 | $CH_3(CH_2)_4CO_2H$ | Caproic acid (hexanoic) | L.*caper*, goat | −3 | 205 | 1.08 | $1.31 \times 10^{-5}$ |
| 7 | $CH_3(CH_2)_5CO_2H$ | Enanthic acid (heptanoic) | Gr.*oenanthe*, vine blossom | −9 | 223 | 0.26 | $1.28 \times 10^{-5}$ |
| 8 | $CH_3(CH_2)_6CO_2H$ | Caprylic acid (octanoic) | L.*caper*, goat | 16 | 238 | 0.07 | $1.28 \times 10^{-5}$ |
| 9 | $CH_3(CH_2)_7CO_2H$ | Pelargonic acid (nonanoic) | pelargonium (geranium family) | 15 | 254 | 0.03 | $1.09 \times 10^{-5}$ |
| 10 | $CH_3(CH_2)_8CO_2H$ | Capric acid (decanoic) | L.*caper*, goat | 32 | 270 | 0.015 | $1.43 \times 10^{-5}$ |
| 12 | $CH_3(CH_2)_{10}CO_2H$ | Lauric acid (dodecanoic) | laurel | 44 | — | 0.006 | — |
| 14 | $CH_3(CH_2)_{12}CO_2H$ | Myristic acid (tetradecanoic) | myristica (nutmeg) | 54 | — | 0.002 | — |
| 16 | $CH_3(CH_2)_{14}CO_2H$ | Palmitic acid (hexadecanoic) | palm oil | 63 | — | 0.0007 | — |
| 18 | $CH_3(CH_2)_{16}CO_2H$ | Stearic acid (octadecanoic) | Gr.*stear*, tallow | 70 | — | 0.0003 | — |
| Miscellaneous carboxylic acids | | | | | | | |
| | $C_6H_5CO_2H$ | Benzoic acid | gum benzoin | 122 | 249 | 0.34 (25°C) | $6.46 \times 10^{-5}$ |
| | $C_6H_5CH{=}CHCO_2H$ | Cinnamic acid (*trans*) | cinnamon | 42 | — | 0.04 | $3.65 \times 10^{-5}$ |
| | $CH_2{=}CHCO_2H$ | Acrylic acid | L.*acer*, sharp | 13 | 141 | soluble | $5.6 \times 10^{-5}$ |
| | (benzene ring with $-CO_2H$ and $OH$) Salicylic acid | Salicylic acid | L.*salix*, willow | 159 | 211 | 0.22 (25°C) | $1.1 \times 10^{-3}$ (19°C) |

The carboxyl group confers considerable polarity to molecules possessing it. A carboxylic acid has a higher boiling point than an alcohol of the same formula weight,

$CH_3CH_2OH$ (f.wt. 46), bp 78°C     $CH_3CH_2CH_2OH$ (f.wt. 60), bp 97°C
$H—CO_2H$ (f.wt. 46), bp 101°C     $CH_3CO_2H$ (f.wt. 60), bp 118°C

partly because pairs of carboxylic acid molecules can be held together by two hydrogen bonds:

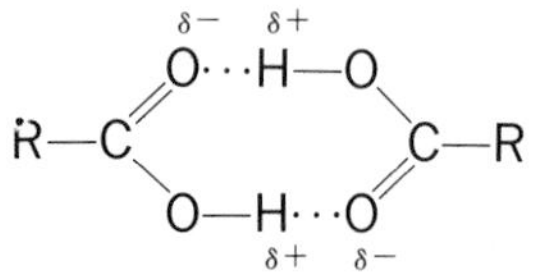

Because a carboxylic acid has these dimeric forms (Greek: *di*-, two; *meros,* part), the effective formula weight is higher than the structural formula implies, and the boiling point is thus higher than otherwise expected. Other hydrogen-bonded polymers are also present.

The first four members of the homologous series of carboxylic acids are soluble in water, for the carboxyl group can both donate and accept hydrogen bonds to and from water molecules. Even so, as the hydrocarbon chain in the acid lengthens, solubility of the acid in water falls off sharply. Since the long-chain fatty acids (those with twelve carbons and more) are normal products of the digestion of fats and oils, the bloodstream, largely aqueous in nature, contains substances that help to dissolve the relatively insoluble fatty acids and then help to transport them.

## ACIDITY OF CARBOXYLIC ACIDS

In an aqueous medium molecules of carboxylic acids interact with water molecules. Collisions that are strong enough occur with a relatively low frequency, and only a small concentration of hydronium ions is present. In general:

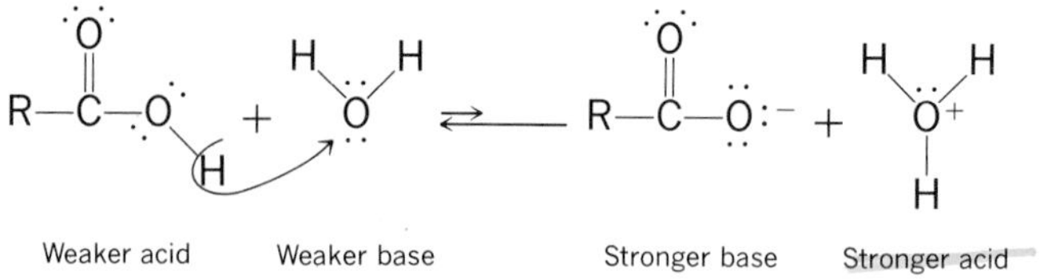

Weaker acid     Weaker base          Stronger base     Stronger acid

A 1 molar solution of acetic acid is ionized only to about 0.5% at room temperature. Evidently water molecules are not very strong proton acceptors in relation to carboxylic acids, which are relatively weak proton donors. Compared with alcohols, however, which also have the hydroxyl group,[1] carboxylic acids are stronger acids by several orders of magnitude.

[1] Here, as always, hydroxyl *group* is not to be confused with hydroxide *ion*.

**Acidity Constants.** Relative acidities of acids can be studied by comparing the acidity constants $K_a$ for their reaction with water:

$$K_a = \frac{[RCO_2^-][H_3O^+]}{[RCO_2H]}$$

*The higher the $K_a$, the stronger the acid.* In Tables 7.1 and 7.2 several such values of $K_a$, most on the order of $10^{-5}$ (0.00001), are recorded. In marked contrast, 1° alcohols have $K_a$ values on the order of $10^{-16}$. To understand this great difference in acidity for two classes possessing hydroxyl groups, we must look to structural differences.

**Structure and Acidity.** The ionization of an alcohol molecule produces an alkoxide ion. A carboxylate ion is produced from an acid.

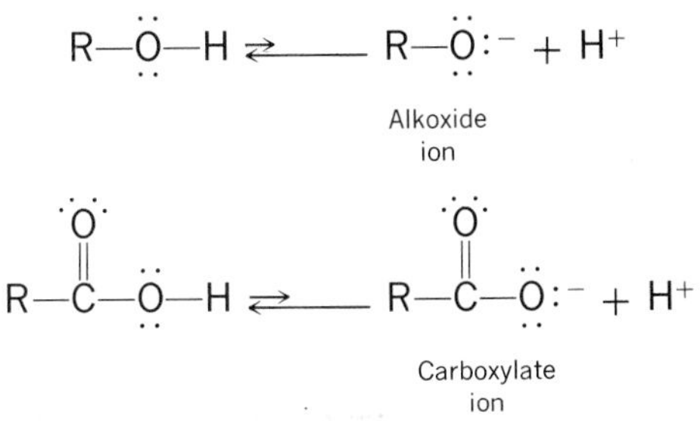

$$R—\overset{..}{\underset{..}{O}}—H \rightleftharpoons R—\overset{..}{\underset{..}{O}}:^- + H^+$$

Alkoxide
ion

$$R—\overset{\overset{\displaystyle \cdot\cdot}{\overset{\displaystyle O}{\|}}}{C}—\overset{..}{\underset{..}{O}}—H \rightleftharpoons R—\overset{\overset{\displaystyle \cdot\cdot}{\overset{\displaystyle O}{\|}}}{C}—\overset{..}{O}:^- + H^+$$

Carboxylate
ion

**Table 7.2  Structure and Acidity of Substituted Carboxylic Acids**

| Structure | $K_a$ (25°C) |
|---|---|
| $\alpha$-Haloacetic acids | |
| F—$CH_2CO_2H$ | $219 \times 10^{-5}$ |
| Cl—$CH_2CO_2H$ | $155 \times 10^{-5}$ |
| Br—$CH_2CO_2H$ | $138 \times 10^{-5}$ |
| I—$CH_2CO_2H$ | $75 \times 10^{-5}$ |
| H—$CH_2CO_2H$ | $1.8 \times 10^{-5}$ |
| | |
| Chloroacetic acids | |
| H—$CH_2CO_2H$ | $1.8 \times 10^{-5}$ |
| Cl—$CH_2CO_2H$ | $155 \times 10^{-5}$ |
| $Cl_2CHCO_2H$ | $5,100 \times 10^{-5}$ |
| $Cl_3CCO_2H$ | $90,000 \times 10^{-5}$ |
| | |
| Monochlorobutyric acids | |
| $\alpha$-chloro  $CH_3CH_2\underset{\underset{Cl}{\mid}}{C}HCO_2H$ | $139 \times 10^{-5}$ |
| $\beta$-chloro  $CH_3\underset{\underset{Cl}{\mid}}{C}HCH_2CO_2H$ | $8.9 \times 10^{-5}$ |
| $\gamma$-chloro  $\underset{\underset{Cl}{\mid}}{C}H_2CH_2CH_2CO_2H$ | $3.8 \times 10^{-5}$ |
| no chloro  $CH_3CH_2CH_2CO_2H$ | $1.54 \times 10^{-5}$ |

The classical structure for an alkoxide ion adequately represents it. Reasonable resonance structures, for example, cannot be drawn for it. For the carboxylate ion two reasonable structures—differing only in locations of electrons—can be drawn, **1** and **2**.

|     |     | (hybrid) |
| :-: | :-: | :------: |
| **1** | **2** | **3** |

According to resonance theory, whenever such a pencil and paper operation is possible, neither structure can be satisfactory for the species. Both **1** and **2** must be used to represent the *one* species present. Blending **1** and **2** into a hybrid produces structure **3**. Since **1** and **2**, although not identical, are still equivalent as far as predicted internal energies are concerned, each must make the same contribution to the hybrid. If they do so, the two carbon-oxygen bonds in **3** must be alike; both carbon-oxygen bond distances must be the same. Spectroscopic examinations do, in fact, confirm this prediction.

    The net effect is that, compared with the molecule from which it forms, the carboxylate ion is more stable than an alkoxide ion. The energy change for the ionization of an acid is more favorable than that for ionization of an alcohol, and the carboxylic acid is a stronger acid than the alcohol.

    **The Inductive Effect and Acidity.** The $K_a$ values for $\alpha$-haloacetic acids given in Table 7.2 reveal that as the halogen substituent changes from $-I$ to $-Br$ to $-Cl$ to $-F$, the acidity *increases*. This order is the same as the order of the relative electronegativities of the halogens. We have learned that one way to stabilize a

If X is a group that withdraws electrons, the anion is stabilized further; hence the acid is a stronger acid.

If X tends to release electrons, the anion is destabilized; negative charge is being forced into a region already negatively charged, making the acid weaker.

charge is to disperse it. Electron-withdrawing substituents such as halogens have this kind of inductive effect, and their presence on the $\alpha$-carbon of acetic acid strengthens the acid. By the same token, an alkyl group, being electron-releasing, weakens the acid. Thus the drop in $K_a$ value between formic acid and acetic acid is reasonable because a methyl group has an electron-releasing effect.

    As more and more electron-withdrawing substituents are placed on the $\alpha$-carbon of acetic acid, its acidity rises sharply. The $K_a$ values for mono-, di-, and trichloroacetic acids (Table 7.2) reveal this. Trichloroacetic acid is almost as strong as a mineral acid and is sometimes used in its place as an acid catalyst.

    As the substituent exerting an inductive effect becomes farther and farther removed from the negatively charged end of the carboxylate ion, its inductive influence on that group drops off markedly. The $K_a$ values for the three mono-chlorobutyric acids in Table 7.2 illustrate this.

## SALTS OF CARBOXYLIC ACIDS

The $K_a$ value of an acid is a measure of the ability of the carboxylic acid group to donate a proton to a water molecule. In this role the neutral water molecule acts as a base, a proton acceptor, and a very weak one at that. In contrast, the hydroxide ion is a far stronger base. Aqueous sodium hydroxide, for example, can quantitatively convert carboxylic acids into their salts. In general:

$$R-\overset{O}{\underset{\|}{C}}-O-H + Na^+OH^- \rightleftharpoons R-\overset{O}{\underset{\|}{C}}-O^-Na^+ + H-OH$$

| Stronger acid | Stronger base | Weaker base | Weaker acid |

Specific examples:

$$CH_3-\overset{O}{\underset{\|}{C}}-O-H + Na^+OH^- \rightleftharpoons CH_3-\overset{O}{\underset{\|}{C}}-O^-Na^+ + H-OH$$

| Acetic acid | Sodium hydroxide | Sodium acetate | Water |

$$CH_3(CH_2)_{16}\overset{O}{\underset{\|}{C}}-O-H + Na^+OH^- \rightleftharpoons CH_3(CH_2)_{16}\overset{O}{\underset{\|}{C}}-O^-Na^+ + H-OH$$

| Stearic acid (insoluble in water) | | Sodium stearate (soluble in water, a soap) | |

$$C_6H_5\overset{O}{\underset{\|}{C}}-O-H + Na^+OH^- \rightleftharpoons C_6H_5\overset{O}{\underset{\|}{C}}-O^-Na^+ + H-OH$$

| Benzoic acid (insoluble in water) | | Sodium benzoate (soluble in water) | |

The salts that form in these reactions are obtained by evaporating the water. All are crystalline solids with relatively high melting points; many decompose before they melt. They are fully ionic substances with the properties to be expected of such materials: solubility in water, insolubility in nonpolar solvents, and relatively high melting points. Table 7.3 lists a few representative examples.

Since they form from substances that are not good proton donors, carboxylate ions ($RCO_2^-$) must be good proton acceptors. In other words, they are quite good bases, especially toward a good proton donor such as hydronium ion:

$$R-\overset{O}{\underset{\|}{C}}-O^- + H-\overset{\overset{H}{+}}{O}-H \rightleftharpoons R-\overset{O}{\underset{\|}{C}}-O-H + H-OH$$

| Carboxylate ion (stronger base) | Hydronium ion (stronger acid) | Carboxylic acid (weaker acid) | Water (weaker base) |

For example,

$$C_6H_5CO_2^-Na^+ + HCl \xrightarrow{\text{water}} C_6H_5CO_2H + Na^+Cl^-$$

Sodium benzoate    Hydrochloric      Benzoic acid    Sodium chloride
(soluble in water)     acid      (insoluble in water)

The ability of a negatively charged carboxylate group to accept a proton from a hydronium ion is important in maintaining control of pH in body fluids. Proteins in the bloodstream, for example, possess such groups in their molecules, and they act to buffer the pH of the blood against any downward change. These same groups in molecules of proteins also help to bring them into solution in water. For helping to make an organic compound soluble in water, the $-CO_2^-$ group is one of the best we shall encounter. Quaternary ammonium ions are also very soluble in water, and such groups are also found in proteins.

## REACTIONS OF CARBOXYLIC ACIDS

The changes of the carboxylic acid group of principal concern in this study are of the type

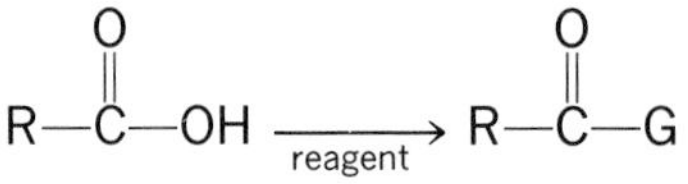

$$R-\overset{\overset{\displaystyle O}{\|}}{C}-OH \xrightarrow{\text{reagent}} R-\overset{\overset{\displaystyle O}{\|}}{C}-G$$

where G is $-Cl$, $-OR'$, $-NH_2$, $-NHR'$, or $-NHR'_2$. By action of some reagent the $-OH$ group may be replaced by one of these groups. We shall study these transformations next.

**Synthesis of Acid Chlorides.** $R-\overset{\overset{\displaystyle O}{\|}}{C}-Cl$. Action of any one of three reagents, thionyl chloride ($SOCl_2$), phosphorus trichloride ($PCl_3$), or phosphorus pentachloride ($PCl_5$), on a carboxylic acid converts it to its acid chloride. In general:

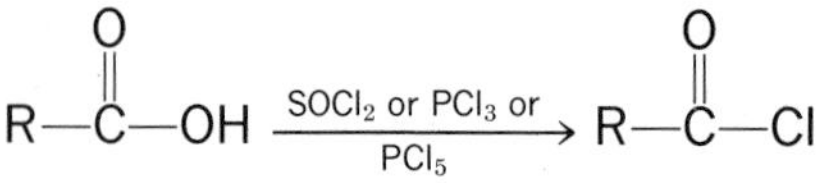

$$R-\overset{\overset{\displaystyle O}{\|}}{C}-OH \xrightarrow[\text{PCl}_5]{\text{SOCl}_2 \text{ or PCl}_3 \text{ or}} R-\overset{\overset{\displaystyle O}{\|}}{C}-Cl$$

**Table 7.3  Salts of Carboxylic Acids**

| Name | Structure | Mp (°C) | Solubility | |
|------|-----------|---------|------------|---|
| | | | Water | Ether |
| Sodium formate | $HCO_2^-Na^+$ | 253 | soluble | insoluble |
| Sodium acetate | $CH_3CO_2^-Na^+$ | 323 | soluble | insoluble |
| Sodium propionate | $CH_3CH_2CO_2^-Na^+$ | | soluble | insoluble |
| Sodium benzoate | $C_6H_5CO_2^-Na^+$ | | 66 g/100 cc | insoluble |
| Sodium salicylate | 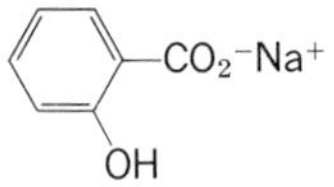 (benzene ring)—$CO_2^-Na^+$ with OH | | 111 g/100 cc | insoluble |

Specific examples:

$$CH_3\overset{O}{\overset{\|}{C}}-OH \xrightarrow{PCl_3} CH_3\overset{O}{\overset{\|}{C}}-Cl \quad (+H_3PO_3)$$

Acetic acid       Acetyl chloride (67%)

$$C_6H_5\overset{O}{\overset{\|}{C}}-OH \xrightarrow{SOCl_2} C_6H_5\overset{O}{\overset{\|}{C}}-Cl \quad (+SO_2 + HCl)$$

Benzoic acid       Benzoyl chloride (91%)

$$CH_3(CH_2)_6\overset{O}{\overset{\|}{C}}-OH \xrightarrow{PCl_5} CH_3(CH_2)_6\overset{O}{\overset{\|}{C}}-Cl \quad (+POCl_3 + HCl)$$

Caprylic acid       Caprylyl chloride (82%)

Acid chlorides are the most reactive of the acid derivatives. Although they do not occur in the chemical inventory of living organisms, they are important to our study in other ways. *In vitro* synthesis of esters and amides are frequently accomplished by first converting the less reactive acid to its acid chloride.

**Synthesis of Acid Anhydrides.** $R-\overset{O}{\overset{\|}{C}}-O-\overset{O}{\overset{\|}{C}}-R$. Although simple anhydrides of carboxylic acids do not occur *in vivo*, very close relatives, the mixed anhydrides between an organic acid and phosphoric acid, do occur in living cells. The reactions of the two types are very similar.

$$R-\overset{O}{\overset{\|}{C}}-O-\overset{O}{\overset{\|}{P}}(OH)_2$$

A mixed phosphoric anhydride

The most important aliphatic anhydride is acetic anhydride, and one of the ways it may be prepared is by the reaction of acetic acid with acetyl chloride:

$$CH_3\overset{O}{\overset{\|}{C}}-OH + Cl-\overset{O}{\overset{\|}{C}}CH_3 \xrightarrow{pyridine} CH_3\overset{O}{\overset{\|}{C}}-O-\overset{O}{\overset{\|}{C}}CH_3 + Pyridine \cdot HCl$$

This method is quite general.

**Synthesis of Esters.** $R-\overset{O}{\overset{\|}{C}}-O-R'$. Esters may be made in a variety of ways. Direct esterification is presented here. In general:

$$R-\overset{O}{\overset{\|}{C}}-OH + H-O-R' \underset{}{\overset{H^+}{\rightleftharpoons}} R-\overset{O}{\overset{\|}{C}}-O-R' + H-OH$$

Specific examples:

$$CH_3\overset{O}{\overset{\|}{C}}OH + HOCH_3 \underset{heat}{\overset{H^+}{\rightleftharpoons}} CH_3\overset{O}{\overset{\|}{C}}OCH_3 + H_2O$$

Acetic acid    Methyl alcohol       Methyl acetate

$$CH_3CH_2\overset{\overset{\displaystyle O}{\|}}{C}OH + HOCH_2CH_3 \underset{\text{heat}}{\overset{H^+}{\rightleftharpoons}} CH_3CH_2\overset{\overset{\displaystyle O}{\|}}{C}OCH_2CH_3 + H_2O$$

Propionic acid  Ethyl alcohol  Ethyl propionate

$$C_6H_5\overset{\overset{\displaystyle O}{\|}}{C}OH + CH_3OH \underset{\text{heat}}{\overset{H^+}{\rightleftharpoons}} C_6H_5\overset{\overset{\displaystyle O}{\|}}{C}OCH_3 + H_2O$$

Benzoic acid  Methyl alcohol  Methyl benzoate

When a straight-chain acid and a 1° alcohol are mixed in a 1:1 mole ratio and heated, about one-third mole percent of acid and alcohol is still present at the eventual equilibrium; about two-thirds mole percent changes to ester and water. The equilibrium may be shifted in favor of the ester by using a large excess either of alcohol (usually the cheaper starting material) or of acid or by removing the water (or ester) as it forms.

**Mechanism of Direct Esterification.** If methanol enriched in $^{18}O$ is allowed to react with benzoic acid, the $^{18}O$ appears in the ester, methyl benzoate, rather than in the water:

$$C_6H_5C\overset{\displaystyle O}{\underset{\displaystyle O\text{—H}}{}} + H\text{—}^{18}O\text{—}CH_3 \xrightarrow{H^+} C_6H_5C\overset{\displaystyle O}{\underset{\displaystyle ^{18}O\text{—}CH_3}{}} + H\text{—OH}$$

This result discloses that direct esterification involves rupture of the C—OH bond in the acid and the H—O bond in the alcohol.

A mineral acid is normally used to catalyze the reaction. Its function is to protonate the carboxyl group to increase the size of the positive charge on the carbonyl carbon. In this condition it will more readily undergo nucleophilic attack by an alcohol molecule.

$$\left[ R\overset{\delta-\;\cdot\ddot{O}\cdot}{\underset{\displaystyle \overset{\delta+}{}\;\overset{..}{O}H}{C}} + H^+ \overset{\text{protonation}}{\rightleftharpoons} R\overset{\displaystyle \ddot{O}\text{--}H}{\underset{\displaystyle \overset{+}{}\;\ddot{O}\text{--}H}{C}} \right]$$

Catalyst

partial plus, mildly reactive       full plus, much more
toward a nucleophile                 reactive toward
(i.e., an electron-rich              a nucleophile
reagent)

After this protonation of the acid, a succession of reactions takes place, all involving equilibria.

$$R\text{—}C\overset{\displaystyle \ddot{O}\cdot + H^+}{\underset{\displaystyle OH}{}} \underset{\text{protonation}}{\rightleftharpoons} \left[ R\text{—}\overset{+}{C}\overset{\displaystyle \overset{..}{O}H}{\underset{\displaystyle \overset{..}{O}H}{}} + H\overset{\displaystyle \ddot{O}\text{—}R'}{} \rightleftharpoons \right] \begin{array}{l} \text{attack by} \\ \text{nucleophile,} \\ R'\ddot{O}H \end{array}$$

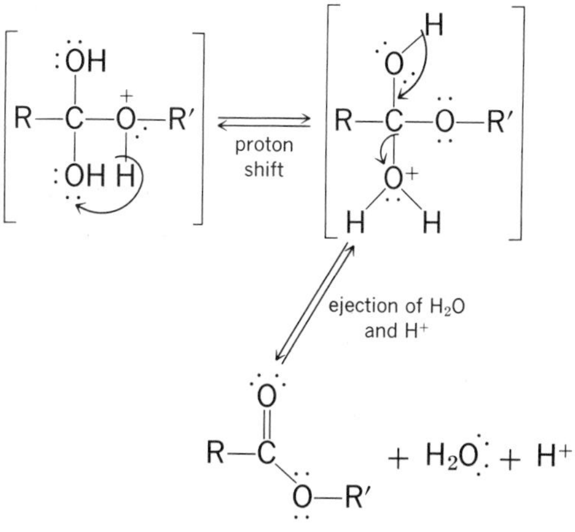

$$\left[ \begin{array}{c} \ddot{\text{O}}\text{H} \\ | \\ \text{R}-\overset{+}{\text{C}}-\overset{..}{\text{O}}-\text{R'} \\ | \\ \ddot{\text{O}}\text{H H} \end{array} \right] \underset{\substack{\text{proton} \\ \text{shift}}}{\rightleftharpoons} \left[ \begin{array}{c} \overset{\text{H}}{\overset{|}{\ddot{\text{O}}}} \\ \text{R}-\overset{|}{\text{C}}-\overset{..}{\text{O}}-\text{R'} \\ | \\ \overset{+}{\ddot{\text{O}}} \\ \overset{\diagup\diagdown}{\text{H}\quad\text{H}} \end{array} \right]$$

ejection of $H_2O$ and $H^+$

$$\text{R}-\overset{\displaystyle\overset{\ddot{\text{O}}}{\|}}{\text{C}}\diagdown\!\!\!\!_{\overset{..}{\text{O}}-\text{R'}} \quad + \; H_2\ddot{\text{O}} + H^+$$

The equilibria are driven from left to right, that is, toward ester formation, by an excess of alcohol or by removing one of the products as it forms. Conversely, an ester can be converted to its original acid and alcohol by the action of excess water in the presence of a strong acid catalyst. The mechanism for the hydrolysis of an ester is the exact reverse of esterification.

From the standpoint of predicting correctly what forms in a specific problem of esterification, the following step-by-step procedure is suggested until experience has been gained. A sample question might ask what ester forms between propionic acid and methyl alcohol.

1. Write the correct structures of the acid and the alcohol,

$$\underset{\text{Propionic acid}}{CH_3CH_2\overset{\displaystyle\overset{O}{\|}}{C}-OH} + \underset{\text{Methyl alcohol}}{H-O-CH_3}$$

2. Blacken out or erase the hydrogen on the OH of the alcohol and the hydroxyl group of the acid.

$$CH_3CH_2\overset{\displaystyle\overset{O}{\|}}{C}-\boxed{OH} + \boxed{H}-O-CH_3 \longrightarrow$$

3. Link the remaining fragments; the oxygen of the alcohol (and everything still attached to it) goes to the carbonyl carbon of the acid:

$$CH_3CH_2\overset{\displaystyle\overset{O}{\|}}{C}\underleftarrow{\qquad}-O-CH_3 \longrightarrow CH_3CH_2\overset{\displaystyle\overset{O}{\|}}{C}-O-CH_3$$

4. Having determined the correct structure of the ester, write the equation for its formation in a neat and orderly manner. The first three steps are done on scratch paper and later, with practice, mentally.

$$\text{CH}_3\text{CH}_2\text{CO}_2\text{H} + \text{CH}_3\text{OH} \xrightarrow{\text{H}^+} \text{CH}_3\text{CH}_2\text{CO}_2\text{CH}_3 + \text{H}_2\text{O}$$

Propionic acid  Methyl
alcohol  Methyl propionate

A mastery of esterification will be of great value in our future study of the chemistry of lipids, for the simple lipids are merely esters of glycerol, a trihydric alcohol.

---

Exercise 7.1    Write equations for direct esterification between the compounds in the following pairs.

(a)  Formic acid and ethyl alcohol.
(b)  Acetic acid and methyl alcohol.
(c)  Propionic acid and isobutyl alcohol.
(d)  Benzoic acid and methyl alcohol.
(e)  Butyric acid and n-butyl alcohol.
(f)  Acetic acid (three moles) and glycerol (one mole).
(g)  Salicylic acid and methyl alcohol.
(h)  Acetic acid and n-pentyl alcohol.
(i)  Valeric acid and isopropyl alcohol.
(j)  Terephthalic acid and ethyl alcohol (two moles).

*Amines : $NH_2$ attached t...*
*Amides : $NH_2$ attached to $C=$...*

---

**Synthesis of Amides.** $R-\overset{\text{O}}{\underset{\|}{\text{C}}}-NH_2$, $R-\overset{\text{O}}{\underset{\|}{\text{C}}}-NHR'$, $R-\overset{\text{O}}{\underset{\|}{\text{C}}}-NR'_2$. Under appropriate conditions ammonia, 1° amines, or 2° amines react with a carboxylic acid and water splits out.

$$R-\overset{\text{O}}{\underset{\|}{\text{C}}}-OH + H-N\diagdown \rightleftharpoons R-\overset{\text{O}}{\underset{\|}{\text{C}}}-N\diagdown + H-OH$$

Carboxylic      $NH_3, NH_2R'$,      Amide
acid            or $NHR'_2$

*In vivo,* the reaction is smoothly controlled by enzyme catalysis, as in protein synthesis (cf. p. 321). *In vitro,* to drive the reaction from left to right and obtain reasonable yields of amides, temperatures must be high, there must be an excess of one of the reactants, and/or water must be removed as it forms.

$$\text{CH}_3\text{CO}_2\text{H} + \text{NH}_3 \rightleftharpoons \text{CH}_3\text{CO}_2^-\text{NH}_4^+ \xrightarrow[\substack{\text{acetic acid,}\\110°\text{C}}]{\text{excess}} \text{CH}_3\overset{\text{O}}{\underset{\|}{\text{C}}}\text{NH}_2 + \text{H}_2\text{O}$$

Acetic acid    Ammonia    Ammonium acetate         Acetamide (90%)

This method of preparing amides is not often used in the laboratory. Usually the acid is first converted to its acid chloride, which is then allowed to react with ammonia (or 1° or 2° amine). The enhanced reactivity of the acid chloride makes up for the inconvenience of using two steps instead of one.

**Summary.** The principal general reactions of carboxylic acids are the following:

$$\xrightarrow{\text{H}_2\text{O}} \quad \text{R}-\overset{\overset{\text{O}}{\|}}{\text{C}}-\text{O}^- + \text{H}_3\text{O}^+ \qquad \text{small percentage ionization in water (weak acids)}$$

$$\xrightarrow{\text{M}^+\text{OH}^-} \quad \text{R}-\overset{\overset{\text{O}}{\|}}{\text{C}}-\text{O}^-\,\text{M}^+ + \text{H}_2\text{O} \qquad \text{formation of metallic salts (M = some metal)}$$

$$\text{R}-\overset{\overset{\text{O}}{\|}}{\text{C}}-\text{OH} \xrightarrow[\text{or PCl}_5]{\text{SOCl}_2,\text{ or PCl}_3,} \text{R}-\overset{\overset{\text{O}}{\|}}{\text{C}}-\text{Cl} \qquad \text{formation of acid chloride}$$

$$\underset{\text{H}^+}{\overset{\text{H}-\text{O}-\text{R}'}{\rightleftharpoons}} \quad \text{R}-\overset{\overset{\text{O}}{\|}}{\text{C}}-\text{O}-\text{R}' + \text{H}_2\text{O} \qquad \text{esterification}$$

$$\overset{\text{H}-\text{N}\langle,\text{ heat}}{\rightleftharpoons} \quad \text{R}-\overset{\overset{\text{O}}{\|}}{\text{C}}-\text{N}\langle + \text{H}_2\text{O} \qquad \text{formation of amide}$$

---

**Exercise 7.2**  Write the structure of the amide that would be expected to form from the following pairs of compounds. Until you have had a little practice, approach each problem in much the same way suggested for similar problems involving esters. (1) Set down the correct structures of the acid and the amine (or ammonia). (2) Erase or blacken out the —OH on the acid and one —H on the amine. (If the amine is a 3° amine, it has no —H, and we write "No amide forms.") (3) Join the remaining fragments—attach the nitrogen to the carbonyl carbon, copying over all other groups attached to each. (4) Write the equation in the conventional way.

(*a*)  Acetic acid and methylamine.
(*b*)  Butyric acid and diethylamine.
(*c*)  Benzoic acid and ammonia.
(*d*)  Propionic acid and trimethylamine.
(*e*)  Valeric acid and aniline.

---

## SYNTHESIS OF CARBOXYLIC ACIDS

The oxidation of 1° alcohols and the oxidation of aldehydes, already studied, produce acids with the same number of carbons as the alcohol or aldehyde. The following methods increase the number of carbons by one.

**The Grignard Synthesis.** In general:

$$\text{R}-\text{X} + \text{Mg} \xrightarrow[\text{step 1}]{\text{ether}} \text{R}-\text{Mg}-\text{X} \xrightarrow[\text{step 2}]{\text{CO}_2} \text{R}-\overset{\overset{\text{O}}{\|}}{\text{C}}-\text{O}^-\text{Mg}^{2+}\text{X}^- \xrightarrow[\text{step 3}]{\text{H}^+} \text{R}-\overset{\overset{\text{O}}{\|}}{\text{C}}-\text{O}-\text{H}$$

This reaction can be presented in another way:

$$R\text{—}X \xrightarrow[\text{3. H}^+ \text{ (e.g., HCl)}]{\substack{\text{1. Mg, ether} \\ \text{2. CO}_2}} R\text{—CO}_2\text{H}$$

The three steps are carried out successively in the same reaction vessel. Specific examples are

$$CH_3CH_2Br \xrightarrow[\text{3. H}^+]{\substack{\text{1. Mg, ether} \\ \text{2. CO}_2 \text{ at } -20°C}} CH_3CH_2CO_2H$$

Ethyl bromide          Propionic acid (72%)

$$(CH_3)_3CCl \xrightarrow[\text{3. H}^+]{\substack{\text{1. Mg, ether} \\ \text{2. CO}_2 \text{ at } 0°C}} (CH_3)_3CCO_2H$$

*t*-Butyl chloride      2,2-Dimethylpropanoic acid (70%)

$$C_6H_5Br \xrightarrow[\text{3. H}^+]{\substack{\text{1. Mg, ether} \\ \text{2. CO}_2 \text{ (cold)}}} C_6H_5CO_2H$$

Bromobenzene        Benzoic acid (90%)

Alkyl halides are usually obtained from corresponding alcohols (cf. 156); aryl halides are made either by direct halogenation of a benzene ring or in a Sandmeyer reaction (cf. 176).

The carbonation of a Grignard reagent is analogous to its addition to any carbonyl group:

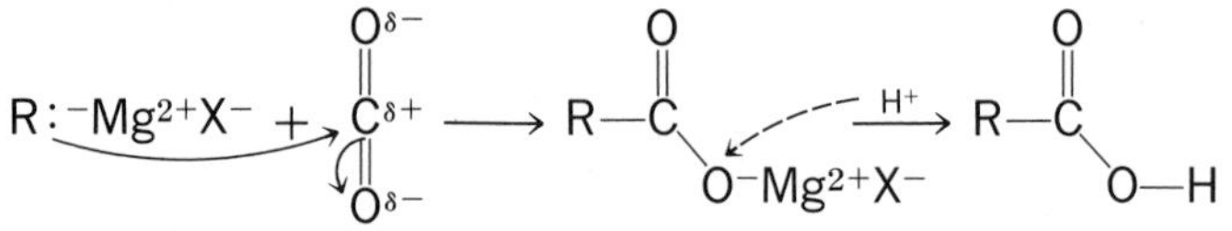

Approximation of         Mixed salt of the acid
the Grignard reagent

**Hydrolysis of Cyano Compounds (Nitriles).** In general:

$$R\text{—C}\equiv N \xrightarrow[\text{heat}]{\text{H}_2\text{O, H}^+, \text{ or OH}^-} R\text{—}\overset{\displaystyle O}{\overset{\|}{C}}\text{—OH} \; (+ \; NH_4^+)$$

A nitrile

Specific examples:

$$CH_3CH_2CH_2CH_2C\equiv N + NaOH(aq) \xrightarrow[\text{followed by H}^+]{\text{H}_2\text{O}}$$

Valeronitrile

$$CH_3CH_2CH_2CH_2\overset{\displaystyle O}{\overset{\|}{C}}OH + NH_3$$

Valeric acid (81%)

$$C_6H_5CH_2C\equiv N + H_2O \xrightarrow{\text{H}^+} C_6H_5CH_2\overset{\displaystyle O}{\overset{\|}{C}}OH + NH_4Cl$$

Phenylacetonitrile        Phenylacetic acid (78%)

The cyano compounds, usually called nitriles, may be prepared by the action of sodium cyanide on an alkyl halide (largely limited to 1° and sometimes 2° halides—3° halides undergo alkene formation too readily). The relationship between methyl cyanide, $CH_3CN$, and acetic acid and between any alkyl cyanide and the carboxylic acid of the same carbon skeleton has long been known. Because the acid is readily formed from the cyanide, the latter is usually named to indicate this relationship. Thus methyl cyanide is acetonitrile, ethyl cyanide, $CH_3CH_2CN$, is called propionitrile, and phenyl cyanide is benzonitrile.

---

Exercise 7.3    Outline the synthesis that converts the given starting material into the acid specified. Some syntheses involve more than one step.
   (a) Ethanol into acetic acid.
   (b) Ethanol into propionic acid.
   (c) t-Butyl alcohol into 2,2-dimethylpropanoic acid.
   (d) Aniline into benzoic acid.
   (e) Isopropyl chloride into isobutyric acid.

---

## DICARBOXYLIC ACIDS

Many important, naturally occurring carboxylic acids have two carboxyl groups per molecule. (Citric acid, a key intermediate in glucose catabolism, has three.) Several are listed in Table 7.4.

These compounds give the same kind of reactions we have already studied, but the proximity of two carboxyl groups is not without its special effects and must therefore be examined. Although knowledge of monofunctional systems helps us to predict the behavior of di- and polyfunctional systems to a considerable extent, some new properties do appear when two or more functional groups are rather close to each other on the molecule.

**Effect of Heat on Dicarboxylic Acids**

1. Oxalic acid:

$$HO_2C\text{—}CO_2H \xrightarrow{150°C} HCO_2H + CO_2$$

   Oxalic acid          Formic acid    Carbon
                                        dioxide

2. Malonic acid (or any $\beta$-dicarboxylic acid):

$$HO_2\overset{\beta}{C}\overset{\alpha}{CH_2}CO_2H \xrightarrow{140°C} CH_3\overset{O}{\overset{\|}{C}}OH + CO_2$$

                             Acetic acid

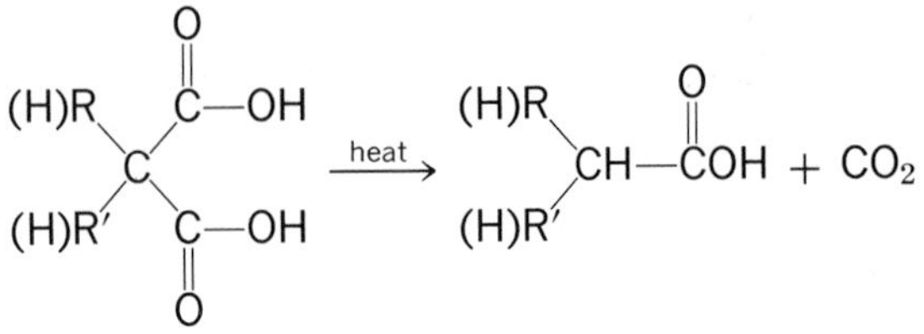

$$(H)R \overset{\displaystyle C-OH}{\underset{\displaystyle C-OH}{\overset{\displaystyle \overset{O}{\|}}{\underset{\displaystyle \underset{O}{\|}}{C}}} (H)R' \xrightarrow{\text{heat}} \begin{matrix}(H)R \\ (H)R'\end{matrix} CH\overset{O}{\overset{\|}{-}}COH + CO_2$$

Substituted
malonic acid

Substituted acetic
acid

3. Succinic acid (or any γ-dicarboxylic acid):

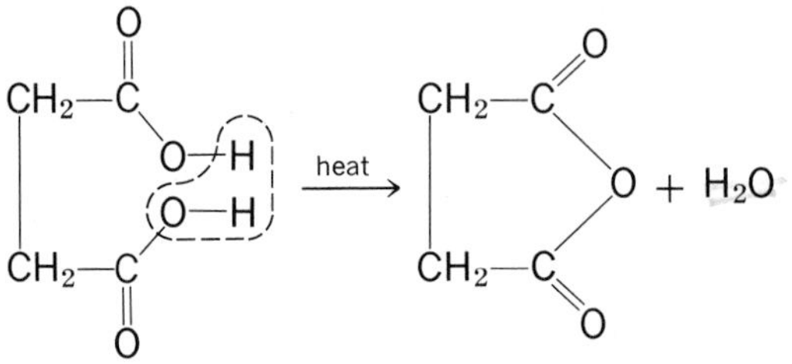

Succinic acid

Succinic anhydride

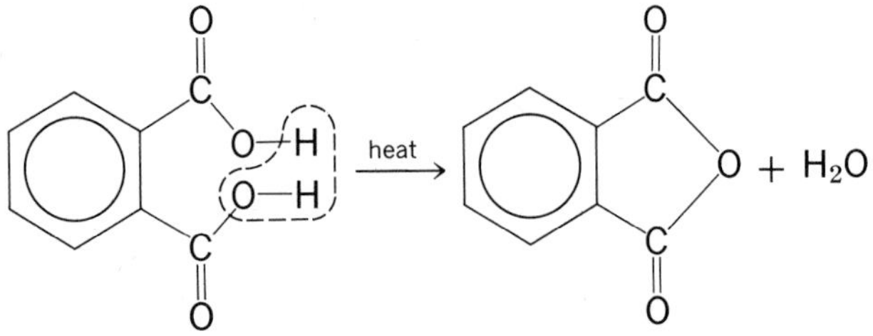

Phthalic acid

Phthalic anhydride

**IMPORTANT INDIVIDUAL ACIDS**

**Acetic Acid.** $CH_3-\overset{O}{\overset{\|}{C}}-OH$. The principal constituent (4 to 5%) of vinegar, acetic acid in its pure state melts at 16.6°C (63°F). At temperatures found in some laboratories and stock rooms, it solidifies in the bottle to an icelike solid; hence the name *glacial* acetic acid. Several methods are used in manufacturing it. In an older process, that formerly used to produce wood alcohol (see p. 134), dried hardwood is heated in a specially designed oven or retort at a temperature that varies from 160 to over 400°C. As stated earlier, this so-called destructive distillation of wood produces gases, which are burned as fuel, a liquid condensate, which consists of a water-soluble mixture and a tarry material, and charcoal. Although mostly water, the liquid contains methyl alcohol (1 to 6%), acetic acid (4 to 10%), acetone (fractional percentage), and trace amounts of other organic compounds. Lime is added to neutralize the acid. When the mixture is distilled, calcium acetate remains as the residue, and from this acetic acid is liberated.

**Table 7.4  Dicarboxylic Acids**

| $n$ | Structure | Name | Origin of Name | Mp (°C) | $K_a$ (25°C) First proton | Second proton |
|---|---|---|---|---|---|---|
| Straight-chain, saturated dicarboxylic acids, $C_nH_{2n-2}O_4$ | | | | | | |
| 2 | $HO_2CCO_2H$ | oxalic acid (ethanedioic) | L.*oxalis*, sorrel | 190 (decom-poses) | $5,900 \times 10^{-5}$ | $6.4 \times 10^{-5}$ |
| 3 | $HO_2CCH_2CO_2H$ | malonic acid (propanedioic) | L.*malum*, apple | 136 | $149 \times 10^{-5}$ | $0.20 \times 10^{-5}$ |
| 4 | $HO_2C(CH_2)_2CO_2H$ | succinic acid (butanedioic) | L.*succinum*, amber | 182 | $6.9 \times 10^{-5}$ | $0.25 \times 10^{-5}$ |
| 5 | $HO_2C(CH_2)_3CO_2H$ | glutaric acid (pentanedioic) | L.*gluten*, glue | 98 | $4.6 \times 10^{-5}$ | $0.39 \times 10^{-5}$ |
| 6 | $HO_2C(CH_2)_4CO_2H$ | adipic acid (hexanedioic) | L.*adipis*, fat | 153 | $3.7 \times 10^{-5}$ | $0.24 \times 10^{-5}$ |
| Miscellaneous dicarboxylic acids | | | | | | |
| | HCCO₂H ‖ HCCO₂H | maleic acid | L.*malum*, apple | 131° | $1,420 \times 10^{-5}$ | $0.08 \times 10^{-5}$ |
| | HO₂CH ‖ HCCO₂H | fumaric acid | fumitory (a garden annual) | sublimes above 200 | $93 \times 10^{-5}$ | $3.6 \times 10^{-5}$ |
| | (phthalic structure) | phthalic acid | na*phthalene* | 231 | $130 \times 10^{-5}$ | $0.39 \times 10^{-5}$ |
| | (isophthalic structure) | isophthalic acid | | 345 | $29 \times 10^{-5}$ | $0.25 \times 10^{-5}$ |
| | (terephthalic structure) | terephthalic acid | | sublimes | $31 \times 10^{-5}$ | $1.5 \times 10^{-5}$ (16°C) |

Another method consists of oxidizing acetaldehyde obtained in the hydration of acetylene, which in turn comes from the hydrolysis of calcium carbide. More recently acetic acid has been produced by the controlled, catalytic oxidation of butane, which is available in natural gas, an abundant, inexpensive source.

Among the many uses of acetic acid may be listed the manufacture of cellulose acetate (p. 295), Paris green (or arsenite, a mixed copper(II) acetate), and white lead. The acetate unit is also one of the most important intermediates in metabolism.

**Oxalic Acid.** HO—C—C—OH. Responsible for the tart taste of rhubarb, this acid in its aqueous solutions will remove stains caused by inks made from iron compounds.

**Citric Acid.**

$$HO-\overset{\overset{\displaystyle CH_2CO_2H}{|}}{\underset{\underset{\displaystyle CH_2CO_2H}{|}}{C}}-CO_2H$$

This acid is responsible for the tart taste of citrus fruits and acts as an intermediate in the metabolism of acetate units (citric acid cycle, p. 457). Sodium citrate is sometimes added to drawn whole blood to prevent it from clotting.

**Lactic Acid.**

$$CH_3\overset{\overset{}{\underset{\underset{\displaystyle OH}{|}}{CH}}}CO_2H$$

The tart taste of sour milk is that of lactic acid. Produced regularly in the body when glucose is metabolized in energy-demanding reactions, it accumulates faster than the body can handle it during severe exercise. The soreness of overworked muscles has been attributed to the presence of excess lactic acid.

## DERIVATIVES OF ORGANIC ACIDS

When the term derivative is used in the context "derivative of an organic acid," we mean that the compound can be prepared from the acid directly or indirectly and can be converted to the acid again, usually by the action of water. In fact, much of the chemistry of acid derivatives consists of converting one into another or into the parent acid. Even though we must deal with several types of compounds, their reactions are so similar that it would be best to note these general patterns first.

For the remainder of this chapter the most noteworthy aspect of acids and their derivatives will be the polarity of the carbonyl group. The carbonyl carbon bears a partial positive charge and is therefore vulnerable to attack by a reagent that is electron-rich. The common theme running throughout the chemistry of these substances is a substitution illustrated by the following:

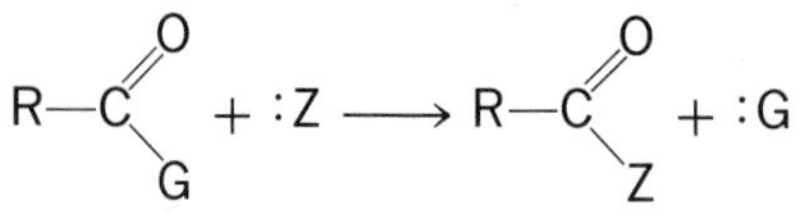

$$R-C\overset{\displaystyle O}{\underset{\displaystyle G}{<}} \; + \; :Z \longrightarrow R-C\overset{\displaystyle O}{\underset{\displaystyle Z}{<}} \; + \; :G$$

where G = —OH (as in carboxylic acids)

—Cl (as in acid chlorides)

$$-O-\overset{\overset{\displaystyle O}{\|}}{C}-R \text{ (as in an acid anhydride)}$$

—O—R' (as in an ester)

—$NH_2$ (as in amides)

—NHR (as in amides)

—$NR_2$ (as in amides)

:Z = some nucleophile which may be a neutral species or negatively charged (e.g., $OH^-$ or $H_2O$, $^-OR$ or HOR, $NH_3$, etc.)

These substitution reactions at the unsaturated carbon of the carbonyl group usually proceed much more readily than analogous substitutions at a saturated carbon (as in an alkyl halide or an alcohol or an ether or an amine).

One reason for the enhanced reactivity of acid derivatives over alkyl halides and alcohols is the fact that the carbonyl carbon goes from an essentially trivalent state to a temporarily tetravalent state as the intermediate forms:

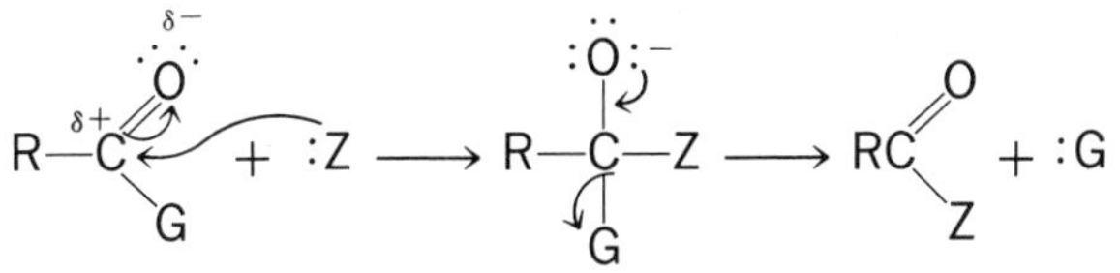

| Carbonyl carbon is trivalent in the sense that it is holding only three groups. | The former carbonyl carbon is tetravalent. | |

In contrast, substitution at a saturated carbon requires, in the concerted mechanism, a shift from a tetravalent carbon to a carbon that is temporarily pentavalent (is holding five groups):

$$G—R + :Z \longrightarrow \left[G\cdots R \cdots Z\right] \longrightarrow G: + R—Z$$

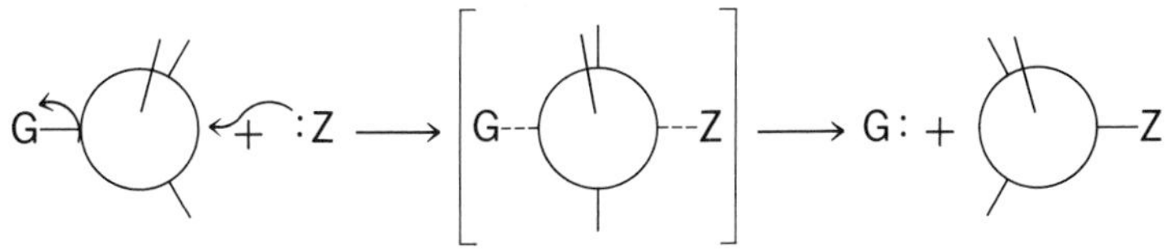

| The carbon holding —G is tetravalent. | In the transition state the carbon is holding five groups. | |

The severe crowding of groups around the temporarily pentasubstituted carbon gives this transition state a higher internal energy than that of a transition state from the carbonyl compound involving a temporarily tetrasubstituted carbon. Substitution at the unsaturated carbon of the carbonyl group therefore usually proceeds with a lower energy of activation, resulting of course in a higher frequency of successful collisions and a more rapid rate.

Another factor making acid derivatives R—C(=O)—G more reactive than their counterparts in saturated systems, R—G, is the fact that in acid derivatives, two groups are acting to exert electron withdrawal from the carbonyl carbon. In R—G, only the group, G, is acting in this way. Hence the relative size of the partial positive charge on a carbonyl carbon may be expected to be greater than that on the carbon in R—G which is directly holding —G. In a sense, then, the carbon to be attacked in acid derivatives is more attractive to the nucleophile than the carbon in saturated systems, R—G.

We can make the following general comparisons between the reactivities of acid derivatives toward nucleophilic substitution and those of their analogs without carbonyl groups.

$$R-\overset{\overset{\displaystyle O}{\|}}{C}-Cl \; > \; R-Cl$$

—Acid chlorides are more reactive than alkyl chlorides.

$$R-\overset{\overset{\displaystyle O}{\|}}{C}-O-R' \; > \; R-O-R'$$

—Esters are more reactive than ethers.

$$R-\overset{\overset{\displaystyle O}{\|}}{C}-NH_2 \; > \; R-NH_2$$

—Amides are more reactive than amines.

Within the series of acid derivatives, relative reactivities toward a common nucleophile (e.g., water) are generally in the following order:

$$\text{Acid chlorides} \gtrsim \text{acid anhydrides} > \text{esters} > \text{amides}$$

Most reactive                                        Least reactive

—————————— decreasing reactivity —————————⟶

Some of the reactions of acid derivatives are catalyzed by mineral acids which render the carbonyl carbon even more positively charged:

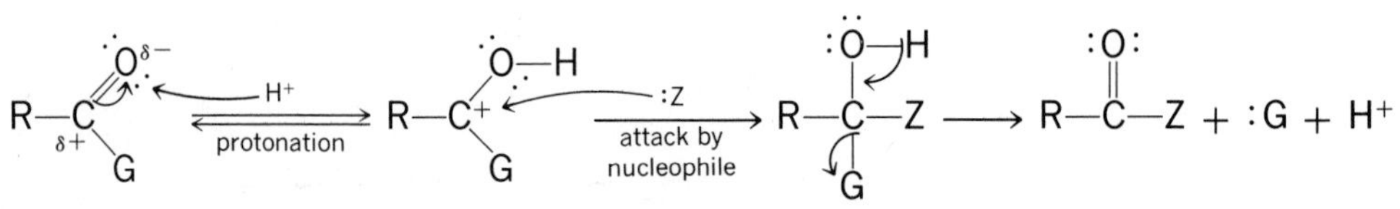

## ACID CHLORIDES AND ANHYDRIDES

Some examples of these substances are given in Table 7.5. Since the reactions of these two types of derivatives are so similar, they will be considered together. The principal reactions are outlined in the following equations.

**Hydrolysis. Conversion to Original Acid**

$$R-\overset{\overset{\displaystyle O}{\|}}{C}-Cl + H_2O \longrightarrow R-\overset{\overset{\displaystyle O}{\|}}{C}-OH + H-Cl$$

Acid chloride

$$R-\overset{\overset{\displaystyle O}{\|}}{C}-O-\overset{\overset{\displaystyle O}{\|}}{C}-R + H_2O \longrightarrow R-\overset{\overset{\displaystyle O}{\|}}{C}-OH + HO-\overset{\overset{\displaystyle O}{\|}}{C}-R$$

Anhydride

**Table 7.5  Acid Chlorides and Anhydrides**

| Name | Structure | Bp (°C) |
|------|-----------|---------|
| (Formyl chloride) | (H—CO—Cl) unknown substance | |
| Acetyl chloride | $CH_3COCl$ | 51 |
| Propionyl chloride | $CH_3CH_2COCl$ | 80 |
| Benzoyl chloride | $C_6H_5COCl$ | 197 |
| Acetic anhydride | $CH_3\overset{O}{\overset{\|}{C}}-O-\overset{O}{\overset{\|}{C}}CH_3$ | 136 |
| Benzoic anhydride | $C_6H_5\overset{O}{\overset{\|}{C}}-O-\overset{O}{\overset{\|}{C}}C_6H_5$ | 42 (mp) |
| Phthalic anhydride | 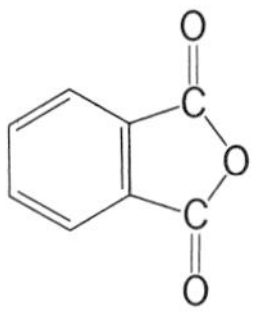 | 131 (mp) |

## Alcoholysis. Ester Formation

$$R-\overset{O}{\overset{\|}{C}}-Cl + H-O-R' \longrightarrow R-\overset{O}{\overset{\|}{C}}-O-R' + H-Cl$$

$$R-\overset{O}{\overset{\|}{C}}-O-\overset{O}{\overset{\|}{C}}-R + H-O-R' \longrightarrow R-\overset{O}{\overset{\|}{C}}-O-R' + H-O-\overset{O}{\overset{\|}{C}}-R$$

## Ammonolysis. Amide Formation

$$R-\overset{O}{\overset{\|}{C}}-Cl + H-N\diagdown \text{ (excess)} \longrightarrow R-\overset{O}{\overset{\|}{C}}-N\diagdown + HCl$$

(as its salt with the excess ammonia or amine)

$H-NH_2$, $H-NHR'$, or $H-NR'_2$

$R-\overset{O}{\overset{\|}{C}}-NH_2$,

$R-\overset{O}{\overset{\|}{C}}-NHR'$, or

$R-\overset{O}{\overset{\|}{C}}-NR'_2$

$$R-\overset{O}{\overset{\|}{C}}-O-\overset{O}{\overset{\|}{C}}-R + H-N\diagdown \text{ (excess)} \longrightarrow$$

$$R-\overset{O}{\overset{\|}{C}}-N\diagdown + HO\overset{O}{\overset{\|}{C}}R$$

(as a salt; see previous reaction)

Exercise 7.4  Complete the following equations by writing the structures of the products that would be expected to form in each. If no reaction is to be expected, write "No reaction."

(a)  $CH_3\overset{\displaystyle O}{\overset{\|}{C}}-Cl + H_2O \longrightarrow$

(b)  $C_6H_5\overset{\displaystyle O}{\overset{\|}{C}}-Cl + CH_3OH \longrightarrow$

(c)  $CH_3\overset{\displaystyle O}{\overset{\|}{C}}-O-\overset{\displaystyle O}{\overset{\|}{C}}CH_3 + CH_3CH_2OH \longrightarrow$

(d)  $CH_3CH_2\overset{\displaystyle O}{\overset{\|}{C}}Cl + NH_3$ (excess) $\longrightarrow$

(e)  $C_6H_5CH_2\overset{\displaystyle O}{\overset{\|}{C}}-Cl + (CH_3)_2NH$ (excess) $\longrightarrow$

(f)  $CH_3\overset{\displaystyle O}{\overset{\|}{C}}-Cl + (CH_3)_3N \longrightarrow$

Exercise 7.5  Write a mechanism for the reaction of water with an acid chloride, for the reaction of an alcohol with an anhydride, and for the reaction of ammonia with an acid chloride.

The reactions of acid chlorides and anhydrides are quite exothermic, and those shown occur readily. The counterpart of anhydride chemistry of particular importance in biochemistry involves derivatives of phosphoric acid.

Phosphoric acid, $H_3PO_4$, is commonly represented by the following structure:

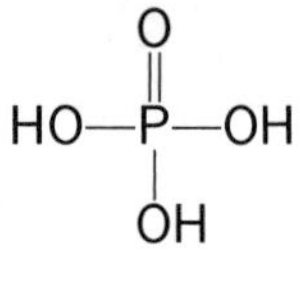

$$HO-\overset{\displaystyle O}{\overset{\|}{\underset{\underset{\displaystyle OH}{|}}{P}}}-OH$$

If a molecule of water is split out between two such molecules, a substance known as pyrophosphoric acid, $H_4P_2O_7$, is the result.

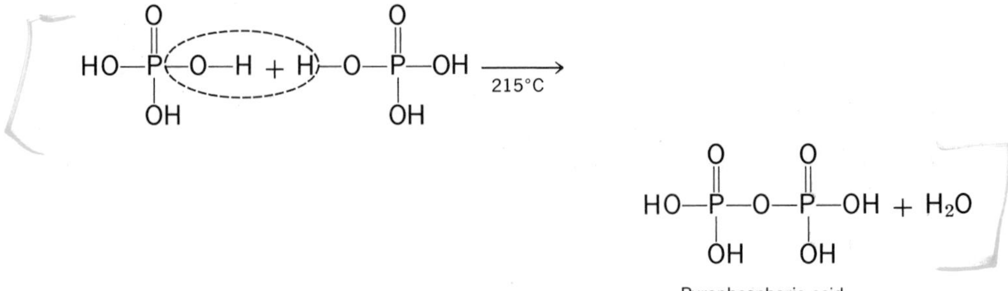

$$HO-\overset{\displaystyle O}{\overset{\|}{\underset{\underset{\displaystyle OH}{|}}{P}}}(O-H + H)O-\overset{\displaystyle O}{\overset{\|}{\underset{\underset{\displaystyle OH}{|}}{P}}}-OH \xrightarrow{215°C}$$

$$HO-\overset{\displaystyle O}{\overset{\|}{\underset{\underset{\displaystyle OH}{|}}{P}}}-O-\overset{\displaystyle O}{\overset{\|}{\underset{\underset{\displaystyle OH}{|}}{P}}}-OH + H_2O$$

Pyrophosphoric acid
(mp 61°C)

This is an acid in which all four hydrogens are replaceable and the successive ionization constants for ionization of the first, second, third, and fourth hydrogens are $K_1 = 1.4 \times 10^{-1}$, $K_2 = 1.1 \times 10^{-2}$, $K_3 = 2.9 \times 10^{-7}$, and $K_4 = 3.6 \times 10^{-9}$. In a sense pyrophosphoric acid is also very much like an anhydride.

$$ -\overset{\displaystyle O}{\overset{\|}{C}}-O-\overset{\displaystyle O}{\overset{\|}{C}}- \qquad -\overset{\displaystyle O}{\overset{\|}{\underset{|}{P}}}-O-\overset{\displaystyle O}{\overset{\|}{\underset{|}{P}}}- $$

Skeleton of a
carboxylic acid
anhydride

Skeleton of a
phosphoric acid
anhydride

The system occurs widely in living cells in the form of compounds of the following general type. They are shown in the acid forms, but usually they are present as singly or doubly ionized particles.

$$ RO-\overset{\displaystyle O}{\overset{\|}{\underset{\underset{\displaystyle OH}{|}}{P}}}-O-\overset{\displaystyle O}{\overset{\|}{\underset{\underset{\displaystyle OH}{|}}{P}}}-OH \qquad \text{for example:} \quad (CH_3)_2C{=}CHCH_2O-\overset{\displaystyle O}{\overset{\|}{\underset{\underset{\displaystyle OH}{|}}{P}}}-O-\overset{\displaystyle O}{\overset{\|}{\underset{\underset{\displaystyle OH}{|}}{P}}}-OH $$

Monoester of pyrophosphoric acid
(a diphosphate)[2]

Isopentenyl pyrophosphate

The R— group is not necessarily a purely hydrocarbon system; in fact, it usually is not. Pyrophosphate esters are simultaneously esters, acids, and anhydrides. As anhydrides they participate in typical reactions of these compounds, for example, hydrolysis:

$$ RO-\overset{\displaystyle O}{\overset{\|}{\underset{\underset{\displaystyle OH}{|}}{P}}}-O-\overset{\displaystyle O}{\overset{\|}{\underset{\underset{\displaystyle OH}{|}}{P}}}-OH + H_2O \longrightarrow RO-\overset{\displaystyle O}{\overset{\|}{\underset{\underset{\displaystyle OH}{|}}{P}}}-OH + HO-\overset{\displaystyle O}{\overset{\|}{\underset{\underset{\displaystyle OH}{|}}{P}}}-OH $$

Diphosphate
ester

Monophosphate
ester

Phosphoric acid

Triphosphate derivatives are also known and are essentially double anhydrides, plus being esters as well as acids. Much more will be studied about these substances, particularly adenosine triphosphate or ATP, in later chapters.

$$ RO-\overset{\displaystyle O}{\overset{\|}{\underset{\underset{\displaystyle OH}{|}}{P}}}-O-\overset{\displaystyle O}{\overset{\|}{\underset{\underset{\displaystyle OH}{|}}{P}}}-O-\overset{\displaystyle O}{\overset{\|}{\underset{\underset{\displaystyle OH}{|}}{P}}}-OH $$

A triphosphate
ester

[2] Organic chemists and biochemists frequently refer to these compounds as diphosphates. Note carefully that in this and similar contexts the term does *not* mean two separate $PO_4{}^{3-}$ groups.

## ESTERS

**Physical Properties.** Since molecules of an ester cannot be hydrogen bond donors, they behave as less polar compounds than carboxylic acids with respect to solubilities and boiling points. The ethyl esters of the fatty acids shown in Table 7.6 all boil at lower temperatures and have lower solubilities in water than the parent acids. Some lower-formula-weight esters have very fragrant odors (Table 7.7), in sharp contrast with those of the parent acids.

**Chemical Properties.** Esters undergo reactions of hydrolysis, alcoholysis, and ammonolysis, all of which involve breaking the carbonyl-to-oxygen bond.

**Hydrolysis.** In general:

$$R-\overset{\overset{\displaystyle O}{\|}}{C}-O-R' + H-OH \quad (\text{excess}) \xrightarrow{H^+} R-\overset{\overset{\displaystyle O}{\|}}{C}-O-H + H-O-R'$$

Specific examples:

$$CH_3\overset{\overset{\displaystyle O}{\|}}{C}-OCH_2CH_3 + H_2O \xrightarrow{H^+} CH_3\overset{\overset{\displaystyle O}{\|}}{C}OH + CH_3CH_2OH$$

Ethyl acetate        Acetic acid    Ethyl alcohol

$$C_6H_5\overset{\overset{\displaystyle O}{\|}}{C}-OCH_3 + H_2O \xrightarrow{H^+} C_6H_5\overset{\overset{\displaystyle O}{\|}}{C}OH + CH_3OH$$

Methyl benzoate       Benzoic acid    Methyl alcohol

**Alcoholysis, transesterification or ester interchange.** In general:

$$R-\overset{\overset{\displaystyle O}{\|}}{C}-O-R' + R''OH \xrightarrow{H^+} R-\overset{\overset{\displaystyle O}{\|}}{C}-O-R'' + H-O-R'$$

         (Large excess)      New ester

Specific examples:

$$CH_2{=}CH\overset{\overset{\displaystyle O}{\|}}{C}OCH_3 + HOCH_2CH_3 \xrightarrow{H^+} CH_2{=}CH\overset{\overset{\displaystyle O}{\|}}{C}OCH_2CH_3 + CH_3OH$$

Methyl acrylate     Ethyl alcohol        Ethyl acrylate (99%)
             (large excess)

$$O{=}C\underset{O-CH_2CH_3}{\overset{O-CH_2CH_3}{\big<}} + \underset{HO-CH_2}{\overset{HO-CH_2}{\big|}} \xrightarrow{K_2CO_3} O{=}C\underset{O-CH_2}{\overset{O-CH_2}{\big<}}\big| + 2HOCH_2CH_3$$

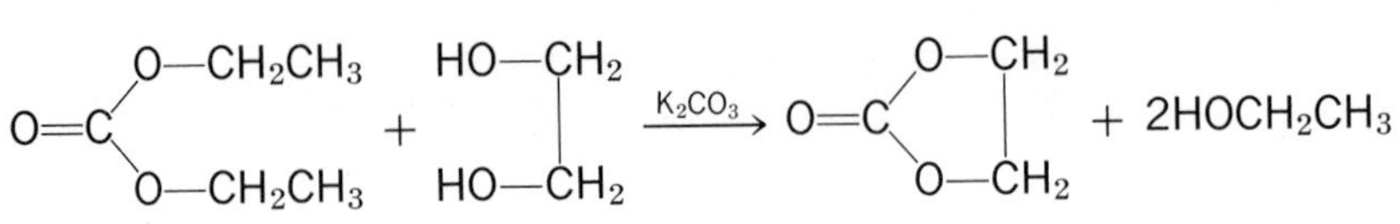

Diethyl carbonate      Ethylene     Ethylene carbonate (55%)
              glycol

**Table 7.6  Esters of Carboxylic Acids**

| Name | Structure | Mp (°C) | Bp (°C) | Solubility (in g/100 g water at 20°C) |
|---|---|---|---|---|
| Ethyl esters of straight-chain carboxylic acids, $RCO_2C_2H_5$ | | | | |
| Ethyl formate | $HCO_2C_2H_5$ | $-79$ | 54 | miscible in all proportions |
| Ethyl acetate | $CH_3CO_2C_2H_5$ | $-82$ | 77 | 7.39 (25°C) |
| Ethyl propionate | $CH_3CH_2CO_2C_2H_5$ | $-73$ | 99 | 1.75 |
| Ethyl butyrate | $CH_3(CH_2)_2CO_2C_2H_5$ | $-93$ | 120 | 0.51 |
| Ethyl valerate | $CH_3(CH_2)_3CO_2C_2H_5$ | $-91$ | 145 | 0.22 |
| Ethyl caproate | $CH_3(CH_2)_4CO_2C_2H_5$ | $-68$ | 168 | 0.063 |
| Ethyl enanthate | $CH_3(CH_2)_5CO_2C_2H_5$ | $-66$ | 189 | 0.030 |
| Ethyl caprylate | $CH_3(CH_2)_6CO_2C_2H_5$ | $-43$ | 208 | 0.007 |
| Ethyl pelargonate | $CH_3(CH_2)_7CO_2C_2H_5$ | $-45$ | 222 | 0.003 |
| Ethyl caproate | $CH_3(CH_2)_8CO_2C_2H_5$ | $-20$ | 245 | 0.0015 |
| Esters of acetic acid, $CH_3CO_2R$ | | | | |
| Methyl acetate | $CH_3CO_2CH_3$ | $-99$ | 57 | 24.4 |
| Ethyl acetate | $CH_3CO_2CH_2CH_3$ | $-82$ | 77 | 7.39 (25°C) |
| $n$-Propyl acetate | $CH_3CO_2CH_2CH_2CH_3$ | $-93$ | 102 | 1.89 |
| $n$-Butyl acetate | $CH_3CO_2CH_2CH_2CH_2CH_3$ | $-78$ | 125 | 1.0 (22°C) |
| Miscellaneous esters | | | | |
| Methyl acrylate | $CH_2{=}CHCO_2CH_3$ | | 80 | 5.2 |
| Methyl benzoate | $C_6H_5CO_2CH_3$ | $-12$ | 199 | insoluble |
| Methyl salicylate | 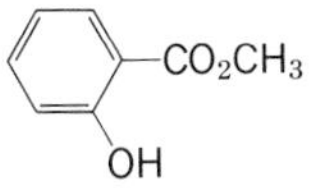 —$CO_2CH_3$, OH | $-9$ | 223 | insoluble |
| Acetylsalicylic acid | 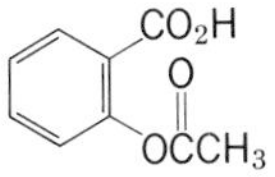 $CO_2H$, O, $OCCH_3$ | 135 | | |
| "Waxes" | $CH_3(CH_2)_nCO_2(CH_2)_nCH_3$ | $n = 23{-}33$: carnauba wax $= 25{-}27$: beeswax $= 14{-}15$: spermaceti | | |

This method is chosen only when circumstances make the more usual synthesis of an ester impractical. Acrylic acid, for example, polymerizes quite readily. Methyl acrylate (see the example just given) is not quite as susceptible to polymerization, and other esters of acrylic acid (e.g., ethyl acrylate in the example) are easily made by alcoholysis of methyl acrylate. In like manner, cyclic esters of carbonic acid are best made from diethyl carbonate, and by distilling the ethyl alcohol as it forms the reaction tends to be forced to completion. This general type of reaction is usually called transesterification to indicate transferral of an ester group from one alcohol portion to another.

**Table 7.7  Fragrances or Flavors Associated with Some Esters**

| Name | Structure | Source or Flavor |
|---|---|---|
| Ethyl formate | $HCO_2CH_2CH_5$ | rum |
| Isobutyl formate | $HCO_2CH_2CH(CH_3)_2$ | raspberries |
| n-Pentyl acetate (n-amyl acetate) | $CH_3CO_2CH_2CH_2CH_2CH_2CH_3$ | bananas |
| Isopentyl acetate (isoamyl acetate) | $CH_3CO_2CH_2CH_2CH(CH_3)_2$ | pears |
| n-Octyl acetate | $CH_3CO_2(CH_2)_7CH_3$ | oranges |
| Ethyl butyrate | $CH_3CH_2CH_2CO_2CH_2CH_5$ | pineapples |
| n-Pentyl butyrate | $CH_3CH_2CH_2CO_2(CH_2)_4CH_3$ | apricots |
| Methyl salicylate | 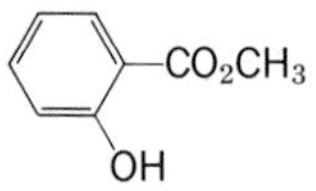 | oil of wintergreen |

**Ammonolysis.** In general:

Specific examples:

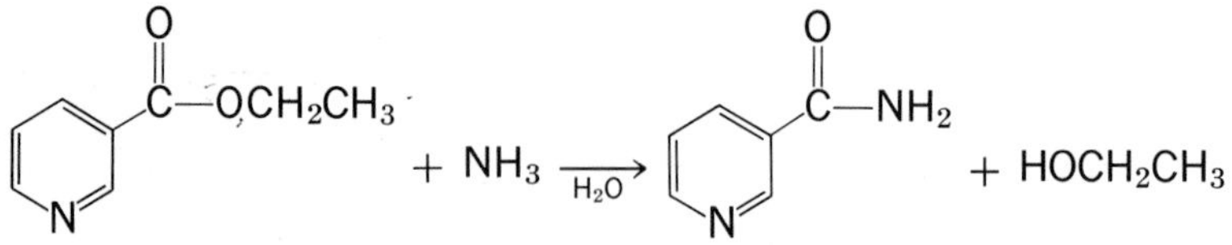

Ethyl nicotinate        Nicotinamide (78%)

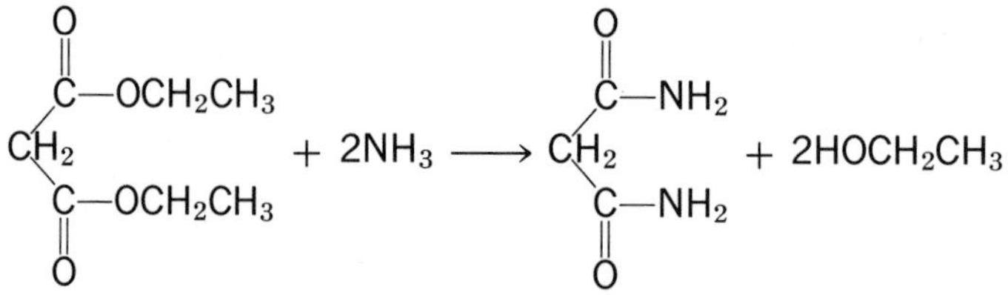

Ethyl malonate          Malondiamide (99%)

Amides are normally prepared by the action of ammonia (or an amine) on an acid chloride or anhydride. Some of these derivatives, however, are not easily made. Ammonolysis of an ester is therefore a useful alternative.

**Saponification.** This important reaction of esters is a slight variation of ester hydrolysis. Ester hydrolysis occurs in the presence of an acid catalyst (or an enzyme); saponification in the presence of aqueous sodium or potassium hydroxide. Essentially the same products form, except that saponification produces not the free acid but its salt. In both hydrolysis and saponification the alcohol portion is liberated as the free alcohol. In general:

$$R-\overset{\overset{\text{O}}{\|}}{C}-O-R' + NaOH \xrightarrow[\text{heat}]{H_2O} R-\overset{\overset{\text{O}}{\|}}{C}-O^-Na^+ + HOR'$$

Specific examples (compare these with the examples of ester hydrolysis):

$$CH_3\overset{\overset{\text{O}}{\|}}{C}OCH_2CH_3 + NaOH \xrightarrow[\text{heat}]{H_2O} CH_3\overset{\overset{\text{O}}{\|}}{C}O^-Na^+ + CH_3CH_2OH$$

Ethyl acetate                    Sodium acetate        Ethyl alcohol

$$C_6H_5\overset{\overset{\text{O}}{\|}}{C}OCH_3 + NaOH \xrightarrow[\text{heat}]{H_2O} C_6H_5\overset{\overset{\text{O}}{\|}}{C}O^-Na^+ + CH_3OH$$

Methyl benzoate                    Sodium benzoate        Methyl alcohol

The term *saponification* comes from the Latin *sapo* and *onis*, "soap" and "-fy"—that is, to make soap. Ordinary soap is a mixture of sodium salts of long-chain carboxylic acids.

**Claisen Ester Condensation.** Esters can also be made to react with themselves, undergoing a reaction very similar to the aldol condensation. In esters as in aldehydes and ketones, the carbonyl carbon is not the only reactive site, for the $\alpha$-position is also reactive. If a hydrogen is attached there, a strong proton acceptor may take it; for example,

$$\overset{\beta}{CH_3}-\overset{\alpha}{CH}-\overset{\overset{\text{O}}{\|}}{C}-O-CH_2CH_3 + {}^-OCH_2CH_3 \rightleftharpoons$$

Ethyl propionate      Ethoxide ion

$$\left[ \begin{array}{c} CH_3-CH-\overset{\overset{\ddot{\text{O}}}{\|}}{C}-O-CH_2CH_3 \\ \updownarrow \\ CH_3-CH=C\overset{\ddot{\text{O}}{}^-}{\underset{OCH_2CH_3}{}} \end{array} \right] + HOCH_2CH_3$$

Anion of ethyl propionate

Resonance stabilization of the anion helps to explain its formation.[3] Even so, it is highly reactive. It can stabilize itself by abstracting a proton from a molecule of ethyl alcohol, regenerating the starting materials. (Hence the equilibrium arrows.) However, the anion of the ester is a new electron-rich species, a nucleophile. What other event might we *expect* to occur? If this anion encounters a molecule of

[3] The reader may ask why the ethoxide ion does not attack the partially positively charged carbonyl carbon. Very likely it does, but no *new* product results. If such an attack causes ejection of the ethoxy group on the ester, the products are identical with the reactants. Such events undoubtedly occur but go nowhere.

unchanged ester, it could easily be attracted to the relatively electron-poor carbonyl carbon:

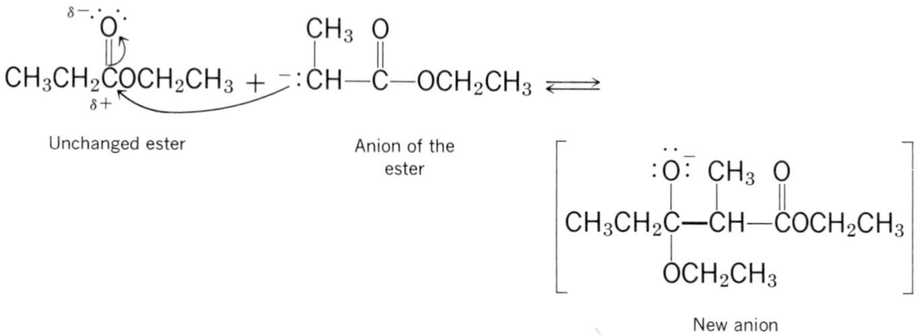

$$CH_3CH_2\overset{\overset{\delta-\,\ddot{\ddot{O}}}{\|}}{\underset{\delta+}{C}}OCH_2CH_3 + {}^-:CH\overset{CH_3}{\underset{}{|}}\overset{O}{\overset{\|}{C}}-OCH_2CH_3 \rightleftharpoons$$

Unchanged ester    Anion of the ester

$$\left[ CH_3CH_2\overset{:\ddot{O}:^-}{\underset{}{C}}-\overset{CH_3}{\underset{OCH_2CH_3}{CH}}-\overset{O}{\overset{\|}{C}}OCH_2CH_3 \right]$$

New anion

The new anion might then stabilize itself by ejecting the ethoxide ion:

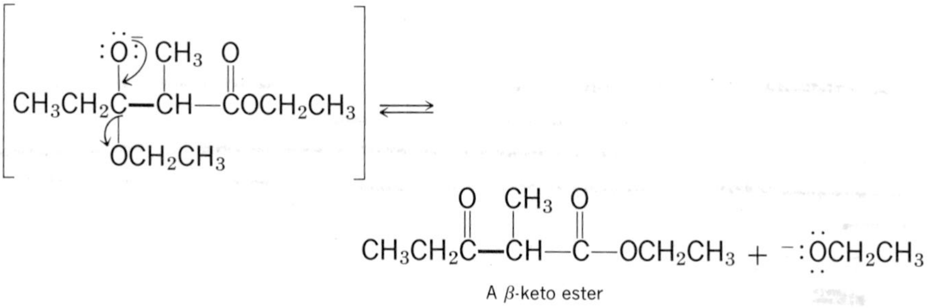

$$\left[ CH_3CH_2\overset{:\ddot{O}:^=}{\underset{OCH_2CH_3}{C}}-\overset{CH_3}{\underset{}{CH}}-\overset{O}{\overset{\|}{C}}OCH_2CH_3 \right] \rightleftharpoons$$

$$CH_3CH_2\overset{O}{\overset{\|}{C}}-\overset{CH_3}{\underset{}{CH}}-\overset{O}{\overset{\|}{C}}-OCH_2CH_3 + {}^-:\ddot{O}CH_2CH_3$$

A β-keto ester

The new carbon-carbon bond is shown in boldface. The product is a **β-keto ester.** Such a molecule has a carbon flanked on *two* sides by carbonyl groups, and the hydrogen attached to it should therefore be more acidic than an α-hydrogen in a monoester. Since the ethoxide ion, a very strong base, is still present, we must *expect* it to stabilize itself by taking that proton:

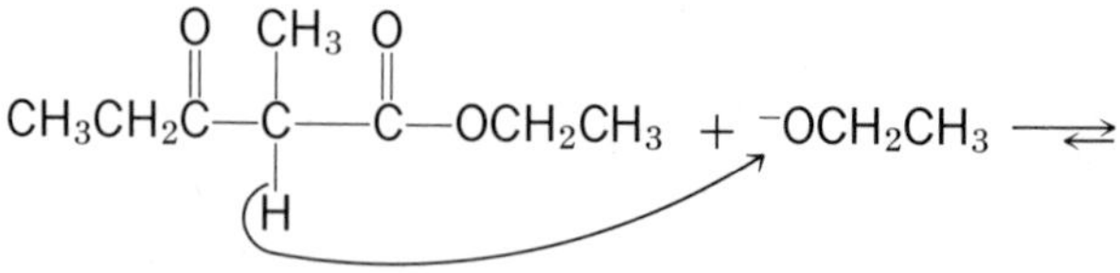

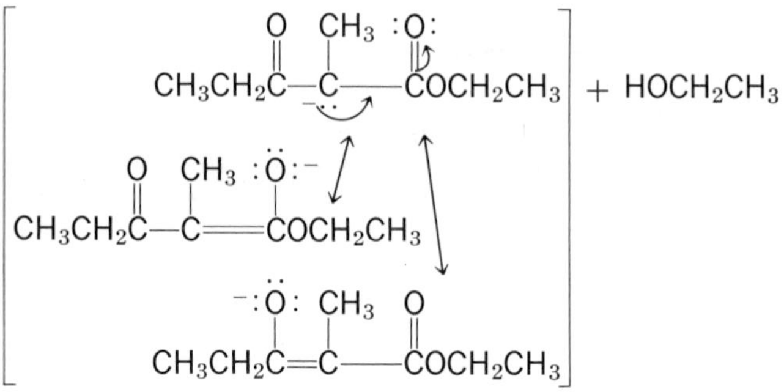

$$CH_3CH_2\overset{O}{\overset{\|}{C}}-\overset{CH_3}{\underset{H}{C}}-\overset{O}{\overset{\|}{C}}-OCH_2CH_3 + {}^-OCH_2CH_3 \rightleftharpoons$$

$$\left[ \begin{array}{c} CH_3CH_2\overset{O}{\overset{\|}{C}}-\overset{CH_3}{\underset{}{C}}-\overset{:\ddot{O}:}{\overset{\|}{C}}OCH_2CH_3 \\ CH_3CH_2\overset{O}{\overset{\|}{C}}-\overset{CH_3}{\underset{}{C}}=\overset{:\ddot{O}:^-}{\underset{}{C}}OCH_2CH_3 \\ CH_3CH_2\overset{O}{\overset{\|}{C}}=\overset{CH_3}{\underset{}{C}}-\overset{:\ddot{O}:}{\overset{\|}{C}}OCH_2CH_3 \end{array} \right] + HOCH_2CH_3$$

The resonance contributors shown help to explain the stability of this new anion. If a relatively strong acid is now added, the electrically neutral β-keto ester will be liberated:

$$CH_3CH_2\overset{\overset{O}{\|}}{C}\overset{CH_3}{\underset{\underset{\cdot\cdot}{|}}{\phantom{C}}}\overset{O}{\overset{\|}{C}}OCH_2CH_3 + H^+ \longrightarrow CH_3CH_2\overset{\overset{O}{\|}}{C}\overset{CH_3}{\underset{|}{-CH-}}\overset{O}{\overset{\|}{C}}-OCH_2CH_3$$

This type of reaction wherein two esters, in the presence of a strong base, react to form a β-keto ester is known as the Claisen ester condensation. In similar reactions in biochemistry enzymes act as the catalysts.

---

Exercise 7.6    Write the structures of the products that would be expected to occur in each of the following situations. If no reaction is to be expected, write "No reaction."

(a) $CH_3CH_2\overset{\overset{O}{\|}}{C}OCH_3 + H_2O \xrightarrow[\text{heat}]{H^+}$
(excess)

(b) $CH_3CH_2O\overset{\overset{O}{\|}}{C}CH_3 + NaOH \xrightarrow[\text{heat}]{H_2O}$

(c) $C_6H_5\overset{\overset{O}{\|}}{C}OCH(CH_3)_2 + NH_3 \xrightarrow[\text{heat}]{H_2O}$
(large excess)

(d) $CH_2{=}\overset{\overset{CH_3}{|}}{C}{-}\overset{\overset{O}{\|}}{C}OCH_3 + CH_3CH_2CH_2OH \xrightarrow[\text{heat}]{H^+}$
(excess)

(e) $CH_3(CH_2)_{14}\overset{\overset{O}{\|}}{C}{-}O{-}CH_2$

$CH_3(CH_2)_{12}\overset{\overset{O}{\|}}{C}{-}O{-}CH + CH_3OH \xrightarrow[\text{heat}]{H^+}$
(large excess)

$CH_3(CH_2)_{16}\overset{\overset{O}{\|}}{C}{-}O{-}CH_2$

(f) $CH_3\overset{\overset{CH_3}{|}}{C}HCH_2\overset{\overset{O}{\|}}{C}{-}O{-}CH_2CH_3 + H_2O \xrightarrow[\text{heat}]{H^+}$

(g) $CH_3CH_2CH_2O\overset{\overset{O}{\|}}{C}{-}\overset{\overset{CH_3}{|}}{C}HCH_3 + H_2O \xrightarrow[\text{heat}]{H^+}$

(h) $CH_3\overset{\overset{O}{\|}}{C}{-}O{-}CH_2$

$CH_3\overset{\overset{O}{\|}}{C}{-}O{-}CH_2 + NaOH \xrightarrow[\text{heat}]{H_2O}$
(excess)

(i) $2CH_3CO_2CH_2CH_3 \xrightarrow[\text{CH}_3\text{CH}_2\text{OH}]{\text{NaOCH}_2\text{CH}_3}$

(j) $2C_6H_5CH_2\overset{\overset{O}{\|}}{C}OCH_2CH_3 \xrightarrow[\text{CH}_3\text{CH}_2\text{OH}]{\text{NaOCH}_2\text{CH}_3}$

---

**Properties of β-Keto Esters and β-Keto Acids. Hydrolysis. Decarboxylation. Acid Cleavage.** β-Keto esters are both ketones and esters, and they exhibit typical reactions of these functional groups. They can be hydrolyzed, for example, to β-keto acids:

$$CH_3\overset{\overset{O}{\|}}{C}CH_2\overset{\overset{O}{\|}}{C}OCH_2CH_3 + H_2O \xrightarrow{H^+} CH_3\overset{\overset{O}{\|}}{C}CH_2\overset{\overset{O}{\|}}{C}OH + CH_3CH_2OH$$

Ethyl acetoacetate                     Acetoacetic acid

Such acids are subject to rather easy decarboxylation, in a manner analogous to that of β-dicarboxylic acids (malonic acid types).

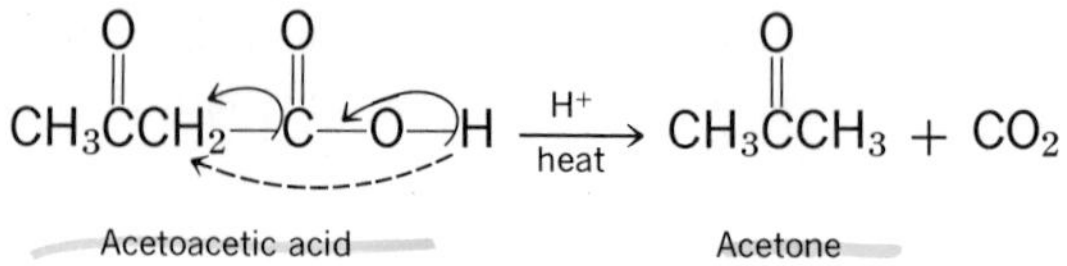

$$CH_3CCH_2{-}C{-}O{-}H \xrightarrow[\text{heat}]{H^+} CH_3CCH_3 + CO_2$$

Acetoacetic acid                    Acetone

The reaction shown occurs in the body, and we shall encounter it again in connection with lipid metabolism. $\beta$-Keto esters undergo another reaction important in lipid metabolism, cleavage of the $\alpha,\beta$-carbon-carbon bond. *In vitro,* the reaction requires strong alkali.

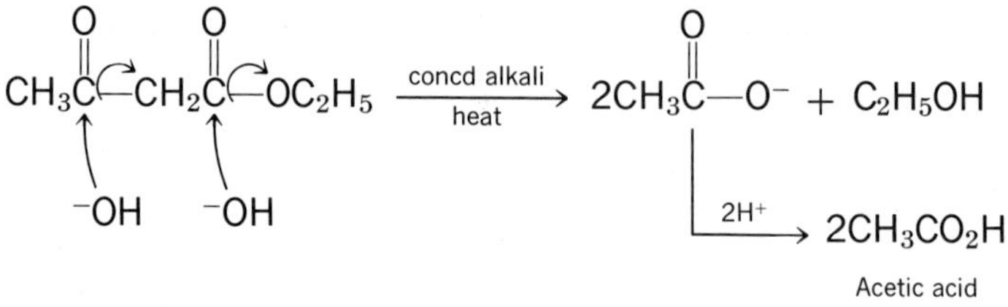

$$CH_3C{-}CH_2C{-}OC_2H_5 \xrightarrow[\text{heat}]{\text{concd alkali}} 2CH_3C{-}O^- + C_2H_5OH$$

$$^-OH \quad\quad ^-OH$$

$$\xrightarrow{2H^+} 2CH_3CO_2H$$

Acetic acid

*In vivo,* an important example of this reaction is the enzyme-catalyzed cleavage of a thio analog of a $\beta$-keto ester. In the following example the symbol CoA—SH represents coenzyme A, which is a thio alcohol.

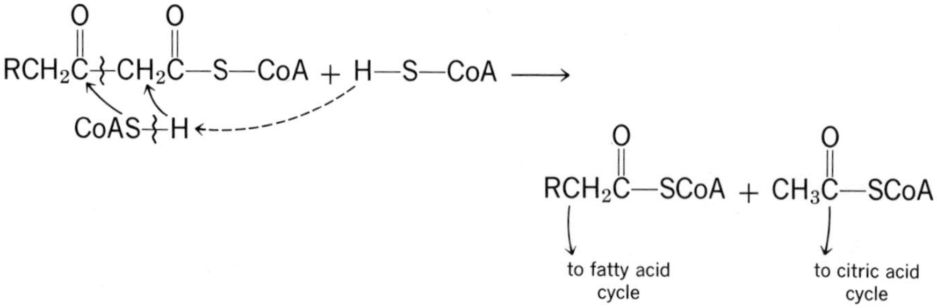

$$RCH_2C{-}CH_2C{-}S{-}CoA + H{-}S{-}CoA \longrightarrow$$

$$CoAS{-}H$$

$$RCH_2C{-}SCoA + CH_3C{-}SCoA$$

to fatty acid                    to citric acid
cycle                                   cycle

**Keto-Enol Tautomerism.** Although some reactions of $\beta$-keto esters are typical of the ketone and ester functions, others are not. Ethyl acetoacetate, for example, reacts instantly with bromine (in water), forming hydrogen bromide as the by-product.

$$CH_3CCH_2COC_2H_5 + Br_2 \xrightarrow{H_2O} CH_3CCHCOC_2H_5 + HBr$$

$$Br$$

This behavior can not be adequately explained in terms of the traditional structures of $\beta$-keto esters. $\beta$-Keto esters are actually *mixtures* of two compounds in mobile equilibrium with each other:

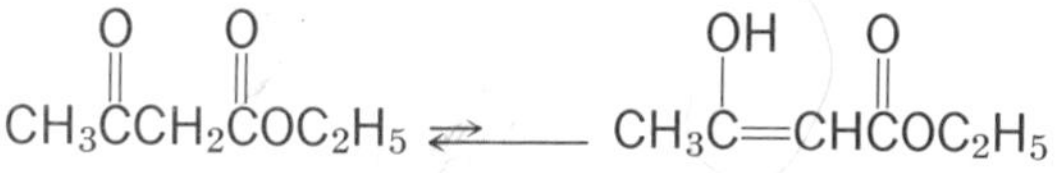

$$\underset{\substack{\text{Keto form}}}{CH_3\overset{O}{\overset{\|}{C}}CH_2\overset{O}{\overset{\|}{C}}OC_2H_5} \rightleftharpoons \underset{\substack{\text{Enol form}}}{CH_3\overset{OH}{\overset{|}{C}}=CH\overset{O}{\overset{\|}{C}}OC_2H_5}$$

Ethyl acetoacetate

The structure with a hydroxyl group attached to an ethylene unit is the *enol* ("ene" plus "ol") form. The rapid bromination of ethyl acetoacetate proceeds by means of the enol. As soon as some of this species is removed, the mobile equilibrium with the keto form shifts to generate additional enol, and the reaction continues:

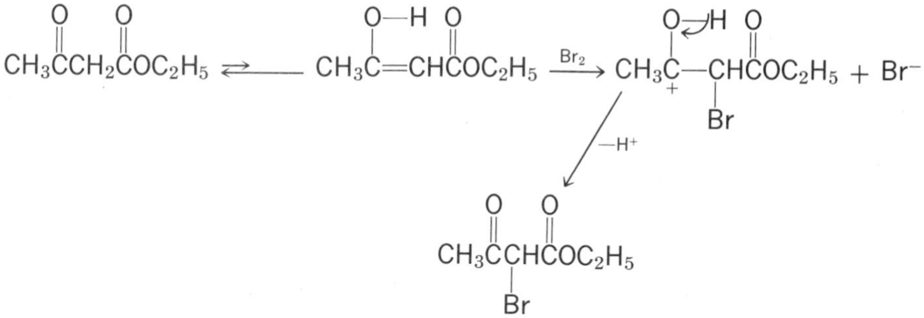

$$CH_3\overset{O}{\overset{\|}{C}}CH_2\overset{O}{\overset{\|}{C}}OC_2H_5 \rightleftharpoons CH_3\overset{O-H}{\overset{|}{C}}=CH\overset{O}{\overset{\|}{C}}OC_2H_5 \xrightarrow{Br_2} CH_3\overset{O-H}{\overset{|}{\underset{+}{C}}}-CH\overset{O}{\overset{\|}{C}}OC_2H_5 + Br^-$$

$$\underset{\substack{-H^+}}{\searrow}$$

$$CH_3\overset{O}{\overset{\|}{C}}CH\overset{O}{\overset{\|}{C}}OC_2H_5$$
$$\underset{Br}{|}$$

This phenomenon in which a substance exists in two molecular forms in dynamic equilibrium with each other is called tautomerism. The two forms are called tautomers, which are simply special examples of isomers so readily interconverted that they are very difficult to separate. All the carbonyl compounds we have studied contain at least trace amounts of corresponding enol tautomers, but normally in such minute quantities (about $10^{-4}\%$ enol in acetone) that their presence may be ignored. When a molecule has two carbonyl groups *beta* to each other, however, the enol form can be stabilized by internal hydrogen bonding. Therefore the percent of enol form present in $\beta$-dicarbonyl compounds is substantially higher than that in other carbonyl compounds, as seen in the data for ethyl acetoacetate and acetylacetone.

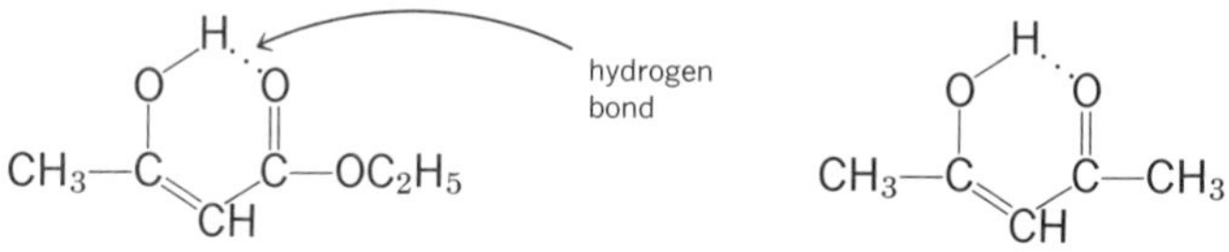

Enol of ethyl acetoacetate
(8% enol:92% keto form)

hydrogen bond

Enol of acetylacetone
(80% enol:20% keto form)

The keto form of most carbonyl compounds is usually so overwhelmingly favored in a keto-enol equilibrium that a reaction which might be regarded as the synthesis of an enol produces, instead, the keto tautomer. Thus the hydrolysis of isopropenyl acetate gives acetic acid and acetone:

$$CH_2{=}\overset{\overset{\displaystyle CH_3}{|}}{C}{-}O{-}\overset{\overset{\displaystyle O}{\|}}{C}{-}CH_3 + H_2O \xrightarrow{H^+} \left[ CH_2{=}\overset{\overset{\displaystyle CH_3}{}}{C}\underset{\underset{\displaystyle H}{O}}{} \right] + \overset{\overset{\displaystyle O}{\|}}{HOCCH_3}$$

Isopropenyl acetate
(a stable ester of an enol)

Acetic acid

An enol,
not isolated
(not stable)

$$\xrightarrow[\text{instantly}]{\text{ketonizes}} CH_3{-}\overset{\overset{\displaystyle CH_3}{\diagup}}{\underset{\underset{\displaystyle O}{\diagdown}}{C}}$$

Acetone

**Properties of Hydroxy Acids and Esters. Lactones.** A substance whose mole-cules possess both an alcohol and a carboxyl function generally gives reactions characteristic of each group. When the two groups are appropriately positioned in relation to each other, some of these properties are modified. We are interested pri-marily in those properties important to our later study of biochemistry.

$\beta$-Hydroxy acids (or esters), like $\beta$-hydroxy aldehydes (and ketones), lose the elements of water very easily to form $\alpha,\beta$-unsaturated acids (or esters) as the major products:

$$CH_3\overset{\overset{\displaystyle }{}}{\underset{\underset{\displaystyle OH}{|}}{C}}H{-}CH_2\overset{\overset{\displaystyle O}{\|}}{C}OH(R) \xrightarrow{H^+} CH_3CH{=}CH\overset{\overset{\displaystyle O}{\|}}{C}OH(R) + H_2O$$

$\alpha,\beta$-Unsaturated acid
(or ester)

$$CH_3(CH_2)_{12}\overset{\overset{\displaystyle OH}{|}}{C}H\overset{\overset{\displaystyle O}{\|}}{C}HCOH \xrightarrow[\substack{\text{(occurs during}\\\text{metabolism of}\\\text{fatty acids)}}]{\text{enzyme}} CH_3(CH_2)_{12}CH{=}CH\overset{\overset{\displaystyle O}{\|}}{C}OH + H_2O$$

$\beta$-Hydroxypalmitic
acid

$$HO{-}\overset{\overset{\displaystyle CH_2COOH}{|}}{\underset{\underset{\displaystyle CH_2COOH}{|}}{C}}{-}COOH \xrightarrow[\substack{\text{(occurs early}\\\text{in the citric}\\\text{acid cycle)}}]{\text{enzyme}} \overset{\overset{\displaystyle CH_2COOH}{|}}{\underset{\underset{\displaystyle CHCOOH}{\|}}{C}}{-}COOH + H_2O$$

Citric acid

In $\gamma$-hydroxy acids and $\delta$-hydroxy acids the two groups are positioned to favor the formation of a cyclic, or internal, ester called a lactone. We know that five- and six-membered rings are stable systems, and have learned that alcohols and acids can be made to form esters. Apparently the hydroxyl group at the "tail" of the $\gamma$-hydroxy or $\delta$-hydroxy acid molecule can more frequently find the carboxyl group at its "head" than a carboxyl group on a neighboring molecule.

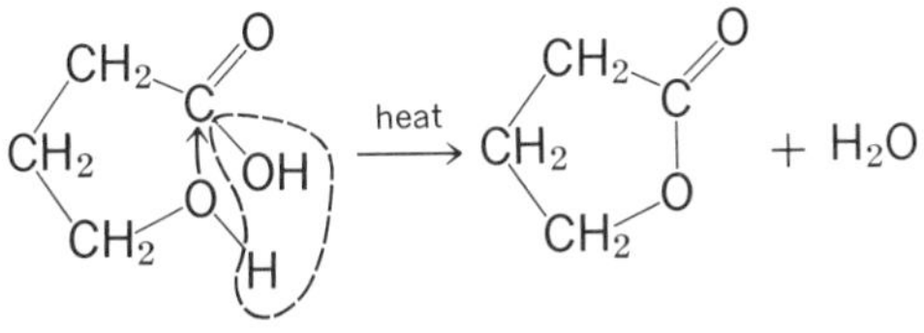

γ-Hydroxybutyric acid      γ-Butyrolactone

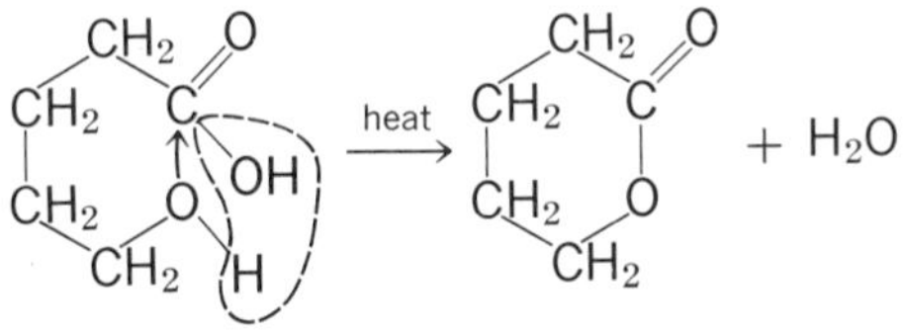

δ-Hydroxyvaleric acid      δ-Valerolactone

**Properties of $\alpha,\beta$-Unsaturated Acids and Other Carbonyl Compounds. 1,4-Addition.** In systems of the type

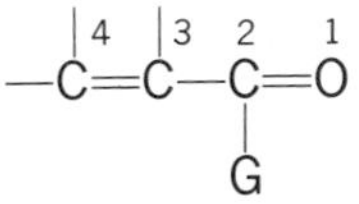

$$-\overset{|}{\underset{4}{C}}=\overset{|}{\underset{3}{C}}-\underset{2}{C}=\underset{1}{O}$$
$$\overset{|}{\underset{G}{}}$$

$\alpha,\beta$-Unsaturated carbonyl compound (G = OH, OR, R)

The numbering is incorrect for nomenclature purposes; here it labels parts of the conjugated system.

the alkene system is joined to an electron-withdrawing group. We might therefore expect the alkene double bond to be less reactive to electron-seeking reagents and possibly somewhat reactive to nucleophiles. We might also expect the direction of addition to the carbon-carbon double bond to be affected by the neighboring carbonyl group. Markovnikov's rule, in fact, is not applicable to conjugated systems. For example,

$$CH_2{=}CHCOH + HCl \longrightarrow ClCH_2CH_2COH \quad \text{NOT} \quad CH_3CHCOH$$

Acrylic acid                     β-Chloropropionic acid

(predicted if Markovnikov's rule is applied)

To understand this system we return to fundamental principles. Reactions, like spontaneous events in general, tend to go through paths of minimum energy. In the addition of hydrogen chloride to an alkene the more stable of two possible carbonium ions forms. Addition of hydrogen chloride to a conjugated system proceeds through the same path. The first step ① is attack by a proton at the site richest in electrons, the oxygen of the carbonyl group. The resonance-stabilized carbonium ion that forms has greater stability than any other that can be fashioned. This ion then picks up a chloride ion ② to produce an unstable enol which rearranges to the keto form ③. The first two steps are essentially identical to 1,4-addition, which we encountered earlier when studying butadiene (cf. p. 102).

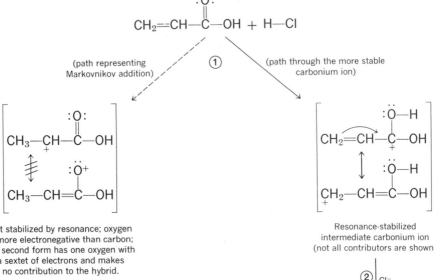

The addition of water to an $\alpha,\beta$-unsaturated carbonyl compound is a similar reaction. Several others occur in molecular biology:

$$RCH{=}CHCOH + H_2O \xrightarrow{\text{enzyme}} RCHCH_2COH \quad \text{(in fatty acid metabolism)}$$

$$HOOCCH{=}CHCOOH + H_2O \xrightarrow{\text{enzyme}} HO{-}CHCOOH \quad \text{(in citric acid cycle)}$$

Fumaric acid                     Malic acid

## IMPORTANT INDIVIDUAL ESTERS

**The Salicylates.** Salicyclic acid can function either as an acid or as a phenol, for it possesses both groups. Salicyclic acid, its esters, and its salts, taken internally, have both an analgesic effect (depressing sensitivity to pain) and an antipyretic action (reducing fever). As analgesics they raise the threshold of pain by depressing pain centers in the thalamus region of the brain. As antipyretics they increase perspiration as well as circulation of the blood in capillaries near the surface of the skin. Both mechanisms have cooling effects. Salicyclic acid itself irritates the moist membranes lining the mouth, gullet, and the stomach because it is too acidic. The sodium or calcium salts, however, are less irritating. In the early history of aspirin[4] the salts had to be administered in aqueous solutions which had a very disagreeable sweetish taste. The search for a more palatable form led to the discovery of acetyl salicyclic acid, or aspirin, first sold by the German company Bayer in 1899 (which still owns the name Aspirin as a trademark in Germany). Aspirin is the most widely used drug in the world. During the 1960s the annual United States production exceeded 27 million pounds from which, annually, over 15 billion aspirin tablets were compounded.

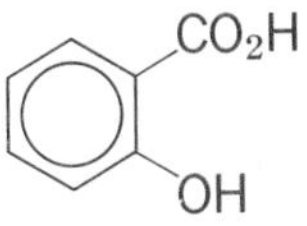

Salicylic acid

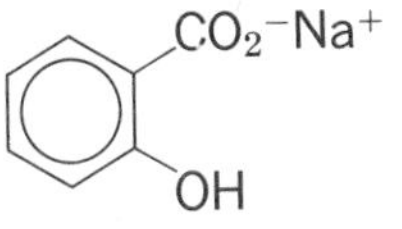

Sodium salicylate

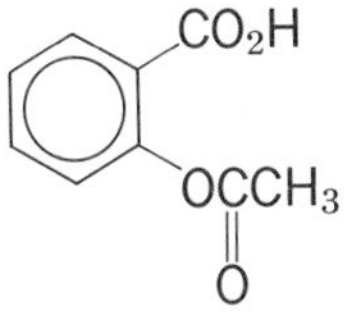

Acetylsalicylic acid ("aspirin")

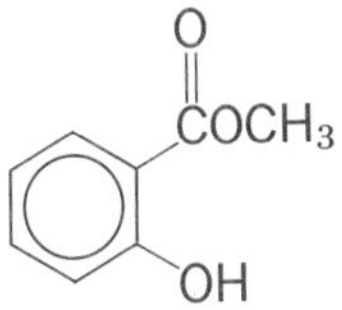

Methyl salicylate ("oil of wintergreen")

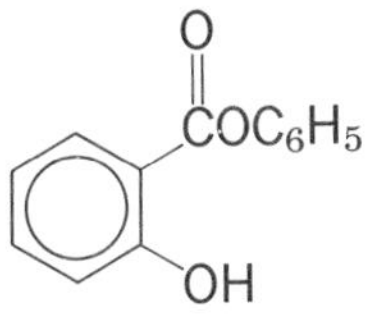

Phenyl salicylate ("salol")

Methyl salicylate is used in liniments, for it is readily absorbed through the skin. Phenyl salicylate is sometimes a constituent of ointments that protect the skin against ultraviolet rays.

**Dacron.** Dacron shares most of the desirable properties of nylon. Especially distinctive is its ability to be set into permanent creases and pleats. It does not tend to become grayish or yellowish with age, as does some white nylon. Chemically, Dacron is a polyester, polyethylene terephthalate, made by ester interchange (transesterification) between ethylene glycol and dimethyl terephthalate.

Dimethyl terephthalate

Ethylene glycol

$-CH_3OH$

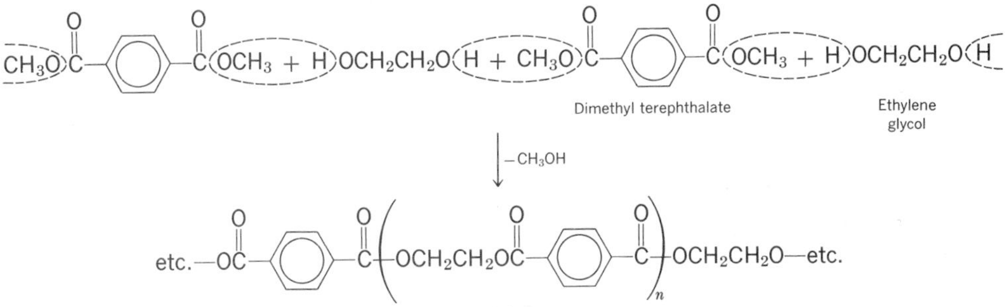

Repeating unit in Dacron

[4] H. O. J. Collier, "Aspirin," *Scientific American*, November 1963, p. 97.

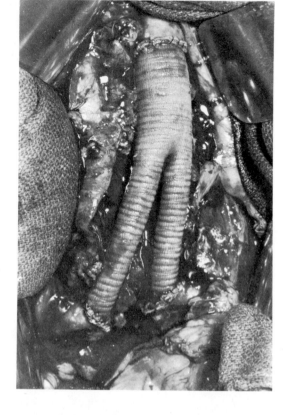

**Figure 7.1** Graft of knitted Dacron tubing in place in an operative site, illustrating an unusual application of this versatile polyester which was pioneered by Dr. Michael E. DeBakey. (Courtesy of Cora and Webb Mading Department of Surgery, Baylor University College of Medicine.)

As is characteristic of all chemicals that can be formed into useful fibers, molecules of Dacron are extremely long and very narrow and symmetrical, properties that resemble those of a natural fiber. A significant use of knitted Dacron is illustrated in Figure 7.1.

Dacron is the trade name for the fiber made from polyethylene terephthalate. To increase the strenth of the fibers, the freshly spun filaments are slowly stretched, allowing the long molecules to become better aligned parallel with the fiber axis. Forces of attraction between neighboring molecules become more effective and the fiber gains considerable strength.

Polyethylene terephthalate may also be made into exceptionally clear, durable films, and in this form it is known as Mylar. Other trade names for polyethylene terephthalate are Terylene and Cronar.

**Glyptal Resins.** Ethylene glycol has but two hydroxyl groups and can form only a linear polymer with terephthalic acid. Glycerol, however, has three hydroxyl groups, the third group making the formation of cross-linked polymers possible. The glyptal resin, prized as a tough, durable surface coating in automobile and other finishes, forms when phthalic anhydride and glycerol interact in a mole ratio of 3:2 (Figure 7.2).

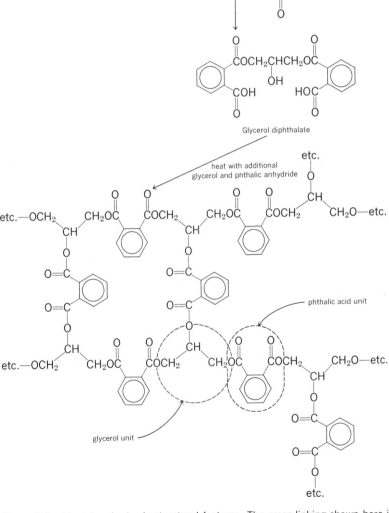

Phthalic
anhydride

Glycerol

Glycerol monophthalate

heat

Glycerol diphthalate

heat with additional
glycerol and phthalic anhydride

etc.

etc.—OCH₂

CH₂O—etc.

phthalic acid unit

etc.—OCH₂

CH₂O—etc.

glycerol unit

etc.

**Figure 7.2** Glyptal resin, basic structural features. The cross-linking shown here is for illustrative purposes only. In the actual resin these links do not occur at every possible position.

Primary alcohols are more reactive in esterification than secondary alcohols. Consequently, in the early stage of making glyptal resin, mixtures of mono- and diphthalate esters of glycerol are present. Actual cross-linking must bring into play glycerol's 2° alcohol function. When the resin is to be used as a surface coating, material that is not for the most part cross-linked is applied to the surface. Baking then completes resinification as the material becomes essentially one gigantic molecule. Understandably, this coating is exceptionally durable.

**Acrylates.** Esters of acrylic acid and methacrylic acid can be made to polymerize in a fashion exactly analogous to vinyl polymerization. The products are well-known plastics.

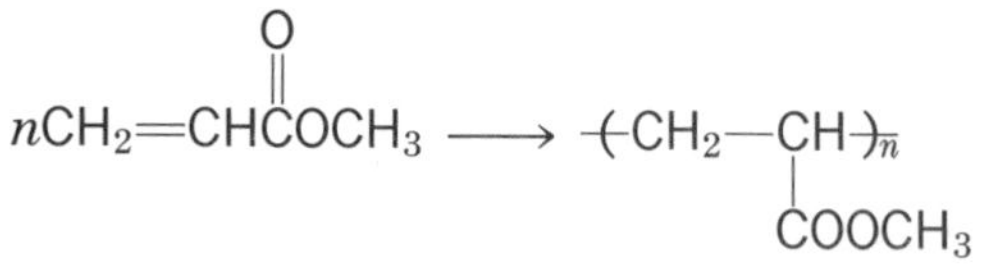

$$n CH_2{=}CH\overset{\overset{\textstyle O}{\|}}{C}OCH_3 \longrightarrow (CH_2{-}\underset{\underset{\textstyle COOCH_3}{|}}{CH})_n$$

Methyl acrylate                    Acryloid

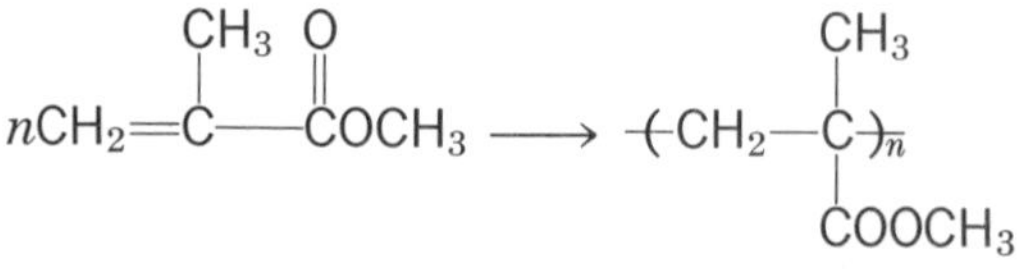

$$n CH_2{=}\overset{\overset{\textstyle CH_3}{|}}{C}{-}\overset{\overset{\textstyle O}{\|}}{C}OCH_3 \longrightarrow (CH_2{-}\underset{\underset{\textstyle COOCH_3}{|}}{\overset{\overset{\textstyle CH_3}{|}}{C}})_n$$

Methyl methacrylate                Plexiglas, Lucite

**Esters of Inorganic Oxygen-Acids.** The one or more —OH groups in the molecule of an inorganic oxygen acid can, like those in the molecule of a carboxylic acid, be replaced by —OR. For this reason several types of esters of inorganic oxygen acids are known, many of which are important and useful materials. Some are formed simply by the reaction of an alcohol with the inorganic acid.

High-molecular-weight alcohols react with sulfuric acid to yield alkyl hydrogen sulfates, as illustrated by the synthesis of a common synthetic detergent, sodium lauryl sulfate:

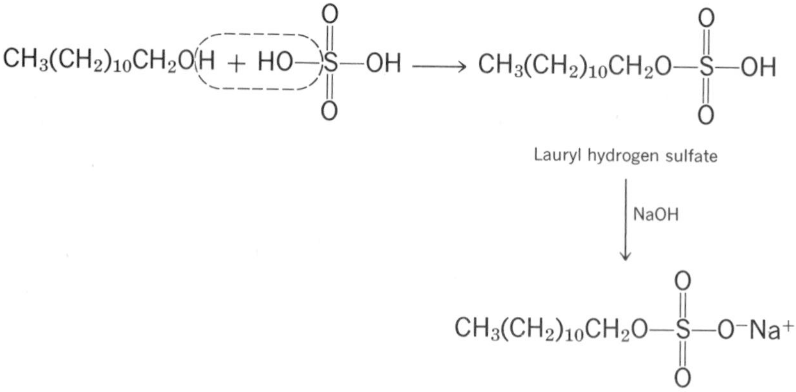

$$CH_3(CH_2)_{10}CH_2O{-}\!\!\left[H + HO\right]\!\!{-}\overset{\overset{\textstyle O}{\|}}{\underset{\underset{\textstyle O}{\|}}{S}}{-}OH \longrightarrow CH_3(CH_2)_{10}CH_2O{-}\overset{\overset{\textstyle O}{\|}}{\underset{\underset{\textstyle O}{\|}}{S}}{-}OH$$

Lauryl hydrogen sulfate

$$\Big\downarrow NaOH$$

$$CH_3(CH_2)_{10}CH_2O{-}\overset{\overset{\textstyle O}{\|}}{\underset{\underset{\textstyle O}{\|}}{S}}{-}O^-Na^+$$

Sodium lauryl sulfate (Dreft)

Esters of nitrous acid, HO—N=O, are made by adding dilute sulfuric acid slowly to a chilled mixture of the alcohol and aqueous sodium nitrite. Isoamyl nitrite has been useful in treating hypertension and is an important drug for relieving the severe pains of angina pectoris. By relaxing smooth muscle, it produces coronary vasodilatation and lowers the blood pressure.

$$(CH_3)_2CHCH_2CH_2OH + Na^{+-}O\!-\!N\!=\!O \xrightarrow{H_2SO_4} (CH_3)_2CHCH_2CH_2O\!-\!N\!=\!O$$

Isoamyl alcohol        Sodium nitrite          Isoamyl nitrite

· Nitric acid esters of glycerol and pentaerythritol are drugs serving the same purposes. These nitrates are also powerful explosives. Nitroglycerine was discovered

$$\begin{array}{c}
CH_2ONO_2 \\
| \\
CHONO_2 \\
| \\
CH_2ONO_2
\end{array}
\qquad\qquad
\begin{array}{c}
CH_2ONO_2 \\
| \\
O_2NOCH_2CCH_2ONO_2 \\
| \\
CH_2ONO_2
\end{array}$$

Glyceryl trinitrate             Pentaerythritol tetranitrate
("nitroglycerin")

in 1847 by an Italian chemist, Sobrero. Alfred Nobel, a Swedish chemist, commenced commercial manufacture of it in 1862, but because it is so sensitive to shock, many fatal accidents occurred. Nobel discovered, however, that the explosive properties of nitroglycerine are not seriously reduced if the liquid is diluted by mixing it with kieselguhr, a powdery "earth" consisting of the siliceous skeletons of diatoms, marine organisms. (Charcoal, powdered wood, and burnt cork also work.) The mixture was called dynamite. In this way the extreme sensitivity to shock of nitroglycerine was brought under control, and the explosive can be handled and shipped much more safely. Nobel became very wealthy from this discovery, and by his will proceeds from his estate fund the annual Nobel prizes.

Part of the success of glycerol trinitrate as an explosive is explained by the fact that each molecule has more than enough oxygen to complete the conversion of its carbon and hydrogen to carbon dioxide and water:

$$4C_3H_5(ONO_2)_3 \xrightarrow{\text{detonation}} 6N_2 + 12CO_2 + 10H_2O + O_2$$

Nitroglycerine

The sudden exothermic conversion of a small volume of oily nitroglycerine into several hundred times as many volumes of hot gases accounts for its blasting properties.

In modern-day dynamites only a very little nitroglycerine is used as a sensitizer for detonating a mixture of ammonium nitrate (up to 55%), sodium nitrate, glycol nitrate, and wood pulp. The importance of dynamite and similar explosives to the peaceful progress of civilization is difficult to overestimate. The opening of canals, the building of dams, the boring of tunnels, the mining of ores, the clearing of land—the list could be extended to a considerable length—remind us that man's quest for controllable power has had its peaceful outlets too.

Esters of phosphoric acid, pyrophosphoric acid, and higher homologs were described earlier as anhydrides. Because they are involved in most reactions in the body, we shall encounter them repeatedly in our study of biochemistry. Some esters, such as tetraethyl pyrophosphate (TEPP) are important insecticides, although this one must obviously be applied with extreme caution because it is highly toxic to animals as well as insects.

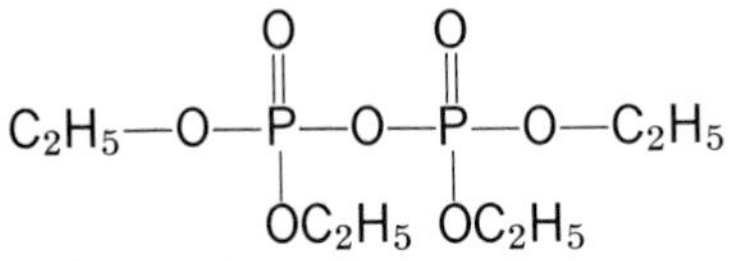

$$C_2H_5-O-\overset{\overset{\displaystyle O}{\|}}{P}-O-\overset{\overset{\displaystyle O}{\|}}{P}-O-C_2H_5$$
$$\underset{\displaystyle OC_2H_5}{|} \quad \underset{\displaystyle OC_2H_5}{|}$$

Tetraethyl pyrophosphate
(TEPP)

## AMIDES

**Physical Properties.** With the exception of formamide, all the simple amides, those derived from ammonia, melt well above room temperature. Their molecules are apparently quite polar and can serve as both hydrogen bond acceptors and donors. Several are listed in Table 7.8.

The amide group is planar. The nitrogen (and any atoms directly linked to it), the carbonyl carbon, and the oxygen all lie in the same plane, apparently because the carbonyl-to-nitrogen bond behaves somewhat as though it were double, as indicated by the following resonance structures:

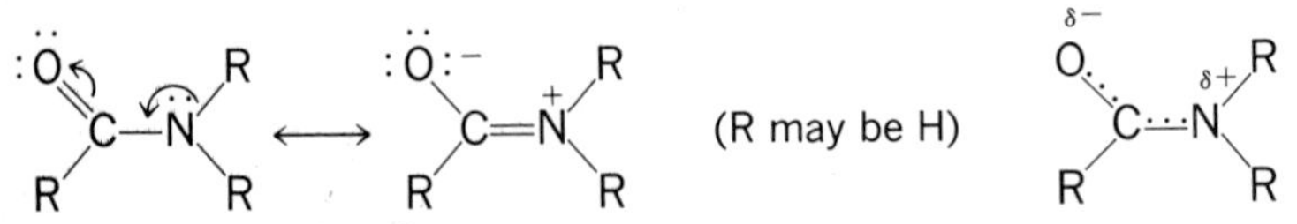

(R may be H)

Hybrid

### Table 7.8   Amides of Carboxylic acids

| Name | Structure | Mp (°C) | Bp (°C) | Solubility in Water |
|---|---|---|---|---|
| Formamide | HCONH$_2$ | 2.5 | 210 (decomposes) | ∞ |
| N-Methylformamide | HCONHCH$_3$ | −5 | 131 (at 90 mm pressure) | very soluble |
| N,N-Dimethylformamide | HCON(CH$_3$)$_2$ | −61 | 153 | ∞ |
| Acetamide | CH$_3$CONH$_2$ | 82 | 222 | very soluble |
| N-Methylacetamide | CH$_3$CONHCH$_3$ | 28 | 206 | very soluble |
| N,N-Dimethylacetamide | CH$_3$CON(CH$_3$)$_2$ | −20 | 166 | ∞ |
| Propionamide | CH$_3$CH$_2$CONH$_2$ | 79 | 213 | soluble |
| Butyramide | CH$_3$(CH$_2$)$_2$CONH$_2$ | 115 | 216 | soluble |
| Valeramide | CH$_3$(CH$_2$)$_3$CONH$_2$ | 106 | sublimes | soluble |
| Caproamide | CH$_3$(CH$_2$)$_4$CONH$_2$ | 100 | sublimes | slightly soluble |
| Benzamide | 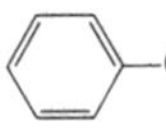 ⬡—CONH$_2$ | 133 | 290 | slightly soluble |

The contribution of the dipolar ionic form to the real state of the molecule may not be great (why?), but if it participates at all, the carbon-to-nitrogen bond will be double in character. Hence the geometry of the system must be like that of an ethylene unit. Resonance theory also predicts considerable polarity for the system. Hydrogen bond networks between amide molecules would account for their relatively high melting points. Amides having no hydrogens on nitrogen ($RCONR'_2$) or only one ($RCONHR'$) will be less capable of donating hydrogen bonds and should have lower melting points. The data in Table 7.8 bear this out in dramatic fashion. The melting point of *N*-methylacetamide, for example, is 54° lower than that of acetamide, and that of *N*,*N*-dimethylacetamide is another 48° lower. We have taken up

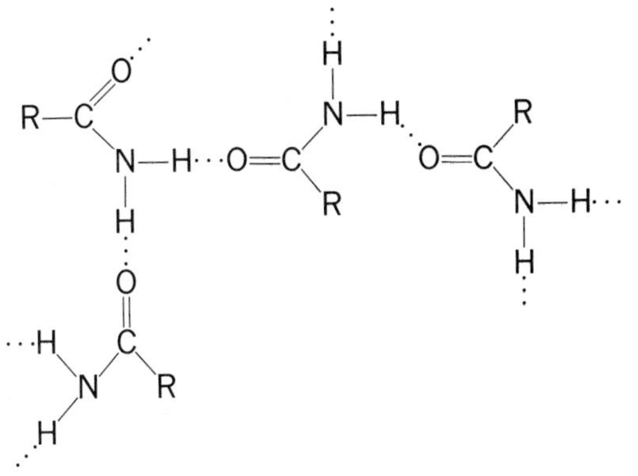

the question of amide structure in this much detail because proteins are polyamides. To understand these most fundamental compounds, we must know the properties of amides.

**Chemical Properties. Hydrolysis.** One chemical property of amides that we shall need to know for later work is simple hydrolysis:

$$R{-}\overset{\displaystyle O}{\overset{\|}{C}}{-}NH_2 + H_2O \xrightarrow[\text{heat}]{\text{acids or bases}} R{-}\overset{\displaystyle O}{\overset{\|}{C}}{-}OH + NH_3$$

$$R{-}\overset{\displaystyle O}{\overset{\|}{C}}{-}NHR' + H_2O \longrightarrow R{-}\overset{\displaystyle O}{\overset{\|}{C}}{-}OH + R'NH_2$$

$$R{-}\overset{\displaystyle O}{\overset{\|}{C}}{-}NR'_2 + H_2O \longrightarrow R{-}\overset{\displaystyle O}{\overset{\|}{C}}{-}OH + R'_2NH$$

It is the carbonyl-to-nitrogen bond that breaks. The reaction usually requires prolonged refluxing of the amide with either acid or base, for they are quite unreactive. *In vivo,* however, enzymes are available for catalyzing the reaction. The digestion of proteins is nothing more than hydrolysis of the amide linkages of proteins.

Exercise 7.7    When acetamide is allowed to react with sodium hydroxide on a $1:1$ mole basis, the products are sodium acetate and ammonia. If hydrochloric acid is used, however, again on a $1:1$ mole basis, the products are acetic acid and ammonium chloride. Explain.

Exercise 7.8    Write the structures of the products that would be expected to form in each of the following. If no reaction is to be expected, write "No reaction."

(a) $CH_3\overset{\displaystyle O}{\overset{\|}{C}}NH_2 + H_2O \xrightarrow{heat}$

(b) $CH_3NH\overset{\displaystyle O}{\overset{\|}{C}}CH_2CH_3 + H_2O \xrightarrow{heat}$

(c) $C_6H_5NH\overset{\displaystyle O}{\overset{\|}{C}}CH_3 + H_2O \xrightarrow{heat}$

(d) $CH_3CH_2\overset{\displaystyle O}{\overset{\|}{C}}N(CH_3)_2 + H_2O \xrightarrow{heat}$

(e) $NH_2CH_2\overset{\displaystyle O}{\overset{\|}{C}}NH\underset{\underset{\displaystyle CH_3}{|}}{C}H\overset{\displaystyle O}{\overset{\|}{C}}OH + H_2O \xrightarrow{heat}$

## IMPORTANT INDIVIDUAL AMIDES

**Nicotinamide** (Niacinamide).    $\overset{\displaystyle O}{\overset{\|}{C}}NH_2$. Nicotinamide, the amide of nicotinic acid (niacin), has the physiological properties of niacin and is the usual form in which it is administered.

**Nylon.**[5] In 1927 Du Pont's chemical director, C. M. A. Stine, persuaded management to allocate 20,000 dollars "to cover what may be called for want of a better name, pure science or fundamental research work." The next year he persuaded a young instructor, Wallace H. Carothers, to be in charge of Du Pont's organic research section. Carothers' principal interests lay in high-molecular-weight compounds, and he was particularly intrigued by the possibilities of allowing substances such as $xAx$ to interact with $yBy$, with $x$ and $y$ designating functional groups able to react with each other. Soon another young man who had just received his doctorate, Julian W. Hill, joined Carothers' team. In 1931 these two men reported that when $xAx$ was $HOCH_2CH_2CH_2OH$ and $yBy$ was $HOOC(CH_2)_{14}COOH$ and the two were strongly heated for days, a hard material formed. When a glass rod was thrust into the molten mass and then pulled out again, long streamers, both hard and

[5] L. E. Alexander, "Nylon—From Test Tube to Counter," *Chemistry*, September 1964, page 8; and *Nylon—The First 25 Years*, E. I. du Pont de Nemours, Wilmington, Del., 1963.

brittle, followed the rod. The filaments were found to stretch to several times their original length, becoming elastic, lustrous, and as strong as real silk. The sub-stance, however, appeared to be too expensive for commercial exploitation. Research efforts turned to a compound of the type $NH_2(CH_2)_nNH_2$ for $xAx$, with $yBy$ remain-ing a dicarboxylic acid, $HOOC(CH_2)_mCOOH$. In 1935 Carothers' team found that hexamethylenediamine ($n = 6$) and adipic acid ($m = 4$), when heated together at high temperatures in a nitrogen atmosphere (to prevent oxidation at these tempera-tures), formed a polymer of unusual properties. Fibers from it, when drawn to about four times their original length, had a tensile strength greater than that of steel, were slightly elastic, lustrous, and insoluble in common solvents, and melted above 260°C (the temperature used in ironing fabrics). In 1936 the Japanese stopped the export from China of swine bristles which were needed for making quality brushes. By the middle of 1938, however, the new fiber, named Exton at that time, was being adopted as a substitute.

During this time an intensive program at Du Pont concentrated on the chemi-cal and engineering problems of converting raw polymer into hosiery. War in the Far East threatened the silk supply and cut it off in 1941. The polymer when made into women's hose was initially scheduled to be called "norun," but the hosiery did "run." A change to "nuron" was abandoned because both "norun" and "nuron" come too close to rhyming with "moron." The $u$ became $y$, the $r$ became $l$, and the word "nylon" was born. This name, generic in the same way that *wood* is, applies to a family of materials, to any synthetic, long-chained, fiber-forming polymer with repeating amide linkages. One member of this family is called nylon 66, suggesting the six carbons in the diamine and the six carbons in the dicarboxylic acid.

$$+ \; NH_2(CH_2)_6NH_2 \; + \; HO\overset{O}{\overset{\|}{C}}(CH_2)_4\overset{O}{\overset{\|}{C}}OH \; + \; NH_2(CH_2)_6NH_2 \; + \; HO\overset{O}{\overset{\|}{C}}(CH_2)_4\overset{O}{\overset{\|}{C}}OH \; +$$

Adipic acid      Hexamethylene-
            diamine

$$\text{etc.}-NH(CH_2)_6NH{\Large(}\overset{O}{\overset{\|}{C}}(CH_2)_4\overset{O}{\overset{\|}{C}}-NH(CH_2)_6NH{\Large)}_n\overset{O}{\overset{\|}{C}}(CH_2)_4\overset{O}{\overset{\|}{C}}-\text{etc.}$$

Repeating unit
in nylon 66

To be useful as a fiber, each molecule in a batch of nylon 66 should contain from fifty to ninety of each of the monomer units. Shorter molecules make weak or brittle fibers.

The nylon woven into most women's hosiery is a monofilament. Raw nylon is first cast as a long ribbon which is chopped into small chips. These are melted on heated grids of metal mesh, and the molten material is carefully strained and then

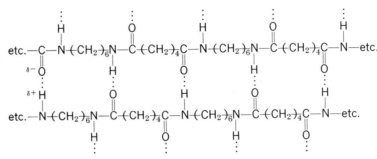

**Figure 7.3** In this somewhat idealized representation of nylon 66, neighboring molecules adhere to each other via hydrogen bonds (dotted lines).

forced through extremely tiny orifices of a spinneret. When the streams emerge, they harden and are caught up on a winding machine and slowly stretched. This drawing brings about a better alignment of the long molecules along the axis of the fiber, permitting a better organization of forces of attraction (hydrogen bonds) between neighboring molecules (Figure 7.3).

Nylon hosiery was on the market long enough before the onset of World War II to familiarize women with its desirable qualities. When the United States entered the conflict, all our nylon production was diverted to the war effort. Parachutes were made of nylon, and for this purpose it proved superior to the unavailable silk. Nylon ropes for towing gliders, climbing mountains, mooring vessels and balloons, and rigging anchors were all twice as strong as those made of Manila hemp. Jungle hammocks, gloves, survival rafts, belts, ponchos, tire cords—the list is very long— when made of nylon were found to be of exceptional quality. Finally the war ended, and with peaceful applications again possible, the demand for nylon so far exceeded the supply that Du Pont was "host" to more than one Congressional investigation, its salesmen to knitting mills were offered bribes, and nylon worth six dollars a pound brought up to thirty-five dollars a pound on the black market. Not until 1952 was the hosiery market sufficiently supplied with nylon to permit expansion into other fields.

The unusual resistance of nylon 66 to breakage by stretching, mechanical abrasion, moisture, action of mild acids and alkalis, light, dry-cleaning solvents, mildew, bacteria, and rotting conditions make it one of the most desirable of all fibers. It is more resistant to burning than wool, rayon, cotton, or silk and is almost as immune to insect attack as Fiberglas. No known insects metabolize nylon molecules, although some will cut their way through a nylon fabric if they are entrapped by it!

At the end of the 1960s annual nylon production was over 1.7 billion pounds. Multimillion dollar plants employing thousands of people in many parts of the world are involved. The initial investment of 20,000 dollars led to a giant molecule that has touched the lives of millions of people.

# REFERENCES AND ANNOTATED READING LIST

## BOOKS

R. T. Morrison and R. N. Boyd. *Organic Chemistry,* second edition. Allyn and Bacon, Boston, 1966. Chapter 18 ("Carboxylic Acids") and Chapter 20 ("Functional Derivatives of Carboxylic Acids") are particularly good references in a standard organic textbook.

L. O. Smith, Jr. and S. J. Cristol. *Organic Chemistry.* Reinhold Publishing Corporation, New York, 1966. Chapter 10 ("Acidity and Basicity in Organic Compounds") is a good survey of organic acids and bases and the structural factors which affect their strengths.

C. A. Vanderwerf. *Acids, Bases, and the Chemistry of the Covalent Bond.* Reinhold Publishing Corporation, New York, 1961. This book has been cited before as an excellent reference to the topics of the title. (Paperback.)

## ARTICLES

L. E. Alexander. "Nylon—From Test Tube to Counter." *Chemistry,* September 1964, page 8.

No author cited. "Nylon—The First 25 Years." Pamphlet published by E. I. du Pont de Nemours, Wilmington, Del., 1963.

H. O. Collier. "Aspirin." *Scientific American,* November 1963, page 97. The history of the salicylates and how they work are discussed.

## PROBLEMS AND EXERCISES

1. Write the structure of each of the following compounds.
   - (*a*)  acetamide
   - (*b*)  methyl propionate
   - (*c*)  acetyl chloride
   - (*d*)  acetic anhydride
   - (*e*)  *N*-methylformamide
   - (*f*)  isopropyl *n*-butyrate
   - (*g*)  glycerol triacetate
   - (*h*)  ethyl acetoacetate
   - (*i*)  malonic acid
   - (*j*)  nylon-66   (Show enough of its structure to illustrate its features.)
   - (*k*)  Dacron   (Give enough of its structure to illustrate its features.)

2. Write common names for each of the following compounds.

   (*a*)  $CH_3CH_2CH_2CH_2O\overset{O}{\overset{\|}{C}}CH_2CH_3$

   (*b*)  $CH_3CH_2CH_2\overset{O}{\overset{\|}{C}}NH_2$

   (*c*)  $CH_3\overset{O}{\overset{\|}{C}}CH_2\overset{O}{\overset{\|}{C}}OH$

   (*d*)  $CH_3CH_2\overset{O}{\overset{\|}{C}}{-}Cl$

   (*e*)  $C_6H_5\overset{O}{\overset{\|}{C}}OCH_2CH_3$

   (*f*)  $CH_3CH_2OCH_2CH_3$

   (*g*)  $CH_3CH_2CH_2CH_2NH_2$

   (*h*)  $CH_3CH_2CH_2CH_2CH_2\overset{O}{\overset{\|}{C}}OH$

   (*i*)  $(CH_3)_2CHO\overset{O}{\overset{\|}{C}}H$

   (*j*)  $CH_3CH_2\overset{O}{\overset{\|}{C}}NH_2$

3. Go back through the chapters on organic chemistry and make a list of all the reactions we have studied, writing a specific example of each, for introducing in one step from some starting compound each of the following functional groups.

(a) —OH  (as in an alcohol)    (b) —Cl  (as in an alkyl halide)

(c) C=C  (as in an alkene)    (d) —CHO  (as in an aldehyde)

(e) $-\overset{\overset{\displaystyle O}{\|}}{C}-$  (as in a ketone)    (f) $-\overset{\overset{\displaystyle O}{\|}}{C}-O-C$  (as in an ester)

(g) $-\overset{\overset{\displaystyle O}{\|}}{C}-N$  (as in an amide)

4. List all the reactions we have studied for making a carbon-carbon single bond.
5. List all the reactions we have studied for breaking a carbon-carbon single bond.
6. List all the reactions we have studied for making a carbon-hydrogen bond.
7. List all the reactions we have studied for breaking a carbon-hydrogen bond.
8. List all the reactions we have studied for making a carbon-oxygen single bond.
9. List all the reactions we have studied for breaking a carbon-oxygen single bond.
10. What functional groups have we studied that are susceptible to oxidation, and what products form when they are oxidized?
11. What functional groups have we studied that are susceptible to reduction by any means, and what are the products they form?
12. Give specific examples of all the *types* of reactions we have studied in which water is a reactant rather than just a solvent. Assume that whatever catalyst is normally used—acid, base, enzyme—is available.
13. Complete the following equations by writing the structure(s) of the principal organic product(s) that are expected to form. If no reaction is expected, write "No reaction."

(a) $CH_3CH_2\overset{\overset{\displaystyle O}{\|}}{C}H \xrightarrow{K_2Cr_2O_7}$

(b) $CH_3OH + CH_3\overset{\overset{\displaystyle O}{\|}}{C}OH \xrightarrow[heat]{H^+}$

(c) $CH_3CH_2CH_3 + H_2SO_4 \longrightarrow$

(d) $CH_3\overset{\overset{\displaystyle OH}{|}}{C}HCH_3 \xrightarrow[heat]{KMnO_4}$

(e) $CH_3\overset{\overset{\displaystyle O}{\|}}{C}NH_2 + H_2O \xrightarrow[heat]{H^+}$

(f) $CH_3\overset{\overset{\displaystyle OH}{|}}{C}HCH_3 \xrightarrow[heat]{H_2SO_4}$

(g) $CH_3O\overset{\overset{\displaystyle O}{\|}}{C}CH_3 + H_2O \xrightarrow[heat]{H^+}$

(h) $CH_3CH=CH_2 + HCl \longrightarrow$

(i) $CH_3\overset{\overset{\displaystyle OCH_3}{|}}{\underset{\underset{\displaystyle OCH_3}{|}}{C}}H + H_2O \xrightarrow[heat]{H^+}$

(j) $CH_3\overset{\overset{\displaystyle O}{\|}}{C}Cl + CH_3CH_2OH \longrightarrow$

(k) $CH_3CH_2OCH_2CH_3 + H_2O \xrightarrow{^-OH}$

(l) $CH_3CH_2\overset{\overset{\displaystyle O}{\|}}{C}H + 2CH_3OH \xrightarrow{HCl}$

(m) $CH_3CH=CHCH_3 + H_2 \xrightarrow[\substack{heat, \\ pressure}]{Ni}$

(n) $CH_3\overset{\overset{\displaystyle OH}{|}}{\underset{\underset{\displaystyle CH_3}{|}}{C}}CH_2CH_3 \xrightarrow{K_2Cr_2O_7}$

(o) $(CH_3)_2CH\overset{\overset{\displaystyle O}{\|}}{C}OH + CH_3OH \xrightarrow{H^+}$

(p) $CH_3Br + NH_3 \longrightarrow$

(q) $CH_3CH_2SH \xrightarrow[\text{agent}]{\text{oxidizing}}$

(r) $CH_3\overset{\underset{\underset{\displaystyle Br}{|}}{}}{C}HCH_3 + KOH \xrightarrow{alcohol}$

(s) $C_6H_6 + Br_2 \xrightarrow{Fe}$

(t) $CH_3CH_2NH\overset{O}{\overset{\|}{C}}CH_3 + H_2O \xrightarrow{heat}$

(u) $CH_3\overset{O}{\overset{\|}{C}}CH_3 \xrightarrow{KMnO_4}$

(v) $NH_2CH_2CH_2CH_3 + HCl(aq) \longrightarrow$

(w) $CH_3CH_2CH_2CH_3 + H_2O \xrightarrow{H^+}$

(x) $CH_3C_6H_5 + HNO_3 \xrightarrow[heat]{H_2SO_4}$

(y) $CH_3OCH_2\overset{O}{\overset{\|}{C}}CH_3 + H_2O \xrightarrow{H^+}{heat}$

(z) $NH_2CH_2\overset{O}{\overset{\|}{C}}NHCH_2\overset{O}{\overset{\|}{C}}OH + H_2O \xrightarrow{heat}$

(aa) $CH_3\overset{O}{\overset{\|}{C}}OH + NaOH(aq) \longrightarrow$

(bb) $CH_3CH_2\overset{CH_3}{\underset{\underset{Cl^-}{H}}{\overset{|}{N}\overset{+}{-}H}} + NaOH(aq) \longrightarrow$

(cc) $CH_3OCH_2\underset{\underset{CH_3O}{|}}{C}HCH_3 + H_2O \xrightarrow{heat}$

(dd) $CH_3NH_2 + CH_3\overset{O}{\overset{\|}{C}}OH \xrightarrow{heat}$

(ee) $CH_3-S-S-CH_3 + H_2 \xrightarrow{Ni}$

(ff) $C_6H_5O\overset{O}{\overset{\|}{C}}CH_3 + H_2O \xrightarrow{H^+}$

(gg) $CH_3\overset{O}{\overset{\|}{C}}OCH_2CH_2O\overset{O}{\overset{\|}{C}}CH_3 + H_2O \xrightarrow[heat]{H^+}$

(hh) $CH_3CH_2Br + Mg \xrightarrow{ether}$

(ii) Product of hh $+ CH_3\overset{O}{\overset{\|}{C}}CH_3 \xrightarrow{then\ H^+}$

(jj) $CH_3CH=CHO\overset{O}{\overset{\|}{C}}CH_3 + H_2O \xrightarrow[heat]{H^+}$

(kk) $CH_3\overset{O}{\overset{\|}{C}}CH_3 + NH_3 + H_2 \xrightarrow[heat,\ pressure]{Ni}$

(ll) ⬡ $\xrightarrow{KMnO_4}$

(mm) $CH_3CH_2OH \xrightarrow[heat]{K_2Cr_2O_7}$

(nn) $C_6H_5Cl + H_2SO_4 \xrightarrow{heat}$

14. Assume that you have available any alcohol of four carbons or fewer plus any needed inorganic reagents. Outline steps for preparing each of the following compounds from these starting materials.
   (a) 2-butene          (b) 1-bromobutane          (c) 1-pentanol
   (d) ethylamine        (e) acetone                (f) acetic acid
   (g) methyl propionate (h) β-hydroxy-n-butyraldehyde (i) ethyl acetoacetate
   (j) n-butyramide      (k) n-octane.

15. Arrange the following compounds in order of increasing boiling point. (You should be able to do this without consulting tables.) Explain your answer.

$CH_3CH_2CH_2CH_2NH_2$        $CH_3CH_2CH_2CH_2OH$        $CH_3CH_2CH_2CH_2CH_3$        $CH_3CH_2\overset{O}{\overset{\|}{C}}OH$
       **A**                           **B**                           **C**                           **D**

16. Arrange the following compounds in order of increasing solubility in water, and outline the reasons for the order you select.

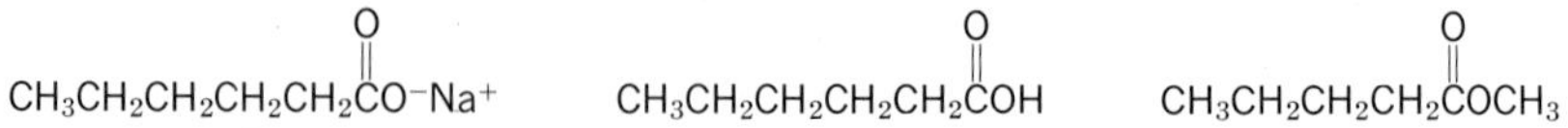

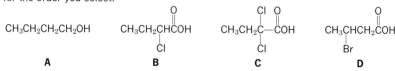

$$CH_3CH_2CH_2CH_2CH_2\overset{\displaystyle O}{\overset{\|}{C}}O^-Na^+$$
A

$$CH_3CH_2CH_2CH_2CH_2\overset{\displaystyle O}{\overset{\|}{C}}OH$$
B

$$CH_3CH_2CH_2CH_2\overset{\displaystyle O}{\overset{\|}{C}}OCH_3$$
C

$$CH_3CH_2CH_2CH_2CH_2OCH_3$$
D

$$CH_3CH_2CH_2CH_2CH_2CH_3$$
E

17. Arrange the following compounds in order of increasing acidity, and outline the reasons for the order you select.

$$CH_3CH_2CH_2CH_2OH$$
A

$$CH_3CH_2\overset{\displaystyle O}{\overset{\|}{C}}HCOH$$
$$\underset{\displaystyle Cl}{\overset{}{|}}$$
B

$$CH_3CH_2\overset{\displaystyle Cl}{\overset{|}{C}}\!-\!\overset{\displaystyle O}{\overset{\|}{C}}OH$$
$$\underset{\displaystyle Cl}{\overset{}{|}}$$
C

$$CH_3\overset{}{C}HCH_2\overset{\displaystyle O}{\overset{\|}{C}}OH$$
$$\underset{\displaystyle Br}{\overset{}{|}}$$
D

18. Arrange the following compounds in their order of increasing basicity, and outline the reasons for the order you select.

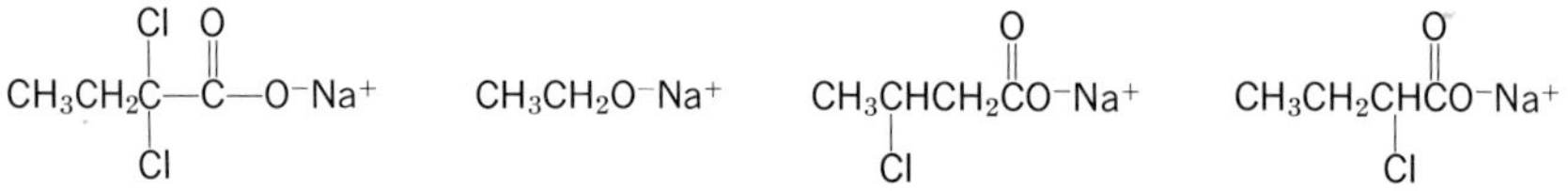

$$CH_3CH_2\overset{\displaystyle Cl}{\overset{|}{C}}\!-\!\overset{\displaystyle O}{\overset{\|}{C}}\!-\!O^-Na^+$$
$$\underset{\displaystyle Cl}{\overset{}{|}}$$
A

$$CH_3CH_2O^-Na^+$$
B

$$CH_3\overset{}{C}HCH_2\overset{\displaystyle O}{\overset{\|}{C}}O^-Na^+$$
$$\underset{\displaystyle Cl}{\overset{}{|}}$$
C

$$CH_3CH_2\overset{\displaystyle O}{\overset{\|}{C}}HCO^-Na^+$$
$$\underset{\displaystyle Cl}{\overset{}{|}}$$
D

19. Write detailed, step-by-step mechanisms for the following reactions: (a) acid-catalyzed hydrolysis of ethyl propionate, (b) aldol condensation of n-butyraldehyde, and (c) Claisen ester condensation of ethyl n-butyrate.

# CHAPTER EIGHT

# Carbohydrates

## PHOTOSYNTHESIS, AN INTRODUCTION

In the sweeping panorama of biological life, man and the animals are parasites, for they ultimately depend on the plant kingdom for food. Plants can take simple inorganic compounds such as carbon dioxide, water, and certain nitrogenous minerals and fashion the molecules that make up the three great classes of foods, carbohydrates, lipids, and proteins. They do this without any direct help from the animal kingdom, depending only on a constant flow of energy from the sun. Plants, not animals, use solar energy to transform the energy-poor and structurally simple substances, carbon dioxide, water, and minerals, into energy-rich and structurally complex materials, including the foods. This overall process is known as photosynthesis, and it is introduced here because its primary products are carbohydrates.

When we use the term *energy-poor compound,* we mean that the particular arrangement of electrons and nuclei that is the structure of the molecules of the compound is quite stable. There is no other structure to which its molecules can spontaneously and quickly rearrange, even when we give the compound an initial boost, as in striking a match to it, and even when it is in the presence of many types of other chemicals, for example, air, water, soil components, etc. An *energy-rich* compound, on the other hand, possesses an arrangement of electrons and nuclei that is relatively unstable. Given the correct initial conditions—activation and other potential reactants—the molecules of an energy-rich compound and those of the other reactants will rearrange in the direction of increasing stability and increasing probability. Released during this chemical event is the energy that is the difference between the energy-rich and the energy-poor arrangements. Often this energy appears as heat, which means simply that the average kinetic energy of the molecules of the products is greater than that of the molecules of the reactants. In an exothermic process molecules of the products move more energetically than those of the reactants. Billions and billions of molecules in violent motion, which give us the sensation of heat, can be made to drive pistons, blast ore deposits, and push jet planes and space rockets.

The energy released when, for example, a piece of wood is burned ultimately came from the sun. A piece of wood, however, is certainly not just a "bottle of sunlight"; it is a complex and highly organized mixture of compounds, mostly organic. The solar energy needed to make these compounds is temporarily stored in the form of distinctive electron-nuclei arrangements of energy-rich molecules that only plants can make from the fundamental materials, carbon dioxide, water, and minerals.

Man has not yet been able to duplicate the plant's remarkable feat of converting the energy of sunlight into the internal energy of chemicals. In fact, the mechanism by which plants perform this conversion, photosynthesis, is not fully understood. The simplest statement of the photosynthesis phenomenon is the "equation"

$$CO_2 + H_2O + \text{solar energy} \xrightarrow[\text{plant enzymes}]{\text{chlorophyll}} \underset{\substack{\text{Basic unit in} \\ \text{carbohydrates}}}{(CH_2O)} + O_2$$

This equation is a compromise between clarity and accuracy, but it is useful as a reference when the balance of nature is considered.

The process of photosynthesis has been estimated yearly to fix about 200 billion tons of carbon from carbon dioxide. Every three centuries all the carbon dioxide found in the atmosphere and in the dissolved state in the waters of the earth goes through the cycle photosynthesis, decay, photosynthesis. All the oxygen in our atmosphere is renewed by this cycle once in approximately twenty centuries. It has been estimated that close to 400 billion tons of oxygen are set free by photosynthesis each year. About 10 to 20% of this activity is carried out on land by familiar land plants. The remaining 80 to 90% is conducted below the surface of the oceans by algae. (The lowly oceanic algae constantly resupply us with oxygen, and yet very few of the thousands and thousands of substances man flushes into the oceans—residues of pesticides and herbicides, detergents, radioactive wastes—have been tested for their effects on algae.) Photosynthesis, of course, can occur only during daylight hours and, over land, only during growing seasons. The great atmospheric patterns of air movements ensure steady supplies of oxygen for people and land animals both at night and during the winter months.

The importance of green plants and the photosynthesis they accomplish is emphasized by the variety of substances they make, as outlined in Table 8.1. Among the important primary products of photosynthesis stand the energy-rich carbohydrates. Before we can consider how the body uses these compounds or any of the foodstuffs, we must first find out more precisely what they are.

## CLASSES OF CARBOHYDRATES

Molecules that comprise the carbohydrates are polyhydroxy aldehydes, polyhydroxy ketones, or substances that by simple hydrolysis yield these. Carbohydrates

**Table 8.1 The Synthesis of Organic Compounds in Green Plants**

| | | | The More Important Secondary Products | |
|---|---|---|---|---|
| Raw Materials | Catalysts | Primary Products | Formula Weight Less Than 1000 | Formula Weight Very High |
| $CO_2$<br>$H_2O$<br>Light energy<br>Inorganic nitrogen compounds | chlorophyll / plant enzymes → | Oxygen<br>Sugars<br>Amino acids<br>↓<br>Reserve materials:<br>Proteins<br>Fats and oils<br>Polysaccharides | Polyene pigments<br>Organic acids<br>Alcohols<br>Terpenes<br>Sterols<br>Waxes<br>Phosphatides (e.g., lecithin)<br>Inositol<br>Aromatic oils and fragrances<br>Alkaloids<br>Pyrrole pigments (e.g., anthocyanine pigments)<br>Nucleic acids | Cellulose<br>Hemicellulose<br>Gums<br>Pectins<br>Resins<br>Rubber<br>Tannins<br>Lignins |

Adapted from J. B. Conant, *Chemistry of Organic Compounds,* revised edition, The Macmillan Company, New York, 1939, page 597.

that cannot be hydrolyzed to simpler molecules are called *monosaccharides* or sometimes simple sugars. They are classified according to

1. How many carbons there are in one molecule:

| If the carbon number is | the monosaccharide is a |
|---|---|
| 3 | triose |
| 4 | tetrose |
| 5 | pentose |
| 6 | hexose, etc. |

2. Whether there is an aldehyde or a keto group present:

   If an aldehyde is present, the monosaccharide is an aldose.
   If a keto group is present, the monosaccharide is a ketose.

These terms may be combined. For example, a hexose that has an aldehyde group is called an aldohexose. A ketohexose possesses a keto group and a total of six carbons.

Carbohydrates that can be hydrolyzed to two monosaccharides are called *disaccharides*. Sucrose, maltose, and lactose are common examples. Starch and cellulose are common *polysaccharides;* their molecules yield many monosaccharide units when they are hydrolyzed.

# MONOSACCHARIDES

## OPTICAL ISOMERISM OF MONOSACCHARIDES

**D- and L-Families. Glyceraldehyde.** The simplest polyhydroxy aldehyde is glyceraldehyde; each molecule has one asymmetric carbon. It can therefore exist in two configurations related as object and a mirror image that cannot be superimposed (Figure 8.1). Until the advent of spectroscopic techniques, scientists could not determine which of two enantiomers was the dextrorotatory form and which the levorotatory; that is, they did not know what the *absolute configuration* of the molecules in samples of each of the two glyceraldehydes is. They noted with considerable interest, however, that a sample of glyceraldehyde that is dextrorotatory can be oxidized to a glyceric acid that is levorotatory. Or if the sample of glyceraldehyde is levorotatory, oxidation produces the dextrorotatory glyceric acid. As we can see from the equation

$$HOCH_2-\overset{\overset{\displaystyle H}{|}}{\underset{\underset{\displaystyle OH}{|}}{C}}-\overset{\overset{\displaystyle O}{\|}}{C}-H \xrightarrow{(O)} HOCH_2-\overset{\overset{\displaystyle H}{|}}{\underset{\underset{\displaystyle OH}{|}}{C}}-\overset{\overset{\displaystyle O}{\|}}{C}-OH$$

Glyceraldehyde          Glyceric acid

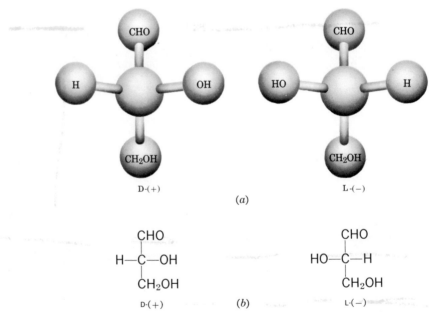

Figure 8.1   Absolute configurations of D- and L-glyceraldehyde, represented (*a*) as ball and stick models, (*b*) in plane projection diagrams.

this oxidation does not touch any of the bonds holding the four groups to the asymmetric carbon. Whatever the absolute configuration of the starting glyceraldehyde, it is retained during the oxidation. The relative configurations of the glyceraldehyde and its oxidation product, glyceric acid, are identical. There is no special correlation between the *sign* of rotation that the *substance* shows and the absolute configuration that its *molecules* have. Whether or not a reaction affects the *configuration* of an asymmetric carbon is of greater interest and importance than any (unpredictable) changes in the direction of optical rotation.

Early in the history of organic chemistry an arbitrary decision with a 50:50 chance of being correct was made. The molecules of the samples of glyceraldehyde that tested dextrorotatory were said to have the absolute configuration indicated in Figure 8.1. It did not matter whether the choice turned out to be wrong, because the only reason for the decision was to have a standard for assigning *relative* configurations to compounds related to glyceraldehyde, such as glyceric acid or the simple sugars. Thus if (+)-glyceraldehyde has the absolute configuration shown, then (−)-glyceric acid will have the same relative configuration. The next step was to devise a simple way to avoid the confusion the signs of rotation might cause. Thus it was agreed that all substances related to (+)-glyceraldehyde, regardless of their specific signs of rotation, would be said to be in the D-*family*. All those related to (−)-glyceraldehyde would be in the L-*family*. The symbols D and L do not stand for any words, nor do they necessarily say anything about the sign of rotation. They are merely small capital letters that give simple names to families of configurations.

D-(+)-Glyceraldehyde        oxidation (configuration does not change) →        D-(−)-Glyceric acid

For the designations D and L to have clear meanings, the route from the standard, D-(+)-glyceraldehyde, to the compound in question must be specified. It is conceivable that a series of steps could convert D-(+)-glyceraldehyde to the dextrorotatory glyceric acid. Instead of simply oxidizing the aldehyde group, we reduce it to a 1° alcohol function; the $HOCH_2$— group at the other end might be the one oxidized to the —COOH group. Conventions must therefore be spelled out to preserve clarity. We say that (−)-glyceric acid is in the D-family because we arbitrarily choose to match the —CHO of the D-(+)-glyceraldehyde and the —COOH of the glyceric acid.

**Plane Projection Formulas.** In writing simple molecules with one asymmetric center, even untalented artists can sketch a perspective stick and ball picture. Drawing complicated systems with perspective, especially those having two or more asymmetric centers, becomes too unwieldy and frustrating. Chemists long ago began to use *plane projection* representations instead of attempting three-dimensional, perspective sketches. Again, certain conventions must be heeded. This method attempts a strictly two-dimensional diagram, creating traps for the unwary. By convention the carbon skeleton that includes the asymmetric carbon is positioned *vertically*. The carbons immediately above and below the asymmetric carbon are imagined to project to the *rear,* as in Figure 8.1. The carbon normally given the lowest number for naming purposes is at the *top*. The two groups positioned horizontally are imagined to project *forward*. The asymmetric carbon is imagined as occupying the intersection of the two perpendicular lines. When more than one asymmetric carbon is present (i.e., two or more such intersections), the vertical

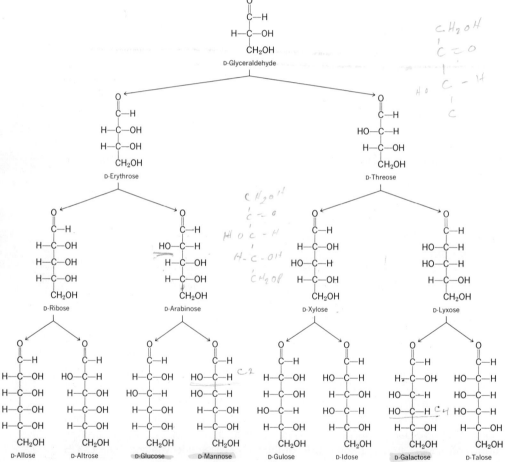

**Figure 8.2** The D-family of aldoses through the aldohexoses. In carbohydrate chemistry, if the —OH on the asymmetric carbon farthest from C-1 projects to the right in a plane projection diagram, the substance is in the D-family. There are eight other aldohexoses, all in the L-family, each being a mirror image of a member of the D-family. The aldoses most important in our study are three, D-ribose, D-glucose, and D-galactose.

bonds *at each intersection* must be imagined to project rearward, the horizontal bonds forward. D- and L-Glyceraldehydes are shown in both their perspective and plane projection formulas in Figure 8.1.

With this brief introduction to D- and L-families, we are in a position to understand what it means that all the nutritionally important sugars are in the D-family. The human animal cannot use members of the L-family of sugars. We can assimilate D-glucose, but we are unable to metabolize L-glucose. Among the simple sugars the configuration of the asymmetric carbon farthest from the carbonyl group determines family membership. If in a plane projection diagram, constructed by the conventions, the hydroxyl at this carbon projects to the right (as in D-glyceraldehyde), the molecule is in the D-family. Figure 8.2 shows the family tree for the D-family of aldoses —from the triose, D-glyceraldehyde, through all the D-aldohexoses, including the all-important glucose. The diagram may be called a family tree because it is possible to start with D-glyceraldehyde and by a series of several reactions make all the compounds shown. In all these plane projection diagrams the asymmetric carbon farthest from the carbonyl has a hydroxyl projecting to the right. Another family tree could be constructed from L-glyceraldehyde.

## CHEMICAL PROPERTIES OF ALDOSES. GLUCOSE

**Evidence for Its Structure.** Given a white powder with the molecular formula $C_6H_{12}O_6$ and several known physical properties, the chemist must still determine how the six carbons, twelve hydrogens, and six oxygens are put together. Are the carbons in a straight chain, a branched chain, or possibly a ring? What is the nature of the oxygens? Are they in hydroxyl groups, keto groups, aldehyde groups, or some other structural arrangement? The following reactions illustrate how information bearing on these questions is gathered.

GLUCOSE—
$C_6H_{12}O_6$

— $NH_2OH$ ⟶ forms a monooxime; therefore $-\overset{\overset{O}{\|}}{C}-H$ or $C-\overset{\overset{O}{\|}}{C}-C$ is indicated.

*anhydride* $(CH_3\overset{\overset{O}{\|}}{C})_2O$ ⟶ forms a pentaacetate; therefore five $-OHs$ are *? all* probably present.

— $Br_2, H_2O$ ⟶ forms a six-carbon carboxylic acid; with no loss of carbon $-\overset{\overset{O}{\|}}{C}-H$ is therefore strongly indicated.

— $HNO_3$ ⟶ forms a six-carbon dicarboxylic acid; with no loss of carbon, not only $-\overset{\overset{O}{\|}}{C}-H$ but also $-CH_2OH$ is therefore present.

— $H_2, Ni$ ⟶ forms $C_6H_{14}O_6$; two $-Hs$ are probably taken up by the aldehyde group.

| vigorous reduction

forms a previously known derivative of *n*-hexane; therefore the carbon skeleton is unbranched.

Of the six carbons in the straight chain, one is used in the aldehyde group, —CHO. The remaining five must accommodate the five hydroxyl groups indicated by the behavior toward acetic anhydride. Because chemists have rarely found a structure in which one carbon holds two hydroxyl groups, we distribute the five hydroxyl groups among the remaining five carbons. The general structure for glucose (without regard for optical isomers) must therefore be

$$CH_2-CH-CH-CH-CH-\overset{\overset{\displaystyle O}{\|}}{C}-H$$
$$\phantom{CH_2-}OH\ \ \ OH\ \ \ OH\ \ \ OH\ \ \ OH$$

Most of the reactions mentioned in this proof of structure have been studied earlier. We must take special note, however, of two oxidizing agents that give oxidized forms of the aldoses of interest in biochemistry as well as of other oxidizing agents that are useful in detecting sugars.

**Oxidation of Aldoses.** A solution of bromine in water acts as a mild oxidizing agent to convert the aldehyde group of an aldose to a carboxylic acid group. The resulting monocarboxylic acid bears the family name, *glyconic acid* (e.g., gluconic acid from glucose, galactonic acid from galactose, etc.). In general:

$$\overset{\overset{\displaystyle O}{\|}}{C}-H \qquad \xrightarrow{\text{Br}_2 \text{ in H}_2\text{O}} \qquad \overset{\overset{\displaystyle O}{\|}}{C}-OH$$
$$(CH-OH)_n \qquad\qquad\qquad (CH-OH)_n$$
$$CH_2OH \qquad\qquad\qquad\quad CH_2OH$$

An aldose       A glyconic acid

The point we shall find helpful later is that the aldehyde group can be oxidized without oxidizing any of the alcohol functions.

Nitric acid, a stronger oxidizing agent than bromine in water, converts aldoses into dicarboxylic acids having the general name, *glycaric acids.* (e.g., glucaric acid from glucose, galactaric acid from galactose, etc.). In general:

$$\overset{\overset{\displaystyle O}{\|}}{C}-H \qquad \xrightarrow{\text{HNO}_3} \qquad \overset{\overset{\displaystyle O}{\|}}{C}-OH$$
$$(CH-OH)_n \qquad\qquad\qquad (CH-OH)_n$$
$$CH_2OH \qquad\qquad\qquad\quad \underset{\underset{\displaystyle O}{\|}}{C}-OH$$

An aldose       A glycaric acid

For quick test tube tests, Tollens' reagent and Benedict's or Fehling's solutions all give positive reactions with the aldoses (cf. p. 196). The reactions are not clean-cut and simple, however, and for this reason they are not of much help in making specific oxidation products of sugars. In contrast with bromine in water and nitric acid, Tollens', Benedict's and Fehling's solutions are all alkaline. The base causes rearrangements through intermediate enols, but the fate of the organic species is too complicated a story to examine in detail. Because all the aldoses and all the ketoses give positive tests with these reagents, they are often called *reducing sugars*.

**Osazone Formation.** On the basis of what we learned on page 206, we might expect aldoses and ketoses to form phenylhydrazones. In acid they do. We would not expect, however, that an excess of phenylhydrazine could react further with the phenylhydrazones until three molecules of phenylhydrazine are consumed. By-products are ammonia and aniline. This behavior is typical of $\alpha$-hydroxy aldehydes and $\alpha$-hydroxy ketones, and the double phenylhydrazones are called *osazones*.

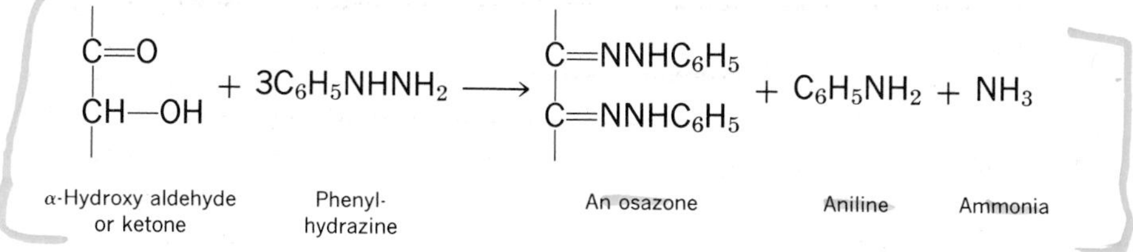

$$
\begin{array}{c}
\text{C}=\text{O} \\
\text{CH}-\text{OH}
\end{array}
+ 3C_6H_5NHNH_2 \longrightarrow
\begin{array}{c}
\text{C}=\text{NNHC}_6\text{H}_5 \\
\text{C}=\text{NNHC}_6\text{H}_5
\end{array}
+ C_6H_5NH_2 + NH_3
$$

| $\alpha$-Hydroxy aldehyde or ketone | Phenyl-hydrazine | An osazone | Aniline | Ammonia |

Specific examples:

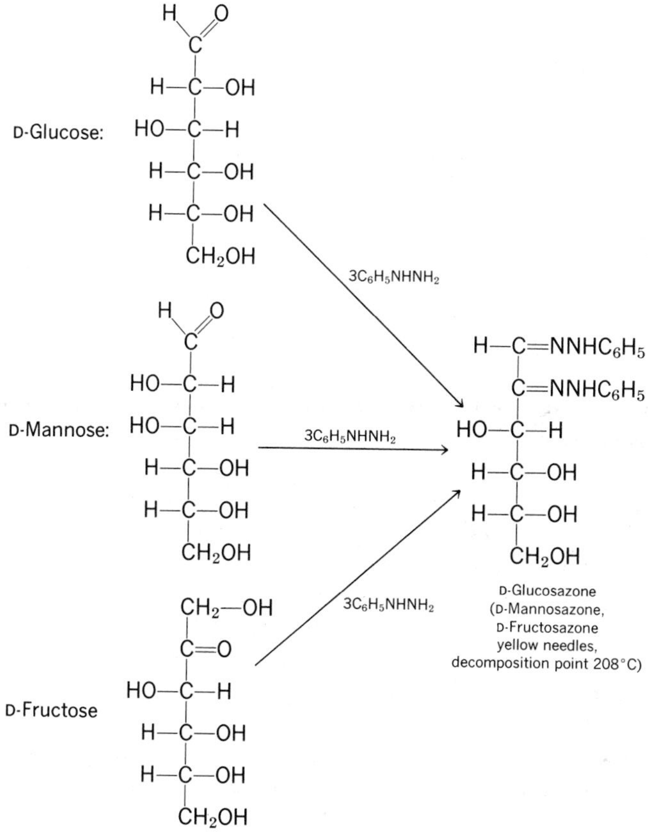

D-Glucose:

3C₆H₅NHNH₂

D-Mannose: 3C₆H₅NHNH₂

H—C=NNHC₆H₅
C=NNHC₆H₅
HO—C—H
H—C—OH
H—C—OH
CH₂OH

D-Glucosazone
(D-Mannosazone,
D-Fructosazone
yellow needles,
decomposition point 208°C)

D-Fructose: 3C₆H₅NHNH₂

The osazones are crystalline solids with characteristic crystalline shapes, decomposition points (rather than melting points), and specific optical rotations. Each aldose or ketose forms its osazone under carefully standardized conditions and at its own characteristic rate. The preparation of an osazone is therefore a useful way of identifying a simple sugar, but, as the examples show, two or more sugars may give the same osazone.

*Thin-layer chromatography* is a more recent technique for identifying sugars. A solution of a mixture of sugars is allowed to move by capillary action across a thin layer of silica, alumina, or some other adsorbing material carefully deposited on glass. The sugars move less rapidly than the solvent, and because of differing abilities to be adsorbed by the layer, each sugar moves at a characteristic rate. In time the sugars are in separate regions of the thin layer, and special sprays can then locate them. The relative rates of movement, easy to measure under standardized conditions, join the list of other physical constants that identify sugars. The members of many other classes of compounds are identified by thin-layer chromatography.

The absolute configuration of D-(+)-glucose, as shown in Figure 8.2, was deduced by a series of reactions and investigation which, although omitted from our study, are discussed in most organic chemistry textbooks. A study of them would make a fine individual or class project.

## CYCLIC FORMS OF GLUCOSE

Some properties of glucose, notably mutarotation and acetal (glucoside) formation, are not consistent with the pentahydroxycaproaldehyde system thus far adopted as its structure.

**Mutarotation.** When ordinary crystalline glucose is freshly dissolved in water, it shows a specific rotation $[\alpha]_D + 113°$. (Here the subscript D refers to the D-line of sodium vapor light.) While the solution remains at room temperature, this value gradually changes, stopping at $[\alpha]_D +52°$. This aged solution can be made to yield the original glucose by suitable conditions of crystallization. Apparently, no deep-seated chemical change has occurred to produce this gradual shift in specific rotation. The change in rotatory power of a solution of an optically active compound is called *mutarotation.*

If glucose is recovered from a solution by evaporating it at a temperature above 98°C, another form is obtained. Its specific rotation, in a freshly prepared solution, is $[\alpha]_D +19°$, but this solution also mutarotates until a final value of $[\alpha]_D +52°$, is obtained, the same final value cited for ordinary glucose. In fact, glucose with a specific rotation, in a fresh solution, of $[\alpha]_D +113°$ can be separated from this second aged solution. The structure of D-(+)-glucose that we have used thus far does not provide any obvious clue to explain this behavior. Moreover, other chemical properties of glucose that do not fit too well with this structure are known.

**Glucosides.** Another peculiarity of glucose is that conversion of its aldehyde group to an acetal requires only one mole of alcohol, instead of two, per mole of glucose (cf. p. 202).

$$C_6H_{12}O_6 + HOCH_3 \xrightarrow{H^+} C_6H_{11}O_5\text{---}OCH_3 + H_2O$$

Glucose     Methyl     Methyl glucoside<br>alcohol

The product, called in the example methyl glucoside, is stable in a basic solution. It does not react positively to Tollens' test. In other words, it does not behave as though it were a potential aldehyde, that is, a hemiacetal. It behaves as though it were an acetal. The reaction shown can in fact be reversed. Like any acetal, methyl glucoside can be easily hydrolyzed. It does not mutarotate, however.

Just as glucose can be prepared in two forms, which are called $\alpha$-glucose and $\beta$-glucose, methyl glucoside can also be made in two forms, $\alpha$-methyl glucoside and $\beta$-methyl glucoside.

**Cyclic Hemiacetal Forms for Glucose.** These observations make sense in terms of structural formulas if glucose can exist in a cyclic hemiacetal form (Figure 8.3). If glucose is already a hemiacetal, then while still being a potential aldehyde capable of showing virtually all the reactions of an aldehyde, it would require only one additional molecule of an alcohol to become an acetal. The two forms of glucose that mutarotate are possible because the hemiacetal —OH group may be on either one side of the ring or the other.

All three forms are present in equilibrium in aged solutions of either $\alpha$-glucose or $\beta$-glucose. The proportions at room temperature are 36% alpha, 64% beta, and 0.02% open form. Thus the phenomenon of mutarotation is explained as resulting from the slow establishment of a dynamic equilibrium between all three forms in solution, regardless of which form in the solid state, the alpha or beta, is used to make the glucose solution. During ring closure of the open form, the newly generated —OH group at C-1 may come out either above the ring or below it, in relation to the —CH$_2$OH group which is our reference. In the closed forms C-1 is now asymmetric. The alpha and beta forms, of course, are not related as object to mirror image. They are not enantiomers but diastereomers. They differ only in the configuration at C-1.

What happens during mutarotation might be more easily understood if we try to isolate only the reacting sites. The original carbonyl carbon is shown in boldface.

$\alpha$-Glucose                  $\beta$-Glucose

Hemiacetal     Alcohol + aldehyde     Hemiacetal<br>(or ketone)

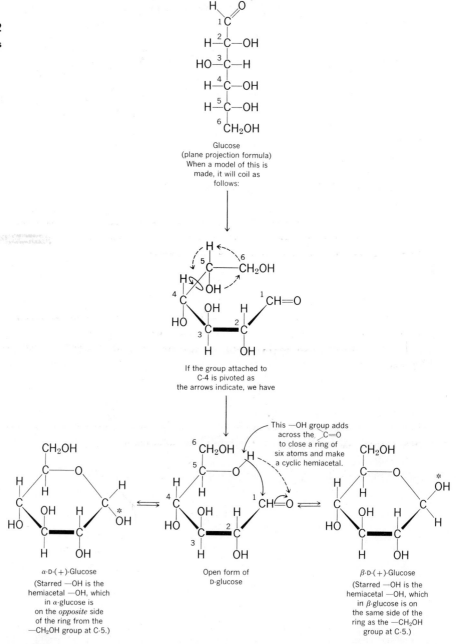

**Figure 8.3** The cyclic hemiacetal forms of D-glucose and their relation to its open-chain polyhydroxy aldehyde structure.

A hemiacetal forms from the simple addition of the elements of an alcohol group across the double bond of a carbonyl group. If the same molecule contains both the —OH and the C=O, the reaction is intramolecular and the hemiacetal will necessarily be cyclic.

The glucosides are C-1 ethers (actually acetals) of the closed forms of the aldoses. Two forms, alpha and beta, are possible because alpha and beta forms for starting materials are possible.

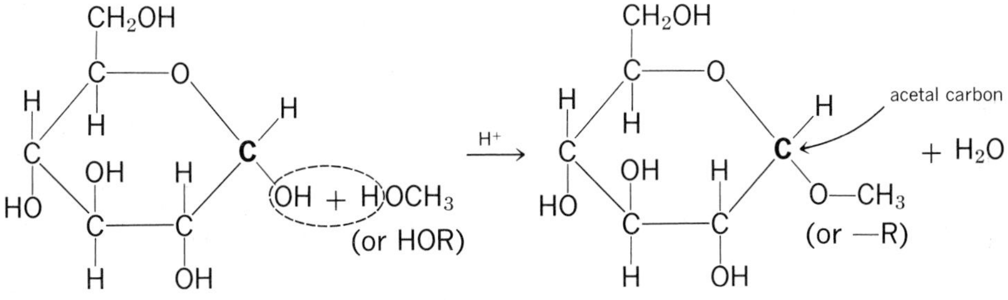

α-Glucose

α-Methyl glucoside

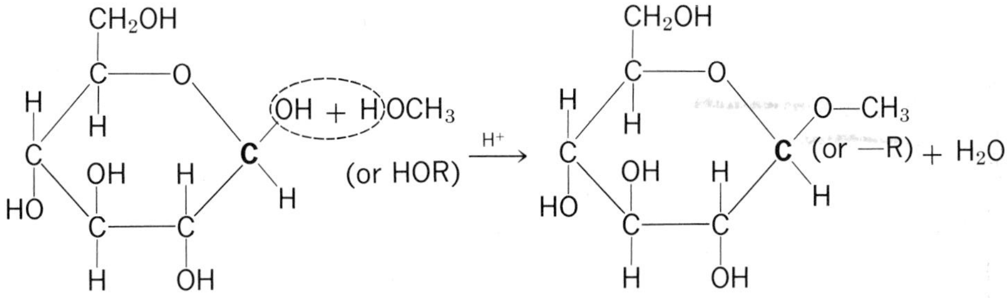

acetal carbon

+ $H_2O$

β-Glucose

β-Methyl glucoside

Since glucose is a hemiacetal, it is a potential aldehyde. The fact that only a trace amount (0.02%) of the open-chain aldehyde is present at equilibrium in an aqueous solution does not reduce the availability of the aldehyde group to any reagent that is capable of demanding it (e.g., oxidizing agents such as Benedict's or Tollens' reagents). The hemiacetal linkage apparently opens easily enough so that when the trace amount of aldehyde form is consumed, the equilibrium shifts to provide a fresh supply. For this reason it is still logical to define the monosaccharides as we did, as polyhydroxy aldehydes or ketones. The central ideas that we shall need to carry forward to our study of the metabolism of glucose are these:

1. Glucose easily forms glucosides at its hemiacetal —OH of the closed forms.
2. Glucose is easily oxidized, especially at its aldehyde group, open form.
3. Glucose has other —OHs that might (a) be oxidized, (b) form esters (especially with phosphoric acid, to anticipate later needs), or (c) form acetals.

The student should learn the structure of glucose, so frequently will we apply it. If we proceed step by step, learning to write the closed forms is not difficult.

1. First write a six-membered ring with an oxygen in the upper right-hand corner.

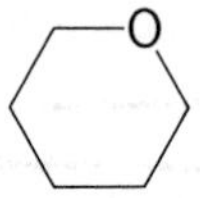

2. Next "anchor" the —$CH_2OH$ on the carbon to the left of the oxygen. (Let all the —Hs attached to ring carbons be "understood.")

or condense to

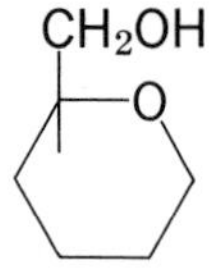

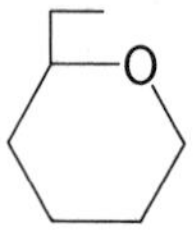

3. Continue in a counterclockwise way around the ring, placing the —OHs first down, then up, then down.

or condense to

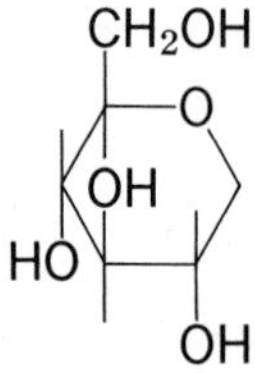

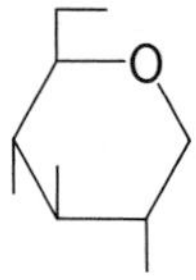

4. Finally, at the last site on the trip, how the last —OH is positioned depends on whether the alpha or the beta form is to be written. The alpha is "down," the beta "up." If this detail is immaterial, or if the equilibrium mixture is intended, the structure may be written as

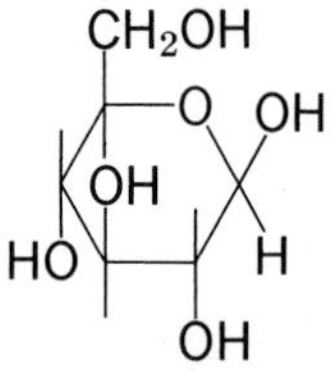

*β*-Glucose

or condense to

or condense to

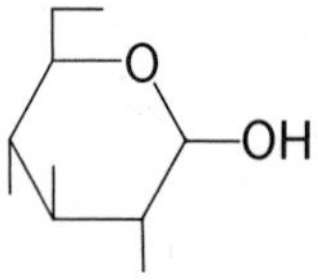

**Occurrence of Glucose.** Glucose, the most important hexose, is found in most sweet fruits, especially in ripe grapes. It is therefore sometimes called grape sugar. A normal constituent of the bloodstream, it is also often called blood sugar. Because glucose is dextrorotatory, a third nickname is dextrose. The energy-rich molecules of glucose constitute one of the chief sources of chemical energy for animals. It is the building block for several other carbohydrates such as maltose, starch, glycogen, the dextrins, and cellulose. It is also a structural unit in lactose (milk sugar) and sucrose (table sugar, cane sugar, beet sugar).

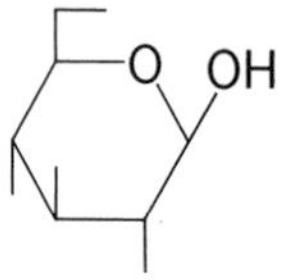

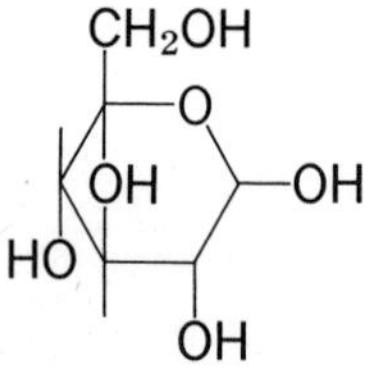

## GALACTOSE

Most galactose occurs naturally in combined forms, especially in the disaccharide, lactose (milk sugar). Galactose, a diastereomer of glucose, differs from it only in the orientation of the C-4 hydroxyl. Like glucose, it is a reducing sugar, it mutarotates, and it exists in three forms, alpha, beta, and open.

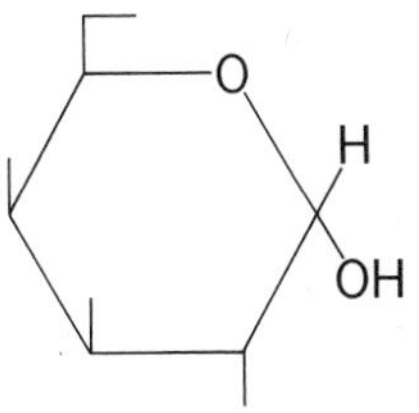

α-Galactose 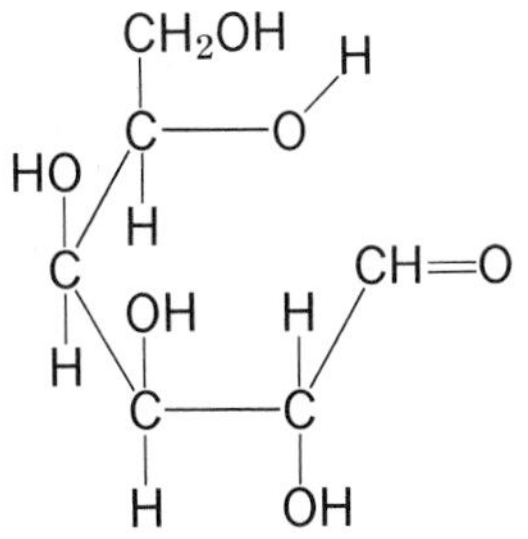      Galactose, open form 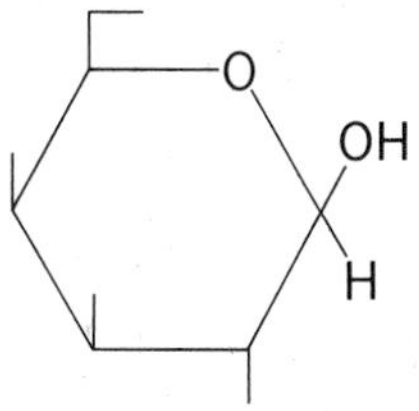      β-Galactose

## KETOHEXOSES. FRUCTOSE

Fructose, the only important ketohexose, is found together with glucose and sucrose in honey and fruit juices. It can exist in more than one form, and as a building unit in sucrose it exists in a cyclic, five-membered hemiketal form.[1]

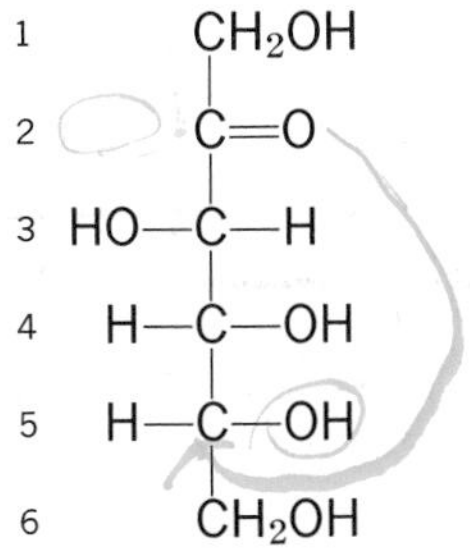

D-(−)-Fructose 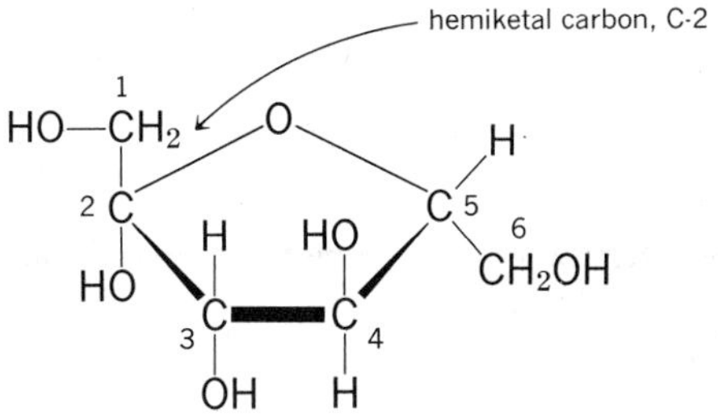      One of the cyclic forms of fructose

Because fructose is strongly levorotatory, we sometimes call it levulose; the specific rotation is $[\alpha]_D -92.4°$.

## AMINO SUGARS

Amino derivatives of glucose and galactose occur widely in nature as building units for some important polymeric substances. Glucosamine is a structural unit in chitin, the principal component of the shells of lobsters, crabs, and certain insects, and in heparin, a powerful blood anticoagulant. Galactosamine is a structural unit of chondroitin, a polymeric substance important to the structure of cartilage, skin, tendons, adult bone, the cornea, and heart valves.

---

[1] When the hemiacetal type of compound is made from a ketone instead of an aldehyde, it is called a hemiketal.

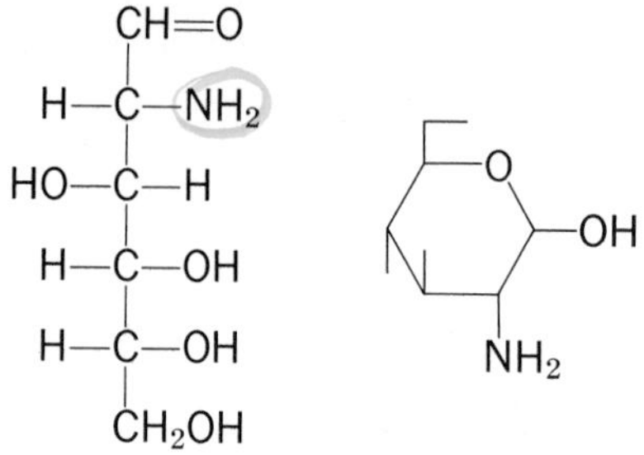

CH=O
H—C—NH₂
HO—C—H
H—C—OH
H—C—OH
CH₂OH

—OH

NH₂

open form       cyclic form

Glucosamine

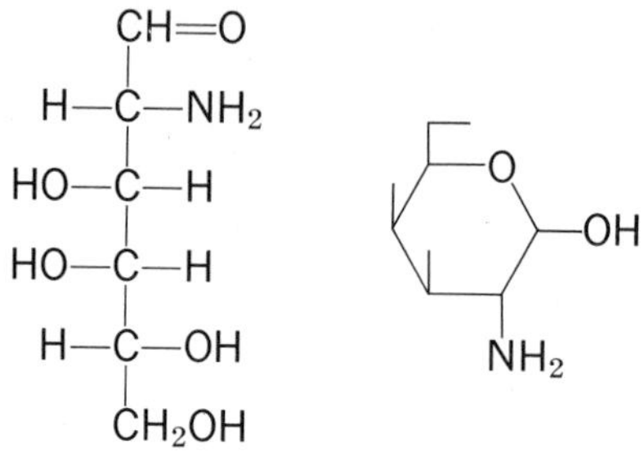

CH=O
H—C—NH₂
HO—C—H
HO—C—H
H—C—OH
CH₂OH

—OH

NH₂

open form       cyclic form

Galactosamine

## DISACCHARIDES

Nutritionally, the three important disaccharides are maltose, lactose, and sucrose. A fourth will be mentioned, cellobiose, which is obtained by partial hydrolysis of cellulose. How these are related to the monosaccharides just discussed may be seen in the following word equations.

$$\text{Maltose} + H_2O \longrightarrow \text{glucose} + \text{glucose}$$
$$\text{Lactose} + H_2O \longrightarrow \text{glucose} + \text{galactose}$$
$$\text{Sucrose} + H_2O \longrightarrow \text{glucose} + \text{fructose}$$
$$\text{Cellobiose} + H_2O \longrightarrow \text{glucose} + \text{glucose}$$

All these disaccharides are glycosides and are formed by one monosaccharide acting as a hemiacetal and the other as an alcohol to make an acetal.

### MALTOSE

Although present in germinating grain, maltose or malt sugar does not occur widely in the free state in nature. It can be prepared by partial hydrolysis of starch. As with glucose, two forms of maltose are known and each mutarotates. Maltose is a reducing sugar and hydrolyzes to yield two molecules of glucose. The enzyme *maltase* catalyzes this reaction, and it is known to act only on α-glucosides, not on β-glucosides. We could imagine maltose being formed by the splitting out of water between the hemiacetal —OH of one glucose unit and the C-4 —OH of the other.

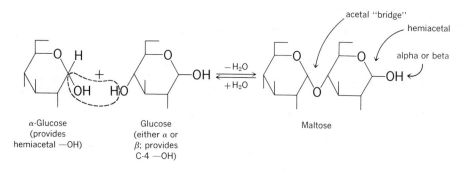

acetal "bridge"

hemiacetal

alpha or beta

−H₂O / +H₂O

α-Glucose
(provides
hemiacetal —OH)

Glucose
(either α or
β; provides
C-4 —OH)

Maltose

## LACTOSE

Lactose or milk sugar occurs in the milk of mammals, 4 to 6% in cow's milk, 5 to 8% in human milk. It is obtained commercially as a by-product in the manufacture of cheese. Like maltose, it exists in alpha and beta forms, is a reducing sugar, and mutarotates. Its hydrolysis yields galactose and glucose, and its formation may be visualized as the splitting out of water between $\beta$-galactose, acting as the hemiacetal, and glucose, furnishing the C-4 —OH group.

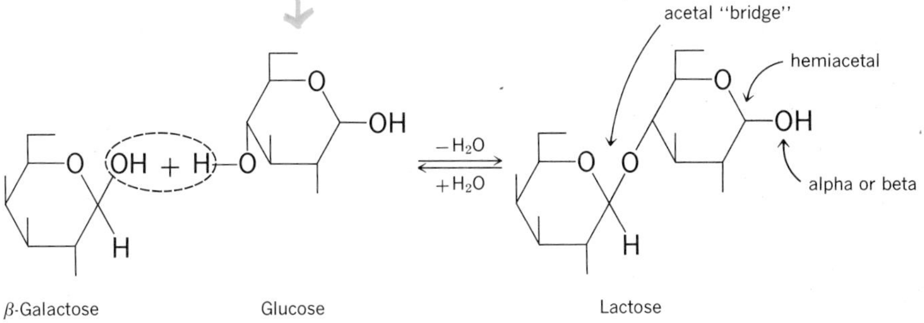

β-Galactose          Glucose                    Lactose

## SUCROSE

The juice of sugarcane, which contains about 14% sucrose, is obtained by crushing and pressing the canes. It is then freed of protein-like substances by the precipitating action of lime. Evaporation of the clear liquid leaves a semisolid mass from which raw sugar is isolated by centrifugation. (The liquid that is removed is *blackstrap molasses.*) Raw sugar (95% sucrose) is processed to remove odoriferous and colored contaminants. The resulting white sucrose, our table sugar, is probably the largest volume *pure* organic chemical produced. Much of our supply of sucrose now comes from sugar beets, and so-called beet sugar and cane sugar are chemically identical. Structurally, a molecule of sucrose is derived from one glucose unit and one fructose unit:

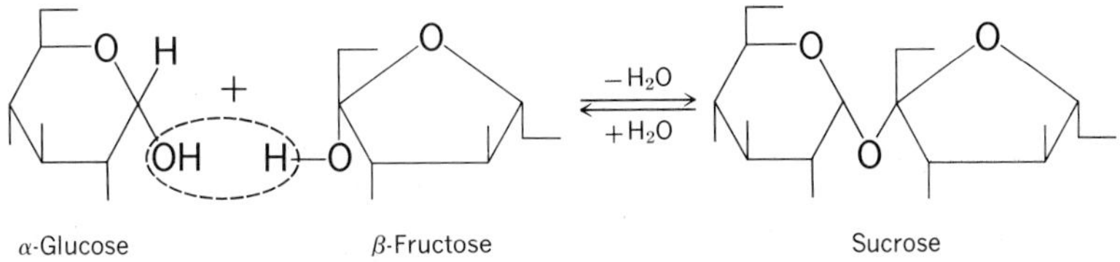

α-Glucose              β-Fructose                  Sucrose

An acetal oxygen bridge links the two units. No hemiacetal group is present in the sucrose molecule. Sucrose is not a reducing sugar.

The 50:50 mixture of glucose and fructose that forms when sucrose is hydrolyzed is often called *invert sugar.* Honey, for example, consists of this sugar for the most part. The term "invert" comes from the change or inversion in sign of optical rotation that occurs when sucrose, $[\alpha]_D$ +66.5°, is converted into the 50:50 mixture, $[\alpha]_D$ −19.9°.

## CELLOBIOSE

Cellobiose ("cello-," cellulose; "bi-," two; "-ose," sugar) is the disaccharide unit in cellulose (e.g., cotton fiber). Like maltose, it delivers two glucose molecules when it is hydrolyzed. It differs structurally from maltose only in the orientation of the acetal oxygen bridge.

β-Glucose          Glucose          Cellobiose

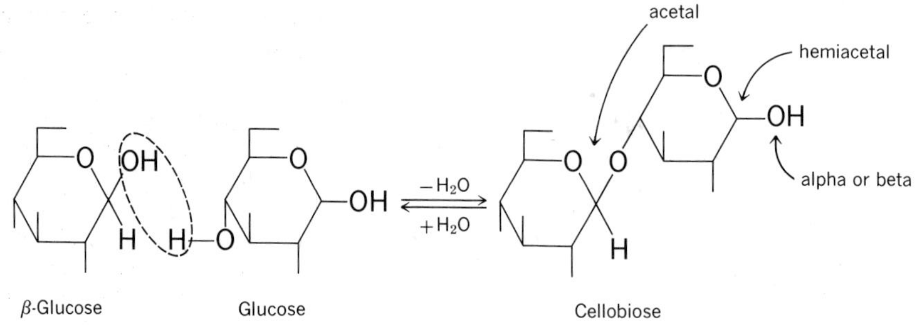

# POLYSACCHARIDES

Much of the glucose produced in a plant by photosynthesis is used to make its cell walls and its rigid fibers. Much is also stored for the food needs of the plant. But instead of being stored as glucose molecules, which are too soluble in water, it is converted to a much less soluble form, starch. This polymer of glucose is particularly abundant in plant seeds. We use these two polymers of glucose in about the same way that plants do. Cellulose and its more woody relatives are made into enclosures ranging from frame houses to shirts and blouses. Starch is used for food. During digestion we break starch down eventually to glucose, and what we do not need right away for energy we store. We do not normally excrete excess glucose but convert it to a starch-like polymer, glycogen, or to fat. In this section we concentrate on the three types of glucose polymers, starch, glycogen, and cellulose.

### STARCH

The skeletal outline of what is believed to constitute the basic structure of starch is shown in Figure 8.4. It is actually a mixture of polyglucose molecules, some linear and some branched. One type, *amylose,* consists of long, unbranched polymers of α-glucose. The other, *amylopectin,* is the branched polymer of α-glucose. Natural starches are about 10 to 20% amylose and 80 to 90% amylopectin.

**Physical Properties.** The huge molecules found in starch, with their many hydroxyl groups, stick to each other quite strongly, but in a disorderly way. These hydrogen bond networks are not so effectively broken up by water molecules that starch is soluble in water. It is only slightly so, and starch "solutions," prepared by grinding starch in a small amount of water and then pouring the slurry into a larger volume of boiling water, are really colloidal dispersions.

**Chemical Properties.** The acetal oxygen bridges linking glucose units together in starch are easily hydrolyzed, especially in the presence of acids or certain

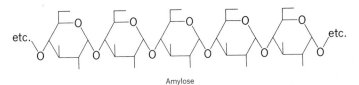

Amylose

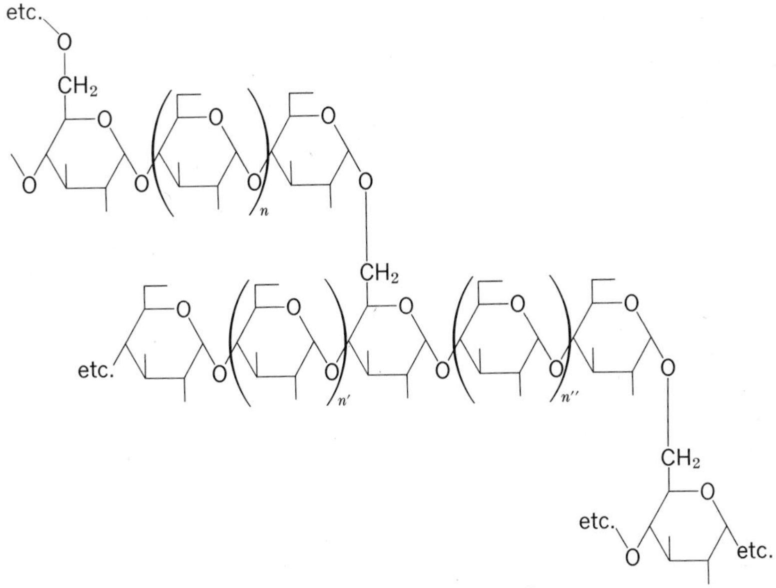

Amylopectin ($n$, $n'$, $n''$ = large numbers)

**Figure 8.4** Basic structural features of the types of glucose polymers found in starches. Molecular weight measurements on various preparations of starches have given results varying from 50,000 to several million. The higher values probably more truly represent the sizes of natural starch molecules, for during the isolation of this substance some partial hydrolysis very likely occurs. An amylose molecule with a molecular weight of 1 million would contain slightly over 6000 gucose units.

enzymes. We have learned to expect this behavior of acetals and, when we digest starch, digestive juices do nothing more than hydrolyze it.

The partial breakdown products of amylopectin are still large molecules called the *dextrins*. They are used to prepare mucilages, pastes, and fabric sizes. The step-by-step hydrolysis of amylopectin to glucose through the dextrins and maltose is described in Figure 8.5.

Starch is not a reducing carbohydrate. The potential aldehyde groups would be at the ends of chains only in percentages too low for detection by Benedict's reagent. Starch does, however, give an intense, brilliant blue-black color with iodine.[2] This *iodine test* for starch can detect extremely minute traces of starch in solution.

[2] The starch-iodine reagent is iodine, $I_2$, dissolved in an aqueous solution of sodium iodide, NaI. Iodine, by itself, is only slightly soluble in water. When iodide ion is present, however, it combines with iodine molecules to form a triiodide ion, $I_3^-$, which liberates iodine easily on demand.

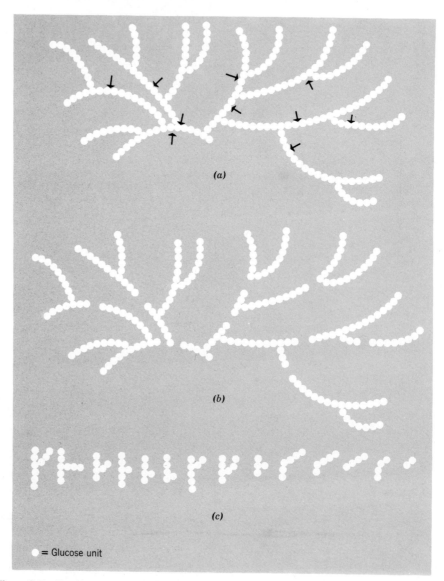

*(a)*

*(b)*

*(c)*

⬤ = Glucose unit

**Figure 8.5** The hydrolysis of amylopectin according to P. Bernfeld. (*a*) Model of amylopectin. (*b*) Dextrins of medium molecular weight that give purple to red colors with iodine. These dextrins result from random hydrolysis of about 4% of the oxygen bridges of amylopectin. See arrows in part *a*. (*c*) Low-formula-weight dextrins (limit dextrins) that are actually hepta-, hexa-, and pentasaccharides. These give a brown color with iodine. (From P. Bernfeld, *Advances in Enzymology*, Vol. 12, 1951, page 390.)

The chemistry of the test is not definitely known, but iodine molecules are believed to become trapped within the vast network of starch molecules. Should this network disintegrate, as it does during the hydrolysis of starch, the test will fail. In fact, the course of the hydrolysis of starch in a test tube can be followed by this test, for as the reaction proceeds the color produced by iodine gradually changes from blue-black to purple to red to no color.

## GLYCOGEN

Liver and muscle tissue are the main sites of glycogen storage in the body. Under the control of enzymes, some of which are in turn controlled by hormones, glucose molecules can be mobilized from these glycogen reserves to supply chemical energy for the body.

Glycogen differs from starch by the apparent absence of any molecules of the unbranched, amylose type. It is branched very much like amylopectin, perhaps even more so. Molecular weights of various glycogen preparations have been reported over a range of 300,000 to 100,000,000, corresponding roughly to 1800 to 60,000 glucose units. During the digestion and absorption of a meal containing carbohydrates, the body builds up its glycogen deposits. Between meals, during fasts, the deposits are made to deliver glucose.

Dispersions of glycogen in water are opalescent, turning a violet-red color with iodine.

## CELLULOSE

Starch and glycogen are polymers of the alpha form of glucose. Cellulose is a polymer of the beta form. Its molecules are unbranched and resemble amylose. A portion of the structure of cellulose is shown in Figure 8.6.

The structural difference between amylose and cellulose is one of the tremendous trifles in nature. In humans amylose is digestible but cellulose is not. Yet the only difference between them structurally is the orientation of the oxygen bridges. Human beings and carnivorous animals do not have the enzyme needed to make the hydrolysis of cellulose rapid enough to be of any use. Many microorganisms, snails, and ruminants (cud-chewing animals) can use cellulose, however. The ruminants have suitable microorganisms in their alimentary tracts whose own enzyme systems catalyze the conversion of cellulose into small molecules the ruminants can use. It is fortunate that *some* systems contain enzymes capable of catalyzing the hydrolysis of cellulose. Otherwise the land would soon be covered by the dead debris of grasses, leaves, and trees that have fallen, and the remains of the annual plants. Soil organisms, fortunately, go to work on these remains and break them down. (Civilization is indeed lucky that DDT does not kill soil bacteria. Its application since World War II has been so widespread that traces of DDT have been found in arctic snows where no one has deliberately brought it. The effect that DDT might have on soil microorganisms, however, was not thoroughly tested before we started using it!)

Cellulose makes up the cell membranes of the higher plants and gives them their rigidity. It is easily the most abundant organic compound in the world. Cotton

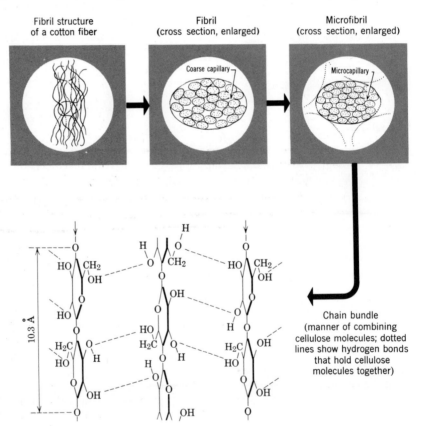

**Figure 8.6** Cellulose, a linear polymer of β-glucose. Note that, in order to indicate the linearity of the structure, every other glucose unit is "flipped over" from the normal way of writing it.

fiber is almost 98% cellulose; each molecule has from 2000 to 9000 β-glucose units, arranged roughly parallel to one another and to the axis of the fiber. Structural details are illustrated in Figure 8.7. Several cellulose molecules twist and overlap to form the next to smallest unit in a cotton fiber, the *chain bundle*. A vast interlacing network of hydrogen bonds provides the necessary forces of attraction. When several chain bundles overlap, they form a microscopic unit of the visible fiber, a *microfibril*. Within it, running between the chain bundles, is a fine capillary network.

Fibril structure
of a cotton fiber

Fibril
(cross section, enlarged)

Microfibril
(cross section, enlarged)

Coarse capillary

Microcapillary

Chain bundle
(manner of combining
cellulose molecules; dotted
lines show hydrogen bonds
that hold cellulose
molecules together)

**Figure 8.7** Details of the cotton fiber. In this series of diagrams successively smaller portions of a cotton fiber are enlarged and depicted. (Adapted from illustrations appearing in H. R. Mauersberger, editor, *Matthews' Textile Fibers,* sixth edition, John Wiley and Sons, New York, 1954, pages 73 and 77.)

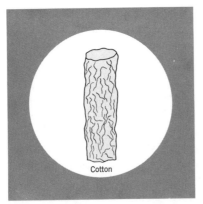 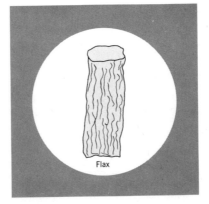

**Figure 8.8** The orientation of cellulose molecules (wavy lines) is better in flax than in cotton.

A collection of intertwined, overlapping microfibrils makes up a visible *fibril,* and within it runs a coarse capillary network. A single cotton fiber is made up of a twisting, intertwining, overlapping aggregation of fibrils.

The manner in which cellulose chains are combined (Figure 8.7) illustrates the three basic requirements a molecule must satisfy if it is to be capable of forming into a fiber.

1. It must be very long and narrow.
2. Its molecular configuration must permit many molecules to be very close to one another all along their lengths. Only with this positioning can any electrical forces of attraction be effective.
3. Electrical features of the molecules must enable them to attract one another. Opportunities for hydrogen bonding are the most important of such features in cellulose.

The ability of a fiber to accept dye often depends on residual electrical forces in fiber molecules which can attract and hold highly polar molecules of the dye.

Much common knowledge about cellulose fibers correlates well with the structure just described. Cellulose fibers from different sources, for example, vary widely in strength. The molecules of cellulose in flax are much better oriented along the fiber axis than those in cotton (cf. Figure 8.8). Since only molecules arranged in this way can take up tension applied to the fiber, linen is stronger than cotton.

Another well-known fact about cellulose is that it readily absorbs water and solutions (e.g., "absorbent cotton"). The vast networks of coarse and fine capillaries in a cellulose fiber provide spaces for water molecules, and the abundance of hydroxyl groups provides means for attracting molecules of water into these spaces. Of course, cellulose fibers are weaker when they are wet, for some of the hydrogen bonds *between* cellulose molecules become reoriented to the water molecules, thus weakening the basis for the strength of the fiber itself.

Cellulose fibers and fabrics are easily weakened by acids or bleaches, but they are quite stable to alkalies. Acid spilled on a shirt or blouse made of cotton quickly produces a hole. Very little happens when a base is spilled, but it should be washed off anyway. The explanation, of course, is that the acetal oxygen bridges in cellulose are susceptible to acid-catalyzed hydrolysis but not to base-catalyzed hydrolysis.

## CELLULOSE IN TECHNOLOGY

The abundance and relative inexpensiveness of cellulose have inspired its adaption to a variety of commercial needs. The three free hydroxyls per glucose unit in cellulose serve as "handles" for chemical modifications. Conversion of these groups to ethers and esters of organic acids and to esters of nitric acid has been especially fruitful. A cellulose obtained from wood, one not suitable to be woven directly into fabrics, is chosen for such modification.

**Cellulose Ethers.** When aqueous sodium hydroxide acts on cellulose and then an alkyl halide is added, one or more of the hydroxyl groups (per glucose unit) are converted into an ether. Methyl ethers (Methocel), ethyl ethers (Ethocel), and benzyl ethers are the commonest. A cellulose ether averaging less than one ether group per glucose unit is soluble in alkali and serves as sizes and finishing agents for textiles, as paper coatings, and as emulsifying and thickening agents for creams, lotions, shampoos, and toothpaste. If the cellulose ether averages more ether groups per glucose unit, it is used in paints, lacquers, varnishes, enamels, films, molded plastics, and packaging sheets.

**Cellulose Nitrates.** The action of a mixture of nitric acid and sulfuric acid on cellulose converts it into cellulose nitrate ("nitrocellulose"). By varying the initial conditions, an average of one, two, or three nitrate ester groups can be put on each glucose unit. The dinitrate is called *pyroxylin,* and the oldest known synthetic plastic, *Celluloid,* is made from it. Pyroxylin, by itself, is brittle, but it can be gelatinized by ethyl alcohol and camphor. If this softened material is molded under pressure and heat and cured at a higher temperature, Celluloid is formed. The immediate incentive for its development over a century ago was a competition to find a substitute for ivory for billiard balls. Even though Celluloid is highly flammable and gives off toxic gases as it burns, it is still used in the manufacture of imitation leather, piano keys, containers, frames for eyeglasses, and quick-drying lacquers.

A material such as camphor which is added to a polymeric material to reduce its brittleness and make it more plastic is called a *plasticizer*. The commercial development and production of plasticizers are a large and important part of the whole plastics industry, and the search for such substances is an active field. Dibutyl phthalate and tricresyl phosphate, for example, have been found superior to camphor as plasticizers for pyroxylin.

The trinitrate of cellulose, or a substance having very nearly three nitrate groups per glucose unit, is called *gun cotton*. It looks something like ordinary cotton but is highly explosive. Alfred Nobel, sometime after his invention of dynamite, discovered that nitroglycerin would mix with gun cotton to form a jelly of high explosive strength but safe enough for handling and shipping. These two substances are mixed in varying proportions for special purposes. A low percentage of gun cotton gives a blasting agent for shattering rocks, a higher percentage a slower-burning, almost smokeless propellant for artillery shells.

**Cellulose Acetate.** The action of acetic acid, sulfuric acid, and acetic anhydride, under carefully controlled conditions, produces cellulose triacetate (i.e., three acetyl groups per glucose unit). Films and threads made from it are brittle, and the substance is otherwise difficult to work. Only expensive solvents will dissolve it. If the triacetate is partially hydrolyzed to approximately the diacetate, it becomes soluble in acetone and gives fibers, the commercially important acetate rayons. Cast as a film, cellulose diacetate is widely used in photography because it is much less flammable than pyroxylin-based film.

In the hydrolysis of the triacetate to the diacetate, some of the acetal oxygen bridges in the cellulose chain are inevitably hydrolyzed, leaving roughly 200 to 300 glucose units per molecule in contrast with 2000 to 9000 in native cotton. Although the acetate rayon fiber is correspondingly weaker, its advantages are enough to support an industry making and selling larger quantities than the total combined production of the purely synthetic fibers, nylon, Orlon, Dacron, saran, etc.

Cellulose esters of propionic and butyric acids and combinations of esters, such as acetate and propionate and acetate and butyrate, also have commercial applications. In addition to forming fibers that are woven into fabrics, cellulose esters have been cast as electrical insulation, eyeglass frames, transparent aircraft enclosures, instrument panels, lacquers, and other molded items.

**Regenerated Cellulose. Viscose Rayon.** Rayon is a generic name for any fiber made by a chemical modification of native cellulose, and in the acetate rayon just described, the modification is permanent. The ester groups are left on the molecule. Viscose rayon, one of the earliest types, is made by temporary modification. Although the cellulose found in wood is not suitable for direct use as a fiber, the action of aqueous alkali and carbon disulfide converts it (or cellulose from any other source) into cellulose xanthate. The solution of the xanthate in aqueous alkali is viscous; hence the name "viscose rayon." This solution is allowed to age or "ripen," during which changes that are not well understood occur. Then it is extruded through spinnerets into dilute acid which hydrolyzes the xanthate units and regenerates cellulose molecules, by now somewhat shortened. The overall reaction may be summarized as

$$R-O-H + S=C=S + NaOH \longrightarrow R-O-\overset{\displaystyle S}{\overset{\|}{C}}-S^-Na^+ \xrightarrow{\ H^+\ }$$

Raw cellulose
[R— is $C_6H_9O_4$
in $(C_6H_{10}O_5)_n$,
cellulose]

Cellulose xanthate

$$R-O-H + CS_2 + Na^+$$

Regenerated cellulose (viscose rayon)

As a filament emerges from the acid bath, it is caught up on a rotating drum. It may simultaneously be twisted together with other filaments to make a thread. If the filaments are stretched as they are spun, the molecules become better oriented, and the product is as strong as silk.

The diameter of a filament or a thread is measured in *denier* units. A denier is the number of grams that 9000 meters (about 5.5 miles) of a filament or a thread weighs. Thus a 15-denier filament weighs 15 grams per 9000 meters and is finer than a 30-denier filament. Filaments are commonly 2 to 3 denier. Threads may be as fine as 50 denier, as coarse as 1000, but are commonly about 150 denier.

If the viscose is extruded into acid through a slit instead of through a spinneret hole, the product emerges as a film called *cellophane*. Synthetic sponges are made by mixing Glauber's salt ($Na_2SO_4 \cdot 10H_2O$) in lumps of various sizes into the viscose, which is then formed into small blocks, coagulated, and leached with warm water. The salt dissolves out of the block, leaving a spongelike mass.

## REFERENCES AND ANNOTATED READING LIST

**BOOKS**

V. M. Ingram. *The Biosynthesis of Macromolecules.* W. A. Benjamin, New York, 1966.

M. Stacey and S. A. Barker. *Carbohydrates of Living Tissues.* Van Nostrand Company, Princeton, N.J., 1962.

C. R. Noller. *Chemistry of Organic Compounds,* third edition. W. B. Saunders Company, Philadelphia, 1965.

**ARTICLES**

E. I. Rabinowitch. "Photosynthesis." *Scientific American,* August 1948, page 34. Although this is an old article, its impressive illustrations of the carbon, hydrogen, and oxygen cycles of nature on this planet make it a fine general introduction to photosynthesis.

R. D. Preston. "Cellulose." *Scientific American,* September 1957, page 156. The structure, technology, and industrial uses of cellulose are discussed.

I. A. Pearl. "Lignin Chemistry, Century-Old Puzzle." *Chemical and Engineering News,* July 6, 1964, page 81. Cellulose in wood is embedded in the natural plastic, lignin. Research departments of the paper industry have toiled long and hard to make something useful from it. This article describes part of this work and its fruits.

## PROBLEMS AND EXERCISES

1. Write the structures (with plane projections) for the products if galactose is allowed to react with each of the following reagents: (*a*) nitric acid, (*b*) bromine in water, and (*c*) excess phenylhydrazine.

2. Write the structure of a monosaccharide that will give the same osazone as galactose.

3. Write structures for α-maltose, β-maltose, and the open form of maltose. (*Hint.* Only one ring is open.)

4. Write equations for the reaction of maltose with (*a*) water (acid- or enzyme-catalyzed), (*b*) bromine in water, and (*c*) excess phenylhydrazine.

5. Define each of the following terms. Illustrative structures may be used.

(*a*) ketohexose      (*b*) disaccharide      (*c*) mutarotation

(*d*) photosynthesis      (*e*) absolute configuration      (*f*) D-family

(*g*) plane projection formula      (*h*) a glycaric acid      (*i*) a glyconic acid

(*j*) a reducing sugar

6. Write equations that illustrate how lactose undergoes mutarotation.

7. If two glucose units are linked together by the reaction

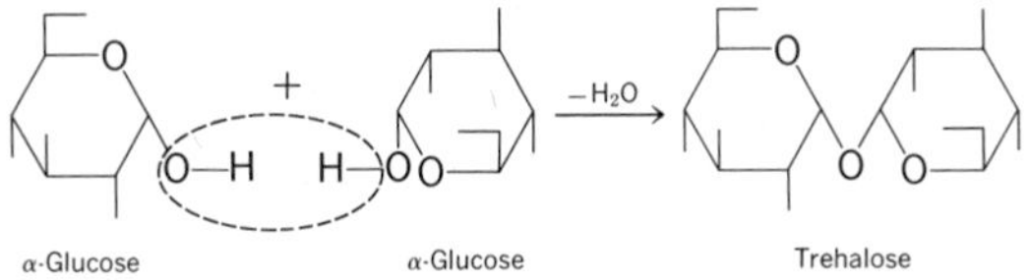

α-Glucose        α-Glucose        Trehalose

the disaccharide formed is trehalose, a disaccharide found in young mushrooms and yeasts and the chief carbohydrate in the blood (hemolymph) of certain insects. On the basis of its structural features, answer the following questions about trehalose. If a reaction is predicted, write it.

(*a*) Is trehalose a reducing sugar? Why?

(*b*) Can it mutarotate? Why?

(*c*) Does it form an osazone? Why?

(*d*) Can it be hydrolyzed?

8. Write enough of the structure of cellulose trinitrate to illustrate all its essential features.

9. Why is cellulose triacetate more soluble in nonaqueous solvents than cellulose?

10. Benedict's test, or some variation of it, is commonly used to detect glucose in urine. What are the structural features in glucose that make it give a positive reaction?

11. In some rare instances galactose appears in the urine. How might an analyst erroneously report the presence of glucose?

# CHAPTER NINE

# Lipids

We defined carbohydrates in terms of their structures, but lipids have an operational definition, one stated in terms of the operation used to isolate them. When plant or animal material is crushed and ground with nonpolar solvents such as benzene, chloroform, or carbon tetrachloride (fat solvents), the portion that dissolves in the solvent is classified as lipid. Carbohydrates and proteins are insoluble. Depending on the plant or animal origin, lipid material may include such a wide variety of compounds that it is difficult to make a precise structural definition. Included in the group could be neutral fats containing only carbon, hydrogen, and oxygen; phosphorus-containing compounds called phospholipids, which usually also contain nitrogen; aliphatic alcohols; waxes; steroids; terpenes; and "derived lipids," those substances resulting from partial or complete hydrolysis of some of the foregoing. Table 9.1 provides a summary of the important lipids.

Some workers and writers in the field have arbitrarily restricted the term lipid or fat to esters of long-chain carboxylic acids and alcohols or closely related substances. This practice will be followed in this chapter, and we shall study primarily the neutral fats, the simple lipids.

## SIMPLE LIPIDS. THE TRIGLYCERIDES

The most abundant group of lipids in plants and animals are the simple lipids or triglycerides. Included in this group are such common substances as lard, tallow, butterfat, olive oil, cottonseed oil, corn oil, peanut oil, linseed oil, coconut oil, and soybean oil. Their molecules consist of esters of the trihydroxy alcohol, glycerol, with various long-chain fatty acids. It is believed that a typical triglyceride molecule contains one unit from each of three different fatty acids rather than three units from a single fatty acid.

Fats and oils of whatever source are mixtures of different glyceride molecules; the differences are in specific fatty acids incorporated, not in the general structural

**Table 9.1 Lipid Classification**

| General Class | Sub-classes | Generic Class Structure | Nature of Hydrolysis Products | Examples and Occurrences |
|---|---|---|---|---|
| Simple Lipids | Waxes | $R-O-\overset{\displaystyle O}{\overset{\|}{C}}-R'$ | Long-chain carboxylic acids Long-chain alcohols | Beeswax, cuticle waxes (on flower petals and fruit skins) |
| | Triglycerides | $CH_2-O-\overset{\displaystyle O}{\overset{\|}{C}}-R$ $CH-O-\overset{\displaystyle O}{\overset{\|}{C}}-R'$ $CH_2-O-\overset{\displaystyle O}{\overset{\|}{C}}-R''$ | Mixture of long-chain acids Glycerol | Animal fats (solids): lard, tallow Vegetable oils (liquids): olive oil, corn oil, peanut oil, cottonseed oil, linseed oil |
| Phospholipids (Phosphoglycerides) | Lecithins | $CH_2-O-\overset{\displaystyle O}{\overset{\|}{C}}-R$ $CH-O-\overset{\displaystyle O}{\overset{\|}{C}}-R'$ $CH_2-O-\overset{\displaystyle O}{\underset{\underset{O^-}{\|}}{\overset{\|}{P}}}-O-CH_2-CH_2-\overset{+}{N}(CH_3)_3$ | Mixture of long-chain acids Glycerol Phosphoric acid Choline (an amino alcohol): $HOCH_2CH_2\overset{+}{N}(CH_3)_3$ $X^-$ | Found in nerve tissue |
| | Phosphatidylethanolamines | $CH_2-O-\overset{\displaystyle O}{\overset{\|}{C}}-R$ $CH-O-\overset{\displaystyle O}{\overset{\|}{C}}-R'$ $CH_2-O-\overset{\displaystyle O}{\underset{\underset{O^-}{\|}}{\overset{\|}{P}}}-O-CH_2-CH_2-\overset{+}{N}H_3$ | Mixture of long-chain acids Glycerol Phosphoric acid Aminoethanol: $HOCH_2CH_2NH_2$ | Found in nerve tissue |
| | Phosphatidylserines | $CH_2-O-\overset{\displaystyle O}{\overset{\|}{C}}-R$ $CH-O-\overset{\displaystyle O}{\overset{\|}{C}}-R'$ $CH_2-O-\overset{\displaystyle O}{\underset{\underset{O^-}{\|}}{\overset{\|}{P}}}-O-CH_2-\underset{\underset{COOH}{\|}}{CH}-\overset{+}{N}H_3$ | Mixture of long-chain acids Glycerol Phosphoric acid Serine: $HOCH_2\underset{\underset{CO_2H}{\|}}{CH}NH_2$ | Found in nerve tissue |
| Sphingolipids | Sphingomyelins | $CH_3-(CH_2)_{12}-CH=CH$ $\qquad O\ \ H\ \ CH-OH$ $R-\overset{\displaystyle O}{\overset{\|}{C}}-\overset{H}{N}-CH$ $\qquad CH_2-O-\overset{\displaystyle O}{\underset{\underset{O^-}{\|}}{\overset{\|}{P}}}-OCH_2CH_2-\overset{+}{N}(CH_3)_3$ | Sphingosine (an unsaturated amino alcohol): $CH_3(CH_2)_{12}CH=CH$ $\qquad CHOH$ $NH_2-CH$ $\qquad CH_2OH$ A long-chain acid Phosphoric acid Choline | Found in brain tissue |
| | Cerebrosides | $CH_3-(CH_2)_{12}-CH=CH$ $\qquad O\ \ H\ \ CH-OH$ $R-\overset{\displaystyle O}{\overset{\|}{C}}-\overset{H}{N}-CH$ $\qquad CH_2-O-$ (galactose ring) | Sphingosine A long-chain acid Galactose | Found in brain tissue (e.g., kerasin, phrenosin, nervon, and oxynervon) |

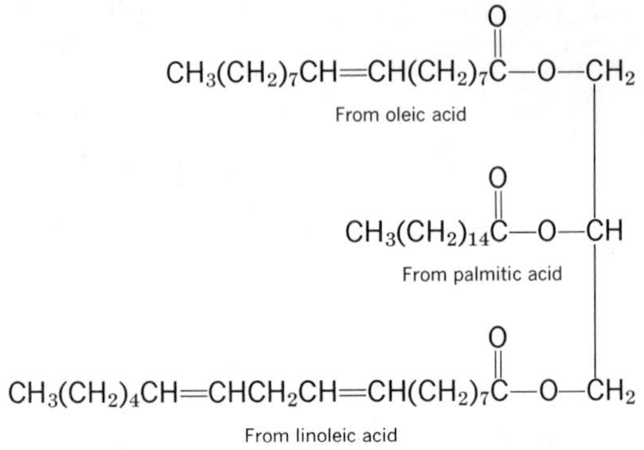

$$CH_3(CH_2)_7CH=CH(CH_2)_7\overset{\displaystyle O}{\overset{\|}{C}}-O-CH_2$$

From oleic acid

$$CH_3(CH_2)_{14}\overset{\displaystyle O}{\overset{\|}{C}}-O-CH$$

From palmitic acid

$$CH_3(CH_2)_4CH=CHCH_2CH=CH(CH_2)_7\overset{\displaystyle O}{\overset{\|}{C}}-O-CH_2$$

From linoleic acid

**Figure 9.1** A typical mixed glyceride showing oleic acid, palmitic acid, and linoleic acid bound by ester linkages to one molecule of glycerol. (The order in which the acid groups are attached to glycerol is arbitrary.)

features. Thus it is not possible to write the structure of, say, cottonseed oil except to describe what a typical molecule is like. One such molecule is represented in Figure 9.1. In a particular fat or oil, certain fatty acids tend to predominate, certain others are either absent or are present in trace amounts, and virtually all the molecules are triglycerides. Data for several fats and oils are summarized in Table 9.2.

**Table 9.2 Composition of the Fatty Acids Obtained by Hydrolysis of Common Neutral Fats and Oils**

| Fat or Oil | Iodine Number | Average Composition of Fatty Acids (%) | | | | | |
|---|---|---|---|---|---|---|---|
| | | Myristic Acid | Palmitic Acid | Stearic Acid | Oleic Acid | Linoleic Acid | Others |
| **Animal fats** | | | | | | | |
| Butter | 25–40 | 8–15 | 25–29 | 9–12 | 18–33 | 2–4 | a |
| Lard | 45–70 | 1–2 | 25–30 | 12–18 | 48–60 | 6–12 | b |
| Beef tallow | 30–45 | 2–5 | 24–34 | 15–30 | 35–45 | 1–3 | b |
| **Vegetable Oils** | | | | | | | |
| Olive | 75–95 | 0–1 | 5–15 | 1–4 | 67–84 | 8–12 | |
| Peanut | 85–100 | — | 7–12 | 2–6 | 30–60 | 20–38 | |
| Corn | 115–130 | 1–2 | 7–11 | 3–4 | 25–35 | 50–60 | |
| Cottonseed | 100–117 | 1–2 | 18–25 | 1–2 | 17–38 | 45–55 | |
| Soybean | 125–140 | 1–2 | 6–10 | 2–4 | 20–30 | 50–58 | c |
| Linseed | 175–205 | — | 4–7 | 2–4 | 14–30 | 14–25 | d |
| **Marine Oils** | | | | | | | |
| Whale | 110–150 | 5–10 | 10–20 | 2–5 | 33–40 | | e |
| Fish | 120–180 | 6–8 | 10–25 | 1–3 | | | e |

[a] Three to four percent butyric acid, 1 to 2% caprylic acid, 2 to 3% capric acid, 2 to 5% lauric acid.
[b] One percent linolenic acid.
[c] Five to ten percent linolenic acid.
[d] Forty-five to sixty percent linolenic acid.
[e] Large amounts of other highly unsaturated fatty acids.

**Table 9.3  Common Unsaturated Fatty Acids**

| Number of Double Bonds | Total Number of Carbons | Name | Structure | Mp (°C) |
|---|---|---|---|---|
| 1 | 16 | palmitoleic acid | $CH_3(CH_2)_5CH{=}CH(CH_2)_7CO_2H$ | 32 |
| 1 | 18 | oleic acid | $CH_3(CH_2)_7CH{=}CH(CH_2)_7CO_2H$ | 4 |
| 2 | 18 | linoleic acid | $CH_3(CH_2)_4CH{=}CHCH_2CH{=}CH(CH_2)_7CO_2H$ | −5 |
| 3 | 18 | linolenic acid | $CH_3CH_2CH{=}CHCH_2CH{=}CHCH_2CH{=}CH(CH_2)_7CO_2H$ | −11 |
| 4 | 20 | arachidonic acid | $CH_3(CH_2)_4CH{=}CHCH_2CH{=}CHCH_2CH{=}CHCH_2CH{=}CH(CH_2)_3CO_2H$ | −50 |

**Fatty Acids.** Fatty acids obtained from the lipids of most plants and animals tend to share the following characteristics.

1. They are usually monocarboxylic acids, $R{-}CO_2H$.
2. The $R{-}$ group is usually an unbranched chain.
3. The number of carbon atoms is almost always even.
4. The $R{-}$ group may be saturated, or it may have one, two, or three double bonds (sometimes four).

The most abundant saturated fatty acids are palmitic acid, $CH_3(CH_2)_{14}CO_2H$, and stearic acid, $CH_3(CH_2)_{16}CO_2H$, having sixteen and eighteen carbons, respectively. Others are included in Table 7.1 (p. 225), the acids above acetic with an even number of carbons, but they are present in only small amounts.

The most frequently occurring unsaturated fatty acids are listed in Table 9.3. The eighteen-carbon skeleton of stearic acid is duplicated in oleic, linoleic, and linolenic acids. Oleic acid is the most abundant and most widely distributed fatty acid in nature.

Among the eighteen-carbon fatty acids, the greater the degree of unsaturation, the lower the melting point:

|  | Mp (°C) |
|---|---|
| stearic acid (saturated) | 70 |
| oleic acid (one double bond) | 4 |
| linoleic acid (two double bonds) | −5 |
| linolenic acid (three double bonds) | −11 |

A double bond produces a stiffened kinkiness in regions of the otherwise flexible chains, thus reducing the number of ways in which neighboring molecules can get close enough to each other that forces of attraction between them operate. With net forces between molecules weakened, they are freer to move about and the material is liquid at lower temperatures than the saturated relatives. For this reason most animal fats are solids and most vegetable oils are liquids at room temperature. The oils are more unsaturated than the fats. There are more double bonds per molecule in the oils and, like the unsaturated acids, they are liquids.

The properties of the fatty acids are those to be expected of compounds having a carboxyl group, double bonds (in some), and long hydrocarbon chains. They are insoluble in water and soluble in nonpolar solvents. They can form salts, be esterified, and be reduced to the corresponding long-chain alcohols. Where present, alkene linkages react with bromine and take up hydrogen in the presence of a catalyst.

**Iodine Number.** The degree of unsaturation in a lipid is measured by its iodine number, defined as the number of grams of iodine that will add to the double bonds present in 100 g of the lipid. Saturated fatty acids, having no alkene linkages, have zero iodine numbers. Oleic acid has an iodine number of 90, linoleic acid 181, and linolenic acid 274. Animal fats have low iodine numbers, vegetable oils (the polyunsaturated oils of countless advertisements) higher values, as the data of Table 9.2 show. The iodine number of the mixed triglyceride of Figure 9.1 is calculated to be 89, in the range of the iodine number of olive oil or peanut oil.

## CHEMICAL PROPERTIES OF TRIGLYCERIDES

**Hydrolysis in the Presence of Enzymes.** Enzymes in the digestive tracts of human beings and animals act as efficient catalysts for the hydrolysis of the ester links of triglycerides. In general:

$$
\begin{array}{l}
\overset{O}{\overset{\|}{RC}}\!\!-\!OCH_2 \\[4pt]
\overset{O}{\overset{\|}{R'C}}\!\!-\!OCH \quad + \ 3H_2O \ \xrightarrow{\text{enzyme}} \\[4pt]
\overset{O}{\overset{\|}{R''C}}\!\!-\!OCH_2
\end{array}
\qquad
\begin{array}{l}
\overset{O}{\overset{\|}{RC}}\!\!-\!OH \quad HOCH_2 \\[4pt]
+ \\[2pt]
\overset{O}{\overset{\|}{R'C}}\!\!-\!OH + HOCH \\[4pt]
+ \\[2pt]
\overset{O}{\overset{\|}{R''C}}\!\!-\!OH \quad HOCH_2
\end{array}
$$

Triglyceride        Fatty acids        Glycerol

Example (using the triglyceride of Figure 9.1):

$$
\begin{array}{l}
CH_3(CH_2)_7CH\!=\!CH(CH_2)_7\overset{O}{\overset{\|}{C}}\!\!-\!OCH_2 \\[6pt]
CH_3(CH_2)_{14}\overset{O}{\overset{\|}{C}}\!\!-\!OCH \quad + \ 3H_2O \ \xrightarrow{\text{enzyme}} \\[6pt]
CH_3(CH_2)_4CH\!=\!CHCH_2CH\!=\!CH(CH_2)_7\overset{O}{\overset{\|}{C}}\!\!-\!OCH_2
\end{array}
$$

$$
CH_3(CH_2)_7CH\!=\!CH(CH_2)_7\overset{O}{\overset{\|}{C}}OH
$$
Oleic acid

+

glycerol + $CH_3(CH_2)_{14}\overset{O}{\overset{\|}{C}}OH$

Palmitic acid

+

$$
CH_3(CH_2)_4CH\!=\!CHCH_2CH\!=\!CH(CH_2)_7\overset{O}{\overset{\|}{C}}OH
$$
Linoleic acid

Dilute acids are not particularly effective as catalysts for this reaction because it is not possible to put much of the lipid into the water phase where the acid is. This problem of solubility exists, of course, in the digestive tract, for the digestive juices are mostly water. One of them, bile, contains detergents, however, the *bile salts*, which emulsify the lipid and thereby facilitate attack by water and enzymes. This enzyme-catalyzed hydrolysis of triglycerides is the only reaction they undergo during digestion.

**Saponification.** When the ester links in triglycerides are saponified by the action of a strong base (e.g., NaOH, KOH), glycerol plus the salts of the fatty acids are produced. These salts are soaps, and how they exert detergent action will be described later in this chapter. In general:

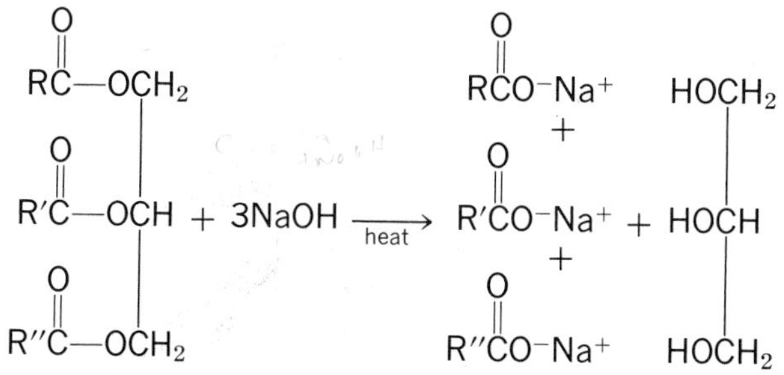

$$
\begin{array}{c}
\underset{\text{O}}{\overset{\text{O}}{\parallel}}\\
\text{RC--OCH}_2\\
\\
\text{R'C--OCH} + 3\text{NaOH} \xrightarrow[\text{heat}]{}\\
\\
\text{R''C--OCH}_2
\end{array}
\quad
\begin{array}{c}
\text{RCO}^-\text{Na}^+ \quad \text{HOCH}_2\\
+\\
\text{R'CO}^-\text{Na}^+ + \text{HOCH}\\
+\\
\text{R''CO}^-\text{Na}^+ \quad \text{HOCH}_2
\end{array}
$$

Mixture of salts     Glycerol

Example (using the triglyceride of Figure 9.1):

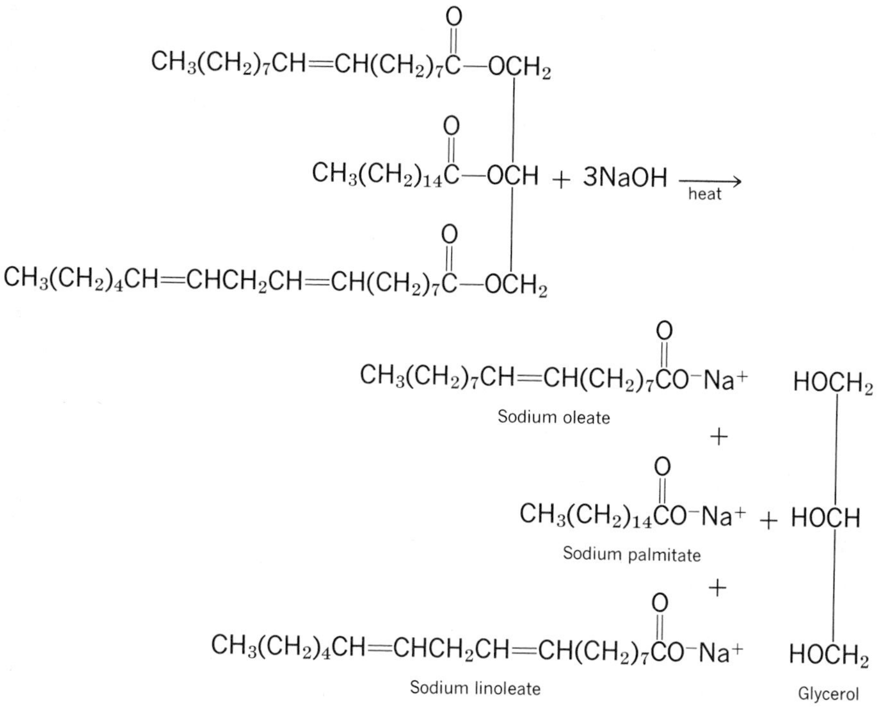

$$
\begin{array}{c}
\text{O}\\
\parallel\\
\text{CH}_3(\text{CH}_2)_7\text{CH}{=}\text{CH}(\text{CH}_2)_7\text{C--OCH}_2\\
\\
\text{O}\\
\parallel\\
\text{CH}_3(\text{CH}_2)_{14}\text{C--OCH} + 3\text{NaOH} \xrightarrow[\text{heat}]{}\\
\\
\text{O}\\
\parallel\\
\text{CH}_3(\text{CH}_2)_4\text{CH}{=}\text{CHCH}_2\text{CH}{=}\text{CH}(\text{CH}_2)_7\text{C--OCH}_2
\end{array}
$$

$$
\begin{array}{c}
\text{O}\\
\parallel\\
\text{CH}_3(\text{CH}_2)_7\text{CH}{=}\text{CH}(\text{CH}_2)_7\text{CO}^-\text{Na}^+ \quad \text{HOCH}_2
\end{array}
$$

Sodium oleate

$$+$$

$$
\begin{array}{c}
\text{O}\\
\parallel\\
\text{CH}_3(\text{CH}_2)_{14}\text{CO}^-\text{Na}^+ + \text{HOCH}
\end{array}
$$

Sodium palmitate

$$+$$

$$
\begin{array}{c}
\text{O}\\
\parallel\\
\text{CH}_3(\text{CH}_2)_4\text{CH}{=}\text{CHCH}_2\text{CH}{=}\text{CH}(\text{CH}_2)_7\text{CO}^-\text{Na}^+ \quad \text{HOCH}_2
\end{array}
$$

Sodium linoleate                 Glycerol

Ordinary soap is a mixture of the sodium salts of long-chain fatty acids. The potassium salts are more soluble in water and are marketed as soft soaps in some shaving creams and in liquid soap. Soaps that float on water have simply had air whipped into them. Perfumes and germicides are usually added. Scouring soaps have had fine sand, pumice, sodium carbonate, or other materials added.

**Hydrogenation.** If some of the double bonds in vegetable oils were hydrogenated, the oils would become like animal fats. They would be solids at room temperature, for example. Complete hydrogenation to an iodine value of zero is not desirable, for the product would be as brittle and unpalatable as tallow. Manufacturers of commercial hydrogenated vegetable oils such as Crisco, Fluffo, Mixo, Spry, etc., limit the degree of hydrogenation. Some double bonds are left. The product has a lower iodine number (e.g., about 50 to 60), a higher degree of saturation, and a melting point that makes it a creamy solid at room temperature, similar to lard or butterfat. The oils of soybean and cottonseed, abundant and inexpensive, are common raw materials for hydrogenated products. If just one molecule of hydrogen were added to the molecule of the mixed triglyceride of Figure 9.1, the iodine number would drop from 89 to 59, from the olive or peanut oil range to that of lard. The peanut oil in popular brands of peanut butter has been partially hydrogenated.

Oleomargarine is made by hydrogenating carefully selected and highly refined oils and fats. One goal is to produce a final product that will melt readily on the tongue, a feature that makes butter so desirable. The product of this hydrogenation is emulsified with about 17% (by weight) of milk that has been cultured with a suitable microorganism to add flavor. Vitamins A and D plus a yellow vegetable dye are usually added as well as biacetyl and acetoin, compounds that give butter its characteristic and highly prized flavor. In some oleomargarines liquid vegetable oils are whipped into the material to make them "more unsaturated," to quote the advertisements. Considerable artifice is expended to provide a cheaper spread that still tastes like butter. The commercial production of oleomargarine exceeds that of butter.

$$CH_3-\overset{\overset{O}{\|}}{C}-\overset{\overset{O}{\|}}{C}-CH_3 \qquad\qquad CH_3-\underset{\underset{OH}{|}}{CH}-\overset{\overset{O}{\|}}{C}-CH_3$$

Diacetyl        Acetoin

**Rancidity.** When fats and oils are left exposed to warm, moist air for any length of time, changes produce disagreeable flavors and odors. That is, the material becomes *rancid.* Two kinds of reactions are chiefly responsible, hydrolysis of ester links and oxidation of double bonds.

The hydrolysis of butterfat would produce a variety of relatively volatile and odorous fatty acids, as the data in the first footnote of Table 9.2 indicate. Water for such hydrolysis is, of course, present in butter, and airborne bacteria furnish enzymes. If the butter is not kept cold (heat accelerates reactions), it will turn rancid. If the mechanism producing rancidity is oxidation, attack by atmospheric oxygen occurs at unsaturated side chains in mixed triglycerides. Eventually, short-chain and volatile carboxylic acids and aldehydes are formed. Both types of substances have extremely disagreeable odors and flavors.

People who exercise much and bathe little develop surface films containing, among many other substances, triglycerides. If these films are not washed off before too long, they turn rancid. Hence the "essence of stale locker room."

**Hardening of Oils. "Drying."** Oxygen attacks unsaturated glycerides not only at their double bonds but also at carbons attached to them, at allylic positions (cf. p. 49). Linolenic acid has four such activated sites, marked by asterisks in the following structure:

$$\overset{*}{C}H_3CH_2CH{=}CHCH_2\overset{*}{C}H{=}CHCH_2\overset{*}{C}H{=}CHCH_2\overset{*}{(}CH_2)_6CO_2H$$

Linolenic acid (asterisks mark allylic sites)

Oxygen attacks these positions to form, first, hydroperoxides (—OOH):

$$-CH{=}CH-\overset{*}{C}H_2- + O_2 \longrightarrow -CH{=}CH-CH- $$
$$\qquad\qquad\qquad\qquad\qquad\qquad\qquad | $$
$$\qquad\qquad\qquad\qquad\qquad\qquad\qquad O-O-H$$

Portion of a side
chain in an
unsaturated
triglyceride

Hydroperoxy group
introduced at an
allylic site

Once these hydroperoxy groups are randomly introduced, they begin to interact with unchanged allylic sites to form peroxide bridges, building a vast interlacing network as illustrated in Figure 9.2. As these chemical changes occur, important

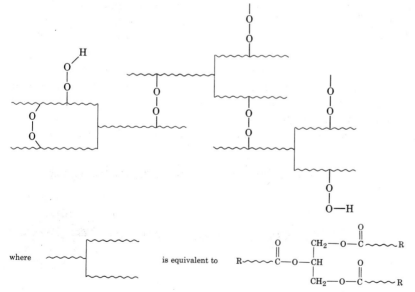

**Figure 9.2** When unsaturated glycerides harden, as in the drying of a paint, peroxide cross-links ultimately develop.

modifications take place in the material. If it has been spread on a surface, a film that is dry, tough, and durable forms, constituting a painted surface. The most highly unsaturated vegetable oils work best, for they have the most allylic sites. Linseed oil, which is quite rich in linolenic acid, is the most widely used drying oil. Fresh (raw) linseed oil still drys too slowly, however, and is made faster drying by heating it with driers, such as lead, manganese, or cobalt salts of certain organic acids. Pigments and dyes are added to give the desired color to the paint. Nearly a billion pounds of drying oils, mostly linseed oil, are used each year in the United States.

Before the widespread introduction of plastics, materials such as oilcloth and linoleum were more widely used than today. Oilcloth is made by coating woven canvas with several layers of a linseed oil paint. To make linoleum, thickened linseed oil is mixed with rosin and cork particles, and the mixture is allowed to harden.

**How Detergents Work.** Triglycerides and other types of "greases," such as hydrocarbon oils, are the binding agents that hold dirt to surfaces. The problem of cleansing reduces itself to finding a way to loosen or to dissolve this "glue." If it can no longer stick to the surface, neither can the dirt particles. Water alone is a poor cleansing agent, for its molecules are so polar that they stick to each other through hydrogen bonds rather than penetrate into a nonpolar region such as a surface film of grease.

Molecules of a typical soap and a typical synthetic detergent are shown in Figure 9.3. Each has a long, nonpolar hydrocarbon "tail" and a very polar, water-insoluble "head." The tail should be easily soluble in nonpolar materials (triglycerides and greases), and the head should be quite soluble in water.

When a detergent, either a soap or a *syndet,* a synthetic detergent, is added to water and the solution is poured onto a surface coated with grease, the tails of the detergent molecules tend to dissolve in the grease layer. The polar head remains in the polar atmosphere of the solution. A little mechanical agitation—stirring, tumbling, or boiling—will help loosen the film (Figure 9.4). As it breaks up into tiny globules, each is "pin-cushioned" by detergent molecules (Figure 9.5). The grease is now in the form of a stable emulsion, because each globule has, in effect, a surface studded with electrical charges, all the same sign. Hence the globules will repel each other. They cannot stick and coalesce.

One of the advantages of the syndets is that they are not precipitated by the

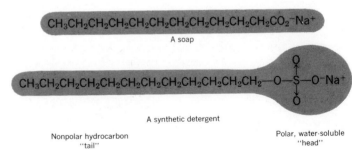

$CH_3CH_2CH_2CH_2CH_2CH_2CH_2CH_2CH_2CH_2CH_2CO_2^-Na^+$

A soap

$CH_3CH_2CH_2CH_2CH_2CH_2CH_2CH_2CH_2CH_2CH_2CH_2-O-\overset{\overset{O}{\uparrow}}{\underset{\underset{O}{\downarrow}}{S}}-O^-Na^+$

A synthetic detergent

Nonpolar hydrocarbon
"tail"

Polar, water-soluble
"head"

**Figure 9.3**  Two organic salts that have detergent properties.

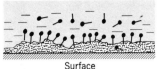 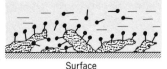

Surface                     Surface

(a)                         (b)

**Figure 9.4** Detergent action. (*a*) Nonpolar "tails" of detergent molecules become embedded in the grease layer. (*b*) Polar "heads" of detergent molecules tend to urge the grease layer away from the surface.

heavier metal ions found in hard water, $Ca^{2+}$, $Mg^{2+}$, and $Fe^{2+}$ or $Fe^{3+}$ or both. The salts of these ions and the fatty acids are insoluble. If ordinary soap is added to hard water, the familiar scum that forms is a mixture of these salts.

**Biodegradability.** Although ordinary soaps have their disadvantages in hard water, they cause virtually no problem when waste water containing them is discharged into sewage disposal systems and eventually into the ground. They are *biodegradable,* that is, microorganisms metabolize them. The search for economical detergents that would not precipitate in hard water led to a variety of types as several companies worked on unique products that they could patent. A few such types that were placed on the market are shown in Table 9.4. Whether these syndets were biodegradable was not considered a serious question until the sewage disposal systems of major metropolitan areas slowly became glutted with the foam and suds of undegraded detergents and homeowners with private wells reported that their tap water foamed.

As tetrapropylene-based alkylbenzene sulfonates (Table 9.4) were more and more widely used, the problem increased because these detergents are the least biodegradable. The enzyme systems of microorganisms are not equipped to degrade the highly branched side chains. But by 1966 the major American syndet makers had switched to *n*-paraffin-based alkylbenzene sulfonates or similar materials. Although these are not completely degradable in groundwater, microorganisms manage to reduce their concentrations to acceptable levels.

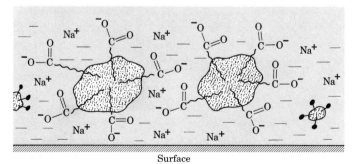

Surface

**Figure 9.5** These two grease globules, pin-cushioned by negatively charged groups, cannot coalesce. Both carry negative charges, and like charges repel. They are now solubilized and are easily washed away.

**Table 9.4  Biodegradability of Representative Types of Synthetic Detergents**

| Types of Synthetic Detergents | Examples to Illustrate Structural Features | Biodegradable? |
|---|---|---|
| Sodium alkyl sulfate | $CH_3(CH_2)_{10}CH_2OSO_2O^-Na^+$ | yes |
| Sodium alkylbenzene sulfonate, tetrapropylene-based | 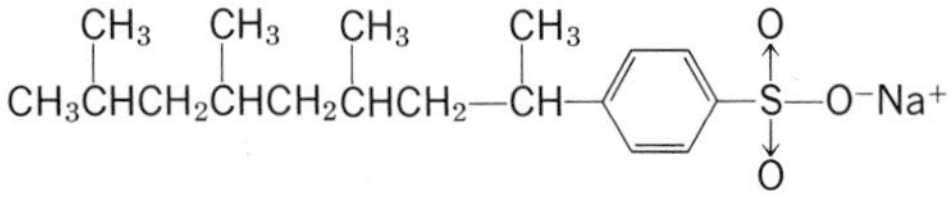 $CH_3CHCH_2CHCH_2CHCH_2$—CH— with $CH_3$ groups, attached to benzene ring—$S$(=O)(=O)—$O^-Na^+$ | no |
| Sodium alkylbenzene sulfonate, $n$-paraffin or $\alpha$-olefin-based | 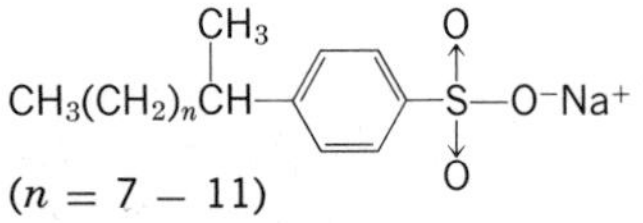 $CH_3(CH_2)_nCH$— with $CH_3$ group, benzene ring—$S$(=O)(=O)—$O^-Na^+$  $(n = 7 - 11)$ | largely so |
| Sodium alkane sulfonate | $C_nH_{2n+1}SO_3^-Na^+$ (chain is largely straight, $n$ is 15 to 18) | almost completely |

# COMPLEX LIPIDS

### GLYCEROL-BASED PHOSPHOLIPIDS

In the simple triglycerides just discussed, all three hydroxyl groups of glycerol are esterified with various fatty acids. In the molecules of the glycerol-based phospholipids, one of these hydroxyl groups is esterified with phosphoric acid which, in turn, is further esterified with a particular amino alcohol. The nature of the groups attached at the remaining two —OHs of glycerol determines the subclass.

**Phosphatides.** In the phosphatides these two —OHs have become esterified with fatty acids. They may be regarded as derivatives of phosphatidic acid. Specific phosphatides are formed by esterifying the phosphate unit with choline, ethanol-amine, or serine.

The phosphatides, and not all types are shown here, are frequently associated with membranes. They apparently have an essential role in the structure of cell walls and of the membranes enclosing cell nuclei, cytoplasmic organelles (e.g., mito-chondria), and the endoplasmic reticulum. Serving this function, they appear to be metabolically more stable than ordinary triglycerides. Thus in animals that are starving the reserves of triglycerides (fats) become depleted as they are withdrawn for energy, but a certain amount of phosphatide material remains, as it must since it is important to holding cells together.

The phosphatides also appear to be essential components of enzyme systems in mitochondria, which are small bodies located in the cytoplasm (see p. 439). Often referred to as the "powerhouse of the cell," a mitochondrion is involved in key energy-producing reactions for which complex systems of enzymes working in co-ordinated teams are necessary. The functional units of such systems consist of a group of enzymes embedded in a network of lipids (phosphatides) and proteins making up the walls of the mitochondria.

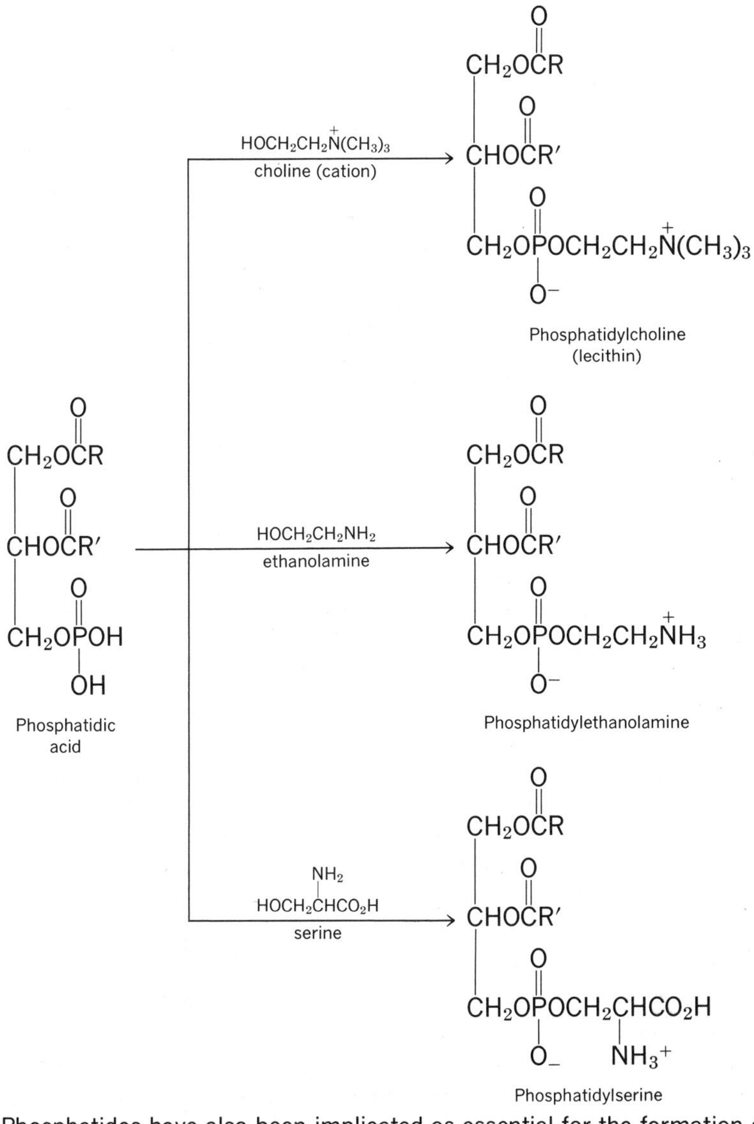

$$CH_2O\overset{\overset{\displaystyle O}{\|}}{C}R$$

$$CHO\overset{\overset{\displaystyle O}{\|}}{C}R'$$

$$\underset{\underset{\displaystyle O^-}{|}}{CH_2O}\overset{\overset{\displaystyle O}{\|}}{P}OCH_2CH_2\overset{+}{N}(CH_3)_3$$

$\underset{\text{choline (cation)}}{HOCH_2CH_2\overset{+}{N}(CH_3)_3}$

Phosphatidylcholine
(lecithin)

$$CH_2O\overset{\overset{\displaystyle O}{\|}}{C}R$$

$$CHO\overset{\overset{\displaystyle O}{\|}}{C}R'$$

$$\underset{\underset{\displaystyle OH}{|}}{CH_2O}\overset{}{P}OH$$

Phosphatidic
acid

$\underset{\text{ethanolamine}}{HOCH_2CH_2NH_2}$

$$CH_2O\overset{\overset{\displaystyle O}{\|}}{C}R$$

$$CHO\overset{\overset{\displaystyle O}{\|}}{C}R'$$

$$\underset{\underset{\displaystyle O^-}{|}}{CH_2O}\overset{\overset{\displaystyle O}{\|}}{P}OCH_2CH_2\overset{+}{N}H_3$$

Phosphatidylethanolamine

$\underset{\text{serine}}{\overset{\overset{\displaystyle NH_2}{|}}{HOCH_2CHCO_2H}}$

$$CH_2O\overset{\overset{\displaystyle O}{\|}}{C}R$$

$$CHO\overset{\overset{\displaystyle O}{\|}}{C}R'$$

$$\underset{\underset{\displaystyle O_-}{|}}{CH_2O}\overset{\overset{\displaystyle O}{\|}}{P}OCH_2\underset{\underset{\displaystyle NH_3^+}{|}}{CH}CO_2H$$

Phosphatidylserine

Phosphatides have also been implicated as essential for the formation of one of the factors needed to bring about the clotting of blood when injury occurs.

**Plasmalogens.** The plasmalogens, another type of glycerol-based phospholipid, are widely distributed in animal tissue, particularly in the myelin sheaths of nerves and in heart and skeletal muscle.

The oxygen link to glycerol's middle carbon in most phospholipids is selectively hydrolyzed by the action of the enzyme *phospholipase A* (or "lecithinase A"), found in snake venom. Since the phosphatides are so intimately associated with the nervous system and with important cell parts, we readily understand why snake poisoning is so serious.

$$CH_2O\overset{\overset{O}{\|}}{C}R$$

$$CHOCH{=}CHR'$$

$$CH_2O\overset{\overset{O}{\|}}{P}OCH_2CH_2{-}\overset{+}{N}(CH_3)_3$$

$$\overset{|}{O_-} \quad or \quad {-}\overset{+}{N}H_3$$

Plasmalogens

## SPHINGOSINE-BASED LIPIDS. SPHINGOLIPIDS

Brain and nerve tissue as well as the lungs and spleen contain a lipid based not on glycerol but on an unsaturated amino alcohol, sphingosine. Figure 9.6 explains the structures of two types of sphingolipids, the sphingomyelins and the cerebrosides. The cerebrosides are interesting lipids in that they contain a sugar unit, usually galactose. Their presence in brain and nerve tissue indicates their importance, but their exact function is not yet fully understood.

**The Myelin Membrane.** Structures called *axons* form the signal-bearing cores of nerve cells. Very long in relation to their diameters, they require protective insulation much like an electrical wire. The material serving this function for axons, called *myelin,* consists of lipids, proteins, polysaccharides, salts, and water and is the most stable membrane known. Once formed about an axon, it endures for the lifetime of the individual. When the formation of myelin is faulty, the outlook is not bright. Niemann-Pick disease, multiple sclerosis, forms of leukodystrophy, and infantile Gaucher's disease are all related to unstable myelin membranes.

The stability of myelin is related, at least in part, to its lipid components. Compared to the proportion of sphingolipid in other membranes, that in myelin is much higher. The sphingolipids contain much longer fatty acid units than the triglycerides, units with from nineteen to twenty-six carbons. Electron microscopy and X-ray diffraction analyses indicate that these long fatty acid side chains probably intertwine to give the molecular equivalent of steel mesh reinforcing concrete. When these side chains are much shorter (eighteen or fewer carbons) or highly unsaturated, studies have shown their intertwining to be less effective. The myelin is less stable, meaning a more vulnerable nervous system. Analysis of the myelin of individuals suffering from the diseases named has shown that for some reason the individuals have not succeeded in incorporating enough of the long-chain fatty acids in their myelin membranes, a significant example of the molecular basis of a disease.

## STEROIDS

The operation of extracting with solvents, which brings out triglycerides and complex lipids from plant and animal material, also brings out a group of nonsaponifiable compounds, the *steroids.* Their structures are generally characterized by a polycyclic carbon skeleton (Figure 9.7). The alcohol group, the keto group, and the

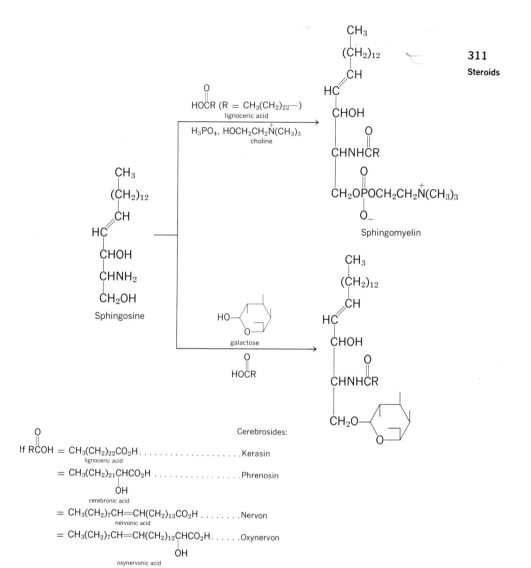

Figure 9.6   The structures of the sphingomyelins and the cerebrosides.

Figure 9.7   (a) Carbon skeleton characteristic of steroids. (b) Most common condensation of the steroid "nucleus."

double bond are common, but the ester linkage is seldom present. Hence steroids are not saponifiable.

The physiological effects of the steroids vary greatly from compound to compound, ranging from vitamin activity to the action of sex hormones. Some steroids are fat emulsifiers found in bile; others are important hormones; one stimulates the heart; another has been implicated in hardening of the arteries; still another ruptures red blood cells. Because of their complexity and importance, it is understandable that research in steroid chemistry and function has earned Nobel prizes for several chemists. The structures, names, and chief physiological properties of a few important steroids are given in the following list.

## Important Individual Steroids

**Sex hormones**

This is one human estrogenic hormone.

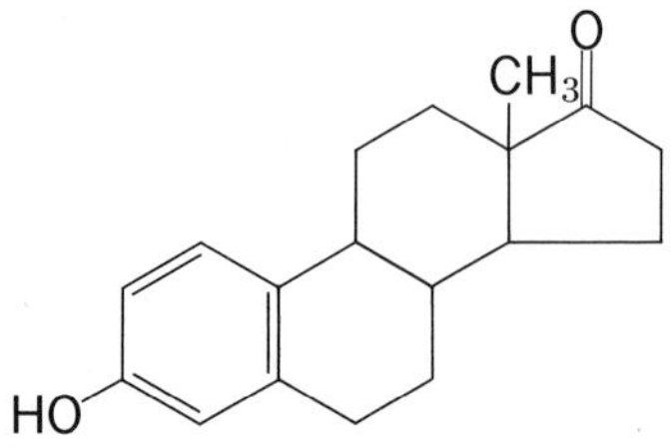

Estrone

This human pregnancy hormone is secreted by the corpus luteum.

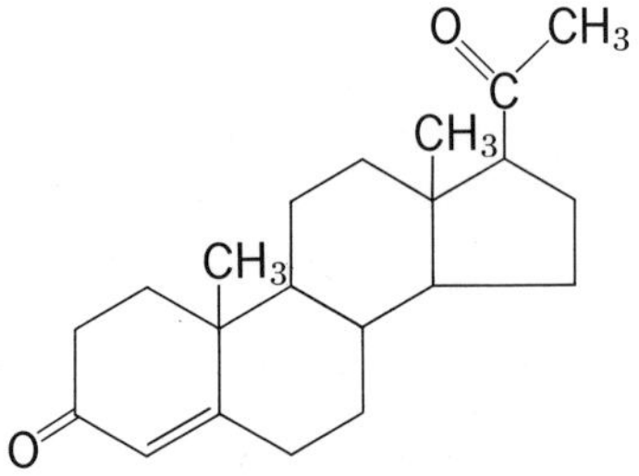

Progesterone

This male sex hormone regulates the development of reproductive organs and secondary sex characteristics.

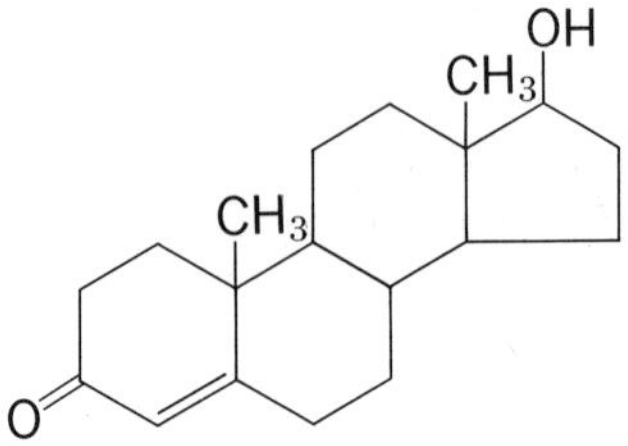

Testosterone

Androsterone, a second
   male sex hormone, is
   less potent than
   testosterone

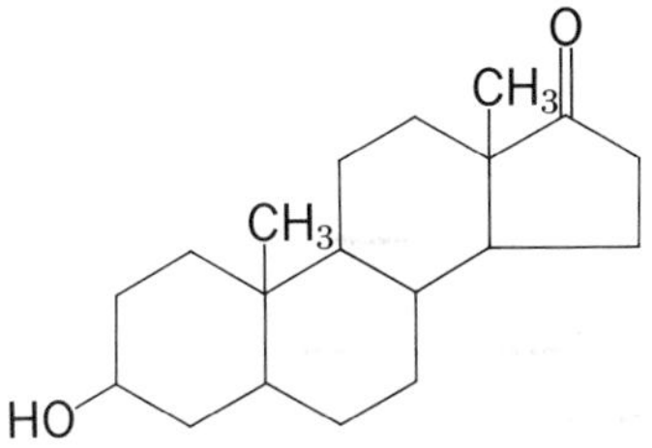

Androsterone

**Vitamin D$_2$ precursor**

Irradiation of this
   hormone, the common-
   est of all plant
   hormones, by ultra-
   violet light opens one of
   the rings to produce
   vitamin D$_2$.

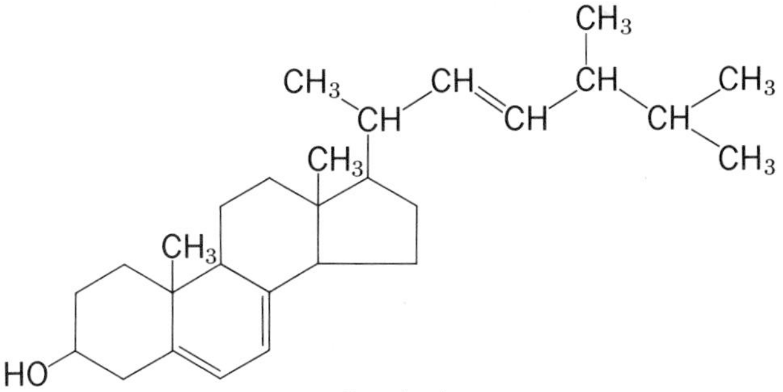

Ergosterol

ultraviolet
light

A dietary deficiency of this
   antirachitic factor
   causes rickets, an infant
   and childhood disease
   characterized by faulty
   deposition of calcium
   phosphate and poor
   bone growth.

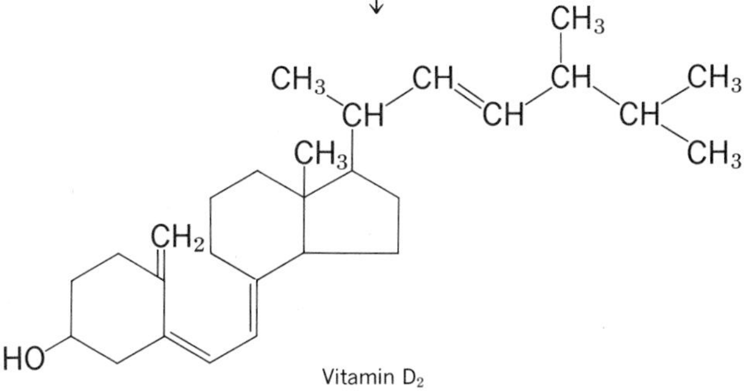

Vitamin D$_2$

**Bile acid**

Cholic acid is found in bile
   in the form of its sodium
   salt. This and closely
   related salts are the bile
   salts that act as powerful emulsifiers of lipid
   material awaiting digestion in the upper intestinal tract. The sodium
   salt of cholic acid is
   soaplike because it has
   a very polar head and a
   large hydrocarbon tail.

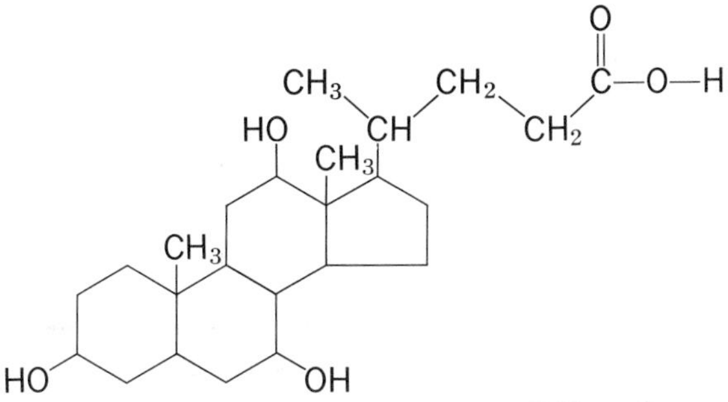

Cholic acid

## Antiarthritic compound

One of the twenty-eight adrenal cortical hormones, cortisone is not only important in the control of carbohydrate metabolism but also effective in relieving the symptoms of rheumatoid arthritis. Hench and Kendall of the Mayo Clinic earned Nobel prizes for demonstrating the antiarthritic properties of cortisone. Their supply of this drug was made available by a research program at Merck and Company, from which came a thirty-two-step synthesis of cortisone from material isolated from ox bile. Cortisone is now made on a fairly large scale from diosgenin, a steroid found in tubers of a Mexican yam, a material more abundant than ox bile.

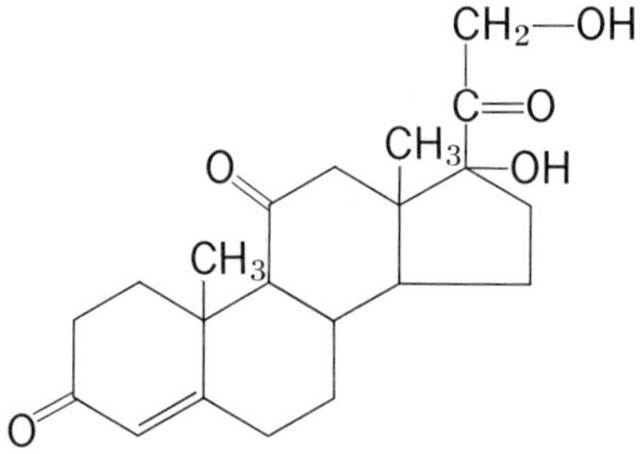

Cortisone

## Cardiac aglycone

Digitoxigenin is found in many poisonous plants, notably digitalis, as a complex glycoside. In small doses it stimulates the vagus mechanism and increases heart tone. In larger doses it is a potent poison.

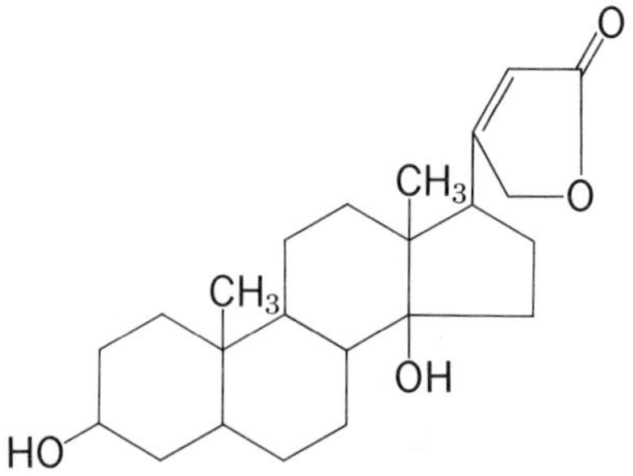

Digitoxigenin

**Cholesterol.** Cholesterol is a steroid alcohol or *sterol*. It is found in nearly all tissues of vertebrates, particularly in the brain and spinal cord, and is the main constituent of gallstones. In recent years it has been associated with circulatory problems such as hardening of the arteries. Its physiological function is still not understood, but we generally assume that the body needs it as a raw material for building other usable steroids.

Human beings and the higher animals can synthesize cholesterol. Most recent evidence indicates that the body can make it from acetate units in about thirty-six steps and in a matter of seconds.

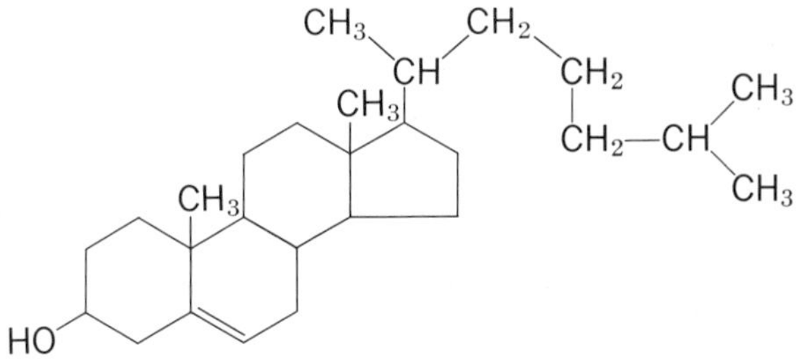

Cholesterol
(Greek: *chole*, bile; *stereos*, solid; plus "-ol," alcohol)

## TERPENES

The pleasant odors and flavors associated with many plants reveal the presence of volatile compounds of ten or fifteen carbons called *terpenes*. Terpenes are isolated by extracting them with ether, or they may be obtained by steam distillation, a process in which live steam is blown through a watery mixture of crushed portions of the plant (or the watery mixture may be boiled to generate the steam within the flask). To perfumers and food flavorers the materials isolated in this way are the *quinta essentia*, the essential parts of the plants, and for this reason they are called the essential oils.

Sandalwood, eucalyptus, peppermint, clove, lavender, rose, citronella, cedar, camphor—the roster of plants yielding essential oils reads like a page from some medieval pharmacopeia or a catalogue of the bottles on grandmother's herb and spice rack.

To the chemist one of the most interesting features of terpenes is the regular occurrence of isoprene units in their molecules. Two isoprene units give the ten-

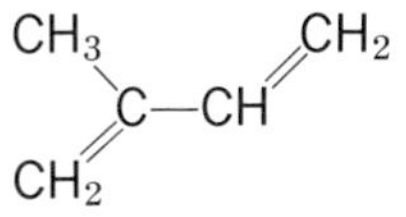

Isoprene
(2-methyl-1,3-butadiene)

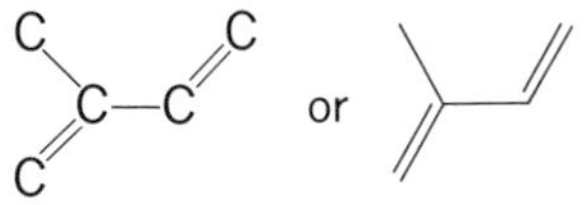

Useful condensations of the
isoprene structure

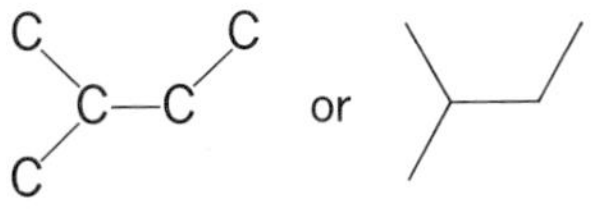

The isoprene "skeleton"

carbon terpenes, the structure to which the term is properly restricted. The *sesqui-terpenes* have three isoprene units making fifteen carbons. *Diterpenes* are twenty-carbon compounds composed of four isoprene units. The thirty-carbon substances (six isoprene units) are *triterpenes*, etc.

The structure of the terpene called *myrcene* (oil of bay)[1] illustrates the connection between isoprene skeletons and terpenes, with the dashed line or lines, here and elsewhere in this section, dividing the molecule into isoprene units. Many terpenes such as *limonene* (oil of peppermint, oil of orange, oil of lemon) and *α-pinene* (oil of turpentine) are cyclic compounds. Important sesquiterpenes (fifteen-carbon compounds) include *α-farnesene* (oil of citronella) and *zingiberene* (oil of ginger).

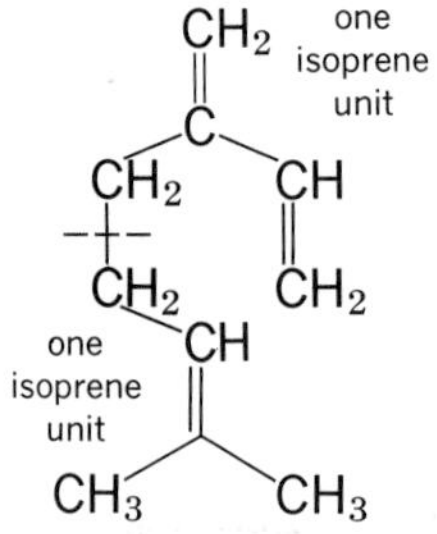

Myrcene

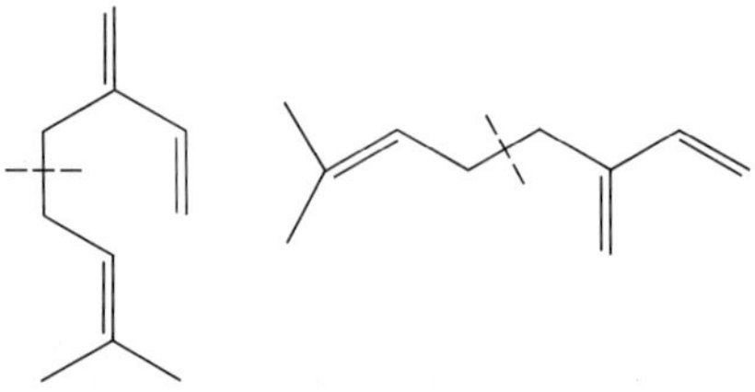

Myrcene (alternative symbols)

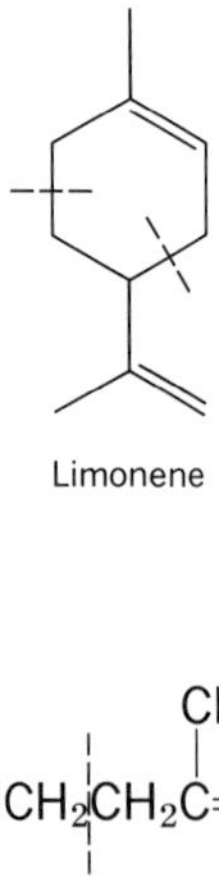

Limonene 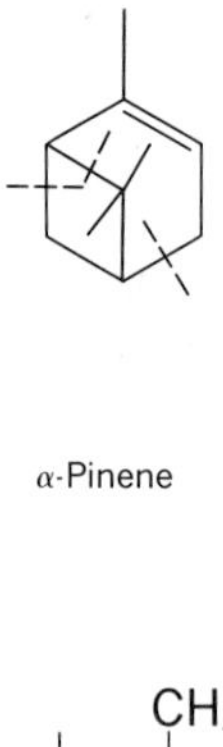        α-Pinene 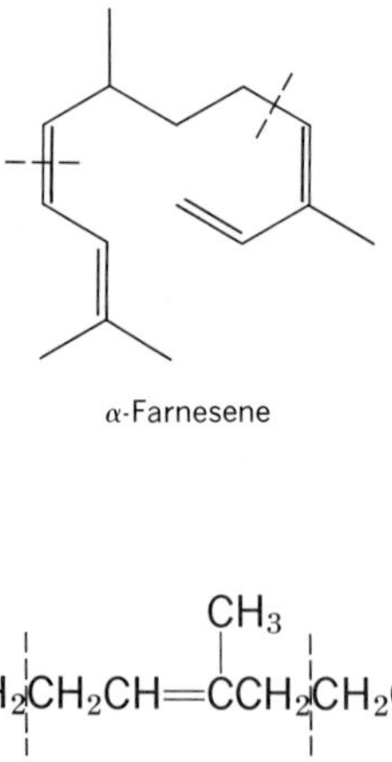        α-Farnesene 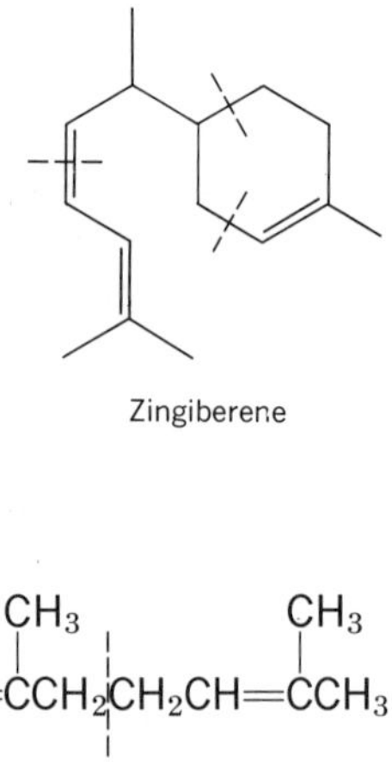        Zingiberene

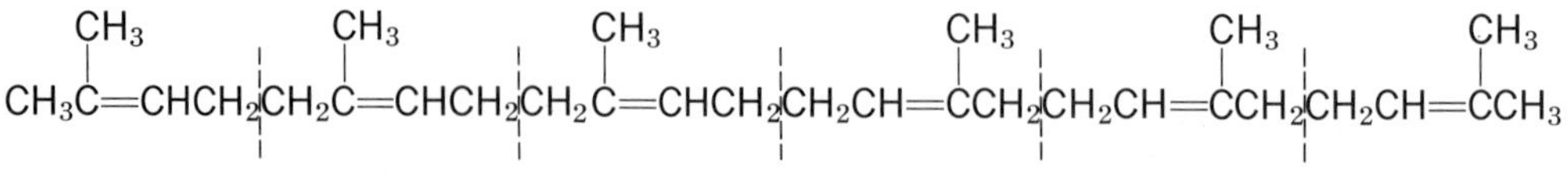

Squalene

*Squalene,* a triterpene, as an unusual hydrocarbon found in shark-liver oil. In the multiple-step conversion of acetate units into cholesterol in the human body, squalene is one of the intermediates.

[1] "Oil of bay" is not a synonym for "myrcene," but oil of bay contains myrcene as one of its most characteristic components. In this section each terpene is named for the particular essential oil most clearly associated with it.

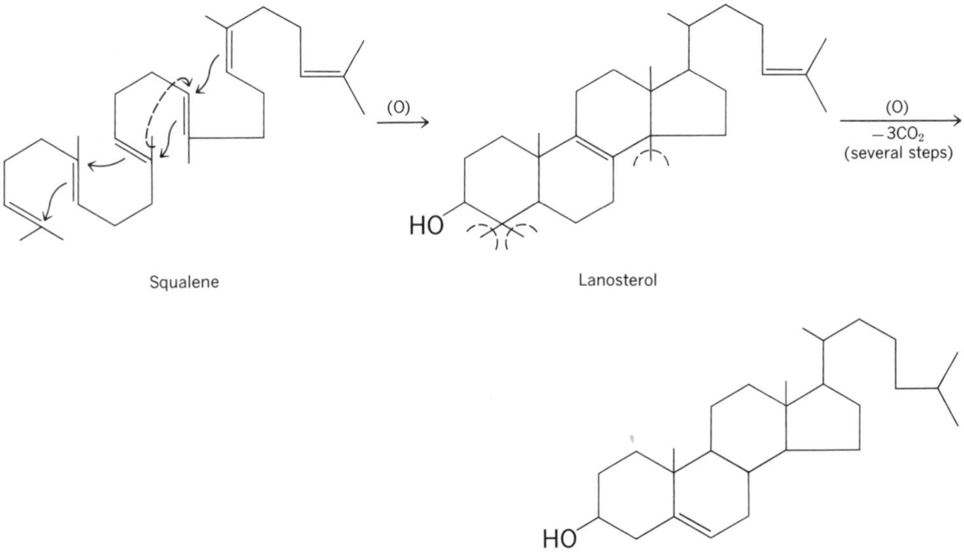

Squalene

Lanosterol

Cholesterol

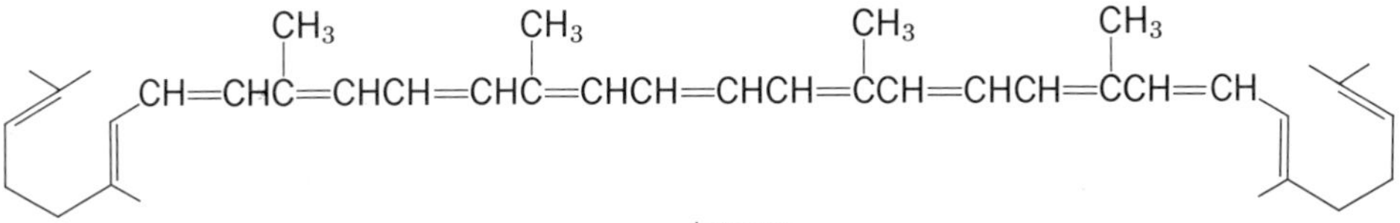

Lycopene

The red-colored matter found in tomatoes is a tetraterpene, *lycopene*. *β-Carotene,* another tetraterpene, is largely responsible for the color of carrots, but it occurs in other plants as well where its color may be masked.

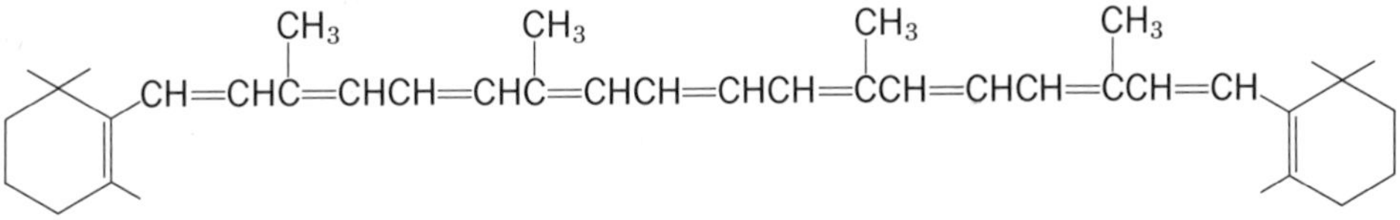

*β*-Carotene

The relation between the terpenoid compounds and isoprene is purely a formalism, for isoprene does not occur naturally and is not directly involved in the biosynthesis of any terpene. *Isopentenyl pyrophosphate* is the five-carbon intermediate through which isoprene units become linked together.

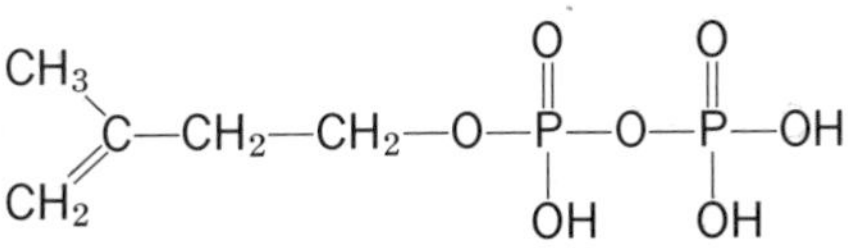

Isopentenyl pyrophosphate

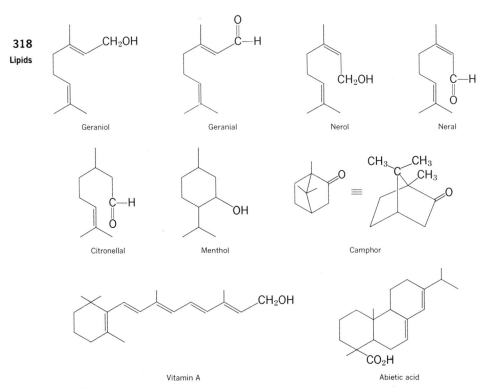

Geraniol          Geranial          Nerol          Neral

Citronellal          Menthol          Camphor

Vitamin A          Abietic acid

**Figure 9.8**   Some oxygen derivatives of terpenes.

**Oxygen Derivatives of Terpenes.** A great variety of alcohol, aldehyde, and ketone derivatives of the terpene hydrocarbons, isolated along with them, are essential contributors to the distinctive flavors and odors of many natural oils and perfumes. The structures of several are given in Figure 9.8. A good perfume, for example, attar of roses, is an extremely complex mixture of substances almost impossible to duplicate by synthetic means. Abietic acid is one of the principal constituents of rosin, a nonvolatile residue obtained when turpentine is extracted from pine. Its sodium salt is used in some laundry soaps.

## REFERENCES

**BOOKS**

C. R. Noller. *Chemistry of Organic Compounds,* third edition. W. B. Saunders Company, Philadelphia, 1965. Chapter 12, "Waxes, Fats, and Oils."

D. J. Hanahan, F. R. N. Gurd, and I. Zabin. *Lipide Chemistry.* John Wiley and Sons, New York, 1960.

L. H. Meyer. *Food Chemistry.* Reinhold Publishing Corporation, New York, 1960. Chapter 2, "Fats and Other Lipids."

**ARTICLES**

J. S. O'Brien, "Stability of the Myelin Membrane." *Science,* Vol. 147, 1965, page 1099.

V. P. Dole. "Body Fat." *Scientific American,* December 1959, page 71. This article is a discussion of obesity.

# PROBLEMS AND EXERCISES

1. Write the structure of a mixed glyceride between glycerol and each of the following three acids: palmitic acid, oleic acid, and linolenic acid.

   (*a*)  Write an equation for the saponification of this triglyceride.

   (*b*)  Write an equation for its digestion.

   (*c*)  Write an equation for a reaction that would reduce the iodine number to zero.

   (*d*)  Calculate the iodine number of this triglyceride. Is it likely to be a solid or a liquid at room temperature?

2. When soap is made, the excess alkali and the glycerol may be washed out by stirring and mixing the crude product with salt water. Explain how salt water will act to inhibit the soap from dissolving.

3. When hydrochloric acid is added to a solution of sodium palmitate, an organic compound precipitates. What is its structure and name?

4. Write structures for a monoglyceride, a diglyceride, and a triglyceride. Use lauric acid, to simplify matters, in all cases. Arrange the structures in order of increasing solubility in water, and explain your reasons for selecting this order.

5. Explain *how* a monoglyceride acts as an emulsifying agent to help disperse a mixture of oil in water.

6. Define or illustrate the following.

   (*a*)   iodine number      (*b*)   steroid            (*c*)   terpene

   (*d*)   phospholipid       (*e*)   saponification     (*f*)   oleomargarine

   (*g*)   syndet             (*h*)   biodegradability

7. Explain how a drying oil "dries."

8. Explain how a detergent emulsifies oily substances.

9. Explain how butter becomes rancid.

# CHAPTER TEN

# Proteins

Proteins are found in all cells and in virtually all parts of cells; of the body's dry weight proteins constitute about half. Proteins hold a living organism together and run it. As skin they give it a shell; as muscles and tendons they provide levers; and as one substance inside bones they, like steel to reinforced concrete, constitute a reinforcing network. Some proteins, in buffers, antibodies, and hemoglobins, serve as policemen and long-distance haulers. Others form the communications network or nerves. Certain proteins, as enzymes, hormones, and gene regulators, direct and control all forms of body repair and construction. No other class of compounds is involved in such a variety of functions, all essential to life.

Diversity of functions might suggest diversity of molecular types. Yet the huge variety of proteins have enough molecular features in common that we can lump them together in one chemical family. One purpose of this chapter is to indicate how a few subtle variations on a theme of *similar structures* might be responsible for the many *different functions* proteins have.

Proteins have one characteristic in common: they are all polymers. The monomer units are classified as α-amino acids; of all the proteins ever analyzed, only about twenty different α-amino acids occur commonly. A few others occur here and there in unusual instances.

We sought to understand certain characteristics of starch, glycogen, and cellulose by studying their monomer unit, glucose. In a similar approach we begin our study of proteins by examining their monomers.

## AMINO ACIDS AS BUILDING BLOCKS FOR PROTEINS

The system common to all the $\alpha$-amino acids is given by the general structure

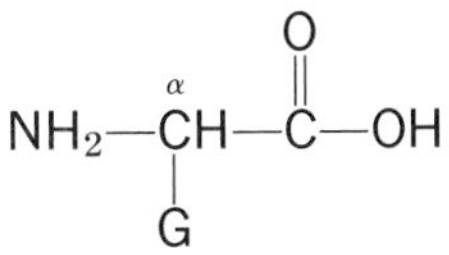

$$NH_2-\overset{\alpha}{\underset{\underset{G}{|}}{C}}H-\overset{\overset{O}{\|}}{C}-OH$$

The G— stands for an organic group or system that is different in each of the various amino acids. The list of twenty-four amino acids in Table 10.1 indicates the nature of these side chains.

**Optical Activity among the Amino Acids and Proteins.** In all the naturally occurring amino acids except the simplest, glycine, the $\alpha$-carbon is asymmetric, and the substances are optically active. Serine is used as the reference for assigning amino acids to configurational families. Naturally occurring serine is in the L-family. Its relation to L-glyceraldehyde shows in the plane projection formulas.

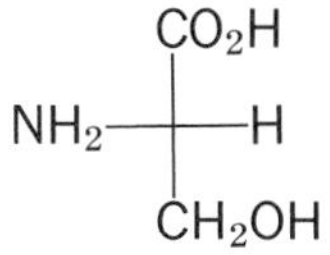

L-Serine
(naturally occurring
enantiomer)

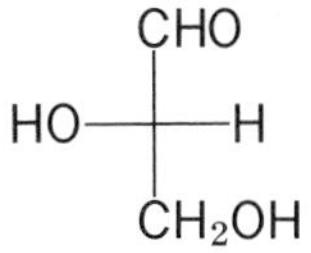

L-Glyceraldehyde

When the other naturally occurring amino acids are compared with L-serine, they are all found to be in the same configurational family. All the optically active, naturally occurring amino acids have the L-configuration. Except in extremely rare instances nature does not use the D-forms. The proteins in our bodies are made only from L-amino acids. If we were fed the D-enantiomers, we could not use them. Our enzymes, which themselves are made from the L-forms of amino acids, work with L-forms only.

**How Amino Acids Are Incorporated into Proteins. Primary Structural Feature of Proteins.** If the $\alpha$-amino group of one amino acid is acted upon by the carboxyl group of another amino acid, an amide link forms between the two. In general:

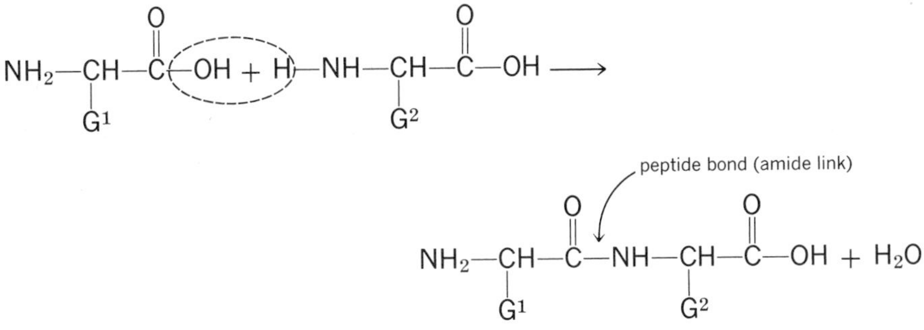

peptide bond (amide link)

A dipeptide

Specific example:

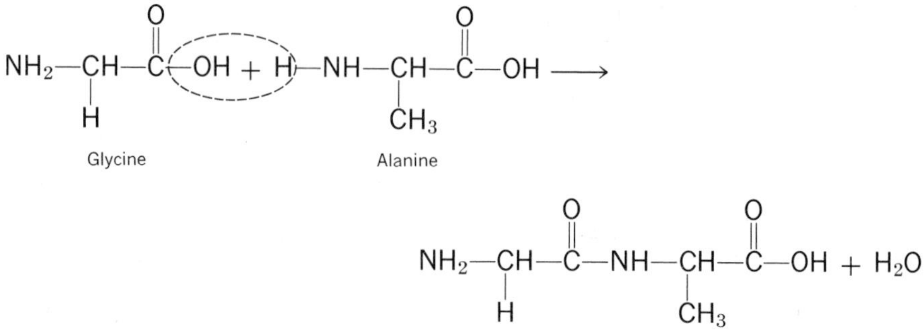

$$NH_2—CH—\overset{\overset{\textstyle O}{\|}}{C}(—OH + H)—NH—CH—\overset{\overset{\textstyle O}{\|}}{C}—OH \longrightarrow$$

(Glycine, with H below the CH; Alanine, with $CH_3$ below the CH)

Glycine                                           Alanine

$$NH_2—CH—\overset{\overset{\textstyle O}{\|}}{C}—NH—CH—\overset{\overset{\textstyle O}{\|}}{C}—OH + H_2O$$

(H below first CH; $CH_3$ below second CH)

Glycylalanine
Gly·Ala[1]

In protein chemistry the amide link is called the *peptide bond*. The product in the example, glycylalanine, is a *dipeptide,* a molecule that can be hydrolyzed to two molecules of amino acid.

The other sequence of the two amino acids could just as well have been used. In general:

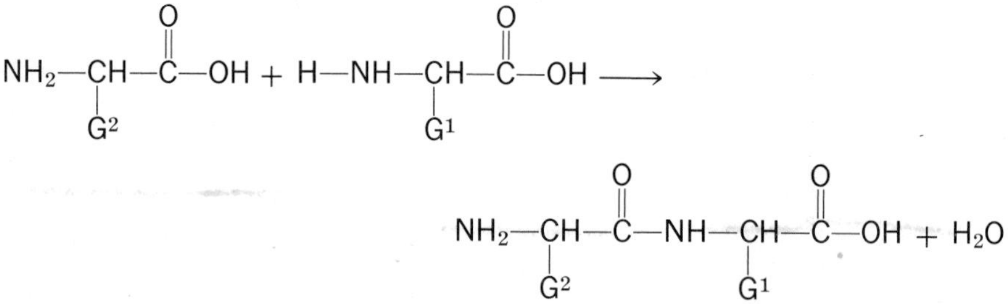

$$NH_2—CH—\overset{\overset{\textstyle O}{\|}}{C}—OH + H—NH—CH—\overset{\overset{\textstyle O}{\|}}{C}—OH \longrightarrow$$

($G^2$ below first CH; $G^1$ below second CH)

$$NH_2—CH—\overset{\overset{\textstyle O}{\|}}{C}—NH—CH—\overset{\overset{\textstyle O}{\|}}{C}—OH + H_2O$$

($G^2$ below first CH; $G^1$ below second CH)

Specific example:

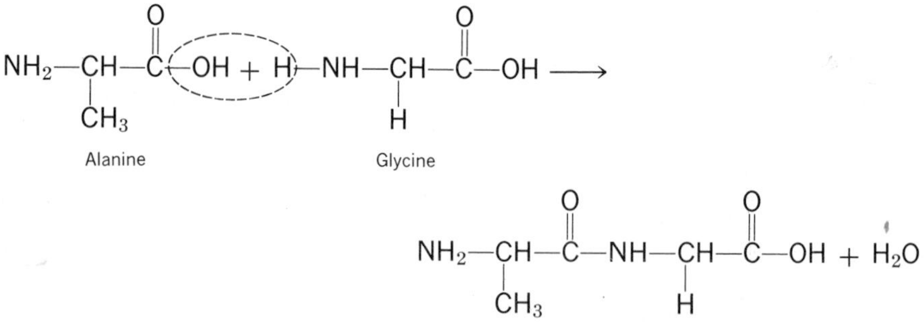

$$NH_2—CH—\overset{\overset{\textstyle O}{\|}}{C}(—OH + H)—NH—CH—\overset{\overset{\textstyle O}{\|}}{C}—OH \longrightarrow$$

($CH_3$ below first CH; H below second CH)

Alanine                                           Glycine

$$NH_2—CH—\overset{\overset{\textstyle O}{\|}}{C}—NH—CH—\overset{\overset{\textstyle O}{\|}}{C}—OH + H_2O$$

($CH_3$ below first CH; H below second CH)

Alanylglycine
Ala·Gly

[1] By convention the symbol Gly·Ala means that the first designated amino acid unit, Gly for glycine, has a free amino group and that the last designated unit, Ala for alanine, has the free carboxyl group. In accordance with this convention amino acids are often written with the —$NH_2$ to the left and the —$CO_2H$ to the right.

**Table 10.1  Amino Acids** HO—C(=O)—CH(—NH$_2$)—G

| G (Side Chain) | Name | Common* | Wellner-Meister Symbol for Side Chain |
|---|---|---|---|
| | | Symbols Used in Representing Amino Acids in Protein Structures | |

**Side chain is nonpolar**

| G (Side Chain) | Name | Common* | Wellner-Meister Symbol |
|---|---|---|---|
| —H | glycine | Gly | |
| —CH$_3$ | alanine | Ala | |
| —CH(CH$_3$)$_2$ | valine | Val | |
| —CH$_2$CH(CH$_3$)$_2$ | leucine | Leu | |
| —CHCH$_2$CH$_3$ (with CH$_3$) | isoleucine | Ile | 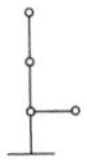 |
| —CH$_2$—C$_6$H$_5$ | phenylalanine | Phe | |
| 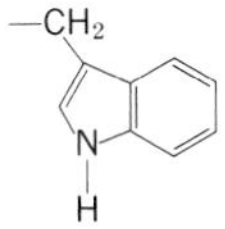 —CH$_2$— (indole) *[handwritten: 2 pka 2.9]* | tryptophan | Trp |  |
| HOC(=O)—CH—CH$_2$ (NH, CH$_2$, CH$_2$) (complete structure) *[handwritten: → 2 pka 2,10]* | proline | Pro | |

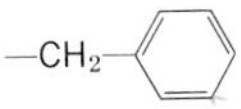

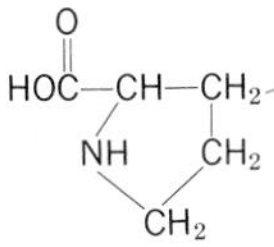

**Side chain has a hydroxyl group**

| G (Side Chain) | Name | Common* | Wellner-Meister Symbol |
|---|---|---|---|
| —CH$_2$OH | serine | Ser | |
| —CHOH (CH$_3$) | threonine | Thr | 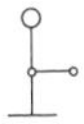 |
| 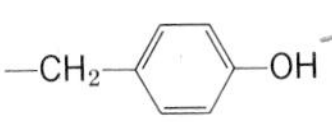 —CH$_2$—C$_6$H$_4$—OH *[handwritten: → 3 pka 2,9,10]* | tyrosine | Tyr |  |
| —CH$_2$—C$_6$H$_2$(I)$_2$—OH | diiodotyrosine | — | — |
| —CH$_2$—C$_6$H$_2$(I)$_2$—O—C$_6$H$_2$(I)$_2$—OH | thyroxine | — | — |
| 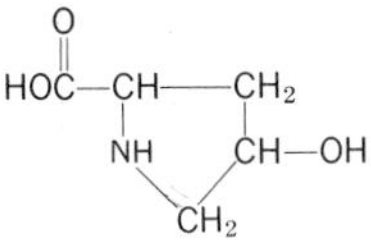 HOC(=O)—CH—CH$_2$ (NH, CH—OH, CH$_2$) (complete structure) | hydroxyproline | Hyp | |

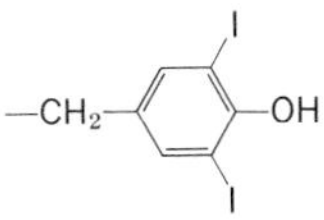

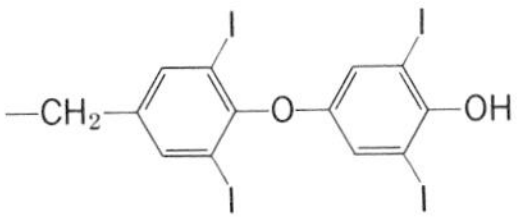

Table 10.1  Amino Acids HO—C—CH—G (continued)

(structure at top: O double bond on C, NH$_2$ below CH)

| G (Side Chain) | Name | Symbols Used in Representing Amino Acids in Protein Structures | |
|---|---|---|---|
| | | Common* | Wellner-Meister Symbol for Side Chain |

**Side chain has a carboxyl group (or the corresponding amide)**

| | | | |
|---|---|---|---|
| —CH$_2$CO$_2$H | aspartic acid | Asp |  |
| —CH$_2$CH$_2$CO$_2$H | glutamic acid | Glu |  |
| —CH$_2$CONH$_2$ | asparagine | Asn |  |
| —CH$_2$CH$_2$CONH$_2$ | glutamine | Gln |  |

**Side chain has a basic amino group**

| | | | |
|---|---|---|---|
| —CH$_2$CH$_2$CH$_2$CH$_2$NH$_2$ | lysine | Lys |  |
| —CH$_2$CH$_2$CH$_2$NH—C—NH$_2$ (with NH) | arginine | Arg | |
| 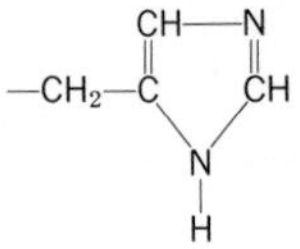 —CH$_2$—C (histidine imidazole ring: CH—N, CH, N—H) | histidine | His |  |

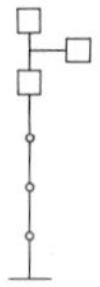

**Side chain contains sulfur**

| | | | |
|---|---|---|---|
| —CH$_2$S—H | cysteine | Cys |  |
| —CH$_2$—S / —CH$_2$—S | cystine | Cys or Cys† |  |
| —CH$_2$CH$_2$SCH$_3$ | methionine | Met |  |

* These three-letter symbols are recommended by the Joint Commission on Biochemical Nomenclature of IUPAC-IUB.
† For half the cystine unit.

[Handwritten annotations in margins: "NH$_2$—CH C—OH" structure with side chain CH$_2$ repeated and NH$_2$; numbers "3", "2" beside side chains; "3 pka 2,9,10" near lysine; "3 pka 2,9,12" near arginine; "3 pka 2,8,10.7" near cysteine.]

Just as the two letters N and O can be arranged to give two different words, NO and ON, so two different amino acids can be joined to give two isomeric dipeptides that have identical "backbones" and differ only in the sequences of side chains.

Dipeptides, of course, are still amino acids, although they are no longer $\alpha$-amino acids. A third $\alpha$-amino acid can react at one end or the other. In general:

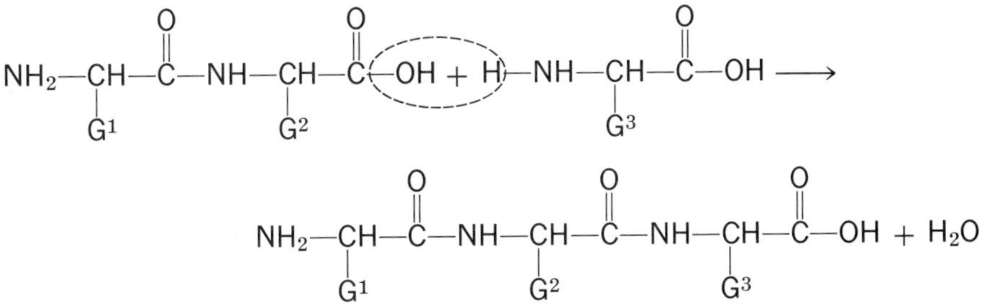

$$NH_2-CH-\overset{\overset{O}{\|}}{C}-NH-CH-\overset{\overset{O}{\|}}{C}-OH + H-NH-CH-\overset{\overset{O}{\|}}{C}-OH \longrightarrow$$

$$\quad\quad\quad | \quad\quad\quad\quad | \quad\quad\quad\quad\quad\quad\quad\quad |$$
$$\quad\quad\quad G^1 \quad\quad\quad\quad G^2 \quad\quad\quad\quad\quad\quad\quad G^3$$

$$NH_2-CH-\overset{\overset{O}{\|}}{C}-NH-CH-\overset{\overset{O}{\|}}{C}-NH-CH-\overset{\overset{O}{\|}}{C}-OH + H_2O$$

$$\quad\quad\quad | \quad\quad\quad\quad\quad | \quad\quad\quad\quad\quad |$$
$$\quad\quad\quad G^1 \quad\quad\quad\quad\quad G^2 \quad\quad\quad\quad G^3$$

A tripeptide

Specific example:

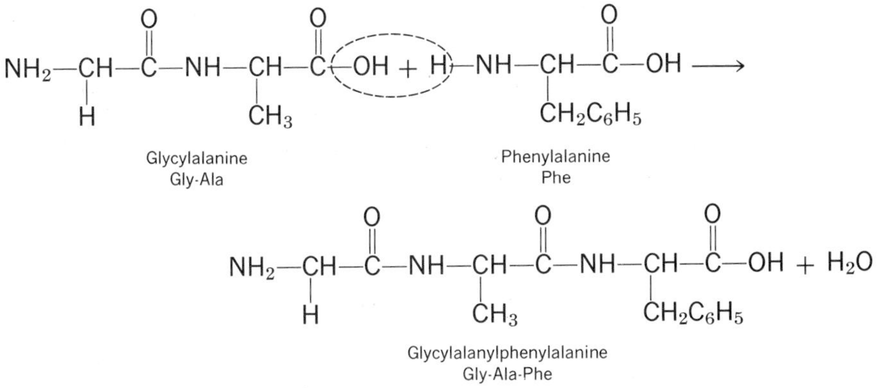

$$NH_2-CH-\overset{\overset{O}{\|}}{C}-NH-CH-\overset{\overset{O}{\|}}{C}-OH + H-NH-CH-\overset{\overset{O}{\|}}{C}-OH \longrightarrow$$

$$\quad\quad\quad | \quad\quad\quad\quad\quad | \quad\quad\quad\quad\quad\quad\quad\quad |$$
$$\quad\quad\quad H \quad\quad\quad\quad CH_3 \quad\quad\quad\quad\quad\quad CH_2C_6H_5$$

Glycylalanine         Phenylalanine
Gly-Ala             Phe

$$NH_2-CH-\overset{\overset{O}{\|}}{C}-NH-CH-\overset{\overset{O}{\|}}{C}-NH-CH-\overset{\overset{O}{\|}}{C}-OH + H_2O$$

$$\quad\quad\quad | \quad\quad\quad\quad\quad | \quad\quad\quad\quad\quad\quad |$$
$$\quad\quad\quad H \quad\quad\quad\quad CH_3 \quad\quad\quad\quad CH_2C_6H_5$$

Glycylalanylphenylalanine
Gly-Ala-Phe

Or, in Wellner-Meister symbols:

Gly-Ala       Phe       Gly-Ala-Phe

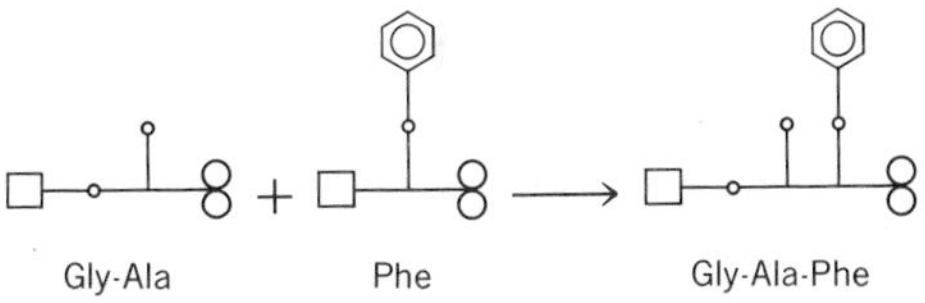

These Wellner-Meister symbols represent protein structures. For the beginner, to whom memorizing twenty or so common names for the amino acids may seem a formidable task, these symbols convey meaning almost immediately, once the very simple system is learned. This chemical shorthand permits rapid recognition of the kind of amino acid units present, their frequency, and their sequences in a polypeptide molecule.

| a square | □ | means nitrogen, $NH_2$, or NH |
|---|---|---|
| a large circle | O | means oxygen, —OH, or C=O (the context together with rules of valence making it clear which is represented) |
| a small circle | o | represents a carbon atom (and whatever —Hs are needed within the rules of valence) |
| two large circles | 8 | means a carboxyl group |
| a large circle and a square | o□ | means an amide group, —$CONH_2$ |

Sulfur atoms are designated by S and aromatic rings by the appropriate, easily recognized geometric figures we have always used for them. Peptide bonds are "understood" to be present.[2]

The tripeptide shown in the last example is only one of the possible isomers involving three different amino acids. If we think in terms of three different letters, A, E, and T, there are six different ways of ordering them:

AET     EAT     TAE

ATE     ETA     TEA

These are not all words in the English language, but then neither are all possible polypeptides found in a particular species. The uniqueness of any tripeptide—or polypeptide, for that matter—lies in the sequence of its side chains. The set of all possible sequences for a tripeptide made from glycine, alanine, and phenylalanine is as follows.

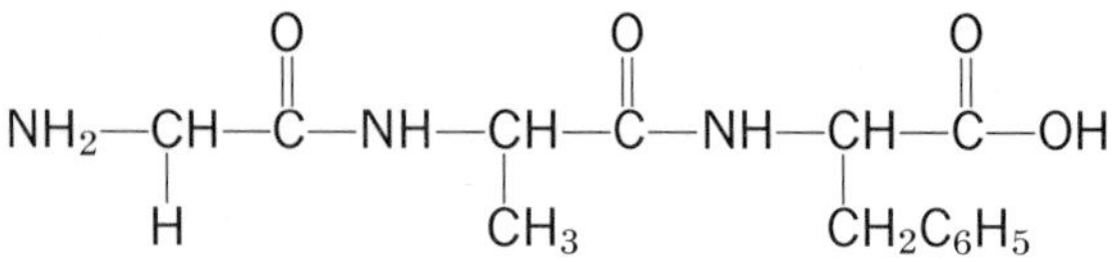

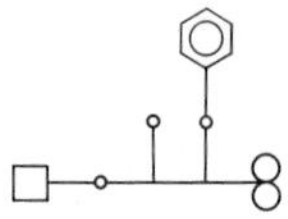

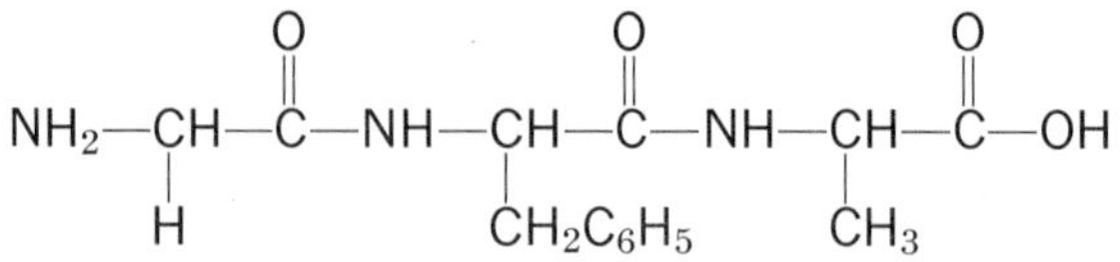

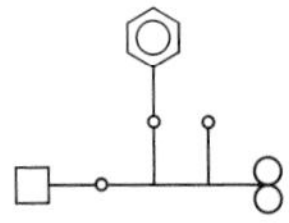

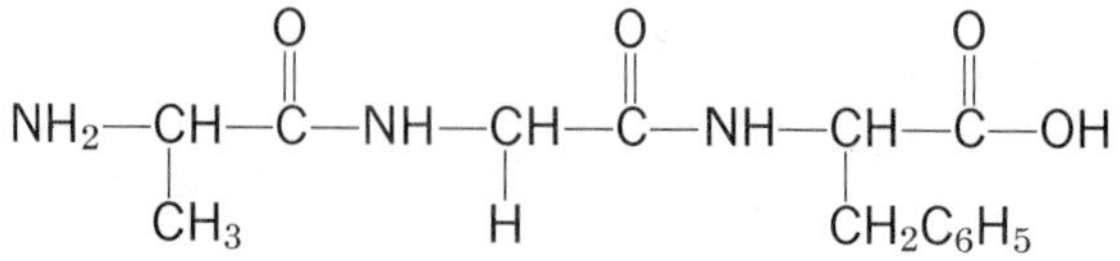

---

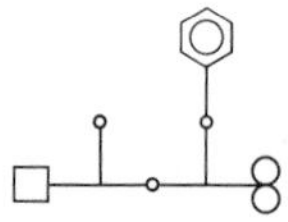

[2] The Wellner-Meister system, not the first of its kind, was introduced in 1966. Whether or not it will be widely accepted remains to be seen. It has so many advantages for an introductory treatment that it seems appropriate to use it here, whatever its future.

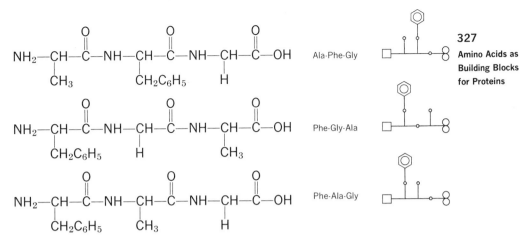

Ala-Phe-Gly

Phe-Gly-Ala

Phe-Ala-Gly

Each of these structures represents a different compound with its own unique set of physical properties. Their chemical properties will be quite similar, however, for the same functional groups are present in all. Each tripeptide is of course an amino acid, although none is an α-amino acid. Each can combine with another amino acid at one end or the other, and with a repetition of this process we can envision how a protein molecule is structured. All protein molecules have the repeating sequence,

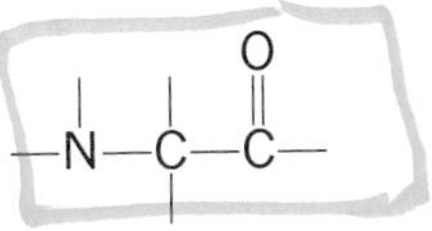

$$ -N-C-C- $$

in the backbone, which together with the sequence of side chains constitutes the *primary structural feature*.

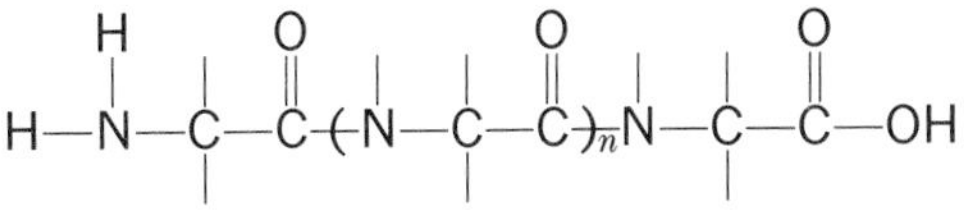

$$ H-N-C-C(N-C-C)_{n}N-C-C-OH $$   (*n* may vary from dozens to thousands)

*N*-terminal
unit

*C*-terminal
unit

In practice, the words *polypeptide* and *protein* are often used interchangeably. We follow this practice, but *polypeptide* will generally mean a smaller protein, one with lower than normal formula weight, perhaps 50 amino acid units or fewer. Many proteins contain upward of 5000 amino acid units. In such systems many amino acids have obviously been used several times. Table 10.2 contains data on the amino acid content of a few representative proteins to illustrate this point.

By long, laborious processes the structures of a few proteins have been determined. Both chemical methods and X-ray diffraction techniques have been used. We shall say nothing about the second technique and give only the general idea for the chemical method. Using letters to represent amino acid units, let us suppose that a protein has the "molecular formula" A, B, C, D, E. That is, analysis shows that a molecule of this protein is a pentapeptide made up of amino acids A, B, C,

**Table 10.2 Amino Acid Composition of Proteins**

Expressed as number of amino acid residues per molecule when the formula weight is known, otherwise as the moles of amino acid residues per 100,000 g protein.

| Amino Acid | Formula weight of protein: | Human Insulin 6000 | Collagen — | Horse Hemoglobin 68,000 | Egg Albumin 45,000 | Silk Fibroin — | Wool Keratin — |
|---|---|---|---|---|---|---|---|
| Glycine | | 4 | 363 | 48 | 19 | 581 | 87 |
| Alanine | | 1 | 107 | 54 | 35 | 334 | 46 |
| Valine | | 4 | 29 | 50 | 28 | 31 | 40 |
| Leucine | | 6 | 28 | 75 | 32 | 7 | 86 |
| Isoleucine | | 2 | 15 | 0 | 25 | 8 | — |
| Phenylalanine | | 3 | 15 | 30 | 21 | 20 | 22 |
| Tryptophan | | 0 | 0 | 5 | 3 | 0 | 9 |
| Proline | | 1 | 131 | 22 | 14 | 6 | 83 |
| Serine | | 3 | 32 | 35 | 36 | 154 | 95 |
| Threonine | | 3 | 19 | 24 | 16 | 13 | 54 |
| Tyrosine | | 4 | 5 | 11 | 9 | 71 | 26 |
| Hydroxyproline | | 0 | 107 | 0 | 0 | 0 | 0 |
| Aspartic acid (or asparagine) | | 3 | 47 | 51 | 32 | 21 | 54 |
| Glutamic acid (or glutamine) | | 7 | 77 | 38 | 52 | 15 | 96 |
| Lysine | | 1 | 31 | 38 | 20 | 5 | 19 |
| Arginine | | 1 | 49 | 14 | 15 | 6 | 60 |
| Histidine | | 2 | 5 | 36 | 7 | 2 | 7 |
| Cysteine | | 0 | 0 | 4 | 5 | — | — |
| Cystine | | 3 | 0 | 0 | 1 | — | 49 |
| Methionine | | 0 | 5 | 4 | 16 | — | 7 |
| Percent hydrocarbon side chains | | 41 | 72 | 47 | 45 | 77 | — |

Data from G. H. Haggis, D. Michie, A. R. Muir, K. B. Roberts, and P. B. M. Walter, *Introduction to Molecular Biology*, John Wiley and Sons, New York, 1964, pages 40–41.

D, and E, but the sequence is to be determined. (There are 125 possible isomers.) To have proceeded this far, the protein must have been hydrolyzed and the mixture of amino acids (plus some di- and tripeptides) separated.

A mixture such as this one is commonly separated by paper chromatography (described in Figure 10.1), which distributes the components of the mixture in various sections of the paper. The paper is then dried and sprayed with chemicals that react to produce visible spots with the components, revealing their positions and sometimes their nature. Using $R_f$ values (Figure 10.1) and color tests, we can identify the components of an unknown mixture of amino acids and di- and tripeptides.

To return to our pentapeptide problem, suppose that hydrolyses were only partial, and that the mixture contains individual amino acids together with dipeptides and tripeptides resulting from random hydrolyses of peptide bonds. Paper chromatography then identifies fragments AD, CB, DC, BE, and DCB, in addition to individual amino acids. The question now is what sequence in the original pentapeptide would give these fragments. If the fragments are lined up as follows, the pattern quickly appears:

AD
DC
DCB
BE
CB
‾‾‾‾
ADCBE = the sequence in the original pentapeptide

Only this sequence could give the fragments observed.

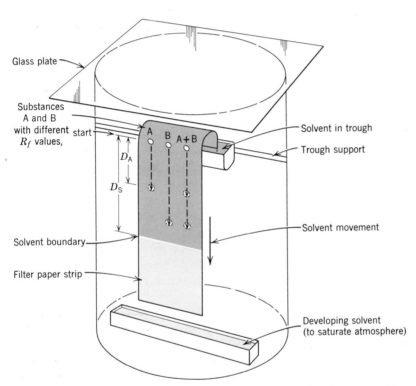

**Figure 10.1** Paper chromatography. A drop of the mixture (A + B) to be separated is spotted on the paper at the starting line. The paper consists largely of cellulose molecules which have water molecules adhering strongly to them by means of hydrogen bonds. (The paper feels dry, however.) The components of the mixture, A + B, are attracted to this stationary phase of water and cellulose, with (as illustrated) molecules of A adsorbing more strongly than those of B. The paper is then positioned to permit a solvent (e.g., a butanol-water mixture) to creep by capillary action past the starting line. As it does, molecules of A and B distribute themselves between the moving phase and the stationary phase. As shown, molecules of B are more soluble in the moving phase, and gradually, as this phase advances, component B pulls away from component A. The $R_f$ value for component A is the ratio of distance traveled by component A ($D_A$) to distance traveled by the moving solvent ($D_S$). When determined under very carefully controlled conditions, $R_f$ values are important physical constants which identify components of unknown mixtures. (Adapted, with permission, from E. C. Conn and P. K. Stumpf, *Outlines of Biochemistry*, second edition, John Wiley and Sons, New York, 1966, page 436.)

Among the proteins whose structures are completely known are the pituitary hormones, oxytocin and vasopressin (Figure 10.2); the adrenocorticotropic hormone (ACTH) (Figure 10.3); the hormone of the pancreas, insulin (Figure 10.4); and an enzyme, ribonuclease (Figure 10.5). The structures of hemoglobin and of the protein associated with the tobacco mosaic virus are also known.

Besides illustrating what protein structures are like, Figures 10.2 through 10.5 teach us other lessons. Among the proteins studied thus far, only proteins performing like functions in different species (e.g., insulins from various animals) have similar amino acid sequences. What seem to be small changes in other sequences result in considerable differences in function. Oxytocin and vasopressin (Figure 10.2) illustrate this. The first is a hormone that stimulates contraction of uterine muscles. The second, in mammals, is able to cause contraction of peripheral blood vessels and a rise in blood pressure. Yet where oxytocin uses leucine (Leu), vasopressin has arginine (Arg); where oxytocin has isoleucine (Ile), vasopressin has phenylalanine (Phe). *Protein function* is highly sensitive to *protein structure* in ways we are just beginning to understand. One of the most dramatic and well-authenticated examples of this sensitivity is the hemoglobin change of sickle-cell anemia.

**Sickle-Cell Anemia and Altered Hemoglobin.** The hemoglobin molecule, as it occurs in adult human beings, has a molecular weight of about 65,000. It consists of four long polypeptide chains, intricately folded, two designated as alpha and two as beta. Each enfolds a large, flat, nonprotein molecule called heme (see Figure 10.6) in which a ferrous ion is held. Each heme unit, in cooperation with its associated globin chain, can bind one oxygen molecule. Hence a complete hemoglobin molecule will carry four oxygen molecules from the lungs. (It is interesting that the cooperation between the globin chains and the heme units is such that when three oxygen molecules have been taken up, the attachment of the fourth and last is made much easier. Thus for the most part only completely oxygenated hemoglobin moves away from the lungs.)

The polypeptide chains of normal hemoglobin contain about 300 amino acid units. In the hemoglobin of those suffering from sickle-cell anemia, *only one of these is changed*. A glutamic acid unit sixth in from the $N$-terminus of the $\beta$-chain is replaced by a valine unit. Glutamic acid has a carboxyl group, —COOH, on its side chain; the side chain of valine is the isopropyl group. The —COOH can ionize, but isopropyl cannot. Normal and sickle-cell hemoglobins are therefore capable of having different electric charges which affect their relative solubilities as well as other properties. (See also page 337.)

When the oxygen supply is high, as in arterial blood, both normal hemoglobin, HHb, and sickle-cell hemoglobin, HbS, have about the same solubility in the bloodstream. When the oxygen supply is lower, as in venous blood, the solubility of HbS is less than that of HHb. Consequently, when circulating blood has delivered its oxygen, the altered hemoglobin tends to precipitate in red cells containing it. This precipitate distorts the shapes of the red cells, the blood becomes more difficult to pump, and a greater strain is placed on the heart. The red cells may also clump together enough to plug a capillary here and there. Sometimes the red cells split open.

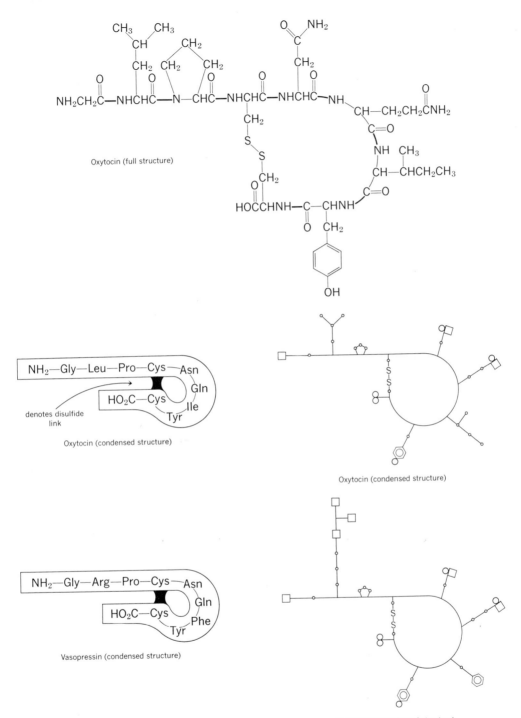

**Figure 10.2** Oxytocin and vasopressin, two low-formula-weight proteins, symbolized in several ways.

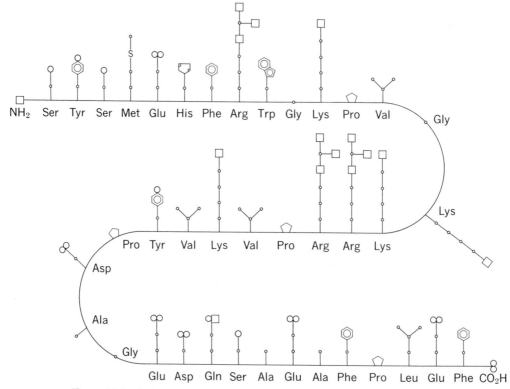

NH₂ Ser Tyr Ser Met Glu His Phe Arg Trp Gly Lys Pro Val Gly

Pro Tyr Val Lys Val Pro Arg Arg Lys — Lys

Asp

Ala — Gly — Glu Asp Gln Ser Ala Glu Ala Phe Pro Leu Glu Phe CO₂H

**Figure 10.3** Amino acid sequence in human adrenocorticotropic hormone (ACTH, adreno-corticotropin, corticotropin). This hormone, made from thirty-nine amino acid units, stimulates the adrenal gland to produce and secrete its steroid hormones. It has been used to treat rheumatoid arthritis, rheumatic fever, and other inflammatory diseases. P. H. Bell and R. G. Shepherd (American Cyanamid) worked out the amino acid sequence of ACTH obtained from pork adrenals. The ACTH molecules from various animal sources are identical up through the first twenty-four amino acid units (counting from the free amino end, upper left). K. Hofmann (University of Pittsburgh) synthesized a polypeptide with a sequence identical to that of the first twenty-three amino-acid units of ACTH; its biological activity is also similar to that of the natural hormone. Apparently the remaining residues from twenty-four through thirty-nine are not essential to biological activity.

The two folds shown in the structure are made only to fit it onto the page. The actual gross shape of the molecule is not indicated by this figure.

A child with the severe form of the disease has received the genetic trait from both parents. A child with the mild form has only one afflicted parent and shows no symptoms except when in an environment with a low concentration of oxygen (e.g., the depressurized cabin of an airplane). The disease is especially widespread in central and western Africa, and children with the severe form often die before they are two years old. The molecular difference responsible for the anemia, a matter of life or death for the victims, is startlingly small. One amino acid in 300 is substituted for another.

The physiological properties of proteins are related not only to the configurations of their amino acid units but also to the kinds of side chains present. We shall therefore examine briefly some generalizations about these side chains that will be useful to our study of proteins.

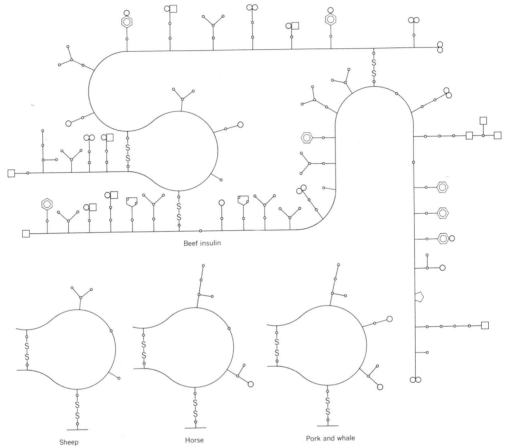

**Figure 10.4** The primary structure of insulin. F. Sanger, British biochemist, and a team of co-workers at Cambridge worked from 1945 to 1955 on the determination of this structure. Sanger received the 1958 Nobel prize in chemistry for this work and the development of the techniques. Insulin from different species have the slight differences noted. Fortunately, these are not serious for diabetics on insulin therapy. (Usually, when an alien protein gets into circulation, the body makes antibodies to combine with it. The rejection of skin grafts from one person to another and the extraordinary problems of transplanting organs are related to this formation of antibodies.) Sheep insulin, however, can be given to human diabetics. Some antibodies form, but the insulin activity is reduced only slightly. The synthesis of human insulin *in vitro* was reported in 1966 at the Brookhaven National Laboratory by a team of biochemists headed by P. G. Katsoyannis. At about the same time Western scientists learned that a team of biochemists in mainland China had succeeded in synthesizing insulin in what was acknowledged to be a much purer state.

**Amino Acids with Nonpolar Side Chains.** The first group of amino acids in Table 10.1 are those with essentially nonpolar side chains.[3] They are said to be *hydrophobic* groups ("hydro-," water; "-phobic," hating); when a huge protein molecule folds into its distinctive shape (cf. tertiary structural features, p. 337),

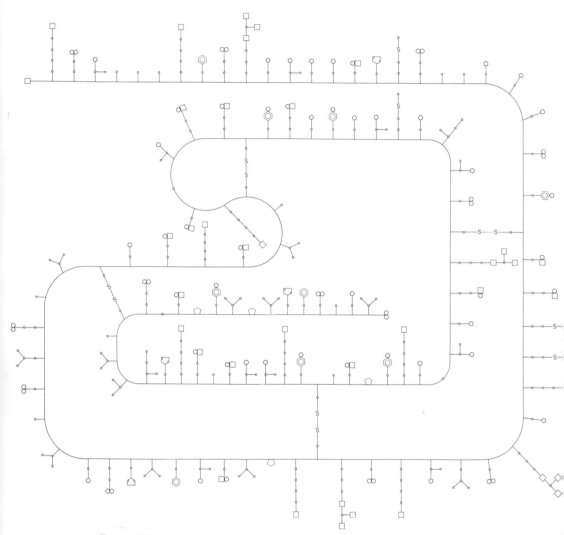

**Figure 10.5** The structure of ribonuclease (RNase), which catalyzes the hydrolysis of ribonucleic acid (RNA) in biological systems. M. Kunitz (Rockefeller Institute) first insolated and crystallized this enzyme in 1940. Other scientists at the Rockefeller Institute, Hirs, Spackman, Smythe, Stein, and Moore, worked out its amino acid sequence. There are 124 amino acid units in the whole molecule. The laboratory synthesis of this enzyme was reported early in 1969 by a Rockefeller University team (led by Merrifield and Gutte) and also by a large group at Merck, Sharp, and Dohme (led by Denkewalter, Hirschmann, Holly, and Verber).

[3] What looks like a 2° amino group in tryptophan is unable to coordinate with a proton. The nitrogen's unshared pair is delocalized into the aromatic system. Hence tryptophan is put in this group because for all practical purposes its side chain is nonbasic and only slightly polar.

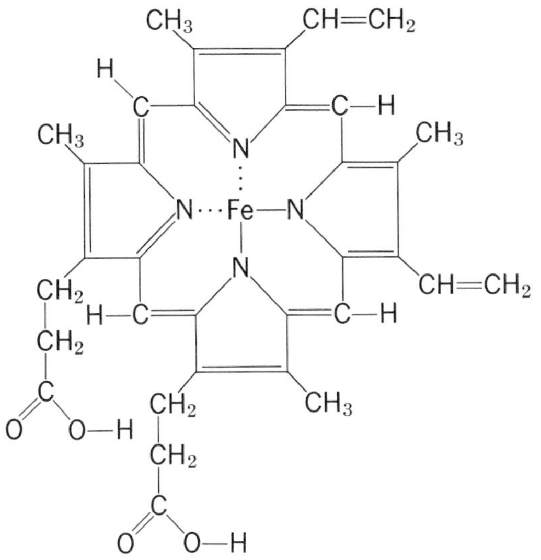

**Figure 10.6** The heme molecule. The iron in heme holds its $Fe^{2+}$ state in *oxygenated* hemoglobin, that is, hemoglobin which has picked up oxygen molecules. In *oxidized* hemoglobin, however, iron is in the $Fe^{3+}$ state, and the substance is brownish in color. When meat is cooked, the color changes from red to brown largely because $Fe^{2+}$ oxidizes to $Fe^{3+}$.

these hydrophobic groups tend to be folded next to each other rather than next to highly polar groups.

**Amino Acids with Hydroxyl-Containing Side Chains.** The second set of amino acids in Table 10.1 consists of those whose side chains carry alcohol or phenol groups. In cellular environments they are neither basic nor acidic, but they are polar and *hydrophilic* (water loving). They can be either hydrogen bond donors or acceptors, and they help bind neighboring groups into ordered structures as the long protein chain folds into its final shape. Amino acids rich in these groups are more soluble in water than otherwise. One of these acids, tyrosine, has a phenolic side chain, which means that proteins containing tyrosine units are vulnerable at these points to oxidizing agents and electrophilic reagents. Proteins in human skin that contain tyrosine are converted to skin pigments, the melanins, by oxidative steps. When nitric acid is spilled on the skin, it turns yellow because the nitro derivatives of the tyrosine units in skin proteins form and are this color.

**Amino Acids with Acidic Side Chains.** Aspartic acid and glutamic acid have an extra carboxyl group each. When these amino acids are included in a protein, their side chains make the protein more acidic than otherwise. They are proton donors and can serve as both hydrogen bond acceptors and donors. Their presence in a protein tends to make it more soluble in water, provided the pH of the medium is right (see p. 343).

In proteins these acids frequently occur in the form of their corresponding amides, arparagine and glutamine. Although still very polar, in this form they can no longer function as acids.

**Amino Acids with Basic Side Chains.** Lysine has an extra primary amino group that makes its side chain basic, hydrophilic, and a hydrogen bond donor or acceptor. The side chain in arginine has the guanidinium group, $-NH-\overset{\overset{\displaystyle NH}{\|}}{C}-NH_2$, represented

as ⌑—⌑—⌑ in the Wellner-Meister symbolism. One of the most powerful proton-accepting groups found in organisms, it exists almost exclusively in its protonated form, $-NH-\overset{\overset{\displaystyle NH_2^+}{\|}}{C}-NH_2$.

**Amino Acids with Sulfur-Containing Side Chains.** Cysteine is a particularly important amino acid because its mercapto group makes it especially reactive toward mild oxidizing agents. Cysteine and cystine, in fact, are interconvertible, a property of far-reaching importance in some proteins.

$$
\underset{\substack{\text{Two molecules of} \\ \text{cysteine}}}{
\begin{array}{c}
\overset{\displaystyle O}{\underset{\displaystyle \underset{\displaystyle NH_2}{|}}{HOCCHCH_2S{-}H}} \\
+ \\
\overset{\displaystyle O}{\underset{\displaystyle \underset{\displaystyle NH_2}{|}}{HOCCHCH_2S{-}H}}
\end{array}}
\quad
\underset{\substack{\text{oxidation} \\ \text{reduction}}}{\overset{\text{oxidation}}{\underset{\text{reduction}}{\rightleftarrows}}}
\quad
\underset{\substack{\text{One molecule of} \\ \text{cystine}}}{
\begin{array}{c}
\overset{\displaystyle O}{\underset{\displaystyle \underset{\displaystyle NH_2}{|}}{HOCCHCH_2S}} \\
| \\
\overset{\displaystyle O}{\underset{\displaystyle \underset{\displaystyle NH_2}{|}}{HOCCHCH_2S}}
\end{array}}
\quad + \ (2H)
$$

The disulfide linkage in cystine is common to several proteins, as we have already seen (Figures 10.2, 10.4, and 10.5). It is especially prevalent in proteins having a protective function, such as those forming hair, fingernails, and shells.

## SECONDARY STRUCTURAL FEATURES OF PROTEINS

If protein molecules were simply long polymeric chains, they might be expected to behave something like pieces of string, clustering together in no particular order and forming randomly tangled bunches similar to sweepings from a sewing room floor. But they do not. Their molecules are usually found in more orderly groups, folded in various ways to make compact and ordered units.

X-ray diffraction analysis is the tool that has revealed this ordering in proteins. Although an explanation of how this technique works is beyond the scope of this book, we can with little trouble accept the results. Highly crystalline materials such as inorganic salt crystals, in which there is a considerable degree of order, give distinctive X-ray diffraction pictures. Solids in which there is little order, *amorphous* substances like waxes, show no particular diffraction pattern. If proteins were clumped together as a tangled brush heap, they would be amorphous. Many proteins, however, give X-ray diffraction patterns indicating the presence of considerable order. Many regions are crystalline, although some parts are amorphous.

The so-called *secondary structural features* of proteins make order in their aggregations possible. (The primary structural feature is the amino acid sequence along the backbone.) We shall discuss these structural features as we examine one of the most characteristic protein shapes, the $\alpha$-helix.

$\alpha$-**Helix.** In many proteins studied thus far, the chains are coiled into a spiral called an $\alpha$-helix. Pauling and Corey, in 1951, were the first to publish extensive evidence for the helix (Figure 10.7). The opportunity for carbonyl oxygens to form hydrogen bonds to amide hydrogens provides the forces that cause the coiling. Thus hydrogen bonds of the type

$$\overset{\diagup}{\underset{\diagdown}{C}}=O\overset{\delta-}{\cdots}\overset{\delta+}{H}-\overset{|}{N}-$$

make up one of the most important of the secondary structural features.

The nature of the side chains affects intrachain hydrogen bonding and determines whether or not an $\alpha$-helix can form. Polylysine, a synthetic polypeptide, does not coil in an acidic medium in which all its side chain groups are in the protonated form $-(CH_2)_4NH_3{}^+$. Repulsions between the rather closely spaced, like-charged groups inhibit any tendency to coil, and in acid the molecule is in a randomly flexing, open form. If now the acid is neutralized and all the side chains become uncharged, $-(CH_2)_4NH_2$, polylysine takes up an $\alpha$-helix configuration. In contrast, polyglutamic acid, another synthetic polypeptide, is coiled in an acid but open in a base. Here the side chains are $-CH_2CH_2COOH$. In a base they are in the ionized form, $-CH_2CH_2COO^-$, and repulsions between like-charged side chains prevent the polymer from assuming an $\alpha$-helix configuration. But in an acidic medium all the side chains are electrically neutral, $-CH_2CH_2COOH$, and coiling is not inhibited.

**Salt Bridges.** When lysine and glutamic acid (or others with similar side chains) are in the same protein, side chains of unlike charge are possible. Depending on the pH of the medium, the $-NH_3{}^+$ and $-CO_2{}^-$ groups may both be present. When sites of opposite full charge are close enough neighbors, forces of attraction called *salt bridges* exist (Figure 10.8). The salt bridge is another important secondary structural feature of proteins.

## TERTIARY STRUCTURAL FEATURES

Helix formation is not always the final shaping, for helices may in turn fold, twist, or assume some other final configuration of minimum energy within the particular environment. The data on the bottom line of Table 10.2 show that the percent of hydrocarbon side chains in proteins is quite high, seldom less than 40%. These groups are not soluble in water. The whole system will be more stable if contact between hydrocarbon side chains and the aqueous medium (if that is what the protein is in) is minimized. Further folding of the molecule, especially common among the globular proteins occurring in organisms in an aqueous environment, minimizes the contact. Once the organism has put together the particular amino acid sequence that makes a protein, folding and coiling apparently follow automatically without any further need for enzymes and other chemicals.

**Myoglobin and Hemoglobin—Tertiary Structural Features.** These two proteins have been extensively examined for their tertiary structural features. John Kendrew

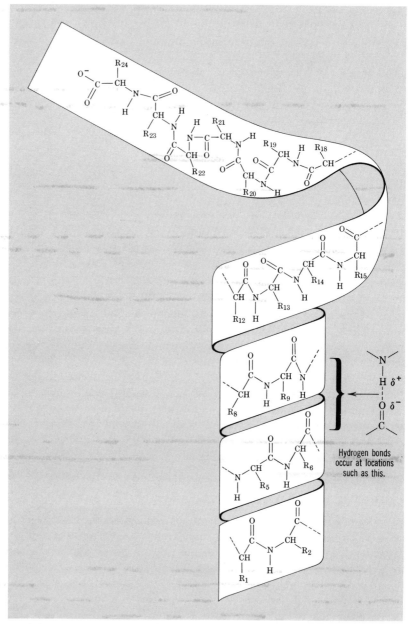

**Figure 10.7** This representation of a section of a protein chain as a ribbon shows the right-handed coiling of an α-helix. Hydrogen bonds exist between carbonyl oxygens and hydrogens on the amide nitrogens. Although each represents a weak force of attraction between the turns of the coil, there are many of them, and the total force stabilizing the helix is more than enough to counteract the natural tendency for the chain to adopt a randomly flexing form. It is understandable that the repeating series of hydrogen bonds up and down the helix is called the "zipper" of the molecule. (From G. H. Haggis, D. Michie, A. R. Muir, K. B. Roberts, and P. M. B. Walker, *Introduction to Molecular Biology*, John Wiley and Sons, New York, 1964, page 51.)

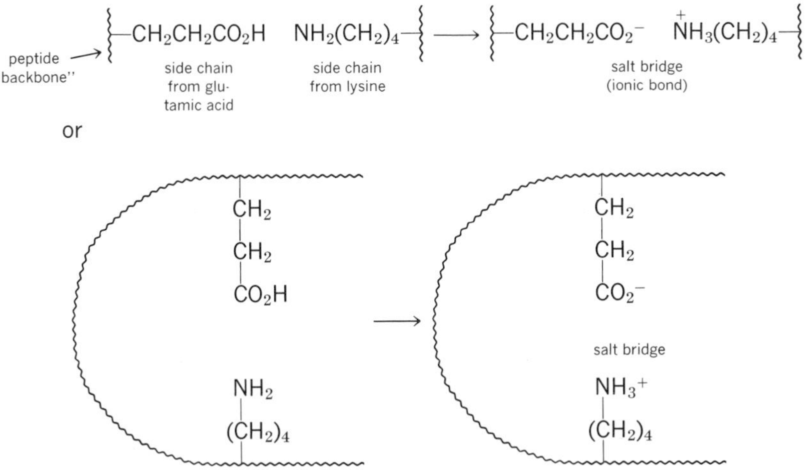

**Figure 10.8** The salt bridge. By acid-base interaction two peptide strands can acquire opposite charges which act to hold them together (top) or stabilize a call within a strand (bottom).

and Max Perutz of England (Cambridge) shared the 1962 Nobel prize in chemistry for their X-ray diffraction analysis of myoglobin and hemoglobin.

Kendrew concentrated on myoglobin, a protein with 153 amino acid residues. He found that its molecules are folded as illustrated in Figure 10.9. Perutz worked on the structure of hemoglobin and found that its $\beta$-chains (146 amino acid residues) are folded very much like the myoglobin system of Figure 10.9. The $\alpha$-chains are folded in a different way. In the whole hemoglobin molecule two $\alpha$-chains and two $\beta$-chains, each folded as shown in Figure 10.10$a$, aggregate as the model in Figure 10.10$b$ indicates.

**A Word of Caution and a Disclaimer.** Protein structures are studied in materials isolated from a living system. The secondary or tertiary structure that a protein has in the isolated crystalline form is not necessarily the same as the structure it has *in vivo*. (The fundamental sequence of the amino acids will be the same, assuming that techniques for isolation are gentle enough.) When they are in an aqueous medium, proteins may have up to 30% by weight of bound water—water molecules attracted to and held by the protein molecule, roughly like water of hydration. Water has an important effect on the activity of proteins in organisms. If it has affected activity, water has very likely affected structure, at least at the secondary and tertiary levels of organization.

This complication serves to remind us, if any reminder is needed, that knowledge of the isolated items in the inventory of a cell is one thing. Translating such information into an understanding of what goes on in a cell is quite another. If the reader ever thought that we shall soon synthesize a living cell *in vitro* from its scrambled molecules without the aid of other living cells, he should by now be disabused of this idea. Such synthesis could happen someday, but not soon. The complexities of dynamic cellular architecture are so vast that biology is a very long way

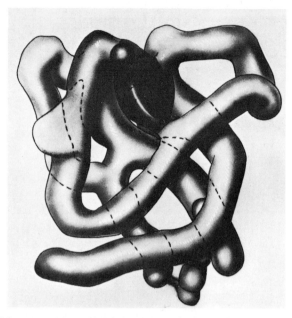

**Figure 10.9** Tertiary structural features of the myoglobin molecule. The sausagelike portion contains the protein chain; where it is relatively straight, the chain is coiled in an α-helix. It is estimated that 70% of the molecule has this secondary feature. The darker, disklike section represents a heme unit. Myoglobin stores and transports oxygen at muscles. (Courtesy of John C. Kendrew, Cambridge University.)

from being reduced to chemistry. The services of all the specialists in biology—anatomists, physiologists, geneticists, microbiologists, pathologists, taxonomists, etc.—will be in demand for a long, long time. Many insights are provided by a study of the molecular basis of life, the principal subject of this book, but the foregoing disclaimer is stated for the sake of a balanced view.

## AMINO ACIDS AND PROTEINS AS BUFFERS

**Dipolar Ionic Character of Amino Acids.** For convenience as well as to conform to common usage, structures of amino acids have been written as having —NH$_2$ and —COOH groups. Neither their chemical properties nor their physical properties agree too well with this picture, however.

$$NH_2—CH—COOH$$
$$|$$
$$G \qquad \text{(glycine, if G = H—)}$$

Conventional way of writing
the structure of
an amino acid

Most carboxylic acids, for example, have acid dissociation constants $K_a$ on the order of $10^{-5}$, but for glycine $K_a$ is $1.6 \times 10^{-10}$. This value is low enough to be in the range of phenols. Similarly, most primary aliphatic amines have basic dissocia-

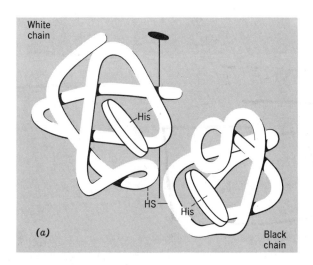

White
chain

His

HS

His

(a)

Black
chain

(b)

**Figure 10.10**   Hemoglobin. ($a$) Tertiary structures of the $\alpha$- and $\beta$-subunits of hemoglobin. The disks represent heme molecules. As with myoglobin (Figure 10.9), the straight sections consist of material in an $\alpha$-helix form. (From G. H. Haggis, D. Michie, A. R. Muir, K. B. Roberts, and P. B. M. Walker, *Introduction to Molecular Biology*, John Wiley and Sons, New York, 1964, page 62.)

($b$) This model, derived from X-ray diffraction data, shows how the most electron-dense regions of hemoglobin are arranged. The white section represents the two $\beta$-subunits; the black has two $\alpha$-subunits. The heme molecules (disks) are for the most part folded inside the system with enough exposure to permit oxygen molecules to attach themselves for transport from lungs to regions requiring them. (Photograph by Edward Leigh, courtesy of *Chemistry in Britain*.)

tion constants $K_b$ on the order of $10^{-5}$, but for glycine $K_b$ is $2.5 \times 10^{-12}$, making it so weak a proton acceptor that it is virtually neutral in this respect.

Amino acids have other peculiarities. Even the smallest ones are nonvolatile, crystalline solids without true melting points. When heated, they decompose and char before melting can occur. Furthermore, the amino acids tend to be insoluble in nonpolar solvents and soluble in water. These facts do not correlate well with the structure we have been using. The *dipolar ionic structure* is more consistent:

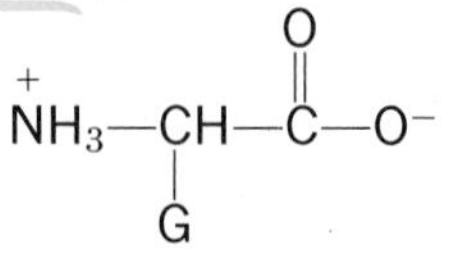

$$\overset{+}{NH_3}-CH-\overset{\overset{\displaystyle O}{\|}}{C}-O^-$$
$$|$$
$$G$$

Dipolar ionic form
of an amino acid

In a crystalline solid made up of such particles, forces of attraction between them will be strong, as in any ionic crystal (e.g., a salt). Before heating supplies enough energy to melt such crystals, covalent bonds are disrupted within the dipolar ions, and decomposition occurs. Ionic substances, of course, are known for their insolubility in nonpolar solvents. As for their low $K_a$ and $K_b$ values, the dipolar ionic structure offers a ready explanation. In an amino acid we are not dealing with a free carboxyl group, —COOH, as the proton donor, but rather with a substituted ammonium ion, —$NH_3^+$. The proton is more strongly bound within this group than it would be on the —COOH group. Similarly, the basic group, the proton-accepting group, in an amino acid is not the free —$NH_2$ group. Rather it is the carboxylate group, COO⁻. The latter is a weaker proton acceptor. The net effect is that amino acids are both very weak acids and very weak bases.

**Amino Acids and Proteins as Buffers.** If a strong base such as hydroxide ion, $OH^-$, is added to a solution of an amino acid, the following reaction will occur:

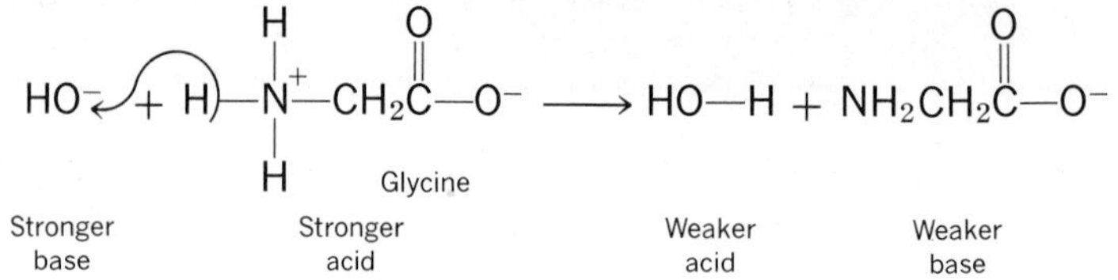

$$HO^- + H-\overset{\overset{\displaystyle H}{|}}{\underset{\underset{\displaystyle H}{|}}{\overset{+}{N}}}-CH_2\overset{\overset{\displaystyle O}{\|}}{C}-O^- \longrightarrow HO-H + NH_2CH_2\overset{\overset{\displaystyle O}{\|}}{C}-O^-$$

Glycine

| Stronger base | Stronger acid | Weaker acid | Weaker base |

If a strong acid (e.g., $H_3O^+$) is added to a solution of an amino acid, it will be neutralized as follows:

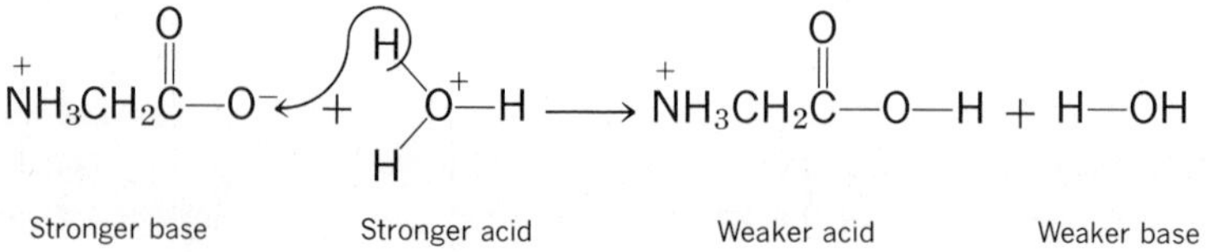

$$\overset{+}{NH_3}CH_2\overset{\overset{\displaystyle O}{\|}}{C}-O^- + \overset{\overset{\displaystyle H}{|}}{\underset{\underset{\displaystyle H}{|}}{O}}-H \longrightarrow \overset{+}{NH_3}CH_2\overset{\overset{\displaystyle O}{\|}}{C}-O-H + H-OH$$

| Stronger base | Stronger acid | Weaker acid | Weaker base |

Amino acids therefore have the ability to neutralize both stronger acids and stronger bases and can thus serve as buffers; they can hold fairly constant the pH of their solutions in water.

To the extent that protein molecules have groups such as those present in dipolar ionic amino acids, they will also act as buffers. Some of the proteins in the bloodstream serve chiefly as buffers, constituting the third buffer system of blood. (The other two are the $H_2CO_3/HCO_3^-$ and the $H_2PO_4^-/HPO_4^{2-}$ pairs.[4])

**Isoelectric Points of Amino Acids and Proteins.** When an amino acid is dissolved in water, proton transfers occur, as illustrated with glycine:

I
$$\overset{+}{NH_3}-CH_2\overset{O}{\overset{\|}{C}}-O \rightleftharpoons + H)-OH \rightleftharpoons \overset{+}{NH_3}CH_2\overset{O}{\overset{\|}{C}}-O-(H +)^-OH$$

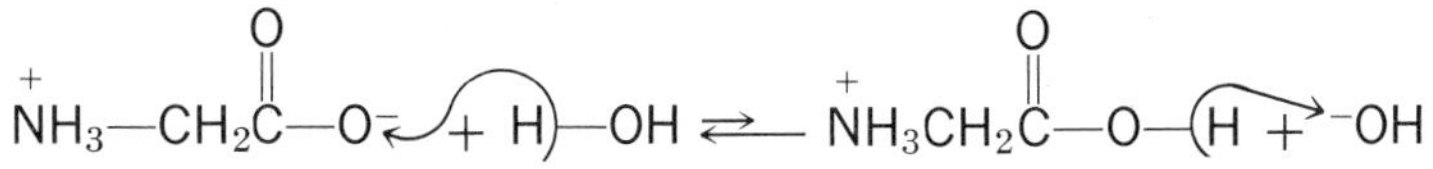

                 1                                                2

II
$$\overset{H}{\underset{H}{>}}O + H)-\overset{H}{\underset{H}{\overset{+}{N}}}-CH_2\overset{O}{\overset{\|}{C}}-O^- \rightleftharpoons H_3O^+ + NH_2CH_2\overset{O}{\overset{\|}{C}}-O^-$$

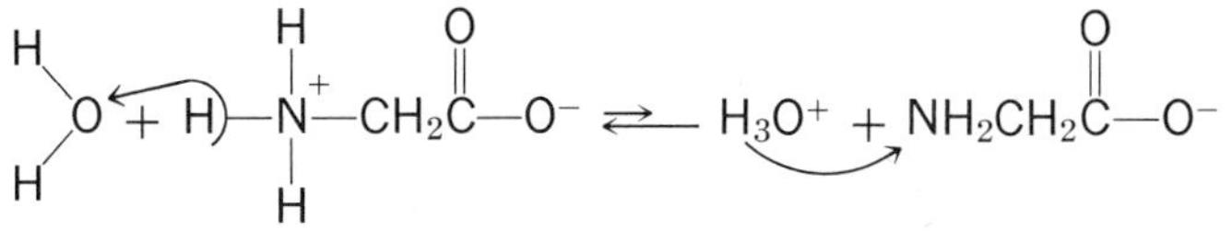

                 1                                               3

In its dipolar ionic form glycine is a slightly better proton donor (equilibrium II) than proton acceptor (equilibrium I). All three forms, **1, 2,** and **3,** coexist at equilibrium, although **2** and **3** are present only in trace amounts. Form **3** is in slight excess of **2.** If a trace of mineral acid were added, equilibrium II would be shifted to the left, and the concentration of **3** could be made equal to that of **2.** At this new pH glycine would exist almost exclusively as **1.** If an electric current were passed through the solution, there would be essentially no migration of glycine units in either direction, toward either electrode. *The pH at which an amino acid exhibits no migration in an electric field is called the isoelectric pH or the isoelectric point.*

Molecules of proteins, because they usually have several amino and carboxyl groups distributed among side chains, have ionization properties like those of simple amino acids. Protein molecules are normally capable of possessing several separate negatively charged and positively charged sites. The *net* charge may be negative, positive, or zero, depending on the specific protein, the pH of the solution, and the presence (or absence) of metallic ions that may complex with the protein and affect the net electric charge.

Protein molecules with no net electric charge are said to be in an isoelectric state.[5] Each soluble protein will have its own unique isoelectric point, the pH at which its net charge is zero, and no migration will occur in an electric field.

---

[4] Buffer systems hold constant the pH of the solution containing them, even when small amounts of strong acids or bases are added. The *carbonate buffer system* works by the following reactions:

    If acid is added: $HCO_3^- + H_3O^+ \longrightarrow H_2CO_3 + H_2O$
    If base is added: $H_2CO_3 + OH^- \longrightarrow HCO_3^- + H_2O$

The phosphate buffer system works by these reactions:

    If acid is added: $HPO_4^{2-} + H_3O^+ \longrightarrow H_2PO_4^- + H_2O$
    If base is added: $H_2PO_4^- + OH^- \longrightarrow HPO_4^{2-} + H_2O$

[5] If metallic ions, by complexing with the protein, make a contribution to the isoelectric state, it is more proper, strictly speaking, to refer to this condition as the isoionic state. Under these circumstances the pH at which no net migration occurs is called the *isoionic point.*

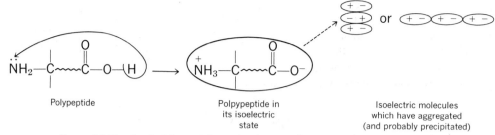

**Figure 10.11** Isoelectric protein molecules can be expected to aggregate in a manner reminiscent of oppositely charged ions collecting together to form ionic crystals. The effect is to convert already large protein molecules into even larger aggregates, reducing their solubility in water.

The purpose of this brief study of isoelectric points is to show that proteins are least soluble at their isoelectric pHs. At other pH values their molecules will each bear some net charge. They must therefore repel each other, and they cannot aggregate and precipitate. But if the molecules are all isoelectric, they are able to clump together (Figure 10.11) until the mass becomes too large to remain in solution and the protein precipitates.

The curdling of milk illustrates this effect. The isoelectric point of casein, the chief protein of cow's milk, is 4.7. It is only slightly soluble in water at this pH. The pH of cow's milk is normally in the range 6.3 to 6.6. When milk sours, bacterial growth is usually taking place and lactic acid is produced. The pH of the milk drops, approaching 4.7. As this happens, the casein molecules become more and more isoelectronic and they precipitate—or in ordinary terms the milk curdles. As long as the pH is something other than the isoelectric pH, however, the casein remains dispersed in the medium.

The importance of maintaining a constant pH of the blood is made clear by these considerations. Many enzymes and hormones circulate in the blood. If the pH of the blood varied too much up or down, presumably the isoelectric points for some of these proteins would be approached. If they were to precipitate, not only would they no longer be able to function but they would tend to plug blood vessels, thereby placing what might be too great a strain on the heart. The ability of hemoglobin to transport oxygen is also closely related to the pH of the blood, a subject we shall study in Chapter 12. The importance of the buffers in the blood cannot be overemphasized.

## DIGESTION OF PROTEINS. HYDROLYSIS

The digestion of proteins is equivalent to their hydrolysis, which is nothing more than the hydrolysis of amide linkages (Figure 10.12). Several digestive enzymes act as catalysts, about which more will be said in Chapter 12.

## DENATURATION

A wide variety of reagents and conditions that do not hydrolyze peptide bonds will destroy the biological nature and activity of the protein. When this happens, the protein is said to have been *denatured*. After denaturation the protein usually coagulates. Several of the more common chemicals and conditions that denature proteins are listed in Table 10.3.

At the molecular level denaturation is a disorganization of the shape of a protein. It can occur as an unfolding or uncoiling of a pleated or coiled structure or as the separation of the protein into subunits which then unfold or uncoil (Figure 10.13).

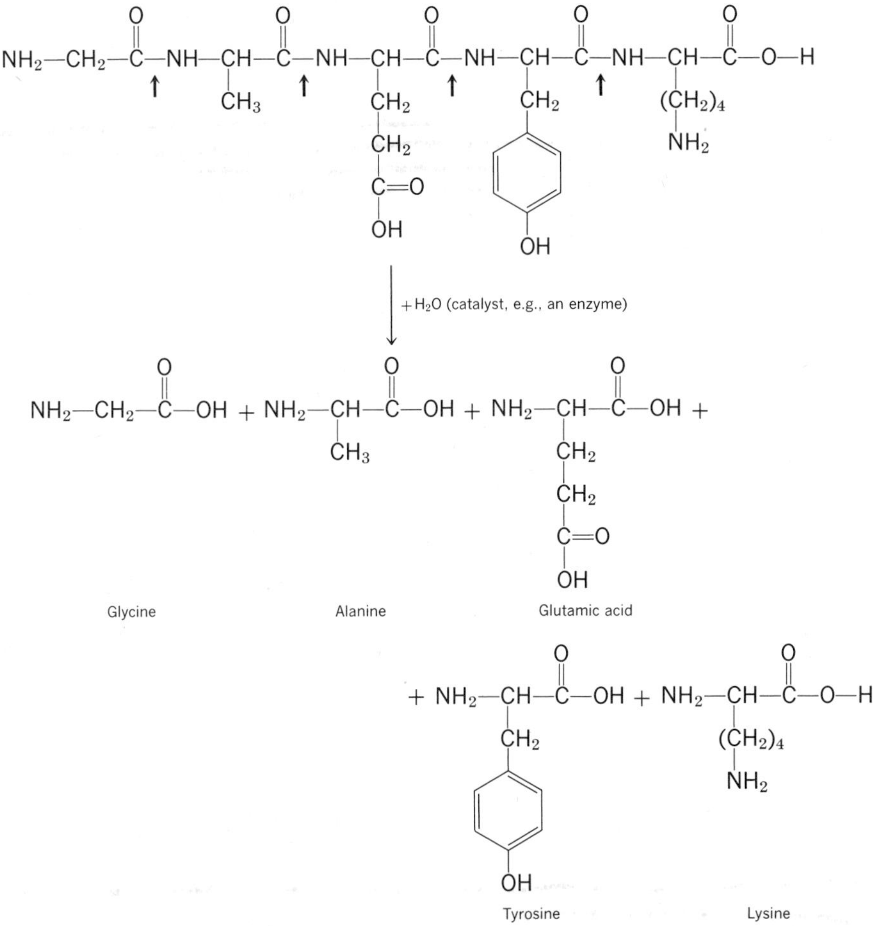

$$NH_2-CH_2-\overset{\displaystyle O}{\overset{\|}{C}}-NH-\overset{\displaystyle \uparrow}{\underset{CH_3}{CH}}-\overset{\displaystyle O}{\overset{\|}{C}}-NH-\overset{\displaystyle \uparrow}{CH}-\overset{\displaystyle O}{\overset{\|}{C}}-NH-\overset{\displaystyle \uparrow}{CH}-\overset{\displaystyle O}{\overset{\|}{C}}-NH-CH-\overset{\displaystyle O}{\overset{\|}{C}}-O-H$$

with side chains:
- $CH_3$
- $CH_2-CH_2-C(=O)-OH$
- $CH_2$—(phenol ring with OH)
- $(CH_2)_4-NH_2$

$$\downarrow +H_2O \text{ (catalyst, e.g., an enzyme)}$$

$$NH_2-CH_2-\overset{O}{\overset{\|}{C}}-OH + NH_2-\overset{CH_3}{\underset{|}{CH}}-\overset{O}{\overset{\|}{C}}-OH + NH_2-CH-\overset{O}{\overset{\|}{C}}-OH +$$

with side chain on the third: $CH_2-CH_2-C(=O)-OH$

| Glycine | Alanine | Glutamic acid |

$$+ NH_2-CH-\overset{O}{\overset{\|}{C}}-OH + NH_2-CH-\overset{O}{\overset{\|}{C}}-O-H$$

with side chains: $CH_2$—(phenol ring with OH) and $(CH_2)_4-NH_2$

| Tyrosine | Lysine |

**Figure 10.12** Protein hydrolysis (digestion). In the complete hydrolysis of this hypothetical polypeptide, only carbonyl-to-nitrogen bonds are ruptured (see arrows). A hydroxyl group, from water, becomes attached to each carbonyl carbon; a hydrogen atom, also from water, becomes attached to each nitrogen:

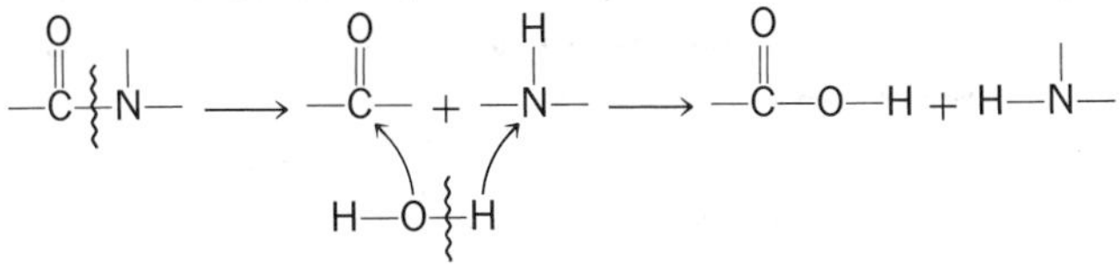

$$-\overset{O}{\overset{\|}{C}}\{N- \longrightarrow -\overset{O}{\overset{\|}{C}}- + -\overset{H}{\underset{}{N}}- \longrightarrow -\overset{O}{\overset{\|}{C}}-O-H + H-N-$$

$$H-O\}H$$

**Table 10.3  Chemicals and Conditions That Cause Denaturation**

| Denaturing Agent | How the Agent May Operate |
| --- | --- |
| Heat | Disrupts hydrogen bonds and salt bridges by making molecules vibrate too violently. Produces coagulation as in the frying of an egg. |
| Solutions of urea $(NH_2-\overset{\overset{O}{\|\|}}{C}-NH_2)$ | Disrupt hydrogen bonds. Being amide-like, urea can form hydrogen bonds of its own. |
| Ultraviolet radiation | Appears to operate the same way that heat operates (e.g., sunburning). |
| Organic solvents (e.g., ethyl alcohol, acetone, isopropyl alcohol) | May interfere with hydrogen bonds in protein, since alcohol molecules are themselves capable of hydrogen bonding. Quickly denatures the proteins of bacteria, thus killing them (e.g., disinfectant action of ethyl alcohol, 70% solution). |
| Strong acids or bases | Can disrupt hydrogen bonds and salt bridges: Prolonged action of aqueous acids or bases leads to actual hydrolysis of proteins. |
| Detergents | May affect salt bridges (by forming new salt bridges of their own), or may affect hydrogen bonds. |
| Salts of heavy metals (e.g., salts of the ions $Hg^{2+}$, $Ag^+$, $Pb^{2+}$) | May disrupt salt bridges (by forming new salt bridges to themselves). These ions usually precipitate proteins (coagulation). |
| Alkaloidal reagents (e.g., tannic acid, picric acid, phosphomolybdic acid) | May affect both salt bridges and hydrogen bonds. These reagents precipitate proteins. |
| Violent whipping or shaking | May form surface films of denatured proteins from protein solutions (e.g., beating egg white into meringue). |

Some denaturations have positive medical advantages. The fact that certain heavy-metal salts denature and coagulate proteins is the basis of a method for treating poisons made of these salts. Mercuric, silver, and lead salts are dangerous primarily because these metallic ions wreak havoc among important proteins of the body, particularly enzymes. When the salts are accidentally taken orally, their ions will be kept from reaching general circulation if they can somehow be precipitated in the stomach. The handiest protein available, raw egg white (albumin), should be swallowed. As the heavy-metal ions denature it, they become tangled in the coagulated mass. Next, *an emetic must be given*. The individual must be made to vomit, thus removing this material from the stomach. Otherwise the digestive juices will go to work on it and eventually release the poisonous ions.

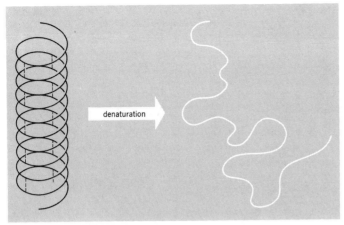

**Figure 10.13** Denaturation of a protein is fundamentally a disorganization of its molecular configuration without necessarily breaking any peptide bonds. It happens whenever the secondary structural features such as hydrogen bonds and salt bridges (represented here as dotted lines) are disrupted.

Another medically important denaturation is the treatment of burns. When a large area of the skin is burned, one of the many serious problems is the loss of water. If the surface proteins are denatured and coagulated, the crust acts to inhibit this loss. In an emergency, if it is simply not possible to get the patient to a doctor, strong tea may be sprayed or dripped onto the burned area. Tea contains tannic acid, a denaturing agent. We must emphasize, however, that this is not a preferred treatment for a burn. A severe burn requires the prompt attention of a doctor. If it is at all possible to get this attention, *nothing* should be placed on the burn that might complicate the doctor's work.

Tannic acid is one of the "alkaloidal reagents," so-called because they have been useful in studying alkaloids. Others are picric acid and phosphomolybdic acid; all denature proteins.

Denaturing agents differ widely in their action, depending largely on the protein. Some proteins (e.g., skin, hide, hair) strongly resist most denaturing actions, as they must.

## HOW PROTEINS ARE CLASSIFIED

Having studied the general principles of protein structure and behavior, let us turn our attention to the myriad types of proteins. They may be classified in the various ways listed.

I. Gross Structure and Solubility

A. *Fibrous proteins.* As the name implies, these proteins consist of fibers. At the molecular level both the α-helix and the pleated sheet (cf. p. 349) are found. These proteins can be stretched, and they contract when the tension is released. They perform important structural, supporting, and protective functions and are insoluble in aqueous media.

1. Collagens. These are the proteins of connective tissue, making up about one-third the body's protein. When acted upon by boiling water, they are converted into more soluble *gelatins*. In contrast to collagen, gelatin is readily digestible. Hence the collagen-to-gelatin conversion that takes place when meat is cooked is an important preliminary to digestion. At the molecular level the change is thought to be simply an unfolding or uncoiling of collagen molecules to expose the peptide bonds to the hydrolytic action of water and digestive enzymes.

2. Elastins. Elastic tissues such as tendons and arteries are elastins. They are similar to collagens, but they cannot be changed into gelatin.

3. Keratins. These proteins make up such substances as wool, hair, hoofs, nails, and porcupine quills. They are exceptionally rich in cystine, which has a disulfide link.

4. Myosins. Muscle tissue is rich in myosin, a protein directly involved in the extension and contraction of muscle.

5. Fibrin. Fibrin is the protein that forms from fibrinogen, its soluble precursor, when a blood vessel breaks. The long fibrin molecules tangle together to form a clot.

B. *Globular proteins.* Members of this broad class are soluble in aqueous media, some in pure water, others in solutions of certain electrolytes. In contrast to the fibrous proteins, the globular proteins are easily denatured. Examples include the following.

1. Albumins. Egg albumin is the most familiar member of this class. Albumins are soluble in pure water (forming, actually, colloidal dispersions), and they are easily coagulated by heat. In the bloodstream albumins contribute to osmotic pressure relations and to the pool of buffers.

2. Globulins. These are soluble in solutions of electrolytes, and they are also coagulated by heat. The γ-globulins in blood are very important elements in the body's defensive mechanisms against infectious diseases.

II. Function

Another way of classifying proteins, including those already discussed, is by the biological functions they serve.

A. *Structural proteins.* Fibrous proteins are examples.

B. *Enzymes.* These are the catalysts of a living organism without which it could not live.

C. *Hormones.* Many, but not all, hormones are proteins.

D. *Toxins.* These proteins produced by bacteria in the living organism act as poisons to that organism.

E. *Antibodies.* The body makes these proteins to destroy foreign proteins that invade it during an attack by an infectious agent.

F. *Oxygen-transporting protein.* Hemoglobin is the name of this important protein.

III. According to the Nature of the Nonprotein Constituent

Not all proteins liberate only amino acids when they are hydrolyzed. Those that do are called *simple proteins*. Others consist of simple proteins bound to a nonprotein group or a *prosthetic group* (Greek *prosthesis,* an addition). The nature of the prosthetic group classifies the protein.

A. *Glycoproteins.* Simple proteins are bound to *carbohydrates.* Mucin in saliva is an example.

B. *Phosphoproteins.* The prosthetic group is some phosphorus-containing substance other than phosphoglycerides or nucleic acids. Casein in milk is a phosphoprotein.

C. *Chromoproteins.* Simple proteins are bound to a pigment. Hemoglobin is an example. The globin is the simple protein, heme the prosthetic group responsible for the characteristic color.

D. *Lipoproteins.* A lipid is the prosthetic group. These complexes are not well understood. They probably are a device for transporting lipids and fatty acids in the bloodstream, and they surely are a part of cellular membranes.

E. *Nucleoproteins.* The prosthetic group is a *nuclei acid,* itself a polymer. Nucleic acids are the chief participants in the chemical events associated with genes and the hereditary messages they bear. Chapter 17 is devoted to their study.

Many other special functions could be listed. With this as a survey of the main classes of proteins, we shall single out one of them, the fibrous proteins, for detailed study. In succeeding chapters many of the other kinds of proteins will be discussed.

## FIBROUS PROTEINS

In our earlier studies of cellulose (p. 291), nylon (p. 264), and Dacron (p. 257) we learned that long molecules are characteristic of fiber-forming materials. It is therefore not surprising that many fibers are proteins. Silk is an example. Wool and hair belong to a family of proteins called the *keratins;* feathers, claws, porcupine quills, horns, hooves, and nails are also in this family. *Collagen,* the reinforcing "rod" material of bones, is also a fibrous protein. So are *myosin* of muscles and *fibrin* of a blood clot.

**Silk Fibroin. The Pleated Sheet as a Tertiary Structural Feature.** $\alpha$-Helices are found among fibrous proteins, but another distinctive structural feature is the *pleated sheet.* Pauling, Corey, and Marsh recognized its existence in silk fibroin through X-ray diffraction analysis. Silk fibroin is unusual in that the two simplest amino acids, glycine and alanine, make up most of the molecule, as the data in Table 10.2 show. The polypeptide strand is coiled to some extent, but the groups that can form hydrogen bonds, C=O and H—N, protrude from the coil at almost right angles to its long axis. This orientation makes it possible for hydrogen bonds to form *between* coils. The coils line up more or less side by side, and this succession of neighboring helices gives an overall sheetlike appearance (Figure 10.14). The individual chains, because they are not fully extended, make the sheet pleated rather than flat. The greatest number of interchain hydrogen bonds can form if adjacent chains run in opposite directions. Details of the packing of protein molecules in silk fibroin are shown in Figure 10.15.

**Keratins.** In hair and wool the keratin molecules appear to have the $\alpha$-helix configuration. If they are soaked in water, they can be stretched to almost twice their lengths, and they then apparently adopt the arrangement found in silk fibroin. Adding considerable strength to wool fibers are interchain disulfide links, for the keratins are especially rich in cystine. In some keratins several $\alpha$-helices are believed to twist together into giant helices much like the several orders of twisting in rope. Salt bridges, hydrogen bonds, and disulfide links are the vulnerable sites in wool at which various reagents and conditions attack and weaken the fibers. Acids and bases, for example, interfere with salt bridges and catalyze the hydrolysis of peptide bonds. Even if only 10 to 15% of these are broken, the loss of fiber strength is severe. Oxidizing and reducing agents weaken wool fibers by attacking disulfide bridges.

Oxidizing agents such as hydrogen peroxide, a common hair bleach, or the combined action of ultraviolet light and oxygen at the beach or on the ski slope generate trace amounts of sulfonic acids and sulfuric acid from the sulfur in hair, causing a certain amount of "frizziness." Reducing agents, on the other hand, convert disulfide links into separate mercaptan groups, weakening the fiber. This reaction is the basis for the first step in some home permanents. The hair is first wetted and then treated with a reducing agent, such as ammonium thioglycollate:

$$\{-CH_2S\!-\!SCH_2-\} + 2H\!-\!SCH_2CO_2{}^-NH_4{}^+ \longrightarrow$$

Disulfide link  
in keratin of  
hair

Ammonium  
thioglycollate

$$\{-CH_2S\!-\!H + H\!-\!SCH_2-\} + \begin{array}{l} SCH_2CO_2{}^-NH_4{}^+ \\ | \\ SCH_2CO_2{}^-NH_4{}^+ \end{array}$$

Two mercaptan  
groups, now on  
separated peptide  
strands

Ammonium  
dithioglycollate

The hair is now curled and clamped in some desired fashion. Next a mild oxidizing agent, such as potassium bromate, is used in the expectation that as disulfide links re-form, they will do so in different patterns that stabilize the "wave":

$$3\{-CH_2S\!-\!H + H\!-\!SCH_2-\} + KBrO_3 \longrightarrow$$

$$3\{-CH_2S\!-\!SCH_2-\} + KBr + 3H_2O$$

**Collagen.** Probably the most abundant protein in the animal kingdom, collagen is the principal fibrous component in cartilage and bone, in tendons and ligaments, and in skin and corneas. In man it accounts for about one-third of body protein. The high content of glycine, proline, and hydroxyproline (cf. Table 10.2) and the fact that the hydroxyl group of hydroxyproline is put in place *after* the peptide has been strung together make collagen different from other proteins. As the structures

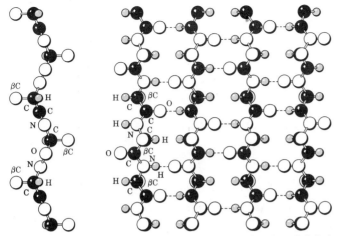

**Figure 10.14** A drawing of a short segment of the antiparallel-chain, pleated-sheet network of protein molecules in silk fibroin. The folding of the "pleats" is horizontal. (From R. E. Marsh, R. B. Corey, and L. Pauling, *Biochimica et Biophysica Acta,* Vol. 16, 1955, page 13.)

of proline and hydroxyproline reveal, when either is incorporated into a protein chain, no hydrogen is left attached to the nitrogen of the amide group. The possibilities for interchain hydrogen bonding are reduced, and the means for stabilizing an $\alpha$-helix are removed. For this reason collagen molecules are incapable of forming $\alpha$-helices, and they adopt a different and unusual arrangement, a three-stranded, twisting system (Figure 10.16) called *tropocollagen.* Collagen fibers are made from the systematic aggregation of these super helices.

In bone tissue collagen fibrils form a matrix within which the mineral hydroxyapatite, $3Ca_3(PO_4)_2 \cdot Ca(OH)_2$, crystallizes. Salts of citric acid and carbonic acid are present, and any remaining spaces are filled by a semiliquid material which provides a means for transporting substances between the bone and the circulatory system.

**Proteins in Skeletal Muscles. Actin, Myosin, and the Molecular Basis of Muscular Work.** Several proteins have been isolated from skeletal muscles, but two are the most directly involved in muscle contraction and mechanical work, *actin* and *myosin.* Actin is known in two forms. Globular actin, or *G-actin,* is a protein with a formula weight of about 45,000. In dilute salt solutions it aggregates into extended linear filaments seemingly without forming any covalent bonds, for the aggregation is easy to break up *in vitro.* If two of these linear filaments intertwine to form an open double helix, the protein is called *F-actin,* the type believed to be present *in vivo.*

Myosin has a very high formula weight, about 500,000, and is largely $\alpha$-helical. Its molecules, rodlike with globular ends, appear to be made of two identical long-chain polypeptide subunits and three much smaller subunits. The two chains are intertwined for most of the length of the myosin molecule, resembling a two-stranded rope.

In mammalian skeletal muscle tissue, myosin and F-actin are arranged as illustrated in Figure 10.17, forming myofibrils. Figure 10.18 is a diagram of some

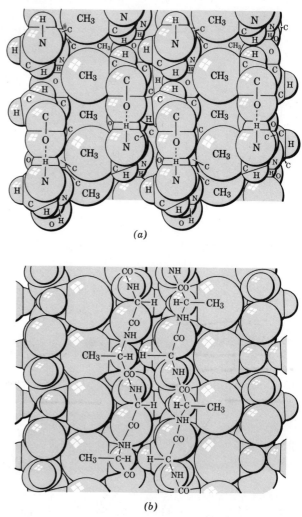

**Figure 10.15** Packing drawings of silk fibroin sections: (*a*) looking down along the axis of the fiber, (*b*) viewing the fiber from an angle perpendicular to its axis and parallel to the plane of the pleated sheets. (From R. E. Marsh, R. B. Corey, and L. Pauling, *Biochimica et Biophysica Acta*, Vol. 16, 1955, page 18.)

of the structural features of a typical striated muscle fiber such as might occur in a leg or arm muscle of a vertebrate. The many mitochondria that are present furnish sources of energy for contraction in ways to be described in Chapters 13 and 14. Interlacing the myofibrils and the mitochondria is the sarcoplasmic reticulum, a fluid-filled space which provides for movement of substances between parts of the cell.

Returning now to the myosin-actin units of Figure 10.17, we note that along the main axis of the myosin unit are projections or side chains. They are distributed around the myosin is a spiral manner and point toward actin units. Just what these side chains are, how they are involved in muscle contraction, and how they interact with the actin units are questions still being actively investigated. And what happens at the molecular level when contraction occurs? Do the individual protein molecules all coil up simultaneously? This would surely cause a shortening of the muscle. Or do they slide past each other in the manner of telescope segments? A British scientist, H. E. Huxley of Cambridge, using the electron microscope, furnished convincing evidence that muscles contract by filaments sliding past each other (Figure 10.19). Just how this sliding is made to happen, in chemical terms, has been the subject of intensive study still in progress. One theory of muscle contraction proposed by R. E. Davies (University of Pennsylvania) is outlined here with the realization that such conjectures are subject to both minor and major overhaul. The reader should keep in mind the fact that what follows, although an excellent theory

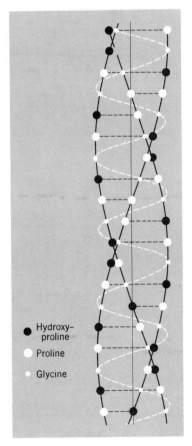

● Hydroxy-
   proline

○ Proline

○ Glycine

**Figure 10.16**  An idealized picture of the helical inter-twining of three protein strands in collagen. Each chain consists of about a thousand amino acid units and has an average formula weight of roughly 120,000. This superhelix formed by these three chains is called tropocollagen. Hydrogen bonds between subunits are believed to stabilize it. The three amino acid units symbolized are not the only ones present, but glycine, which occurs at about every fourth position, is followed immediately by proline and hydroxyproline. (From G. H. Haggis, D. Michie, A. R. Muir, K. B. Roberts, and P. B. M. Walter, *Introduction to Molecular Biology,* John Wiley and Sons, New York, 1964, page 93.)

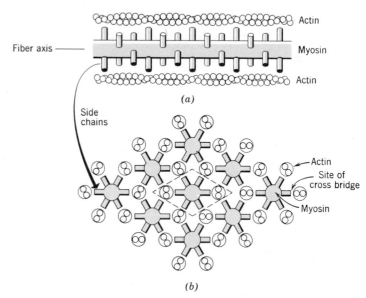

Figure 10.17 Arrangement of filaments of actin and myosin in fixed striated muscle: (a) looking perpendicular to the fiber axis, (b) looking down the fiber axis. (From R. E. Davies, *Nature*, Vol. 199, 1963, page 1068, as modified by information in H. E. Huxley and W. Brown, *Journal of Molecular Biology*, Vol. 30, 1967, page 394.)

well founded on experimental data, is a theory nonetheless, one that is still being tested.

**The Davies Theory of Muscle Contraction.** The side chains on the myosin filaments, described in greater detail in Figure 10.20, are important in this theory. According to Davies, each of these side chains terminates in a protein-bound molecule of adenosine triphosphate, universally called ATP. (ATP will be discussed in greater detail in Chapters 13 and 14. For the present all that interests us is its triphosphate system, which is simultaneously an ester, an anhydride, and an acid, with the acid groups probably being ionized as shown in Figure 10.20. The an-

hydride system, $-\overset{\overset{\displaystyle O}{\|}}{\underset{|}{P}}-O-\overset{\overset{\displaystyle O}{\|}}{\underset{|}{P}}-$, was discussed on page 245.) Near the base of the

side chain is a fixed negative charge, probably $-COO^-$ of the side group of glutamic or aspartic acid. Repulsion between this fixed charge and the net negative charge at the ATP tip is thought to keep the side chain extended in the resting molecule. Otherwise the side chain polypeptide would tend to coil up. Should it do this, according to the theory, the phosphoric anhydride system would be drawn into a region of the myosin that actually provides a local catalyst for hydrolyzing such anhydrides. In Figure 10.20 this region is labeled ATPase. (The names of enzymes end in -ase, and ATPase is an enzyme that acts on ATP.)

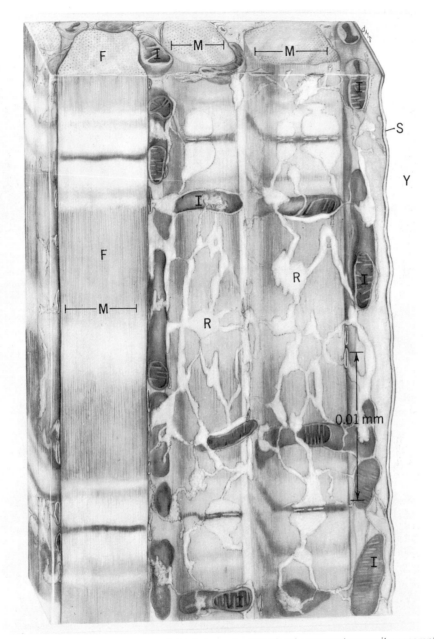

**Figure 10.18** Structural features of striated muscle. This sketch represents a semitransparent block cut out of the muscle: Y, extracellular space; S. the plasma membrane or sarcolemma of the muscle cell; M, myofibrils; F, myofilaments of molecular dimensions that make up a myofibril; I, mitochondria scattered among the myofibrils; R, sarcoplasmic reticulum (or sarcotubules) interlacing the myofibrils. (From H. Stanley Bennett, "Structure of Muscle Cells," in *Biophysical Science—A Study Program,* edited by J. L. Oncley, John Wiley and Sons, New York, 1959, page 396.)

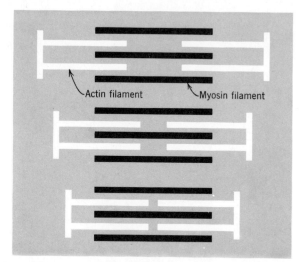

**Figure 10.19**   The relations between filaments of actin and myosin: at the top, the extended state; in the middle, the resting state; at the bottom, a partly contracted state. By examining many electron microscope pictures of muscles, H. E. Huxley (*Endeavour,* Vol. 15, 1956, page 177) deduced this sliding filament model for protein contraction.

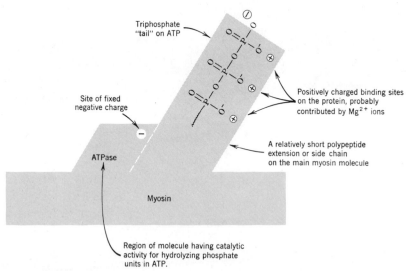

**Figure 10.20**   The cross-bridge in muscle myosin is depicted here in its resting or extended state. The net charge at the ATP tip is $-1$. Repulsion between it and the fixed charge keeps the side chain extended. The side chain is called H-meromyosin. (Adapted, by permission, from R. E. Davies, *Nature,* Vol. 199, 1963, page 1068.)

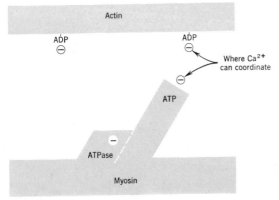

**Figure 10.21** Detail of how actin and myosin are arranged in the resting state. When the muscle is activated, calcium ions are released from the sarcoplasmic reticulum, and they move to the negatively charged sites, as shown. Then a cyclic series of events (Figure 10.22) makes the actin and myosin filaments slide past each other. (ADP stand for adenosine diphosphate, which is identical with ATP except that it has one less phosphate unit.) (Adapted, by permission, from R. E. Davies, *Nature*, Vol. 199, 1963, page 1068.)

The side chain on myosin is called a cross-bridge because it can be part of an ionic bond or bridge to a neighboring actin molecule. Relative arrangements of actin and myosin in the resting state are shown in greater detail in Figure 10.21. This bridge is completed in the moment of activation when calcium ions are released from a region of the cell called the sarcoplasmic reticulum. This ion with two plus charges can fit between the two sites with minus charges, one on myosin and one on actin. Once this happens, the minus charge at the ATP tip is neutralized. The force of repulsion between this tip and the fixed negative charge at the base of the side chain is removed. The side chain at once coils into a helix (Figure 10.22). It cannot do this, of course, without pulling on the actin molecule, for as long as the calcium ion is there the actin and myosin are for all practical purposes bound together. Referring back to Figure 10.17, which shows actin units and myosin neighbors surrounded on all sides, we see that this "pull" can only cause the actin and myosin filaments to slide by each other, as Huxley said they must. This sequence of events is described further in Figure 10.23, which carries the story through the relaxation stage.

When the side chain is in its coiled phase, the ATP tip is exposed to the action of the ATPase (Figure 10.24) and a phosphate unit is clipped off by hydrolysis. Now comes a key step, putting a phosphate unit back on. The resynthesis of ATP is the major metabolic function of glucose and fatty acids. We shall devote three chapters to this story. It is enough for the present simply to note that ATP is resynthesized and the ATP tip, now regenerated with its net negative charge, has this charge repelled by the fixed negative charge. The side chain uncoils and shoots back

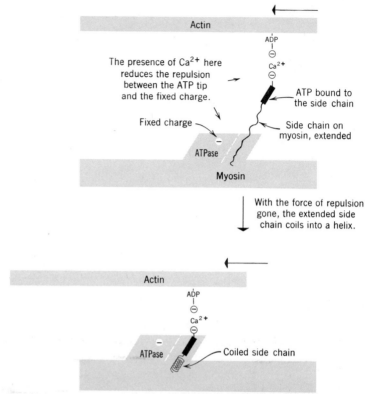

**Figure 10.22** The calcium ion removes the force that keeps the side chain extended, and it coils back toward the main chain of myosin. As it does so, the actin filament is pulled in the direction of the arrows, causing it to slip along the myosin chain. (Adapted, by permission, from R. E. Davies, *Nature*, Vol. 199, 1963, page 1068.)

out ready for another cycle. Many rapid repetitions of these stages produce the overall effect of contraction. The presence of calcium in the sarcoplasm activates a relaxing mechanism which pumps the calcium ions back into the sarcoplasmic reticulum. When the calcium has been removed, the muscle is no longer able to contract.

The Huxley model of muscle contraction (Figure 10.19) is thus given a molecular interpretation by the theory of R. E. Davies. The modified and amplified forms are shown in Figure 10.25. The Z-lines in this figure are believed to be continuous membranes which bisect bundles of filaments, and individual filaments probably terminate at these Z-lines. The cross-bridges are indicated by the thin lines between the actin and myosin units, and the successive formation and breaking of these cross-bridges produce the telescoping effect.

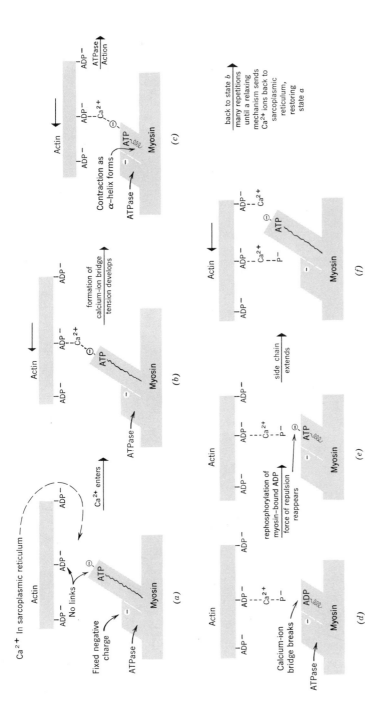

**Figure 10.23** The contraction of a skeletal muscle according to the theory of R. E. Davies. (*a*) Resting state (cf. Figure 10.21). (*b*) When Ca$^{2+}$ enters, the negative charge at the ATP tip on the myosin side chain is neutralized (cf. Figure 10.22). (*c*) The side chain rapidly and automatically coils into a helix. Because of the calcium-ion bridge to actin, the actin filament is pulled along.

(*d*) When the helix forms, the ATP tip is dragged into the part of the myosin molecule, the ATPase region, that has catalytic activity for hydrolyzing a phosphate unit. For details see Figure 10.24. The phosphate (represented here simply as P$^-$) stays with the calcium ion, and the bridge between actin and myosin is broken at this point. Until now the actin has been given one tiny jerk. For continued work the system must be recharged, which translates at the molecular level into the need to put a phosphate unit back on the ATP tip. (*e*) The negative charge at the ATP tip reappears, and since it is quite close to the fixed negative charge, the charges repel each other and the side chain becomes reextended. (*f*) A phosphate ion is displaced from the vicinity of calcium ion, and the calcium-ion bridge is reestablished. This action of course again removes the negative charge at the ATP tip, the side chain coils again, and the whole cycle starts over. Many repetitions all up and down the actin and myosin systems result in muscular work. (From R. E. Davies, *Nature*, Vol. 199, 1963, page 1068. Used by permission.)

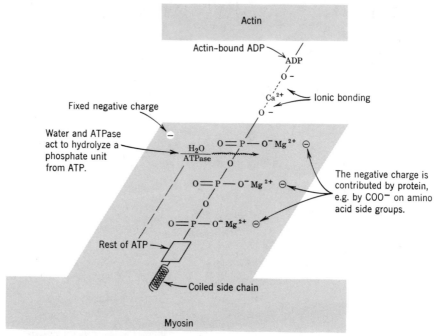

**Figure 10.24**    A site on the myosin molecule acts as an enzyme to catalyze the hydrolysis of a phosphate unit from the myosin-bound ATP, breaking the calcium-ion bridge to the actin at this point. It will be re-formed to a different actin-bound ADP in the next step (see Figure 10.23). (Adapted, by permission, from R. E. Davies, *Nature*, Vol. 199, 1963, page 1068.)

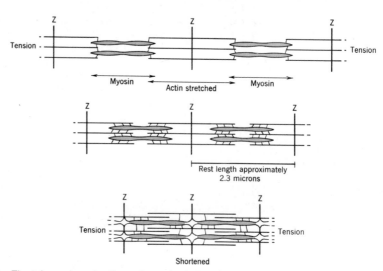

**Figure 10.25**    The telescoping of actin and myosin during the contraction of skeletal muscle. (Hanson-Huxley model as amplied by R. E. Davies, *Nature*, Vol. 199, 1963, page 1068. Used by permission.)

# REFERENCES AND ANNOTATED READING LIST

## BOOKS

R. E. Davies. "On the Mechanism of Muscular Contraction," in *Essays in Biochemistry*, edited by P. N. Campbell and G. D. Greville, Vol. 1. Academic Press, New York, 1965, page 29.

S. W. Fox and J. F. Foster. *Introduction to Protein Chemistry*. John Wiley and Sons, New York, 1957.

G. H. Haggis, D. Michie, A. R. Muir, K. B. Roberts, and P. B. M. Walker. *Introduction to Molecular Biology*. John Wiley and Sons, New York, 1964.

J. L. Oncley, editor in chief. *Biophysical Science—A Study Program*. John Wiley and Sons, New York, 1959.

R. F. Steiner. *The Chemical Foundations of Molecular Biology*. Van Nostrand, Princeton, N.J., 1965.

A. White, P. Handler, and E. L. Smith. *Principles of Biochemistry,* third edition. McGraw-Hill Book Company, New York, 1964.

H. R. Mauersberger, editor. *Matthews' Textile Fibers,* sixth edition. John Wiley and Sons, New York, 1954.

K. D. Kopple. *Peptides and Amino Acids*. W. A. Benjamin, New York, 1966.

## ARTICLES

D. Wellner and A. Meister. "New Symbols for the Amino Acid Residues of Peptides and Proteins." *Science,* Vol. 151, 1966, page 77.

M. F. Perutz. "The Hemoglobin Molecule." *Scientific American,* November 1964, page 64. In an easy-to-understand style, the 1962 Nobel prize winner in chemistry discusses the work that earned this award.

E. Zuckerkandl. "The Evolution of Hemoglobin." *Scientific American,* May 1965, page 110. In this interesting discussion of how evolution might proceed at the molecular level, the evolution of a specific chemical, human hemoglobin, is reconstructed by comparing its amino acid sequences with those in hemoglobin from other animals.

Choh Hao Li. "The ACTH Molecule." *Scientific American,* July 1963, page 46. One of the leaders in the successful isolation and purification of ACTH writes about this work and about how the function of this compound is related to its structure.

J. C. Kendrew. "The Three-Dimensional Structure of a Protein Molecule." *Scientific American,* December 1961, page 96. The work on myoglobin described in this article earned for Kendrew a share (with Perutz) of the 1962 Nobel prize in chemistry.

R. B. Merrifield. "The Automatic Synthesis of Proteins." *Scientific American,* March 1968, page 56. To make a specific polypeptide of predetermined amino acid sequence, one end of the first amino acid to be used has to be protected or tied down while at the other end the second amino acid is joined to it by a peptide bond. The result is a specific dipeptide and not its isomer (e.g., Gly·Ala and not Ala·Gly). While the new dipeptide is still tied down at one end, a third amino acid can be put on at the other end, and so on. In principle, the polypeptide chain can in like manner be extended in any desired sequence, and a specific protein can be made. Merrifield tells in this article how he automated this procedure and used it to make insulin. (He also used his automatic procedure to make ribonuclease.)

K. R. Porter and C. Franzini-Armstrong. "The Sarcoplasmic Reticulum." *Scientific American,* March 1965, page 73. Electron micrographs and elegant drawings clarify the function of the complex network of tubules and sacs in a striated muscle fiber, the sarcoplasmic reticulum.

J. Gross. "Collagen." *Scientific American,* May 1961, page 121. Studies of the molecular nature of this important protein are described. The article contains some beautiful electron photomicrographs.

F. C. McLean. "Bone." *Scientific American,* February 1955, page 84. Bone is an active tissue. This article describes its structure and functions.

H. E. Huxley. "The Mechanism of Muscular Contraction." *Scientific American,* December 1965, page 18. The electron microscope techniques upon which were based the sliding filament model of muscle contraction are described by the chief architect of this theory.

R. E. Davies. "A Molecular Theory of Muscle Contraction: Calcium-Dependent Contractions with Hydrogen Bond Formation Plus ATP-Dependent Extensions of Part of the Myosin-Actin Cross-Bridges." *Nature,* Vol. 199, 1963, page 1068.

## PROBLEMS AND EXERCISES

1. Describe in structural terms and symbols what all proteins have in common.

2. What are the structural factors that are the basis of differences among proteins?

3. Write the structures of the isomeric dipeptides that could be hydrolyzed to a mixture of serine and methionine.

4. Draw the structures of the following pentapeptides: (*a*) Ala-Gly-Phe-Val-Leu, (*b*) Glu-Lys-Tyr-Thr-Ser. Make the best judgment about the placement of protons and the location of charged sites. (*c*) Which of the two compounds would tend to be more soluble in a nonaqueous medium? Why? (*d*) Which would tend to be more soluble in water? Why? (*e*) Which would be better able to participate in salt bridges? Why? (*f*) If in the pentapeptide of part *b* the serine residue (Ser) were replaced by arginine, would the new compound be more or less basic?

5. The side chain in arginine is derived from guanidine, $\ddot{N}H_2$—$\overset{\overset{\displaystyle :NH}{\|}}{C}$—$\ddot{N}H_2$. It is a very strong base, having a $K_b$ value of about 1, in comparison with aliphatic amines having $K_b$ values on the order of $1 \times 10^{-5}$. This strength suggests that the protonated form of guanidine, $\ddot{N}H_2$—$\overset{\overset{\displaystyle +NH_2}{\|}}{C}$—$\ddot{N}H_2$, is especially stable compared to the nonprotonated form. As an exercise in resonance theory, offer an explanation for this unusual stability.

6. Write equations to show how valine in its dipolar ionic form can act as a buffer.

7. The following peptides were subjected to partial hydrolyses giving the fragments shown. Deduce the amino acid sequences in the peptides. (Commas in the "formulas" indicate the sequence is unknown.)

   (*a*) Glu, Gly, His, Phe, Tyr gave Phe·Glu + Gly·Tyr + Glu·His + Tyr·Phe.

   (*b*) Asp, Ile, Met, Pro, Tyr gave Tyr·Asp + Met·Ile + Asp·Pro + Pro·Met.

8. Describe what would happen, chemically, to the following pentapeptide under these conditions: (*a*) digestion, (*b*) action of a reducing agent, (*c*) addition of hydroxide ion (not intending hydrolysis of peptide bonds).

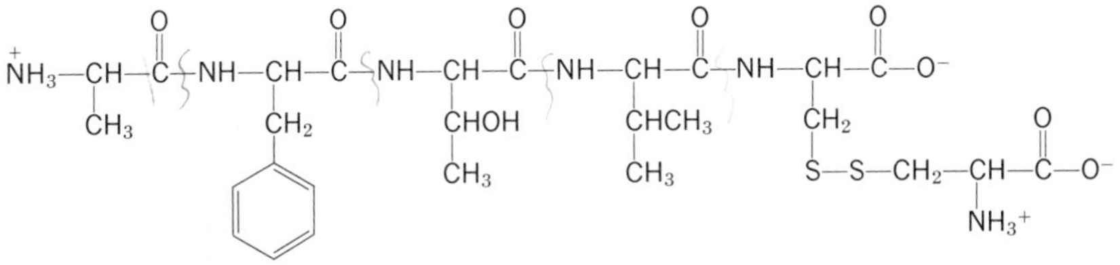

$$\overset{+}{N}H_3-CH-\overset{\overset{O}{\|}}{C}-NH-CH-\overset{\overset{O}{\|}}{C}-NH-CH-\overset{\overset{O}{\|}}{C}-NH-CH-\overset{\overset{O}{\|}}{C}-NH-CH-\overset{\overset{O}{\|}}{C}-O^-$$

with side chains: $CH_3$; $CH_2$—(phenyl ring); $CHOH$—$CH_3$; $CHCH_3$—$CH_3$; $CH_2$—$S$—$S$—$CH_2$—$CH(\overset{+}{N}H_3)$—$\overset{\overset{O}{\|}}{C}$—$O^-$

9. Discuss the role of the hydrogen bond in protein structure.
10. Explain why there is no net migration of amino acid molecules in an electric field when they are in a medium at their isoelectric point.
11. Why is a protein least soluble in a medium at its isoelectric point?
12. What is an isoionic point?
13. In terms of events at the molecular level, what happens during denaturation?
14. What is the relation between collagen and gelatin?
15. Discuss functions for the following: (*a*) elastins, (*b*) albumins, (*c*) keratins, (*d*) fibrin, (*e*) collagen, (*f*) keratin.
16. Discuss in molecular terms what might be responsible for the fact that collagen molecules do not form $\alpha$-helices.
17. What is the difference between F-actin and G-actin?
18. In your own words and by means of your own drawings, discuss the Davies theory of muscle contraction and how it fits into the Huxley model.

# CHAPTER ELEVEN

# Biochemical Regulation and Defense

Within both the plant and animal kingdoms much is shared in common from species to species. The higher organisms in the animal kingdom have, for example, carbohydrates, lipids, and proteins as part of their composition and their diets. Yet any thoughtful person who has walked anywhere on the face of the earth is sooner or later struck by the bewildering variety of living things. Because a mere handful of the chemical elements are involved, both the similarities among species as well as their differences must be accounted for in terms of these elements. One of the outstanding triumphs of studies in the molecular basis of life and health has been the success in understanding both sameness and variety in common terms. The practice of medicine would hardly be possible without a great deal of sameness in people, but neither Olympic games nor great literature would exist without their differences.

In a sense this chapter is Part I of the "molecular basis of human uniqueness." It is about enzymes and hormones and vitamins. At one level of understanding the uniqueness of a species is related to the uniqueness of its set of enzymes and hormones. The survival of the species requires that each new offspring somehow gain possession of substantially the same set of enzymes and hormones that the parents have. The description of how this happens molecularly could be called Part II of the molecular basis of human uniqueness and is the subject of Chapter 17.

Survival also requires defensive mechanisms against a host of enemies, the most dangerous of which attack the body's biochemical regulatory systems. This topic is also considered in this chapter.

## CHEMICAL NATURE OF ENZYMES

Life as we know it would be unthinkable if there were no catalysts to aid and control the processes of metabolism. These catalysts are the enzymes.

All the enzymes that have thus far been studied consist of large polypeptide or protein units plus some other type of substance called a *cofactor*. The protein portion is designated the *apoenzyme* (Greek *apo*, away from). Without its cofactor it is catalytically inactive. The cofactor is sometimes a simple, divalent metallic ion (e.g., $Mg^{2+}$, $Ca^{2+}$, $Zn^{2+}$, $Co^{2+}$, or $Mn^{2+}$), sometimes a nonprotein organic compound. Some enzymes require both kinds of cofactors. If the cofactor is firmly bound to the apoenzyme, it is called a *prosthetic group* (Greek *prosthesis*, an addition). If, instead of being more or less permanently bound to the apoenzyme, the organic cofactor is brought into play during the act of catalysis, it is called a *coenzyme*. The fully intact enzyme is sometimes referred to as the *holoenzyme*. The relation expressed in a word equation is

Cofactor + apoenzyme $\longrightarrow$ holoenzyme   (or enzyme, for short)

**Classifying and Naming Enzymes.** Before continuing our examination of how enzymes work, it will be helpful to note how names for enzymes are derived. The names of most end in "-ase," for example, maltase, oxidase, esterase, peptidase, transferase, reductase. Prefixes to this ending indicate either the substance acted upon or the kind of reaction performed. For example, maltase is an enzyme that acts as a catalyst for a reaction, the hydrolysis, of maltose; an esterase acts on an ester; a transferase assists the transfer of a group from one substrate to another. The names of some very common enzymes, such as several in the digestive system —pepsin, trypsin, and chymotrypsin, for example—stand as exceptions to the rules.

The following classification of enzymes includes only the major types together with specific examples.

A. *Enzymes that catalyze hydrolysis. Hydrolases*

1. Esterases and lipases. These act on esters in general or lipids in particular.

$$R-\overset{\overset{\displaystyle O}{\|}}{C}-O-R' + H_2O \underset{}{\overset{1}{\rightleftharpoons}} R-\overset{\overset{\displaystyle O}{\|}}{C}-OH + HO-R'$$

2. Carbohydrases (or glycosidases). There are several of these, and each has the responsibility for catalyzing the hydrolysis of a specific kind of acetal

---

[1] Double arrows indicate equilibria in all these examples because a catalyst does not affect the *direction* of a reaction. It affects *how rapidly an equilibrium between reactants and products will be established.* A catalyst acts to lower the energy of activation for a change, and the lowering of this energy barrier for the forward reaction simultaneously means a lowering for the reverse reaction. If the forward reaction is to be favored, some other event (e.g., removal of a product as it forms, perhaps by a second reaction) must occur, and in cells it does. Therefore the discussions are written with the left-to-right reaction as the forward reaction.

oxygen bridge (glycosidic link) in a di- or polysaccharide. Examples are maltase, lactase, sucrase, and amylase.

$$H_2O + sugar'—O—sugar'' \rightleftharpoons sugar' + sugar''$$

3. **Proteolytic enzymes.** These assist hydrolysis of peptide linkages in proteins. Many specialized proteolytic enzymes are known, each being very effective for catalyzing hydrolysis of a peptide bond between two particular amino acids or between two types of amino acids. Examples include pepsin, trypsin, and chymotrypsin which participate in the digestion of proteins.

$$R—\overset{\overset{O}{\|}}{C}—NH—R' + H_2O \rightleftharpoons R—\overset{\overset{O}{\|}}{C}—OH + NH_2—R'$$

4. **Phosphatases.** Many metabolic intermediates are esters or anhydrides of phosphoric acid (or its anhydrides), and several reactions consist of the hydrolysis of such groups.

$$R—O—\underset{\underset{OH}{|}}{\overset{\overset{O}{\|}}{P}}—OH + H_2O \rightleftharpoons R—OH + HO—\underset{\underset{OH}{|}}{\overset{\overset{O}{\|}}{P}}—OH$$

B. *Respiratory enzymes.* Respiration in this usage is not confined to events in the lungs. Tissue cells are the main sites for reactions with oxygen, and the body can oxidize a wide variety of chemicals that are stable in air. Such reactions are usually quite productive of energy, and any enzyme participating in them may be classified as a respiratory enzyme.

1. **Oxidases.** An enzyme that catalyzes the transfer of hydrogen from a molecule directly to molecular oxygen is classified as an oxidase.

2. **Hydroperoxidases.** Hydrogen peroxide is a (temporary) product in certain metabolic pathways, and hydroperoxidases are enzymes that catalyze its further breakdown.

Catalase:    $2H_2O_2 \rightleftharpoons 2H_2O + O_2$

Peroxidase:   $H_2O_2$ + some acceptor $\rightleftharpoons H_2O$ +  oxidized form of
                     molecule                              acceptor molecule

3. **Dehydrogenases.** A dehydrogenase catalyzes the transfer of hydrogen from some donor molecule to an organic acceptor (in contrast to oxidases, which transfer hydrogen directly to oxygen). Although numerous examples have been studied, only general types of hydrogen transfer processes will be given.

Alcohol dehydrogenation:

$$\underset{R'}{\overset{R}{\diagdown}}\underset{\diagup}{C}\overset{\diagdown}{\underset{(H\ H)}{O}} + acceptor \rightleftharpoons \underset{R'}{\overset{R}{\diagdown}}C{=}O + acceptor\text{-}H_2$$

Aldehyde oxidation:

$$R-\overset{\displaystyle O}{\underset{H}{\overset{\|}{C}}} + \; OH \underset{\text{acceptor}}{\rightleftharpoons} R-\overset{\displaystyle O}{\overset{\|}{C}}-OH + \text{acceptor·}H_2$$

The most important dehydrogenases are enzymes whose coenzymes are pyridine nucleotides, NAD and NADP (cf. p. 440).

C. *Transferases.* Many metabolic events consist of the transfer of a group from one molecular species to another with the preservation of the energies of the bonds associated with the reactants. (The energy is not liberated as heat, for example.) The principal transferases are

—Transphosphorylases: catalyzing the transfer of a phosphate unit from one molecule to another.

—Transglycosidases: catalyzing the transfer of a saccharide unit as, for example, when a glucose unit is transferred to fructose in the formation of sucrose.

—Transaminases: catalyzing the transfer of an amino group, making it possible for an organism to make some of its needed amino acids internally from molecular fragments of nonprotein origin.

—Transacylases: catalyzing the transfer of an acyl group, for example, acetyl.

D. *Isomerases.* Several metabolic reactions result only in an isomerization of a particular molecule, either structural isomerization or racemization. Enzymes catalyzing such changes are classified as isomerases or sometimes as mutases.

E. *Carboxylases.* Enzymes responsible for stripping carbon dioxide from an organic reactant or for inserting it are called carboxylases.

**Apoenzymes.** Much of the information we need to understand the circumstances under which enzymes do or do not work can be deduced from the fact that they are substantially protein in nature. They can be expected, for example, to be sensitive to any or all of the denaturing agents. They retain maximum catalytic activity over very narrow ranges of temperature and pH. In the previous chapter the influence of these conditions on secondary and tertiary structural features of proteins was described. Apparently, if a particular protein is to function as an apoenzyme, it must have both the correct sequence of amino acids and a particular final gross shape that leaves just the right groups and polar sites exposed in just the right pattern. It is in terms of this pattern that we seek to understand one of the most remarkable properties of enzymes, their specificity. As far as test tube experiments are concerned, we know that both acids and bases catalyze a wide variety of reactions. In contrast, an enzyme that catalyzes the hydrolysis of an ester may be (and usually is) ineffective in catalyzing the hydrolysis of an amide. Various digestive juices have different enzymes for hydrolyzing terminal amino acid groups, internal amino acid groups, ester linkages in lipids, and acetal linkages in amylose and separate enzymes for hydrolyzing maltose, sucrose, and lactose. None of these enzymes may be at all efficient in catalyzing the chemical breakdown of the tissues lining the digestive tract.

**Coenzymes. Vitamins.** Not all diseases are caused by germs. If the diet lacks vitamins or if the system cannot use the vitamins it receives, some very debilitating maladies result, for example, scurvy, beriberi, and pernicious anemia. Science is a long way from a complete knowledge of how vitamins are involved in health. In fact, all the vitamins may not have been discovered yet. But certain of the vitamins are known to function as parts of enzyme systems. The term vitamin is applied to an organic compound if

1. It cannot be synthesized by the host, and it must therefore be provided by the diet.
2. Its absence causes a specific disease, a "vitamin deficiency disease."
3. Its presence is required for normal growth and health.
4. It is present in ordinary foods in very small concentrations and is not a carbohydrate, a lipid, or a protein.

The list of vitamins needed by man includes between sixteen and twenty independent types, with some variations within a type. Each species apparently has its own vitamin requirements; the list for man is not the same as the list for, say, microorganisms,[2] and some vitamins may appear on more than one list.

Since vitamins are required in very small daily amounts, we have generally thought that they are somehow involved in the catalysis of biochemical reactions. This assumption has not been shown to be true for all vitamins, but a large number of coenzymes either are identical with certain of the vitamins or are simple derivatives of them. Table 11.1 contains a partial list of vitamins that have been implicated in enzymic activity. Names and structures of one or more coenzymes derived from the vitamins are usually given. In some structures parts have been shaded to draw attention to the way the vitamin unit is fitted into the coenzyme. Most of the coenzymes are esters of phosphoric acid.

**Relation of Vitamin Activity to Vitamin Structure.** It has long been known that some diseases are caused by deficiencies in the diet and that they can be cured by changing the diet. The value of eating liver to prevent night blindness has been known from antiquity. As early as the seventeenth century it was recognized by some that scurvy does not develop if the daily food intake includes citrus fruit. By the nineteenth century the Japanese had found that beriberi can be prevented by a proper diet, and that its incidence rises when polished rice is the principal food consumed. And about this time it was realized that rickets does not occur or can be cured if the daily diet includes some cod liver oil. What substance in liver, citrus fruit, cod liver oil, and a generally balanced diet was essential for good health? Whatever it was, it should have a name, and what could be more natural than to use such terms as "antiberiberi factor," "antiscurvy factor," "antirickets factor,"

---

[2] Vitamin designates those compounds required by the higher animals. The term nutrilite is applied to substances required only by microorganisms.

# Table 11.1 Vitamins in Coenzyme Systems

| Vitamin | Structure | Coenzyme System | General Area of Need |
|---|---|---|---|
| Thiamine | | Thiamine pyrophosphate (cocarboxylase) | Removing $CO_2$ from acids |
| Riboflavin | | Riboflavin phosphate (flavin mononucleotide "yellow coenzyme," FMN) | Oxidation-reduction reactions |
| Pyridoxal | | Pyridoxal phosphate | Utilization of amino acids |

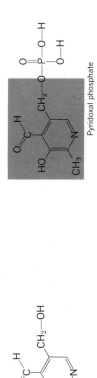

(continued on next page)

**Table 11.1  Vitamins in Coenzyme Systems** (continued)

| Vitamin | Structure | Coenzyme System | General Area of Need |
|---|---|---|---|
| Nicotinic acid Nicotinamide | 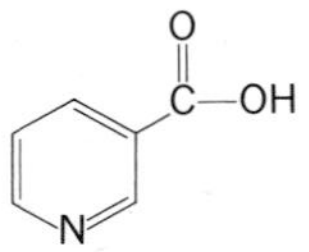 Nicotinic acid (niacin) 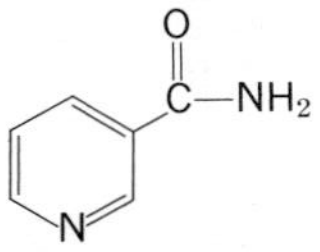  Nicotinamide (niacinamide) | 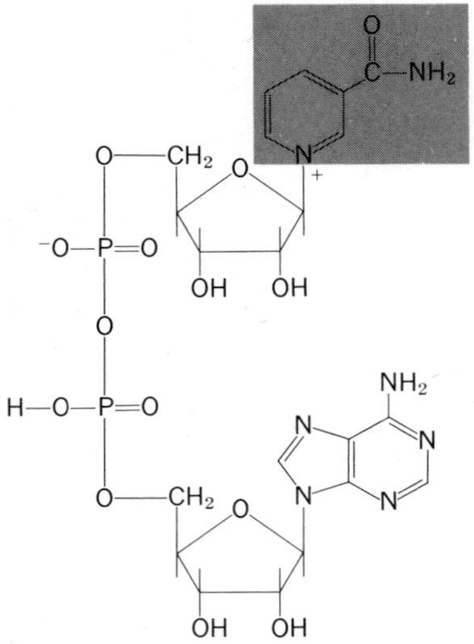 Nicotinamide adenine dinucleotide (NAD), older usage: diphosphopyridine nucleotide (DPN) | Oxidation-reduction reactions; hydrogen transfers; occurs in several important enzymes |

**Table 11.1  Vitamins in Coenzyme Systems** (continued)

| Vitamin | Structure | Coenzyme System | General Area of Need |
|---|---|---|---|
| | | 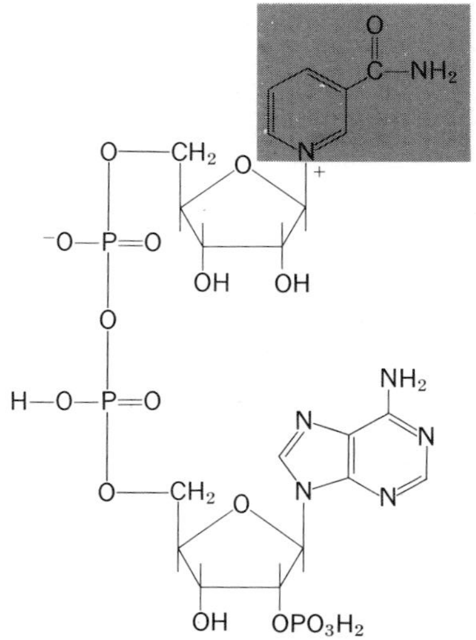 | Oxidation-reduction reactions; hydrogen transfers; occurs in several important enzymes |

Nicotinamide adenine dinucleotide phosphate (NADP), older usage: triphosphopyridine nucleotide (TPN)

(continued on next page)

**Table 11.1  Vitamins in Coenzyme Systems** (continued)

| Vitamin | Structure | Coenzyme System | General Area of Need |
|---|---|---|---|

**Pantothenic acid**

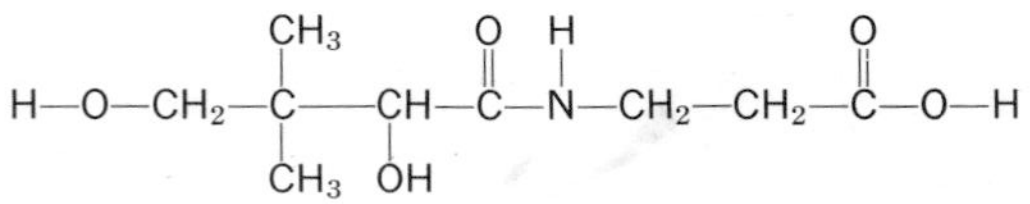

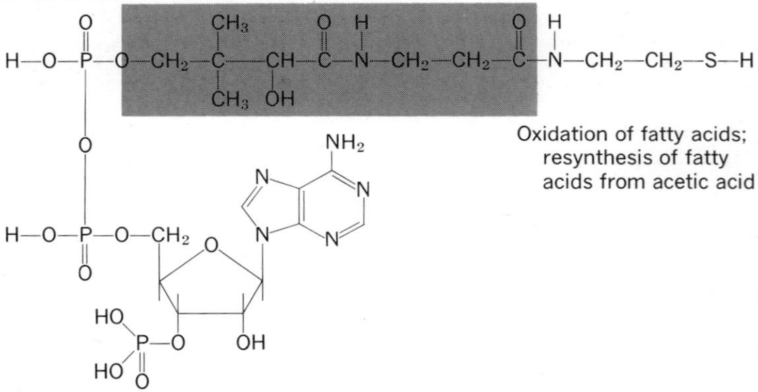

Structure:

$$CH_3 \quad O \quad H$$
$$H-O-CH_2-C-CH-C-N-CH_2-CH_2-C-O-H$$
$$CH_3 \quad OH$$

Coenzyme System:

$$CH_3 \quad O \quad H$$
$$H-O-CH_2-C-CH-C-N-CH_2-CH_2-C-N-CH_2-CH_2-S-H$$
$$CH_3 \quad OH \quad O$$

with adenine (NH$_2$) ring, ribose, and phosphate groups:

$$H-O-P-O-P-O-CH_2$$

Coenzyme A (CoA—SH)

General Area of Need: Oxidation of fatty acids; resynthesis of fatty acids from acetic acid

---

**p-Aminobenzoic acid (a nutrilite)**

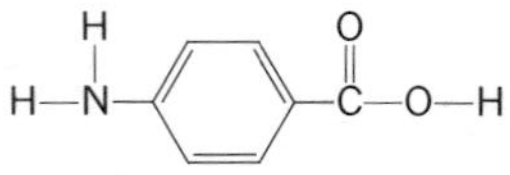

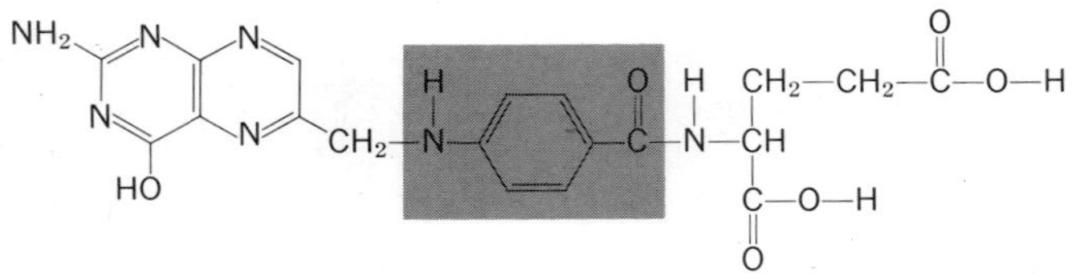

Structure: (benzene ring with H—N—H and —C(=O)—O—H)

Coenzyme System:

Folic acid (pteroylglutamic acid; also classed as a B vitamin by some)

General Area of Need: Utilization of amino acids

**Table 11.1  Vitamins in Coenzyme Systems** (continued)

| Vitamin | Structure | Coenzyme System | General Area of Need |
|---|---|---|---|
| Lipoic acid (thioctic acid) | 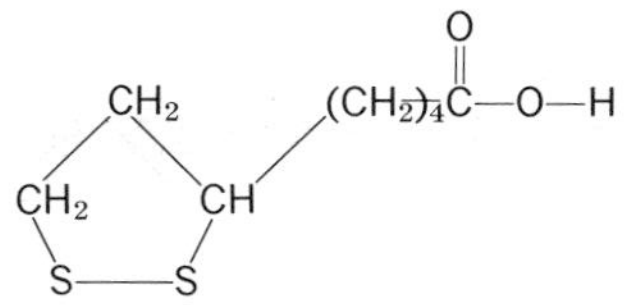 | Lipoic acid itself appears to be a coenzyme. | Biological oxidation-reductions (note the easily reduced disulfide linkage) |
| Biotin | 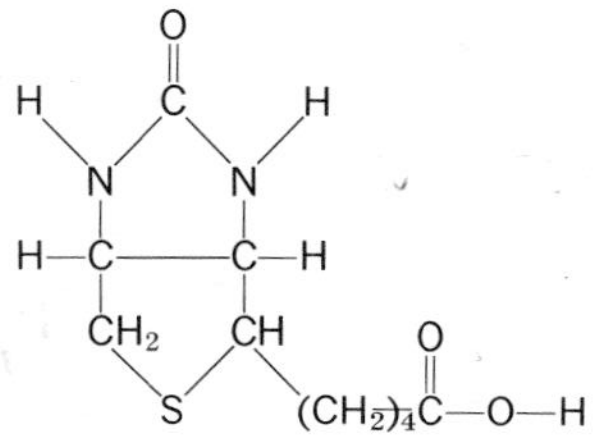 | This vitamin is a coenzyme for at least two enzyme systems. | May be needed to make other enzymes; may be needed in fatty acid synthesis; may be involved in the metabolism of carbohydrates and proteins<br>(continued on next page) |

373

**Table 11.1 Vitamins in Coenzyme Systems** (continued)

| Vitamin | Structure | Coenzyme System | General Area of Need |
|---|---|---|---|
| Vitamin B$_{12}$ (cyanocobalamin) | | | Anti-pernicious anemia factor |

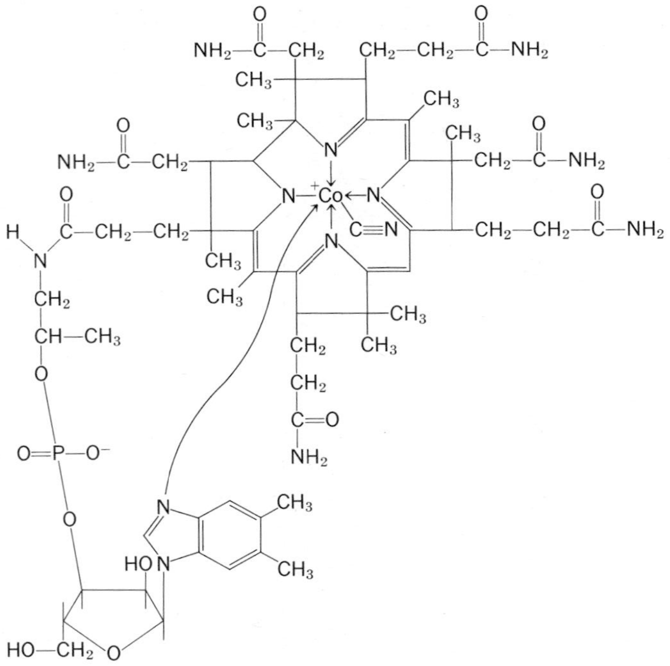

(The 1964 Nobel prize in chemistry went to British scientist Dorothy Crowfoot Hodgkin for determining this structure, and others, with X-ray diffraction techniques.)

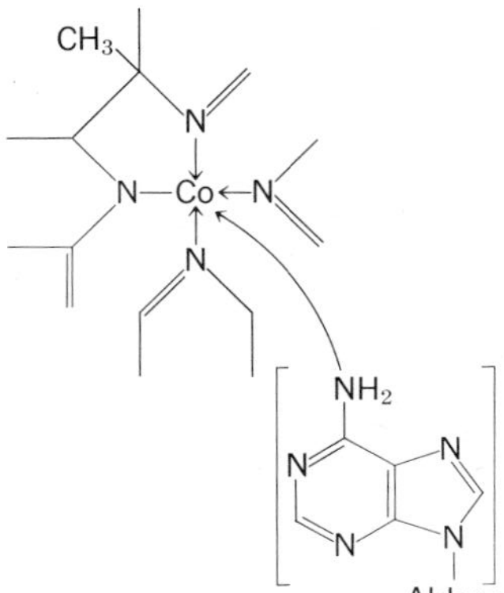

This unit (structure uncertain) replaces the $-C\equiv N$ group of the vitamin in the coenzyme.

Aldose unit

and "antipellagra factor"? Casimer Funk, a Polish biochemist, in 1912 coined the word "vitamine" as a general name for all these factors, believing that all these sub-stances were amines (*vita* is the Latin word for life). When it was found that not all the factors contained the amino group, the "-e" was dropped and the generic term became "vitamin."

In the earlier work, before actual chemical compounds were isolated and identified as vitamins, various foods were said to have this or that *vitamin activity*. In addition, it came to be recognized (especially with the pioneering work by Sir Frederick Gowland Hopkins) that foods generally contain trace amounts of sub-stances necessary to normal growth, and these were named *accessory growth factors*. In studies with rats, for example, it was found that the animals needed two accessory growth factors. The first proved to be relatively nonpolar because it could be extracted by nonpolar solvents and was called the "fat-soluble A factor." The second, soluble in water, was named the "water-soluble B factor." Soon these factors that had been thought to affect only growth were found to act (A factor) against xerophthalmia[3] and (B factor) against beriberi. In this example, at least, "growth factors" and "vitamin activities" merged and so did the nomenclature, the "fat-soluble A factor" becoming "vitamine A," the "water-soluble B factor" becom-ing "vitamine B." (The "-e's," as we have just explained, were later dropped.) This development and other similar discoveries led to the practice of defining a vitamin in terms of both its chemical structure and its activity. Definitive rules for the no-menclature of vitamins were adopted by the International Union of Pure and Applied Chemistry in 1957, but it will take some time for them to be universally accepted. The IUPAC rules are followed in this textbook, but the reader should be aware that many references follow other systems. For example, niacin is sometimes placed in the vitamin B complex, but the IUPAC rules list it separately under the name "nico-tinic acid." As another example, "vitamin $B_6$" is defined as an *activity* that pro-motes growth and prevents anemia and other conditions in animals, that prevents certain kinds of skin lesions in human beings, and that suppresses teeth decay, nausea, and vomiting in pregnant women. In this context, then, "vitamin $B_6$" refers to an *activity*, but what compound constitutes vitamin $B_6$? *One* compound cannot be singled out; the activity is produced by any one of three, pyridoxol, pyridoxal, and pyridoxamine, which are collectively called "pyridoxine." Thus a specific vita-min activity can sometimes be produced by more than one compound.

## ENZYME ACTION

One of the most rapidly advancing fields of research is concerned with the question how enzymes work. Before examining the general features of current theory, we must look at some of the facts with which the theory makers must deal.

1. *Enzyme molecules not only are largely protein in nature but are very often much, much larger than the molecules of the chemical or chemicals whose reactions they catalyze.*

[3] Xerophthalmia is a lusterless, dry, thickened condition of the eyeball.

2. *The catalysis produced by enzymes is usually selective.* Some enzymes, for example, possess *absolute specificity.* They catalyze one specific reaction of one specific compound. The enzymes urease and fumarase are examples. Urease catalyzes the hydrolysis of urea—only the hydrolysis and only that for urea:

$$NH_2-\overset{\displaystyle O}{\overset{\displaystyle \|}{C}}-NH_2 + H_2O \xrightarrow{\text{urease}} 2NH_3 + CO_2$$

Although it is essentially nothing more than the hydrolytic cleavage of two amide linkages, urease has no catalytic effect on the hydrolysis of biuret, closely related to urea, or on that of any other amide. This specificity is in striking contrast with

$$NH_2-\overset{\displaystyle O}{\overset{\displaystyle \|}{C}}-NH-\overset{\displaystyle O}{\overset{\displaystyle \|}{C}}-NH_2$$

Biuret

the fact that mineral acids will catalyze the hydrolysis of virtually any amide. Similarly, the enzyme fumarase catalyzes the addition of water to the double bond of fumaric acid but has no catalytic activity for this kind of reaction on even such a closely related compound as maleic acid (the cis isomer of fumaric acid).

$$\begin{array}{c} H-C-CO_2H \\ \| \\ HO_2C-C-H \end{array} + H_2O \xrightarrow[\text{fumarase}]{} \begin{array}{c} HO-CH-CO_2H \\ | \\ HO_2C-CH_2 \end{array}$$

Fumaric acid             Malic acid

$$\begin{array}{c} H-C-CO_2H \\ \| \\ H-C-CO_2H \end{array} + H_2O \xrightarrow[\text{fumarase}]{\ \ \ //////\ \ \ } \text{no reaction}$$

Maleic acid
(cis isomer of fumaric acid)

Most enzymes are characterized by *group specificity.* They catalyze reactions involving only a certain functional group, for example, an ester linkage. Such group specificity may be absolute or relative. An enzyme with absolute group specificity acts on only one kind of functional group. An enzyme with relative group specificity acts predominately on one type of linkage, such as an amide, but it could act on another, for example, an ester. The digestive enzyme trypsin appears to be an example of the second type.

3. *The rates of enzyme-catalyzed reactions are extraordinarily more rapid than the same or similar reactions subject to nonenzymic catalysis.* This fact about enzyme-catalyzed reactions has stubbornly resisted explanation and has been one of the major subjects of research in this field. Reactions catalyzed by enzymes frequently proceed from 100,000 to 10,000,000 times as rapidly as they do under ordinary catalysis. The digestion of amylose (cf. p. 288) is catalyzed by the enzyme $\beta$-amylase, and it involves the hydrolysis of the acetal-oxygen bridges between glucose units. $\beta$-Amylase works at a rate equivalent to the hydrolysis of 4000 oxygen bridges per *second* per *molecule* of enzyme! If the $\beta$-amylase, in a separate experi-

ment, is hydrolyzed to a mixture of its constituent amino acids, this mixture has no detectable catalytic effect on the hydrolysis of starch.

4. *Enzymes promote reactions under relatively mild temperatures.* The healthy existence of a human being requires that his body temperature be maintained fairly constantly at about 98.6°F, the temperature of a hot but not intolerable summer day. Enzymes must therefore do their work at this temperature. Normally we can accelerate reactions by simply applying heat, but too much heat applied to an enzyme-catalyzed reaction, even if it occurs in a test tube, may do nothing more than denature the enzyme.

5. *Enzymes promote reactions at nearly neutral pHs.* The variations in the pHs of body fluids are not large. Neither strongly acidic conditions nor strongly basic conditions prevail in the fluids of the body,[4] but enzymes will catalyze reactions that might, in a test tube in the absence of the enzyme, require either a fairly high pH (strongly basic) or a fairly low one (strongly acidic). Of course, both acids and bases can denature some proteins. Furthermore, at its isoelectric pH an enzyme is likely to have its lowest solubility and quite likely its lowest activity. Not all enzymes, however, have to be in solution. Many are incorporated into the molecular matrix making up the wall of, for example, a mitochondrion, a type of small granule present in cells (see p. 439).

## THEORY OF ENZYME ACTION

We know almost intuitively that enzyme action must involve a coming together of the enzyme molecule and the molecule(s) of reactant(s). With the name *substrate* (or substrates) given to what the enzyme acts upon, the enzyme (E) and the substrate (S) are thought to form an unstable intermediate called the *enzyme-substrate complex* (E-S). During the very brief existence of this complex, the enzyme activates the substrate for further reaction (E-S*),[5] and the end product (P) soon starts to form (E-P) and disengages, freeing the enzyme for more work.

$$E + S \rightleftharpoons E\text{-}S \rightleftharpoons E\text{-}S^* \rightleftharpoons E\text{-}P \rightleftharpoons E + P$$

(The shifting of the equilibrium from left to right could be mediated by a variety of factors we need not consider here.) In at least one example, reported in 1964, the enzyme-substrate complex for a particular reaction was actually isolated.

The first fact cited in the previous section, concerning the relative sizes of the enzyme and the substrate molecules, strongly indicates that only a small segment of the apoenzyme is directly involved. It also implies that the substrate does not take up just any position on the enzyme molecule but rather selects some one segment. These sites on the enzyme molecule that act as substrate binders or activators or both are called *active sites*.

The second fact in the earlier section, relating to specificity, requires that active substrate-binding sites be able to discriminate between very similar substrate molecules. The high specificity makes almost irresistible the analogy between

---

[4] The stomach is something of an exception to this rule. The pH of gastric juice is about 2, and the enzyme pepsin works in this medium.

[5] The asterisk means that the substrate is activated.

enzyme-substrate interaction and the fit of a key to a tumbler lock. In fact, the theory is often called the *lock and key theory* (Figure 11.1). An enzyme-substrate complex of reasonable stability cannot be expected to form unless the two can fit very closely and comfortably together. A left foot can be forced into a right shoe, a right hand into a left-hand glove, although the results are uncomfortable and not long endured. But a right hand will slip easily into a right-hand glove. Moreover, the process of fitting involves some flexing of the glove. It is believed that the act of a substrate fitting to an active binding site of an enzyme may sometimes involve some flexing of the enzyme molecule. Figure 11.2 illustrates how the process of binding between enzyme and substrate may induce a flexing in the enzyme to bring its active catalytic sites into proper alignment. (Some scientists have suggested that a few hormones may exert their action by causing protein chains of enzymes to flex into configurations capable of entering into otherwise unlikely processes.) Thus the specificity of an enzyme is rationalized in terms of fitting together enzyme

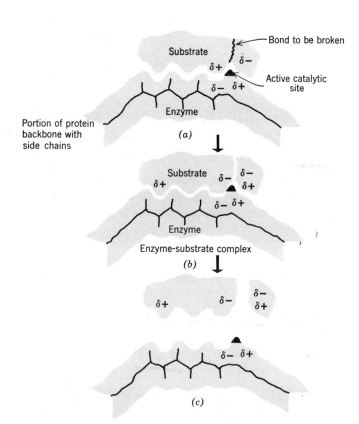

**Figure 11.1** The "lock and key" theory of enzymic action. (*a*) Enzyme and substrate have complementary shapes, and electric forces assist their association. (*b*) While they are locked together in the enzyme-substrate complex, the reaction proceeds. (*c*) A change in relative polarities accompanying the reaction causes the pieces of the substrate to leave the enzyme.

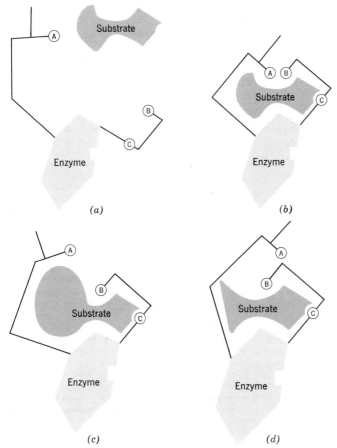

**Figure 11.2** The concept of a flexible active site. As a substrate of proper shape and size nestles to the binding site of the enzyme (*a*), it causes other portions of the enzyme to flex (*b*) and bring about a proper alignment of catalytic groups A and B. A substrate that is either too large (*c*) or too small (*d*) may be bound to the enzyme, but neither can induce the proper flexing, and the reaction therefore fails. (From D. E. Koshland, Jr., *Science*, Vol. 142, 1963, page 1539. Copyright 1963 by the American Association for the Advancement of Science. Used by permission.)

and substrate. If anything happens to alter the geometry of an active binding site of an enzyme, it cannot function. The binding site can be altered by incorrect synthesis of the enzyme by the organism, for example, by mutation of a gene (discussed in Chapter 17), or by some denaturing action that changes the shape of the protein or blocks its active site, the effect of some poisons.

The enormously fast rates of enzyme-catalyzed reactions are more difficult to explain. Suppose, for example, that an enzyme is to catalyze some reaction between two substrates, A and B, as in Figure 11.3. Suppose, furthermore, that three active sites, R, S, and T, are needed on the enzyme. As drawn in the figure, the reaction

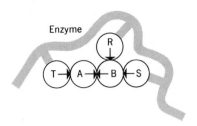

**Figure 11.3** An active site in an enzyme may involve two or more side chains on the protein, or one may be part of the coenzyme. Shown here are three catalytic side chains, R, S, and T. If they have the correct shapes and sites of partial charge, they will force the reactants A and B to become oriented in just one way, and a reaction between them will occur. (From D. E. Koshland, Jr., *Science,* Vol. 142, 1963, page 1538. Copyright 1963 by the American Association for the Advancement of Science. Used by permission.)

will require a carefully oriented coming together of five pieces. If five particles were to come together by random collision in solution, with R, T, and S on physically separated particles, the probability of their colliding in exactly one orientation (as in Figure 11.3) would be vanishingly remote. Just to make A and B collide as indicated and not in some other way (e.g., ⟨←A⟩⟨←B⟩ or ⟨↖A⟩⟨B→⟩ or ⟨A⟩⟨B↓⟩, etc.) is of low enough probability. A three-body collision of precise orientation is even more remote; and a five-body collision by a chance coming together of all five in exactly the right orientation and right collision energy is statistically almost impossible. But if the three active sites are fixed and are from the start properly oriented, and if A and B not only can fit as shown but may also be attracted to these sites by electric forces, the probabilities for the reaction are vastly improved. In fact, Koshland has calculated that the rate of a reaction involving five bodies like those in Figure 11.3 would be accelerated by a factor of $10^{17}$ to $10^{23}$ compared to a random five-body collision in solution. Incredible as it may seem, even this acceleration is not enough to explain some known velocities of enzyme-catalyzed reactions.

When molecules of substrate and enzyme approach each other, electric forces of attraction are believed sometimes to assist the swift formation of the enzyme-substrate complex. These forces are quite likely those of electrically polar regions, partially charged groupings, or ionic sites—all of which are possible in the side chains made available by the amino acid units in proteins. The formation of the complex is illustrated schematically in Figure 11.1. The reaction that the substrate undergoes may then introduce new functional groups with radically changed polarities. After the reaction the enzyme and the chemically changed substrate, which still adheres momentarily to the enzyme's surface, may repel each other. The enzyme therefore "peels" away quite naturally and rapidly and is freed for further action.

Thus far the theory has not explained how coenzymes and, indirectly, vitamins are involved. In many complexes studied, the coenzyme contributes the active catalytic site, illustrated schematically in Figure 11.4. There is good evidence, for example, that the coenzyme *cocarboxylase* (made from thiamine; cf. Table 11.1) catalyzes decarboxylations of $\alpha$-keto acids by temporarily transforming them to resemble a $\beta$-dicarboxylic acid. This type of acid, we have learned (p. 25), undergoes decarboxylation rather readily. Figure 11.5 illustrates how cocarboxylase might accomplish this.

Nicotinamide (niacinamide) is a vitamin used by the body in making the coenzyme nicotinamide adenine dinucleotide (NAD).[6] Its structure is given in Table 11.1, but we need pay direct attention only to one small portion of it, the nicotinamide unit. Figure 11.6 illustrates how NAD acts to catalyze a biological oxidation by removing the elements of hydrogen. This is one coenzyme for which the associated apoenzyme is known to impart *substrate binding* specificity and the cofactor (NAD) is known to provide the active *catalytic* site.

Numerous other examples illustrating how coenzymes and their vitamin units participate in biological reactions could be discussed. Our purpose, however, is to be illustrative rather than comprehensive. These rather complicated examples make one point that should not be overlooked: an understanding of life at the molecular level does not appear to require general physical principles any different from those used successfully to account for other natural processes. We have not always believed this to be true. In spite of the commonplaceness of the idea now, it can still inspire awe in the minds of those who, as students or scientists, have examined it in depth. The power of this idea for inspiring further research in all areas pertaining to the processes of life is immense. The field of chemotherapy (p. 385) is but one of many living monuments to the idea. Endocrinology has also profited, in spite of the fact that molecular explanations for the activities of hormones have stubbornly resisted discovery.

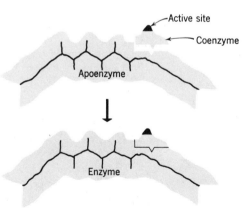

**Figure 11.4** The formation of a complete enzyme by the union of an apoenzyme and a coenzyme.

[6] Still called in many sources diphosphopyridine nucleotide (DPN).

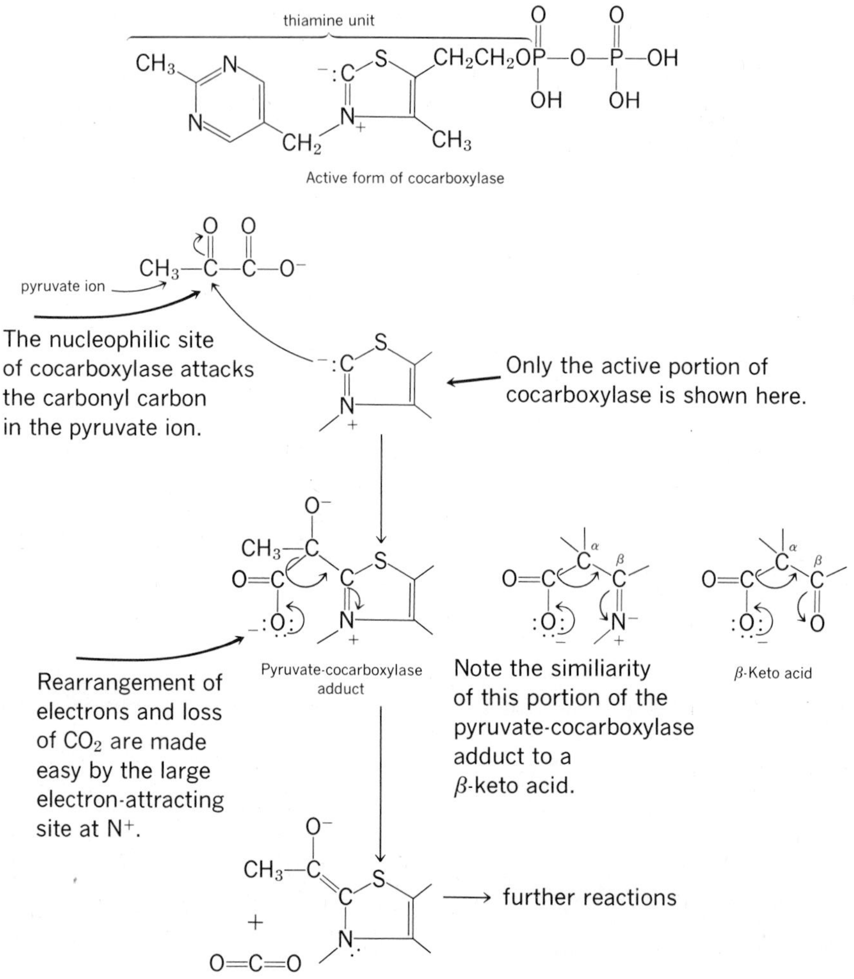

Active form of cocarboxylase

pyruvate ion

The nucleophilic site
of cocarboxylase attacks
the carbonyl carbon
in the pyruvate ion.

Only the active portion of
cocarboxylase is shown here.

Pyruvate-cocarboxylase
adduct

Rearrangement of
electrons and loss
of $CO_2$ are made
easy by the large
electron-attracting
site at $N^+$.

Note the similiarity
of this portion of the
pyruvate-cocarboxylase
adduct to a
$\beta$-keto acid.

$\beta$-Keto acid

$\longrightarrow$ further reactions

**Figure 11.5** Cocarboxylase catalyzes the decarboxylation of pyruvic acid. When the pyruvate ion becomes attached to the cocarboxylase as shown in this figure, the new system is analogous to a $\beta$-keto acid. These easily lose the elements of carbon dioxide (cf. p. 251).

## HORMONES

In higher forms of life, as in complex societies, communication between highly specialized sections is essential. Although specialization creates its own unique problems, without it we would have difficulty imagining what we call "higher living forms," with their specialized organs and tissues. In a living human body, as one example, communication between tissues and organs is accomplished in two ways, neural and humoral. Neural transmission occurs in the nervous system. Humoral communication takes place via the circulating fluids of the body, notably the bloodstream. Certain organs of the body are so specialized toward sending "chemical signals" via humoral circulation that they are considered a specific group, the *endocrine glands,* the glands of internal secretion. (Endocrine comes from the Greek: *endon,* within; *krinein,* to separate.) These glands specialize in the synthesis and secretion of compounds called *hormones* (Greek *hormon,* arousing, exciting). For each hormone there is a certain organ or a certain kind of cell whose biochemical reactions are most affected by it—a *target organ* or a *target cell.* The overabundance or the lack of various hormones has been related to such dramatic gross effects as cretinism, goiter, diabetes, dwarfism, and giantism. Sexual development is especially dependent on normal hormonal action. The sex hormones are steroids (p. 312), although most (but not all) other hormones are proteins.

In spite of decades of research and hundreds of reported investigations by hundreds of scientists, a detailed biochemical picture of *how* hormones act is still in its early stages of development. We shall limit our discussion to the theories

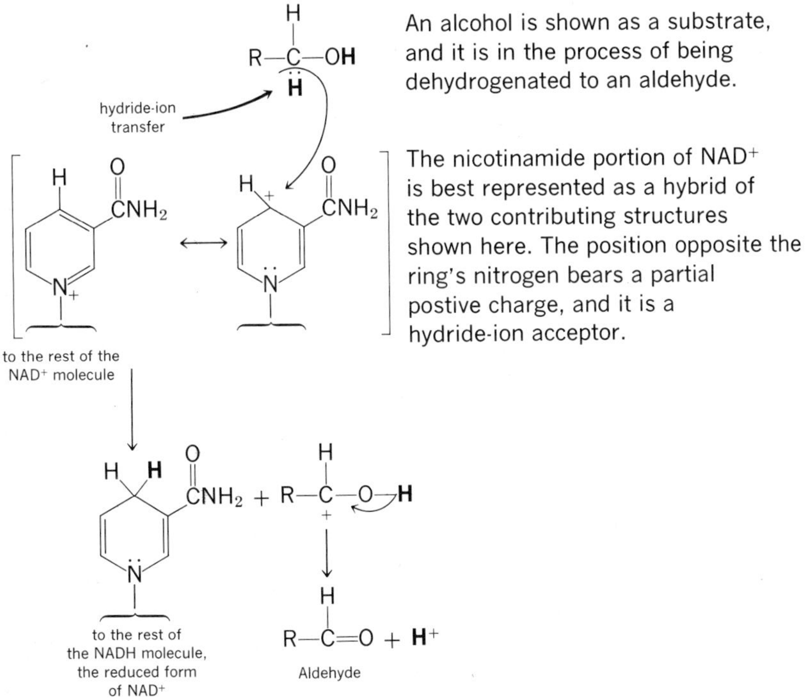
An alcohol is shown as a substrate, and it is in the process of being dehydrogenated to an aldehyde.
hydride-ion transfer
The nicotinamide portion of NAD+ is best represented as a hybrid of the two contributing structures shown here. The position opposite the ring's nitrogen bears a partial postive charge, and it is a hydride-ion acceptor.
to the rest of the NAD+ molecule
to the rest of the NADH molecule, the reduced form of NAD+
Aldehyde

**Figure 11.6** Removal of the elements of hydrogen ($H:^-$ and $H^+$) from an alcohol is catalyzed by an enzyme whose coenzyme is $NAD^+$ (nicotinamide adenine dinucleotide). Only the nicotinamide unit is shown here. The first step is the transfer of a hydride ion from the alcohol to the coenzyme, which is structured to be a good hydride-ion acceptor.

that have been proposed to explain the effects of hormones and to serve as the basis for devising further research. Then in succeeding chapters the action of specific hormones will be mentioned from time to time, reflecting the intimate relation between hormonal action and a wide variety of metabolic reactions.

Consider a very generalized metabolic event. Compound A reacts with compound B to establish an equilibrium with compound C; a specific enzyme E is required:

$$A + B \overset{E}{\rightleftharpoons} C$$

The rate at which this equilibrium is established will depend on (1) the availability of A, (2) the availability of B, (3) the availability of E, (4) the specific activity of E, and (5) the concentration of C already present. Hormones do not appear to *initiate* reactions between substances that would not otherwise react at all, without an enzyme. Instead, hormones affect the *rates* of metabolic events. In the example a specific hormone could therefore control the overall rate at which the equilibrium is established by altering any one (or more) of the factors. Not all hormones alter factors in the same way.

Since 1950 the evidence has increasingly pointed to hormonal control of the availability of reactant A or B or both, at least in a wide variety of (but not all) instances. If the reaction occurs inside a cell, to inhibit it the hormone need merely deny reactant A or B or both access to the cell's interior. Even when enzyme E is already present inside the cell in a highly active form, if A or B or both cannot get through the cell wall, the reaction obviously will not occur. The same argument would apply to a reaction occurring inside a subcellular particle such as a mitochondrion; these have their own specialized membranes.

Suppose a so-called "growth hormone" were needed for normal development. Cells grow by dividing. Before they divide a great deal of chemical synthesis must occur to provide for the new daughter cells. Amino acids, for example, must be allowed into any cell before it can divide in a normal way; amino acids are required for protein synthesis. But suppose the absence of the growth hormone means that amino acids cannot be transported through the cell membrane (or at least not in sufficient amounts). Then growth will be retarded. But if the growth hormone is available, it can exert an effect on the cell membrane that will allow amino acids through. The molecules of the hormone itself need not penetrate the cell wall to have this effect. Many of the hormones are proteins or polypeptides, that is, large molecules. This theory of hormonal action, as affecting transport properties of cell membranes, is attractive because it does not require large hormone molecules to move inside the cells of the target organ.

One protein hormone, insulin, has been studied particularly well; evidence indicates that its principal but not exclusive action is altering the permeability of the walls of its target cells for glucose molecules. The five most important hormones involved in regulating secretion of digestive juices all appear to exert their effects by modifying the transport properties of cellular membranes. (We shall return to these hormones in Chapter 12.) The hormones that regulate secretion and reabsorption in the tubules of the kidneys also appear to work by affecting permeabilities of cell walls.

This theory of hormonal action has another intriguing aspect, its *possible* relation to embryonic cell differentiation. The fertilized egg is one cell. After one division there are two cells. When these divide there are four, and the embryo is on its way to development. Sooner or later, however, some of the new cells must begin to take on specialized characteristics. Some must eventually become brain tissue, others bones, still others various internal organs, etc. Such cell differentiation must imply *selective* admission or exclusion of some nutrients rather than others from the outside medium. In fact, in the fully developed organism, each kind of cell must have a way of picking and choosing from the fluid in which it is bathed just those nutrients it needs. Without a mechanism for this selection, there can be no differentiation, development, or even long-term maintenance. It appears that in many instances hormones must be considered vital factors in this mechanism. We do not mean to imply that hormones exert their influence in this one way only. We shall briefly examine another theory that probably applies to other hormonal actions.

Most individual reactions in the body are but single steps in a long *metabolic pathway,* a sequence of chemical events wherein the product of one step is a reactant for the next. Each step requires an enzyme. The speed of the overall sequence cannot be any faster than the speed of the slowest step in the pathway, the "bottleneck step." If the pathway is to be accelerated, the hormone may exert its action by increasing the activity, the catalytic "ability," of the enzyme for the slowest step. Alternatively, if the hormone functions to slow down a particular pathway, it may do so by inhibiting one of the enzymes in the sequence to the point that this step becomes the slowest of all.

The hormone may exert its effect on the enzyme by interacting with the apoenzyme and changing its molecular shape, the nature of the folding or coiling of its protein chain. The hormone may also affect the concentration of the enzyme, perhaps by influencing the gene that directs its synthesis (cf. p. 561). Or the hormone and the cofactor may interact in some way. Conceivably the hormone is itself a cofactor, or perhaps it competes with a cofactor for binding to the apoenzyme, inhibiting the enzyme.

Thus there are several possibilities, several ways in which hormonal action may be explained. The theory postulating that hormones affect transport of nutrients through cellular membranes appears to apply to the majority of hormones, but in no instance do we yet understand the detailed chemistry. Some of the most sophisticated research is being directed to this problem. The profound overall effects that hormones can have make this research particularly important. Control of cancer could conceivably require complete information about what regulates the transport of nutrients through cell membranes. Rapidly growing cancer cells are obviously able to let in more than their share of nutrients.

## CHEMOTHERAPY

Chemotherapy is the use of chemicals, or drugs, to destroy infectious organisms without seriously harming human protoplasm. Chemicals that simply inhibit the growth of a microbe are called *antimetabolites* and are classified as chemotherapeutic agents. Excluded from this class of chemotherapeutic substances are such general bactericidal agents as iodine, phenol (carbolic acid), and silver compounds which, although good germicides outside the body, destroy most kinds of tissue and cannot be administered internally.

Chemotherapy consists of using chemicals unrelated to antibodies or any other weapon produced by the organism. The basic principles of the field are discernible in the story of Paul Ehrlich's discovery of a drug for syphilis. This scientist, who invented the term "chemotherapy," conceived the idea that chemicals extremely toxic to an infecting microbe but harmless to the host must exist. The injection of one chemical, one "magic bullet" as he called them, would spell death to one infectious microbe. Each infectious disease would presumably require a special chemical. This idea led Ehrlich to a cure for syphilis, and his theory dominated the new field of chemotherapy for decades.

Ehrlich's fight against syphilis began as a search for a drug to cure African sleeping sickness. Atoxyl was effective, but it was decidedly unsafe because it blinded too many users. Ehrlich reasoned that if the structure of atoxyl was altered a trifle, it might become harmless to people but remain deadly to the trypanosomes of African sleeping sickness.

Atoxyl

"606" (Salvarsan or arsphenamine)

Searching among organic compounds of arsenic, Ehrlich investigated many series of compounds. The six hundred and sixth compound he studied proved to be effective against trypanosomes in mice and horses. In the meantime Ehrlich had read a report of the discovery of the causative organism of syphilis, the syphilis spirochete. Its discoverer declared it to be a close relative to the trypanosome. He was mistaken, but Ehrlich did not know this at the time and concluded that what would kill a trypanosome would also kill any of its close relatives. He tried his newly discovered "magic bullet" for trypanosomes on the syphilis spirochetes. It worked. Although not the ideal cure, when carefully administered it proved effective. Named Compound "606," it appears to work by reacting with free —SH groups on proteins, and the proteins in the syphilis spirochetes are rich in these groups. Penicillin has replaced 606 as the drug of choice in treating syphilis.

## SULFA DRUGS AND ANTIMETABOLITES

Between 1909 and 1935 thousands and thousands of chemicals were prepared and tested as "magic bullets" for other diseases. And thousands and thousands of these experiments failed. So few compounds were found to have any value that Ehrlich's ideas fell from favor. Then, in 1935, from the efforts of the I. G. Farbenindustrie of Germany to find better dyes, there exploded into the history of chemotherapy an event that changed medical history.

Because dye molecules had been studied as "magic bullets" by Ehrlich, it became routine to test all dyes as chemotherapeutic agents. In 1935 Gerhard Domagk, a doctor employed by I. G. Farbenindustrie, reported to the world the results of a desperate gamble to save the life of his little girl. She had contracted a streptococcal infection from a pin prick, and she was dying. As a last resort Domagk gave his daughter an oral dose of a brick-red dye called prontosil, which in his tests with mice had proved to be an inhibitor of the growth of streptococci. His daughter recovered. The modern era of chemotherapy was born. Domagk later won the Nobel prize.

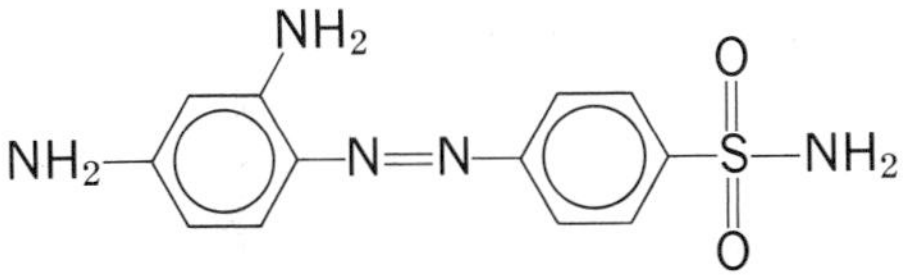

Prontosil 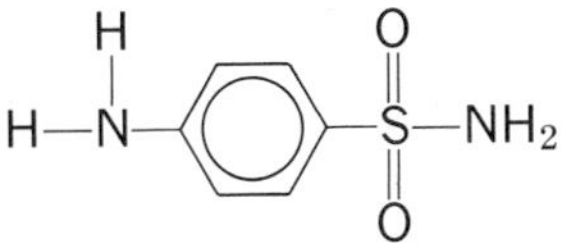                                          Sulfanilamide

French chemists reasoned that prontosil is reduced in the body to what is the active agent, sulfanilamide. Within a year of Domagk's discovery, Fourneau, of the Pasteur Institute in Paris, synthesized sulfanilamide, and it was found to be as effective as prontosil.

Sulfanilamide is not completely harmless to the human body. In efforts to improve it, chemists, following Ehrlich's lead, concentrated on changing the molecule by small bits. Literally thousands of variations of the basic sulfanilamide molecule were synthesized. The names and structures of a few of the better ones are shown in Figure 11.7. The successful variations are those with changes in the groups attached to the sulfur atom.

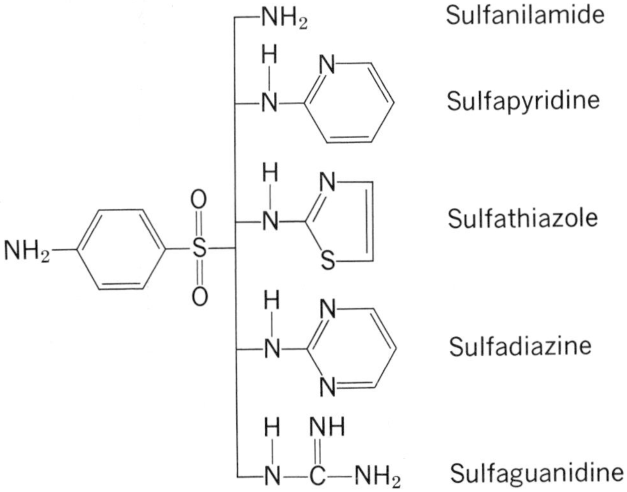

Figure 11.7   Some important sulfa drugs. Sulfapyridine was shown to be effective against pneumonia by Whitby, an English scientist, in 1938. A scourge which had brought sudden, premature death to countless people was removed. Sulfadiazine is credited with saving the lives of thousands of battlefront casualities in World War II and later wars.

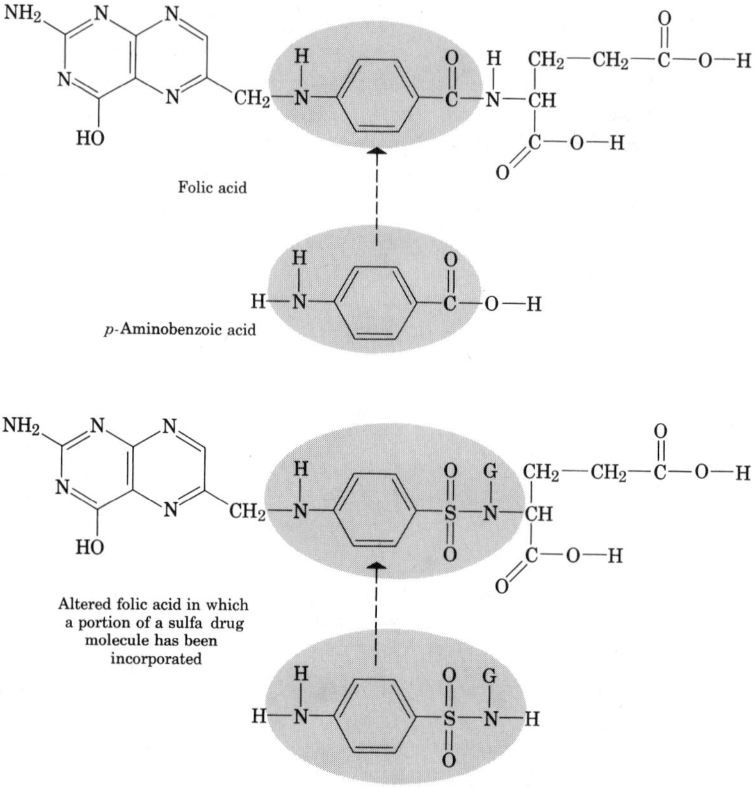

Folic acid

*p*-Aminobenzoic acid

Altered folic acid in which
a portion of a sulfa drug
molecule has been
incorporated

Sulfa drug (G = a group attached to nitrogen)

**Figure 11.8**   Sulfa drugs as antimetabolites. When a microbe selects a sulfa drug to synthe-size folic acid, it fashions an altered folic acid that is ineffective as a coenzyme. Reactions re-quiring it do not occur, and the growth of the microbe is inhibited. It suffers, in effect, from a vitamin deficiency disease.

Sulfa drugs are antimetabolites and work by producing deficiency diseases in bacteria. According to D. D. Woods, bacteria sensitive to sulfa drugs cannot dis-criminate between a molecule of sulfanilamide and a molecule of *p*-aminobenzoic acid, which it needs to make the coenzyme folic acid for one of its enzyme systems. A molecule of a sulfa drug resembles this *p*-aminobenzoic acid enough to permit its in-corporation into the coenzyme (Figure 11.8). The altered coenzyme fails to function as required. An important metabolic process in the microbe is not initiated, its growth is inhibited, and it may die. In the meantime the normal defensive mechanisms of the body are able to cope with the infection before it spreads too far.

## ANTIBIOTICS. PITTING MICROBE AGAINST MICROBE

In 1928 the British bacteriologist Sir Alexander Fleming, experimenting with cultures of *Micrococcus pyogenes* var. *aureus,* noticed that one culture had become contaminated by a mold, *Penicillium notatum,* similar to common bread mold.

Spores of the mold, borne by the breeze, had evidently alighted on the petri dish in which the micrococci were multiplying. Both the mold spores and the micrococci used the nutrient medium in the dish as their food and competed for it. The mold spores, however, secreted a chemical that was fatal to the micrococci. Micrococcal colonies failed to grow in the vicinity of the mold, or those that had started there showed signs of dissolution. One living thing, the mold, had apparently killed the other living organism. Fleming named the active chemical produced by the mold *penicillin*. Although Fleming's chance observation was made in 1928, he could not find anyone who would or could isolate and purify the mold's active substance. Not until 1940 did Florey and Chain succeed in isolating it.

The term *antibiotic* means "against life." Antibiotics may be defined as chemicals that are produced by microorganisms and that can, in small concentrations, inhibit the growth of or even destroy other microorganisms. Microbes fruitful in producing antibiotics have been found chiefly in the soil. Different soil samples contain different antibiotic-producing microbes. Tens of thousands of soil samples from all over the world have been examined.

In 1944, in a handful of New Jersey soil, Dr. Selman Waksman found a strain of actinomycetes that produces the antibiotic *streptomycin*, effective against many microbes not attacked by penicillin, including the tough tubercle bacillus. In 1947 Dr. Paul Burkholder discovered another strain of actinomycetes in soil sent to him from Caracas, Venezuela. It produces *chloramphenicol* (Chloromycetin), effective against typhus, spotted fever, typhoid fever, and certain types of dysenteries.

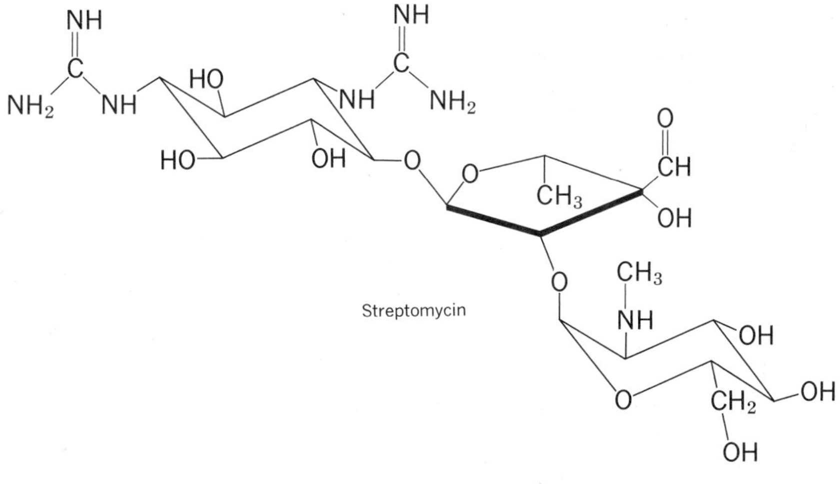

Streptomycin

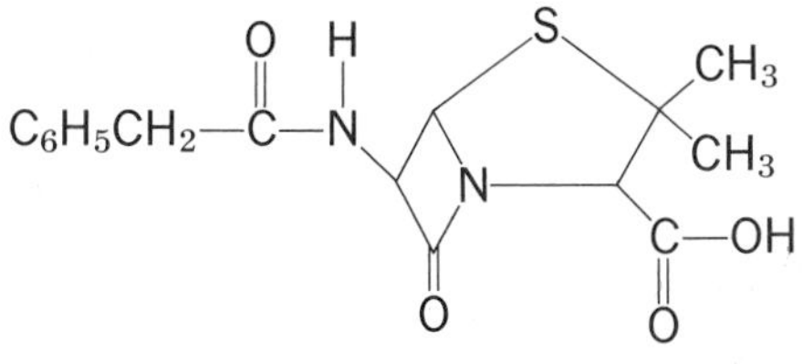

Penicillin G

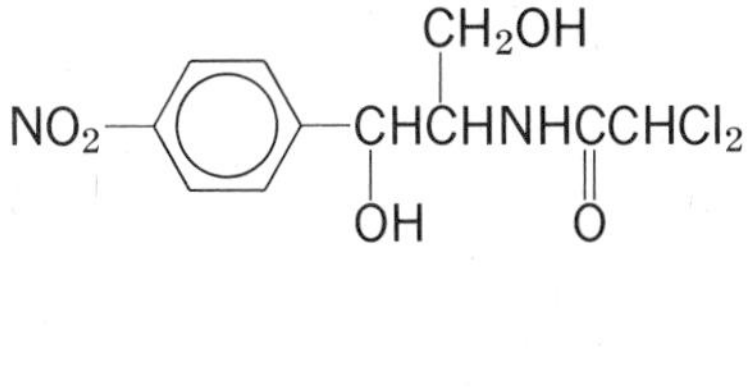

Chloramphenicol

Aureomycin

Terramycin

Dr. Benjamin Duggar found a strain of actinomycetes in Missouri River mud that produces *Aureomycin,* useful against a broad spectrum of infectious agents, especially Asiatic cholera and brucellosis. In Indiana soil was found still another strain that produces Terramycin, an extremely efficient, broad-spectrum antibiotic.

How do antibiotics work? No single answer can be given, for each works differently. Many act as antimetabolites. Penicillin, for example, inhibits the passage of an important amino acid across the cell wall of bacteria. Deprived of this compound, they cannot properly maintain their cell walls, and the walls deteriorate, spilling the contents of the cell. Chloromycetin inhibits protein synthesis in bacteria, quite likely by interfering with important enzymes or even by interfering with the synthesis of apoenzymes.

The chemical aspects of chemotherapy just studied illustrate how *struggles between the invading microbe and the host are, at root, interactions between chemicals.* These interactions must involve the ordinary laws of chemical behavior, even though they occur among uncommonly complicated molecules. This last statement amounts to a declaration of faith. Belief in it impels scientists to seek *chemical* answers and explanations to the problems of infections. The more that is learned about the chemistry of diseases, the better they can be controlled.

## INHIBITION OF ENZYME ACTIVITY. POISONS

Anything that prevents formation of an enzyme from its apoenzyme and cofactor(s) will clearly inhibit or prevent the reaction for which the enzyme is needed. Anything that inhibits an enzyme and its substrate from forming their complex will give the same negative results. In these inhibiting effects the connection between poisons and enzymes can be seen. *The most powerful poisons known wreak their devastation by inhibiting key enzymes.* All are effective in very small doses, for only catalytic amounts of poisons, after all, are needed to inhibit catalytic amounts of enzyme.

**Cyanides.** The cyanide ion reacts with many metal-ion activators of enzymes and makes these ions unavailable to an apoenzyme. Cells that consume oxygen require the presence of the enzyme cytochrome oxidase, which contains iron. Cyanide ion inhibits this enzyme, so that cellular respiration ceases instantly. Ironically, the individual is stimulated to breathe more deeply in an effort to rush oxygen to the cells. Under the usual conditions of poisoning by the gas hydrogen cyanide, this deeper breathing simply brings in more poison. Antidotes (e.g., photographer's hypo: sodium thiosulfate) have been worked out in animal experiments, but they are effective only when given simultaneously with the cyanide, rendering them of little help to an actual victim.

**Arsenic Compounds.** Arsenic poisons such as sodium arsenate, $Na_3AsO_4$, block enzymes that contain phosphate units. Such units are present in many of the coenzymes listed in Table 11.1. Arsenate ions, $AsO_4^{3-}$, closely resemble phosphate ions, $PO_4^{3-}$, and can take their place. If that happens, however, the coenzyme or the enzyme does not function properly.

**Nerve Poisons.** A variety of deadly poisons act by attacking enzymes of the nervous system. In certain parts of the system, the so-called cholinergic nerves, arrival of a nerve impulse at one nerve cell stimulates formation of a trace amount of acetylcholine all along its length. This compound then reacts with a *receptor protein* on the next cell, making it possible for that cell to send the signal along. Once the signal has been sent, the acetylcholine must be hydrolyzed to leave the cell in readiness for the next signal. Otherwise excess acetylcholine forms, resulting in overstimulation of muscles, glands, and nerves; choking, convulsions, and paralysis cause death. The hydrolysis of acetylcholine must occur within a few millionths of a second after it appears. The equation for this reaction, the hydrolysis of an ester, is

$$CH_3-\overset{\overset{\displaystyle CH_3}{|+}}{\underset{\underset{\displaystyle HO^-\quad CH_3}{}}{N}}-CH_2-CH_2-O-\overset{\overset{\displaystyle O}{||}}{C}-CH_3 + H_2O \xrightarrow{\text{cholinesterase}}$$

Acetylcholine

$$CH_3-\overset{\overset{\displaystyle CH_3}{|+}}{\underset{\underset{\displaystyle ^-OH\quad CH_3}{}}{N}}-CH_2-CH_2-OH + CH_3-\overset{\overset{\displaystyle O}{||}}{C}-O-H$$

Choline · · · · · · · · · · · · · · · · · · · · · · · · · · · · Acetic acid

Cholinesterase is the enzyme. Without it the reaction is too sluggish and for all practical purposes does not take place at all. The excess acetylcholine produces a prolonged, unwanted, and therefore dangerous stimulation of the nerves and the glands and muscles they control. Eyes, salivary glands, heart muscles, the digestive tract, the adrenal glands, and other parts of the body are all affected.

"Nerve gases," such as diisopropyl fluorophosphate (DFP), function by reacting with cholinesterase at its active site. This mechanism and those of other nerve poisons are shown schematically in Figure 11.9. The anticholinesterase activity of nerve gases deactivates one key enzyme in the body. Because one vital reaction does not occur, the victim dies within a few minutes.

$$F-\overset{\overset{\displaystyle O}{||}}{P}\left(O-\overset{\overset{\displaystyle CH_3}{\diagup}}{\underset{\underset{\displaystyle CH_3}{\diagdown}}{CH}}\right)_2$$

Diisopropyl fluorophosphate

The standard treatment for nerve gas poisoning has been the injection of counterpoisons, atropine, and certain curare alkaloids. The curare alkaloids are the

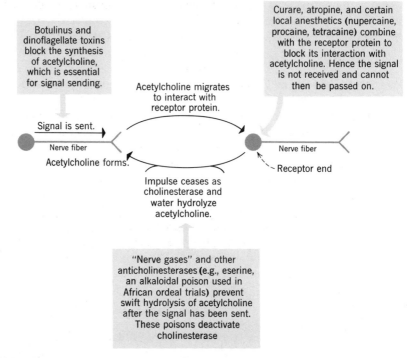

**Figure 11.9** The acetylcholine cycle and the action of nerve poisons.

active principles of "brews" concocted by South American natives for their poison arrows and darts. Whereas nerve gases overstimulate signal sending along the nervous system, atropine and curare alkaloids render nerve cells less sensitive to signal sending. A newer and better antidote, pyridine aldoxime methiodide, can actually remove the nerve gas molecule from the enzyme.

**Food Poisoning.** The toxic proteins secreted by the botulinus bacillus, a microorganism that multiplies in spoiled food, are the most powerful food poisons yet discovered, surpassing even the nerve gases. The toxin of *Clostridium botulinum* is so noxious that its lethal dose is estimated to be only eight *molecules* per nerve cell. One milligram would be enough to kill 1200 *tons* of mice, and half a pound would destroy all the human beings on earth. As indicated in Figure 11.9, these toxins act to block the synthesis of acetylcholine. When this compound fails to form, signals are not sent, paralyzing key organs such as the heart.

**Dinoflagellate Poisons.** At various unpredictable periods dinoflagellates, microscopic marine organisms, multiply almost explosively. Their secretions kill fish by the millions in what people living along coastal areas know all too well as "red tides." Dinoflagellate poisons act in the same way as the botulinus toxins.

**Cancer Therapy and Enzymic Inhibition.** The hope that selective enzymic inhibition may some day be helpful in the treatment of cancer accounts for much of the intense interest that scientists direct to the mechanism of enzyme activity.

Our brief study has revealed that enzyme activity and enzyme inhibition (poisoning) are virtually two sides of the same coin. Until both are thoroughly understood, any *selectivity* in enzymic inhibition will be difficult and risky to manipulate.

## ENZYMES IN MEDICAL DIAGNOSIS

Enzymes in the body are normally confined to intracellular fluids, with only minute traces present in such extracellular fluids as urine, plasma, cerebrospinal fluid, and bile. In certain diseases, however, the concentration of a particular enzyme in one or another of these extracellular fluids is known to increase greatly. Since samples of such fluids are easily withdrawn from a patient, and since enzymes are highly specific in the reactions they catalyze, analysis of a fluid for enzyme content is relatively easy. This procedure amounts to a *biochemical biopsy* of a tissue.

**Human Infectious Hepatitis.** In this disease the concentration of the enzyme glutamic pyruvic transaminase in blood serum increases tenfold *even before the symptoms appear*. Assay of blood serum for the enzyme is a much simpler way of detecting the disease or its carriers than microscopic examination of a sample of liver tissue.

**Coronary Occlusion with Damage to Heart Muscle.** An electrocardiograph is generally used to diagnose coronary occlusion and to measure the extent of damage to heart muscles. With patients suffering from a second attack, however, the electrocardiograph gives data that are very difficult to interpret. Coronary occlusions followed by injury to heart muscles are accompanied by a rise in the concentration of the enzyme glutamic oxaloacetic transaminase in blood serum. Serum analysis for this enzyme assists the physician not only in his diagnosis but also in his assessment of the possible degree of disability or recovery.

# REFERENCES AND ANNOTATED LIST

### BOOKS

T. P. Bennett and E. Frieden. *Modern Topics in Biochemistry*. The Macmillan Company, New York, 1966. Chapter 4, "Biological Catalysts: The Enzymes," is a brief but good summary of how enzymes work. (Paperback.)

F. R. Jevons. *The Biochemical Approach to Life*. Basic Books, New York, 1964. Chapter 6 introduces vitamins and coenzymes with an historical treatment.

G. Litwack and D. Kritchevsky. *Actions of Hormones on Molecular Processes*. John Wiley and Sons, New York, 1964. Chapter 1, written by T. R. Riggs, is a discussion of the influence of hormones on the transportation of nutrients across cell membranes.

A. F. Wagner and K. Folkers. *Vitamins and Coenzymes*. Interscience Publishers, New York, 1964. Chapter 1 of this comprehensive and advanced treatise covers the evolution of the vitamin theory.

J. B. Neilands and P. K. Stumpf. *Outlines of Enzyme Chemistry,* second edition. John Wiley and Sons, New York, 1958. Portions of this textbook, such as those dealing with the classes of enzymes and the mechanism of enzymic activity, can be understood with the background in organic chemistry presented in this text.

ARTICLES

D. E. Koshland, Jr. "Correlation of Structure and Function in Enzyme Action." *Science,* Vol. 142, 1963, page 1533. This paper was an important reference for the writing of this chapter.

D. C. Phillips. "The Three-Dimensional Structure of an Enzyme Molecule." *Scientific American,* November 1966, page 78. The spatial arrangement of the groups in the enzyme lysozyme has been worked out; this article reports the structure and how it correlates with the activity of the enzyme.

I. E. Frieden. "The Enzyme-Substrate Complex." *Scientific American,* August 1959, page 119. "Life is essentially a system of cooperating enzyme reactions." Thus the author closes this article, which traces the idea of the enzyme-substrate complex from the first insights of Jons Jakob Berzelius (1835), to the first isolation of an enzyme in crystalline form (urease, by J. B. Summer of Cornell University, in 1926), through a succession of experiments by several teams of scientists who have fixed the concept firmly in science.

Sir Solly Zuckerman. "Hormones." *Scientific American,* March 1957, page 77. For a general survey of the field of hormones, this article is both well written and well illustrated.

A. Csapo. "Progesterone." *Scientific American,* April 1958, page 40. The dramatic history of the study of progesterone is related in this article. As an account that illustrates the definition, "Science is what scientists do," this article is first-rate.

L. Wilkins. "The Thyroid Gland." *Scientific American,* March 1960, page 119. The thyroid gland and its secretions regulate nearly all the basic cellular reactions in the body. In an article that is splendidly illustrated, the history of the discoveries about this important part of the endocrine system is related.

R. H. Levey. "The Thymus Hormone." *Scientific American,* July 1964, page 66. The role of the thymus gland in the synthesis of antibodies is discussed.

J. D. Woodward. "Biotin." *Scientific American,* June 1961, page 139. Although no cases of natural biotin deficiency have been known, biotin is required by all cells. The article presents the interesting history of this potent vitamin and modern theory concerning it.

E. Adams. "Poisons." *Scientific American,* November 1959, page 76. Some of the most lethal poisons are also important drugs. Adams discusses this relationship as part of a general description of poisons and the way they work. Nerve poisons receive special treatment.

A. Tyler. "Fertilization and Antibodies." *Scientific American,* June 1954, page 70. The chemical basis for the union of an egg and a sperm is compared with the antigen-antibody reaction. The two events are quite similar.

Sir Macfarlane Burnet. "How Antibodies Are Made." *Scientific American,* November 1954, page 74. "The Mechanism of Immunity." *Scientific American,* January 1961, page 58. The 1960 Nobel laureate in medicine discusses his theory of antibody production.

M. Yoeli. "Animal Infections and Human Diseases." *Scientific American,* May 1960, page 161. The pattern of infection that involves animal carriers and insect transmitters is described in this article, and the ever-present threat of epidemic hovering over much of mankind is graphically portrayed.

F. Wroblewski. "Enzymes in Medical Diagnosis." *Scientific American,* August 1961, page 99. The detection of cancer before it is a malignancy may be possible by assaying extracellular fluids for enzymes. Wroblewski's article explains how.

# PROBLEMS AND EXERCISES

1. Define the following terms.

   (*a*) apoenzyme      (*b*) cofactor      (*c*) prosthetic group

   (*d*) coenzyme      (*e*) holoenzyme      (*f*) esterase

   (*g*) lipase      (*h*) proteolytic enzyme      (*i*) dehydrogenase

   (*j*) vitamin      (*k*) absolute enzyme specificity      (*l*) substrate

   (*m*) enzyme-substrate complex      (*n*) hormone      (*o*) endocrine glands

   (*p*) target organ      (*q*) chemotherapy      (*r*) antimetabolite

2. Outline the relations of apoenzymes, coenzymes, vitamins, minerals, and enzymes.

3. Explain the specificity of enzymes.

4. Explain why enymes have optimum pHs and temperatures.

5. Discuss the factors that help to account for the high rates of enzyme-catalyzed reactions.

6. How might certain hormones influence chemical events inside cells, even though their own molecules are not able to get inside?

7. How do the most dangerous poisons usually work?

# CHAPTER TWELVE

# Important Fluids of the Body

## THE INTERNAL ENVIRONMENT

To live and move and have our being in this world, we must be able to cope with wide fluctuations in outside temperature, humidity, wind velocity, amount of sunshine, and other factors. We also have an internal environment, a *milieu interieur*. Claude Bernard, the great French physiologist of the nineteenth century, was the first to call attention to its importance and to the fact that higher animals are able to exert nearly perfect control over their composition, regardless of the external conditions. The internal environment consists of the extracellular fluids, all the fluids that are not inside cells. They amount to about 20% of the weight of the body. Three-fourths of this is the *interstitial fluid* within which most cells are constantly bathed. Virtually all the remaining one-fourth is blood plasma. Its circulation provides the main link for chemical communication between the internal and the external environments. (Other forms of communication are provided by eyes, ears, nose, and the nerves that are sensitive to heat and touch.) The digestive juices, cerebrospinal fluid, synovial fluid, aqueous and vitreous humor, and the lymph are also "compartments" of the extracellular fluid.

In this chapter we study only certain selected aspects of the internal environment, concentrating for the most part on the digestive juices and the bloodstream. The processes of digestion dismantle the large molecules of the diet into small molecules which are distributed by the bloodstream wherever they are needed. Since we do not eat constantly, traffic from the digestive tract into general circulation has its rush hour periods. Through feast and famine, however, the composition of the bloodstream fluctuates narrowly, for means of storing excess nutrient molecules are available. The concentration of electrolytes and the pH of the blood are kept remarkably constant. For this service we are very much dependent on the work of the kidneys, which we shall study; the formation of urine (diuresis) is an important part of the chemistry of blood. Then we shall study how other chemicals are exchanged between cells and the bloodstream; here we must look at the lymphatic system, which is also important to blood chemistry. Because oxygen is one of the important chemicals delivered to cells, we study oxygen–carbon dioxide transport. Finally, we examine the clotting mechanism that stands guard over tissues enclosing the bloodstream.

# DIGESTION

Food is something we seldom take for granted, even when we have plenty. The digestion of food, however, we always take for granted, until something goes wrong with the process. No small amount of contentment depends on the efficiency of chemical systems to digest almost any combination of food and drink. In this section we study the molecular basis of this life process.

## THE DIGESTIVE TRACT

The digestive tract is essentially a tube running through the body; its principal parts and organs are located and named in Figure 12.1. The major subdivisions are the mouth, throat, and alimentary canal, which consists of the esophagus, stomach, and intestines. Connected to these at various places are the salivary glands, liver, and pancreas. These manufacture juices that are released into the digestive tract and are needed for the chemical processes of digestion. Other digestive fluids are made in and released from cells in the walls of the stomach or intestines.

When food is taken into the mouth, chewing shreds it to small particles with greater total surface area. Digestive fluids are therefore given larger areas on which to make their chemical attacks. During chewing saliva lubricates the food while launching a preliminary digestive attack on one substance in food, starch. The tongue then pushes the food back toward the throat where powerful muscles in the throat and esophagus send it downward to the stomach. Protein digestion starts here.The well-churned contents of the stomach are eventually released, portion by portion, through the pyloric valve (Greek *pylorus,* a gate guard) into the duodenum,[1] the first section of the small intestine. The digestion of lipids, carbohydrates, and proteins is completed in the small intestine. As the undigested material moves on, most of the water is reabsorbed into the body through the walls of the first section of the large intestine. This part of the alimentary tract contains a multitude of beneficial bacteria which act on residues passing through and produce some vitamin K and some B-complex vitamins. They also synthesize or modify compounds that give the characteristic odor to the final residue of undigested food, the feces.

The overall processes of digestion are principally those of breaking large molecules into small ones that can be readily absorbed through the intestinal walls into the bloodstream or the lymph system. The chemical changes are quite simple.

---

[1] Duodenum is from the Latin for "twelve." Herophilus measured this section as the width of twelve fingers. Actually it is about 10 inches long; the total length of the small intestine is about 21 feet.

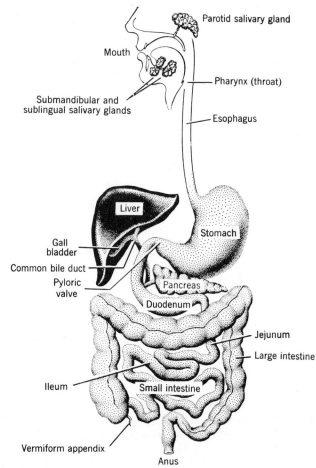

**Figure 12.1** Organs of the digestive tract. (Adapted from Eva D. Wilson, Katherine H. Fisher, and Mary E. Fuqua, *Principles of Nutrition,* second edition, John Wiley and Sons, New York, 1965.)

Ester linkages in lipids are hydrolyzed, and so too are amide (peptide) bonds in proteins and acetal-oxygen bridges in di- and polysaccharides. Each of these reactions requires specific enzymes which it is the business of the digestive juices to provide. With the exception of bile, all the digestive juices are dilute solutions of enzymes together with some simple inorganic ions.

### DIGESTIVE PROCESSES. DIGESTIVE JUICES AND ENZYMES

We now study the various chemical events of digestion by examining each digestive juice, its components, and its action on foodstuff molecules.

**Saliva.** The flow of saliva, controlled by nerves, is stimulated by the sight, smell, taste, or even the thought of food. Its flow, on the average, is 1 to 1.5 liters

**Table 12.1  Components of Saliva**

| Nonenzymic substances: | 1. Water (99.5%)<br>2. Mucin, a glycoprotein that gives saliva its characteristic consistency<br>3. Inorganic ions: $Ca^{2+}$, $Na^+$, $K^+$, $Mg^{2+}$, $H_2PO_4^-$, $Cl^-$, and $HCO_3^-$; average pH of saliva, 6.8<br>4. Miscellaneous molecules found also in blood and urine: urea, ammonia, cholesterol, amino acids, uric acid |
|---|---|
| Enzyme: | $\alpha$-Amylase (ptyalin) |

per person per day. Table 12.1 shows its principal components. Mucin functions to lubricate the food. $\alpha$-Amylase, the principal enzyme in saliva, begins the hydrolysis or digestion of amylose.

**Gastric Juice.** The chief components of gastric juice, secreted into the stomach, are listed in Table 12.2. It is a mixture of several secretions originating in various glands or areas of the stomach wall. Its hydrochloric acid comes from one group, the parietal cells, its enzymes (or enzyme precursors) from another, the chief cells. An average adult's stomach receives about 2 to 3 liters of gastric juice per day from the ducts of several million gastric glands located in the stomach walls.

The secretion from the parietal cells of gastric glands is a solution that contains $0.16M$ HCl and $0.007M$ KCl plus trace amounts of other inorganic ions and virtually no organic substances. Its flow is stimulated in part by the action of a hormone, *gastrin*, which apparently consists of two active peptides, gastrin I and gastrin II. Its mechanism of action is unknown, but the hormone is released into circulation by mechanical distension of the stomach as well as by the appearance of chemicals (of the foods). When this hormone reaches its target cells (parietal cells), they are made to release their particular contribution to gastric juice. The ability of parietal cells to make hydrochloric acid is quite a remarkable feat, since the final hydrogen-ion concentration of gastric juice is about $10^6$ times that of

**Table 12.2  Components of Gastric Juice**

| Nonenzymic substances: | 1. Water (99%)<br>2. Mucin<br>3. Inorganic ions: $Na^+$, $K^+$, $Cl^-$, $H_2PO_4^-$<br>4. HCl (0.5%); average pH, 1.6–1.8 |
|---|---|
| Enzymes: | 1. Pepsin, a proteolytic enzyme<br>2. Gastric lipase* |

* A lipase is known to be present in the stomach contents. Whether this enzyme is produced as part of gastric secretions is an unanswered question. Lipases are secreted into the duodenum, and the lipase in the stomach may result from regurgitation of lipase-containing intestinal fluids back through the pylorus. In any event, the optimum pH for this lipase is near neutrality, and it would not be capable of much activity in the relatively high acidity of the stomach. (It may be functional in infants, however, for they have a lower gastric acidity and their important lipid is the highly emulsified fat in milk, a form readily attacked.)

blood plasma. This acid is apparently made by an energy-demanding process such as

$$NaCl + H_2CO_3 \rightleftharpoons HCl \text{ (to gastric juice)} + NaHCO_3 \text{ (to blood plasma)}$$

*In vitro* the right-to-left reaction is spontaneous. Parietal cells appear to be able to force the left-to-right change.

The chief cells of gastric glands synthesize a polypeptide, *pepsinogen,* in a slightly alkaline medium. This polypeptide is the enzymically inactive precursor for the enzyme *pepsin.* Pepsinogen is converted to pepsin when it encounters the hydrochloric acid from the parietal cells. Once some pepsin forms, it acts to catalyze a more rapid conversion of unchanged pepsinogen to more pepsin. The reaction is believed to consist of a simple hydrolytic splitting of a small segment of the polypeptide chain. The remainder automatically takes up a new secondary or tertiary shape (cf. p. 337) in which the active site of the enzyme is now exposed. No cofactor is involved; therefore pepsinogen is not really a apoenzyme. Rather, it is classified as a *zymogen.*

Pepsin, a proteolytic enzyme, catalyzes the hydrolysis of peptide bonds in proteins, the digestion of which commences in the stomach. The optimum pH for its activity is in the neighborhood of 2, and at pHs higher than 6 it is readily denatured. Thus when pepsin together with the contents of the stomach moves into the duodenum (pH = 6.0 to 6.5 during digestive activity), its enzymic work ceases and other proteolytic enzymes take over. Pepsin has its best specificity for amide linkages when an amino acid with an aromatic side group furnishes the amino group for the bond. In contrast with some of the proteolytic enzymes, pepsin can help cleave peptide bonds in the interior of a protein chain. Others, to be studied shortly, favor terminal amino acid groups only. The action of pepsin therefore generates intermediate breakdown products of proteins with varying chain lengths; their names are given in the following word equation:

Proteins → proteoses → peptones → polypeptides → amino acids

➤— Fragments become progressively smaller —→

Proteoses, peptones, and polypeptides are the principal products of peptic digestion. A theory proposing to explain how pepsin works is described in Figure 12.2.

Because the membranes constituting the digestive tract are themselves made up of proteins, we may wonder why they are not digested. In most people digestion of the stomach or intestinal walls obviously does not occur. But for a few it does, and they have ulcers. The defensive mechanism has several features. In addition to the chief and parietal cells, the gastric glands possess a third type, the *mucous cells.* These make and secrete a thick viscous fluid containing a mucoprotein, a complex between a polysaccharide and a protein. Other cells in the inner surface layer of the stomach also secrete a "mucus" which acts to coat the stomach walls. Since the mucus is constantly replenished, it is only slowly digested and has both acid-binding and peptic-inhibitory abilities. It therefore helps to keep enzymes away from the walls of the stomach. (A similar mucus is found in the intestines.) Should the concentration of hydrochloric acid in the stomach become and remain too high

Step 1. Two neighboring carboxyl groups on side chains of the pepsin molecule iteract to form the more reactive anhydride system (p. 241).

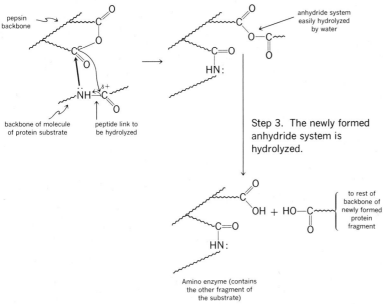

Step 2. A molecule of protein substrate, assisted by the geometrical and electrical factors described in the lock and key theory of enzymic action, collides with the anhydride unit of the enzyme.

Step 3. The newly formed anhydride system is hydrolyzed.

Step 4. The "amino enzyme" unit has a free carboxyl group as a close neighbor. Cleavage of the amino group is relatively easy in such systems, and the mechanism might be something like the following.

**Figure 12.2**   Proposed mechanism for the action of pepsin. (From M. L. Bender and F. J. Kezdy, *Annual Review of Biochemistry,* Vol. 34, 1965, page 63.)

(*hyperacidity*), the mucus itself is more rapidly digested, and the stomach walls are exposed to digestive action. If this goes too far, the wall is perforated. Perforation may happen to occur at a blood vessel, producing serious hemorrhaging which can be fatal.

The secretion of hydrochloric acid is controlled in part by the action of another hormone, *enterogastrone*. When food leaves the stomach and enters the duodenum, cells in its walls are stimulated to release this substance (which may consist of two polypeptides) into the bloodstream. When circulation has brought enterogastrone to the gastric glands, it acts to shut down their secretion of hydrochloric acid. Thus a regulatory mechanism controls the secretion of gastric juice. When it is needed, it is produced. When the need diminishes, its flow is reduced.

The churning, semidigested mixture of food and fluids present in the digestive tract is called *chyme* (Latin *chymus,* juicelike). Over a two- to five-hour period the gastric chyme is released in portions through the pyloric valve into the duodenum where several enzyme-rich secretions complete the digestion of food.

**Clinical Importance of Gastric Analysis.** After stimulating the secretion of gastric juice by feeding the subject a meal or alcohol or by injecting histamine, samples of gastric juice are withdrawn at about fifteen-minute intervals and analyzed for both free and combined acid. The results may reveal disorders that are very remote from the stomach. A few are listed.

1. Achlorhydria: the absence of free acid. Although this condition is rare, if it is discovered and if pepsin is also absent, pernicious anemia or stomach cancer is indicated.

2. Hypoacidity (hypochlorhydria): too low a concentration of acid. Only infrequently is this condition serious, but it can indicate stomach cancer, chronic constipation, or inflammation of the stomach (chronic gastritis).

3. Hyperacidity (hyperchlorhydria): an excess of acid. Relatively more common, this condition is associated with chronic "heartburn," chronic "indigestion," gastric or duodenal ulcers, or even inflammation of the gallbladder (cholecystitis).

Figure 12.3 shows how the concentration of free acid varies with time in the normal individual, in the patient with pernicious anemia, and in the patient who has a duodenal ulcer.

**Pancreatic Juice.** The pancreas produces two secretions. The internal secretion which contains insulin is discharged into the bloodstream. The external secretion, pancreatic juice, empties into the duodenum. Its principal components are listed in Table 12.3.

The flow of pancreatic juice and its composition are regulated by two hormones, *secretin* and *pancreozymin*. Both are released into the bloodstream from cells of the upper intestinal mucosa whenever it is exposed to certain chemicals in the chyme. Release of secretin is prompted primarily by the presence of an acid that comes into the duodenum from the stomach. When molecules of this hormone reach the pancreas, they stimulate an increase in the volume flow and the electrolyte content (particulary $HCO_3^-$) of pancreatic juice. Secretin does not itself stimulate the release of either digestive enzymes or zymogens. Their release requires

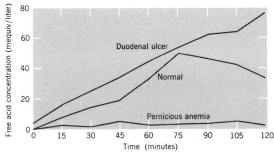

Free acid concentration (mequiv/liter) vs Time (minutes)

**Figure 12.3** Gastric analysis. The graph shows for three situations how the concentration of free acid varies with the time lag between the initial stimulation of gastric secretion and the withdrawal of a sample for analysis. (From A. White, P. Handler, and E. L. Smith, *Principles of Biochemistry*, third edition, McGraw-Hill Book Company, New York, 1964, page 712. Used by permission.)

the cooperation of the second hormone, pancreozymin, which is secreted when chemicals originating from ingested food appear in the duodenum.

Part of the enzymic activity of pancreatic juice requires the preliminary action of substances in bile and intestinal juice. Our discussion of the enzyme-catalyzed reactions in the upper intestinal tract will therefore be continued after these other digestive juices have been described.

**Intestinal Juice (*Succus entericus*).** Mucosal glands of the duodenum secrete this digestive juice whenever chyme enters from the stomach. The regulation of its flow is not very well understood, for it appears to be stimulated in more than one way, including hormonal action. The principal components of intestinal juice are listed in Table 12.4.

The enzyme-rich intestinal juice is important to the digestion of all foods. Various peptidases complete the hydrolysis of proteins to amino acids. Sucrase, maltase, and lactase handle conversion of the important disaccharides to glucose, fructose, and galactose. Simple lipids are acted upon by a lipase; more complex

**Table 12.3 Components of Pancreatic Juice**

| | |
|---|---|
| Nonenzymic substances: | 1. Water; average daily volume, 500–800 cc |
| | 2. Inorganic ions: $Na^+$, $K^+$, $Ca^{2+}$, $Cl^-$, $HCO_3^-$, $HPO_4^{2-}$; average pH, 7–8* |
| Enzymes: | 1. Pancreatic lipase ("steapsin") |
| | 2. $\alpha$-Amylase |
| | 3. Maltase |
| | 4. Ribonuclease |
| Zymogens: | 1. Trypsinogen |
| | 2. Chymotrypsinogen |
| | 3. Procarboxypeptidase |

* Under physiological conditions the contents of pancreatic juice will not function in a slightly alkaline fluid. The mixing of acidic chyme from the stomach with this juice plus bile and intestinal juice produces a neutral or slightly acidic material.

**Table 12.4    Components of Intestinal Juice**

| Nonenzymic substances: | 1. Water |
| | 2. Inorganic ions: $Na^+$, $K^+$, $Ca^{2+}$, $Cl^-$, $HCO_3^-$, $HPO_4^{2-}$ |
| Enzymes: | 1. Aminopeptidase |
| | 2. Dipeptidase |
| | 3. Nucleases, nucleotidase, nucleosidase |
| | 4. Maltase |
| | 5. Sucrase |
| | 6. Lactase |
| | 7. Intestinal lipase |
| | 8. Lecithinase |
| | 9. Phosphatase |
| | 10. Enterokinase |

lipids are hydrolyzed by the catalytic action of lecithinase and phosphatase. Nucleic acids (cf. Chapter 17) respond to enzymes peculiar to their hydrolytic needs.

One of the most important enzymes found in intestinal juice is *enterokinase,* whose major function is to convert trypsinogen to trypsin. Once trypsin is formed, it serves to activate the other zymogens in pancreatic juice. Figure 12.4 presents

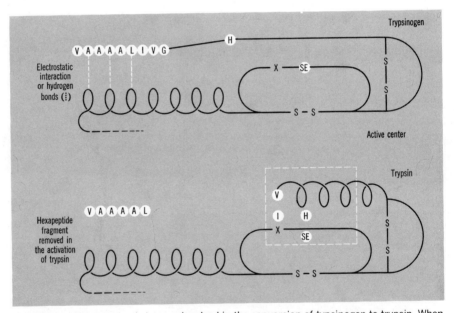

**Figure 12.4**    The structural changes involved in the conversion of typsinogen to trypsin. When the small hexapeptide is split off, the remainder of the chain at this end of trypsinogen automatically adopts a more helical configuration, permitting the histidine unit (H) to come close to the serine unit (SE), a juxtaposition required for enzymic activity. A, aspartic acid unit; G, glycine unit; H, histidine unit; I, isoleucine unit; SE, serine unit; V, valine unit; X, active site. (From H. Neurath, "Protein Structure and Enzyme Action," in *Biophysical Science—A Study Program,* edited by J. L. Oncley, John Wiley and Sons, New York, 1959, page 189.)

a theory to explain the activation of trypsin. Thus for pancreatic juice to be effective in the duodenum, the participation of intestinal juice is needed.

Processes.
Digestive Juices
and Enzymes

| *Zymogen* | | *Enzyme* |
|---|---|---|
| trypsinogen | $\xrightarrow[H_2O]{enterokinase}$ | trypsin |
| chymotrypsinogen | $\xrightarrow[H_2O]{trypsin}$ | chymotrypsin |
| procarboxypeptidase A | $\xrightarrow[H_2O]{trypsin}$ | carboxypeptidase A |
| procarboxypeptidase B | $\xrightarrow[H_2O]{trypsin}$ | carboxypeptidase B |

Some of the enzymic influences of the intestinal juice are exerted within the cells of the duodenal mucosa rather than in the duodenum itself. If nutrient molecules that are not completely hydrolyzed begin to dialyze through the intestinal walls, their hydrolysis is completed during the journey by the action of the intestinal juice enzymes.

**Bile.** Bile, the third juice that empties into the duodenum, contains no digestive enzymes. Nevertheless, it is vitally important to digestion, for it contains *bile salts* whose fat-emulsfying action is essential to rapid digestion of lipids. The chief problem in lipid digestion is that lipids are notoriously water-insoluble. They tend to form large globules, and any digestive attack on their molecules can occur only at the surface of the globule. The bile salts, structures for two of which are shown in Figure 12.5, have all the structural features of soaps, that is, large hydrocarbon-like portions joined to highly polar (ionic) ends (cf. p. 306). Emulsification of an oil, of course, means that large globules are broken into many times their number of smaller globules. Although the total amount of oil is obviously the same, the total surface area of all the globules is enormously increased, and lipases plus water can more readily hydrolyze the lipid material. Some scientists believe that the bile soaps act primarily to emulsify fats, although this subject is a matter of controversy in a few circles. Bile soaps may also be needed for washing fatty coatings from partly digested particles of other foods.

Bile also contains *bile pigments* (from the breakdown of hemoglobin) and cholesterol. If the concentrations of cholesterol and bile salts in bile become too high while they are in the gallbladder, they may precipitate to form gallstones. When these become large enough, they can plug the duct from the gallbladder to the duodenum. Usually the only remedy is to remove the gallbladder, and those who have had such an operation must henceforth exercise careful control over the lipid content of their diet.

The flow of bile is regulated by a hormone, *cholecystokinin*, which is synthesized in cells of the upper intestinal mucosa. Whenever chyme that contains fats or fatty acids enters the duodenum, this hormone is released into general circulation. When its molecules arrive at the gallbladder, they cause a powerful contraction of its muscles, forcing bile out of it. Details of the mechanism are not known.

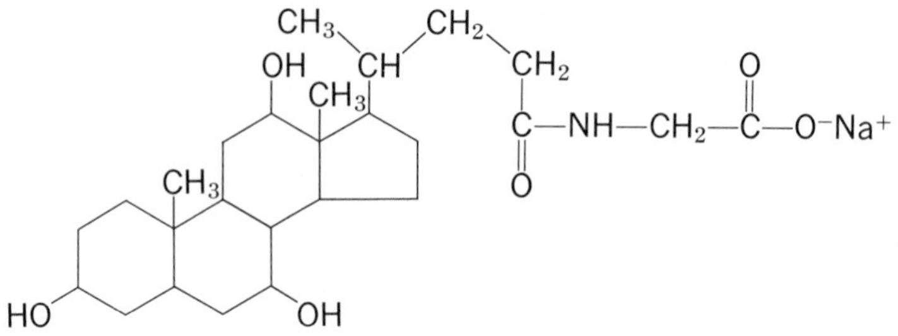

Sodium salt of glycocholic acid

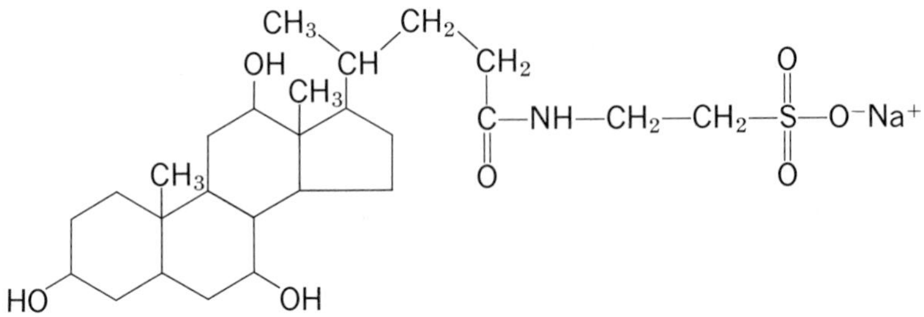

Sodium salt of taurocholic acid

**Figure 12.5** Two of the more common bile salts that act as lipid-emulsifying soaps in the duodenum.


### Digestion in the Upper Intestinal Tract

1. **Proteolytic digestion.** The digestion of proteins begins in the stomach, but gastric chyme that enters the duodenum can still contain an abundance of large protein molecules. Trypsin and chymotrypsin are the two enzymes chiefly responsible for handling such materials. The smaller fragments of proteins that they produce are further degraded by several specialized proteolytic enzymes, carboxypeptidases, tripeptidase, and dipeptidase. Table 12.5 contains a summary of the specificity of these enzymes. With a variety of proteolytic enzymes available, it is apparent that almost any protein of whatever unique or unusual amino acid sequence can be digested, provided the amino acids belong to the L-family. The principal lack in human beings is an enzyme for digesting keratins. These are rich in disulfide linkages for which man has no digestive enzyme.

2. **Carbohydrate digestion.** Pancreatic juice contains an amylase that acts on acetal linkages in amylose and amylopectin molecules that have escaped hydrolysis in the mouth. The end product is maltose, which is hydrolyzed to glucose by the action of maltase. (Mentioned earlier, p. 403, were the actions of lactase and sucrase.)

3. **Lipid digestion.** The pancreatic lipase, steapsin, is the most important fat-splitting enzyme in the digestive tract. It catalyzes the hydrolysis of the ester linkages in triglyceride molecules that are at the opposite ends of the glycerol portions.

$$R'-\overset{O}{\overset{\|}{C}}-O-\underset{\displaystyle \underset{\displaystyle CH_2-O-\overset{O}{\overset{\|}{C}}-R''}{\overset{\displaystyle CH_2-O-\overset{O}{\overset{\|}{C}}-R}{|}}{CH}} \quad \xrightarrow[\text{lipase}]{2H_2O} \quad R'-\overset{O}{\overset{\|}{C}}-O-\underset{\displaystyle \underset{\displaystyle CH_2-OH + H-O-\overset{O}{\overset{\|}{C}}-R''}{\overset{\displaystyle CH_2-OH + H-O-\overset{O}{\overset{\|}{C}}-R}{|}}{CH}}$$

A monoester of       Two molecules<br>glycerol       of fatty acids

One of the ester linkages survives, and there is evidence that the successful mobilization of lipid material out of the intestinal tract does not require further digestion of this monoester. The fatty acids plus the monoester may be absorbed into cells of

**Table 12.5 Specificity of Proteolytic Enzymes**

| Enzyme | Preferred Site of Action* | |
| --- | --- | --- |
| **Endopeptidases (act on nonterminal sites)** | | |
| Pepsin | $-\overset{O}{\overset{\|}{C}}-NH-\underset{\displaystyle Ar'}{CH}-\overset{O}{\overset{\|}{C}}\wr NH-\underset{\displaystyle Ar}{CH}-$ | Ar' or Ar = side chain from a tyrosine or phenylalanine unit |
| Trypsin | $-\overset{O}{\overset{\|}{C}}-NH-\underset{\displaystyle \underset{G}{NH_3^+}}{CH}-\overset{O}{\overset{\|}{C}}\wr NH-$ | Positively charged side chain from arginine or lysine |
| Chymotrypsin | $-\overset{O}{\overset{\|}{C}}-NH-\underset{\displaystyle Ar}{CH}-\overset{O}{\overset{\|}{C}}\wr NH-$ | Ar = an aromatic side chain |
| **Exopeptidases (act on terminal amino acid units)** | | |
| Carboxypeptidase A | $-\overset{O}{\overset{\|}{C}}\wr NH-\underset{\displaystyle Ar}{CH}-\overset{O}{\overset{\|}{C}}-OH$ <br> <u>A C-terminal amino acid unit</u> | Ar = an aromatic side chain |
| Carboxypeptidase B (aminopeptidase) | $NH_2-\underset{\displaystyle \underset{G}{NH_3^+}}{CH}-\overset{O}{\overset{\|}{C}}\wr NH-$ <br> <u>An N-terminal amino acid unit</u> | Positively charged side chain from arginine or lysine |
| Prolidase | $-\underset{\displaystyle}{CH}-\overset{O}{\overset{\|}{C}}\wr\underset{\displaystyle \underset{CH_2}{CH_2}}{N}-\underset{\displaystyle CH_2}{CH}-\overset{O}{\overset{\|}{C}}-$ | |
| Tripeptidase | $NH_2-\underset{}{CH}-\overset{O}{\overset{\|}{C}}\wr NH-\underset{}{CH}-\overset{O}{\overset{\|}{C}}-NH-\underset{}{CH}-\overset{O}{\overset{\|}{C}}-OH$ | |
| Dipeptidase | $NH_2-\underset{}{CH}-\overset{O}{\overset{\|}{C}}\wr NH-\underset{}{CH}-\overset{O}{\overset{\|}{C}}-OH$ | |

* A vertical wavy line shows where the hydrolysis occurs.

the intestinal mucosa where another lipase acts rapidly to finish the hydrolysis of the last ester linkage, provided it is of a short-chain fatty acid. If it is not, lipid metabolism in these cells takes what might seem to be a surprising turn. Reesterification occurs. The newly formed glycerides enter circulation via the lymph system, and when they or fatty acids reach the bloodstream, they are believed to be picked up by circulating protein molecules and are transported as lipoprotein complexes.

## BRIEF SUMMARY OF THE DIGESTION OF THE THREE BASIC FOODS

**Carbohydrates.** Starch and the three disaccharides, maltose, lactose, and sucrose, are the chief carbohydrates in nutrition. Digestion of cooked starch begins in the mouth, where α-amylase is present, and continues for a time in the stomach until the hydrochloric acid in the gastric juice inactivates the α-amylase that has accompanied swallowed food. α-Amylase is not considered active toward uncooked starch. Dextrins and maltose are the end products of the work of this enzyme.

The disaccharides probably undergo some nonenzymic hydrolysis in the stomach, where the action of hydrochloric acid hydrolyzes acetal linkages in these molecules. This hydrolysis is not of great importance, however.

The amylase in pancreatic juice completes the hydrolysis to maltose of both raw and cooked starch. The three carbohydrases, maltase, lactase, and sucrase, that appear in the duodenum complete the digestion of the disaccharides. *The end products of carbohydrate digestion then are fructose, galactose, and glucose.*

Cellulose in the diet (e.g., in leafy vegetables) passes through essentially unchanged, furnishing much of the bulk needed for efficient operation of the excretory system in the lower intestinal tract.

**Lipids.** For all practical purposes lipids undergo no digestive action until they reach the duodenum where bile salts rapidly emulsify them. Once emulsified, pancreatic lipase smoothly hydrolyzes lipids to *glycerol, fatty acids, and monoglycerides, the end products of the digestion of the neutral fats and oils.* The lipase of the intestinal juice acts on lipids that escape hydrolysis catalyzed by steapsin. Reesterification of glycerol begins during absorption.

**Proteins.** Simple proteins are not acted on in the mouth. In the stomach pepsin catalyzes the conversion of proteins to proteoses and peptones. Trypsin and chymotrypsin, in the duodenum, act rapidly on proteoses and peptones (and, to a certain extent, on native proteins), converting them to simple peptides and amino acids. Peptidases in intestinal juice and pancreatic juice complete the digestion of proteins. *The end products of protein digestion are amino acids and, perhaps, some simple soluble di- and tripeptides.*

## BLOOD

### FUNCTIONS

A schematic diagram of the circulatory system is given in Figure 12.6. The functions of the bloodstream are quite well known. It transports oxygen, other nutrients, and waste products, thereby establishing communication between the external and the internal environments. It also transports substances from one site to another within the body, for example, moving fatty acids from adipose tissue to the liver, and moving freshly made glucose and ketone bodies, from the liver to tissues that use them. The bloodstream is important in heat exchange. By bringing some of the flowage through capillaries near the skin, heat can be pumped out of the system. The body's defensive mechanisms, which guard against infectious diseases as well as physical damage to the tissues that enclose the moving blood,

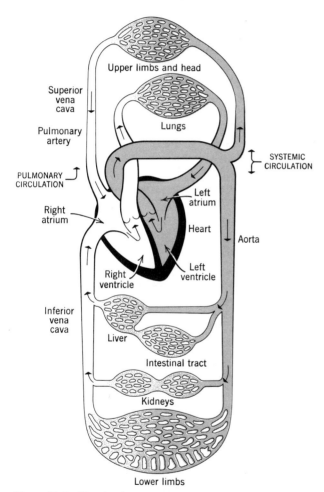

**Figure 12.6** The circulatory system in man. Shaded areas designate oxygenated blood. Blood received from the veins at the right atrium of the heart passes into the right ventricle, and from there it is forced through the spongy capillary networks of the lungs. In the capillaries carbon dioxide is removed and oxygen is taken up. The blood then goes to the left atrium for delivery to the left ventricle and on to the rest of the body.

depend greatly on substances carried by it. To meet all these needs, the blood contains a host of specialized chemicals and cells.

## COMPOSITION

Blood consists of a solution called *plasma* in which are dispersed *formed elements* or cellular bodies: erythrocytes (red blood cells), leukocytes (white blood cells), and thrombocytes (blood platelets). *Blood serum* is what is left of the plasma after the formed elements and fibrinogen (the clotting protein) have been separated. So-called *defibrinated blood* is whole blood from which this clotting protein has been removed; the cellular bodies are still present.

**Formed Elements.** Oxygen transport is the most important function of erythrocytes. The blood of a normal adult contains about five million of these red cells per cubic millimeter of blood. (A drop of blood has a volume of about 20 to 30 cubic millimeters.) Variations in the concentration of red cells can be measured ("red cell count") and have diagnostic value. In anemia, for example, the red cell count is abnormally low.

Leukocytes, or white blood cells, number from 5 to 10,000 per cubic millimeter of blood. Several varieties are recognized. One, the polymorphonuclear leukocyte, attacks invading bacteria and other foreign bodies. Another, the lymphocyte, is involved in the synthesis of antibodies and the fixation of toxins. The fatal disease of the blood-forming organs, leukemia, is accompanied by huge increases in the number of leukocytes in the blood. Thus the white cell count is also an important diagnostic tool.

Thrombocytes or platelets are important in blood clotting.

**Blood Plasma.** The average adult has about 5 liters of blood, 55% of which is plasma. The erythrocytes make up another 42 to 43%. The white cells and the platelets are each present in trace percentages. The other principal components of blood are various proteins; upward of seventy have been identified, and more are believed to be present. The proteins have been classified according to certain solubility relations, certain properties that make rough separations of these proteins possible. When plasma, for example, is mixed with a saturated solution of ammonium sulfate, some of the proteins precipitate, and they are called the *globulins*. After these have been removed, addition of solid ammonium sulfate causes another batch of proteins to precipitate, the *albumins*. It may seem altogether arbitrary to base a system for classifying proteins on such a procedure, but the proteins in each class do share common functional purposes.

One of the important functions of albumins, which account for about 55% of the total protein in blood, is the regulation of osmotic pressure. Another general function is to act as transporting agents for other relatively insoluble chemicals.

Globulins constitute about 40% of the total protein in blood. Three general types, $\alpha$-globulin, $\beta$-globulin, and $\gamma$-globulin, are recognized; each type consists of several different proteins. The $\gamma$-globulins are important in the body's defense against infectious disease. The other globulins appear to be necessary for the transport of metal ions such as $Fe^{2+}$ and $Cu^{2+}$, which would otherwise be insoluble in the slightly basic medium of blood.

Fibrinogen, the third major protein found in plasma, makes up about 5 to 6% of the total protein. It is the immediate precursor of fibrin, a plasma-insoluble protein that precipitates as a tangled brush heap of molecules in a developing blood clot.

Electrolytes constitute another general type of plasma solute. Inorganic ions are of course found in all body fluids; Figure 12.7 shows the kinds and concentrations present in plasma, interstitial fluid, and cell fluid. The distribution of ions in cell fluid is by no means uniform, however. Localized pockets of higher or lower concentrations are found in the cell. The most significant difference to be noted in the data of Figure 12.7 is that inside the cell the ions $K^+$, $Mg^{2+}$, and $SO_4^{2-}$ and various phosphate ions predominate. Outside the cell in plasma or interstitial fluid the ions $Na^+$, $Cl^-$, and $HCO_3^-$ predominate. Furthermore, the concentration of large molecules, proteins, is much greater inside a cell than outside.

Osmotic pressure, you will recall, is related to particle concentration without substantial regard to the sizes of these particles or the charges on them. Since cells have walls that are freely permeable to water molecules, and since cells normally neither burst nor shrink, it must be true that the osmotic pressure inside a cell is the same as that outside. Therefore any water that migrates out of a cell is replaced by water that moves in. The effective particle concentration is the same inside a cell as ouside, but the kinds of particles (sodium or potassium ions, etc.) are different. If the fluid is locally diluted either inside or outside a cell (the second being the more common problem), ions are moved from one side to another to minimize the effects of the imbalance. Which ions are moved is kept under control. Normally, for example, sodium ions would not move from the inside of a cell (low concentration) to the outside (high concentration). We should expect the reverse. A similar statement could be made about potassium ions. To make the separation of these ions possible, as indicated by the data in Figure 12.7, the cell must do work. Energy-demanding chemical processes take place, in effect to pump sodium ions in one direction. If the sodium ions are *actively* transported by chemical processes that use energy to pass them to the outside, some positive ions must take their place. Potassium ions do, and they need not be actively transported. Passive diffusion will accomplish it. For example, if red cells, at the temperature of an ice bath, are kept in contact with saline solution containing glucose, they lose their potassium ions and sodium ions move in until a general distribution of these ions has occurred. If the system is warmed to body temperature (37°C), sodium ions are pumped out of the red cells and potassium ions diffuse back until the two are for the most part separated. If the glucose is omitted, when the system is warmed to body temperature, the *active transport* of sodium ions outside and potassium ions inside does not occur. Then if the glucose is added, sodium ions move out and potassium ions diffuse back. Glucose is apparently necessary to the process, probably as a source of ATP or some other high-energy compound needed to supply the energy for moving and keeping the system away from the ionic equilibrium. What we have described about red cells is true for most types of cells. Significant portions of their energy needs are for active transport of certain substances against an *electrochemical gradient*—from a region of low to high concen-

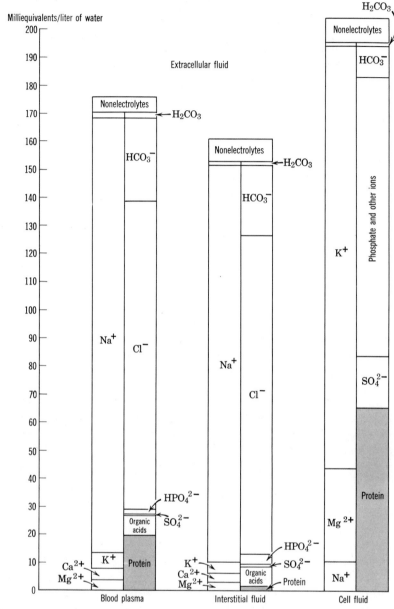

**Figure 12.7** Electrolyte composition of body fluids. (Adapted, by permission, from J. L. Gamble, *Chemical Anatomy, Physiology and Pathology of Extracellular Fluid,* sixth edition, Harvard University Press, Cambridge, Mass., 1954.)

tration of charged particles. It is apparent that cells do a great deal of work not directly connected with such obvious functions as muscular exercise, thinking, and the like. It is also obvious that the delivery of a fresh supply of chemicals to the bloodstream from digested food could alter the sensitive balance of the concentrations of electrolytes. The cells of one organ, the kidneys, therefore perform specialized tasks in regulating the volume of water left in the bloodstream and its concentrations of ions and other small molecules. Whether or not the bloodstream does its job depends in large measure on the work of the kidneys, which we shall describe next.

# URINE

## DIURESIS

The whole process of urine formation is called diuresis (Greek *ouresis,* to make water or urine). The kidneys, in cooperation with certain hormones, exert control over the concentrations of electrolytes, including those affecting the acid-base balance of the blood. It will be helpful to our understanding of what kidneys do to know something about what they are.

The kidneys are located slightly above the waist, behind the other organs on either side of the spinal column. Into each a single large artery and a single large vein are connected. While in the kidneys the blood undergoes a filtering process of immense proportions. Liquid leaves the bloodstream and returns at a rate of 130 ml per minute or about 190 liters per day. Only about 1 to 2 liters is converted into urine. The blood fluid that temporarily leaves the bloodstream carries with it sodium chloride (at the rate of over 1 kg per day), sodium bicarbonate (about 0.5 kg per day), glucose (150 g per day), amino acids, other ions (notably phosphates), and other small molecules. Virtually all this material reenters the bloodstream. Very little remains in the urine being formed. To accomplish such feats of filtration, enormous reabsorption surface areas are required, and these are provided by the successive branching of the circulatory system into smaller and smaller tubes until the individual filtration and reabsorption units, the *nephrons,* are reached. Each kidney has about a million of them, and their principal parts are described in Figure 12.8. Arterial blood finally reaches a tuft of parallel capillaries called the *glomerulus;* here the blood water together with dissolved ions and small molecules leaves the bloodstream and enters the *renal capsule.* This liquid is now called the *glomerular filtrate.* The cellular elements, plasma proteins, and lipids remain behind in the blood. The glomerular filtrate passes first through the *proximal tubule* where about seven-eighths is reabsorbed by the capillary beds interlacing the nephron. In this manner nearly all the electrolytes, all the blood sugar, and most of the water are soon returned to circulation. In the region of the proximal tubule *obligatory reabsorption* is said to occur. Urea, uric acid, and creatinine tend to remain in the urine being formed. The remaining one-eighth of the glomerular filtrate may or may not be reabsorbed farther along the line, in the *distal tubule.* In other words, distal reabsorption is *optional reabsorption.* The exercise of this option is regulated by an antidiuretic hormone called *vasopressin.*

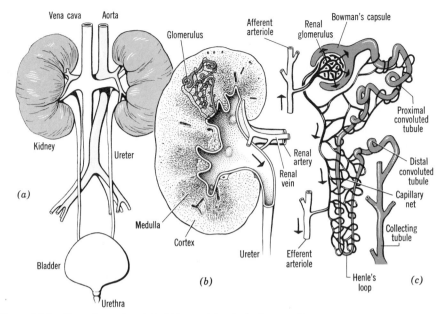

**Figure 12.8**  Operating units of the kidneys: (*a*) principal parts, (*b*) cross section of the kidney, (*c*) nephron. (Adapted, with permission, from G. E. Nelson, G. G. Robinson, and R. A. Boolootian, *Fundamental Concepts of Biology,* John Wiley and Sons, New York, 1967, page 131.)

When vasopressin is present, optional absorption occurs, as much as 100%. The volume of urine made is therefore small. When vasopressin is absent, optional absorption does not occur and relatively larger amounts of urine are made. Vasopressin is secreted by cells of the posterior, or neural, region of the hypophysis, a small endocrine gland lying under the brain. These cells respond to the osmotic pressure of the blood in the following ways.[2]

If the osmotic pressure of the blood goes up, the hypophysis secretes vasopressin, which means that optional reabsorption occurs and the water initially in the blood is kept there. If the osmotic pressure of the blood is high, the system acts to keep it from going higher, as it would if water were removed. The thirst mechanism is now also stimulated to encourage drinking, thus bringing in more water.

If the osmotic pressure of the blood goes down, little or no vasopressin is secreted. Hence optional reabsorption tends not to occur, and at least some of the water taken out of the bloodstream in the glomerulus is kept out and voided as urine. By removing this water from the blood, the concentration of the remaining blood rises. These changes in osmotic pressure of the blood need be no more than 2% up or down for the hypophysis to act to restore the osmotic pressure to normal.

[2] Since the concept of osmotic pressure will be used frequently in this chapter, it would be well to consult any textbook in general college chemistry (or an encyclopedia) to review it. In general, a concentrated solution has a high osmotic pressure, a dilute solution a low osmotic pressure.

If the pressure goes up, the kidneys do not make as much urine, water is conserved, and the thirst mechanism is stimulated to bring in more water. If the osmotic pressure goes down, water is removed. The hypophysis acts as the "thermostat" for this regulation, relaying signals not by electrical means but by the chemical messenger vasopressin.

In a rare disease called diabetes insipidus, the secretion of vasopressin is blocked. Virtually no optional reabsorption occurs, and the volume of the urine increases rather dramatically. Upward of 5 to 12 liters of urine may be voided per day. In contrast, with average intake of fluids a normal person eliminates about 1.5 liters of urine each day.

Another hormone, *aldosterone*, participates in diuresis by regulating the reabsorption of specific ions. As noted earlier, major portions of $Na^+$, $Cl^-$, $K^+$, and $HCO_3^-$ leave the bloodstream and return in the obligatory reabsorption of the proximal tubule. In the distal tubule final adjustments in the concentrations of these ions are made, with aldosterone acting in a regulatory manner.

**Regulation of the Acid-Base Balance of the Blood.** As measured at room temperature, the pH of the blood must be kept within the narrow range of 7.0 to 7.9; for venous plasma it is normally in the extremely narrow range of 7.38 to 7.41. A drop in pH, that is, a shift toward the acid side, is called *acidosis*. *Alkalosis* is a rise in pH, a shift to greater basicity. Acidosis is the more common tendency because ordinary metabolism produces acids. Two agents, the kidneys and the buffers of the blood, remove this acid and maintain a steady pH. The kidneys "pump" hydrogen ions from the blood to the urine being formed; in their place (to preserve electrical neutrality) sodium ions are returned via optional reabsorption.

The buffers of the blood were discussed on pages 342 and 343. The most important is the system $HCO_3^-/H_2CO_3$. In neutralizing acids, bicarbonate ions in blood combine with hydrogen ions:

$$H^+ \quad + \quad HCO_3^- \quad \rightleftharpoons \quad H_2CO_3$$

| Hydrogen ions of developing acidosis | Bicarbonate ions in the blood | Carbonic acid, a weak acid with little tendency to dissociate |
|---|---|---|

Too much carbonic acid can be handled by decomposing it to water and carbon dioxide:

$$H_2CO_3 \xrightarrow{\text{carbonic anhydrase}} H_2O + CO_2 \quad \text{(excreted at the lungs)}$$

Removal of carbon dioxide in the lungs means a net replacement of a hydrogen ion by a water molecule at the expense of a bicarbonate ion, thus depleting part of the buffer system. The kidneys come to the rescue, however, for they are able to generate a fresh supply of bicarbonate ions for the bloodstream and at the same time produce a more acidic urine.

**Acidification of the Urine.** R. F. Pitts proposed the following mechanism to explain how cells of the distal tubules increase the supply of bicarbonate ion in blood and feed hydrogen ions to the urine. In this way they restore the acid-base balance and prevent acidosis from developing further. Pitts's scheme may best be studied by referring to Figure 12.9, with steps indicated by encircled numbers.

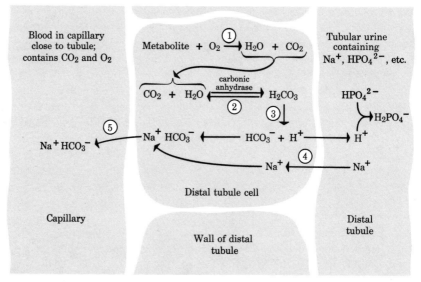

**Figure 12.9**  The acidification of the urine, according to R. F. Pitts. Circled numbers are explained in text. (From *American Journal of Medicine*, Vol. 9, 1950, page 356.)

The oxidation of some nutrient molecule in a distal tubule cell produces carbon dioxide ①. Action of the enzyme carbonic anhydrase causes it to combine with water to form carbonic acid ②, which ionizes to produce bicarbonate ions and hydrogen ions ③. In a sense the ionization is forced, for sodium ions from tubular urine are made to exchange themselves for hydrogen ions ④. Aldosterone appears essential to the promotion of this exchange. The urine thereby becomes more acidic. Bicarbonate ions leave the distal cell and enter the bloodstream ⑤, replacing bicarbonate ions used earlier to neutralize hydrogen ions. Freshly voided urine normally has a pH of about 6. In severe acidosis the kidneys, as they battle to control it, may produce urine with a pH as low as 4.6. Such urine has a hydrogen-ion concentration several hundred times that of normal blood.

When the kidneys fail to function and the formation of urine falls off or stops, a condition called *uremia* exists. (Uremia means urine in the blood, in the sense that the substances normally removed from the bloodstream by the kidneys stay instead in the blood. The term uremic poisoning is sometimes used in this connection.) Clinical evidence for uremia is obtained by measuring the rise in nonprotein nitrogen (NPN)—chiefly urea with smaller quantities of uric acid and creatinine. The blood urea nitrogen (BUN) can also be measured, as can the pH and the concentrations of several ions. Symptoms of uremic poisoning include nausea, vomiting, drowsiness, a tendency to bleed, and a developing coma. In some instances of renal failure the kidneys have not been permanently injured, and the body can recover. Treatment varies, but artificial kidneys have been developed to clean the blood until normal recuperative processes have regenerated kidney function. In principle, the device by dialysis selectively removes dissolved substances from the blood shunted through it. Cellophane has served as a dialyzing membrane in these instruments, which have saved many lives. Because their cost and scarcity have limited their availability, hospitals providing them have had to set up committees of doctors, clergymen, professional men, and others to make the agonizing decisions of who will have a place in the time schedule of the artificial kidney and who will not.

**Proteinuria (Albuminuria).** Protein molecules do not normally leave the bloodstream in the glomerulus. When protein is detected in urine, it usually indicates some injury or disease in the kidneys. Pores in the glomerulus are believed to become enlarged to permit the passage of the larger protein molecules. Since the albumins are the major proteins in blood, the condition is frequently called albuminuria, but other proteins also escape. Loss of the blood proteins upsets osmotic pressure relations, affecting functions throughout the body.

# LYMPH

## THE LYMPHATIC SYSTEM

The bloodstream is one of the circulatory systems in the body. The second one, the lymphatic, performs the following functions: (1) recirculates interstitial fluid to the bloodstream, (2) transports some chemicals from where they are made to general circulation and conveys freshly absorbed, resynthesized lipid material from the intestinal tract to the bloodstream, and (3) participates in the defensive mechanisms of the body.

The lymphatic system (Figures 12.10 and 12.11) is made up of ducts that branch successively to terminate in a spongy mesh of tiny, thin-walled capillaries, all with closed ends, bedded within most of the soft tissue of the body. The larger lymph vessels, which have valves, receive lymph fluid from the branches and convey it to large blood vessels near the neck. The thoracic duct drains the lower limbs and the tissues of all the organs except the lungs, the heart, and the upper part of the diaphragm. The *right lymphatic duct* takes care of these organs, and the *cervical ducts* take lymph from the head and neck. Spaced along the larger lymphatic vessels are *lymph nodes,* specialized capillary beds which filter out solid matter in lymph before it reaches the bloodstream. More importantly, these nodes contain both white cells that destroy bacteria and other foreign substances and special cells that make antibodies. The lymph nodes are very important to the defenses of the body. Conversely, however, when cancer strikes the lymph ducts are avenues for its spread from one tissue to another, a phenomenon known as metastasis.

# TRANSPORT OF CHEMICALS

Having learned something about the bloodstream and some of the ways in which its composition is controlled, and having looked briefly at the second circulatory system, the lymphatics, we are now in a position to discuss how these two systems bring nutrients to cells and carry waste products away.

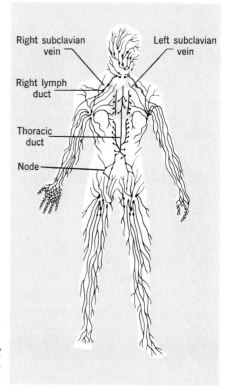

**Figure 12.10** The relations of the larger vessels of the lymphatic system to one another and the bloodstream. (From G. E. Nelson, G. G. Robinson, and R. A. Boolootian, *Fundamental Concepts of Biology,* John Wiley and Sons, New York, 1967, page 126.)

## EXCHANGE OF CHEMICALS AT TISSUE CELLS

The walls of capillaries of the bloodstream may be treated as selectively permeable membranes or filters. They allow water with dissolved nutrient molecules, including oxygen, to pass in one direction and water with dissolved waste chemicals to return in the other. Fluids that leave the bloodstream should return in equal volume; two mechanisms cooperate to make this possible while ensuring that the direction of the flow is correct. The pumping action of the heart creates a pressure which is higher on the arterial side of the capillary than on the venous side. Since the fluids of the blood, the cells, and the interstitial spaces have dissolved materials, osmosis and dialysis also occur. The cooperation of these two factors plus the involvement of the lymph system are illustrated in Figure 12.12.

Plasma differs from interstitial fluid primarily in being more concentrated in proteins and should therefore more effectively induce osmosis and dialysis. The natural *net* direction for dialysis would be from the interstitial compartment into the bloodstream, on both the arterial and the venous ends of the capillary constriction. But on the arterial end the higher blood pressure reverses this tendency, making the net direction of fluid flow be from the capillary into the interstitial space. Nutrients are brought into contact with the tissue cells which selectively abstract those they need and send out waste chemicals such as carbon dioxide and excess water. The *net concentration* of dissolved solutes does not change, but the solutes themselves vary as nutrients are consumed and by-products are produced. The foregoing description constitutes a hypothesis formulated by E. H. Starling (1866–1927), one of the great British physiologists. He did not allow for "leakage" of proteins and lipids from the capillaries, but later work using radioactively labeled proteins and other techniques (notably by the Americans H. S. Mayerson, K. Wasserman, and S. J. LeBrie, who built on hypotheses by C. K. Drinker) has shown that these larger molecules do, in fact, routinely leave the bloodstream. Because it is very difficult for them to reenter the bloodstream directly, they return via lymph fluid.

Enormous amounts of fluid diffuse in the various capillary beds of the body within a short period of time. It has been estimated that in the entire capillary network of a 160-pound man, fluid diffusion proceeds at a *rate* of about 400 gallons

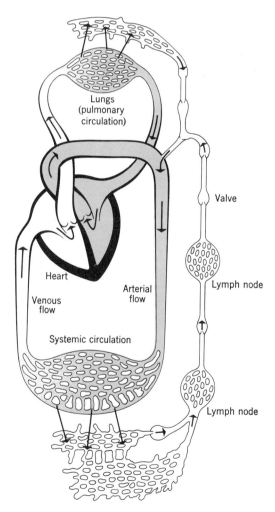

Lungs
(pulmonary
circulation)

Valve

Lymph node

Heart

Arterial
flow

Venous
flow

Systemic circulation

Lymph node

Figure 12.11  General features of the relations between lymph flow and the bloodstream. Oxygenated blood is shaded. Water and other substances leave the bloodstream at its capillaries, circulating around cells and seeping through cell walls. Fluids, unused nutrients, waste molecules, and the like return to circulation, partly by means of the lymph system and largely by passing directly through the wall of a blood capillary.

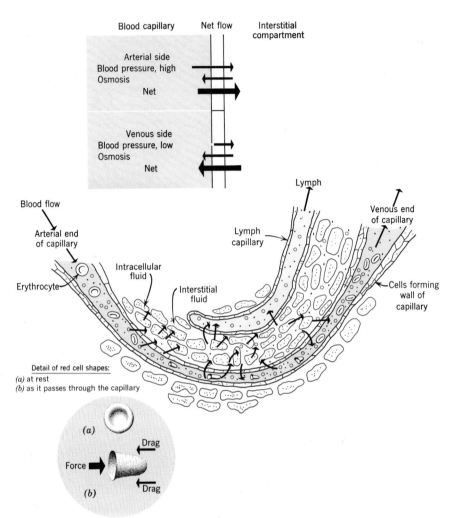

**Figure 12.12**  Fluid exchange at capillaries. Dots represent proteins, circles lipids. Erythrocytes (red cells) have the shape of a disk dented on opposite sides, but during their passage through the most constricted region of a capillary, their shapes are distorted as shown. This distortion is believed to aid somehow in the release of oxygen. As much as half the total protein in the blood leaves the bloodstream each day and returns via the lymph.

The schematic diagram at the top summarizes Starling's hypothesis, explaining how fluids leave the capillaries at their arterial sides and reenter at their venous sides. On the arterial side blood pressure counterbalances osmosis and dialysis, and the net movement is from capillary to interstitial space. Blood pressure drops as it goes through the constricted capillary; on the venous side it no longer overpowers osmosis and dialysis, and the net movement is reversed.

per minute. Delicate balances in fluid exchange at cells may be upset by important breakdowns in function.

**Shock.** The amount of proteins leaked through capillaries appears to increase when an individual has been subjected to sudden, severe injury, extensive burns, or even major surgery. Loss of protein from the blood upsets osmotic-pressure relations. Fluids forced out on the arterial side of a capillary loop do not return as they should. The blood volume drops, and the circulatory system no longer brings enough oxygen and nutrients to the cells. Those in the central nervous system are the most seriously affected, and a condition known as shock ensues. Shock may also result from extensive hemorrhage, for any failure of the circulatory system will cause it. Restoration of blood volume is mandatory for recovery from shock.

**Edema.** Fluids accumulating in the interstitial and cellular regions in abnormally large amounts produce a condition called edema. It may have one of several causes, only a few of which can be mentioned here.

If proteins leak through the kidneys into the urine being formed (albuminuria, p. 417), osmotic pressure relations may be upset and enough fluid retained in the interstitial compartment to produce noticeable swelling or edema in some tissues, notably those of the lower limbs. The swelling that usually accompanies malnutrition has similar causes. Prolonged malnutrition amounting to starvation will eventually result in less concentrated blood fluid, with the same result, edema.

Localized edema or swelling occurs around an injured area. The injury may have damaged either blood vessels or lymph capillaries, preventing normal drainage, and the fluid that accumulates produces the swelling. In a particularly ugly tropical disease, elephantiasis, small worms (filari) enter the body and live in the lymphatics, thereby reducing drainage through them. The swelling assumes grotesque proportions.

Obstructions in veins, as in certain cancers or in varicose veins, may cause a back pressure on the venous side of the capillary loops, hindering dialysis of interstitial fluids into veins. Again not quite as much fluid returns as enters the interstitial compartment, and swelling occurs.

## RESPIRATORY FUNCTIONS OF THE BLOOD

Moving oxygen from air to tissue cells is a task of enormous magnitude. If an adult generates 3000 kcal per day from his food, he requires 600 liters of oxygen, and he generates about 480 liters of carbon dioxide. The movement of these gases to and from air and cells is explained in Figure 12.13. The rate of this movement requires large surface areas across which contact between the blood flow and the inhaled air can be rapidly made. To provide this surface, the trachea ("wind pipe") undergoes a succession of branchings until the final, smallest functional unit, the individual *alveolus*, is reached. Incoming air is distributed to about 300 million alveoli, each one of which is little more than a bundle of fine capillaries suspended in air. An estimated 1500 miles of such exposed capillaries provide about 800 square feet of diffusing surface. Through these capillaries blood flows at a rate of about 5 liters per minute during resting periods; during severe exercise the rate is from 20 to 30 liters per minute.

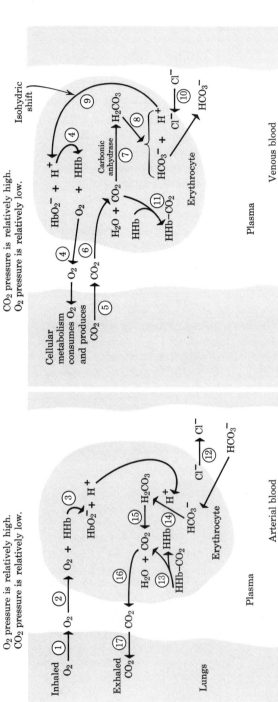

**Figure 12.13** Oxygen–carbon dioxide transport by the blood. Starting at the lungs with oxygen transport, the mechanics of breathing make oxygen pressure relatively high in the lungs. Oxygen is forced to diffuse into the bloodstream ①. Although blood may carry some dissolved gaseous oxygen, virtually all is drawn into erythrocytes ②, where it combines with hemoglobin, HHb, to form oxyhemoglobin, $HbO_2^-$, and $H^+$ ③:

$$HHb + O_2 \rightleftharpoons HbO_2^- + H^+$$

Two factors help shift this equilibrium from left to right in the lungs. The oxygen pressure is relatively high, and the hydrogen ion is neutralized by bicarbonate ion ⑭, something that must happen anyway if carbon dioxide is to be released. Thus arrival of bicarbonate ion helps draw oxygen into the cell. In cellular areas needing oxygen, two factors help in its release from $HbO_2^-$. Oxygen pressure is now relatively low, and carbon dioxide is poised to enter at ⑤ and ⑥. To release oxygen from $HbO_2^-$ ④, a hydrogen ion is needed ⑨; the influx of carbon dioxide makes it available in ⑦ and ⑧. If carbon dioxide is not available, the cell presumably does not need oxygen anyway. This shift of hydrogen ion ⑨ is called the *isohydric shift*. Bicarbonate ions leave the red cell during transport, and to replace their charge chloride ions move in ⑩. (This switch is called the *chloride shift*.) Most carbon dioxide is taken to the lungs as bicarbonate ion, but some combines with hemoglobin to be carried as carbamino hemoglobin ⑪, written as HHb–$CO_2$ for short. Very little moves simply as the gas dissolved in the plasma. When the erythrocyte returns to the lungs, a reversal of events, ⑫ to ⑰, sends carbon dioxide out into exhaled air and the cyclical process repeats itself.

**Oxygen Toxicity.** If an individual breathes pure oxygen, enough simply dissolves in plasma to supply nearly all the needs of tissues. Consequently, oxyhemoglobin has difficulty in dissociating at cells (④ in Figure 12.13). The isohydric shift ⑨ cannot therefore be used, and the hydrogen ions produced at ⑧, instead of being buffered by a hemoglobin molecule, tend to force the pH of the blood down. The resulting acidosis may be severe and very dangerous.

**Oxygen Deficiency. Anoxia.** Up to an altitude of about 18,000 feet, air pressure is high enough to provide the air needed, when inhaled into the lungs, to saturate erythrocytes with oxygen, provided little exercise is engaged in. At higher altitudes the oxygen pressure is not enough, and symptoms of oxygen deficiency—restlessness, confused thinking, and eventually unconsciousness ("blackout")—become evident. Portable oxygen or air tanks and pressurized cabins are familiar solutions to the problem.

Oxygen deficiency occurs in pneumonia, when the alveoli fill with fluids that hinder rapid exchange of gases. *Fibrosis* in alveoli, a condition which causes tissue to harden and become fibrous, also interferes with this exchange. Miners who breathe ore dusts and farmers who breathe organic dusts are taking risks that may eventually lead to lung failure. Certain organic substances in cigarette smoke have been implicated in lung cancer, the only known cure for which is surgical removal of the affected area. Only about 10% of such patients survive more than five years, unless the cancer is caught in its earliest stages. The lungs have some capacity for repair, and medical experts insist that those who smoke excessively should cut down or quit, however long they have been smoking.

Carbon monoxide is a poison that affects oxygen transport by combining with hemoglobin about 200 times more firmly than oxygen. In such a situation nearly pure oxygen must be administered to displace the molecules of carbon monoxide and to make up for oxygen deficiency.

## BLOOD CLOTTING

Blood must clot when it is shed, and it must not clot when it is not shed. These simply stated rules are a matter of life and death, and a complicated apparatus controls whether or not clotting occurs. Because the details are imperfectly understood, we confine our study to a broader view.

When blood clots, the soluble protein *fibrinogen* is converted to an insoluble protein *fibrin*, whose fibers aggregate in a matlike way to seal off the sites where the blood vessel has been broken or cut. The formation of a clot can be studied *in vitro*. In fact, special precautions must be taken to prevent freshly drawn blood from clotting. To change fibrinogen to fibrin, an enzyme is needed, but it can hardly be expected to be present in blood at all times. The key to the control of blood clotting is restricting the formation of this enzyme to the times when it is needed.

The clotting enzyme is called *thrombin,* and it exists in the bloodstream in an inactive form called *prothrombin.* To convert prothrombin to the active enzyme requires at least ten factors—calcium ion, antihemophilic factors, a coenzyme, and a group of molecules collectively called thromboplastin. Some are present in blood,

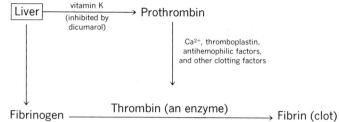

**Figure 12.14**   Probable mechanism of blood clotting. Both fibrinogen and prothrombin are made in the liver. Some clotting factors occur in blood, some in platelets, and some in cells that release them when cut or injured.

and others are released when tissues are cut. Thus the clotting mechanism is not thrown into operation unless a cut or injury occurs. When blood is removed from the body, it is not easy to prevent any and all contact with tissues slightly injured by the act of inserting a needle. Clotting under these circumstances can be prevented simply by tying up the calcium ions through the administration of sodium oxalate or sodium citrate. With the first sodium salt calcium precipitates as calcium oxalate; with the second salt calcium ions complex with citrate ions and are effectively removed.

Figure 12.14 illustrates the principal features of a probable mechanism for clot formation. Vitamin K deficiency, as in some hemorrhagic diseases, reduces the formation of the inactive prothrombin in the liver. The same effect can be achieved with dicumarol, a chemical formed in decaying sweet clover and sometimes given to prevent thrombosis, the unwanted formation of a clot within a blood vessel. Heparin, an extremely potent anticoagulant, is found in cells located near the walls of the tiniest capillaries. It acts by slowing the conversion of prothrombin to thrombin.

## BIOCHEMICAL INDIVIDUALITY

Many facets of nutrition have been studied in this chapter, but we shall leave to other references such matters as specific nutritional requirements and the best sources of nutrients. One topic, however, is inadequately covered in the literature. Roger J. Williams (University of Texas) deserves most of the credit for calling our attention to it, biochemical individuality.

Understanding of the molecular basis of life and health or of illness and disease is not the same thing as applying this knowledge to the control and cure of a metabolic disorder. One major emphasis of this book has been and will be the likenesses and similarities at the molecular level of the members of any given species. But obvious differences exist among human beings. External differences are completely apparent, but few people are aware of internal differences.

In an effort to simplify the study of any subject, we try to generalize. Such expressions as "a typical cell," "a normal adult," "a normal child," and an "average appetite" are concepts of some usefulness in the beginning stages of a serious study. Generalizations too often become dogmas, however, and when practical application is attempted to an *individual* case, any deviation from the normal is greeted with surprise at best or outrage at worst. Students of anatomy are very familiar with the huge variations in sizes and shapes of major internal organs among individuals who are outwardly of similar height and weight; the organs themselves may be quite healthy. A few examples of variations in whole organs point to the possibility of more fundamental biochemical differences.

The cross-sectional area of the esophagus can vary as much as fourfold. Some people are able to bolt their food; others cannot. Some people can swallow large pills with ease; others tend to choke on them. Some people easily admit a stomach tube; in others it is inserted with great difficulty. How pointless to blame the patient's uncooperativeness (although it may sometimes be a factor).

Human stomachs exhibit considerable variety. In fact, from an anatomical point of view, there is no such thing as a normal stomach. Stomachs vary so much in size and shape that some can hold from six to eight times as much food as others. (One wonders how many children have had to endure anatomically unrealistic stomach stuffings by well-meaning but uninformed mothers!)

The liver participates in several important metabolic pathways. Because its shape, its size, and its location vary from individual to individual, we suspect that its molecular processes may not be the same in each person.

Williams cites organ after organ whose anatomical variations can be quite marked in different individuals. He notes, for another example, that there are so many variations in the circulatory system—the heart, the arteries, the veins, and the capillaries—that it is difficult to draw a line between "congenital abnormalities" and "unimportant differences." From individual to individual the same specific artery can vary as much as three times in cross-sectional area, which means that blood-carrying ability has the same range. Considering the importance of an adequate circulation, such variations may have a profound influence on events at the molecular level in every tissue of the body, and hence on the health and feeling of well-being of every individual involved. People who are prone to fainting and dizziness may just be experiencing poor circulation to the brain, which may in turn be caused by arteries of small cross-sectional area, a matter of heredity.

The variations cited and documented by Williams are of hereditary origin. As a tentative theory, a provisional principle, Williams suggests the *principle of genetic gradients:* whenever an extreme genetic character is discovered in an individual, a less extreme manifestation of this character probably exists in other individuals. Thus if one person is found to have virtually no sensitivity to pain, we should expect this characteristic to range in a large population from very little sensitivity, to moderate sensitivity, all the way to extreme sensitivity. If one person has very low thyroid activity, the principle of genetic gradients leads us to suspect that in a large enough population thyroid activity from individual to individual varies gradually from one extreme (low activity) to the other. Too often we tend to think that only three possibilities exist, abnormally low, normal, and abnormally high. Williams considers an entire continuum of activity much closer to the truth.

The applicability of Williams' principle of genetic gradients to molecular biology is evident in the field of nutrition. Although there are many body likenesses, to be sure, each human being has a very distinctive chemistry. Williams warns that differences are ignored only to our peril. We tend too to think that malformation, vulnerability to infection and disease, and rate of general degeneration are all set irrevocably during the genetic lottery that occurs at conception. But the genes and the enzymes they beget cannot function by themselves. Chemical nutrients are needed and must be provided in the diet. Attention to individual needs by means of nutrition and diet (in other words, environmental factors as opposed to genetic factors) can yield important health-giving results. Williams has postulated a further principle incorporating this idea, the *genetotrophic principle:* each individual organism, because of its distinctive genetic background, has distinctive nutritional needs that must be satisfied for maximum well-being.

Heredity cannot be ignored; we simply would not *be* without it. Environment cannot be ignored; we simply could not *become* without it. The genetrophic principle implies that unless we know the uncommon nutritional needs imposed by our heredity we cannot do anything about them, which seems perfectly obvious. But all too few are even aware that uncommon nutritional needs can be imposed by heredity, and those who are aware seldom realize how many people have these special needs. A great deal of research in nutrition has focused on what *all* people require. Such research is clearly important, and more is needed. Williams' work should alert us to the fact that research in nutrition must also be directed to identifying individual needs.

# REFERENCES AND ANNOTATED READING LIST

### BOOKS

I. N. Kugelmass. *Biochemistry of Blood in Health and Disease.* Charles C Thomas, Springfield, Ill., 1959.

J. L. Gamble. *Chemical Anatomy, Physiology and Extracellular Fluid,* sixth edition. Harvard University Press, Cambridge, Mass., 1954.

A. White, P. Handler, and E. L. Smith. *Principles of Biochemistry,* third edition. McGraw-Hill Book Company, New York, 1964.

R. J. Williams. *Biochemical Individuality: The Basis for the Genetotrophic Concept.* John Wiley and Sons, New York, 1956.

### ARTICLES

H. Neurath. "Protein-Digesting Enzymes." *Scientific American,* December 1964, page 68. By studying the molecular structures of certain digestive enzymes, Neurath and his group at the University of Washington hope to discover clues to the actions of all enzymes.

D. M. Surgenor. "Blood." *Scientific American,* February 1954, page 54. The author describes the functions of the blood, discusses its preservation and handling, and states the problems attendant on separating its components.

J. E. Wood. "The Venous System." *Scientific American,* January 1968, page 86. Veins are more than tubes. They are flexible reservoirs for blood.

H. S. Mayerson. "The Lymphatic System." *Scientific American*, June 1963, page 80. After discussing the anatomy of the lymphatic system and its general functions, Mayerson describes experiments he and others have performed to prove that it is normal for capillaries to "leak" proteins and lipids.

J. H. Comroe, Jr. "The Lung." *Scientific American*, February 1966, page 57. The anatomy of the respiratory system is described, and pulmonary circulation and the mechanics of breathing are discussed.

H. W. Smith. "The Kidney." *Scientific American*, January 1953, page 40. After a well-illustrated discussion of the structure and function of the kidney, the author speculates on its evolution.

J. P. Merrill. "The Artificial Kidney." *Scientific American*, July 1961, page 132. While discussing the problems of devising an artificial kidney, the author explains normal kidney function.

K. Laki. "The Clotting of Fibrinogen." *Scientific American*, March 1962, page 60. How the enzyme thrombin works is discussed in detail.

L. Pauling. "Orthomolecular Psychiatry." *Science*, Vol. 160, 1968, page 265. Biochemical individuality is probably nowhere more significant than in the brain. As Pauling marshals evidence for this supposition, he proposes that the methods used in treating patients with mental disease—psychotherapy, chemotherapy, shock therapy—be augmented by another general method, orthomolecular psychiatric therapy. He defines this as treatment that provides the optimum molecular environment for the mind, especially the optimum concentrations of chemicals normally present in the human body.

# PROBLEMS AND EXERCISES

1. Define each of the following terms.

| | | |
|---|---|---|
| (*a*) internal environment | (*b*) extracellular fluids | (*c*) interstitial fluids |
| (*d*) plasma | (*e*) serum | (*f*) defibrinated blood |
| (*g*) red cell count | (*h*) leukocytes | (*i*) lymphocytes |
| (*j*) thrombocytes | (*k*) isohydric shift | (*l*) thrombin |
| (*m*) prothrombin | (*n*) anoxia | (*o*) α-amylase |
| (*p*) parietal cells | (*q*) chief cells | (*r*) mucous cells |
| (*s*) gastrin | (*t*) pepsinogen | (*u*) zymogen |
| (*v*) pepsin | (*w*) proteolytic enzymes | (*x*) proteoses |
| (*y*) peptones | (*z*) mucoprotein | |

| | | |
|---|---|---|
| (*aa*) gastric hyperacidity | (*bb*) enterogastrone | (*cc*) chyme |
| (*dd*) pyloric valve | (*ee*) duodenum | (*ff*) gastric achlorhydria |
| (*gg*) gastric hypoacidity | (*hh*) secretin | (*ii*) pancreozymin |
| (*jj*) enterokinase | (*kk*) cholecystokinin | (*ll*) steapsin |
| (*mm*) active transport | (*nn*) diuresis | (*oo*) glomerular filtrate |
| (*pp*) diabetes insipidus | (*qq*) acidosis | (*rr*) alkalosis |
| (*ss*) BUN | (*tt*) NPN | (*uu*) uremia |
| (*vv*) albuminuria | (*ww*) edema | (*xx*) alveolus |
| (*yy*) thrombin | (*zz*) prothrombin | |

2. Explain how proteins of the stomach wall are protected from being digested.

3. Describe the function of each of the following.

(a) trypsin       (b) bile salts       (c) serum globulins
(d) serum albumins       (e) fibrinogen       (f) vasopressin
(g) aldosterone       (h) acidification of the urine       (i) lymph
(j) lymph nodes

4. Explain how the flow of each of the following is controlled.

(a) saliva       (b) gastric juice       (c) pancreatic juice
(d) bile       (e) vasopressin

5. How does carbon monoxide act as a poison?

6. Alcohol in blood suppresses the secretion of vasopressin. How does this affect diuresis?

7. In anemia the supply of hemoglobin is low. How does this affect respiration?

8. Referring to this chapter and to previous ones, what are the three main buffers in blood? Show equations that illustrate *how* they work. Explain *why* they *must* work.

9. Explain how malnutrition will upset osmotic pressure relations between blood and the interstitial fluids.

10. Explain how the kidneys act to reduce developing acidosis.

11. Explain how fluid exchange occurs between plasma and cells? (How does arterial blood become venous in composition?)

12. Describe ways in which edema may originate.

13. What is the principle of genetric gradients?

14. What is the genetotrophic principle?

15. Explain how the $H^+$ needed to help release $O_2$ from $HbO_2{}^-$ is produced precisely where it is needed.

16. Explain how $H^+$ is tied up at precisely the right point in order to assist the uptake of $O_2$ as $HbO_2{}^-$.

17. Explain in your own words how a blood clot forms.

18. Why is it unwise to breathe pure oxygen if your respiratory organs are functioning properly?

19. How does the administration of vitamin K assist the clotting mechanism?

# CHAPTER THIRTEEN

# Energy for Living

"In the sweat of your face you shall eat bread until you return to the ground" said an ancient writer, "for out of it you were taken; you are dust, and to dust you shall return" (Genesis 3:19).

This gloomy commentary on the human predicament recognized our kinship with nature and its laws. All our ancestors, including our parents, dined on plants and on animals which fed on plants. Indirectly we are indeed made of dust, born out of stable, randomly dispersed chemicals such as carbon dioxide, water, and a few mineral substances which plants can convert into foods with the indispensable help of energy from the sun.

Sooner or later the stuff of our bodies will become these common things again, but there is a marvelous interlude between dust and dust, and we call it life. However long or briefly it endures, it is possible only as long as the organism can maintain itself in a condition quite remote from the ultimate equilibrium. The unrelenting tendency in nature to achieve universal equilibrium is easily life's most implacable foe. Universal equilibrium is the impossibility, or the extreme improbability, of any further *net* change—the existence of maximum disorder and maximum chaos, the universal "sea level" of existence, dust and ashes scattered to the four winds. A living organism, by comparison, is a highly organized, a finely ordered, and a relatively high-energy state of affairs. To keep it that way, even temporarily, requires information, energy, and "spare parts." We obtain energy and spare parts from the chemicals, from the foods in our diet and the air that we breathe. We obtain information genetically, as discussed in Chapter 17, and environmentally, as studied systematically in other courses. The emphasis in this chapter will be on the *energy* we receive, how we obtain it, how we store it, how we transform it, and how we use it.

By now we should be somewhat used to the notion that we can obtain energy from chemicals. We know that certain combinations or mixtures undergo changes that are accompanied by the evolution of energy. Sometimes the change is little more than an explosion, but if it is made to occur in certain ways, objects are use-fully pushed—ore in a mine, a cylinder in a car engine, a rocket engine. We are also accustomed to the idea that chemical changes are rearrangements or redistributions of electrons in relation to atomic nuclei. With a little clever engineering we can sometimes make this electron "flow" occur through circuits, through motors or heaters or light bulbs, and useful work is done.

Because chemicals can be made to do work for us, whatever in chemicals makes this work possible is important enough to have a name. We call it chemical energy. We do not say that the chemicals have *work,* because we use this word when we wish to talk about energy actually being expended *in a purposeful way.* The chemicals simply have *a potential for doing work,* and thus chemical energy is a special kind of potential energy. By a series of chemical changes which we are about to study, we use the chemical energy in the foods we eat and the air we breathe to synthesize members of a special family of energy-rich substances, the organophosphates. The most important member of this family is adenosine triphosphate or ATP.

### ADENOSINE TRIPHOSPHATE. ATP

Adenosine triphosphate is simultaneously an ester, an acid, and an anhydride. In the anhydride part of the structure, the system of primary interest to us, are two bonds customarily drawn as squiggles, $\sim$. These are the bonds most directly involved in reactions that convert chemical energy into other forms within living cells.

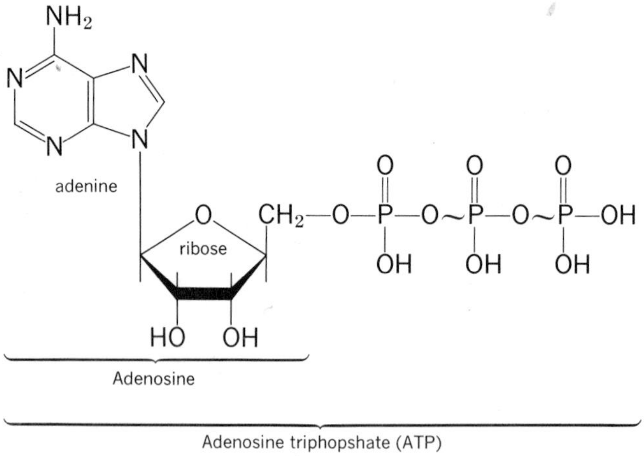

Adenosine triphopshate (ATP)

Thus the energy needed for muscular work comes from a reaction of ATP bound to protein molecules in a muscle. Although a detailed mechanism was discussed earlier (pp. 351 to 360), a very simplified statement of muscle contraction is given by the equation

$$\text{Relaxed muscle—ATP} \longrightarrow \text{contracted muscle—ADP} + P_i$$

Some heat is also evolved, but not nearly as much as would have been formed had ATP been converted to ADP and $P_i$ by reacting directly with water.

**The Reaction of ATP with Water.** Although the hydrolysis of ATP is not a useful reaction in a living cell, it can be made to occur. The products produced by hydrolysis, adenosine diphosphate (ADP) and inorganic phosphate ($P_i$), are the same as those produced when ATP reacts to release its chemical energy for cellular work. Therefore studies of energy changes for the hydrolysis of ATP *in vitro* provide data for estimating energy changes for other reactions of ATP *in vivo*. Of the several ways by which the ATP molecule can be hydrolyzed, the following are important as sources of energy:

Adenosine diphosphate (ADP)          Phosphoric acid ($P_i$)

Adenosine monophosphate (AMP)          Diphosphoric acid ($PP_i$)

Hydrolysis at position 1 is by far the most common. The exact amount of energy this reaction makes available depends on several factors, the pH of the medium, its temperature, and the concentrations of reactants.

Because the organophosphates have ionizable hydrogens, the pH factor is important. In fact, the exact state of ionization of these compounds depends on the pH of the medium. Phosphoric acid, $H_3PO_4$, in a medium of pH 5, for example, exists almost entirely in the form of the dihydrogen phosphate ion, $H_2PO_4^-$. In a slightly alkaline medium, pH 8, phosphoric acid has become mostly the monohydrogen phosphate ion, $HPO_4^{2-}$. Between pH 5 and 8 there are virtually no concentrations of the fully ionized phosphate ion, $PO_4^{3-}$, or of the unionized acid, $H_3PO_4$. In this range of pH inorganic phosphate exists as a mixture of $H_2PO_4^-$ and $HPO_4^{2-}$. Thus at physiological pH (7.3 as measured at room temperature) phosphoric acid and its relatives, diphosphoric acid, AMP, ADP, and ATP (and many others), will be in various states of ionization. To simplify the writing of equations, all these considerations will be assumed, and such symbols as $P_i$ (for inorganic phosphate), $PP_i$ (inorganic diphosphate), AMP, ADP, and ATP will stand for the substances, whatever their states of ionization. In a solution of pH 7 having a temperature of 37°C (body temperature) and concentrations of reactants approximately those within cells, the estimated amount of energy available for *useful* chemical work from the conversion of ATP into ADP and $P_i$ is 9000 calories per mole of ATP.

**Free Energy.** In an energy-releasing reaction the total energy liberated is not all convertible into useful work. Some is wasted, unavoidably so. To describe how much energy is made available for *useful* work by a spontaneous reaction, scientists have developed the concept of *free energy,* so-called because it is just that, free for our use if we can harness it. This concept is most easily developed for reactions taking place in two sets of circumstances, at constant temperature and constant volume and at constant temperature and constant pressure. The second set of conditions, constant temperature and pressure, is very closely approximated within living systems; and for events taking place under these conditions the *change* in free energy in going from reactants to products is given the symbol $\Delta G$. (The Greek capital delta, $\Delta$, is a symbol for "change in"; $G$ honors Josiah Willard Gibbs (1839–1903), one of the greatest scientists in American history. The free energy change at constant temperature and pressure is very commonly called the *Gibbs free energy,* or sometimes simply the *Gibbs function.*[1])

By the almost universally accepted convention for all thermodynamic properties, the value of $\Delta G$ is given a *minus* sign when the free energy is *released* by a system undergoing a change. If a system is made to undergo a change by applying energy from the outside, the system's change in free energy is given a plus sign. For any spontaneous natural process occurring under conditions of constant temperature and pressure, the change in the Gibbs free energy is negative. Moreover, the value of $\Delta G$ under such conditions is the *maximum* amount of useful energy we could conceivably obtain, free, from that change. How successfully or how completely the maximum yield of free energy will be harnessed depends on the *process* whereby the change occurs. For example, a great deal of energy can be obtained from the *change* of glucose and oxygen to carbon dioxide and water by the *process* of direct combustion, but it is mostly heat energy and some energy produced by the expansion of gases. The body cannot harness either form of energy, for it is not an internal combustion engine. Causing this same change in glucose and oxygen to take place by another process—in a long series of successive chemical changes, each one representing quite minor molecular surgery—the body traps about 50% of the energy released, using some of it for the synthesis of organic triphosphates such as ATP and a few other energy-rich compounds. Before studying how this is done, we shall examine some speculations about the remarkable position occupied by the triphosphates, particularly ATP, in the metabolism of nearly all forms of life.

**The Energy Richness of ATP and Other Triphosphates.** Adenosine triphosphate, by far the most widely distributed of the high-energy phosphates, is found in the lowest and highest forms of both plant and animal life. For the changes that are believed to have occurred during evolution, ATP was at a very early stage selected as the almost universal medium of energy transactions. It has a prominence not shared by other high-energy compounds. Carboxylic acid anhydrides (as well as acid chlorides) are also high-energy compounds. Fritz Lipmann, one of the pioneers in focusing the attention of scientists on free energy factors in biochemical changes, suggested that the phosphate system, rather than anhydrides of carboxylic acids (or acid chlorides), is "selected" by living organisms, but not because it is more (or

---

[1] The student is warned that another symbol is commonly used for the Gibbs free energy, namely $\Delta F$. The symbol $\Delta G$ is recommended by the International Union of Pure and Applied Chemistry.

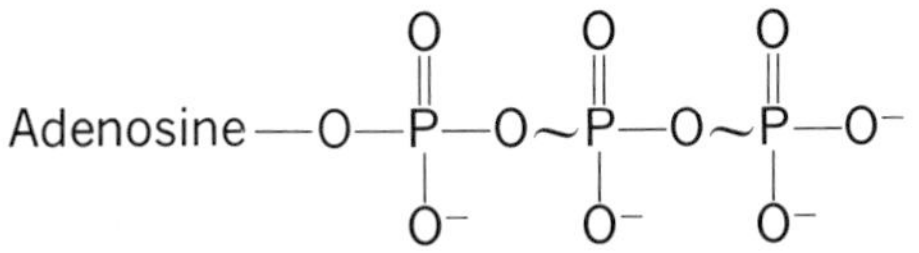

Adenosine—O—P—O~P—O~P—O⁻         R—C—O~C—R

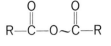

Phosphoric anhydride system
in ATP

Anhydride of a
carboxylic acid

less) energy-rich. It is not. Rather the phosphate system reacts less rapidly with water at physiological pH conditions. The reaction of ATP with water under physiological conditions proceeds slowly enough that there is no serious leakage of free energy.

Fritz Lipmann (1899–      ). United States biochemist. In 1953 he shared the Nobel prize in physiology and medicine with Hans Krebs for his discovery of coenzyme A. He was born in Germany, but he came to the United States and eventually became a citizen in 1944.

Bernard and Alberte Pullman of France have provided insights into the source of the energy wealth of the triphosphate system. By quantum mechanical calculations they have been able to assign numerical values to the partial charges in this network.[2]

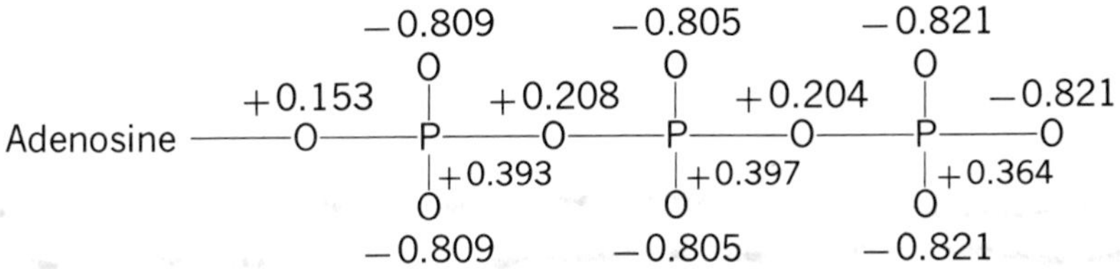

```
                    −0.809        −0.805        −0.821
              +0.153   O    +0.208   O    +0.204   O    −0.821
Adenosine ———————O————P————O————P————O————P————O
                      |+0.393       |+0.397       |+0.364
                      O             O             O
                    −0.809        −0.805        −0.821
```

ATP

The net charge, the algebraic sum of the partial charges, is −3.972 or, roughly, −4. The structure shown therefore refers to the quadruply ionized or maximally ionized ATP molecule. The significant feature of these results is that like charges are nearest neighbors. All along the chain O—P—O—P—O—P there is a succession of partial positive charges, and like charges repel. Around the oxygen-populated outskirts is a series of partial negative charges. As long as this molecule hangs together, there is an internal tension. Some relief can be expected by shortening that chain. Even by losing one phosphate unit, some reduction in repulsive forces and some net gain in stability will be expected—and not just for the reason already suggested. Total internal energy will be further reduced by the increased resonance stabilization possible in the products (ADP and phosphate) and by increased

[2] A. Pullman and B. Pullman, "π Molecular Orbitals and the Process of Life," in *Molecular Orbitals in Chemistry, Physics, and Biology,* edited by P. Löwdin and B. Pullman, Academic Press, New York, 1964, page 561.

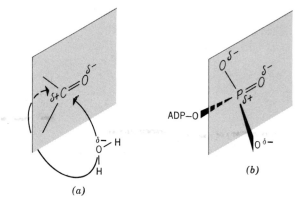

(a)

(b)

**Figure 13.1** Speculation about the lower reactivity of ATP toward water when compared with those of anhydrides and acid chlorides derived from carboxylic acids. Even though these three are "high-energy" systems, the phosphoric acid anhydride network is less reactive. (a) Attack by a nucleophile such as water at the carbonyl carbon of an anhydride or acid chloride derived from a carboxylic acid is made relatively easy by the openness of the approach from either side and by the partial positive change at this site. (b) Attack by a nucleophile at the phosphorus of a terminal phosphate unit of ATP is hindered because less space is available through which to make an approach, and the neighboring oxygens have partial negative charges that tend to repel an electron-rich nucleophile.

opportunities for the hydration of the ions. Yet, as we have said, ATP does not react rapidly with water under physiological conditions, especially when compared with other high-energy systems such as acid chlorides and anhydrides. To account for this lower rate we might seek a steric explanation. As depicted in Figure 13.1, the "least-resistance" or lowest-energy approach by a nucleophile such as water to the carbonyl carbon of a carboxylic acid anhydride or chloride is along a line somewhat perpendicular to the plane in which the carbonyl carbon and its three attached groups lie. The four groups attached to the phosphorus of the terminal phosphate unit in ATP, however, do not lie in a plane. The arrangement is tetrahedral as in methane and is therefore more crowded. Moreover, for the nucleophile to approach the phosphorus from almost any angle, it must pass through a region of negative charge density created by the oxygens.

We are faced with the obvious conclusion that the *energy of activation* for hydrolysis of ATP must be *high* if the *rate* for it is *low*. And this conclusion may be interpreted as meaning that of all the collisions or potential collisions between water molecules and the terminal phosphorus in ATP, only a very small fraction, those with potentially the highest collision energy, are successful. This fraction of fruitful collisions is dependent on obstacles in the path of the two collision centers, the terminal phosphorus in ATP and the oxygen in water. There are more obstacles in the ATP system, but the energy payoff for a successful collision is still high. Thus ATP is apparently uniquely adapted to be the coinage for a large number of energy exchanges in the cellular realm. Foreign molecules bearing free energy (e.g., glucose, fatty acids) must have it converted into this form. We are now in a better position to study how this conversion takes place.

# BIOENERGETICS

## RESPIRATION

In everyday usage respiration means breathing, but scientists have found it convenient to give the word other meanings. In this section the word describes the chemical events most closely connected with the reduction of molecular oxygen to water. Reference will be made, for example, to "respiratory enzymes" and to the "respiratory chain," a series of chemical events leading to oxygen consumption. We intend to illustrate in some depth the connection between the synthesis of ATP and the uptake of atmospheric oxygen.

Our study of the molecular basis of muscular work began in Chapter 10, and proceeded as far as connecting mechanical work with the conversion of ATP to ADP and $P_i$. The rest of the story is the use of oxygen and some nutrient (e.g., glucose or fatty acids) to restore the ATP. The reduction of oxygen to water requires for each oxygen atom a pair of hydrogen ions and a pair of electrons, or in its barest statement,[3]

$$:\overset{..}{O} + : + 2H^+ \longrightarrow H:\overset{..}{O}:H$$

Viewed this way, the problem is threefold: (1) how to obtain a pair of electrons from an organic molecule and transport them to oxygen, (2) how to obtain the hydrogens, and (3) how to deliver the large free energy associated with the reduction of oxygen at least partly into storage as ATP.

## MAJOR OXIDATIVE PATHWAYS. AN OVERVIEW

Two of our foods, the carbohydrates and lipids, provide us with virtually all our energy for living. Each follows its own pathway for several steps, and then they converge into two of the most important metabolic sequences in the body, the *citric acid cycle* and the *respiratory chain* (Figure 13.2). Before studying each in detail, let us first describe them in a general way.

**Glycolysis.** The word may be thought of as derived from "glycogen loosening" (Greek *lysis*, act of loosening) or from "glycose loosening" (glycose is a generic name for any of the isomeric aldohexoses, including glucose). Glycolysis is a series of reactions in the body by which glucose is broken down to lactic acid. Because oxygen is not required, the reactions are sometimes called the *anaerobic sequence*. The sequence accomplishes two things: it powers the synthesis of a small amount of ATP, and it yields a chemical that can be passed eventually to the citric acid cycle (leading to more ATP). Leaving the details for a later chapter, we find that the over-

---

[3] We do not believe of course that electrons are really available in the way indicated by this purely hypothetical equation. The equation merely serves to trim the problem to its minimum statement.

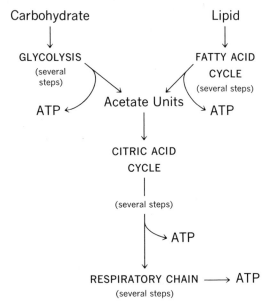

**Figure 13.2** Major oxidative pathways in the body.

all change balances as follows:

$$C_6H_{12}O_6 + 2ADP + 2P_i \xrightarrow{\text{glycolysis}} 2CH_3\underset{\underset{OH}{|}}{C}HCO_2H + 2ATP$$

Glucose           Lactic acid

**Fatty Acid Cycle.** In this series of reactions a fatty acid molecule has its chain shortened two carbons at a time. The two-carbon fragments, acetyl units, are fed into the citric acid cycle leading to ATP production. The shortened fatty acid goes around the fatty acid cycle again and again until it has no more two-carbon units to lose. The cycle itself produces considerable ATP in its many turns, but it does so indirectly through the participation of the respiratory chain. The details will be given in Chapter 15. The provision of a raw material, acetyl units, for the citric acid cycle interests us here.

**Citric Acid Cycle.** In this sequence a four-carbon carrier molecule picks up a two-carbon acetyl unit from either glycolysis or the fatty acid cycle. The resulting six-carbon molecule, citric acid, is successively degraded until two carbons have been broken off as carbon dioxide, hydrogen and electrons have been fed to the respiratory chain for ATP production, and the four-carbon carrier molecule has been regenerated. The cycle is then ready for another "turn."

**Respiratory Chain.** The chain may be regarded as a series of reactions, as a series of carrier molecules, or as both, depending on the context of the discussion. It is the mechanism for delivering the elements of hydrogen (H:⁻ and H⁺) to molecular oxygen taken in by breathing. Water, of course, is the end product. The chief purpose of the chain is to tap some of the energy potentially available from reducing oxygen to water and to use this energy in powering the production of ATP.

**ATP Production from the Major Oxidative Pathways. A Summary.** The overall equation for the oxidation of glucose in the body is

$$C_6H_{12}O_6 + 6O_2 + 36ADP + 36P_i \longrightarrow 6CO_2 + 42H_2O + 36ATP$$

This may be regarded as the sum of the following reactions:[4]

| | | |
|---|---|---|
| $C_6H_{12}O_6 + 6O_2 \longrightarrow 6CO_2 + 6H_2O$ | $\Delta G = -680,000$ cal | |
| $36ADP + 36P_i \longrightarrow 36ATP + 36H_2O$ | $\Delta G = +324,000$ cal | |
| Net | $\Delta G = -356,000$ cal | |

Considering the amount of free energy available (680,000 cal) and the amount actually used (324,000 cal), the efficiency of the operation, in terms of conserving free energy, is

$$\frac{324,000}{680,000} \times 100 = 48\%$$

Thus when the body uses glucose to obtain energy (as ATP), roughly 48% of what is potentially available is actually conserved.

The complete oxidation of a typical fatty acid, palmitic acid, and the simultaneous production of ATP proceeds by the equation

$$CH_3(CH_2)_{14}CO_2H + 23O_2 + 130ADP + 130P_i \longrightarrow$$
$$16CO_2 + 146H_2O + 130ATP$$

This breaks down as follows:

| | | |
|---|---|---|
| $CH_3(CH_2)_{14}CO_2H + 23O_2 \longrightarrow 16CO_2 + 16H_2O$ | $\Delta G = -2,340,000$ cal | |
| $130ADP + 130P_i \longrightarrow 130H_2O + 130ATP$ | $\Delta G = +1,170,000$ cal | |
| Net | $\Delta G = -1,170,000$ cal | |

The efficiency is

$$\frac{1,170,000}{2,340,000} \times 100 = 50\%$$

That is, of all the free energy potentially available from the oxidation of palmitic acid, the body taps 50% of it to make ATP.

**Regulation of ATP Production. Homeostasis.** Many factors must be present to make ATP production possible. Organic nutrients are needed to provide electrons and hydrogens for the reduction of oxygen; oxygen, of course, is needed; ADP and $P_i$ must be available; all the enzymes must be functioning. Let us therefore assume that the subject is healthy, his diet has been adequate, nutrients are available, and he is breathing normally. Suppose that he has been resting and is completely

---

[4] Values of $\Delta G$ are for the changes occurring under physiological conditions of temperature, pressure, pH, and concentrations. Because we do not know with certainty what these conditions are within cells and tissues, the values of $\Delta G$ given throughout this book are estimates that have been proposed in some references but not in all.

relaxed. It is at least probable that his ATP production will not be as rapid as that of an identical subject who happens to be exercising, for the synthesis of ATP is tied to the demand for it.

For the reaction

$$ADP + P_i \longrightarrow ATP$$

we must think in terms of the ratio

$$\frac{[ATP]}{[ADP][P_i]}$$

When this ratio is high, there is probably not much tendency to make it even higher, that is, not much tendency to make more ATP. Conversely, when the ratio is small, the time is ripe to make fresh ATP. When this ratio drops, because ATP has been used in some energy-demanding process, reactions start spontaneously to bring it back to normal. The response works like a thermostat, a device for translating information (room temperature has dropped) into action (the furnace is turned on) until new information (the temperature comes back up) shuts the system off again. In engineering this reponse is called an inverse feedback mechanism, in molecular biology *homeostasis*. Walter Cannon coined this word for the behavior of an organism toward stimuli (e.g., an energy-demanding event uses up ATP and the ratio changes) that commences a series of metabolic events (e.g., respiratory chain) to restore the system to normal (e.g., produce ATP to replace what was used). Several conditions in the human body are so constant that diseases can be diagnosed if they change measurably. Body temperature is perhaps the most obvious. The pH of the blood, its white cell count and red cell count, and concentrations of various dissolved substances are others. Mechanisms that operate to keep these constant are called homeostatic.

---

Walter Bradford Cannon (1871–1945). Professor of physiology at Harvard from 1906 to 1942, Cannon was one of the greatest physiologists of the United States.

---

Cellular respiration is stimulated not so much by the presence of oxygen as by some activity (e.g., muscular work) that uses up ATP and generates ADP plus $P_i$. A rise in concentration of ADP and $P_i$ triggers reactions that use them, reactions that lead to ATP. We therefore begin our more detailed study of bioenergetics with the sequence that stands closest to actual oxygen consumption, the respiratory chain.

## THE RESPIRATORY CHAIN

Of the many major metabolic sequences in the body, our knowledge of the respiratory chain is still the most sketchy. The following discussion is a reasonable approximation of the state of our knowledge in this area at the present time.

The respiratory enzymes are quite well known and understood. The sequence of electron transport from enzyme to enzyme is also believed to be fairly well known, although our discussion will not include all the details. (There will be, in fact, considerable simplification, and it is acknowledged that what is simplification for one is oversimplification for another or vexatious difficulty for still another.) The coupling of the flow of electrons down the respiratory chain of enzymes to the conversion of ADP and $P_i$ into ATP is the most speculative part of the explanation. We shall be talking about coupling enzymes and transfer enzymes, but thus far they are as much idea as hard fact. Evidence exists, to be sure, but a goodly portion of what follows is just one plausible theory that explains many aspects of *oxidative phosphorylation*—that is, the making of ATP by using energy from oxidation. We shall begin with the aspect of the respiratory chain that is best understood, its enzymes.

## RESPIRATORY ENZYMES AND COENZYMES

**Mitochondria.** The enzymes whose coenzymes are to be described are organized primarily in membranes of mitochondria (Greek: *mitos,* a thread; *chondros,* a grain). Frequently called the "powerhouses of cells," they are the primary sites for oxidative phosphorylation. Individual cells can have hundreds, sometimes thousands, of these cigar- or sausage-shaped mitochondria (Figure 13.3). The inside surface of the inner membrane and the outside surface of the outer membrane of the mitochondrion are covered with thousands of tiny particles, each one an elementary unit of mitochondrial activity. A single mitochondrion in the flight muscle of a wasp, for example, has as many as 100,000 such units—a reflection of the exceedingly intense energy demands within this particular tissue. The particles on the outside membrane of the mitochondrion accept electrons and hydrogens from suitable donors which we generally call metabolites.[5] These particles also catalyze reactions powered by ATP. The inner particles convey the electrons along the chain of respiratory enzymes until ATP is made and atmospheric oxygen is reduced to water. These enzymes make up about 15% of the proteins of the membranes. They are therefore probably important not only to cell *function* but also to mitochondrial structure.[6] The enzymes of the citric acid cycle are also associated with mitochondrial structure. Thus mitochondria can receive metabolites, ADP, $P_i$, and oxygen, and they can discharge carbon dioxide, water, and ATP.

**Coenzymes of the Respiratory Chain.** Only four types of coenzymes or prothestic groups—pyridine nucleotides, flavin nucleotides, ubiquinones, and cytochromes—participate intimately in the enzymes of the respiratory chain. We may ignore the

---

[5] We use this word to mean an intermediate that is produced by chemical action on some *nutrient* (any product of digestion) and will eventually be used for another chemial change. The word metabolite therefore arbitrarily excludes enzymes and final products such as urea, water, and carbon dioxide.

[6] This fact accounts for the major difficulty in unraveling the workings of the respiratory chain. Experiments designed to isolate its enzymes and coenzymes disrupt their cooperative existence. When separated from each other, they sometimes fail to give the reactions of the intact mitochondrion. Furthermore, deducing what happens *in vivo* from activity *in vitro* is always open to question.

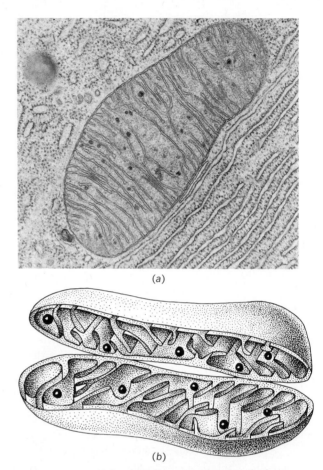

(a)

(b)

**Figure 13.3** A mitochondrion. (*a*) An electron micrograph of a mitochondrion in a pancreas cell of a bat (×53,000). (Courtesy of Dr. Keith R. Porter.) (*b*) Perspective drawing of an "opened" mitochondrion. The dots represent units of mitochondrial activity. Larger granules of uncertain composition are also present. (Courtesy of G. E. Nelson, G. G. Robinson, and R. A. Boolootian, *Fundamental Concepts of Biology,* John Wiley and Sons, New York, 1967, page 40.)

protein portions of these enzymes, the apoenzymes that give binding site uniqueness to the individual catalysts, and study only the respiratory coenzymes.

**Pyridine Nucleotides. NAD$^+$ and NADP$^+$.** A *nucleotide* is a compound made from a molecule of phosphoric acid, plus a five-carbon sugar, plus one of about half a dozen heterocyclic amines. In addition to being building blocks for genes (cf. p. 544), they are parts of several coenzymes. Adenosine monophosphate (AMP) is a typical nucleotide (next page).

One of the important coenzymes of the respiratory chain is a derivative of AMP, *nicotinamide adenine dinucleotide* (NAD$^+$). Its structure consists of two nucleotides joined together at the phosphate groups. One nucleotide has adenine as its side chain amine, the other nicotinamide, one of the vitamins. Only the shaded portion of the complicated structure, where the catalytic activity occurs, concerns us now, although the rest is essential in ways that are not very well understood.

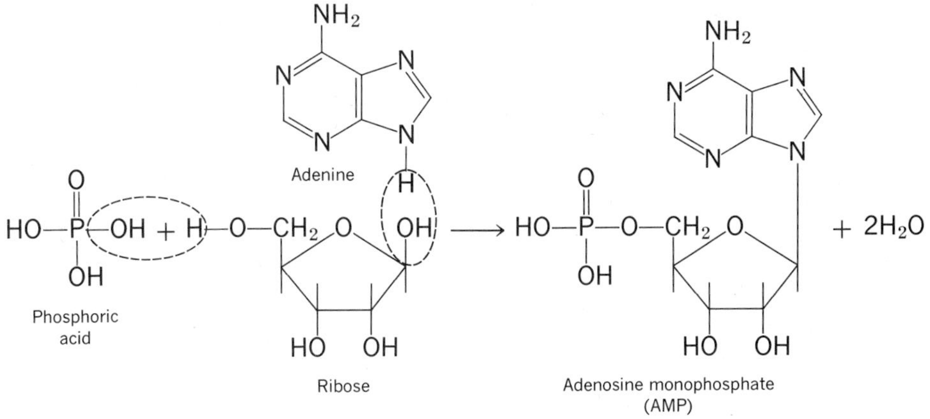

Phosphoric acid

Adenine

Ribose

Adenosine monophosphate (AMP)

$+ 2H_2O$

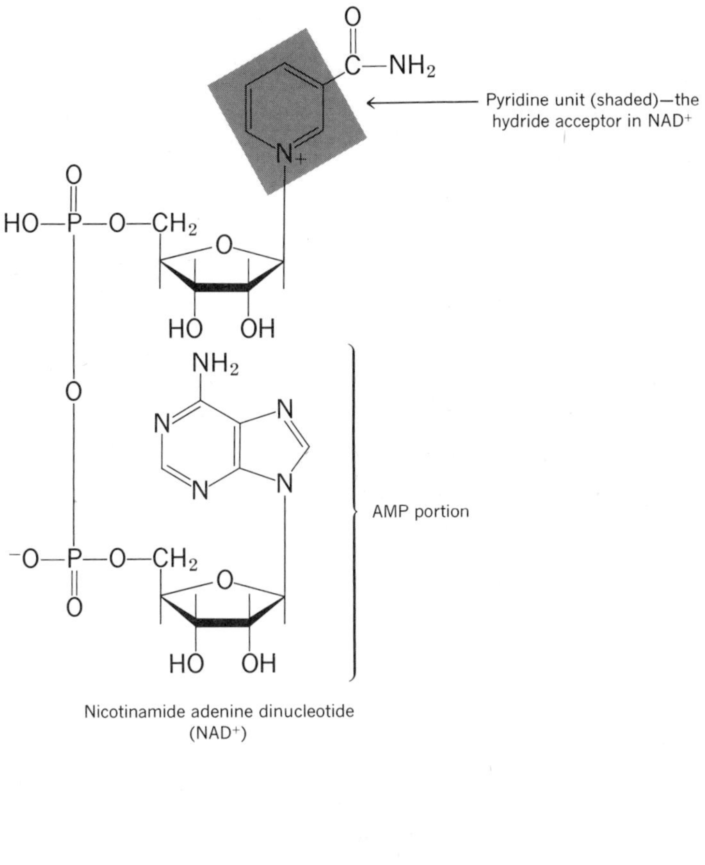

Pyridine unit (shaded)—the hydride acceptor in NAD⁺

AMP portion

Nicotinamide adenine dinucleotide (NAD⁺)

At its pyridine unit $NAD^+$ is a hydride acceptor. The following equation shows how it accepts the hydride:

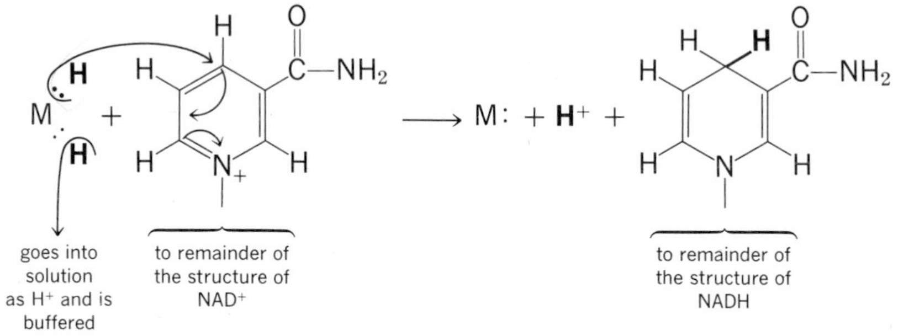

goes into solution as $H^+$ and is buffered

to remainder of the structure of $NAD^+$

to remainder of the structure of NADH

Boldface indicates the pair of electrons and the two hydrogens that are about to depart from the metabolite, $MH_2$. Stated in more simplified form, the equation may be written as

$$MH_2 + NAD^+ \longrightarrow M: + H^+ + NADH$$

Or, to use a method of organizing biochemical events that we shall frequently employ,

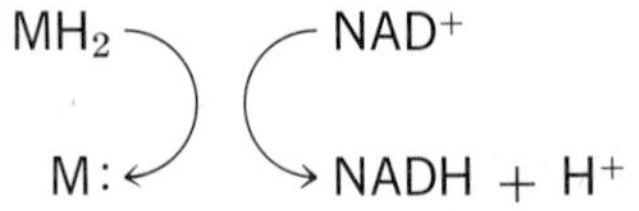

where $MH_2$    is shorthand for any intermediate or metabolite that can lose the elements of hydrogen (i.e., $H:^-$ and $H^+$)

$NAD^+$  will henceforth be our only symbol for the *oxidized* form of nicotinamide adenine dinucleotide (read as "en–ay–de–plus")[7]

M:      represents what is left of the metabolite after it has lost the elements of hydrogen

NADH will henceforth be our only symbol for the *reduced* form of nicotinamide adenine dinucleotide.

For many metabolites the equation given represents the first step in a journey that a pair of electrons from their molecules makes down the respiratory chain for final delivery to molecular oxygen. The hydrogens trail along, and water is the final product. Our bodies do not have little metallic wires to conduct electrons. Electrons must be passed from enzyme to enzyme, but we still consider the process electron flow. We find $NAD^+$ not only in enzymes bound to the surfaces of mitochondria but also in soluble enzymes.

A close relative of $NAD^+$, *nicotinamide adenine dinucleotide phosphate* ($NADP^+$),

---

[7] In some contexts it is simply written NAD. The plus sign is useful in writing balanced equations for hydride transfers, but the actual *net* electric charge on the molecule will be a function of the pH of the immediate environment. There is, after all, a diphosphate system present; in a statement on page 431 the policy followed in this text for such systems is given.

another of the pyridine nucleotides,[8] is encountered less often. It functions in the same way as NAD+; at least the equation for its action is very similar:

$$NADP^+ + MH_2 \longrightarrow M\text{:} + H^+ + NADPH$$

When the body converts excess sugar to fat, a considerable amount of chemical reduction is necessary. A molecule of a fatty acid has many more carbon-hydrogen bonds than does a molecule of glucose. The principal supplier of hydrogen for this reduction is NADPH.

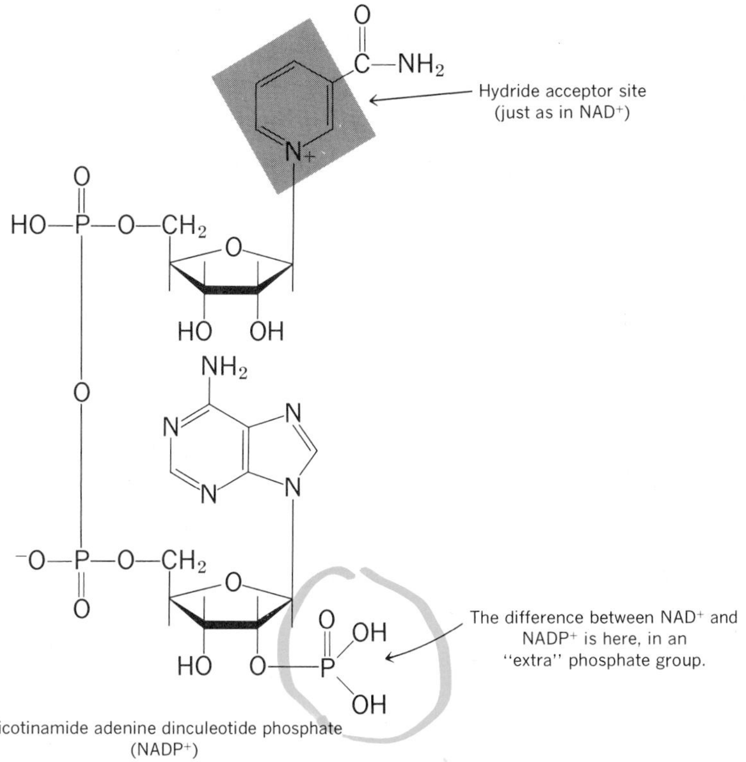

Nicotinamide adenine dinucleotide phosphate
(NADP+)

**Flavin Nucleotides. Flavoproteins.** A large group of respiratory enzymes use as their cofactor one of two derivatives of riboflavin, one of the vitamins. They are *flavin mononucleotide* (FMN) and *flavin adenine dinucleotide* (FAD). We shall be concerned for the most part with FAD, particularly the shaded portion of its struc-

[8] The symbols introduced in this section for the pyridine nucleotides conform to the recommendations of the Commission on Enzymes of the Joint Commission on Biochemical Nomenclature of the International Union of Pure and Applied Chemistry and the International Union of Biochemistry, 1965. These symbols have been recommended as replacements for older ones which will undoubtedly continue to be used for some time in many texts and references. Thus NAD+ was formerly DPN+ (for diphosphopyridine nucleotide) and NADP+ was formerly TPN+ (for triphosphopyridine nucleotide). The older terms are considered objectionable because in neither compound is a "phospho-" group attached to a pyridine ring.

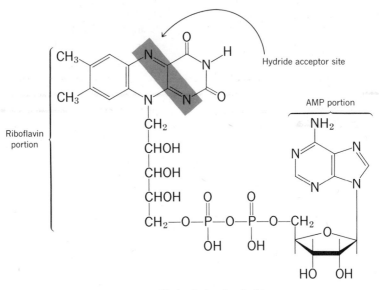

Flavin adenine dinucleotide
(FAD)

ture. The combination of either FMN or FAD with an apoenzyme is called a flavoprotein (FP). The flavoproteins catalyze the removal of hydride ion, $H:^-$, and hydrogen ion, $H^+$, from a metabolite:

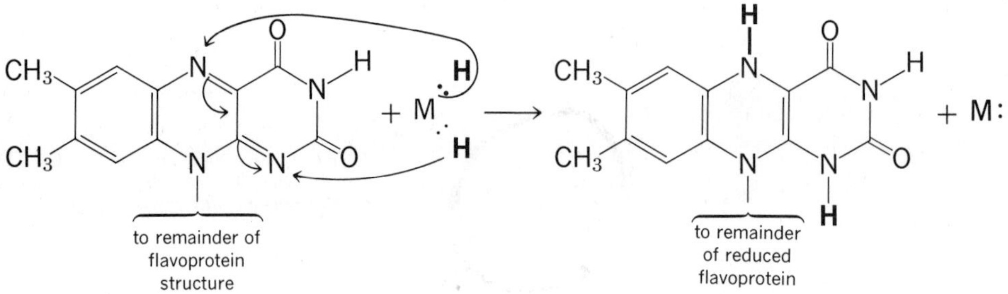

to remainder of
flavoprotein
structure

to remainder
of reduced
flavoprotein

or $$FAD + MH_2 \longrightarrow FADH_2 + M:$$

or

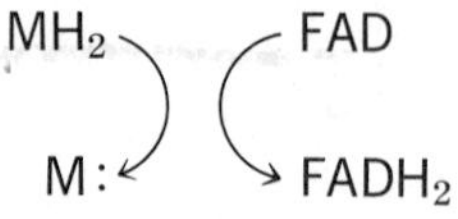

$$MH_2 \qquad FAD$$
$$M: \qquad FADH_2$$

An important source of hydrogen for this reaction is a reduced pyridine nucleotide:

$$H^+ + NADH + FAD \longrightarrow NAD^+ + FADH_2$$

This reaction is a key part of the respiratory chain and the second stage in the trip of a pair of electrons and a pair of protons from a metabolite to oxygen. Putting

the first two stages together, we have

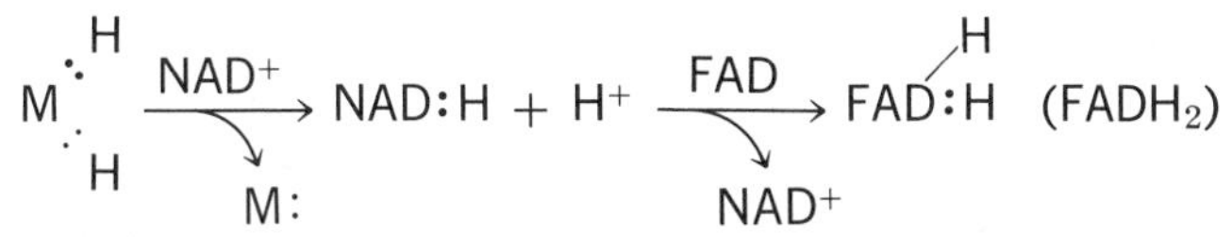

$$M \xrightarrow[\quad H\quad]{NAD^+} NAD:H + H^+ \xrightarrow[\quad NAD^+\quad]{FAD} FAD:H \quad (FADH_2)$$

or

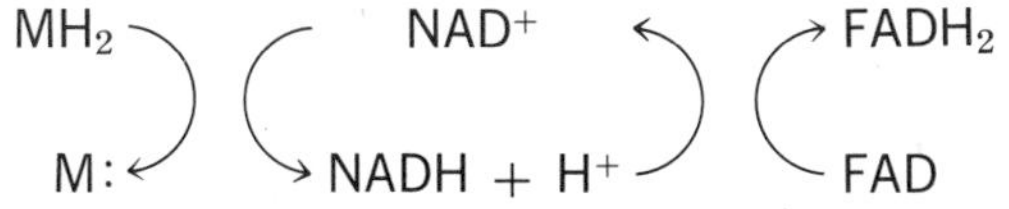

**Ubiquinones. Coenzyme Q.** The ubiquinones are a group of coenzymes having the general structure shown. The name ubiquinone reflects both structure and occurrence. The coenzymes are quinones, and they are found not just in mitochondria but also in cell nuclei, microsomes, and elsewhere. Hence the contraction of the term "ubiquitous quinones" to ubiquinones. We use the synonym coenzyme Q (CoQ).

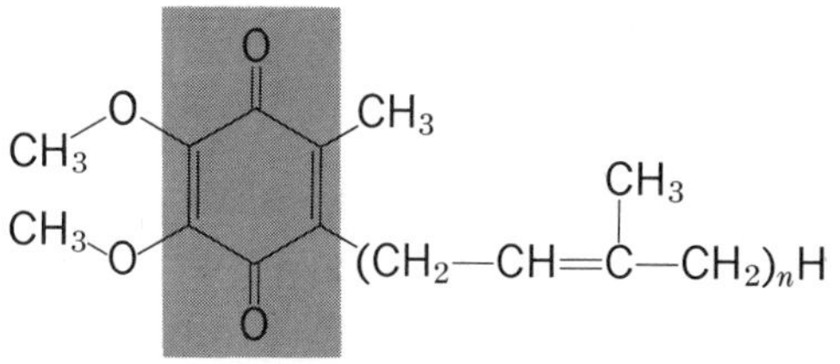

Ubiquinone, $Q_n$     $(n = 6, 7, \ldots, 10)$     $p$-Quinone
(coenzyme Q, CoQ)                                   (mp 116°C)

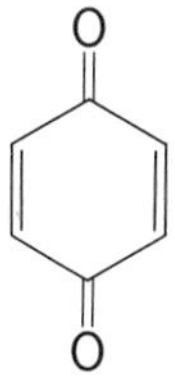

The shaded portion of the structure, the quinone system itself, is of interest here. *One* of the ways by which coenzyme Q *appears* to be involved is in hydrogen transfer from FADH₂, a reaction which may be represented as

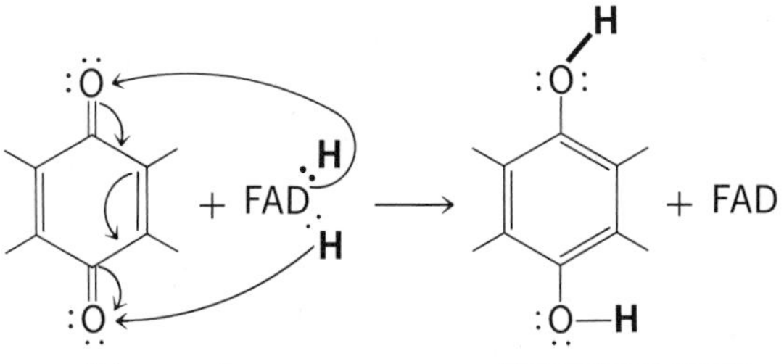

Quinone system                Hydroquinone system
                             (or reduced quinone)

or $$CoQ + FADH_2 \longrightarrow CoQH_2 + FAD$$

To keep current the log of the electron and proton travels, this is the third stage in those sequences involving the pyridine nucleotides. Putting the three stages together, we have the equation

$$\text{M} \overset{\text{H}}{\underset{\text{H}}{:}} \xrightarrow[\text{M:}]{\text{NAD}^+} \text{NAD:H} + \text{H}^+ \xrightarrow[\text{NAD}^+]{\text{FAD}} \text{FAD:H} \xrightarrow[\text{FAD}]{\text{CoQ}} \text{CoQ:H}$$

MH$_2$ ⟶ M:    NAD$^+$ ⟶ NADH + H$^+$    FADH$_2$ ⟶ FAD    CoQ ⟶ CoQH$_2$

The pair of electrons has been passed from MH$_2$ to NAD$^+$ and then to FAD and then to CoQ. Hydrogens (more correctly, protons) have gone along, but the trip is only about half complete. From here on the electrons are handed successively through a series of cytochromes, the last group in the respiratory chain.

**Cytochromes.** Several members of this family are known, but only general characteristics concern us. Cytochromes belong to the family of *hemoproteins*, which includes hemoglobin (cf. p. 330). As their name implies (Greek: *kyto-*, cell;

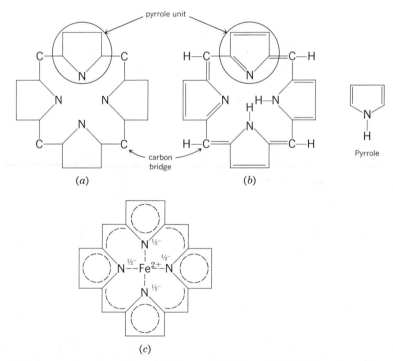

**Figure 13.4** Cytochrome structure, general features. (*a*) Tetrapyrrole skeleton of porphin; (*b*) porphin, showing the extensive conjugated system; (*c*) the Fe$^{2+}$–porphin complex as it occurs in heme and in at least some of the cytochromes. By the loss of an electron Fe$^{2+}$ can become Fe$^{3+}$ and still remain complexed within the porphin skeleton. Not all the complexing potential of the iron ion is used in what is shown here. In cytochrome c (Figure 13.5) two more groups, provided by the protein portion, are complexed to the central iron by means of their nitrogens.

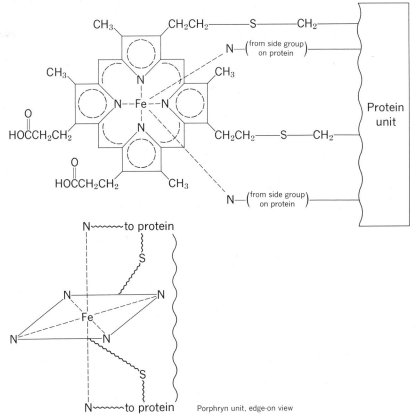

Porphryn unit, edge-on view

**Figure 13.5** Cytochrome c. The complete amino acid sequence of this compound, as isolated from heart cells of several species, is known but is not shown here. The porphyrin portion is attached to the protein chain by two sulfide links (via side chains of the amino acid cysteine) and by complexing with nitrogens from the side chains of two histidine units.

*chroma,* pigment), they are colored. The molecular unit responsible for the color is the *porphin* skeleton, made of four pyrrole units linked by —CH— bridges (Figure 13.4).

Substituted porphins, called *porphyrins,* are capable of forming organometallic complexes with several metal ions, notably $Fe^{2+}$ and $Fe^{3+}$. The structure of one of the cytochromes, *cytochrome c,* is shown in Figure 13.5. The iron-porphyrin unit is partially wrapped by a protein coat. Just how electrons can shuttle in and out of these units is not understood, and the structures are so complicated that some simplification is obviously necessary. We therefore refer only to the iron ions. One $Fe^{3+}$ ion is able to accept one electron and become $Fe^{2+}$, rather than becoming $Fe^+$ or $Fe^0$ by taking more. Hence, if a cytochrome is to accept a *pair* of electrons, we must visualize two $Fe^{3+}$ ions being present. Not enough is known about the cytochromes, however, to say categorically how many metallic ions are part of a single enzyme. To strike a chemical balance, we imagine the involvement of *two* iron ions in one enzyme. No effort will be made to analyze current speculations and the evidence for them. To complicate matters further, *cytochrome a* also con-

tains $Cu^+$ and $Cu^{2+}$ ions; we henceforth ignore them at some sacrifice of accuracy.
  *Cytochrome b* is believed to be capable of accepting electrons from the reduced form of coenzyme Q, $CoQH_2$:

$$CoQH_2 + 2Fe^{3+} \longrightarrow CoQ + 2H^+ + 2Fe^{2+}$$

To represent this reaction in terms of our running log on the travels of a pair of electrons from some metabolite to oxygen, we have

$$\overset{\textbf{H}}{\underset{\textbf{H}}{CoQ\!:}} \;+\; \overset{Fe^{3+}}{\underset{Fe^{3+}}{}} \;\longrightarrow\; CoQ + 2H^+ + \overset{Fe\cdot^{2+}}{\underset{Fe\cdot^{2+}}{}}$$

| Coenzyme Q | Cytochrome b | Coenzyme Q | Cytochrome b |
|---|---|---|---|
| (reduced form) | (oxidized form) | (oxidized form) | (reduced form) |

Once the pair of electrons has entered the cytochrome system, it is passed from cytochrome to cytochrome until it finally reaches one that can catalyze its transfer to oxygen, while hydrogen ions come in and water forms. We can now put together the sequence of electron flow in the respiratory chain. Then we shall backtrack and discuss how this electron flow is made to do work, putting ADP and $P_i$ back together to form ATP.

## ELECTRON TRANSPORT IN THE RESPIRATORY CHAIN

  Piecing together material from the previous section, we summarize the electron transport chain in Figure 13.6. Some metabolites can deliver the elements of hydrogen directly to FAD. Others pass them first to $NAD^+$, which accounts for the "branching" in the figure. Only one coenzyme Q is indicated, but probably two or more are involved. The circled numbers in Figure 13.6 identify reactions that may be written as follows:

① $MH_2 + NAD^+ \longrightarrow M\!: + H^+ + NADH$

② $NADH + FAD + H^+ \longrightarrow NAD^+ + FADH_2$

③ $FADH_2 + CoQ \longrightarrow FAD + CoQH_2$

④ $CoQH_2 + 2Fe_b{}^{3+} \longrightarrow CoQ + 2H^+ + 2Fe_b{}^{2+}$

⑤ $2Fe_b{}^{2+} + 2Fe_{c_1}{}^{3+} \longrightarrow 2Fe_b{}^{3+} + 2Fe_{c_1}{}^{2+}$

⑥ $2Fe_{c_1}{}^{2+} + 2Fe_c{}^{3+} \longrightarrow 2Fe_{c_1}{}^{3+} + 2Fe_c{}^{2+}$

⑦ $2Fe_c{}^{2+} + 2Fe_a{}^{3+} \longrightarrow 2Fe_c{}^{3+} + 2Fe_a{}^{2+}$

⑧ $2Fe_a{}^{2+} + \frac{1}{2}O_2 + 2H^+ \longrightarrow 2Fe_a{}^{3+} + H_2O$

Sum: $MH_2 + \frac{1}{2}O_2 \longrightarrow M\!: + H_2O$

①' $Succinate + FAD \longrightarrow$ $fumarate + FADH_2$

②' $FADH_2 + CoQ \longrightarrow FAD + CoQH_2$

then to step ④

$Succinate + \frac{1}{2}O_2 \longrightarrow$ $fumarate + H_2O$

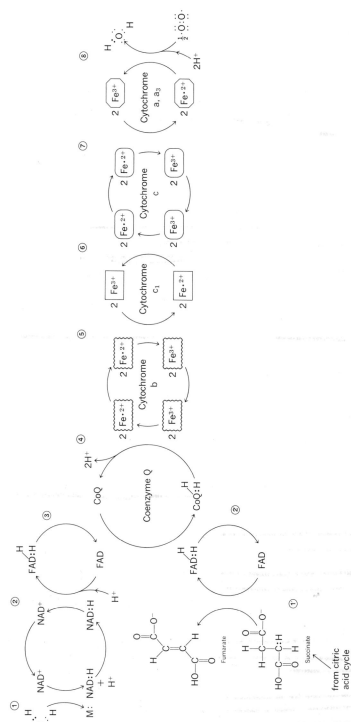

**Figure 13.6** The respiratory chain, tentative pathway for electron transfer. In this schematic outline not all postulated cytochromes are indicated, and the positioning of coenzyme Q is for convenience. The succinate ion, for example, may feed to one CoQ system (later tied to cytochrome b), and NADH may feed to another CoQ unit. The NAD$^+$/NADH system is not often bypassed. The circled numbers identify reactions summarized in another way in the textual discussion. (References: D. R. Sanadi, *Annual Review of Biochemistry*, Vol. 34, 1965, page 22; A. L. Lehninger, *The Mitochondrion*, W. A. Benjamin, New York, 1964, pages 60 and 115; and D. A. Green, "The Mitochondrion," *Scientific American*, January 1964, page 63.)

The net effect, or sum, represents a considerable fall down the free energy hill. The exact change in free energy depends on the metabolite involved and the concentrations, but whatever its amount, it powers the phosphorylation of ADP to form ATP.

## OXIDATIVE PHOSPHORYLATION

What follows is an outline of one theory attempting to explain the most speculative part of ATP synthesis. One of the problems in oxidative phosphorylation is the large *positive* free energy needed for the job:

$$ADP + P_i \longrightarrow ATP + H_2O \qquad \Delta G = +9000 \text{ cal/mole}$$
$$\text{(under } in \ vivo \text{ conditions)}$$

The free energy fall in the respiratory chain is ample enough to provide for ATP synthesis, but to use the free energy made available by one reaction to drive another, the two reactions must be coupled mechanistically. It is not enough that they happen in the same solution. There must be a common intermediate, and its identity is our next problem. How do we coordinate enzymic actions in electron transport with enzymic actions in ATP synthesis? How do we divert energy released by electron transport to do the chemical work of making ATP?

At least two *types* of enzymes are believed to intervene between ADP and ATP. One is called a *coupling enzyme* (CE), the other a *phosphate transfer enzyme* (TE). These enzymes are believed to be located physically adjacent to the group of respiratory enzymes, and all of them together constitute one unit of respiration, a *respiratory assembly*. These are the units that spangle the membrane surfaces of a mitochondrion.

The coupling of electron transport to ATP synthesis is believed to occur in at least three places along the respiratory chain. We illustrate what has been postulated to happen at one such place, the region in which NADH interacts with FAD (step 2 of Figure 13.6). Analogous reactions for other sites are summarized as part of Figure 13.7.

According to the theory, then, for ATP synthesis at the NADH-FAD zone of the respiratory assembly:

1. A coupling enzyme attaches itself temporarily to NADH.

$$NADH + CE \longrightarrow NADH—CE$$

2. When FAD pulls H:⁻ from this complex, a high-energy bond is left behind.

$$NADH—CE + FAD + H^+ \longrightarrow$$
$$NAD^+{\sim}CE + FADH_2$$

3. The coupling enzyme is now activated to the point that it can be transferred to phosphate ion.

$$NAD^+{\sim}CE + P_i \longrightarrow NAD + CE{\sim}P$$

4. CE~P is a high-energy phosphate and as such transfers phosphate to the transfer enzyme.

$$CE{\sim}P + TE \longrightarrow CE + P{\sim}TE$$

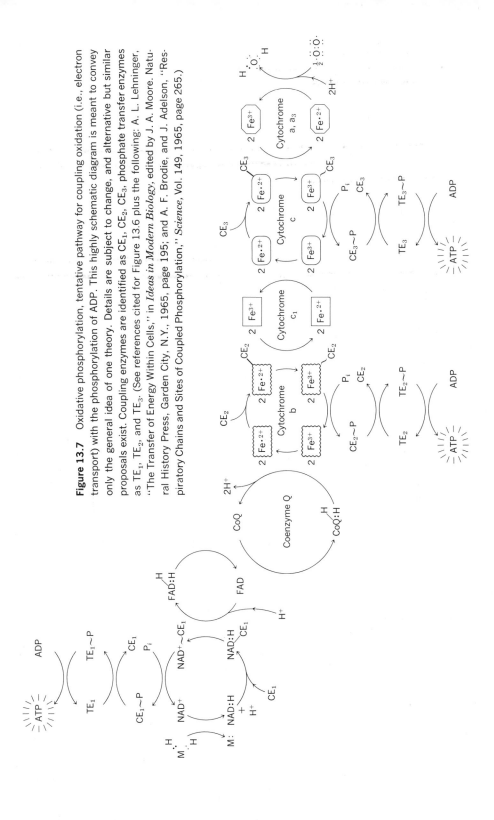

**Figure 13.7** Oxidative phosphorylation, tentative pathway for coupling oxidation (i.e., electron transport) with the phosphorylation of ADP. This highly schematic diagram is meant to convey only the general idea of one theory. Details are subject to change, and alternative but similar proposals exist. Coupling enzymes are identified as $CE_1$, $CE_2$, $CE_3$; phosphate transfer enzymes as $TE_1$, $TE_2$, and $TE_3$. (See references cited for Figure 13.6 plus the following: A. L. Lehninger, "The Transfer of Energy Within Cells," in *Ideas in Modern Biology*, edited by J. A. Moore. Natural History Press, Garden City, N.Y., 1965, page 195; and A. F. Brodie, and J. Adelson, "Respiratory Chains and Sites of Coupled Phosphorylation," *Science*, Vol. 149, 1965, page 265.)

5.  P~TE is a higher-energy phosphate than ATP, and it can therefore transfer its phosphate to ADP.

$$P\sim TE + ADP \longrightarrow TE + ATP$$

Net effect:

$$NADH + FAD + H^+ + ADP + P_i \longrightarrow NAD^+ + FADH_2 + ATP$$

Step 2 is the key step. Instead of a simple transfer of H:⁻ from NADH to FAD (which releases free energy), *the transfer results in the formation of a new high-energy bond* in what is left behind. Free energy is therefore conserved in that bond,[9] and it powers the remaining sequences. Thus in step 2 energy available from one sequence, namely electron transport, is tapped to drive another sequence, oxidative phosphorylation. The events are organized in a more schematic way in Figure 13.7.

The respiratory chain cannot function, of course, unless many conditions are met, one of which is obviously an adequate supply of the metabolites that can furnish the elements of hydrogen. The citric acid cycle is the principal source of this supply.

# CITRIC ACID CYCLE

## A GENERAL VIEW

The series of reactions sometimes referred to as the *Krebs cycle,* sometimes as the tricarboxylic acid cycle, but more usually as the citric acid cycle is outlined in Figure 13.8. Only the major intermediates are shown, and all details concerning enzymes and possible mechanisms are omitted. Strictly speaking, the cycle itself does not include pyruvic acid, but by showing it we indicate the connection between this cycle and glycolysis. From pyruvic acid the overall balanced equation is

$$2CH_3-\overset{O}{\overset{\|}{C}}-\overset{O}{\overset{\|}{C}}-OH + 5O_2 \longrightarrow$$

$$6CO_2 + 4H_2O \qquad \Delta G_{est} = -560,000 \text{ cal} \quad \text{(for two moles pyruvic acid)}$$

---

Hans Krebs (1900–     ), German-born British biochemist who shared the 1953 Nobel prize in medicine and physiology with Fritz Lipmann (p. 433). Forced out of Germany by Nazi pressures in the early 1930s, Krebs eventually joined the faculty of Oxford University. Using evidence he had discovered himself and piecing together the work of others, Krebs together with W. A. Johnson postulated much of the citric acid cycle in 1937. Since then he has been spontaneously honored by scientists the world over who very often refer to the cycle as the Krebs cycle. So many people have contributed to it that citric acid cycle, the name Krebs used, will be employed in this text.

---

[9] In a manner of speaking only. Free energy, of course, is a property of the entire molecule, not just the bond.

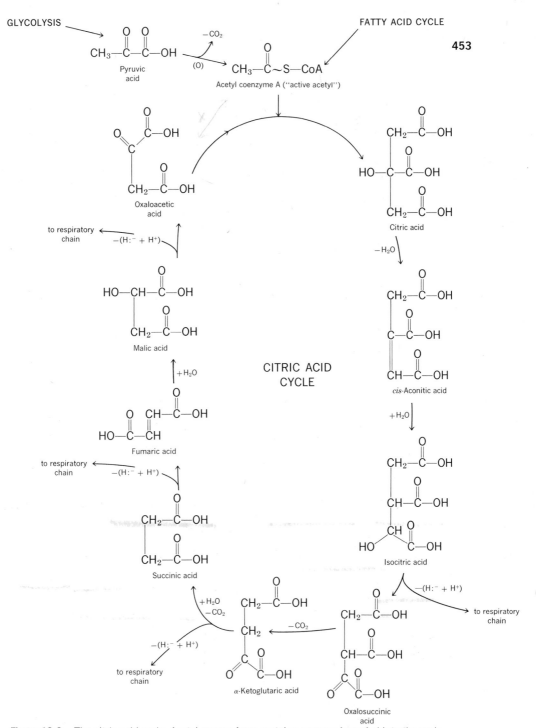

**Figure 13.8** The citric acid cycle. Acetyl groups from acetyl coenzyme A are fed into the cycle, and by a succession of steps two carbons are broken off as carbon dioxide and four units of the elements of hydrogen (H:⁻ + H⁺) are sent on to the respiratory chain.

Oxygen, shown as a reactant even though it does not react directly with any intermediate in the citric acid cycle, is included because this cycle does not occur unless the respiratory chain operates. Intermediates in the cycle deliver hydrogens and electrons to the chain, and eventually these combine with oxygen to form water. The respiratory chain is not called into action unless ATP has for some reason been used up. If we were to ignore the respiratory chain, the equation just given would read as follows:

$$2CH_3\!-\!\overset{\overset{\displaystyle O}{\|}}{C}\!-\!\overset{\overset{\displaystyle O}{\|}}{C}\!-\!OH + 6H_2O \longrightarrow 6CO_2 + \underbrace{10(H\!:^- + H^+)}_{\substack{\text{delivered to the} \\ \text{respiratory chain}}} \longrightarrow$$

The large free energy drop for the oxidation of pyruvic acid powers the formation of ATP, and for every two molecules[10] of pyruvic acid ATP production[11] is

$$30ADP + 30P_i \longrightarrow 30ATP + 30H_2O \qquad \Delta G_{\text{est}} = +270{,}000 \text{ cal}$$

Figure 13.9 shows how the citric acid cycle couples to the respiratory chain and at what points the elements of hydrogen are transferred.

Many of the individual reactions in the citric acid cycle are very much like the ordinary organic reactions we have already studied. The principal difference is that enzymes rather than acids or bases intervene as catalysts. Each step in the cycle causes rather modest changes: water is put in or taken out of a molecule; the elements of hydrogen are removed; carbon dioxide breaks off. To a beginner the cycle may appear formidable because there seem to be so many complicated compounds. In the next section we shall therefore take a slow, careful look at each step and attempt to make the difficult seem a little less so.

## CITRIC ACID CYCLE, A DETAILED VIEW OF ITS INDIVIDUAL STEPS

**Sources of Acetyl Coenzyme A.** As indicated in Figures 13.8 and 13.9, the citric acid cycle must start with a supply of acetyl coenzyme A. Several substances serve as sources of the acetyl group, and acetyl coenzyme A is at one of the major "crossroads" of metabolic pathways in the body.

Proteins are one source. Several amino acids can be degraded by enzyme systems of various cells to produce acetyl coenzyme A. Some of them do so by way of pyruvic acid, which is also a source of acetyl groups.

Fatty acids are a rich and abundant source of acetyl groups. By steps to be described in Chapter 15, fatty acids are degraded eventually to the coenzyme A derivative of the β-keto acid, acetoacetic acid. This structure has two potential acetyl

---

[10] The data are for *two* molecules of pyruvic acid because that many are obtainable from one molecule of glucose via glycolysis.

[11] Based on an estimated +9000 cal for each mole of ATP when it is formed under physiological conditions rather than standard conditions.

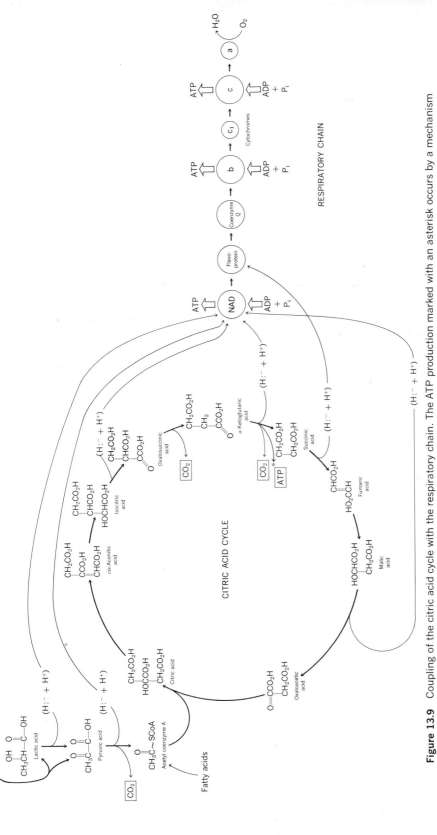

**Figure 13.9** Coupling of the citric acid cycle with the respiratory chain. The ATP production marked with an asterisk occurs by a mechanism discussed in Appendix II. (See references cited for Figures 13.6 and 13.7.)

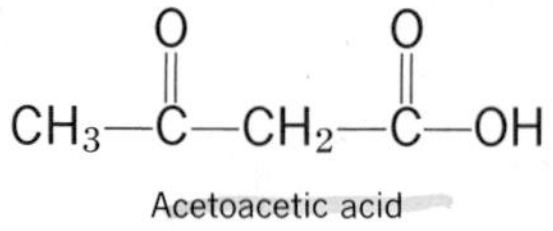

$$CH_3-\overset{\overset{\displaystyle O}{\|}}{C}-CH_2-\overset{\overset{\displaystyle O}{\|}}{C}-OH$$

Acetoacetic acid

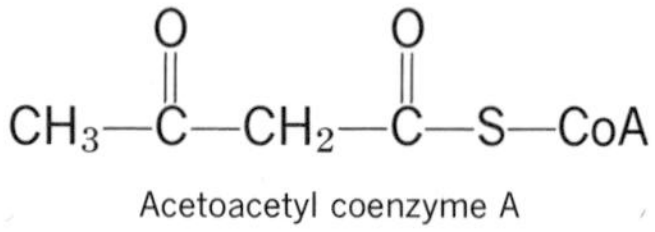

$$CH_3-\overset{\overset{\displaystyle O}{\|}}{C}-CH_2-\overset{\overset{\displaystyle O}{\|}}{C}-S-CoA$$

Acetoacetyl coenzyme A

groups. In the presence of more coenzyme A it can split in two, giving two molecules of acetyl coenzyme A.

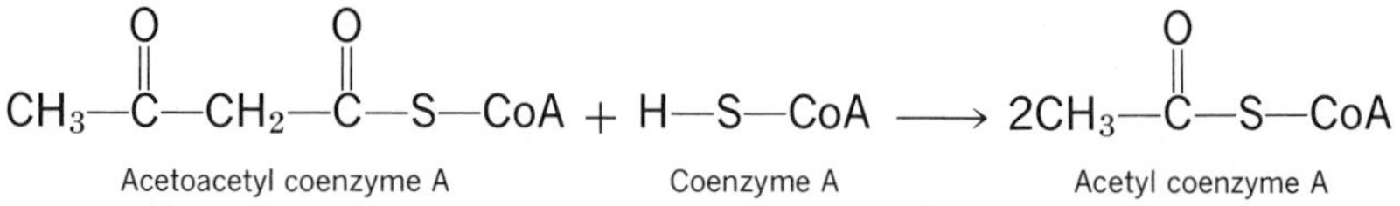

$$CH_3-\overset{\overset{\displaystyle O}{\|}}{C}-CH_2-\overset{\overset{\displaystyle O}{\|}}{C}-S-CoA + H-S-CoA \longrightarrow 2CH_3-\overset{\overset{\displaystyle O}{\|}}{C}-S-CoA$$

Acetoacetyl coenzyme A    Coenzyme A    Acetyl coenzyme A

This source of acetyl coenzyme A is particularly important for many different kinds of muscles.

Carbohydrates are another major supply of acetyl groups. The pathways in the metabolism of fructose, galactose, and glucose soon converge, and we may deal solely with glucose. One molecule of glucose is converted by glycolysis to two molecules of pyruvic acid. This source of acetyl groups is particularly important to brain and nerve tissues.

The conversion of pyruvic acid to acetyl coenzyme A is both a decarboxylation and an oxidation.

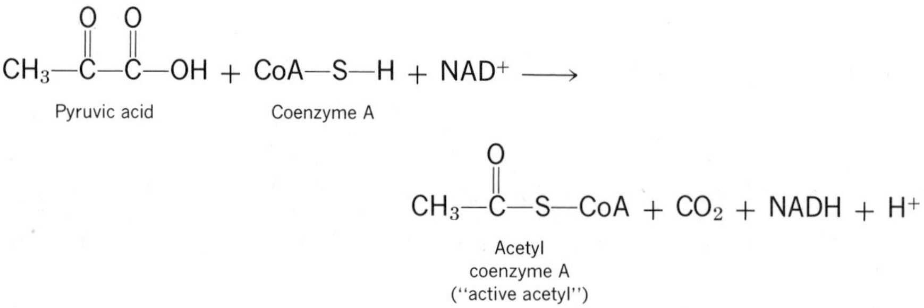

$$CH_3-\overset{\overset{\displaystyle O}{\|}}{C}-\overset{\overset{\displaystyle O}{\|}}{C}-OH + CoA-S-H + NAD^+ \longrightarrow$$

Pyruvic acid    Coenzyme A

$$CH_3-\overset{\overset{\displaystyle O}{\|}}{C}-S-CoA + CO_2 + NADH + H^+$$

Acetyl
coenzyme A
("active acetyl")

Several intermediate steps not yet fully understood are still being investigated. The theory given in Appendix I is another instructive example of how a complicated chemical event can be coordinated by teams of enzymes, of how essential vitamins are (for coenzyme systems), and of how extremely important is the *organization* of chemicals within a cell or a subcellular particle. It is not enough to consider the chemical basis of cellular activity; the cellular basis of chemical activity requires equal attention.

Having examined the principal sources of active acetyl, we are now able to feed one into the citric acid cycle and study what happens to it.[12]

---

[12] Just as we have done with phosphoric acid derivatives, we shall ignore the state of ionization of any carboxylic acid. If the medium is slightly alkaline, they would exist largely as ions, $R-CO_2^-$.

**Oxaloacetatic Acid to Citric Acid**

$$
\begin{array}{c}
\underset{\displaystyle \overset{+}{\underset{\displaystyle O=C-CO_2H}{}}}{H-CH_2-\overset{\displaystyle O}{\overset{\|}{C}}\sim SCoA} \\
CH_2-CO_2H
\end{array}
\longrightarrow
\begin{array}{c}
CH_2-\overset{\displaystyle O}{\overset{\|}{C}}\sim SCoA \\
HO-C-CO_2H \\
CH_2-CO_2H
\end{array}
\xrightarrow{H_2O}
\begin{array}{c}
CH_2-\overset{\displaystyle O}{\overset{\|}{C}}OH + HSCoA \\
HO-C-CO_2H \\
CH_2-CO_2H
\end{array}
$$

Oxaloacetic acid + acetyl CoA                                                     Citric acid

Catalyzed by a complex condensing enzyme, citrate synthase, the mechanism of this two-step (or more) reaction is not well understood. The first stage, however, is analogous to an aldol condensation (cf. p. 209). On paper, at least, it is nothing more than the addition of one carbonyl compound having an $\alpha$-hydrogen across the double bond of the carbonyl of another:

$$
R-\overset{\displaystyle O}{\overset{\displaystyle \curvearrowright}{\overset{\|}{C}}} \overset{\displaystyle H}{\underset{}{+}} \;CH_2-\overset{\displaystyle O}{\overset{\|}{C}}- \longrightarrow R-\overset{\displaystyle O-H}{\overset{\displaystyle |}{\underset{\displaystyle |}{C}}}-CH_2-\overset{\displaystyle O}{\overset{\|}{C}}-
$$

The second step, again on paper at least, is simply the hydrolysis of a thio ester. With the completion of this step the two-carbon acetyl group is locked into the citric acid cycle. In the next step the hydroxyl group will be moved over one carbon.

**Citric Acid to Aconitic Acid to Isocitric Acid**

$$
\begin{array}{c}
CH_2-CO_2H \\
HO-C-CO_2H \\
H-CH-CO_2H
\end{array}
\underset{+H_2O}{\overset{-H_2O}{\rightleftharpoons}}
\begin{array}{c}
CH_2-CO_2H \\
C-CO_2H \\
\| \\
CH-CO_2H
\end{array}
\underset{-H_2O}{\overset{+H_2O}{\rightleftharpoons}}
\begin{array}{c}
CH_2-CO_2H \\
H-C-CO_2H \\
HO-CH-CO_2H
\end{array}
$$

Citric acid (90%)                      *cis*-Aconitic acid (4%)                 Isocitric acid (6%)

We have learned that alcohols can be dehydrated and that alkene linkages can add the elements of water. These are the changes in this series of reactions. One enzyme, aconitase, controls them, and at equilibrium the proportion of each isomer is given by the percentages shown. The equilibrium, however, is normally kept off balance by the next step which is irreversible.

**Isocitric Acid to Oxalosuccinic Acid to $\alpha$-Ketoglutaric Acid**

to respiratory chain

$$
\begin{array}{c}
CH_2CO_2H \\
CHCO_2H \\
H-O-C-CO_2H \\
| \\
H
\end{array}
\xrightarrow{(H:^- + H^+)}
\left[
\begin{array}{c}
CH_2CO_2H \\
(\alpha)\;CH-\overset{\displaystyle O}{\overset{\|}{C}}-O \\
(\beta)\;C \\
\overset{\displaystyle O}{}\quad CO_2H \quad H
\end{array}
\right]
\xrightarrow{Mn^{2+}}
\begin{array}{c}
CH_2CO_2H \\
CH_2 \\
C \\
\overset{\displaystyle O}{}\quad CO_2H
\end{array}
+ CO_2
$$

Isocitric acid                             Oxalosuccinic acid                     $\alpha$-Ketoglutaric acid

The first installment of the delivery of the elements of hydrogen from the citric acid cycle to the respiratory chain is made here where a 2° alcohol group in isocitric acid is oxidized. The product, oxalosuccinic acid, has several functional groups, and one of the combinations is that of a $\beta$-keto acid. These easily decarboxylate (cf. p. 251), and the second step occurs before the intermediate oxalosuccinic acid has a chance to leave the surface of the enzyme. Manganese ion is required as a cofactor.

The overall reaction as shown from left to right is favored by an estimated free energy change of $\Delta G = -5000$ cal/mole under physiological conditions. For all practical purposes the reaction is irreversible.

### $\alpha$-Ketoglutaric Acid to Succinic Acid

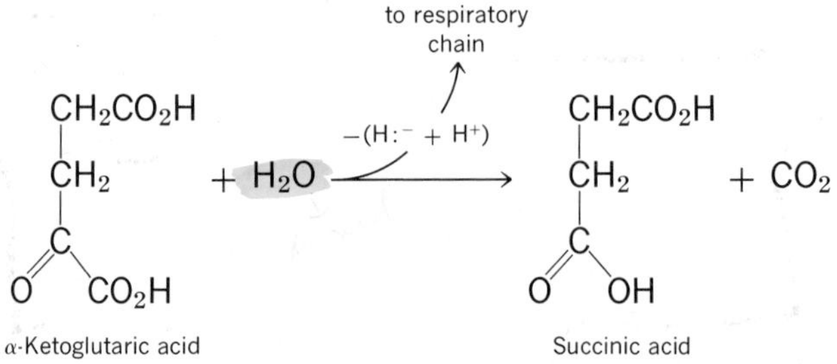

This step is an oxidative decarboxylation, a simultaneous oxidation and a loss of carbon dioxide.[13] We have seen an oxidative decarboxylation happen before, to pyruvic acid. Several enzymes and coenzymes participate. A postulated mechanism for this transformation is given in Appendix II, which also discusses how ATP is made at this stage without using the respiratory chain.

### Succinic Acid to Fumaric Acid

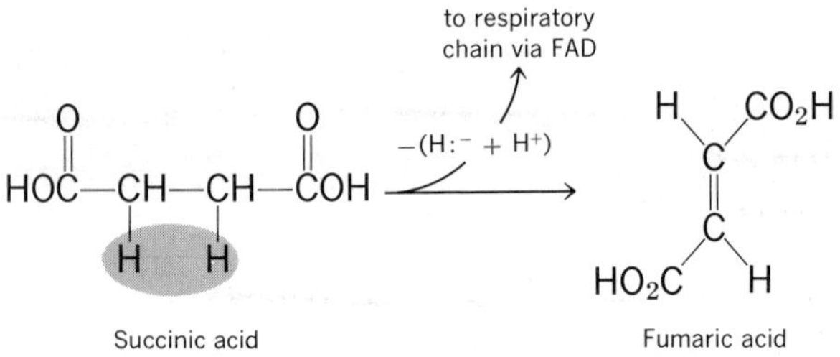

We have not studied any simpler example of a dehydrogenation (loss of $H_2$) such as this, but such reactions are known, particularly when the product is a conjugated system (alternating double and single bonds). Because the reaction illustrates

[13] If an $\alpha$-keto acid were to undergo no more change than loss of carbon dioxide, an *aldehyde* would be left behind:

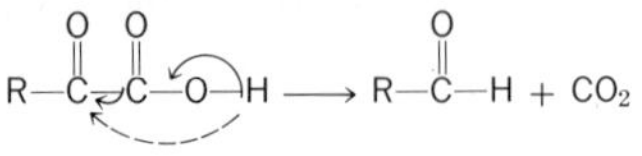

To obtain a carboxyl group instead, oxidation must also occur.

**Table 13.1    Yield of ATP from Pyruvic Acid in the Citric Acid Cycle and the Respiratory Chain**

| Steps | | Receiver of (H:⁻ + H⁺) in the Respiratory Chain | Molecules ATP Formed |
|---|---|---|---|
| Pyruvic acid | ⟶ acetyl CoA | NAD⁺ | 3 |
| Isocitric acid | ⟶ α-ketoglutaric acid | NAD⁺ | 3 |
| α-Ketoglutaric acid | ⟶ succinyl-CoA* | NAD⁺ | 3 |
| Succinyl-CoA | $\xrightarrow{\text{GDP}}$ succinic acid* | — | 1 |
| Succinic acid | ⟶ fumaric acid | FAD | 2 |
| Malic acid | ⟶ Oxaloacetic acid | NAD⁺ | 3 |
| Total ATP per molecule pyruvic acid: | | | $\overline{15 \text{ molecules}}$ |

* Details of how ATP forms in this step are given in Appendix II.

enzyme specificity in more than one way, it is interesting that the cis isomer of fumaric acid, maleic acid, does not form.

$$\underset{\text{Maleic acid}}{\overset{\displaystyle \text{H} \quad \text{CO}_2\text{H}}{\underset{\displaystyle \text{H} \quad \text{CO}_2\text{H}}{\text{C}=\text{C}}}}$$

**Fumaric Acid to Malic Acid**

$$\underset{\text{Fumaric acid}}{\overset{\displaystyle \text{H} \quad \text{CO}_2\text{H}}{\underset{\displaystyle \text{HO}_2\text{C} \quad \text{H}}{\text{C}=\text{C}}}} \quad + \quad \underset{}{\overset{\displaystyle \text{H}}{\text{O—H}}} \quad \longrightarrow \quad \underset{\text{Malic acid}}{\overset{\displaystyle \text{H—CHCO}_2\text{H}}{\underset{\displaystyle \text{HO—CHCO}_2\text{H}}{|}}}$$

This reaction is simply the addition of water to a double bond.

**Malic Acid to Oxaloacetic Acid**

$$\underset{\text{Malic acid}}{\overset{\displaystyle \text{H—O}}{\underset{\displaystyle \text{H CH}_2\text{—CO}_2\text{H}}{\text{C—CO}_2\text{H}}}} \quad \xrightarrow[\text{(H:}^-\text{ + H}^+\text{)}]{\text{to respiratory chain}} \quad \underset{\text{Oxaloacetic acid}}{\overset{\displaystyle \text{O} \quad \text{CO}_2\text{H}}{\underset{\displaystyle \text{CH}_2\text{—CO}_2\text{H}}{\text{C}}}}$$

By this oxidation of a 2° alcohol, the cycle has returned to the starting point, the four-carbon carrier that is able to pick up an acetyl group.

Table 13.1 summarizes the yield of ATP obtained from pyruvic acid via the citric acid cycle and the respiratory chain. For every hydrogen unit (H:⁻ + H⁺) delivered to NAD⁺, three ATPs are produced. If NAD⁺ is bypassed, as at the succinic acid–to–fumaric acid stage, and delivery is to FAD, only two ATPs form.

**REFERENCES AND ANNOTATED READING LIST**

**B O O K S**

A. L. Lehninger. *Bioenergetics.* W. A. Benjamin, New York, 1965. This outstanding book will serve as a reference for much of the rest of this text. Written for college undergraduates, it begins with a survey of elementary thermodynamic principles and discusses the flow of energy in the biological world, the importance of ATP, and most of the important sources and uses of ATP. (Paperback.)

A. L. Lehninger. *The Mitochondrion.* W. A. Benjamin, New York, 1964. Although the physiological aspects of bioenergetics receive more emphasis in this book, the principal topic is still energy for living.

D. E. Griffiths. "Oxidative Phosphorylation," in *Essays in Biochemistry,* edited by P. N. Campbell and G. D. Greville, Vol. 1. Academic Press, New York, 1965. This chapter is a summary discussion of theories (as of about 1964) pertaining to the coupling of electron flow in the respiratory chain to the synthesis of ATP from ADP and $P_i$.

*Annual Review of Biochemistry* (published by Annual Reviews, Palo Alto, Calif.) has been consulted in the preparation of this chapter.

A. White, P. Handler, and E. L. Smith. *Principles of Biochemistry,* third edition. McGraw-Hill Book Company, New York, 1964.

**A R T I C L E S**

A. L. Lehninger. "Energy Transformation in the Cell." *Scientific American,* May 1960, page 102. Lehninger writes for a more general audience in this article, and much work has appeared since 1960.

E. Racker. "The Membrane of the Mitochondrion." *Scientific American,* February 1968, page 32. This article describes studies on how chemicals making up the folded inner membrane of mitochondria contribute to ATP synthesis.

D. E. Green. "Biological Oxidation." *Scientific American,* July 1958, page 56. The role of mitochondria is emphasized, and fewer biochemical details are given.

D. E. Green. "The Mitochondrion." *Scientific American,* January 1964, page 63. Particular attention is given to the molecular structure of a mitochondrion and its serviceability in performing its functions. Green has taken vigorous exception to many aspects of the most widely held views on oxidative phosphorylation. For a report see *Scientific Research,* May 13, 1968, page 33.

M. A. Amerine. "Wine." *Scientific American,* August 1964, page 46. This interesting, relaxing article on the history and technology of wine making includes a brief summary of the chemistry of fermentation.

## PROBLEMS AND EXERCISES

1. Define the following terms. (Neither equations nor full structures, if they are especially complicated, are called for.)

| | | |
|---|---|---|
| (a) fermentation | (b) zymase | (c) respiration |
| (d) glycolysis | (e) anaerobic sequence | (f) respiratory chain |
| (g) homeostasis | (h) oxidative phosphorylation | (i) mitochondrion |
| (j) metabolite | (k) NAD | (l) NADP |
| (m) DPN | (n) FAD | (o) CoQ |
| (p) cytochromes | (q) coupling enzyme | (r) respiratory assembly |
| (s) Krebs cycle | (t) "active acetyl" | |

2. If a mixture of glucose and oxygen is less stable than a mixture of carbon dioxide and water (the combustion products of glucose), explain how glucose can be stored almost indefinitely in air.

3. In the reduction of oxygen to water in the body, what are some sources of $H^+$ and electrons? List the compounds (their structures) that supply these.

4. Discuss how ATP production is geared to ATP need.

5. Write equations to show how $NAD^+$ abstracts the elements of hydrogen from (a) lactic acid, (b) isocitric acid, and (c) malic acid. Use as a symbol for $NAD^+$ the following partial structure:

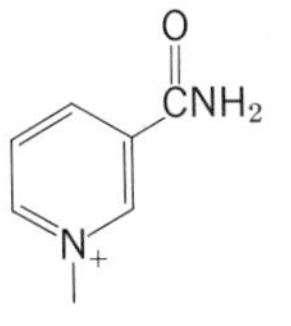

6. Write the following in equation form:

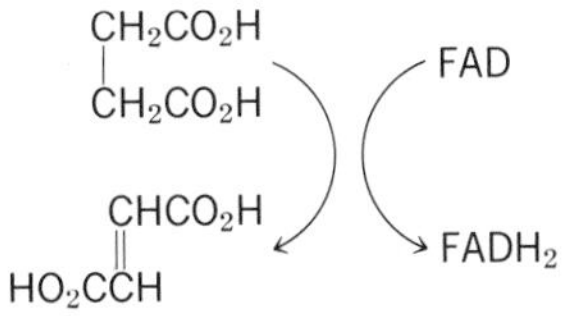

7. In this chapter the expression "homeostatic mechanism" was introduced. Even though this name was not used, we encountered an example of homeostasis in Chapter 12. It had to do with vasopressin. What was it? Describe how it illustrates homeostasis.

# CHAPTER FOURTEEN

# Metabolism of Carbohydrates

No one eats continually. Even the most compulsive nibbler goes without while he sleeps. The body has remarkable resources for storing nutrients in various forms and for releasing them as needed. These needs of course vary considerably from hour to hour. Everyone has known periods of severe exercise when breathing becomes increasingly vigorous, and sooner or later he must stop to "catch his breath." At times the system simply cannot take in and distribute oxygen as fast as it is needed, and yet the person keeps going, at least for a while. At other times ATP demand in certain tissues is higher than the supply available from their respiratory chains. If oxygen does not arrive at respiratory units of mitochondria rapidly enough, ATP production by the respiratory chain slows down and its rate adjusts to the rate of oxygen supply. Assuming adequate supplies of stored fat and glycogen, and assuming a body in shape, the energy output of a long-distance runner is limited primarily by the rate at which oxygen can enter at the lungs and be distributed.

Sometimes, when emergencies arise, the body needs a brief, very intense burst of energy, as in a short sprint or a hundred-yard dash. The body then relies on a mechanism for generating ATP in the absence of oxygen—an *anaerobic sequence* for making ATP known as *glycolysis*. In glycolysis glucose units are broken down to lactic acid while a small amount of ATP is made.

Glucose is not the only carbohydrate that can undergo glycolysis. We concentrate our attention on it because the major metabolic pathways of all the dietary carbohydrates—starch, sucrose, maltose, lactose, fructose, galactose, and glucose—converge (Figure 14.1).

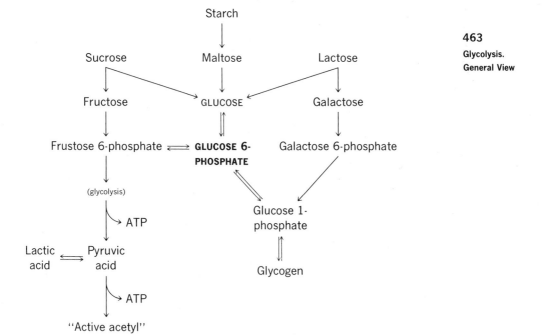

**Figure 14.1**  Convergence of pathways of dietary carbohydrates.

## GLYCOLYSIS

### GENERAL VIEW

**ATP Production.** The amount of ATP made by glycolysis depends on the starting point. If it is glycogen, three ATPs per glucose unit are made; if it is glucose, only two ATPs per glucose molecule are produced. Glucose 6-phosphate can be made from glycogen without sacrificing any ATP, but when it is made from glucose one ATP has to be used, lowering the net yield. The following equations represent the overall effects of glycolysis.

Glycolysis from glycogen:

Glycolysis from glucose:

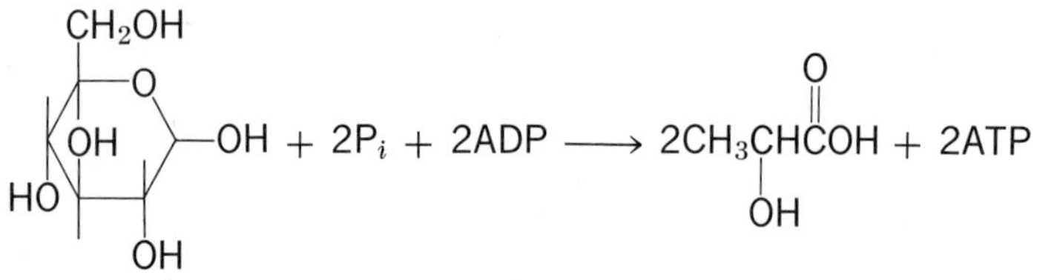

Glucose Lactic acid

Between the left and right sides of these equations, several steps occur (Figure 14.2). From beginning to end, two ATPs are consumed while four are produced for a net of two ATPs per initial glucose molecule (Table 14.1). The yield of ATP may not seem much, but it can be produced quickly without oxygen. Moreover, lactic acid can be catabolized further when oxygen becomes available, or it may migrate to other tissues and cells where the supply of oxygen is favorable for running the respiratory chain. As indicated in Figure 13.9, we obtain eighteen ATPs from each lactic acid molecule when it is processed by the citric acid cycle and the respiratory chain. Therefore glycolysis does not waste the free energy available from glucose; it merely taps some of it in an anaerobic way while leaving its end product, lactic acid, available for further ATP production.

**The Role of Lactic Acid in Glycolysis.** Even though glycolysis does not require oxygen, an oxidation does occur, the conversion of glyceraldehyde 3-phosphate to 1,3-diphosphoglyceric acid. In Figure 14.2 the conversion is shown as a dehydrogenation with hydrogen transferred to $NAD^+$.

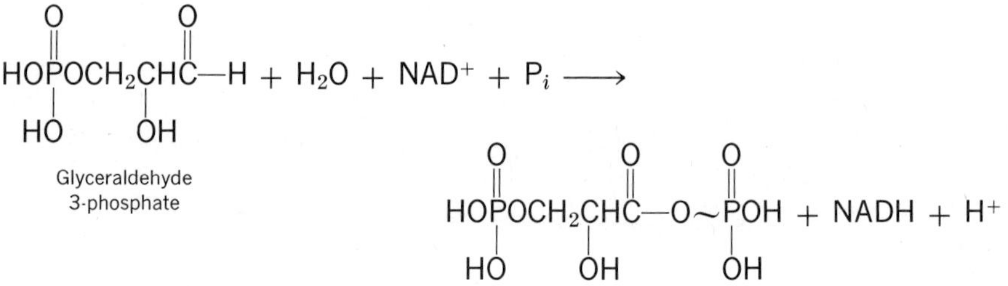

Glyceraldehyde 3-phosphate

1,3-Diphosphoglyceric acid

If the hydrogen stays there (i.e., as NADH + $H^+$), the next molecule of glucose to be processed this far can go no further. The enzyme (having $NAD^+$ as its coenzyme) is out of service; glycolysis and production of ATP will stop, for this is an anaerobic sequence. The respiratory chain is not working, meaning that the elements of hydrogen in NADH + $H^+$ cannot be passed to it, thereby regenerating the $NAD^+$-dependent enzyme. The problem here is the regeneration of this $NAD^+$. In animals the problem is solved by using something produced later in glycolysis, pyruvic acid, as the acceptor for the "$H_2$" in NADH + $H^+$. Thus the last step in glycolysis is regeneration of this $NAD^+$-dependent enzyme, and the lactic acid that forms becomes at least a temporary reservoir for hydrogen. This key aspect of glycolysis is discussed again in Figure 14.3.

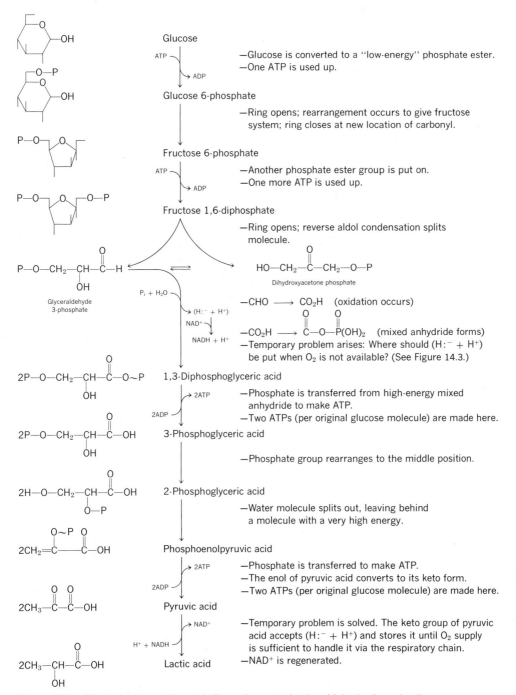

**Figure 14.2** Glycolysis. Anaerobic catabolism, glucose to lactic acid. In the formulas P represents a phosphate ester group.

**Table 14.1   Yield of ATP from Glucose in Glycolysis—Anaerobic Catabolism, Glucose to Lactic Acid**

| Steps | | Molecules of ATP Gained (+) or Lost (−) per Initial Glucose Unit |
|---|---|---|
| Glucose + ATP $\longrightarrow$ glucose 6-phosphate + ADP | | −1* |
| Fructose 6-phosphate + ATP $\longrightarrow$ fructose 1,6-diphosphate + ADP | | −1 |
| 1,3-Diphosphoglyceric acid (2 molecules) + 2ADP $\longrightarrow$ 3-phosphoglyceric acid (2 molecules) + 2ATP | | +2 |
| Phosphoenolpyruvic acid (2 molecules) + 2ADP $\longrightarrow$ pyruvic acid (2 molecules) + 2ATP | | +2 |
| Net gain in ATP via glycolysis | | +2 |

* If the starting point is glycogen instead of glucose, this consumption of ATP is unnecessary, and the net ATP production is three per glucose unit.

**Reversal of Glycolysis.** When lactic acid forms during anaerobic muscular exercise, it migrates from muscle cells. Circulation carries it to other tissues, many of which can use it as a metabolite by changing it back to pyruvic acid and from that to acetyl coenzyme A. At this point it can be catabolized by the citric acid cycle and the respiratory chain for ATP synthesis. At the liver (or the kidneys), however, lactic acid will encounter enzymes that could change it back to glycogen instead of oxidizing it. In other words, the effect of glycolysis can be reversed. In fact, the steps for accomplishing this are, for the most part, the exact reverse of steps in glycolysis. Since we wish here to gain only a general conception of what happens, details are reserved for a later section. Two points should be made at this stage, however. At least two steps in glycolysis are not readily reversible, and methods for bypassing them are available, making the control of glycolysis easier. By exerting special control over the enzymes of the bypasses, the system can determine whether or not glycolysis will be reversed.

Since glycolysis *produces* ATP, its reverse must *require* some. If glycolysis is going downhill, the reverse is going uphill. To provide energy for this, some of the lactic acid is channeled into the citric acid cycle to generate the ATP needed to carry the remaining back to glycogen. Only a small fraction of lactic acid (one molecule out of six) need be oxidized further, for the aerobic sequence (citric acid cycle and respiratory chain) is particularly productive of ATP. Starting from one molecule of *lactic acid,* this pathway yields eighteen molecules of ATP. One-sixth of this is three molecules. Glycolysis *from glycogen* (rather than from glucose) yields a net three molecules of ATP. Its reverse will therefore need at least this much.

## A DETAILED VIEW

We shall start a detailed study with glucose and carry one molecule through to lactic acid, rationalizing each step in terms of organic reactions already studied.

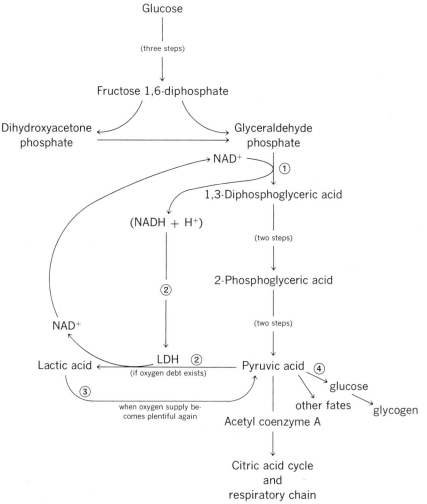

**Figure 14.3** Lactic acid as a reservoir of hydrogen. Principal fates of pyruvic acid. During gly-colysis the NAD⁺-dependent enzyme at ① is reduced and, unless it is regenerated, glycolysis will stop. The enzyme can be regenerated in two ways. If the oxygen supply is sufficient, the elements of hydrogen removed at ① by NAD⁺ can be passed on down the respiratory chain. If the oxygen supply is insufficient (a state of oxygen debt), the enzyme lactic acid dehydrogenase (LDH) catalyzes the transfer ② of (H:⁻ + H⁺) to pyruvic acid, and lactic acid becomes the storage depot. The enzyme for ① is restored. Once the oxygen supply is rebuilt, the hydrogen stored in lactic acid is removed ③ and delivered to the respiratory chain. The pyruvic acid that is made simultaneously may go into the citric acid cycle, or it may be converted ④ to glucose. (Reference: C. L. Markert, "Mechanisms of Cellular Differentiation," in *Ideas in Modern Biology*, edited by J. A. Moore, Natural History Press, Garden City, N.Y., 1965, page 234.)

Then we shall examine variations as they occur in special systems such as muscles or microorganisms. Equilibrium constants are for pH 7.3 and are only approximate. Free energy values are also approximate.

### Glucose to Glucose 6-Phosphate

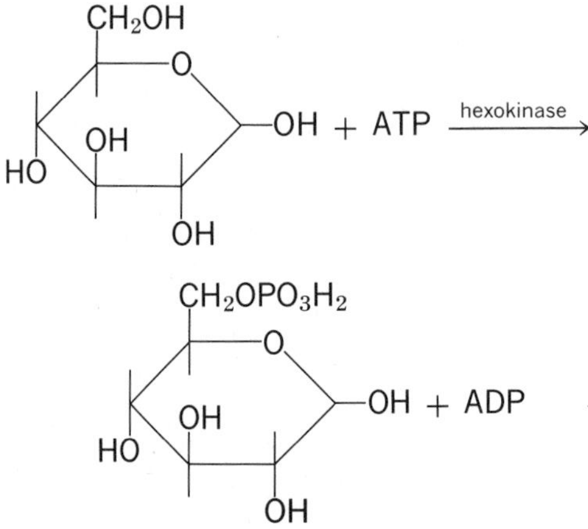

$$\frac{[\text{glucose 6-phosphate}][\text{ADP}]}{[\text{glucose}][\text{ATP}]} \approx 6300$$

$$\Delta G \approx -5000 \text{ cal/mole}$$

The forward reaction is obviously strongly favored. This phosphorylation of glucose is the principal way by which a cell captures a glucose molecule. Molecules of glucose 6-phosphate are not free to migrate out of cells, but glucose molecules have much more of such freedom. The product is simply an ester of phosphoric acid, and the reaction is analogous to the formation of an ordinary ester from an alcohol and an acid anhydride. Here the alcohol is glucose, the acid anhydride a phosphoric acid anhydride, ATP.

### Glucose 6-Phosphate to Fructose 6-Phosphate

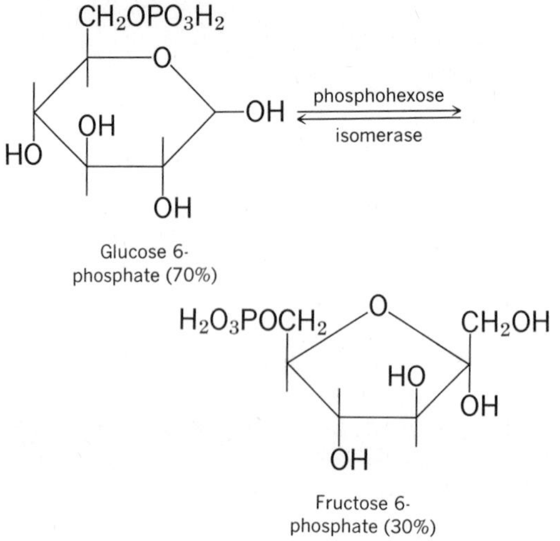

Glucose 6-
phosphate (70%)

$$\frac{[\text{fructose 6-phosphate}]}{[\text{glucose 6-phosphate}]} = 0.4$$

Fructose 6-
phosphate (30%)

In this fairly mobile equilibrium one isomer converts into another, and to understand what is happening we need only recall that

1. Cyclic forms of sugars exist in equilibrium with open-chain forms.
2. Carbonyl compounds exist in equilibrium with enol forms. With these facts in mind, the following sequence indicates how the isomerization might take place. Although this mechanism is speculative, there is evidence for the *ene-diol* bound to the surface of the enzyme.

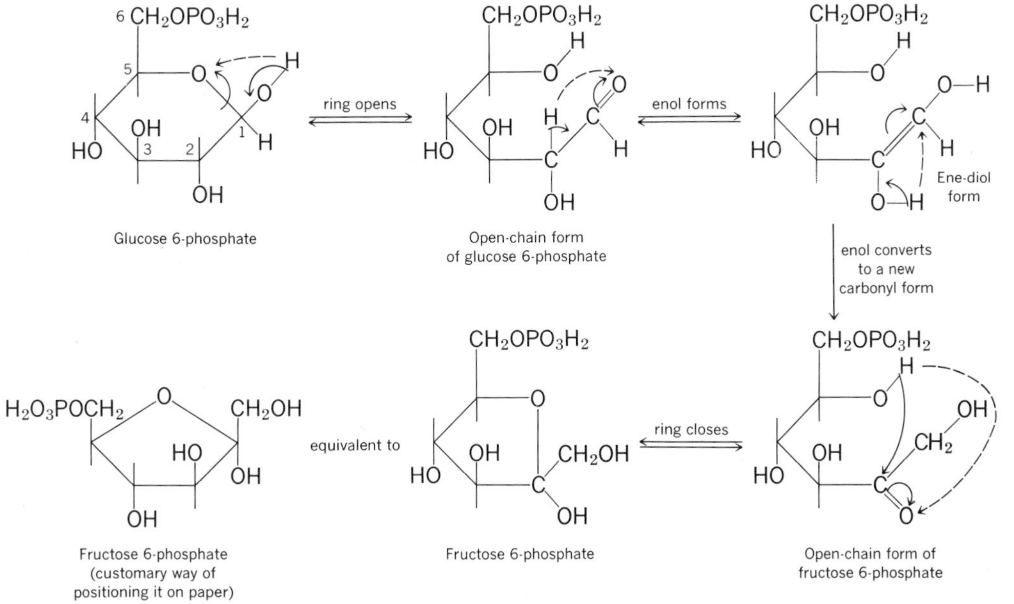

$6\ CH_2OPO_3H_2$

Glucose 6-phosphate

ring opens

$CH_2OPO_3H_2$

Open-chain form
of glucose 6-phosphate

enol forms

$CH_2OPO_3H_2$

Ene-diol
form

enol converts
to a new
carbonyl form

$H_2O_3POCH_2$  $CH_2OH$

Fructose 6-phosphate
(customary way of
positioning it on paper)

equivalent to

$CH_2OPO_3H_2$

Fructose 6-phosphate

ring closes

$CH_2OPO_3H_2$

Open-chain form of
fructose 6-phosphate

Since the ene-diol is simultaneously an enol for both the aldehyde group (of glucose) and the ketone group (of fructose), one or the other can readily form as equilibrium is established.

### Fructose 6-Phosphate to Fructose 1,6-Diphosphate

$H_2O_3POCH_2$  $CH_2OH$

Fructose 6-phosphate

$+\ ATP\ \xrightarrow{\text{phosphofructokinase}}$

$H_2O_3POCH_2$  $CH_2OPO_3H_2$

$+\ ADP$

Fructose 1,6-diphosphate

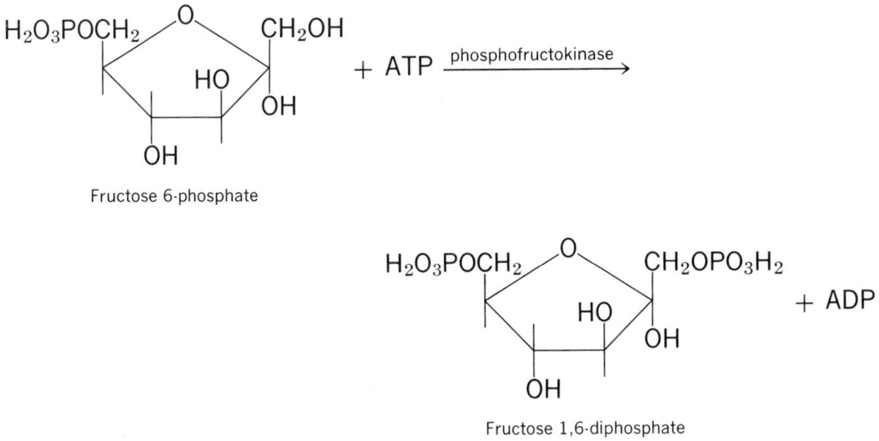

$$\Delta G \approx -5000 \text{ cal/mole} \quad \text{(a rough estimate)}$$

Another molecule of energy-rich ATP is invested in this essentially irreversible reaction. From here on, however, glycolysis will be largely downhill. This reaction is exactly analogous to the formation of a phosphate ester from glucose, the first step studied (from which the estimated $\Delta G$ was taken).

**Fructose 1,6-Diphosphate to Glyceraldehyde 3-Phosphate and Dihydroxyacetone Phosphate**

$H_2O_3POCH_2$ ... $CH_2OPO_3H_2$

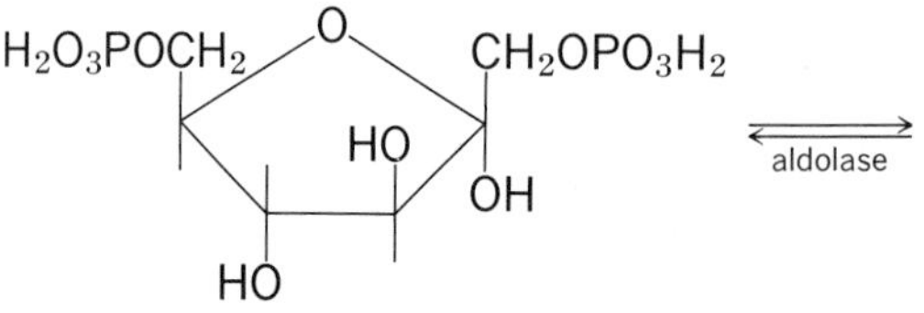

$\xrightleftharpoons[\text{aldolase}]{}$

Fructose 1,6-diphosphate

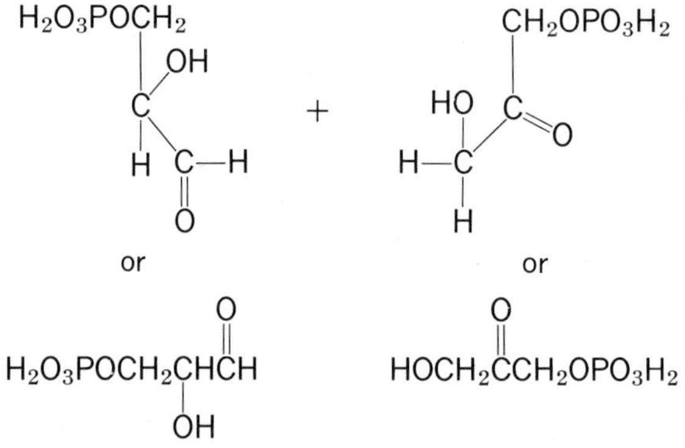

$H_2O_3POCH_2$
$\quad$ OH
$\quad$ C
$\quad$ H C—H
$\qquad$ O

$+$

$CH_2OPO_3H_2$
HO C$=$O
H—C
$\quad$ H

or

or

$$H_2O_3POCH_2CHCH$$
$$\quad\quad\quad\quad OH \quad\quad O$$

$$HOCH_2CCH_2OPO_3H_2$$

Glyceraldehyde 3-
phosphate

Dihydroxyacetone
phosphate

$$\frac{\begin{bmatrix}\text{glyceraldehyde 3-}\\\text{phosphate}\end{bmatrix}\begin{bmatrix}\text{dihydroxyacetone}\\\text{phosphate}\end{bmatrix}}{[\text{fructose 1,6-diphosphate}]} \approx 1 \times 10^{-5}$$

The position of equilibrium certainly does not favor the products, but succeeding steps remove them as they form and the equilibrium constantly shifts. The falling apart of the fructose 1,6-diphosphate molecule is really quite simple if we remember two facts from organic chemistry.

1. Closed-ring forms of sugars can exist in equilibrium with open-chain forms.
2. The aldol condensation can be reversed (cf. p. 213). This segment in the process of glycolysis can therefore be regarded as occurring in the following two steps.

Step 1. Ring opening and closing:

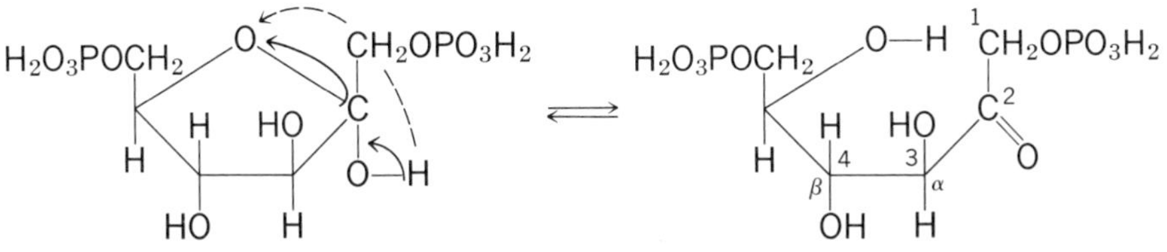

Fructose 1,6-diphosphate
(closed-ring form)

Fructose 1,6-diphosphate
(open-chain form)

The —OH group at position 4 in the open form is located *beta* to the carbonyl group at position 2. This $\beta$-hydroxy carbonyl system is precisely what is characteristic of aldols (cf. p. 209).

Step 2. Reverse aldol condensation:

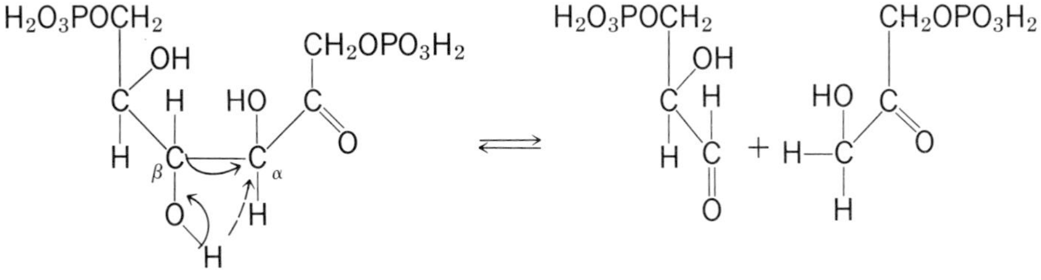

Fructose 1,6-diphosphate
(open-chain form)

Glyceraldehyde 3-
phosphate

Dihydroxyacetone
phosphate

**Dihydroxyacetone Phosphate to Glyceraldehyde 3-Phosphate.** These two phosphates are isomers and can interconvert in much the same way that glucose 6-phosphate and fructose 6-phosphate isomerize, via an ene-diol.

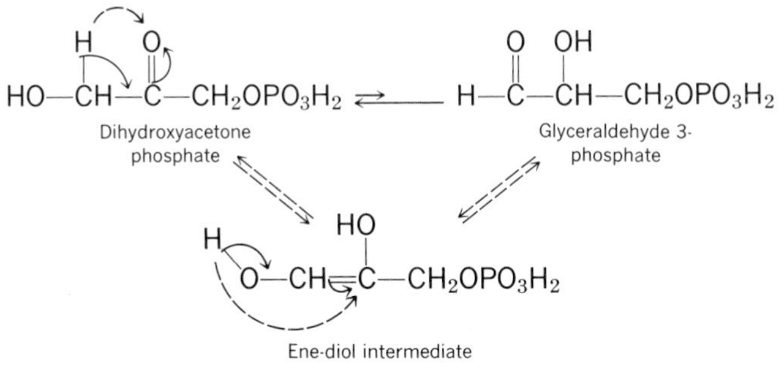

Dihydroxyacetone
phosphate

Glyceraldehyde 3-
phosphate

$$\frac{\left[\begin{array}{c}\text{glyceraldehyde 3-}\\\text{phosphate}\end{array}\right]}{\left[\begin{array}{c}\text{dihydroxyacetone}\\\text{phosphate}\end{array}\right]} \approx 0.04$$

Ene-diol intermediate

Of these two compounds the jumping-off point for further events in glycolysis is glyceraldehyde 3-phosphate, even though it is not the favored species in the equilibrium. Additional reactions, of course, remove it and the equilibrium shifts to produce more. The fact that glyceraldehyde 3-phosphate can be made from dihydroxyacetone phosphate means that this second compound is not wasted, that all the original glucose molecule is used, and that, in effect, each glucose molecule yields two of glyceraldehyde 3-phosphate.

## Glyceraldehyde 3-Phosphate to 1,3-Diphosphoglyceric Acid

$$H_2O_3POCH_2\overset{\overset{\displaystyle HO}{|}}{CH}\overset{\overset{\displaystyle O}{\|}}{CH} + NAD^+ + P_i \xrightarrow{\text{phosphoglyceraldehyde dehydrogenase}}$$

Glyceraldehyde 3-phosphate

$$H_2O_3POCH_2\overset{\overset{\displaystyle HO}{|}}{CH}-\overset{\overset{\displaystyle O}{\|}}{C}-O\sim\overset{\overset{\displaystyle O}{\|}}{\underset{\underset{\displaystyle OH}{|}}{P}}-OH + NADH$$

1,3-diphosphoglyceric acid

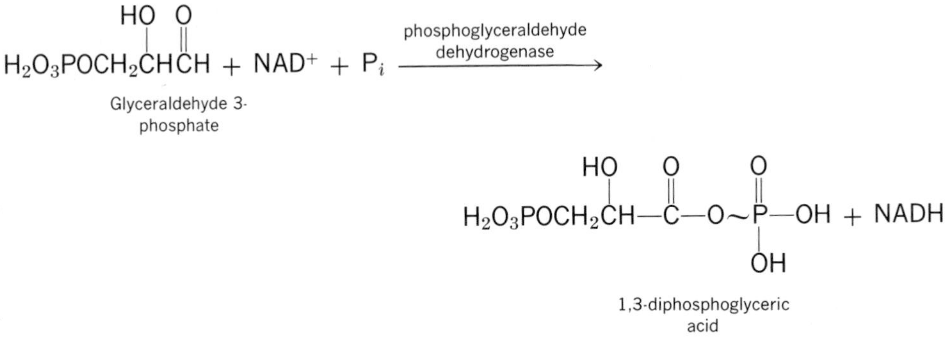

$$\frac{[\text{1,3-diphosphoglyceric acid}][\text{NADH}]}{[\text{glyceraldehyde 3-phosphate}][\text{NAD}^+][P_i]} \approx 1$$

This step is an oxidation, actually a dehydrogenation; $NAD^+$ accepts $H{:}^-$ from the aldehyde group and NADH forms. The last step in glycolysis will regenerate the $NAD^+$, which is, of course, the coenzyme for this step. Although the mechanism for this reaction is not fully known, there is evidence that the enzyme contains an

—S—H group and that it participates somewhat as follows ($R = H_2O_3POCH_2\overset{\overset{\displaystyle HO}{|}}{CH}-$):

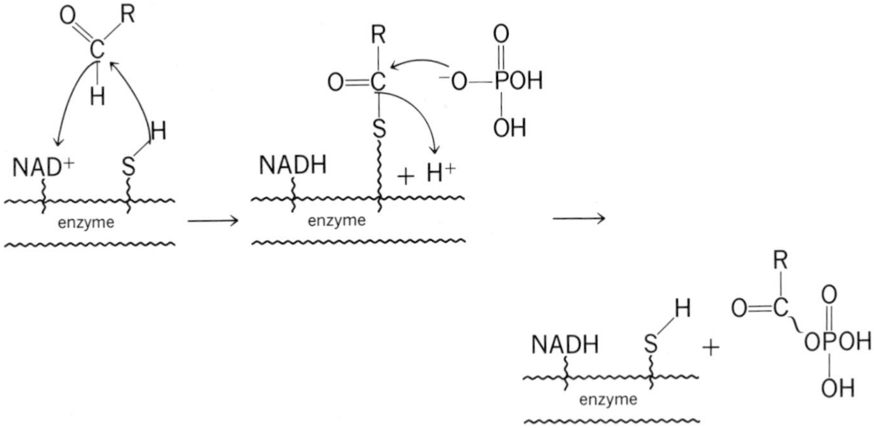

Thus attack by sulfur (a powerful nucleophile) on the carbonyl carbon of the aldehyde group helps "push" $H{:}^-$ out where the $NAD^+$ is positioned to accept it. Then phosphate ion attacks the same carbonyl carbon and pushes the sulfur out. The final product, 1,3-diphosphoglyceric acid, has at one end of its molecule a mixed acid anhydride system. Like all anhydrides it is energy-rich. In the next step this energy is finally caught in the form of ATP. The glycolysis pathway is now functioning as it must, converting some of the free energy of glucose into biologically usable energy, ATP, without requiring oxygen.

## 1,3-Diphosphoglyceric Acid to 3-Phosphoglyceric Acid

$$\underset{\text{1,3-Diphosphoglyceric acid}}{H_2O_3POCH_2\overset{\overset{\displaystyle HO}{|}}{C}H\overset{\overset{\displaystyle O}{\|}}{C}O\sim\overset{\overset{\displaystyle O}{\|}}{\underset{\underset{\displaystyle OH}{|}}{P}}OH} + ADP \xrightarrow[\text{kinase}]{\text{3-phosphoglyceric acid}} \underset{\text{3-Phosphoglyceric acid}}{H_2O_3POCH_2\overset{\overset{\displaystyle HO}{|}}{C}H\overset{\overset{\displaystyle O}{\|}}{C}OH} + ATP$$

$$\frac{[\text{3-phosphoglyceric acid}][\text{ATP}]}{[\text{1,3-diphosphoglyceric acid}][\text{ADP}]} \approx 5000$$

The previous step was an oxidation. As so often happens in the body, such a reaction is the prelude to the synthesis of a high-energy phosphate such as ATP. The position of equilibrium in this step lies far to the right, and the formation of ATP is favored.

## 3-Phosphoglyceric Acid to 2-Phosphoglyceric Acid

$$\underset{\substack{\text{3-Phosphoglyceric acid} \\ \text{or 3-PGA}}}{H_2O_3POCH_2\overset{\overset{\displaystyle HO}{|}}{C}H\overset{\overset{\displaystyle O}{\|}}{C}OH} \xrightarrow{\text{phosphoglyceromutase}}$$

$$\underset{\substack{\text{2-Phosphoglyceric acid} \\ \text{or 2-PGA}}}{HOCH_2\overset{\overset{\displaystyle H_2O_3PO}{|}}{C}H\overset{\overset{\displaystyle O}{\|}}{C}OH} \qquad \frac{[\text{2-phosphoglyceric acid}]}{[\text{3-phosphoglyceric acid}]} \approx 0.25$$

The mechanism for this isomerization is believed to involve an enzyme phosphate (abbreviated Enz-P) which donates its phosphate group to the middle position in 3-PGA (3-phosphoglyceric acid) and takes a phosphate unit back from the end position to produce 2-PGA (2-phosphoglyceric acid).

$$\text{3-PGA} + \text{Enz-P} \rightleftharpoons H_2O_3POCH_2\overset{\overset{\displaystyle H_2O_3PO}{|}}{C}H\overset{\overset{\displaystyle O}{\|}}{C}OH + \text{Enz} \rightleftharpoons \text{2-PGA} + \text{Enz-P}$$

## 2-Phosphoglyceric Acid to Phosphoenolpyruvic Acid

$$\underset{\text{2-Phosphoglyceric acid}}{\overset{\overset{\displaystyle H_2O_3PO}{|}}{C}H_2-\underset{\underset{\displaystyle OH}{|}}{C}-\overset{\overset{\displaystyle O}{\|}}{C}OH} \xrightarrow[\phantom{x}]{\text{enolase}} \underset{\text{Phosphoenolpyruvic acid}}{CH_2=\overset{\overset{\displaystyle H_2O_3PO}{|}}{C}-\overset{\overset{\displaystyle O}{\|}}{C}-OH} \qquad \frac{[\text{phosphoenolpyruvic acid}]}{[\text{2-phosphoglyceric acid}]} \approx 0.7$$

The starting material is a relatively low-energy phosphate ester, but by the simple loss of the elements of water the most energy-rich phosphate known, phosphoenolpyruvic acid, is formed.

## Phosphoenolpyruvic Acid to Pyruvic Acid

$$\underset{\substack{\text{Phosphoenolpyruvic}\\\text{acid}}}{H_2O_3PO\;\; CH_2{=}\overset{\displaystyle O}{\underset{\displaystyle}{C}}{-}\overset{\displaystyle O}{\underset{\displaystyle}{C}}{-}OH} + ADP \xrightarrow{\text{pyruvic acid kinase}}$$

$$\underset{\text{Pyruvic acid}}{CH_3{-}\overset{\displaystyle O}{C}{-}\overset{\displaystyle O}{C}{-}OH} + ATP \qquad \frac{[\text{pyruvic acid}][\text{ATP}]}{[\text{phosphoenolpyruvic acid}][\text{ADP}]} \approx 5100$$

There are two ATP payoffs in glycolysis, and this is the second. The tendency for an enol to isomerize to its keto form (p. 253) is at least part of the driving force for this reaction.

$$\underset{\substack{\text{Phosphoenolpyruvic}\\\text{acid}}}{H_2O_3PO\;\; CH_2{=}\overset{O}{C}{-}\overset{O}{C}{-}OH} \xrightarrow{H_2O} \underset{\substack{\text{Enol form of}\\\text{pyruvic acid}}}{CH_2{=}\overset{H-O}{C}{-}\overset{O}{C}{-}OH} \longrightarrow \underset{\substack{\text{Keto form of}\\\text{pyruvic acid}}}{\overset{H}{CH_2}{-}\overset{O}{C}{-}\overset{O}{C}{-}OH}$$

To write a balanced algebraic sum of the reactions of glycolysis, we have

Glucose + 2NAD$^+$ + 2ADP + 2P$_i$ $\longrightarrow$
2 pyruvic acid + 2NADH + 2H$^+$ + 2H$_2$O + 2ATP

The NAD$^+$ was part of an enzyme system, and if all enzymes are to be restored to normal under anaerobic conditions, the next step must occur or else glycolysis cannot continue with another glucose molecule.

### Pyruvic Acid to Lactic Acid

$$\underset{\text{Pyruvic acid}}{CH_3{-}\overset{O}{C}{-}\overset{O}{C}{-}OH} + H^+ + NAD{:}H \longrightarrow$$

$$\underset{\text{Lactic acid}}{CH_3{-}\overset{OH}{CH}{-}\overset{O}{C}{-}OH} + NAD^+ \qquad \frac{[\text{lactic acid}][\text{NAD}^+]}{[\text{pyruvic acid}][\text{NADH}]} \approx 30{,}000$$

When hydrogen must be transferred to pyruvic acid instead of to the respiratory chain, the system is said to be running an *oxygen debt*, at least at the tissue involved. Lactic acid accumulates under these circumstances.

## GLYCOLYSIS IN SPECIAL SYSTEMS

**Glycolysis in Muscles.** Although muscle cells possess a specific enzyme, *hexokinase*, for converting glucose to glucose 6-phosphate, the major source for this starting material is glucose 1-phosphate, which is made from the glycogen stored by the muscle cells. Action of the enzyme, *glycogen phosphorylase*, and of the phosphate ion on terminal glucose units in glycogen produces glucose 1-phosphate directly, a way of making a phosphate ester of glucose without sacrificing ATP. A second enzyme, *phosphoglucomutase*, catalyzes the rearrangement of the phosphate group from position 1 to position 6. From this point on we have glycolysis.

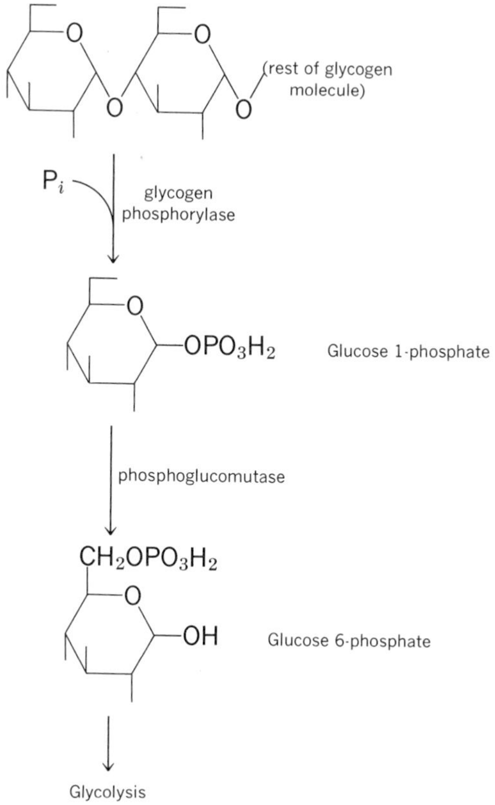

The supply of ATP in a relaxed, resting muscle is only enough to sustain muscular work for about half a second. The mechanisms for regenerating ATP must indeed work quickly and smoothly. As long as oxygen supplies are sufficient, the citric acid cycle with the respiratory chain fills the needs. If the cells become anaerobic, glycolysis takes over. For a few seconds, however, ATP (at least *actin-bound* ATP) can be regenerated from another high-energy phosphate, one with more energy than ATP. Called phosphocreatine, the resting muscle has from four to six times as much of it as ATP. Apparently phosphocreatine can donate a phosphate group to ADP, and muscles have the necessary enzyme:

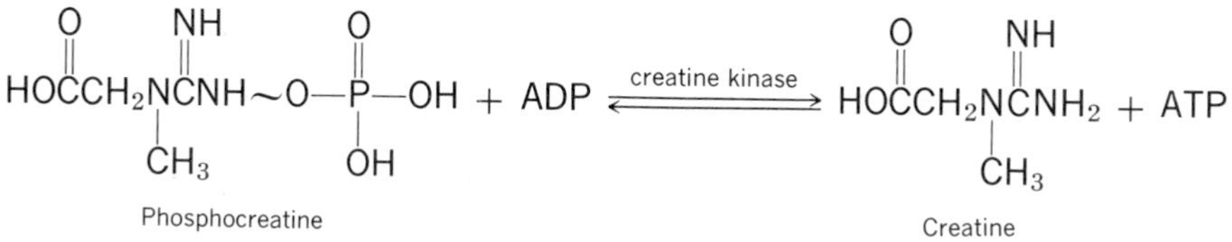

If the ADP is actin-bound (cf. p. 354), so too is the resulting ATP and the muscle is set to go on working. The supply of phosphocreatine can be regenerated from 1,3-diphosphoglyceric acid, an intermediate of very high energy produced during glycolysis (muscle cells possess the requisite enzyme):

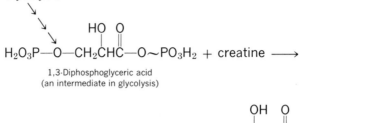

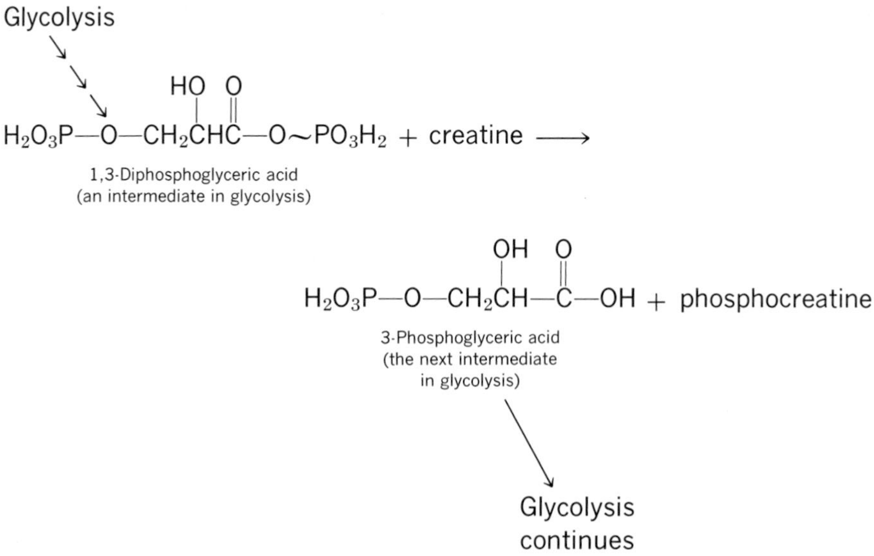

Glycolysis

$$\text{H}_2\text{O}_3\text{P}-\text{O}-\text{CH}_2\overset{\overset{\displaystyle\text{HO}}{|}}{\text{CH}}\overset{\overset{\displaystyle\text{O}}{\|}}{\text{C}}-\text{O}\sim\text{PO}_3\text{H}_2 + \text{creatine} \longrightarrow$$

1,3-Diphosphoglyceric acid
(an intermediate in glycolysis)

$$\text{H}_2\text{O}_3\text{P}-\text{O}-\text{CH}_2\overset{\overset{\displaystyle\text{OH}}{|}}{\text{CH}}-\overset{\overset{\displaystyle\text{O}}{\|}}{\text{C}}-\text{OH} + \text{phosphocreatine}$$

3-Phosphoglyceric acid
(the next intermediate
in glycolysis)

Glycolysis
continues

The net effect of the intervention of phosphocreatine at this stage is summarized in the following diagram:

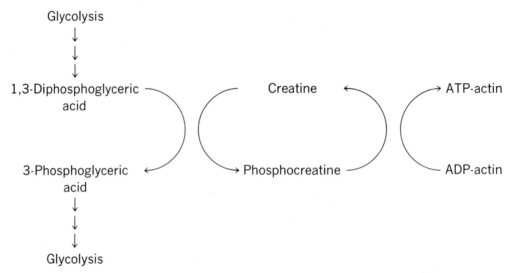

Glycolysis

1,3-Diphosphoglyceric acid — Creatine ← → ATP-actin

3-Phosphoglyceric acid ← → Phosphocreatine — ADP-actin

Glycolysis

Apparently there are several ways to keep the supply of ATP at a steady level, and the organism is poised to respond quickly to demands for it. Figure 14.4 provides a brief summary.

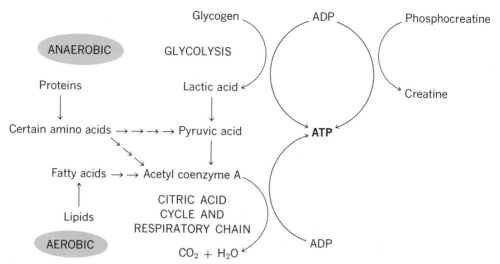

**Figure 14.4** Major sources of chemical energy for making ATP.

**Glycolysis in Microorganisms. Alcohol Fermentation.** When certain yeasts act to convert sugars to alcohol, they employ essentially the same reactions of glycolysis that we have already studied. The major difference is the way in which they handle the anaerobic storing of "$H_2$," a problem considered in Figure 14.2 and particularly in Figure 14.3. In muscle cell glycolysis this "$H_2$" is transferred to the keto group of pyruvic acid to make lactic acid, the temporary reservoir. In alcohol fermentation, once yeasts have made pyruvic acid, they decarboxylate it to form a new "$H_2$" acceptor, acetaldehyde:

$$CH_3-\overset{\overset{\displaystyle O}{\|}}{C}-\overset{\overset{\displaystyle O}{\|}}{C}-O \xrightarrow{\text{in certain yeasts}} CH_3-\overset{\overset{\displaystyle O}{\|}}{C} + O=C=O\uparrow$$

Pyruvic acid            Acetaldehyde    Carbon dioxide

When it accepts "$H_2$" from reduced $NAD^+$, ethyl alcohol is the final product:

$$CH_3-\overset{\overset{\displaystyle O}{\|}}{C} + NADH + H^+ \xrightarrow{\text{in certain yeasts}} CH_3-CH_2-O + NAD^+$$

Acetaldehyde                Ethyl alcohol

The liberated carbon dioxide causes the frothing or foaming action that accompanies fermentation.

## PHOSPHOGLUCONATE OXIDATIVE PATHWAY OF GLUCOSE CATABOLISM

A second way of catabolizing glucose in the absence of oxygen is by the phosphogluconate pathway or the "hexose monophosphate shunt." Our discussion of this pathway is brief, for even indirectly it is not important for ATP production in skeletal muscle. It is, however, a way of converting glucose ultimately to carbon dioxide and water without going through the citric acid cycle. *Its most important function is generating NADPH* (the reduced form of NADP, nicotinamide adenine dinucleotide phosphate—cf. p. 443), which is needed to furnish hydrogen for reduction steps whenever the body synthesizes fatty acids and steroids (referred to again in Chapter 15).

Starting from glucose 6-phosphate, the balanced equation representing the net effect of this pathway is

$$6 \text{ glucose 6-phosphate} + 12\text{NADP}^+ \xrightarrow{\text{hexose monophosphate shunt}}$$

$$5 \text{ glucose 6-phosphate} + 6\text{CO}_2 + 12\text{NADPH} + 12\text{H}^+ + \text{P}_i$$

There are of course several intermediate steps, some of them using enzymes of the glycolysis pathway. Among the intermediates are certain five-carbon sugars (e.g., ribose) needed by the body to make nucleotides and nucleic acids. For our future studies, however, we should remember that this shunt is an important way to make NADPH.

## REVERSAL OF GLYCOLYSIS

Although Figure 14.2 does not indicate whether the steps of glycolysis are reversible, most of them are. By referring to Figure 14.5, we see which ones are not. Certain tissues in the body, notably the liver (but kidney tissue too), possess enzymes for circumventing the bottlenecks to the complete reversal of glycolysis. One of these limiting steps is the conversion of pyruvic acid back to phosphoenolpyruvic acid. In glycolysis this step (in the forward sense) is essentially irreversible. To avoid it for the reversal of glycolysis, the body temporarily uses carbon dioxide. By converting pyruvic acid to oxaloacetic acid, the —CH$_2$— group becomes flanked by carbonyls, presumably making it energetically easier for the group to form an enol and become phosphorylated. We might envision the reactions as follows:

Step 1. $O{=}C + CH_2C{-}COH \xrightarrow[\text{ADP} + \text{P}_i]{\text{ATP}} HO{-}C{-}CH_2C{-}COH$

Pyruvic acid   (pyruvate carboxylase)   Oxaloacetic acid

$$P{-}O{-}\{\text{GTP system}\}$$

Step 2. $H{-}O{-}C{-}CH_2{-}C{-}C{-}OH \xrightarrow{\text{phosphoenolpyruvate}}$

Oxaloacetic acid

carboxykinase
(Activity of this enzyme
is sharply increased in
acute insulin insufficiency.)

$$O{\sim}P + \{\text{GDP system}\}$$

$$H^+ + O{=}C + CH_2{=}C{-}C{-}OH$$

Phosphoenolpyruvic acid

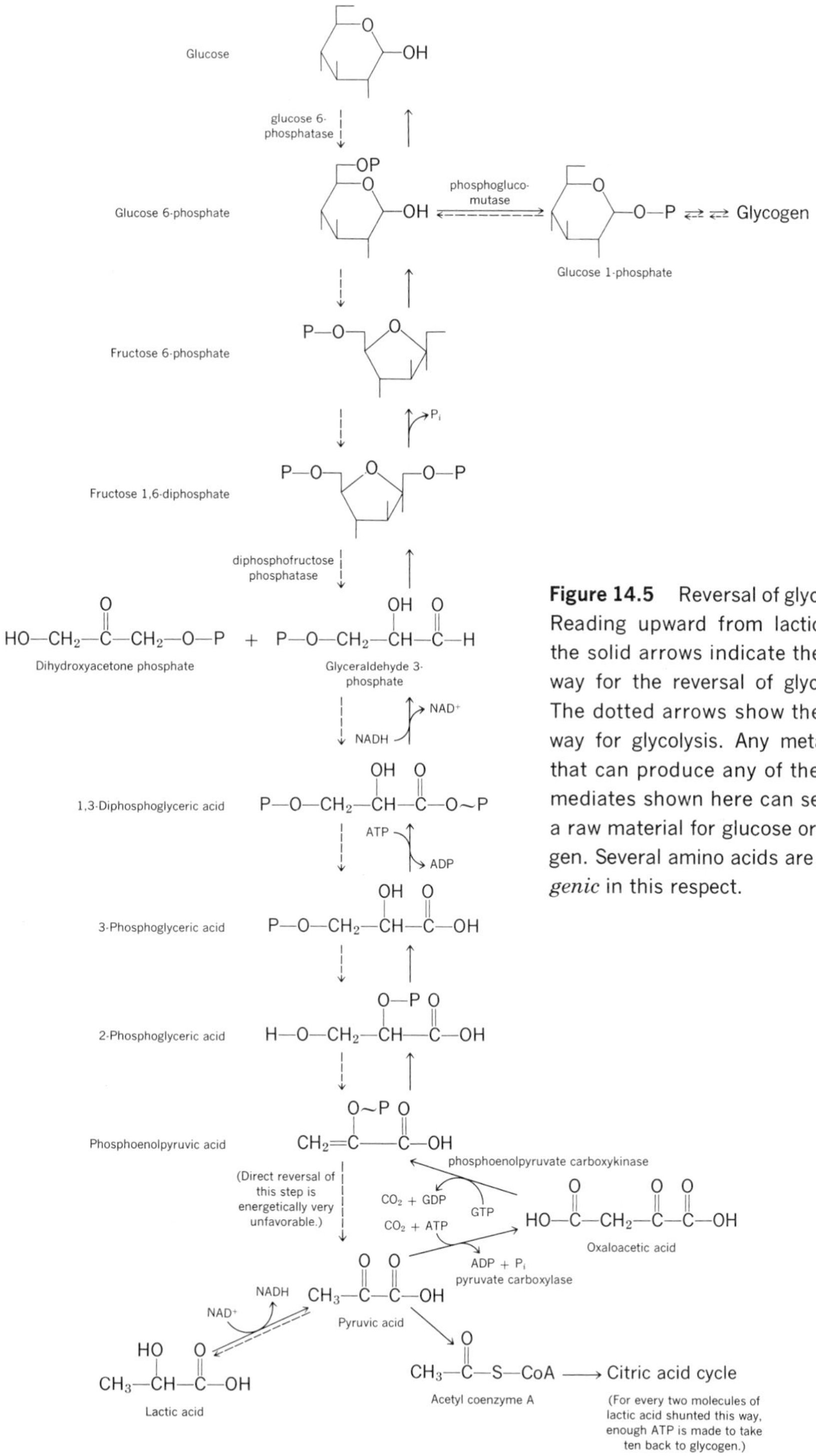

**Figure 14.5** Reversal of glycolysis. Reading upward from lactic acid, the solid arrows indicate the pathway for the reversal of glycolysis. The dotted arrows show the pathway for glycolysis. Any metabolite that can produce any of the intermediates shown here can serve as a raw material for glucose or glycogen. Several amino acids are *glucogenic* in this respect.

$$\overset{\displaystyle H}{\underset{\displaystyle |}{}}$$

Step 1, reminiscent of an aldol condensation, is the addition of an $\alpha - C$ system across the carbonyl of another molecule, here carbon dioxide. Just how ATP powers this reaction is not known, but it is consumed at this point. The rough mechanism indicated in step 2 is pure conjecture. A second high-energy phosphate is required as a phosphate donor. Instead of ATP, a close relative, GTP (guanosine triphosphate) or ITP (inosine triphosphate), is used. (We need not be concerned about their structures since they behave like ATP.) Thus two molecules of high-energy phosphate are required to circumvent this bottleneck and to make phosphoenolpyruvic acid from pyruvic acid.

Another bottleneck is the conversion of fructose 1,6-diphosphate to fructose 6-phosphate. Because only certain kinds of cells, notably those of the liver and kidneys, have the requisite enzyme, diphosphofructose phosphatase, glycolysis can be reversed only in them. When the glucose 6-phosphate stage is reached, the next step depends on the presence and activity of still another enzyme, glucose 6-phosphatase. Again liver cells contain it. Its activity is controlled by hormones, a topic discussed in greater detail in our study of diabetes mellitus. In any event, at this stage either glucose or glycogen is made.

Two important points have been made in this section. First, glycolysis can be reversed; that is, glucose can be made from noncarbohydrate sources. Second, the reversal of a metabolic pathway is not simply the reversal of each separate step. In at least one place, usually more, a slightly different pathway, a bypass involving enzymes not needed for the forward reactions, is chosen. Wherever such bypasses occur, the direction of the pathway can be controlled by managing the activities of the enzymes for the bypass.

## GLUCOSE DELIVERY AND TRANSPORT

### ABSORPTION

Glucose, fructose, and galactose, the principal end products of the digestion of carbohydrates, readily move across the intestinal barrier, partly by simple diffusion and partly by active transport. In migration by active transport enzyme-catalyzed reactions occur, consuming energy (ATP) and converting the substances temporarily into forms that themselves diffuse. Interlacing the walls of the various parts of the gastrointestinal tract are capillaries that are tributaries of larger and larger blood vessels which finally join the hepatic portal vein[1] (Figure 14.6). This network provides intimate access between the digestive tract and the circulatory system. Nutrient molecules formed during digestion are therefore carried first by portal circulation through the liver. The blood passes through innumerable

---

[1] *Hepatic* is from the Greek: *hepat-*, liver; *-ikos*, akin to—that is, of, relating to, or affecting the liver. *Portal* is from the Latin *porta*, gate—that is, relating to a place of entry, a communicating part or region of an organ. The portal region of the liver, the *porta hepatis*, is the transverse fissure on its underside, where a number of vessels enter.

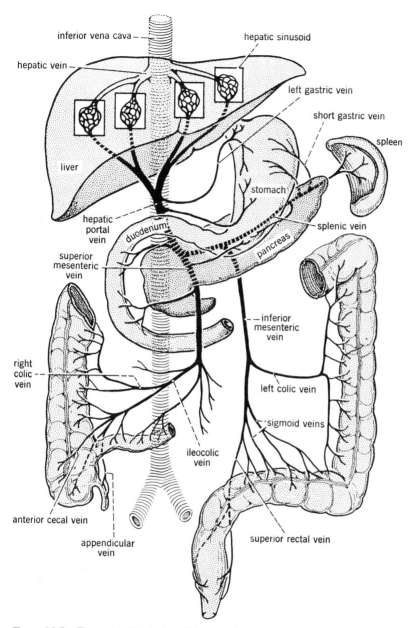

**Figure 14.6** The major tributaries of the hepatic portal vein. (From A. J. Berger, *Elementary Human Anatomy,* John Wiley and Sons, New York, 1964, page 372.)

capillary networks (hepatic sinusoids in Figure 14.6) before it emerges into the inferior vena cava and travels from there to the right atrium of the heart. While the blood percolates through the liver, nutrients may or may not be removed, depending on several factors. This largest organ of the body has an extraordinary number of important functions involving all classes of foods. We are concerned here about what it does with glucose and how it participates in regulating glucose metabolism elsewhere. In describing these activities, we shall find certain terms useful. Since they are compounded from reasonably familiar word parts, they are almost self-defining.

**Glucogenesis** (glucose genesis or glucose making). When the body synthesizes glucose from any product of glycolysis (e.g., lactic acid or pyruvic acid) or from other carbohydrate sources, glucogenesis is said to have occurred. The liver is the principal site for it (Figure 14.7).

**Gluconeogenesis** (gluco-neo-genesis; "neo-" is a word part meaning "new"). Noncarbohydrate materials such as proteins and lipids can be converted into glucose by certain cells, notably in the liver. Glucose can also be made from the skeletons of *glucogenic* amino acids (sometimes trimming may be necessary). Fatty acids must normally participate to provide hydrogen. When glucose is made from noncarbohydrates, a new source, gluco*neo*genesis is said to take place. Specific details must be left to Chapter 16.

**Glycogenesis** (glycogen genesis or glycogen making). Whatever the source of glucose molecules, some of them may be taken out of circulation and converted into glycogen, a process called glycogenesis which occurs principally in the liver and in muscles. Such glycogen is in a dynamic state, meaning that its glucose units enter and leave constantly. Figure 14.8 outlines the principal steps in glycogenesis from glucose.

**Glycogenolysis** (glycogen-lysis; i.e., glycogen breakdown or hydrolysis). When glycogen reserves are depleted because glycogen breaks down or is hydrolyzed faster than glucose units are added, the process is called glycogenolysis. The glycogen content of muscles is rapidly used up in severe exercise, and fresh glucose molecules must be taken in from the bloodstream. To replenish this supply, liver

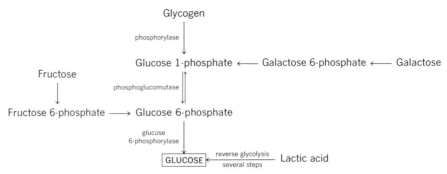

**Figure 14.7**  Glucogenesis. Sources of glucose other than direct absorption of dietary glucose. Phosphorylase is activated by epinephrine. The activity of glucose 6-phosphorylase appears to be enhanced in diabetes mellitus.

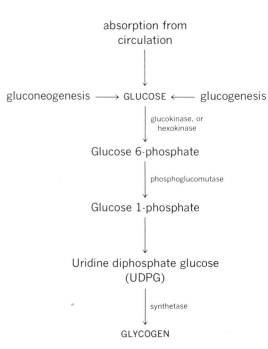

absorption from
circulation
|
↓

gluconeogenesis ──→ GLUCOSE ←── glucogenesis

| glucokinase, or
| hexokinase
↓

Glucose 6-phosphate

| phosphoglucomutase
↓

Glucose 1-phosphate

|
↓

Uridine diphosphate glucose
(UDPG)

| synthetase
↓

GLYCOGEN

**Figure 14.8** Glycogenesis. The activity of the enzyme *synthetase* is normally stimulated by an excess of glucose 6-phosphate, which means that the extra glucose 6-phosphate will be removed and made into glycogen. The activity of the enzyme *glucokinase* drops in diabetes. (UDPG is a complex derivative of glucose whose structure we shall not examine.)

glycogen can break down. Several factors initiate it: physical exercise, fasting, epinephrine (a hormone released during periods of fright), failure of the insulin supply, and other hormonal activities. Glucose released from glycogen reserves in the liver reaches muscles by the circulatory system.

Glucose found in the bloodstream is called *blood sugar*. Like the circulation of money in a healthy economy, the circulation of blood sugar in a healthy body is controlled by routine supply and demand and must be able to respond to emergencies. The connections between liver glycogen, muscle glycogen, blood sugar, and lactic acid are displayed by means of a modified Cori cycle in Figure 14.9.

Carl and Gerty Cori, who shared the Nobel prize in physiology and medicine with B. Houssay in 1947, made their greatest contributions to the chemistry of glycogenesis and glycogenolysis. They succeeded in synthesizing glycogen *in vitro* from glucose, ATP, and certain enzymes.

## BLOOD SUGAR LEVEL

The word *level* in this term means concentration, and it is expressed in milligrams of glucose per 100 ml of blood. After eight to twelve hours of fasting, the blood sugar level in the venous blood of an average adult is usually in the range of 60 to 100 mg per 100 ml, called the *normal fasting level*.

Special words, describing levels outside this range, are constructed from word

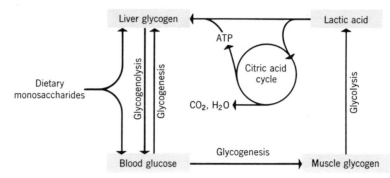

**Figure 14.9** Modified Cori cycle, showing relations of glycogenesis, glycogenolysis, glycolysis, and the citric acid cycle to one another.

parts that make the terms almost self-defining.[2] When concentrations of blood sugar are above the normal fasting level, the condition is called *hyperglycemia*. *Hypoglycemia* is a condition with blood sugar below the normal fasting level. If the blood sugar level goes high enough, the kidneys remove some of the glucose and put it into the urine. The blood sugar level above which this happens, called the *renal threshold* (Latin *renes,* kidneys) for glucose, is normally about 140 to 160 mg of glucose per 100 ml of blood, sometimes higher. When detectable quantities of glucose appear in the urine, the condition is called *glucosuria*. These conditions are not in themselves diseases, but they are usually symptoms of a disorder. Anyone who eats a meal extremely rich in carbohydrate may temporarily become hyperglycemic, even to the point of being glucosuric. Anyone who fasts or is starving will sooner or later become hypoglycemic. People with diabetes mellitus are usually hyperglycemic and glucosuric; if they give themselves an overdose of insulin, they will rather promptly become hypoglycemic and go into "insulin shock." A very low blood sugar level is obviously serious. Hyperglycemia, on the other hand, is not too serious *in itself;* events happen in the body to reduce it (Figure 14.10).

**Tissue Dependence on Blood Sugar Level.** Different tissues depend to various degrees on direct access to glucose from the bloodstream as a source of energy. Heart muscle, for example, can use a variety of metabolites including lactic acid and fatty acids. Skeletal muscles use fatty acids and certain of their breakdown products to maintain themselves during resting periods, and they draw upon stored glycogen in times of stress. This glycogen must eventually be replenished, but there will be no swift muscle failure if the blood sugar level drops. The brain and the central nervous system are more vulnerable.

[2] The word parts are as follows:

    *hyper-*, over, above, beyond, excessive; e.g., *hyper*sensitive.

    *hypo-*, under, beneath, down; e.g., *hypo*dermic (beneath the skin).

    *glyco-* or *glyc-*, may refer to a "glycose" (generic name for a monosaccharide) or to glycogen; usually refers to glucose, as in hyper*glyc*emia.

    *gluco-* or *gluc-*, refers to glucose.

    *-emia*, refers to blood or a condition of having a specific substance in the blood.

    *-uria*, refers to urine or a condition of having a specific substance in the urine.

The glycogen content of the brain is about 0.1% by weight, a reserve supply of glucose units far too low to sustain brain function for more than a short time. Under normal conditions the brain obtains virtually all its required energy from glucose taken directly from the bloodstream. Mitochondria of the brain and the central nervous system are particularly active in glycolysis. The citric acid cycle also occurs. Any interference in the smooth operation of this cycle will soon appear as malfunction of the brain and the central nervous system. Thus if the blood sugar level drops for any reason to very hypoglycemic levels, the brain loses its main source of energy. For a short while it can manage by catabolizing amino acids and lipids, but severe hypoglycemia usually means convulsions and coma. Damages to the brain may be permanent. Even a temporary, mild hypoglycemia may cause dizziness and fainting spells. Regulation of the blood sugar level is obviously a matter of considerable importance. A healthy liver and a smoothly functioning endocrine system are indispensible. Figure 14.11 shows the general means by which the endocrine system participates in controlling the blood sugar level. Proper control of the blood sugar level by the body is a measure of its *glucose tolerance,* that is, its ability to use glucose in a normal way.

## THE LIVER IN GLUCOSE METABOLISM

For a detailed study of the molecular basis of a disease, diabetes mellitus, we need to know more about glucose metabolism in the liver. In the following discussion the circled numbers refer to those in Figure 14.12.

In Figure 14.12 we find glucose in circulation ① in a capillary within the liver.

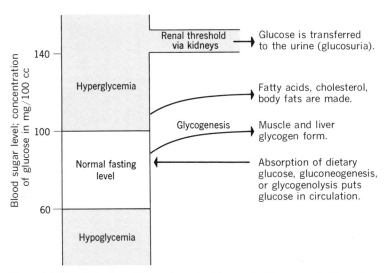

**Figure 14.10** Reduction of hyperglycemia. Glucose that is not needed to replenish glycogen reserves in muscles or the liver or to supply the brain and central nervous system is for the most part converted to fat. Normally the renal threshold for glucose is not exceeded, and glucose is not found in the urine. In diabetes mellitus, however, glucosuria is standard.

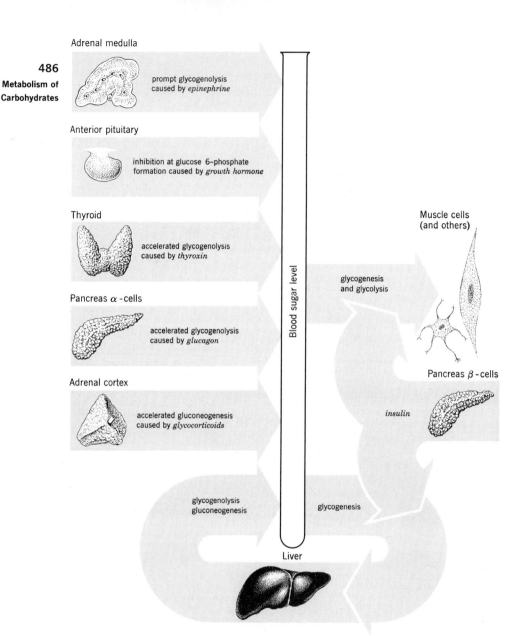

**Figure 14.11** The blood sugar level as influenced by hormones. Arrows pointing *in* signify factors that tend to make the blood sugar level rise. Arrows pointing *out* signify factors causing lower blood sugar levels. Some of the hormones exert their action within the liver, which gives this organ a central role in the regulation of the blood sugar level. (Adapted from I. N. Kugelmass, *Biochemistry of Blood in Health and Disease,* Charles C Thomas, Springfield, Ill., 1959. Courtesy of Charles C Thomas.)

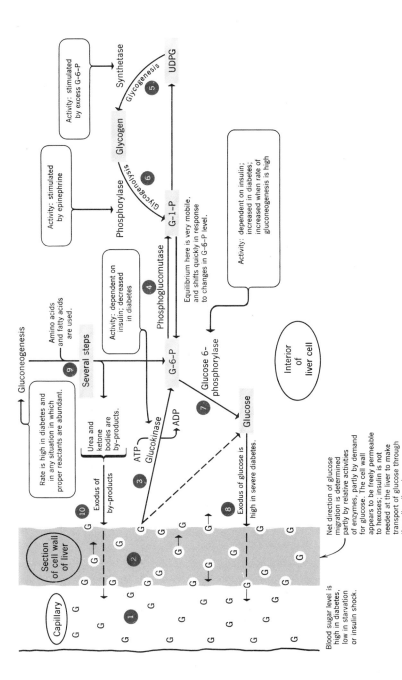

**Figure 14.12** Some aspects of glucose metabolism in liver cells. Circled numbers are explained in text. G, glucose; G-6-P, glucose 6-phosphate; G-1-P, glucose 1-phosphate; UDPG, uridine diphosphate glucose (a derivative of glucose).

Some glucose molecules are leaving this capillary and migrating through a wall of a liver cell ②. Once inside the liver cell a glucose molecule will normally be phosphorylated ③, that is, converted to glucose 6-phosphate (G-6-P) at the expense of ATP. The enzyme for this is *glucokinase,* and insulin somehow affects its activity, apparently by controlling the rate of its synthesis and therefore its concentration. The presence of another enzyme, *phosphoglucomutase,* makes possible the rearrangement of G-6-P to glucose 1-phosphate (G-1-P) at ④. By steps we shall not study, G-1-P can be made into glycogen at ⑤. But if the glucose supply is low (e.g., developing starvation), glucose can be removed from storage ⑥ through the action of the enzyme *phosphorylase*—followed by reversal of ④, followed by hydrolysis of G-6-P at ⑦, followed by migration out of the cell at ⑧. Epinephrine, a hormone, makes the enzyme phosphorylase at ⑥ especially active, causing rapid depletion of the reserves of glycogen.

The enzymes at ③ and ⑦, *glucokinase* and *glucose 6-phosphorylase,* have activities that adjust to the concentrations of glucose and G-6-P. In a *fed* animal the rate at ③ will normally equal or slightly exceed that at ⑦, and a net intake of glucose occurs. In the *fasted* animal the rate at ⑦ moderately exceeds that at ③ to help take glucose out of storage and put it into circulation, especially to meet the needs of the brain and the central nervous system.

In diabetes mellitus the rate at ③ is much slower than at ⑦. In other words, this mechanism for "trapping" glucose inside the cell by converting it to a phosphate ester is no longer effective. Furthermore, the enzymes for gluconeogenesis at ⑨ become much more active in diabetes. At least the liver makes even more glucose by this method. In so doing it also makes urea and the so-called "ketone bodies," byproducts of gluconeogenesis which migrate out of the cell at ⑩. *This overproduction of "ketone bodies" would make unchecked diabetes mellitus fatal if some other complication did not cause death.* Exactly what the ketone bodies are and why their overproduction is serious we shall discuss in Chapter 15. For the present it is sufficient to say that two of these compounds are acids, and that they lower the pH of the blood.

In summary, the liver can take glucose out of circulation, store it, catabolize it, synthesize it (glucogenesis or gluconeogenesis), or release it back into circulation. What happens at any particular moment depends on concentrations of the metabolites, including glucose, or hormones, and on the relative activities of many enzymes. As we have seen, hormones frequently act by moderating or activating enzymes. In diabetes mellitus, even though the blood sugar level is already high, liver cells make even more glucose to send the blood sugar level higher. Very harmful by-products are formed at the same time. The most important factor in regulating what the liver does to glucose is the hormone insulin.

## DIABETES MELLITUS AND INSULIN

Diabetes comes from the Greek *diabetes,* to pass—here meaning to pass urine in greater than normal amounts. Mellitus signifies "honey-sweet." Diabetes mellitus therefore signifies passage of urine containing significant quantities of dissolved sugar (glucose), which is the condition glucosuria.

The primary defect in diabetes mellitus is the failure of glucose to enter into specific tissues because effective insulin is absent. Insulin is a protein manufactured and stored in the $\beta$-cells of the islets of Langerhans in the pancreas. Located as shown in Figure 14.6, the pancreas controls the release of insulin. When the blood sugar level rises, insulin is put into circulation; when the blood sugar level drops, the rate of insulin release falls off.

The primary action of insulin is to promote the transfer of glucose from circulation, across cell walls, and into the cells of several kinds of tissues. Insulin produces hundreds of changes all through the body, but most, if not all of them, are the result of its primary action, helping to take glucose out of circulation and put it into various cells. Insulin also directly enhances the activity of the enzyme glucokinase in the liver (cf. Figure 14.12). At least the activity of this enzyme is related to the presence of insulin, and there is evidence that insulin in some unknown way stimulates its synthesis.

Not all types of tissues are sensitive to insulin. Insulin is not needed, for example, to help transport glucose into cells in the brain, the kidneys, and gastrointestinal tract, and into red blood cells. It is needed to get glucose across the cell barriers in skeletal and cardiac muscle and in adipose tissue (fatty tissue), among others.

If effective insulin is absent, the blood sugar level rises much higher than normal, even in a fasting condition. In addition, liver cells are stimulated to manufacture even more glucose (by reactions to be studied later). The blood sugar level is eventually so high that the renal threshold for glucose is exceeded, and the kidneys transfer some of it to the urine.

Most cases of diabetes are discovered when a patient sees a doctor about some other condition (e.g., arteriosclerosis, persistent drowsiness, loss of appetite, skin infections, and many other ailments). Routine urine analysis then reveals the presence of glucose, and even if this is not observed, the doctor may use other more sensitive tests. Analysis of the glucose in the blood may reveal that the fasting blood sugar level (in blood taken from a vein, i.e., venous blood) is above 120 (but not necessarily above the renal threshold), indicating diabetes; even a level between 100 and 120 is suggestive. Further tests are made. The blood sugar level may be analyzed two hours after the patient has ingested a meal containing 100 g of glucose. By this time, in a normal person, the blood sugar level will be back below 100. If it is not, diabetes is indicated.

**Glucose Tolerance Test.** Perhaps the best and most useful test for diagnosing diabetes is the glucose tolerance test. For about three days before the test the patient ingests upward of 300 g of carbohydrate per day in addition to the normal components of the diet. Then after a fast of about ten hours (overnight) the subject drinks a solution of 100 g of glucose in 400 ml of water flavored with lemon. A specimen of venous blood is immediately analyzed for blood sugar level, and further analyses are made at intervals of 0.5, 1, 1.5, 2, 3, and sometimes 4 and 5 hours. Urine specimens are also analyzed. In a person without diabetes the following values are usually observed.

| | |
|---|---|
| Fasting blood sugar level: | not above 100 mg per 100 cc of venous blood |
| Maximum or peak value: | less than 160 |
| Value at two-hour interval: | less than 120 |

In the diabetic subject the fasting and peak levels are higher and, of special significance, the rate at which the level returns to "normal" is much slower. Typical *glucose tolerance curves* plotted from the data for diabetic and nondiabetic persons are shown in Figure 14.13.

**Hyperinsulinism. "Insulin Shock."** When too much insulin appears in the blood, a condition called hyperinsulinism, an excessive amount of glucose is removed from the bloodstream by any or all of the several processes of glucose utilization, and the blood sugar level drops. The ensuing hypoglycemia may cause malnutrition in the brain. Since its cells depend directly on blood sugar, hyperinsulinism may produce convulsions and "insulin shock." The remedy is immediate ingestion of some readily digestible carbohydrate such as fruit juice, candy, sugar lumps, etc.

Severe hypoglycemia can occur in diabetic people if too much insulin is inadvertently injected. Moreover, their insulin needs can vary if they become ill in other ways. If a nondiabetic individual ingests a relatively large amount of carbohydrate, the sharp rise in blood sugar level may cause a small overstimulation of the pancreas. As the normal glucose tolerance curve of Figure 14.13 indicates, about two to three hours later the blood sugar level is slightly below the normal fasting level. If it goes too low, the brain cells experience a mild scarcity of glucose, and the subject may feel irritable, bleary-eyed, and sleepy. Many whose breakfast consists almost entirely of fruit juice, a heavily sugared roll or doughnut, and sweetened coffee feel drowsy by midmorning. The coffee break becomes a welcome relief, but

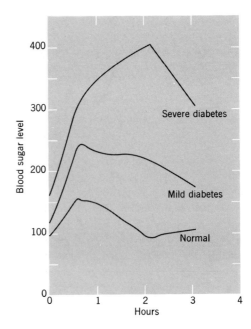

**Figure 14.13** Glucose tolerance curves. (From I. N. Kugelmass, *Biochemistry of Blood in Health and Disease*, Charles C Thomas, Springfield, Ill., 1959. Courtesy of Charles C Thomas.)

of course it is only another round of pastry and sugared coffee. The blood sugar level rises and the brain cells perk up, but the pancreas dutifully pours out insulin to restore the level to normal—again a bit more insulin than needed, and two or three hours later grogginess reappears. Many do not achieve any balance in their diet until the evening meal.

The most important meal of the day is breakfast, and it should be rich in proteins with some fat and carbohydrate. Proteins and fat tend to moderate the speed with which glucose leaves the digestive tract. Instead of pouring suddenly into the bloodstream, glucose enters over a longer period, and overstimulation of the pancreas is avoided.

## EPINEPHRINE

Of the several hormones besides insulin that regulate glucose catabolism, epinephrine (adrenaline) is particularly important in emergencies. Relatively simple in

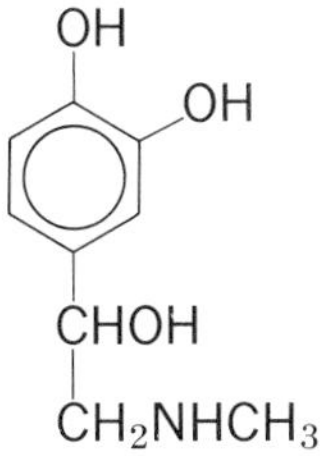

Epinephrine

structure, this hormone is made by cells in the adrenal medulla, and in times of danger or stress it is discharged into the bloodstream in minute amounts. It activates the enzyme phosphorylase in the liver (cf. Figure 14.12) and in muscles, leading to prompt breakdown of glycogen to glucose. Glucose thus formed in the liver migrates quickly into the bloodstream, and the blood sugar level may rise abruptly, very often to the renal threshold. This sudden appearance of blood glucose makes available to the brain its chief nutrient at the precise moment when it may be required to react the most sharply to a threatening situation. This hormone also affects lipid metabolism by stimulating the breakdown of stored triglycerides and causing the concentration of free fatty acids in the bloodstream to rise quickly.

## REFERENCES

E. T. Bell. *Diabetes Mellitus.* Charles C Thomas, Springfield, Ill., 1960.

I. N. Kugelmass. *Biochemistry of Blood in Health and Disease.* Charles C Thomas, Springfield, Ill., 1959.

A. L. Lehninger. "The Transfer of Energy Within Cells," in *Ideas in Modern Biology,* edited by J. A. Moore. Natural History Press, Garden City, N.Y., 1965.

A. L. Lehninger. *Bioenergetics.* W. A. Benjamin, New York, 1965.

A. L. Lehninger. *The Mitochondrion.* W. A. Benjamin, New York, 1964.

G. Litwack and D. Kritchevsky, editors. *Actions of Hormones on Molecular Processes.* John Wiley and Sons, New York, 1964.

A. Marble and G. F. Cahill, Jr. *The Chemistry and Chemotherapy of Diabetes Mellitus.* Charles C Thomas, Springfield, Ill., 1962.

E. F. Neufeld and V. Ginsburg. "Carbohydrate Metabolism." *Annual Review of Biochemistry,* Vol. 34, 1965, page 297.

A. White, P. Handler, and E. L. Smith. *Principles of Biochemistry,* third edition. McGraw-Hill Book Company, New York, 1964.

R. H. Williams. *Disorders in Carbohydrate and Lipid Metabolism.* W. B. Saunders, Philadelphia, 1962.

H. G. Wood and M. F. Utter. "The Role of $CO_2$ Fixation in Metabolism," in *Essays in Biochemistry,* edited by P. N. Campbell and G. D. Greville, Vol. 1. Academic Press, New York, 1965.

## PROBLEMS AND EXERCISES

1. Discuss the meaning of each of the following terms.

   (a) glycolysis  (b) anaerobic sequence  (c) an ene-diol
   (d) oxygen debt  (e) active transport  (f) glucogenesis
   (g) gluconeogenesis  (h) glycogenesis  (i) glycogenolysis
   (j) blood sugar  (k) blood sugar level  (l) normal fasting level
   (m) portal vein  (n) hyperglycemia  (o) hypoglycemia
   (p) renal threshold  (q) glucosuria  (r) glucose tolerance
   (s) glucose tolerance test  (t) hyperinsulinism

2. Describe what each of the following do. Use equations where appropriate.

   (a) glucokinase  (b) glucose 6-phosphorylase  (c) phosphoglucomutase
   (d) phosphorylase  (e) insulin  (f) epinephrine

3. What is the difference between glucogenesis and gluconeogenesis?

4. Explain how pyruvic acid serves to regenerate an enzyme needed in glycolysis.

5. Describe by equations two events in glycolysis that involve an ene-diol.

6. Write the equations for the two "ATP payoffs" in glycolysis.

7. Explain by equations how glycolysis by yeast cells differs from that accomplished in human muscle cells.

8. Explain why hypoglycemia is serious.

9. Discuss the relative activities of glucokinase and glucose 6-phosphorylase in (a) a fed animal, (b) a fasted animal, and (c) a person with diabetes mellitus.

10. Describe how the body maintains a fairly constant blood sugar level, even though it ingests glucose only two or three times a day.

11. In periods of fasting, would you expect the amount of glycogen in the liver to increase or to decrease? Why?

12. Why does severe hyperinsulinism lead quickly to "insulin shock," whereas the absence of insulin, at least for a short period, is not so serious?

13. Is glucosuria always caused by diabetes mellitus? If not, how else might it arise?

# CHAPTER FIFTEEN

# Metabolism of Lipids

People who, for whatever reason, eat much and exercise little gain weight. The body has a remarkable capacity for converting carbohydrates and proteins into fat and storing it. In this chapter we study how the body makes and uses lipids, particularly the simple triglycerides.

## ABSORPTION, DISTRIBUTION, AND STORAGE OF LIPIDS

The complete digestion of simple lipids produces glycerol and a mixture of long-chain fatty acids. Such complete hydrolysis, however, is not necessary for absorption to occur, for monoesters of glycerol, the monoglycerides, can also leave the intestinal tract.

As molecules of monoglycerides, fatty acids or their anions, and glycerol migrate across the intestinal barrier, triglyceride molecules are extensively resynthesized until what is delivered for circulaton is essentially reconstituted lipids. Only a very small fraction is delivered directly into portal circulation, however. Most of the absorbed lipid emerges into the lymph system. Lymph coming from regions of the digestive system is ultimately delivered to the bloodstream via the thoracic duct; most of the lipid is present as microdroplets or chylomicra averaging one micron in diameter.

Triglycerides found in the bloodstream, whether as microdroplets or as individual molecules, are associated with plasma-soluble proteins, and thus the strong tendency for lipids to coalesce and to remain out of solution in water is defeated. Triglycerides and fatty acids are normally transported with no difficulty. The plasma is known to contain a lipase, an enzyme that can catalyze the hydrolysis of the ester linkages in triglycerides. Thus fatty acids are formed, and these too are bound to plasma-soluble proteins for ready transport. Certain tissues can remove these fatty acids and use them for ATP synthesis. Those not so used are eventually stored in the specialized connective tissue, adipose tissue.

Cells of adipose tissue can remove fatty acids from the bloodstream, and they can synthesize triglycerides not only from fatty acids but also from glucose. The fat collects as larger and larger droplets in the cytoplasm of each cell until the cell's nucleus is pushed to the edge and becomes flattened. This fat-laden tissue is found principally around abdominal organs and beneath the skin, and it constitutes the body's chief reserve of chemical energy, far exceeding the normal carbohydrate reserves. Lipid deposits around internal organs serve to cushion and protect them from mechanical shock when the jars and bumps of living are experienced. Adipose tissue found subcutaneously, that is, beneath the skin, serves also as insulation against excessive heat loss to the environment (or as some protection against cold). Once thought to be simply a passive insulation, something like an ordinary blanket, subcutaneous adipose tissue is now believed to act more like an electric blanket. Metabolic activity in the cells of this tissue generates a small amount of heat to offset heat loss when the outside temperature drops. Adipose tissue, wherever it is found, is quite active metabolically. The cells have large numbers of mitochondria, nerves extend in and among them, and the bloodstream pushes its capillary network deep into this tissue. In fact, one of the problems associated with obesity is the extra load placed on the heart not only to assist in moving the extra weight during exercise but also to pump blood throughout a larger network.

The average adult human male receiving an adequate diet has enough lipid in reserve to sustain his life for thirty to forty days, assuming he has enough water. For an individual in a modern, developed country this reserve is obviously much more than he needs, although in more primitive times it may have been essential to the survival of the race. The body need not be without food too long before the fat reserves are tapped. After a few hours the body's glycogen reserves virtually disappear. There is only enough lipid material or glucose in the bloodstream to sustain average metabolic needs for several minutes. The fat deposited in the liver provides the energy needed for slightly more than an hour. Obviously other, larger reserves are needed and the body has two. One is the unabsorbed food in the digestive tract. Although a temporary reserve, it is an important one and quite plainly much used by people who are fortunate enough to eat regularly. The other reserve is the adipose tissue. In the steady state, that is, in the normally fed individual, about 30% of the carbohydrate taken in is temporarily changed into triglyceride, and virtually all this conversion takes place within the cells of this tissue. In an area of very active research virtually no one dissents from this conclusion. (Earlier researchers believed that the liver was the main site for fatty acid synthesis, and that newly made fatty acid molecules were released into circulation, picked up by adipose cells, and converted into triglycerides.)

The advantages of storing chemical energy as triglyceride can be understood in terms of the *energy densities,* grams per calorie, of various storage materials. When glucose is stored as wet glycogen in tissues, it requires about 0.6 g of such material to yield 1 calorie. Essentially no water is involved in the storage of triglycerides, however, and only about 0.13 g of the substance is needed to yield 1 calorie.

Energy density can be thought of in other terms. A molecule of a fatty acid can deliver over a hundred molecules of ATP, depending on the chain length, whereas a molecule of glucose can make available only about thirty-eight. Almost half the hydrogens in glucose are already attached to oxygen, but hydrogens attached to *carbon* are by and large the ones that fuel the respiratory chain (from which they ultimately emerge bound to oxygen). Fatty acid molecules have all but one of their hydrogens attached to carbon.

The conversion of glucose to fatty acids and thence to triglycerides is not without some cost. Energy is required, and about a quarter of the glucose supply must be sacrificed to furnish it. If 30% of carbohydrate taken in is changed into lipids, and if another 25% is used to power this change, a significant portion of all oxygen consumption and heat production in the body goes into just this operation.

If food intake (more correctly, chemical energy intake) exceeds energy expenditure, the newly synthesized triglyceride becomes part of a relatively inactive "compartment" of adipose tissue. Considerable indirect evidence indicates that there are two compartments for storing fat. Just what or precisely where they are or how they are distributed is not known, anatomically, but it is known that a portion of the deposited lipid stays put a long time, and that another portion has a more rapid turnover. Hence the *idea* of two compartments, even though the word itself may misleadingly suggest walled-off regions. The half-life of lipid in the one compartment is from 350 to 500 days, meaning that 50 g of an initial 100 g have after this length of time been replaced through the release of stored lipid and the substitution of a fresh supply. In the active "compartment" the half-life is on the order of 30 to 40 days, as determined by radioactive labeling studies. A great deal of metabolic activity occurs in this compartment. In succeeding sections we shall study in greater detail how fatty acids are made, how glycerides are put together, how fatty acids are taken from storage, and how they are oxidized to produce ATP. The events that take place in the turnover of triglyceride in adipose tissue are outlined in Figure 15.1.

## SYNTHESIS OF TRIGLYCERIDES FROM GLUCOSE

**General View.** Glucose is in a more oxidized state than a fatty acid. To convert glucose into fatty acid therefore requires a reducing agent, that is, a source of the elements of hydrogen, $(H:^- + H^+)$. To make triglycerides we also need a source of glycerol, or its equivalent. And we obviously need carbon parts for the fatty acids. Glucose provides essentially all these services (Figure 15.2).

Some of the glucose taken in by adipose tissue is oxidized by the phosphogluconate pathway, the hexose monophosphate shunt described briefly in Chapter 14 (p. 477) and called here simply the *shunt*. The net result of this series of reactions, which we did not study in detail, was the oxidation of glucose and the reduction of $NADP^+$, the coenzyme described on page 443 as a close relative of $NAD^+$. The shunt was summarized as follows:

$$6 \text{ glucose 6-phosphate} + 12NADP^+ \longrightarrow$$
$$5 \text{ glucose 6-phosphate} + 12NADPH + 12H^+ + 6CO_2 + P_i$$

Cells of adipose tissue have the necessary enzymes for this shunt, and one of its main purposes appears to be furnishing NADPH for fatty acid synthesis.

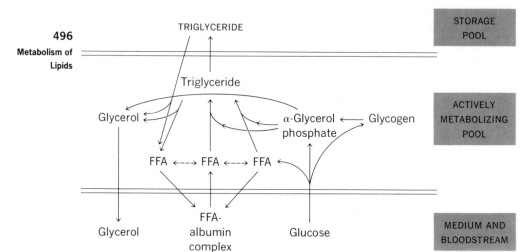

**Figure 15.1** Triglyceride turnover in adipose tissue (FFA stands for free fatty acids). In this diagram the concept of compartments for triglyceride storage represents the *idea* rather than implying that they have been identified anatomically. (From B. Shapiro, "Triglyceride Metabolism," in *Handbook of Physiology, Section 5: Adipose Tissue,* edited by A. E. Renold and G. F. Cahill, Jr., American Physiological Society, Washington, D.C., 1965, page 222. Used by permission.)

Most of the rest of the glucose taken in by adipose tissue is catabolized by glycolysis to furnish, finally, two compounds essential to the formation of triglycerides. One is needed directly, $\alpha$-glycerol phosphate, for this form of glycerol, its monoester with phosphoric acid, is required to make triglycerides. Glycerol itself will not do. It must be the $\alpha$-phosphate ester. The other product of glycolysis, pyruvic acid, is needed for forming not triglyceride but acetyl coenzyme A. Fatty acids are made from acetyl coenzyme A (Figure 15.3) and in turn interact with $\alpha$-glycerol phosphate to furnish triglycerides.

**Detailed View. The Lipogenesis Cycle.** Since the generation of acetyl coenzyme A from glucose has been described in the preceding chapters, we may begin our study with this intermediate. We shall limit our discussion to the synthesis of palmitic acid, $CH_3(CH_2)_{14}COOH$, a typical long-chain fatty acid, about whose formation much is known. Although its synthesis consists of several steps, they may be classified as two types—first, the *activation of acetyl coenzyme A* by ATP and carbon dioxide; second, attachment of another acetyl coenzyme A to *build up the length of the carbon chain.*

Step 1. Activation:

$$CH_3\overset{O}{\overset{\|}{C}}SCoA + ATP + CO_2 \xrightarrow[\text{(a biotin enzyme)}]{\text{acetyl CoA–carboxylase–Mn}^{2+}} \quad \underset{\underset{O}{\overset{\|}{C}-SCoA}}{\overset{\overset{O}{\overset{\|}{C}OH}}{\underset{\displaystyle CH_2}{}}} + ADP + P_i$$

Malonyl coenzyme A

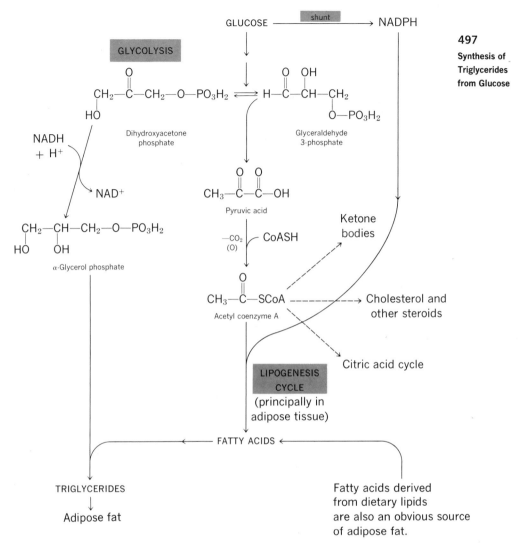

**Figure 15.2** Triglyceride synthesis in adipose tissue with glucose as the principal mate-rial. Glucose, via the *shunt,* produces the needed reducing agent, NADPH. Via glycolysis, glucose produces acetyl units for fatty acids. Dihydroxyacetone phosphate, an intermediate in glycolysis, is also a source of $\alpha$-glycerol phosphate, the form glycerol must be in if it is to be used to make triglycerides. Dashed arrows indicate pathways from acetyl coenzyme A that are alternatives to triglyceride synthesis.

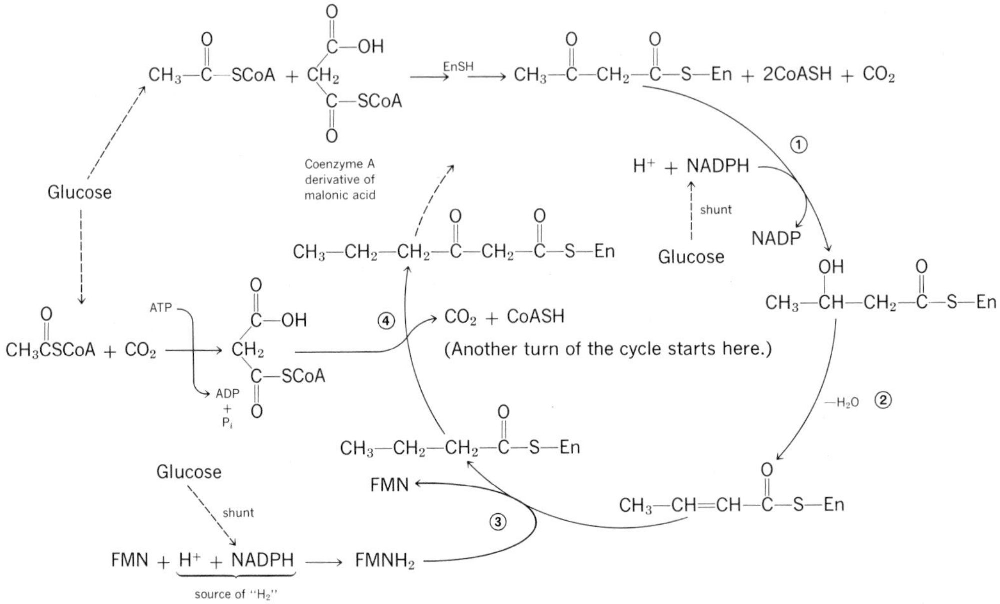

$$CH_3-\overset{\overset{\textstyle O}{\|}}{C}-SCoA + \underset{\underset{\textstyle \overset{\|}{O}}{\overset{\textstyle |}{C-SCoA}}}{\overset{\overset{\textstyle O}{\|}}{\overset{\textstyle C-OH}{\underset{}{CH_2}}}} \xrightarrow{\text{EnSH}} CH_3-\overset{\overset{\textstyle O}{\|}}{C}-CH_2-\overset{\overset{\textstyle O}{\|}}{C}-S-En + 2CoASH + CO_2$$

Coenzyme A derivative of malonic acid

Glucose

$$CH_3-CH_2-CH_2-\overset{\overset{\textstyle O}{\|}}{C}-CH_2-\overset{\overset{\textstyle O}{\|}}{C}-S-En$$

$H^+ + NADPH$ ①

shunt

NADP

Glucose

$$CH_3-\underset{\underset{\textstyle OH}{|}}{CH}-CH_2-\overset{\overset{\textstyle O}{\|}}{C}-S-En$$

$$CH_3\overset{\|}{C}SCoA + CO_2 \xrightarrow[\substack{ATP \\ ADP \\ + \\ P_i}]{} \underset{\underset{\textstyle \overset{\|}{O}}{C-SCoA}}{\overset{\overset{\textstyle O}{\|}}{\overset{\textstyle C-OH}{CH_2}}}$$

④ → $CO_2 + CoASH$

(Another turn of the cycle starts here.)

$-H_2O$ ②

$$CH_3-CH_2-CH_2-\overset{\overset{\textstyle O}{\|}}{C}-S-En$$

Glucose

FMN ←

shunt

③

$$CH_3-CH=CH-\overset{\overset{\textstyle O}{\|}}{C}-S-En$$

$FMN + \underbrace{H^+ + NADPH}_{\text{source of ``H}_2\text{''}} \longrightarrow FMNH_2$

**Figure 15.3** Lipogenesis cycle. Biosynthesis of fatty acids from acetyl units. The reactions taking place at the circled numbers are discussed in the text. An enzyme having an —SH group, EnSH, part of the palmitate synthetase complex, locks the growing chain to a multiple-enzyme system. The growing fatty acid unit swings from specialized enzyme to specialized enzyme, pivoting about the S—En bond as the steps of the cycle take place. (FMN is flavin mononucleotide, FMNH$_2$ its reduced form.)

Step 2. Chain lengthening, a summary:

$$CH_3\overset{\overset{\textstyle O}{\|}}{C}SCoA + 7\underset{\underset{\textstyle \overset{\|}{O}}{CSCoA}}{\overset{\overset{\textstyle O}{\|}}{\overset{\textstyle COH}{CH_2}}} + 14NADPH + 14H^+ \xrightarrow[\text{(several steps)}]{\text{palmitate synthetase}}$$

$$CH_3(CH_2)_{14}\overset{\overset{\textstyle O}{\|}}{C}OH + 7CO_2 + 14NADP^+ + 8CoASH + 6H_2O$$

Palmitic acid

The carbon dioxide is furnished by the decomposition of bicarbonate ion. In step 1, by affixing a carboxyl group on the $\alpha$-position of the acetyl unit, this position is activated. In step 2 we see that enzymes can apparently direct the *acetyl*ation of the $\alpha$-position in the system. At least the first of the several steps in chain lengthening is believed to be something like the following:

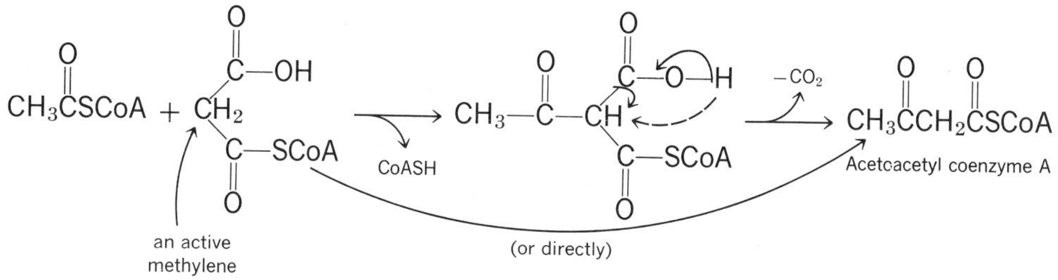

The initial phase of this reaction is very much like a Claisen ester condensation (cf. p. 249). From here on the percentage of conjecture or educated guessing about the pathway increases. The rest of the discussion refers to Figure 15.3, and the circled numbers indicate steps in that figure. The next step ①, believed to be reduction of the keto group in the acetoacetyl portion of the last product, requires NADPH (cf. p. 495). Then the new hydroxyl group is lost ② as the molecule is dehydrated, producing an unsaturated acid derivative. Reduction of this acid derivative ③, which requires more NADPH, gives an enzyme-bound derivative of butyric acid, the four-carbon fatty acid. The net effect, is therefore to convert a two-carbon acid into a four-carbon acid. Thus in one "turn" of the lipogenesis cycle the chain is lengthened by two carbons.

In the next turn the four-carbon acid derivative tacks on another two-carbon unit, starting with the following step ④, the exact analog of the reaction of acetyl coenzyme A with malonyl coenzyme A:

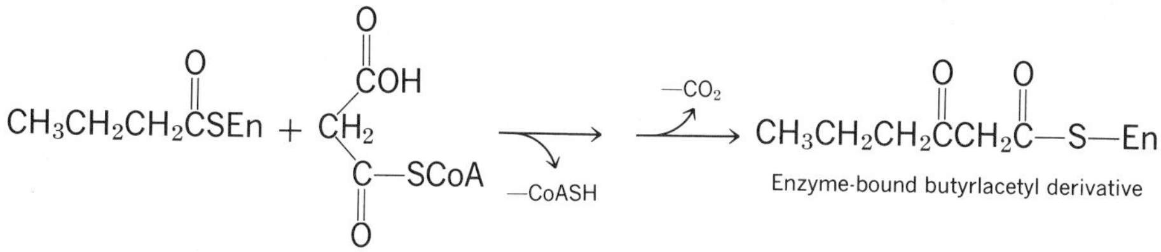

In the next three steps (reduction of the keto group, dehydration, reduction of the alkene linkage, as shown in Figure 15.3) the keto group is reduced to a —CH$_2$— unit. Thus an enzyme-bound, six-carbon acid derivative is prepared. Further turns of the lipogenesis cycle give the eight-carbon acid, then the ten-, twelve-, fourteen-, and finally the sixteen-carbon acid, palmitic acid. The enzyme for the cycle, palmitate synthetase, is actually a complex of enzymes and is not well understood. The synthesis of fatty acids of other chain lengths as well as the synthesis of unsaturated fatty acids involves other enzyme systems, some of which are similar to palmitate synthetase, or portions of it. In later sections we shall see that the oxidation of a fatty acid involves steps very similar to those described in Figure 15.3. In fact, the synthesis of fatty acids from two-carbon fragments was once believed to be the *exact* reverse of the steps in the breakdown of these acids to two-carbon units, acetyl coenzyme A. Synthesis, however, occurs primarily in adipose tissue and decomposition elsewhere, notably the liver. Moreover, the breakdown of fatty acids does not involve malonyl coenzyme A and carbon dioxide.

**Conversion of Fatty Acids into Neutral Triglycerides.** $\alpha$-Glycerol phosphate, its principal source an intermediate in glycolysis as already indicated in Figure 15.1, is needed instead of free glycerol to convert fatty acids to triglycerides. The following reactions indicate how it participates in the synthesis of triglycerides, using fatty acids made by various lipogenesis cycles.

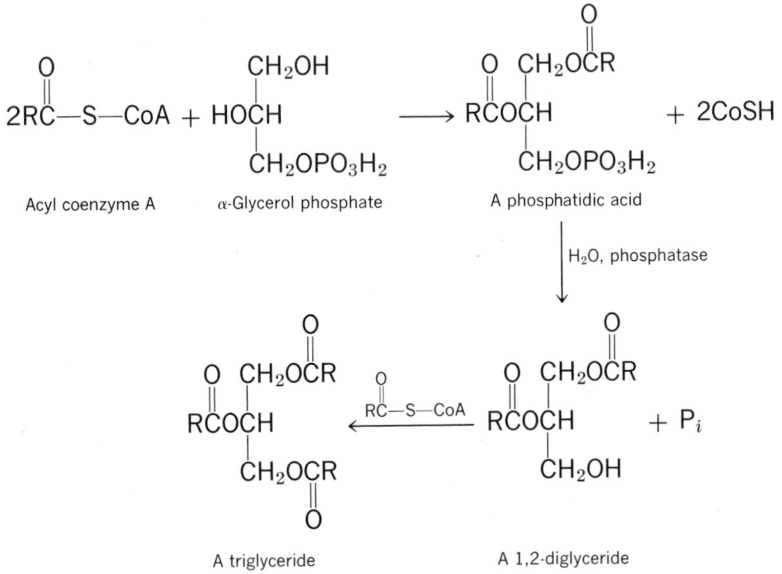

Acyl coenzyme A    $\alpha$-Glycerol phosphate     A phosphatidic acid

A triglyceride     A 1,2-diglyceride

(The question of having different R groups at different positions in the triglyceride is ignored here.)

Having now discussed how triglycerides are formed, we shall study how they are used to make ATP.

## CATABOLISM OF TRIGLYCERIDES

To mobilize the energy reserves in adipose lipids, several successive steps must occur (Figure 15.4).

1. Triglycerides in adipose tissue (or other depots) must be hydrolyzed to give free fatty acids (FFA) and glycerol.
2. These products must be transported in the bloodstream to the sites of oxidation. Glycerol can enter the glycolysis pathway after it has been converted into glyceraldehyde 3-phosphate.
3. Fatty acids are broken down two carbons at a time to release acetyl coenzyme A units.
4. These units are fed into the citric acid cycle to furnish hydrogen donors.
5. Hydrogen donors fuel the respiratory chain that produces ATP and water.

For all these steps to occur, there must of course be a need for ATP. The principal site of fatty acid catabolism is the liver, but some of the smaller fragments produced from the fatty acids can be utilized by muscles. In fact, we have already noted that

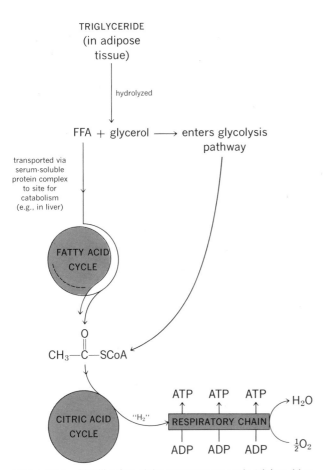

**Figure 15.4** Mobilization of the energy reserves in triglycerides.

most of the energy requirements of resting muscle are met by intermediates from fatty acid catabolism. The degradation of fatty acids occurs by a repeating series of steps known as the fatty acid cycle.

**Fatty Acid Cycle. Fatty Acid Spiral.** Before a molecule of a fatty acid can yield ATP, it must be activated, consuming ATP. The price is small, for the return is great.

$$\underset{O}{\overset{\displaystyle O}{\|}}$$

R—C—OH + CoASH + ATP ⟶ R—C—SCoA + AMP + PP$_i$

In this rare reaction of ATP its molecule is cleaved to produce not simple inorganic phosphate, P$_i$, but the diphosphate (or pyrophosphate) ion, here symbolized simply as PP$_i$. Once this initial activation is achieved, a series of reactions clips two carbons at a time from the fatty acid unit. When two carbons have been removed, as an acetyl unit, the remainder of the original fatty acid goes through the cycle again to have another acetyl group removed. These reactions continue until the original fatty acid has been cut down to the size of a final acetyl group. The acetyl groups normally enter the citric acid cycle as diagramed in Figure 15.4.

The steps in the fatty acid cycle, outlined in Figure 15.5, are as follows; the circled numbers refer to this figure.

① Dehydrogenation. FAD is the hydrogen acceptor.

$$CH_3(CH_2)_{12}CH_2-CH_2-\overset{\overset{\displaystyle O}{\|}}{C}-SCoA + FAD \longrightarrow$$

Palmitic acid CoA, from a $C_{16}$ acid

$$CH_3(CH_2)_{12}CH=CH-\overset{\overset{\displaystyle O}{\|}}{C}-SCoA + FADH_2$$

An $\alpha,\beta$-unsaturated
acid derivative
of CoASH

FAD ⤶

2ATP ⟵ Respiratory chain

Several flavoproteins with FAD as the cofactor act as enzymes, each one differing in the size of the fatty acid unit it acts upon. Each $FADH_2$ produced presumably yields two ATPs via the respiratory chain, as implied in Figure 13.9.

② Addition of water to the double bond.

$$CH_3(CH_2)_{12}CH=CH-\overset{\overset{\displaystyle O}{\|}}{C}-SCoA + H_2O \longrightarrow CH_3(CH_2)_{12}\overset{\overset{\displaystyle OH}{|}}{CH}-CH_2-\overset{\overset{\displaystyle O}{\|}}{C}-SCoA$$

A $\beta$-hydroxy acid derivative of CoASH

③ Dehydrogenation. The 2° alcohol group is oxidized to a keto group, with $NAD^+$ the hydrogen acceptor.

$$CH_3(CH_2)_{12}\overset{\overset{\displaystyle OH}{|}}{CH}-CH_2-\overset{\overset{\displaystyle O}{\|}}{C}-SCoA + NAD^+ \longrightarrow$$

$$CH_3(CH_2)_{12}\overset{\overset{\displaystyle O}{\|}}{C}-CH_2-\overset{\overset{\displaystyle O}{\|}}{C}-SCoA + H^+ + NADH$$

A $\beta$-keto acid derivative of CoASH

$NAD^+$ ⤶

3ATP ⟵ Respiratory chain

Each NADH produced presumably yields three ATPs via the respiratory chain (cf. again Figure 13.9).

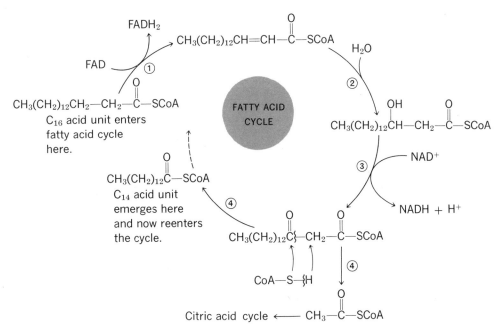

**Figure 15.5**  Fatty acid cycle. The reactions taking place at the circled numbers are discussed in the text. In this cycle the same intermediates do not recur; only types of them do, and types of reactions recur. The $C_{14}$ acid unit produced at ④ will next be dehydrogenated to give a $C_{14}$ unsaturated acid. This acid will be hydrated ②, and oxidized ③, and then cleaved ④ to produce a $C_{12}$ acid unit and another acetyl coenzyme A. Seven turns of the cycle are needed to break the original palmitic acid ($C_{16}$) into eight acetyl coenzyme A units.

④ Cleavage of the β-keto acid system.

$$CH_3(CH_2)_{12}\overset{O}{\underset{\|}{C}}-CH_2\overset{O}{\underset{\|}{C}}-SCoA + CoASH \longrightarrow$$

$$CH_3(CH_2)_{12}\overset{O}{\underset{\|}{C}}-SCoA + CH_3\overset{O}{\underset{\|}{C}}-SCoA$$

A $C_{14}$ acid derivative

$$12ATP \longleftarrow \underset{\text{chain}}{\text{Respiratory}} \longleftarrow \underset{\text{cycle}}{\text{Citric acid}}$$

A summary of the energy (ATP) yield for the complete oxidation of one molecule of palmitic acid is given in Table 15.1.

The intermediates in this catabolism do not normally accumulate to any significant extent, for as soon as they form they react further. When the acetyl coenzyme A stage is reached, several options are available. Hints of these have appeared in a few figures (e.g., Figure 15.2), and we are now at the point that we can study them. One of them has an important bearing on the synthesis of cholesterol; another is particularly ominous in diabetes.

## METABOLIC FATES OF ACETYL COENZYME A

**General View.** As we have often seen, acetyl coenzyme A stands at one of the most important metabolic intersections in all body chemistry. Figure 15.6 serves as a reminder and also summarizes some features not yet noted. The ready interconversion of acetyl coenzyme A to acetoacetyl coenzyme A leads to another important crossroad. The formation of $\beta$-hydroxy-$\beta$-methylglutaryl coenzyme A (we call it HMG-CoA for our limited needs) makes possible the subsequent syntheses of cholesterol and conceivably of the steroid hormones and the bile salts. It also makes possible the production of three compounds, somewhat inaccurately called the ketone bodies, that are important to our study of the molecular basis of diabetes mellitus.

**Reactions Leading to the Ketone Bodies. Formation of acetoacetyl coenzyme A.** This reaction (Figure 15.6) resembles a Claisen ester condensation (cf. p. 249):

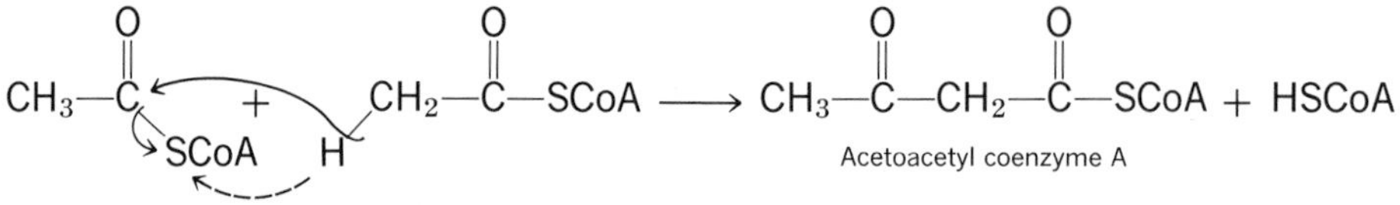

Claisen ester condensation:

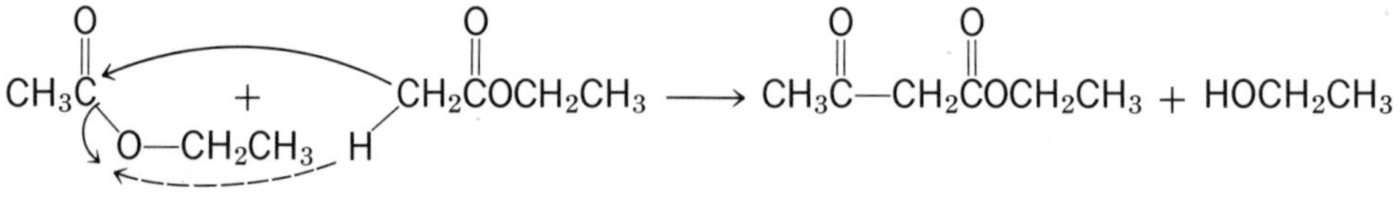

Table 15.1   Yield of ATP During Complete Oxidation of Palmitic Acid

| Step | ATP Made or Used |
|---|---|
| $CH_3(CH_2)_{14}CO_2H \longrightarrow CH_3(CH_2)_{14}\overset{O}{\overset{\|}{C}}-SCoA$ | $-1$ |
| $CH_3(CH_2)_{14}\overset{O}{\overset{\|}{C}}-SCoA \longrightarrow 8CH_3\overset{O}{\overset{\|}{C}}-SCoA$ | |
| 1. $7FADH_2 \longrightarrow 7FAD + 7(H\!:^- + H^+)$ to chain | $+14 \ (7 \times 2)$ |
| 2. $7NADH + H^+ \longrightarrow 7NAD^+ + 7(H\!:^- + H^+)$ to chain | $+21 \ (7 \times 3)$ |
| $8CH_3\overset{O}{\overset{\|}{C}}-SCoA \longrightarrow 16CO_2 + 8H_2O + 8CoASH$ | $+96 \ (8 \times 4 \times 3)$ |
| Net ATP yield | $+130$ |

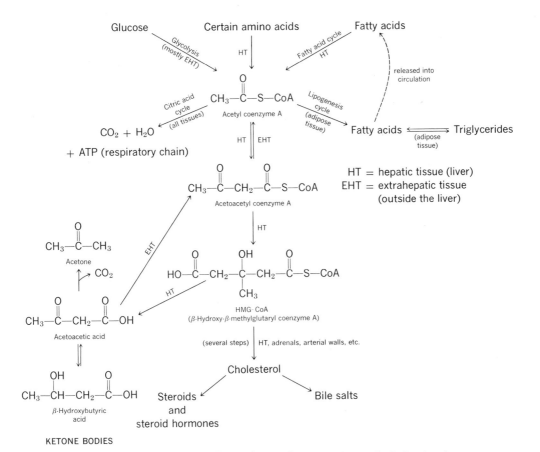

**Figure 15.6** The HMG-CoA crossroads. The various pathways occur usually, but not exclusively, in the tissues indicated.

**Formation of HMG-CoA.** Having joined two acetyl units to make acetoacetyl coenzyme A, the system can next attach a third acetyl unit in a step that resembles the aldol condensation (cf. p. 209):

HMG–CoA
(β-hydroxy-β-methylglutaryl coenzyme A)

This sequence is not meant to show all the details of the mechanism. Having come this far, we can look back and see that putting together three acetyl coenzyme A units has produced a sizable carbon skeleton. Figure 15.7 is a general outline of how the body goes on to make cholesterol. For more immediate needs we shall study how HMG-CoA breaks down to give the ketone bodies.

**Formation of acetoacetic acid.** If forming HMG-CoA is like an aldol condensation, this next step is like a reverse aldol condensation:

OH is in the $\beta$-position to carbonyl

$$\underset{\text{HMG–CoA}}{\text{HOCCH}_2\overset{\displaystyle CH_3}{\underset{\displaystyle |}{C}}CH_2C-SCoA} \longrightarrow \underset{\text{Acetoacetic acid}}{\text{HOCCH}_2CCH_3} + CH_3C-SCoA$$

**Formation of $\beta$-hydroxybutyric acid.** The keto group in the product of the last reaction is reducible by NADH. The reaction is reversible, and both the starting acid and the product are found in circulation in a ratio that varies widely, depending on other factors.

$$\underset{\text{Acetoacetic acid}}{CH_3CCH_2COH} + NADH + H^+ \rightleftharpoons \underset{\beta\text{-Hydroxybutyric acid}}{CH_3CHCH_2COH} + NAD^+$$

**Formation of acetone.** This reaction has its direct counterpart in *in vitro* reactions. $\beta$-Keto acids, in general, easily lose the elements of carbon dioxide (cf. p. 251). One of the few reactions in the body for which a specific enzyme catalyst has not been established, it is unimportant except in diabetes. With this condition its frequency rises, and the odor of acetone may even be detected on the breath of a person with severe diabetes.

$$CH_3-C-CH_2-C-O \longrightarrow \underset{\text{Acetone}}{CH_3-C-CH_3} + CO_2$$

**The Ketone Bodies.** The three compounds acetoacetic acid, $\beta$-hydroxybutyric acid, and acetone are traditionally called the *ketone bodies*. Collective analysis of the three of them is possible, and the ketone body concentration is usually expressed in terms of $\beta$-hydroxybutyric acid. Formed slowly but relatively continuously in the liver and released into general circulation, these compounds are normally present. Yet the ketone body level in venous circulation in an adult is roughly only 1 mg per 100 ml. This low level is maintained partly because two of the three ketone bodies, the two acids, serve as sources of energy in many tissues. Muscles, for example, obtain

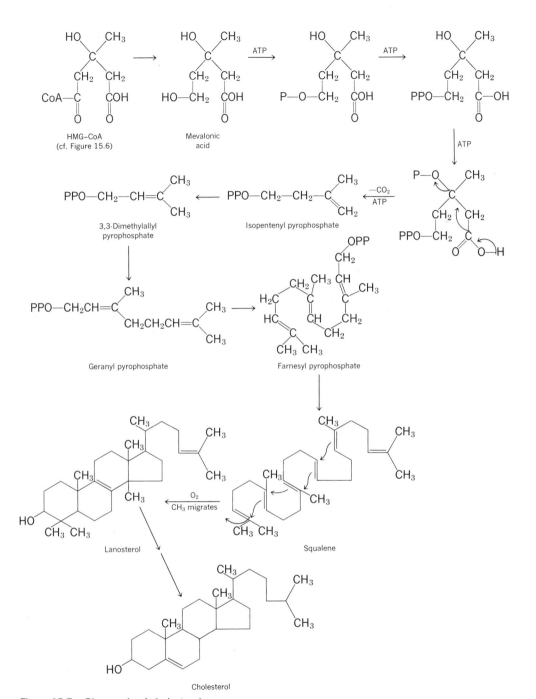

**Figure 15.7**   Biogenesis of cholesterol.

much of their ATP for resting functions from acetoacetic acid, a supplier of acetyl groups for acetyl coenzyme A and the citric acid cycle as indicated in Figure 15.6. The kidneys also remove ketone bodies, excreting about 20 mg of these substances per day via the urine. When the ketone body level in the blood rises, a condition of *ketonemia* exists. When, as a result of high ketonemia, significant quantities of these compounds are found in the urine, the condition is called *ketonuria*. Usually "ketone breath" is by now noticeable. These three conditions—ketonemia, ketonuria, and ketone breath—are collectively called *ketosis*. The situation is dangerous. Unchecked, ever-rising ketosis means death by means we shall soon study.

## KETOSIS AND ACIDOSIS

Two of the ketone bodies are carboxylic acids. Persistent overproduction of them will therefore eventually overtax the buffer systems of the blood. The kidneys can remove hydrogen ions, and the pH of urine can become as low as 4 (meaning a hydrogen-ion concentration about 1000 times that of normal blood). The kidneys can also remove the two acids of the ketone bodies as their negative ions. But if an excessive output of ketone bodies continues, the kidneys must eventually put increasing amounts of sodium ion into the urine being formed. (For every negative ion from a carboxylic acid put into the urine, a positive ion must be placed there too. Electric neutrality requires this.) Loss of sodium ion means, indirectly, loss of buffer capacity in the blood. The pH of the blood slowly starts to drop. We have learned that this drift downward in the pH of the blood, from about 7.4 to a value of 7 or lower (as measured at room temperature), is called acidosis. As the pH drops, buffers in the blood act to neutralize the acids. One of these buffers, the $H_2CO_3/HCO_3^-$ system, is the most affected. Bicarbonate ions are neutralized, and the excess carbonic acid decomposes to water and carbon dioxide which is expelled via the lungs. Normally the bicarbonate level in blood is 22 to 30 millimoles/liter. This level drops to 16 to 20 millimoles in mild acidosis; to 10 to 16 millimoles in moderate acidosis; and below 10 millimoles in severe acidosis.

Acidosis is serious for many reasons; one disruption, that of the mechanism for transporting oxygen, was studied in Chapter 12. In moderate to severe acidosis the difficulties in taking in oxygen at the lungs have become so great that the person experiences severe "air hunger"; breathing is very painful and laborious (a condition called dyspnea). By the time the bicarbonate level has dropped to 6 to 7 millimoles/liter in the adult, he is in a coma.

In even moderate acidosis the body loses an excessive amount of fluids via the kidneys. To put the salts of the carboxylic acids (from two of the ketone bodies) into the urine being formed, large quantities of water must also be removed. The person suffers from a general dehydration. This coupled with disruption of his ability to transport oxygen will normally depress the central nervous system. Even in mild acidosis a person experiences fatigue, a desire to stay in bed, lack of appetite, nausea, headache, reduced power of concentration, an indisposition to converse, and difficulty in making even simple decisions. Acidosis has been described in some detail because it is instructive to realize how much of human well-being depends on control of the pH of the blood. And acidosis is much more common than a healthy person might realize. Everyone experiences it briefly after severe physical exercise.

*Lactic acid acidosis* develops during strenuous muscular work. We learned in Chapter 14 that glycolysis becomes an important source of ATP under anaerobic conditions, producing lactic acid. When it forms faster than it can be removed (be converted to glycogen at the expense of some of it), its level in the blood will rise, the level of bicarbonate ion will drop, and so will the pH. The lactic acid level in blood may rise from its resting level of 1 to 2 millimoles/liter to 10 to 12 milli-moles/liter after hard work. This will cut the bicarbonate level roughly in half, to a level of 12 to 14 millimoles/liter, meaning that the moderate form of acidosis results. All athletes have experienced the violent dyspnea, the painful, gulping air hunger that accompanies maximum effort in a contest. When the lactic acid level is about 10 millimoles/liter, further work is rendered virtually impossible regardless of will power. Lactic acid acidosis is obviously a controlling factor in the improvement of athletic performance.

Lactic acid acidosis may develop during milder exercise if the supply of oxygen drops. Men living in lower altitudes who fancy themselves rugged individuals may return from even brief hikes at high altitudes with severe pains in the chest and the ego. Neither is serious. To recover from lactic acid acidosis, the person who has overexerted need merely rest.

**Causes and Prevention of Ketosis.** We have seen that ketosis may lead to coma and death; we need to know now how it can be prevented or controlled. The best evidence indicates that overproduction of the ketone bodies is caused by overproduc-tion of acetyl coenzyme A. At the acetyl coenzyme A crossroads near the top of Figure 15.6, three pathways are available for removing acetyl coenzyme A as it forms. The citric acid cycle, however, will not remove it unless there is a demand for ATP, leaving to nonactive people the production of fat and acetoacetyl coenzyme A as ways of handling extra acetyl coenzyme A. In diabetes mellitus and starvation acetoacetyl coenzyme A is the more important drain for excess acetyl coenzyme A, leading in turn to the overproduction of HMG-CoA from which the ketone bodies form. With this connection between ketone bodies and acetyl coenzyme A established, we must determine how acetyl coenzyme A may be synthesized too rapidly in the body.

The two main sources of acetyl coenzyme A are glucose and fatty acids. Glu-cose reserves, we have noted (p. 494), do not last long before they are depleted. After this depletion the body switches to fatty acids as the almost exclusive source of acetyl coenzyme A. Although glucose can be utilized directly out of the blood-stream by essentially all tissues, fatty acids must by and large be collected and first processed at the liver. This organ then sends out acetoacetate units which can be picked up in lieu of glucose in extrahepatic tissues and used to make acetyl coenzyme A. Glucose reserves obviously disappear in starvation. Less obvious is their disappearance in diabetes.

In starvation the supply of glucose runs out. In diabetes the glucose supply is more than adequate, but the system's ability to use it has deteriorated. We have learned that the primary defect in diabetes mellitus is the failure of the system to produce and circulate effective insulin. The pancreas may have lost its ability to make it, or the insulin made may be so tightly bound to proteins as it circulates in the blood that for all practical purposes it might as well not be there. Insulin may be present, but it is not *effective insulin*. A small amount of evidence, recently reported, indicates that the insulin produced in some individuals is structurally not quite right. A genetic defect, an inborn error issuing from a mutation, is responsible. Whatever the reason for the inadequate supply or the ineffectiveness of the insulin, the result is approximately the same.

The tissue of the body most sensitive to the action of insulin is adipose tissue. Its cells exhibit a response to insulin even when the concentration of this hormone drops to one-twentieth its usual fasting level. Insulin appears to exert its action by controlling the entry of glucose into the cells. When insulin is absent, glucose does not enter; when insulin is present, glucose enters. Once inside, glucose is phosphorylated and trapped. From then on roughly half of it is catabolized by the shunt to make NADPH (cf. p. 477); the other half undergoes glycolysis to furnish acetyl coenzyme A, most of which is converted, with the help of NADPH, into fatty acids and then to glycerides. All these reactions were discussed earlier in this chapter. In effect, when insulin is present, cells of adipose tissue tend to make triglycerides. A rise in blood sugar level normally signals the pancreas to release insulin to speed the removal of glucose and its conversion to lipid. In diabetes, except for the so-called *obese diabetic,* glucose is not converted to lipid.

Triglyceride is normally not only made in adipose tissue but also broken down. The *net* effect, uptake of glucose and lipogenesis *or* lipolysis and release of free fatty acids, obviously depends on the supply of effective insulin as well as other factors. *In the nonobese diabetic the release of free fatty acid predominates.* This free fatty acid is picked up by protein molecules in the bloodstream and carried to the liver which catabolizes the fatty acids as fast as it can.[1] Thus acetyl coenzyme A is produced at a rate faster than normal. Obviously acetyl coenzyme A is not needed to make fatty acid. Even if it could reach adipose cells again, it would not be taken in, converted back to fatty acids, and stored as triglyceride because to make triglyceride $\alpha$-glycerol phosphate is needed. Glucose is essential to supply the $\alpha$-phosphate ester, but since insulin is not present, adipose tissue is starved for glucose. The superabundance of acetyl coenzyme A cannot all be removed via the citric acid cycle. Demand for ATP simply cannot be sustained at that high level, and without ATP demand this cycle is not in action. The only pathway left for the excess acetyl coenzyme A is toward HMG-CoA and the production of ketone bodies and cholesterol (Figure 15.6).

In summary, when adipose tissue does not take in glucose, it releases fatty acids faster than it can make them. These are catabolized in the liver (primarily) at a more rapid rate than usual. The excess acetyl coenzyme A thus produced is con-

---

[1] When the concentration of lipid-protein complex in the blood rises, a condition of lipemia ("lipid-emia") exists, and the liver is presented with an abundance of this metabolite. The liver swells as it becomes fat-logged.

verted into the ketone bodies faster than they can be catabolized in extrahepatic tissue. Ketosis results. If it goes unchecked long enough, acidosis ensues and leads to dehydration, disturbances in the central nervous system, and eventually coma and death. Acidosis produced by insulin deficiency is managed with insulin therapy. In some cases, when the patient is in a coma, isotonic sodium bicarbonate (1.5%) is administered intravenously to maintain life until insulin therapy becomes effective for the long-range treatment. When starvation causes acidosis, the remedy is obvious but is not always available. Starvation, defined here as total fasting whether voluntary or involuntary, is distinguished from undernutrition in which acidosis is seldom observed. Ancel Keys found during the course of the Minnesota Experiment that ketosis was essentially absent in volunteers who for six months were each given a daily diet of only 1500 calories (in contrast with an estimated daily per capita intake in the United States of over 3000 calories). During this time they lost a fourth of their weight.

**The Obese Diabetic.** At the time of diagnosis about 40% of all diabetics are overweight compared to a 10% incidence in nondiabetics. Quite obviously adipose tissue is overly successful in making and storing triglycerides in these individuals. In most other forms of diabetes the individual loses weight as the fats in adipose tissue are mobilized for energy. Several factors must be remembered in considering this paradox. First, diabetes is defined *clinically* in terms of symptoms and clinical measurements of impaired glucose tolerance. It is not defined in terms of any particular explanation for this impairment at the molecular level. The best, most widely accepted explanation of faulty glucose tolerance is *relative lack* of effective insulin, or at least delays and poor timing in the release of effective insulin. Conditions of glucosuria and ketosis (and subsequent acidosis) are explained at the molecular level in terms of what this primary defect does to carbohydrate-lipid-protein metabolism. Second, diabetes is not just one disease. The clinical definition leaves open the possibilities that the symptoms and the impaired glucose tolerance may have a variety of causes, that several organs and tissues may be involved. Insulin therapy is not the complete solution to diabetes, as clinically defined, even though it has been of immeasurable help in permitting diabetics to lead fuller, more useful lives. Finally, much about diabetes must still be explained. We have been studying theories that do not necessarily account for all aspects of clinically defined diabetes.

The obese diabetic, the overweight individual with clinically defined diabetes, appears able to make insulin and release it. But either much or all of it is *atypical insulin* or it is largely *typical insulin* so tightly bound to serum proteins that it is ineffective at most tissues. At least these are two theories that have recently attracted considerable attention. Adipose tissue, we have learned, is very sensitive even to low concentrations of insulin. Thus glucose can get into cells of adipose tissue, where it becomes fat, but not into other cells. Net effects are always balances of opposing tendencies. Although the cells of adipose tissue in the obese diabetic are active in making fat, they are also active in releasing fatty acids. Fat making exceeds fatty acid release, but the rates of both are high. Weight rises (helped, no doubt, by too much food intake), and fatty acid catabolism also increases. These

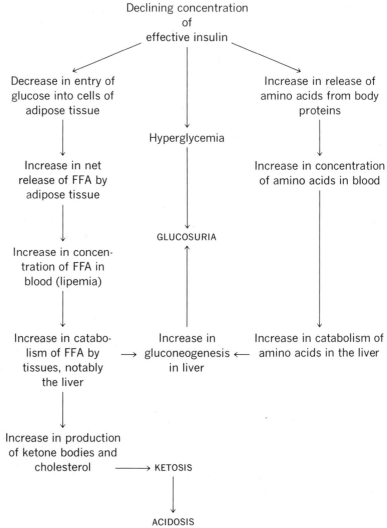

**Figure 15.8**  General outline of the main sequences of events in diabetes mellitus. (From A. Marble and G. F. Cahill, Jr., *The Chemistry and Chemotherapy of Diabetes Mellitus,* Charles C Thomas, Springfield, Ill., 1962. Courtesy of Charles C Thomas.)

fatty acids compensate for the glucose that is not available to other cells for energy, and they are also used to make even more glucose.

To compensate for the difficulty with which glucose penetrates cells of other tissues, the system works to increase even further the glucose concentration of the blood. This activity is carried out primarily in the liver. We have learned that the synthesis of glucose from essentially noncarbohydrate sources, called gluconeogenesis (p. 482), requires amino acids (primarily for carbon skeletons) and fatty acids (for hydrogen). Thus fatty acids are again mobilized and catabolized faster than normal; ketosis may result. Figure 15.8 outlines these general events. Ketosis is seldom as severe in obese diabetes as it is in other forms of the disease.

Associated with diabetes, especially with relatively severe cases, is the tendency to have problems with the circulatory system. Arteriosclerosis increases in what appears to be an independent phenomenon; at least it is not well controlled by insulin. Although the course of illness is imperfectly understood, vascular diseases are now the major cause of death in diabetes. Before Banting and Best introduced insulin therapy in the 1920s, acidosis was the major cause.

**Tolbutamide.** In many diabetics the pancreas retains some ability to make and secrete insulin, although not enough. Frequently in such cases the oral administration of the drug tolbutamide (or chlorpropamide) is sufficient, for this compound apparently stimulates insulin secretion in the pancreas.

Tolbutamide
(Orinase, Rastinon, Diabuton,
Mobenol, Toluina, Diaben,
Ipoglicone, Orabet, Oralin,
Artosin, Dolipol, U-2043, D-860)

Chlorpropamide
(Diabinese, Catanil, P-607)

# REFERENCES

A. White, P. Handler, and E. L. Smith. *Principles of Biochemistry,* third edition. McGraw-Hill Book Company, New York, 1964.

W. E. M. Lands. "Lipid Metabolism." *Annual Review of Biochemistry,* Vol. 34, 1965, page 313.

P. R. Vagelos. "Lipid Metabolism." *Annual Review of Biochemistry,* Vol. 33, 1964, page 139.

A. E. Renold and G. F. Cahill, Jr., section editors. *Handbook of Physiology, Section 5: Adipose Tissue.* American Physiological Society, Washington, D.C., 1965.

R. H. Williams. *Disorders in Carbohydrate and Lipid Metabolism.* W. B. Saunders Company, Philadelphia, 1962.

G. Litwack and D. Kritchevsky, editors. *Actions of Hormones on Molecular Processes.* John Wiley and Sons, New York, 1964.

H. G. Wood and M. F. Utter. "The Role of $CO_2$ Fixation in Metabolism," in *Essays in Biochemistry,* edited by P. N. Campbell and G. D. Greville, Vol. 1. Academic Press, New York, 1965.

A. L. Lehninger. *The Mitochondrion.* W. A. Benjamin, New York, 1964.

K. Bloch, editor. *Lipid Metabolism.* John Wiley and Sons, New York, 1960.

A. Marble and G. F. Cahill, Jr., *The Chemistry and Chemotherapy of Diabetes Mellitus.* Charles C Thomas, Springfield, Ill., 1962.

E. Kirk. *Acidosis.* William Heinemann Medical Books, Ltd., London, 1946.

A. Keys, J. Brozek, A. Henschel, O. Mickelsen, and H. L. Taylor. *The Biology of Human Starvation.* University of Minnesota Press, Minneapolis, 1950.

M. P. Cameron and M. O'Connor, editors. *Aetiology of Diabetes Mellitus and Its Complications,* Vol. 15, Ciba Foundation Colloquia on Endocrinology. Little, Brown, Boston, 1964.

R. M. C. Dawson and D. N. Rhodes, editors. *Metabolism and Physiological Significance of Lipids.* John Wiley and Sons, New York, 1964.

H. S. Mayerson. "The Lymphatic System." *Scientific American,* June 1963, page 80.

## PROBLEMS AND EXERCISES

1. Define each of the following terms.
   - (a) ketonemia
   - (b) ketonuria
   - (c) chylomicra
   - (d) ketosis
   - (e) energy density
   - (f) lactic acid acidosis
   - (g) lipemia
   - (h) adipose tissue

2. List the functions of adipose tissue.

3. Write equations for the steps in the conversion of $CH_3CH_2CH_2\overset{O}{\underset{\|}{C}}$—S—En to

$$CH_3CH_2CH_2CH_2CH_2\overset{O}{\underset{\|}{C}}—S—En$$

4. Write equations for the steps in the conversion of $CH_3CH_2CH_2CH_2CH_2\overset{O}{\underset{\|}{C}}$—SCoA to

$$CH_3CH_2CH_2\overset{O}{\underset{\|}{C}}—SCoA \text{ and } CH_3\overset{O}{\underset{\|}{C}}—SCoA$$

5. How does removal of the two acidic ketone bodies as their negatively charged ions at the kidneys deplete the amount of sodium ion in the blood?

6. Referring to Figure 12.9, how does depletion of $Na^+$ interfere with the transfer of $H^+$ into the urine?

7. Referring to Figure 12.13, at what point does acidosis interfere with the uptake of $O_2$ and why?

8. How does a condition of being overweight place an extra burden on the heart?

9. Explain, step by step, how lack of effective insulin will lead (a) to overproduction of acetyl coenzyme A, (b) thence to overproduction of ketone bodies, (c) thence to acidosis, and finally (d) to coma and death.

10. Suppose molecules of the following compound are included in the diet of an individual:

$$CH_3CH_2CH_2\overset{O}{\underset{\|}{C}}{}^*—OH$$

   Assume that the carbon with the asterisk is the radioactive carbon-14 isotope. Write a flow chart to explain how each of the following results might be observed.
   - (a) Carbon dioxide containing the carbon-14 isotope is exhaled.
   - (b) Carbon-14 appears in a molecule of cholesterol.

11. Construct a flow chart that explains how a carbohydrate-rich diet will cause a person to become fat.

# CHAPTER SIXTEEN

# Metabolism of Proteins

## NITROGEN FIXATION. ITS IMPORTANCE TO ANIMAL METABOLISM

Nitrogen nuclei are essential in the makeup of all living things, but we as human beings, who live and move in an atmosphere that is roughly 80% nitrogen gas, have no ability to appropriate nitrogen directly from the air. To be of use to us, the nitrogen nuclei must be a part of derivatives of ammonia such as proteins and amino acids. We simply lack the enzymes needed to catalyze the reduction of elemental nitrogen to ammonia.

Fortunately, among the scores of millions of microbes to be found in each gram of most topsoils, there are many nitrogen-fixing types. Some, like *Clostridium pasteurianum* and *Azotobacter vinelandii,* are free-living bacteria. Others, such as those in the genus *Rhizobium,* are parasites of certain plants. By steps not yet fully understood, *Clostridium* and *Azotobacter* can reduce nitrogen to ammonia; and once this compound forms, the routes are open to amino acids. In Chapter 6 (p. 207) we studied a reaction called reductive amination:

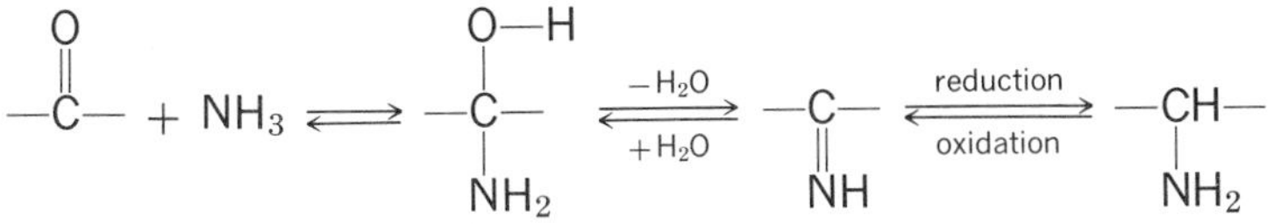

$$\underset{\substack{\text{Carbonyl in an} \\ \text{aldehyde or ketone}}}{-\overset{\displaystyle O}{\underset{\displaystyle \parallel}{C}}-} + NH_3 \rightleftharpoons -\overset{\displaystyle O-H}{\underset{\displaystyle \underset{NH_2}{|}}{C}}- \underset{+H_2O}{\overset{-H_2O}{\rightleftharpoons}} \underset{\text{An imine}}{-\overset{\displaystyle}{\underset{\displaystyle \underset{NH}{\parallel}}{C}}-} \underset{\text{oxidation}}{\overset{\text{reduction}}{\rightleftharpoons}} \underset{\text{Amine}}{-\overset{\displaystyle}{\underset{\displaystyle \underset{NH_2}{|}}{C}}H-}$$

The reductive amination, the conversion of the carbonyl group of an aldehyde or ketone to an amino group, proceeds from left to right. The reverse process, oxidative deamination, is also possible and is an important reaction in metabolism. The formation of glutamic acid from one of the intermediates in the citric acid cycle, namely α-ketoglutaric acid, is a principal route for converting ammonia into an important amino acid. The enzyme for this reaction, *glutamic acid dehydrogenase,*

$$NH_3 + HOCCH_2CH_2C-C-OH + NADH + H^+ \longrightarrow$$

α-Ketoglutaric acid

$$HOCCH_2CH_2CHCOH + NAD^+ + H_2O$$
$$NH_2$$

L-Glutamic acid

which is widely distributed among plants and animals, catalyzes only the formation of the L-enantiomer. Thus in this reaction an intermediate from carbohydrate (or lipid) metabolism can be used to make a compound of importance in protein metabolism.

Bacteria of the genus *Rhizobium* have a relationship with legumes such as alfalfa, clover, soybeans, and peas by which, working together, they "fix" nitrogen from the air. The bacteria alone cannot do this, nor can the legume acting without the bacteria. Working together through the plant's root system, they reduce atmospheric nitrogen to hydroxylamine, $NH_2OH$, instead of to ammonia. But this compound can form oximes with aldehyde or keto groups (cf. p. 206), and reduction of oximes gives amino groups. Thus the overall effect is the same as having ammonia the reduction product. The living together of widely differing organisms in a mutually beneficial way, as illustrated by *Rhizobium* and legumes, is called *symbiosis.* With favorable conditions legumes and *Rhizobium* can put as much as 200 pounds of fixed nitrogen per year into an acre of soil.

Atmospheric nitrogen is converted to other forms, to nitrates and nitrites, by oxidation. The daily worldwide production of nitric acid by electric discharges during storms and possibly by solar radiation is estimated to be hundreds of thousands of tons. (This is still only a small percentage of the nitrogen fixation accomplished by organisms.) Upon entering the soil, nitrates and nitrites become raw materials for many microorganisms and higher plants which can convert these ions into ammonia and thence to amino acids.

The higher animals, including man, are absolutely dependent on the soil bacteria and the plants to provide amino acids. By eating plants or by eating animals which feed on plants, the higher animals obtain the dietary proteins which, when digested, give the inventory of amino acids the particular species must have for life. When plants and animals die, decay processes return their nitrogen to the soil and atmosphere. This conservation of nitrogen is necessary, for it must be used again and again.

## DYNAMIC STATE OF NITROGEN METABOLISM

The mixture of dietary amino acids produced by digestion of proteins is rapidly absorbed from the intestinal tract, although not by simple dialysis. The process consumes energy (ATP), and it appears to require vitamin $B_6$ (pyridoxal) and manganese ion, $Mn^{2+}$. There is evidence that the absorptive process discriminates against the D-amino acids. Virtually all naturally occurring amino acids are in the L-family; the body cannot utilize the D-forms, at least not at rates useful enough for survival.

Reference compounds, carbohydrates:

$$\underset{\text{D-Glyceraldehyde}}{\overset{\displaystyle \overset{O}{\underset{\text{CH}}{\|}}}{\underset{\text{CH}_2\text{OH}}{\underset{H}{\diagup}\bigcirc\underset{OH}{\diagdown}}}} \qquad \underset{\text{L-Glyceraldehyde}}{\overset{\displaystyle \overset{O}{\underset{\text{CH}}{\|}}}{\underset{\text{CH}_2\text{OH}}{\underset{HO}{\diagup}\bigcirc\underset{H}{\diagdown}}}}$$

Reference compounds, proteins:

$$\underset{\text{D-Serine}}{\overset{\displaystyle \overset{O}{\underset{\text{COH}}{\|}}}{\underset{\text{CH}_2\text{OH}}{\underset{H}{\diagup}\bigcirc\underset{NH_2}{\diagdown}}}} \qquad \underset{\substack{\text{L-Serine}\\\text{(naturally occurring}\\\text{enantiomer)}}}{\overset{\displaystyle \overset{O}{\underset{\text{COH}}{\|}}}{\underset{\text{CH}_2\text{OH}}{\underset{NH_2}{\diagup}\bigcirc\underset{H}{\diagdown}}}}$$

Virtually all the dietary amino acids enter circulation via the portal vein (p. 48). They do not remain in the bloodstream long before they are removed and used in one or more processes. The following pathways are the options.

1. Synthesis of protein—the repair of tissue, the formation of new tissue, the synthesis of enzymes and some hormones.

2. Synthesis of nonprotein, nitrogenous compounds—substances needed internally such as nucleic acids, heme, creatine, some hormones, and complex lipids.

3. Synthesis of certain other amino acids that might temporarily be in short supply.

4. Conversion into nonnitrogenous compounds—eventually glycogen and/or triglyceride, with the elimination of nitrogen in the form of urea.

5. Catabolism leading to ATP, carbon dioxide, water, and urea.

When the amount of nitrogen excreted as urea equals the amount ingested in other forms in the diet, the person is in a state of *nitrogen equilibrium.* Growing infants and children as well as those recovering from a wasting disease have a *positive nitrogen balance;* more nitrogen is taken in than excreted because the body conserves nitrogen nuclei to make and repair tissue. The processes of aging, starvation, and suffering from a debilitating disease produce a *negative nitrogen balance;* what nitrogen goes out exceeds intake. We find it useful to think in terms of a nitrogen pool (Figure 16.1). Although nothing in protein metabolism resembles

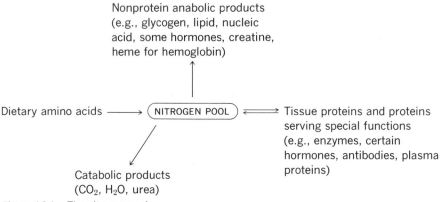

**Figure 16.1** The nitrogen pool.

the lipid reserves in adipose tissue or the glycogen reserves in the liver and muscles, we may still regard any tissue (i.e., all tissue) in which amino acids are located, whether as proteins or as dissolved amino acids, as a depot or pool for amino acids. Most tissue proteins are in a dynamic state, constantly undergoing degradation and resynthesis. This turnover is fairly rapid among proteins of the liver and blood plasma, very slow among muscle proteins. Enzyme proteins do not last indefinitely; they are broken back down to amino acids and must be replaced as needed.

The body has no mechanism for the temporary storage of a chance excess of dietary amino acids, as it has for glucose (glycogen) and lipids (depot fat). It does not lay in a supply for some emergency. Amino acids not used internally are catabolized. The carbon fragments enter one of a variety of pathways including gluconeogenesis, lipogenesis, and the citric acid cycle. Nitrogen is excreted in the form of urea.

**The Synthesis of Urea.** Urea, the major end product of nitrogen metabolism in human beings, is made in the liver. From this organ it is carried in the bloodstream to the kidneys which remove it and place it in the urine being formed. Urea is made from carbon dioxide and ammonia. The net equation is as follows, although a few steps are required to achieve synthesis (Figure 16.2).

$$2NH_3 + CO_2 \longrightarrow NH_2{-}\overset{\overset{\displaystyle O}{\|}}{C}{-}NH_2 + H_2O$$
Urea

**Essential Amino Acids.** Although the body has the capacity to use some amino acids to make others, a few cannot be synthesized this way. Unless they are present in the proteins of the diet, nitrogen equilibrium is upset and the nitrogen balance becomes negative. Presumably human metabolism cannot form these amino acids from the carbon skeletons and the amino groups of other intermediates, and for this reason they are said to be *essential* (see Table 16.1). The *nonessential amino acids,* also listed in Table 16.1, can be made in the body. (Nonessential, in this context, means *temporarily* dispensable. Obviously all the amino acids listed

**Table 16.1   Amino Acids, Classified as Nutritionally Essential or Nonessential in Maintaining Nitrogen Equilibrium in an Adult Man**

| Essential | Nonessential |
| --- | --- |
| Isoleucine | Alanine |
| Leucine | Arginine |
| Lysine | Aspartic acid |
| Methionine | Cystine |
| Phenylalanine | Glutamic acid |
| Threonine | Glycine |
| Tryptophan | Histidine |
| Valine | Hydroxyproline |
| | Proline |
| | Serine |
| | Tyrosine |

Data from W. C. Rose, *Federation Proceedings,* Vol. 8, 1949, page 546.

are used by the body, and protein synthesis could not proceed without them.) Dietary proteins that contain all the essential amino acids are called *adequate proteins*. Gelatin without tryptophan and zein (protein in corn) without lysine are inadequate proteins.

**Kwashiorkor.** In parts of Latin America, Asia, and Africa the death rate among children is several times that in developed, industrialized societies. Children by the

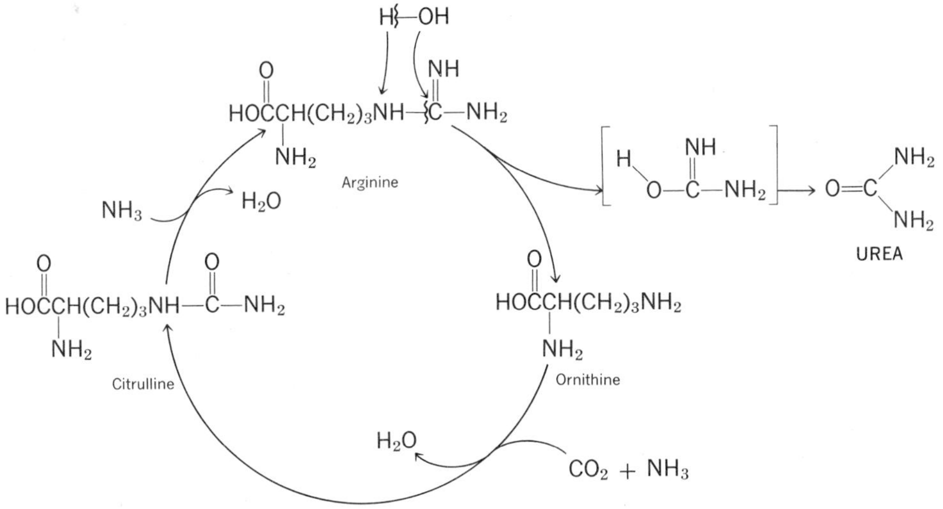

Net effect:   $2NH_3 + CO_2 \longrightarrow NH_2-\overset{\overset{\text{O}}{\|}}{C}-NH_2 + H_2O$

Urea

**Figure 16.2**   Urea synthesis by the Krebs ornithine cycle. The ornithine needed as a carrier molecule can be made from glutamic acid as shown in Figure 16.3.

thousands are doomed to short lives with bloated bellies, patchy skin, and discolored hair. As long as they are nourished at their mother's breast, they enjoy health. When the second child comes and displaces the first, the symptoms appear in the first child. The disease is called kwashiorkor, a name taken from two words of an African dialect meaning "first" and "second"—the disease that the first child contracts when the second one is born. The diet of the firstborn, instead of milk, is now starchy and contains inadequate protein. Hardly recognized until the 1940s, the ailment is now known to be a protein deficiency disease. Both undernutrition and malnutrition are responsible.[1] The initial symptoms are a loss of appetite and diarrhea—both of which lead the mother to reduce the amount of food she gives the child, thus hastening the onset of additional complications. In the weakened state the child is even more susceptible to the diseases that are a constant hazard in the tropics. Efforts to improve the quality of protein in the diet of people in these regions have been intensive. The Institute of Nutrition of Central America and Panama (INCAP) has developed a protein-rich flour from cottonseed. In areas where the major cash crop is cotton, this flour plus vitamins and minerals is added to local corn meal, which alone is an inadequate protein. Intensive research is being conducted to develop a hybrid corn that produces an adequate protein. The sea is a vast, untapped reservoir of protein-rich fish, and groups in several countries have developed fish meal flour. Plankton, seaweed, and algae are also being investigated. At the present time the population growth in Latin America is 3% per year, the growth in food production 2.5%. What may seem a small fraction of a percent actually means death from starvation (or from diseases brought on during the weakened condition of developing starvation) for millions of people. Between the 1980s and the turn of the century, this planet may well see the greatest catastrophe in recorded history, and history has recorded some monstrous famines.[2] At the present rate of population growth, by the 1980s the developed countries with high-yield agriculture will no longer have the capacity, regardless of intentions and maximum effort, to make up the food deficits of underdeveloped countries. Children are the first to suffer, for their dietary needs are greater and they must have a positive nitrogen balance to grow. Yet they are weaker than adults, and in the competition for food the battle is particularly grim.

When a body is on a starvation diet, its resources are mobilized to stave off the crisis. The glucose shortage is one of the serious consequences, for it means that gluconeogenesis must take place in an effort to supply what blood sugar is needed. Gluconeogenesis requires certain of the amino acids as well as fatty acids. Thus demand for this metabolic pathway is high at a time when the body has precious few amino acids to spare. To supply them, degradation of tissue protein must occur,

[1] Undernutrition is a general inadequacy of the diet; malnutrition is a serious imbalance in the necessary components of the diet, for example, vitamin deficiency or deficiency in the essential amino acids.

[2] For example, the Great Famine in Bengal, 1769–1770, ten million people or a third of the population perished; the Irish famine, 1846–1847, over a million people died when the potato crops failed; the famine of 1877–1878 in North China, over nine million people perished; the Russian famine of 1921–1922, three million died; World War II famines deliberately induced in the Warsaw ghetto and Nazi concentration camps; the Bengal famine of 1943, one and a half million died; and the Nigerian war famine in Biafra in 1968.

partly accounting for the wasting-away aspects of starvation. The lipid reserves also disappear. Fatty acid catabolism and gluconeogenesis generate ketone bodies which create serious problems of their own (already discussed, Chapter 15). We have mentioned that the body can make glucose and lipids from amino acids, and it can make some amino acids from other amino acids and from glucose and lipids. We shall now examine in greater detail just how these processes are carried out.

## SYNTHESIS OF NONESSENTIAL AMINO ACIDS

Multiple lines of evidence indicate that the body can make some amino acids from molecular parts contributed by carbohydrates, lipids, and the amino acids that it cannot make, the essential amino acids. Much of the research work has been done with rats, but most of the results appear to apply to man. When young animals do not grow on a diet deliberately made deficient in an amino acid, that amino acid probably cannot be made internally. When the diet is made to include amino acids labeled with $^{15}N$ in the $\alpha$-position and this radioactive form of nitrogen soon appears in all the amino acids (except lysine), it is obvious that pathways for passing the nitrogen of an amino group from amino acid to amino acid exist.

**Transamination.** The body contains a family of enzymes called *transaminases* that catalyze the following type of reaction:

$$R\!-\!\overset{\overset{O}{\|}}{C}\!-\!\overset{\overset{O}{\|}}{C}\!-\!OH + R'\!-\!\overset{\overset{NH_2}{|}}{CH}\!-\!\overset{\overset{O}{\|}}{C}OH \rightleftharpoons R\!-\!\underset{\underset{NH_2}{|}}{CH}\overset{\overset{O}{\|}}{C}OH + R'\!-\!\overset{\overset{O}{\|}}{C}\!-\!\overset{\overset{O}{\|}}{C}\!-\!OH$$

| $\alpha$-Keto acid | $\alpha$-Amino acid | New $\alpha$-amino acid | New $\alpha$-keto acid |

Transferring an amino group from one molecule to another is called *transamination*. Several intermediate steps occur, and vitamin $B_6$, pyridoxal, is apparently the essential coenzyme for all the transaminases.

Glutamic acid serves as a common provider of amino groups; most of the nonessential amino acids can be made from it as indicated in Figure 16.3.

## CATABOLISM OF AMINO ACIDS

By a variety of experimental techniques (use of radioactive isotopes as labels, effects of certain diets) we now know that the majority of the amino acids can be converted into glycogen (with excretion of urea), and are therefore called the glycogenic amino acids (Table 16.2). The catabolism of a few amino acids leads to ketone bodies, called the ketogenic amino acids (Table 16.2). Since glycogen can be converted into lipids and since amino acids can be made into glycogen (gluconeogenesis), molecular parts can obviously undergo considerable shuffling in the body.

With over twenty amino acids, the discussion of their catabolism could fill books. Since our purpose is to illustrate rather than to document, the plan for this section is as follows. A few of the simpler examples of the conversion of amino acids to precursors of glycogen will be presented. Some of the more involved metabolic pathways will be placed in figures with legends that summarize briefly the overall results. Depending on the scope of the course, these figures may be used or ignored. After describing amino acid catabolism, including the net formation of glucose units, the synthesis of lipids, and the channeling of units into the citric acid cycle to make ATP, we shall give a few examples of how amino acids are used to make nonprotein nitrogenous compounds such as diamines and heme. Comparatively few of the metabolic reactions of amino acids are described, however.

**Methods for Removing Groups.** Transamination and direct oxidative deamination are two of several ways for removing the $\alpha$-amino group of an amino acid.

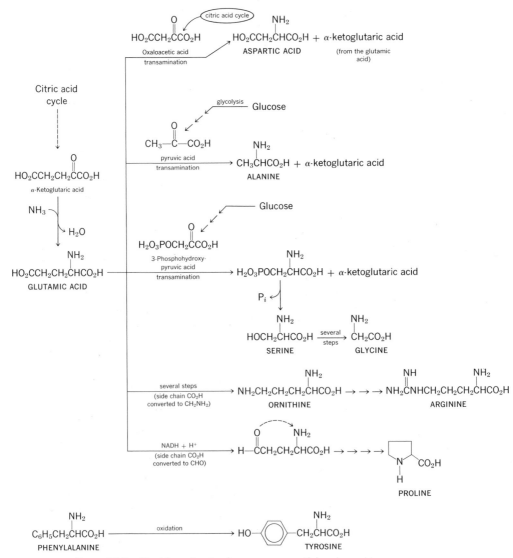

**Figure 16.3** The biosynthesis of some nonessential amino acids.

**Table 16.2 Amino Acids as Raw Materials for Glycogen and Fatty Acids**

| | |
|---|---|
| Glycogenic amino acids | |
|   Alanine | Hydroxyproline |
|   Arginine | Methionine |
|   Aspartic acid and asparagine | Proline |
|   Cysteine | Serine |
|   Glutamic acid and glutamine | Threonine |
|   Glycine | Tryptophan |
|   Histidine | Valine |
| | |
| Ketogenic amino acids* | |
|   Isoleucine | Phenylalanine |
|   Leucine | Tyrosine |
|   Lysine | |

* All but leucine are also considered to be glycogenic.

1. **Transamination** (indirect oxidative deamination). The amino group is transferred to $\alpha$-ketoglutaric acid which becomes glutamic acid and is oxidized.

(a) $\quad \underset{\underset{NH_2}{|}}{RCHCO_2H} \; + \; HO_2CCH_2CH_2\overset{\overset{O}{\|}}{C}CO_2H \; \rightleftharpoons$

$\qquad\qquad\qquad$ $\alpha$-Amino acid $\qquad\qquad\qquad$ $\alpha$-Ketoglutaric acid

$\qquad\qquad\qquad\qquad\qquad\qquad R\overset{\overset{O}{\|}}{C}CO_2H \; + \; HO_2CCH_2CH_2\underset{\underset{NH_2}{|}}{CHCO_2H}$

$\qquad\qquad\qquad\qquad\qquad\qquad\qquad$ $\alpha$-Keto acid $\qquad\qquad\qquad\qquad$ Glutamic acid

(b) $\quad HO_2CCH_2CH_2\underset{\underset{NH_2}{|}}{CHCO_2H} \; + \; NAD^+ \; + \; H_2O \longrightarrow$

$\qquad\qquad\qquad$ Glutamic acid

$\qquad\qquad\qquad\qquad\qquad\qquad HO_2CCH_2CH_2\overset{\overset{O}{\|}}{C}CO_2H \; + \; NH_3 \; + \; NADH \; + \; H^+$

$\qquad\qquad\qquad\qquad\qquad\qquad\qquad$ $\alpha$-Ketoglutaric acid

Sum $\quad \underset{\underset{NH_2}{|}}{RCHCO_2H} \; + \; NAD^+ \; + \; H_2O \longrightarrow R\overset{\overset{O}{\|}}{C}CO_2H \; + \; NH_3 \; + \; NADH \; + \; H^+$

$\qquad\quad$ $\alpha$-Amino acid $\qquad\qquad\qquad\qquad\qquad\qquad$ $\alpha$-Keto acid

$\qquad\qquad\qquad\qquad\qquad\qquad\qquad\qquad\qquad$ Urea $\qquad$ NAD$^+$ $\searrow$ (H:$^-$ + H$^+$)

$\qquad\qquad\qquad\qquad\qquad\qquad\qquad\qquad\qquad\qquad\qquad\qquad\qquad$ Electron transport chain

2. **Direct oxidative deamination.** Step $b$ in the previous method is an oxidation (NAD$^+$ is reduced) and a deamination. Being essentially irreversible, this step over-

comes the unfavorable equilibrium in the first step. There is some evidence that flavoprotein enzymes (FAD enzymes) can catalyze this type of reaction directly on amino acids other than glutamic acid.

$$(a) \quad R{-}\underset{\underset{NH_2}{|}}{CH}{-}CO_2H + H_2O + FAD \rightleftharpoons R{-}\overset{O}{\overset{||}{C}}{-}CO_2H + NH_3 + FADH_2$$

$-H_2$

$-NH_3$

(mechanism?)

$$\underset{+}{\underset{R{-}\overset{\overset{NH}{||}}{C}{-}CO_2H}{}} \xrightarrow{\text{H---OH}} R{-}\overset{\overset{N{\,}H{\,}H}{|}}{\underset{\underset{H}{\overset{||}{O}}}{C}}{-}CO_2H$$

FADH$_2$

$$(b) \quad FADH_2 + O_2 \longrightarrow FAD + H_2O_2$$

$$\xrightarrow{\text{catalase}} H_2O + \tfrac{1}{2}O_2$$

Figure 16.4 is a general summary of how some of the amino acids fit into the metabolic picture of the citric acid cycle and gluconeogenesis. The major entry points are pyruvic acid and oxaloacetic acid, the second a key intermediate in glu- coneogenesis because it serves as a way station for converting pyruvic acid back to the very high-energy phosphoenolpyruvic acid. (See Figure 14.5 and the related discussion for a review of this problem.) In the normal functioning of the citric acid cycle oxaloacetic acid is also an intermediate. A four-carbon "carrier" molecule which picks up an acetyl group to form citric acid, it is later regenerated. Any *extra* oxaloacetic acid that the system may produce, such as from the catabolism of amino acids, is a net gain and may be used to make "new" glucose. That is, gluco- neogenesis may occur. As indicated in Figure 16.4, aspartic acid can be converted to oxaloacetic acid, and valine, proline, and glutamic acid also lead to this interme- diate. Several other amino acids can be converted to pyruvic acid, which can also be made into glucose.

The ketogenic amino acids lead to acetoacetic acid or acetyl coenzyme A. Before we describe in greater detail how some of the amino acids in Figure 16.4 are catabolized, the student is reminded that these are by no means the only reactions of the amino acids. Amino acids are used principally to make or to replace tissue proteins, enzymes, some hormones, and other nitrogenous substances needed for life and health. Gluconeogenesis is important primarily in periods of inadequate

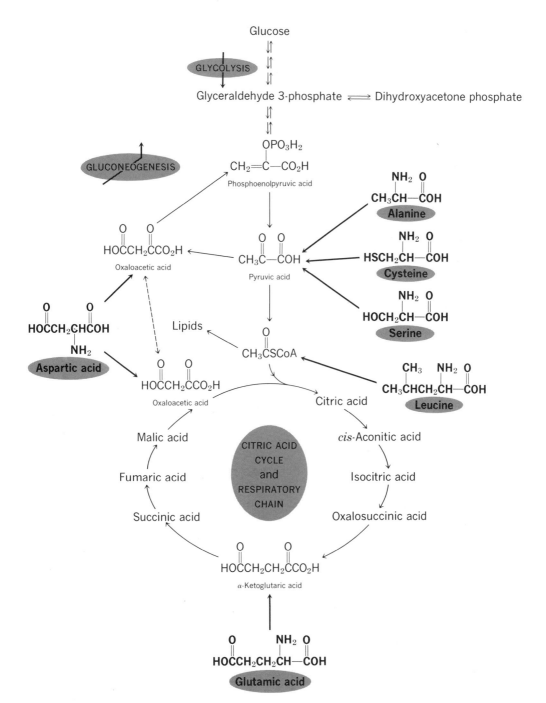

**Figure 16.4** Catabolism of some amino acids, illustrative examples. Several amino acids can enter the gluconeogenesis pathways. A few produce acetyl coenzyme A, and from it lipids, including cholesterol and steroids, can be made. Individual steps are discussed in greater detail in the text.

diet, fasting, starvation, diabetes, and wasting diseases. It is now clear that if the body is to make "new" glucose, it must sacrifice some of its proteins (as well as some of its fatty acids).

The following have been selected to illustrate the catabolism of amino acids.

**Alanine.** Oxidative deamination produces pyruvic acid.

$$\underset{\text{Alanine}}{\underset{\overset{\displaystyle NH_2}{|}}{CH_3CHCO_2H}} \xrightarrow[\text{deamination}]{\text{oxidative}} \underset{\text{Pyruvic acid}}{CH_3\overset{\displaystyle O}{\overset{||}{C}}COH} \begin{array}{l} \longrightarrow \text{Gluconeogenesis} \\ \\ \searrow \text{Citric acid cycle} \end{array}$$

**Aspartic acid.** Again, oxidative deamination directly produces oxaloacetic acid, an intermediate in gluconeogenesis.

$$\underset{\text{Aspartic acid}}{HO\overset{\displaystyle O}{\overset{||}{C}}CH_2\underset{\underset{\displaystyle NH_2}{|}}{\overset{\displaystyle O}{\overset{||}{C}}H}COH} \xrightarrow[\text{deamination}]{\text{oxidative}} \underset{\text{Oxaloacetic acid}}{HO\overset{\displaystyle O}{\overset{||}{C}}CH_2\overset{\displaystyle O}{\overset{||}{C}}COH} \begin{array}{l} \longrightarrow \text{Gluconeogenesis} \\ \\ \searrow \text{Citric acid cycle} \end{array}$$

**Cysteine.** The mercaptan group is first oxidized to a sulfinic acid, $HO_2S—$, which is very much like a carboxyl group, $HO_2C—$. Then the amino group is lost. In the product, $\beta$-sulfinylpyruvic acid, the keto group is in the $\beta$-position to the sulfinyl group. If it were a $\beta$-keto carboxylic acid, we would expect loss of carbon dioxide (cf. p. 25). Being instead a $\beta$-keto sulfinic acid, it loses sulfur dioxide. The sulfur dioxide is oxidized to sulfate ion, but pyruvic acid is the organic product. The initial intermediate, cysteinesulfinic acid, is also used to make many biologically important compounds of sulfur. The bile acid, taurocholic acid (p. 406), is one of them. Sulfate ions produced by cysteine catabolism are excreted via the urine.

$$\underset{\text{Cysteine}}{HSCH_2\underset{\underset{\displaystyle NH_2}{|}}{\overset{\displaystyle O}{\overset{||}{C}}H}COH} \xrightarrow{(O)} \underset{\substack{\text{Cysteine sulfinic}\\\text{acid}}}{HOSCH_2\underset{\underset{\displaystyle NH_2}{|}}{\overset{\displaystyle O}{\overset{||}{C}}H}COH} \xrightarrow{\text{deamination}}$$

$$\underset{\beta\text{-Sulfinylpyruvic acid}}{H{-}O{-}\overset{\displaystyle O}{\overset{||}{S}}{-}CH_2\overset{\displaystyle O}{\overset{||}{C}}COH}$$

$$\downarrow [SO_2] \xrightarrow{(O)} SO_4{}^{2-}$$

$$\underset{\text{Pyruvic acid}}{CH_3\overset{\displaystyle O}{\overset{||}{C}}COH} \begin{array}{l} \longrightarrow \text{Gluconeogenesis} \\ \\ \searrow \text{Citric acid cycle} \end{array}$$

**Glutamic acid.** Oxidative deamination of this amino acid gives $\alpha$-ketoglutaric acid. In three more steps of the citric acid cycle (Figure 16.4) this compound is converted into oxaloacetic acid.

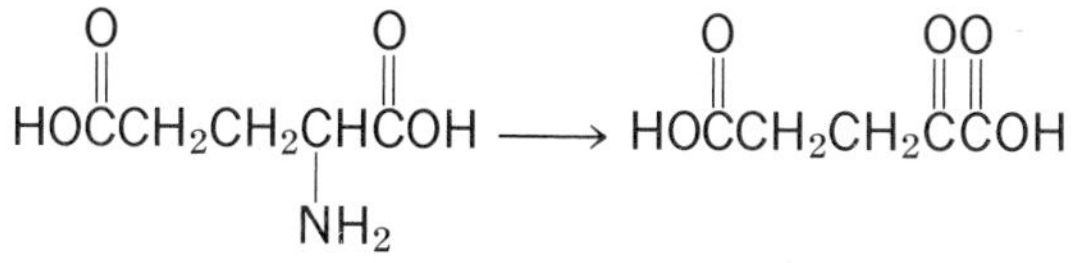

$$\underset{\text{Glutamic acid}}{\overset{\displaystyle O \qquad\qquad O}{HO\overset{\|}{C}CH_2CH_2\underset{\underset{\displaystyle NH_2}{|}}{C}H\overset{\|}{C}OH}} \longrightarrow \underset{\text{$\alpha$-Ketoglutaric acid}}{\overset{\displaystyle O \qquad\qquad OO}{HO\overset{\|}{C}CH_2CH_2\overset{\|\;\|}{C}COH}}$$

$$\rightarrow\rightarrow\rightarrow \quad \underset{\text{acid}}{\text{Oxaloacetic}} \begin{array}{c} \nearrow \text{Gluconeogenesis} \\ \searrow \text{Citric acid cycle} \end{array}$$

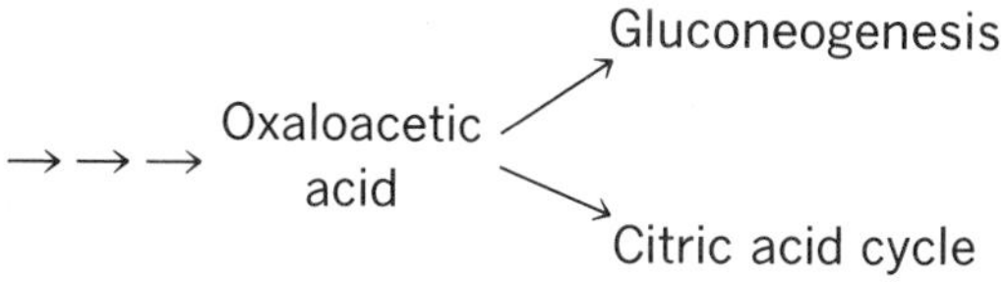

**Serine.** Dehydration of this amino acid leads to unstable intermediates which eventually become pyruvic acid.

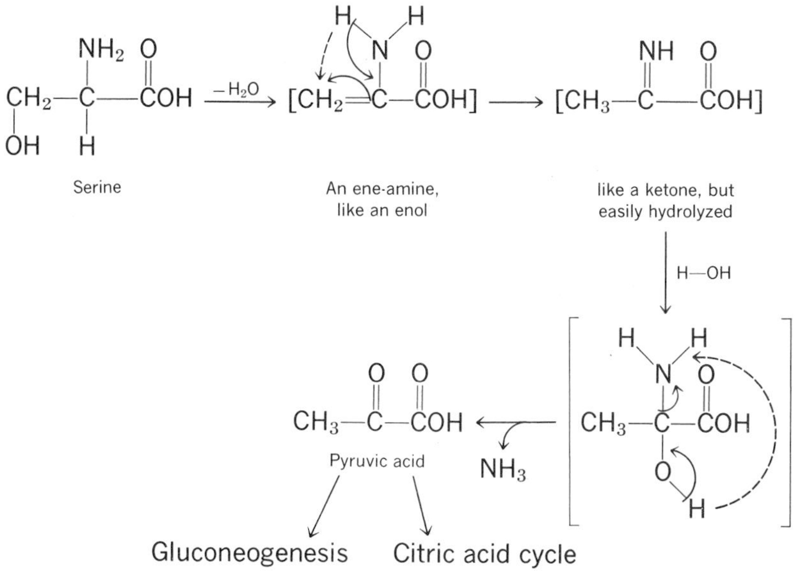

$$\underset{\text{Serine}}{\overset{\displaystyle NH_2 \; O}{\underset{\underset{\displaystyle OH \;\; H}{|\qquad |}}{CH_2{-}C{-}\overset{\|}{C}OH}}} \xrightarrow{-H_2O} \underset{\substack{\text{An ene-amine,}\\\text{like an enol}}}{[CH_2{=}\overset{\displaystyle \overset{H \quad H}{N} \quad O}{C}{-}\overset{\|}{C}OH]} \longrightarrow \underset{\substack{\text{like a ketone, but}\\\text{easily hydrolyzed}}}{[CH_3{-}\overset{\displaystyle \overset{NH}{\|}}{C}{-}\overset{\displaystyle O}{\overset{\|}{C}OH}]}$$

$$\Big\downarrow \text{H—OH}$$

$$\underset{\text{Pyruvic acid}}{\overset{\displaystyle O \;\; O}{CH_3{-}\overset{\|}{C}{-}\overset{\|}{C}OH}} \longleftarrow \quad NH_3 \qquad \Big[ CH_3{-}\overset{\displaystyle \overset{H \quad H}{N} \quad O}{C}{-}\overset{\|}{C}OH \Big]$$

$$\swarrow \qquad\qquad \searrow$$
$$\text{Gluconeogenesis} \qquad \text{Citric acid cycle}$$

**Threonine.** Threonine does not undergo loss of its amino group by transamination. Rather, a special enzyme found principally in the liver and the kidneys catalyzes the following change.

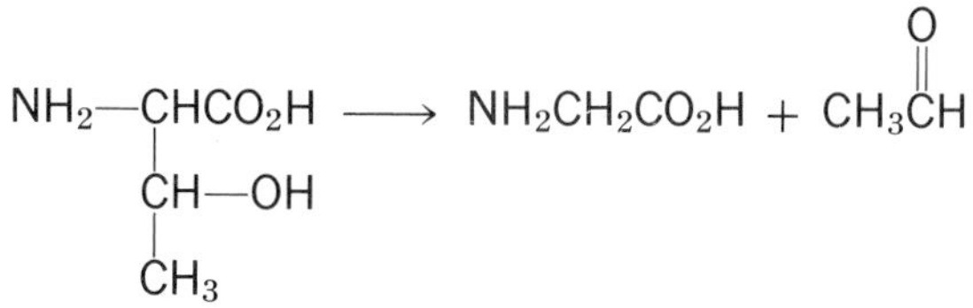

$$\underset{\text{Threonine}}{NH_2{-}\underset{\underset{\displaystyle CH_3}{|}}{\underset{\underset{\displaystyle CH{-}OH}{|}}{C}}HCO_2H} \longrightarrow \underset{\text{Glycine}}{NH_2CH_2CO_2H} + \underset{\text{Acetaldehyde}}{CH_3\overset{\displaystyle O}{\overset{\|}{C}H}}$$

Acetaldehyde is oxidized to acetic acid which is changed to acetyl coenzyme A.

**Leucine.** The catabolism of leucine illustrates the fate of a ketogenic amino acid, one that can furnish carbon fragments for the synthesis of lipids and steroids. In the reactions that follow we have ① transamination, ② oxidative decarboxylation, ③ dehydrogenation, ④ carboxylation, ⑤ hydration of a double bond, and ⑥ a reaction resembling a reverse aldol condensation or a reverse Claisen ester condensation (cf. p. 213 or p. 252). In the last stages we have the interrelations described in Figure 15.6 (p. 505).

$$\underset{\text{CH}_3\text{CHCH}_2\text{CH}-\text{COH}}{\overset{\text{CH}_3\quad \text{NH}_2\ \text{O}}{}} \xrightarrow[\textcircled{1}]{\text{transamination}} \underset{\text{CH}_3\text{CHCH}_2\text{CCOH}}{\overset{\text{CH}_3\quad \text{OO}}{}}$$

$$\xrightarrow[\text{CoASH}]{\text{NAD}^+} \quad \textcircled{2}\ \text{oxidative decarboxylation}$$

CO$_2$

(This resembles the conversion of pyruvic acid to acetyl CoA: $\underset{\text{CH}_3\text{CCOH}}{\overset{\text{OO}}{}} \longrightarrow \underset{\text{CH}_3\text{CSCoA}}{\overset{\text{O}}{}} + \text{CO}$

$$\underset{\text{CH}_3\text{C}=\text{CHCSCoA}}{\overset{\text{CH}_3\quad \text{O}}{}} \xleftarrow[-2\text{H}]{\overset{\textcircled{3}}{\text{dehydrogenation}}} \underset{\text{CH}_3\text{CHCH}_2\text{CSCoA}}{\overset{\text{CH}_3\quad \text{O}}{}}$$

CO$_2$ ↘ ④

$$\underset{\text{HOCCH}_2\text{C}=\text{CHCSCoA}}{\overset{\text{O}\quad \text{CH}_3\quad \text{O}}{}} \xrightarrow[\text{H}_2\text{O}]{\text{addition of}\ \textcircled{5}} \underset{\text{HOCCH}_2\text{CCH}_2\text{CSCoA}}{\overset{\text{O}\quad \text{CH}_3\quad \text{O}}{}}$$

OH

⑥ / HMG—CoA

several steps

$$\underset{\text{HOCCH}_2\text{CCH}_3}{\overset{\text{O}\quad \text{O}}{}} + \underset{\text{CH}_3\text{CSCoA}}{\overset{\text{O}}{}}$$

Acetoacetic acid    Acetyl coenzyme A

Cholesterol

in extra-hepatic tissue    CoASH

$$\underset{\text{CoASCCH}_2\text{CCH}_3}{\overset{\text{O}\quad \text{O}}{}}$$

Acetoacetyl coenzyme A

Steroids and steroid hormones    Bile salts

Lipogenesis cycle ⤡   ⤢ Citric acid cycle

## GLYCINE AND THE SYNTHESIS OF HEME

We shall learn later in this chapter about the catabolism of heme; here we shall simply state that the body must replenish its supply constantly and routinely. The heme molecule (Figure 16.5) is of such size and complexity that its routine

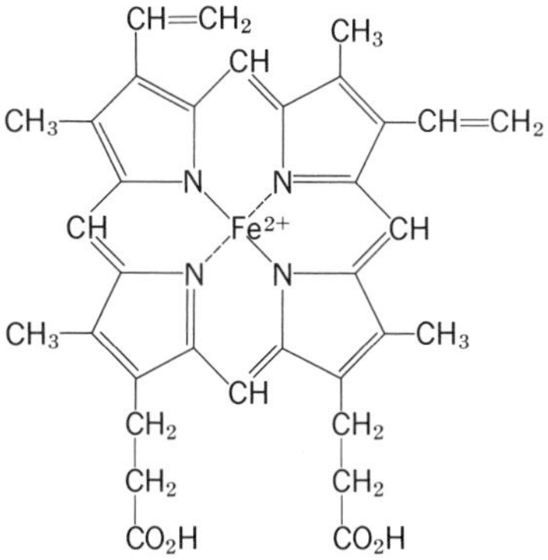

Heme

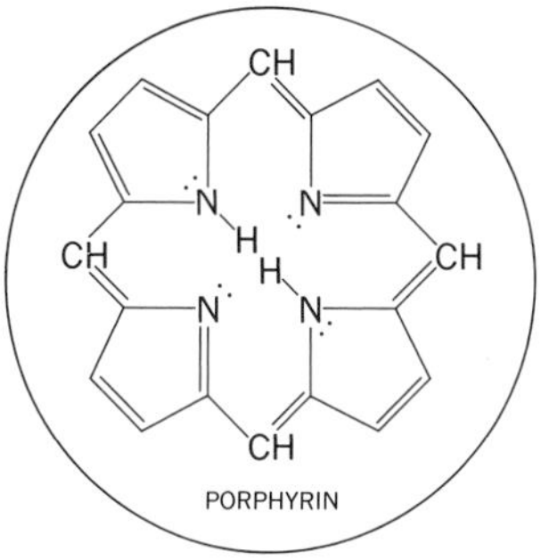

PORPHYRIN

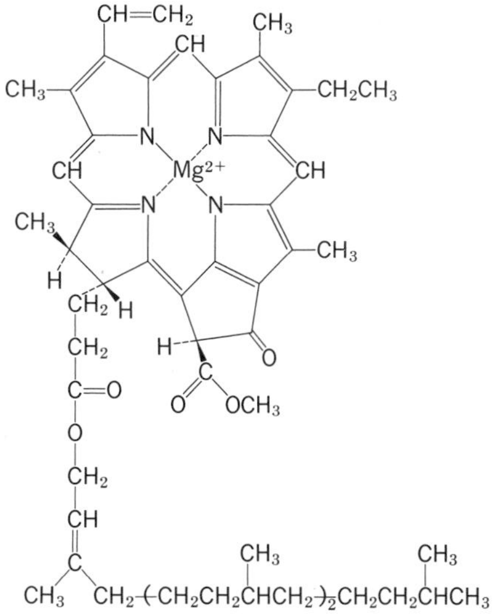

Chlorophyll a

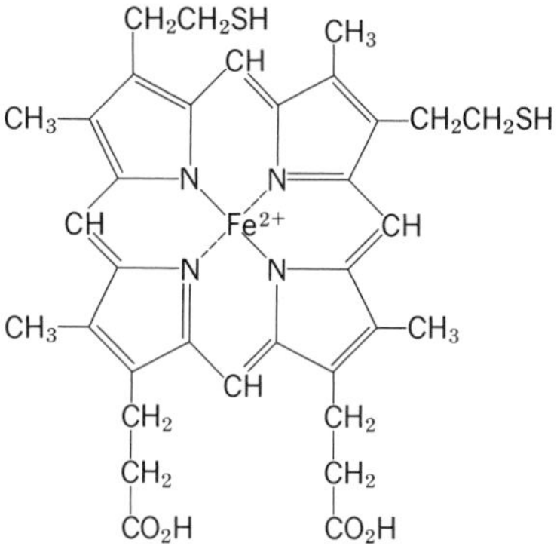

Porphyrin group present in cytochrome c

(Cytochrome b has the same porphyrin c group, the heme group, as hemoglobin, but it is attached to the protein in a different way.)

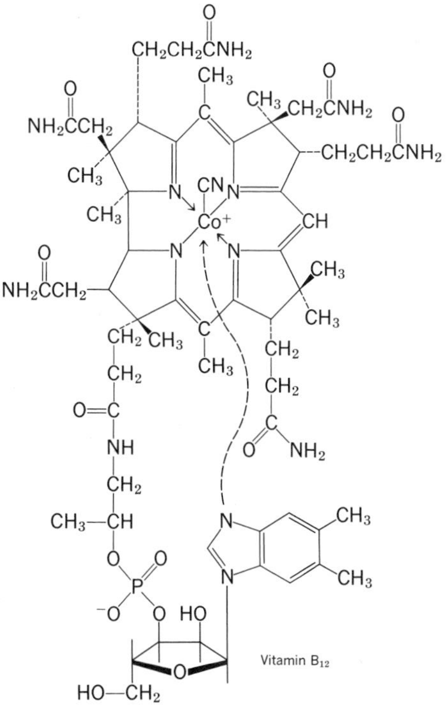

Vitamin $B_{12}$

**Figure 16.5** The porphyrin nucleus is a part of many important substances. All red-blooded animals have heme. All green leafy plants have chlorophyll. The cytochromes are essential to respiratory chains of enzymes. Vitamin $B_{12}$, sometimes called cyanocobalamin, is the anti-pernicious anemia factor.

synthesis would seem to be a very formidable job. Yet the body starts with the simplest amino acid, glycine, and almost the simplest dicarboxylic acid, succinic acid, to put together this awesome molecule. The same organic skeleton found in heme is widely distributed in nature, for it is a part not only of hemoglobin but also of the cytochromes and chlorophyll, as structures in Figure 16.5 show. The porphyrin molecule does not occur by itself in nature.

In the early stages of the synthesis of heme, reactions very much like ordinary organic reactions occur. We cannot forget that special enzymes are needed to make them possible, but we can look upon one of the steps, the step that makes the pyrrole ring, with a view informed by our earlier experiences with ordinary organic chemistry.

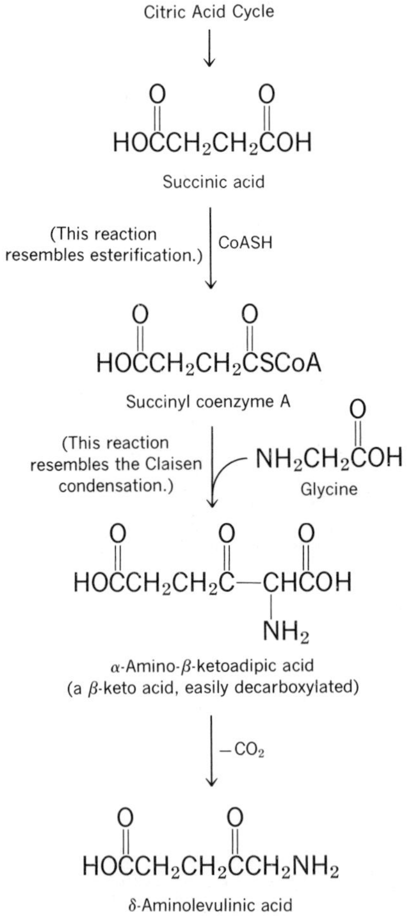

Citric Acid Cycle

$\downarrow$

$$\underset{\text{Succinic acid}}{\text{HO}\overset{\displaystyle O}{\overset{\|}{C}}\text{CH}_2\text{CH}_2\overset{\displaystyle O}{\overset{\|}{C}}\text{OH}}$$

(This reaction resembles esterification.) | CoASH

$$\underset{\text{Succinyl coenzyme A}}{\text{HO}\overset{\displaystyle O}{\overset{\|}{C}}\text{CH}_2\text{CH}_2\overset{\displaystyle O}{\overset{\|}{C}}\text{SCoA}}$$

(This reaction resembles the Claisen condensation.) | $\underset{\text{Glycine}}{\text{NH}_2\text{CH}_2\overset{\displaystyle O}{\overset{\|}{C}}\text{OH}}$

$$\text{HO}\overset{\displaystyle O}{\overset{\|}{C}}\text{CH}_2\text{CH}_2\overset{\displaystyle O}{\overset{\|}{C}}-\overset{\displaystyle \underset{\displaystyle NH_2}{|}}{C}\text{H}\overset{\displaystyle O}{\overset{\|}{C}}\text{OH}$$

α-Amino-β-ketoadipic acid
(a β-keto acid, easily decarboxylated)

$\downarrow -CO_2$

$$\underset{\text{δ-Aminolevulinic acid}}{\text{HO}\overset{\displaystyle O}{\overset{\|}{C}}\text{CH}_2\text{CH}_2\overset{\displaystyle O}{\overset{\|}{C}}\text{CH}_2\text{NH}_2}$$

In the next step two molecules of δ-aminolevulinic acid condense to close the five-membered pyrrole ring of an intermediate called porphobilinogen:

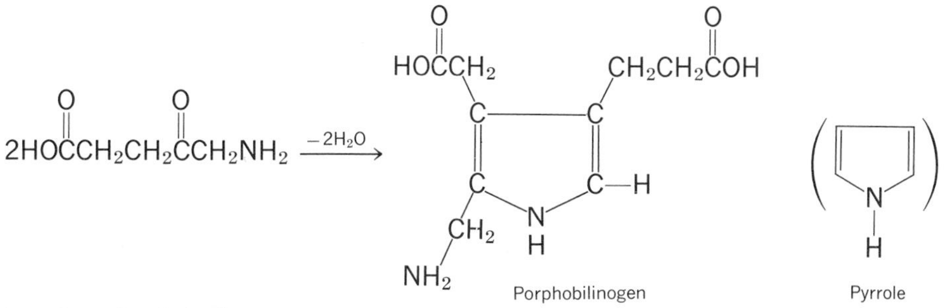

$$2HO\overset{O}{\overset{\|}{C}}CH_2CH_2\overset{O}{\overset{\|}{C}}CH_2NH_2 \xrightarrow{-2H_2O}$$

Porphobilinogen

Pyrrole

Or, viewed mechanistically,

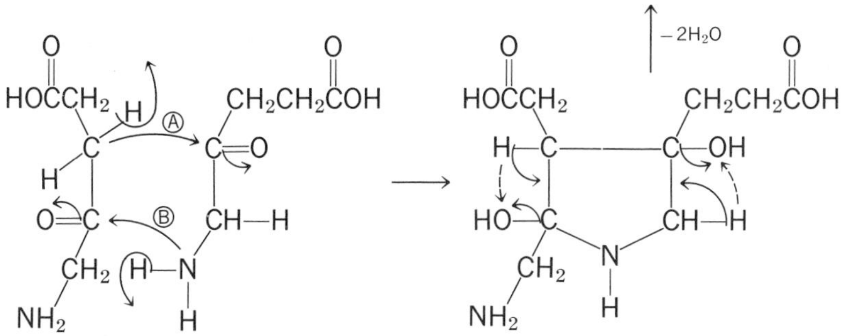

(Circled letters Ⓐ and Ⓑ index two steps to be described in greater detail.)

This mechanistic view of the ring closure is conjecture, for the timing of the steps and other details of what happens when the enzyme or enzymes bring it about are not known. Yet it is interesting that the process can be rationalized to a certain extent in terms of Ⓐ, a step very much like the initial phase of an aldol condensation, and Ⓑ, a step that resembles the beginning of the formation of an imine.

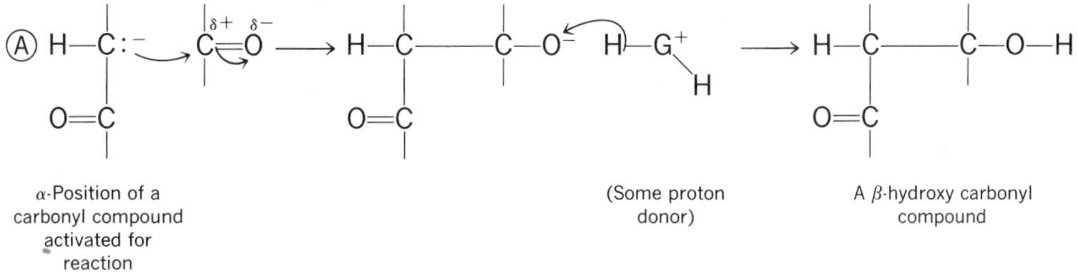

Ⓐ 

α-Position of a carbonyl compound activated for reaction

(Some proton donor)

A β-hydroxy carbonyl compound

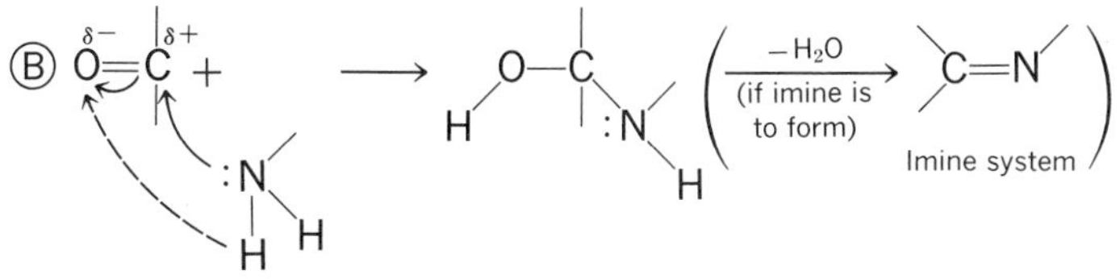

Ⓑ 

$$\xrightarrow[\text{to form}]{-H_2O \atop \text{(if imine is}}$$

Imine system

With one of the rings needed for the heme molecule already formed, we next visualize how four such rings might be put together. Two steps are probably involved. The first produces a linear tetrapyrrole system, and the second ties one end of it back to the other end to form the huge network of pyrrole rings. To simplify these steps we use the following symbols.

| | | |
|---|---|---|
| A = | $-CH_2CO_2H$ | for a side chain *acetic* acid unit |
| P = | $-CH_2CH_2CO_2H$ | for a side chain *propionic* acid unit |
| M = | $-CH_3$ | for a side chain *methyl* unit |
| V = | $-CH=CH_2$ | for a side chain *vinyl* unit |

Furthermore, we visualize the side chain acetic acid unit as being converted to a methyl group by decarboxylation:

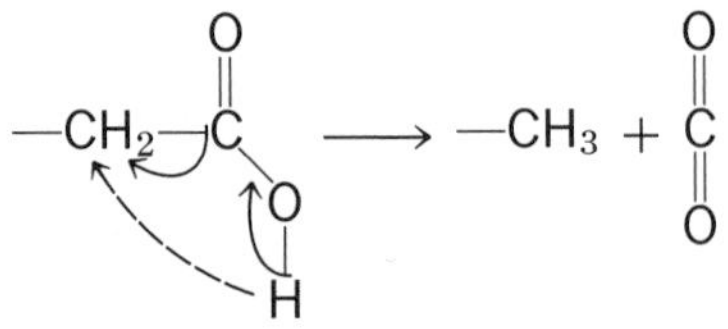

$$-CH_2-\overset{\overset{O}{\|}}{C}\overset{O}{\underset{H}{}} \longrightarrow -CH_3 + \overset{\overset{O}{\|}}{\underset{\overset{\|}{O}}{C}}$$

or

$$A \longrightarrow M$$

In a similar manner some of the propionic acid side chains can be converted to vinyl groups. Both decarboxylation and dehydrogenation must occur.

$$-CH_2CH_2\overset{\overset{O}{\|}}{C}-OH \longrightarrow -CH=CH_2 + CO_2 + (H:^- + H^+)$$

or

$$P \longrightarrow V$$

Figure 16.6 outlines the steps from porphobilinogen to heme.

## CATABOLISM OF HEME

Hemoglobin circulates inside red blood cells or erythrocytes, cells which have a life-span of only about 120 days. After traveling through the bloodstream for that period, they split open and the hemoglobin spills out. The hemoglobin is then catabolized, and its breakdown products are eliminated through the bile, in the feces, and, to a slight extent, in the urine. Some of the characteristics of bile, feces, and urine are caused by partly degraded heme molecules.

In the catabolism of heme, which begins before the globin portion of hemoglobin breaks away, one of the carbon bridges in the large tetrapyrrole ring is re-

**Figure 16.6** General pathway for the formation of heme from porphobilinogen. The synthesis ▶ of the latter from glycine and succinic acid is described in the text. Symbols: A = $-CH_2CO_2H$ (acetic acid unit), P = $-CH_2CH_2CO_2H$ (propionic acid unit), M = $-CH_3$ (methyl), V = $-CH=CH_2$ (vinyl).

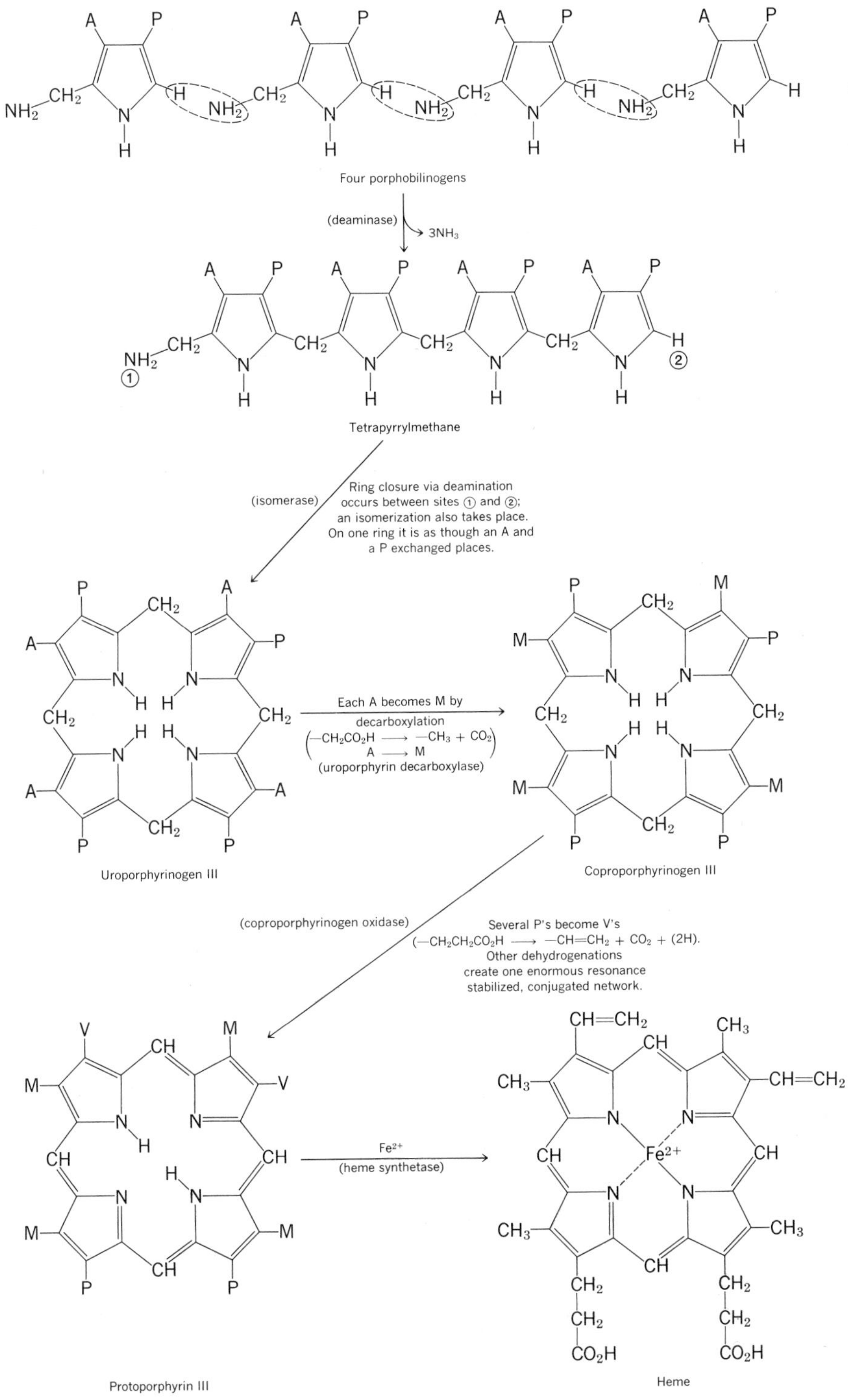

Four porphobilinogens

(deaminase) → 3NH₃

Tetrapyrrylmethane

(isomerase)    Ring closure via deamination
occurs between sites ① and ②;
an isomerization also takes place.
On one ring it is as though an A and
a P exchanged places.

Uroporphyrinogen III

Each A becomes M by
decarboxylation
$$\left(-CH_2CO_2H \longrightarrow -CH_3 + CO_2\right)$$
$$A \longrightarrow M$$
(uroporphyrin decarboxylase)

Coproporphyrinogen III

(coproporphyrinogen oxidase)

Several P's become V's
$$(-CH_2CH_2CO_2H \longrightarrow -CH=CH_2 + CO_2 + (2H).$$
Other dehydrogenations
create one enormous resonance
stabilized, conjugated network.

Protoporphyrin III

$$\xrightarrow[\text{(heme synthetase)}]{Fe^{2+}}$$

Heme

moved to open it to a linear form. This and succeeding steps are shown in Figure 16.7.

The slightly broken hemoglobin molecule, now called *verdohemoglobin*, then splits into globin, ferrous ion, and a greenish tetrapyrrole pigment called *biliverdin* (Latin: *bilis,* bile; *viridus,* green). Globin enters the nitrogen pool, and the iron is conserved by the body in the form of a storage protein, ferritin, to be reused. Biliverdin undergoes further reactions as outlined in Figure 16.7.

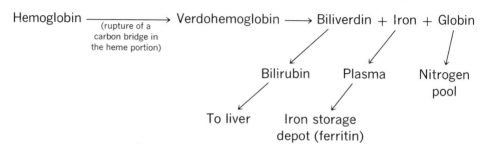

In human beings biliverdin is converted by enzymes in the liver to a reddish-orange pigment called bilirubin (Latin: *bilis,* bile; *rubin,* red), the principal pigment in human bile. The reaction is nothing more than the addition of two hydrogens.

Bilirubin is not only made by the liver but is also removed from circulation by the liver, which transfers it to bile. In this fluid it finally enters the intestinal tract where bacterial action reduces it to a colorless substance, *mesobilirubinogen.* When the structure of this compound, shown in Figure 16.7, is examined and compared with that of biliverdin or bilirubin, we see that the conjugated system has by now become broken up and limited to four short and separate units in the rings. For a compound to be colored, its conjugated chain must normally be longer than four atoms.

Mesobilirubinogen is then reduced further to yield *bilinogen* which goes by two names acquired before the identity of this material obtained from different sources was discovered. Bilinogen that stays in the intestinal tract and leaves the body in feces is called *stercobilinogen* (Latin *stercus,* dung). *Urobilinogen,* the same compound, exits with the urine.

Slight oxidation of bilinogen yields *bilin* which also has two names for the same reason. *Stercobilin* is bilin that leaves the body in the feces, and *urobilin* leaves via the urine. The characteristic brown color of feces and urine is produced by various amounts of bilin.

Figure 16.8 shows the various breakdown products of heme in terms of where they form and where they are excreted. Most of the bilinogen leaves via the feces. Only relatively small amounts are reabsorbed and eventually placed into the urine. On the average a normal adult excretes 250 mg of bilin each day in the feces and 1 to 2 mg in the urine.

**Jaundice.** Jaundice (French *jaune,* yellow) is a condition symptomatic of malfunction somewhere along the pathway of heme metabolism. If bile pigments accumulate in the plasma in concentrations high enough to impart a yellowish colora-

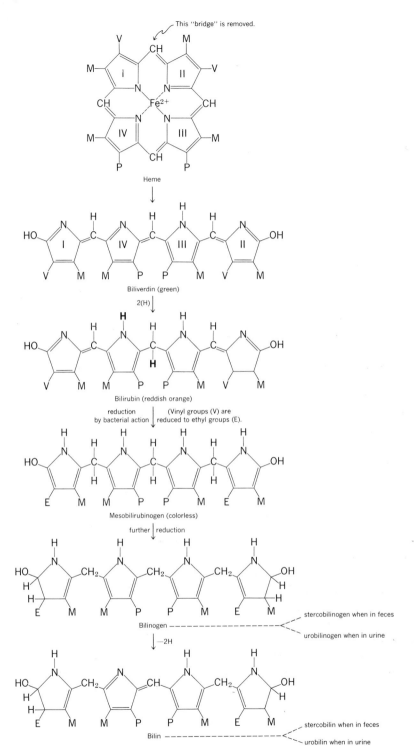

**Figure 16.7** Pathway of heme catabolism.

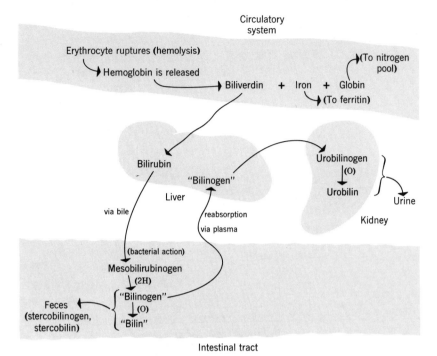

**Figure 16.8**   Elimination of hemoglobin breakdown products.

tion to the skin, the condition is diagnosed as jaundice. Three kinds of malfunction produce it.

**Hemolytic jaundice.** In this condition hemolysis takes place at an abnormally fast rate, and bile pigments, particularly bilirubin, form faster than the liver can clear them.

**Hepatic diseases.** If the liver itself is not able to remove bilirubin, it remains in circulation and the body becomes jaundiced. Hepatic diseases such as *infectious hepatitis* and *cirrhosis* may be responsible. The stools are usually clay-colored, since the tetrapyrrole pigments do not reach the intestinal tract.

**Obstruction of bile ducts.** Should the ducts that deliver bile to the intestinal tract become obstructed, the tetrapyrrole pigments in bile cannot be eliminated and tend to reenter general circulation. The kidneys remove large amounts of bilirubin, but the stools are usually clay-colored. As the liver works harder and harder to handle its task of removing excess bilirubin, it may weaken and become permanently damaged.

## REFERENCES

S. W. Fox and J. F. Foster. *Introduction to Protein Chemistry.* John Wiley and Sons, New York, 1957.

A. White, P. Handler, and E. L. Smith. *Principles of Biochemistry,* third edition. McGraw-Hill Book Company, New York, 1964.

E. D. Wilson, K. H. Fisher, and M. E. Fuqua. *Principles of Nutrition,* second edition. John Wiley and Sons, New York, 1965.

## PROBLEMS AND EXERCISES

1. Define each of the following terms.
   - (*a*) nitrogen fixation
   - (*b*) symbiosis
   - (*c*) nitrogen equilibrium
   - (*d*) positive nitrogen balance
   - (*e*) negative nitrogen balance
   - (*f*) nitrogen pool
   - (*g*) essential amino acid
   - (*h*) nonessential amino acid
   - (*i*) kwashiorkor
   - (*j*) adequate protein
   - (*k*) glycogenic amino acid
   - (*l*) ketogenic amino acid
   - (*m*) jaundice
   - (*n*) hemolysis
   - (*o*) ferritin
   - (*p*) verdohemoglobin
   - (*q*) biliverdin
   - (*r*) bilirubin
   - (*s*) "bilinogen" and its relation to stercobilinogen and urobilinogen
   - (*t*) "bilin" and its relation to stercobilin and urobilin

2. What is the difference between undernutrition and malnutrition?

3. How are we dependent on soil bacteria?

4. In general terms, how does the body try to meet its needs for glucose during developing starvation?

5. What are the principal end products of amino acid catabolism?

6. Outline the sequences of reactions that show how alanine might be used to make (*a*) glucose and (*b*) palmitic acid.

7. Write the sequence of reactions that show how aspartic acid might be used to make "new glucose."

8. In Guatamala corn makes up about 70% of the human diet. The protein in corn is deficient in lysine. Discuss the implications of this.

9. What is the difference between gluconeogenesis and glucogenesis, and in what ways are proteins and lipids used in the former? Be specific, writing specific reactions.

# CHAPTER SEVENTEEN

# The Chemistry
# of Heredity

## HEREDITY AND ENZYMES

The physical links between one generation and the next are the sperm cell of the male and the egg cell of the female. After the union of these two cells, a series of chemical events unfolds. All the information that is needed to ensure that the fertilized egg will develop physically into an authentic member of the parents' species must be present in that tiny and wonderful unit of matter. If the baby animal is to have fur rather than feathers, at its conception a sequence of chemical reactions is started that will lead to fur making rather than feather making. In spite of the basic similarity in diets, especially after digestion has done its work, radically different pathways are taken by various species.

We have learned that nearly every reaction occurring in a living organism requires special enzymes which act uniquely on fairly common molecules such as glucose, fatty acids and glycerol, and amino acids. Each species possesses its own peculiar set of enzymes and hormones, even though some remarkable similarities exist, especially at the coenzyme level. Fur-bearing animals have enzyme systems that catalyze the ordering of amino acid sequences to produce fur. Feathered creatures have different enzyme systems that generate the proteins of feathers from basically the same set of amino acids. Whatever the genetic message may be, whether a set of instructions to develop as a cowbird or a set to develop as a cow, it very likely concerns the generation of a distinctive enzyme system. We therefore expect any theory concerning the chemical basis of heredity to explain how a species acquires and reproduces its special set of enzymes.

## GENERAL FEATURES OF THE PHYSICAL BASIS OF HEREDITY

In a typical animal, union of a sperm cell with an egg cell produces a new cell called a *zygote,* which proceeds to multiply by a process of cell division called *mitosis.* The daughter cells of the zygote themselves divide, and so on, as an embryo takes form. Early in this stage of development two fundamentally different kinds of cells can be distinguished: *germ cells,* which will give rise to either sperm or eggs, and *somatic cells,* from which will form all the myriad tissues and organs unique to the body of the species. The germ cells are cells set apart, protected from change and unaffected by the tremendous variations taking place among the somatic cells. When the somatic cells have proceeded far enough in their development that the gonads are fully elaborated and sexual maturity is reached, the germ cells become active. They develop sperm or eggs, depending on the sex of the individual. If the sperm and the egg of the parents contained the essentials to produce a unique enzyme system, the germ cells of the children must also possess these essentials, in order that they, in turn, may pass them to the next generation.

**Chromosomes and Genes.** The cell is the structural unit of life, and chemicals associated with living things are, by and large, organized in these units. All the material comprising the cells is called the *protoplasm.* Discrete "bodies," such as mitochondria and the cell nucleus, exist in the protoplasm. Figure 17.1 depicts a generalized animal cell, with several parts labeled. The cell nucleus contains a fluid in which twisted and intertwined filaments called chromonemata exist; these apparenty bear strings of the basic units of heredity, the *genes.* Chromonemata and their gene strings constitute individual *chromosomes.*

During mitosis division of the nucleus precedes division of the cell as a whole. In preliminary stages of nucleus division the chromonemata and the genes normally produce exact duplicates of themselves. The final cell division, then, separates duplicated chromonemata and gene strings into the individual nuclei of two new

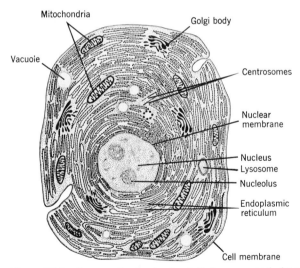

Mitochondria

Golgi body

Vacuole

Centrosomes

Nuclear
membrane

Nucleus
Lysosome
Nucleolus

Endoplasmic
reticulum

Cell membrane

**Figure 17.1** A generalized animal cell. Cells vary greatly from tissue to tissue, but most have the common features shown here. In this chapter we are particularly concerned with the nucleus and the ribosome-studded endoplasmic reticulum. (From Jean Brachet, "The Living Cell," *Scientific American,* September 1961, page 50. Copyright © 1961 by Scientific American, Inc. All rights reserved.)

6. New Interphase.

Two new daughter cells emerge with
sets of chromosomes and genes identical
with each other and the "parent" cell.

Daughter Cells

Late Telophase

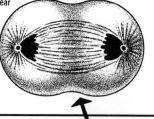

5. Telophase.

Cell begins to divide. New nuclear
membranes begin to form.

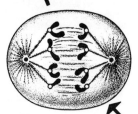

Early Telophase

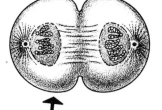

Early Anaphase

Anaphase

4. Anaphase.

Centromeres now divide and "daughter"
chromosomes are pulled apart as the
centromeres move toward opposite poles
of the spindle.

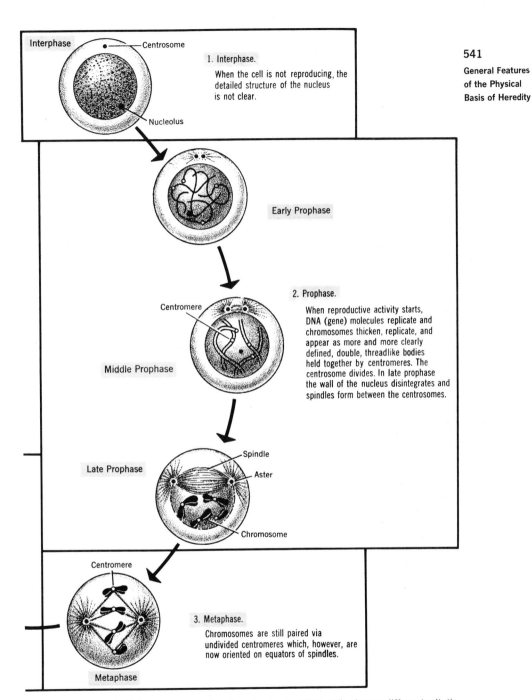

Interphase

Centrosome

1. Interphase.

When the cell is not reproducing, the detailed structure of the nucleus is not clear.

Nucleolus

Early Prophase

2. Prophase.

When reproductive activity starts, DNA (gene) molecules replicate and chromosomes thicken, replicate, and appear as more and more clearly defined, double, threadlike bodies held together by centromeres. The centrosome divides. In late prophase the wall of the nucleus disintegrates and spindles form between the centrosomes.

Centromere

Middle Prophase

Late Prophase

Spindle

Aster

Chromosome

Centromere

3. Metaphase.

Chromosomes are still paired via undivided centromeres which, however, are now oriented on equators of spindles.

Metaphase

**Figure 17.2** Mitosis. This sequence applies to animal cells. Plant cells show a different mitotic sequence, especially in the telophase. (Cell structures from E. J. Gardner, *Principles of Genetics*, third edition, John Wiley and Sons, New York, 1968, page 28.)

daughter cells. The division of one parent cell involves one duplication of each gene; this reproductive duplication is often called *replication*. Figure 17.2 shows the chief stages in mitosis. To understand how genes might be responsible for carrying heredity messages, we must study the chemistry of these heredity units.[1]

## NUCLEIC ACIDS AS HEREDITY UNITS

We have long known that cell nuclei are rich in a polymeric material called deoxyribonucleic acid (DNA). All available evidence supports the conclusion that DNA is the actual chemical constituting genes. At the University of Cambridge in 1953 F. H. C. Crick and J. D. Watson proposed a structure for DNA that provides a means for correlating its physical and chemical properties with its properties as the apparent chemical of genes. For this work these two men shared with Maurice Wilkins the 1962 Nobel prize in medicine and physiology. Much about DNA was known before 1953, and good evidence suggested that this chemical was the actual constituent of genes. Data from X-ray diffraction studies even suggested that the polymeric molecules making up DNA were aligned side by side in a twisted helix. But until the work of Crick and Watson the picture was not clear enough for scientists to see just *how* the structure of DNA had something to do with genetic properties. Moreover, other polymeric substances chemically similar to DNA existed in cells, and how might these be involved?

**Nucleic Acids.** Deoxyribonucleic acid is a member of a family of polymers called *nucleic acids.* Their monomer units are called *nucleotides,* and these, in turn, are built from simpler parts. Figure 17.3 shows the relation of the nucleic acids to their building blocks by indicating how hydrolysis splits the polymer successively into smaller and smaller units. If the original nucleic acid is of the DNA type, one product of its hydrolysis (Figure 17.3) will be deoxyribose ("de-" means "lacking"; "deoxy-" means "lacking in an oxygen atom found in the close structural relative, ribose"). The other major type of nucleic acid, ribonucleic acid (RNA), will hydrolyze to form ribose instead of deoxyribose.

The heterocyclic bases obtained from nucleic acids are derived from either purine or pyrimidine. In Figure 17.3 all the bases but adenine are shown in a keto form (really a lactam form) and an enol form (actually lactim). In brief, the following equilibria are analogous.

| Keto form | Enol | Lactam (cyclic amide) | Lactim |

In their enol forms the systems are fully aromatic and derive stabilization via resonance. As isolated chemicals they are believed to exist largely in such forms, but when these purines and pyrimidines are bound into nucleic acids, they are in

---

[1] It is beyond the intended scope of this book to examine the development of the gene concept. Some of the numerous outstanding references available are cited at the end of this chapter.

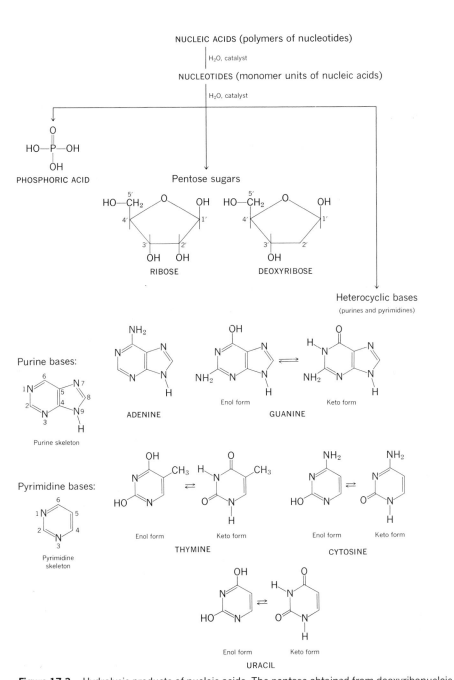

**Figure 17.3** Hydrolysis products of nucleic acids. The pentose obtained from deoxyribonucleic acid (DNA) is deoxyribose. From ribonucleic acids (RNA) ribose is obtained. The five principal heterocyclic bases are shown here, although others are known. As separate chemicals they exist largely in their enol forms. Bound into nucleic acids, they exist largely in the keto forms.

their keto forms. What might appear a loss of stabilization is recovered through the greatly increased availability of sites for possible hydrogen bonding, and stabilization is achieved in a different way.

Of the three pyrimidine bases shown, uracil is found almost exclusively in RNA. The complete hydrolysis of DNA yields the following four bases: adenine (A), guanine (G), thymine (T), and cytosine (C).

The monomer units of nucleic acids, the nucleotides, are made from one molecule each of phosphoric acid, a pentose, and a heterocyclic base. The manner of assembly of a typical nucleotide is shown in Figure 17.4. The heterocyclic amine is always attached at C-1′ of the pentose; the phosphate ester always forms at the C-5′ position.

Each nucleotide has at least two places where new ester groups may form, at hydroxyls on the pentose ring and at the phosphate terminus, making the polymerization of nucleotides and the formation of nucleic acids possible (Figure 17.5). The phosphoric acid unit on one nucleotide splits out the elements of water with an alcohol unit of the pentose of the next nucleotide. A phosphate ester linkage or bridge forms between the two nucleotides. The process is visualized as continuing until hundreds of nucleotides are incorporated into the polymer.

The backbone of any nucleic acid from any of the forms of life investigated thus far appears to be the simple alternating system phosphate-pentose-phosphate-pentose-phosphate-pentose-etc. Because each pentose unit has a heterocyclic amine attached, a more complete picture of a nucleic acid is

$$\text{etc.—phosphate—pentose—phosphate—pentose—phosphate—pentose—etc.}$$

with amine′, amine″, amine‴ attached to the respective pentose units.

A more useful condensation of a nucleic acid structure is given in Figure 17.6.

If this structure represents what nucleic acids have in common, how they are different becomes apparent. The lengths of the backbones and the sequences

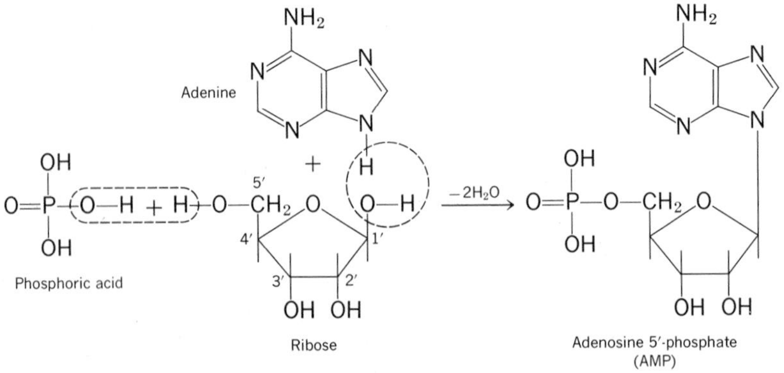

Figure 17.4  A typical nucleotide, a monomer for RNA. If the —OH group at C-2′ were replaced by —H, the nucleotide would be one of the monomers for DNA. By convention, the ring position numbers are primed for the ribose ring to distinguish them from the unprimed numbers of the heterocyclic ring.

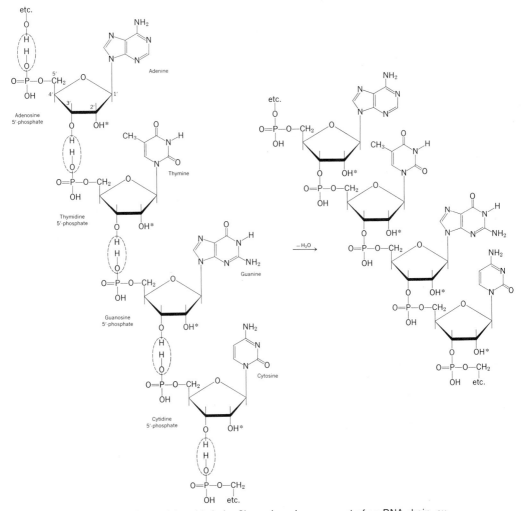

**Figure 17.5** Formation of a nucleic acid chain. Shown here is a segment of an RNA chain, except that uracil replaces thymine in RNA. If the —OHs marked by asterisks were replaced by —Hs, this would be a segment of a DNA chain. The sequence of the heterocyclic amines is purely arbitrary in this drawing, but one each of the four amines common to DNA has been included. J. Cairns has found a molecular weight of $2.8 \times 10^9$ for the DNA of one species (*Escherichia coli*). If we use a value of 325 as the average formula weight of each nucleotide, this DNA will be made of 8,600,000 nucleotide units. Such a DNA molecule would probably make up a collection of genes rather than just one. Genetic studies indicate that the average gene size is 1500 nucleotide pairs (of a double helix).

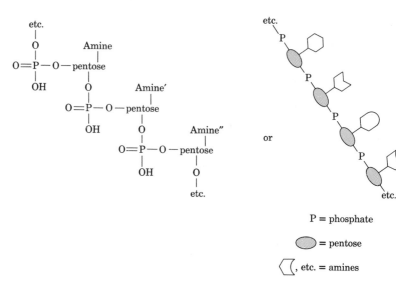

P = phosphate

= pentose

, etc. = amines

**Figure 17.6**  Condensed structural representations of nucleic acids.

in which the amines are strung along the backbones can both be different. In fact, we shall shortly identify this sequence with the genetic code itself. The facts of heredity will be correlated, in principle, with this aspect of the structure of a chemical.

**Watson-Crick Theory.** The DNA samples from a variety of species have been analyzed for proportions of the heterocyclic bases; Chargaff called the attention of scientists to some regularities that proved to be of great theoretical importance. To illustrate what these regularities are, some of the analytical data have been compiled in Table 17.1. The purines, we recall, are adenine (A) and guanine (G), the pyrimidines thymine (T) and cytosine (C).[2] Chargaff noted that for all the species analyzed

1. The purines equaled the pyrimidines in total percentages; that is, the purine-pyrimidine ratio was $(G + A)/(C + T) = 1$.

2. The percentage of adenine (a purine) was always very close to the percentage of thymine (a pyrimidine), that is, $A/T = 1$.

3. The percentage of guanine (a purine) was always very close to the percentage of cytosine (a pyrimidine), that is, $G/C = 1$.

4. The *characteristic composition* that was unique for a species was the ratio $(A + T)/(G + C)$.

This characteristic ratio for a particular species is not affected by factors such as the age of the species or its manner of growth. Different organs of the same species (e.g., thymus and liver of human beings, Table 17.1) exhibit substantially the same characteristic ratios. This constancy indicates something important about the kind of chemical that was, even before Crick and Watson's theory, believed to be the stuff of genes. Whatever the genetic material may be, and however the genetic code may

---

[2] 5-Methylcytosine, one of the rarer pyrimidines, is present in some species and the percents given for cytosine include this compound.

**Table 17.1   DNA Composition of Various Species**

| Species | Base Proportions (mole %) Purines G | A | Pyrimidines C | T | $\frac{G+A}{C+T}$ (i.e., purine/ pyrimidine) | $\frac{A}{T}$ | $\frac{G}{C}$ | $\frac{A+T}{G+C}$ (species characteristic) | |
|---|---|---|---|---|---|---|---|---|---|
| *Sarcina lutea* | 37.1 | 13.4 | 37.1 | 12.4 | 1.02 | 1.08 | 1.00 | 0.35 | |
| *Brucella abortus* | 20.0 | 21.0 | 28.9 | 21.1 | 1.00 | 1.00 | 1.00 | 0.73 | |
| *Escherichia coli* K12 | 24.9 | 26.0 | 25.2 | 23.9 | 1.08 | 1.09 | 0.99 | 1.00 | |
| Wheat germ | 22.7 | 27.3 | 22.8 | 27.1 | 1.00 | 1.01 | 1.00 | 1.19 | |
| *Staphylococcus aureus* | 21.0 | 30.8 | 19.0 | 29.2 | 1.07 | 1.05 | 1.11 | 1.50 | |
| Human thymus | 19.9 | 30.9 | 19.8 | 29.4 | 1.03 | 1.05 | 1.01 | 1.52 | |
| Human liver | 19.5 | 30.3 | 19.9 | 30.3 | 0.99 | 1.00 | 0.98 | 1.54 | |

Data from A. White, P. Handler, and E. L. Smith, *Principles of Biochemistry*, third edition, McGraw-Hill Book Company, New York, 1964, page 172. Used by permission.

be structured, some things about it must be uniform, at least within a species. That there are some constant aspects was apparent from Chargaff's data. It remained for Watson and Crick to exploit these observations as well as results of X-ray analysis and the known general features of the DNA polymer. Watson and Crick's first paper on this topic, "A Structure for Deoxyribose Nucleic Acid," appeared on April 25, 1953, in the British publication *Nature*. Soon after, on May 30, 1953, a second paper, "Genetic Implications of the Structure of Deoxyribonucleic Acid," appeared in the same journal. These are two landmarks in the history of science. They were not, as many have written, documents of a revolution in science, nor were they turning points or radical departures. They were major advances. In a revolution old ideas are discarded, old institutions destroyed, and new ideas, new institutions are built on their remains. The accomplishment of Watson and Crick was built on what had gone before. They were not the first to propose that DNA might be the chemical of genes. The basic manner of hooking together the nucleotides had been established, and many proposed models for DNA gave the polymer a particular configuration in space. Pauling and Corey of $\alpha$-helix fame (in connection with protein structure) had earlier in 1953 proposed a model in which three DNA molecules wrapped around each other. Fraser, also in 1953, had proposed a three-stranded structure and Furberg, in 1952, a single-stranded helix. None of these models, however, was successful in explaining how the chemical could do what the gene was known to do—become duplicated exactly. This was the singular contribution of Watson and Crick. The model of a gene molecule that they proposed not only best fit the X-ray data—the best data appeared in a paper by M. H. F. Wilkins, A. R. Stokes, and H. R. Wilson published immediately after Watson and Crick's first paper—but could direct the essential operation required of genetic material, namely self-duplication. For this work Watson, Crick, and Wilkins shared the 1962 Nobel prize.

To begin our discussion of the Watson-Crick model for a gene, several portions of their two papers will be quoted (with permission). The first paper opens as follows:

"We wish to suggest a structure for the salt of deoxyribose nucleic acid (D.N.A.). This structure has novel features which are of considerable biological

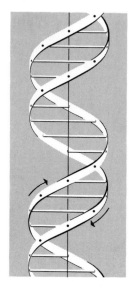

**Figure 17.7**    The DNA double helix. "This figure is purely diagrammatic. The two ribbons symbolize the two phosphate-sugar chains,
and the horizontal rods the pairs of bases holding the chains together. The vertical line marks the fibre axis." (Figure and quotation from J. D. Watson and F. H. C. Crick, *Nature,* Vol. 171, 1953,
page 737. Used by permission.)

interest. [A brief discussion of structures that have been proposed by others, and
the objections that Crick and Watson see in them, follows.] We wish to put forward
a radically different structure for the salt of deoxyribose nucleic acid. This structure
has two helical chains each coiled round the same axis (see diagram). [Their diagram
is reproduced in Figure 17.7 together with their legend.] The novel feature of the
structure is the manner in which the two chains are held together by the purine and
pyrimidine bases. The planes of the bases are perpendicular to the fibre axis. They
are joined together in pairs, a single base from one chain being hydrogen-bonded
to a single base from the other chain, so that the two lie side by side with identical
$z$ coordinates. One of the pair must be a purine and the other a pyrimidine for bonding
to occur . . . .

"If it is assumed that the bases only occur in the structure in the most
plausible tautomer forms (that is, with the keto rather than the enol configurations),
it is found that only specific pairs of bases can bond together. These pairs are:
adenine (purine) with thymine (pyrimidine), and guanine (purine) with cytosine
(pyrimidine).

"In other words, if an adenine forms one member of a pair, on either chain,
then on these assumptions the other member must be thymine; similarly for
guanine and cytosine. The sequence of bases on a single chain does not appear to
be restricted in any way. However, if only specific pairs of bases can be formed, it
follows that if the sequences of bases on one chain is given, then the sequence on
the other chain is automatically determined. [Crick and Watson next refer to ratios
of adenine to thymine and guanine to cytosine as given, for example, by Chargaff's
work and in Table 17.1.]

"It has been found experimentally that the ratio of the amounts of adenine to thymine, and the ratio of guanine to cytosine, are always very close to unity for deoxyribose nucleic acid. [Their postulated structure, of course, *requires* just such a ratio. They next comment on the provisional character of their postulated structure and the need for more exact X-ray studies. They are aware that Wilkins' paper, which was to give more accurate data, will follow their own in the journal, but they note that they were unaware of the details of Wilkins' results while they were devising their model. The next paragraph of their first paper opens the window.]

"It has not escaped our notice that the specific pairing we have postulated immediately suggests a possible copying mechanism for the genetic material."

In their second paper they describe the view from this window in richer detail. Before we can assimilate it ourselves, we should learn more about "base pairing" and how it is important to the structure of DNA.

The horizontal rods shown in Figure 17.7 symbolize base pairing in a very general way; details are given in Figure 17.8. Base pairing is the mutual attraction of a purine base for a pyrimidine base by means of hydrogen bonds. The geometries, the bond angles, and the availability of hydrogen-bond-donating and hydrogen-bond-accepting groups are apparently just right for this attraction when thymine and adenine are paired and cytosine and guanine are paired. (5-Methylcytosine can be substituted for cytosine.) Thus, of the four common bases in DNA, A pairs with T and C pairs with G. This peculiar specificity is described and explained by Crick and Watson in their second paper.

"The bases are joined together in pairs, a single base from one chain being hydrogen-bonded to a single base from the other. The important point is that only certain pairs of bases will fit into the structure. One member of a pair must be a purine and the other a pyrimidine in order to bridge between the two chains. If a pair consisted of two purines, for example, there would not be room for it.

"We believe that the bases will be present almost entirely in their most probable tautomeric forms. If this is true, the conditions for forming hydrogen bonds are more restrictive, and the only pairs of bases possible are

adenine with thymine;
quanine with cytosine.

The way in which these are joined together is shown in Figures 4 and 5. [Figure 17.8 of this text corresponds to these Figures 4 and 5.] A given pair can be either way round. Adenine, for example, can occur on either chain; but when it does, its partner on the other chain must always be thymine."

If two purines paired, "there would not be room." The X-ray data revealed these limitations on the space available between the two spiral backbones (the ribbons in Figures 17.7 and 17.9). If two pyrimidines (recall that these are narrower molecules with only one ring each) paired, the space could not be filled to the point that hydrogen bonds would be possible. A purine with a pyrimidine, however, works out just right. Figure 17.9 is a fuller but still schematic representation of the Watson-Crick DNA double helix. Figure 17.10 pictures a scale model of a short section of the double helix.

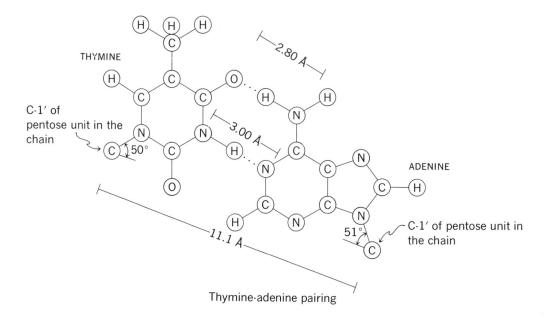

Thymine-adenine pairing

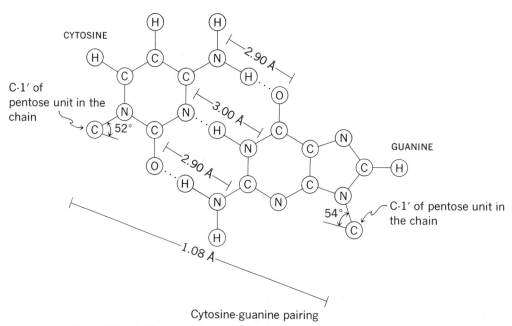

Cytosine-guanine pairing

**Figure 17.8** Pairing of purine-pyrimidine bases in DNA. The dimensions and geometries of these molecules are such that in their keto forms they can fit to each other via hydrogen bonds, two between thymine and adenine and three between cytosine and guanine. These pairs, which are almost entirely coplanar, form the "rungs" of the "spiral staircase" which the Watson-Crick model resembles. The perpendicular axis of Figure 17.7 would come up through the regions of the hydrogen bonds. The dimensions shown were determined by Pauling and Corey.

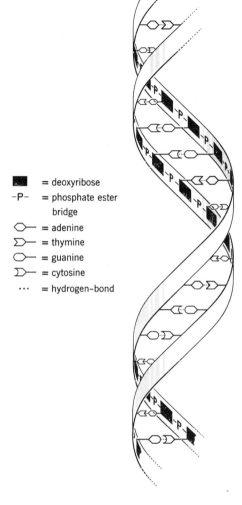

= deoxyribose
-P- = phosphate ester
      bridge
= adenine
= thymine
= guanine
= cytosine
··· = hydrogen–bond

**Figure 17.9**   A DNA double helix showing base pairing.

## REPLICATION OF DNA

Speculation about how chemicals might become duplicated in preparation for cell division has generally included the idea of a chemical mold or a template. In industry a template is a pattern or guide used in laying out and scribing a workpiece. In shipbuilding it is a full-sized wooden mold or paper pattern used to make hull parts. A template can be a gauge or a pattern for checking contours and dimensions. The idea of a chemical serving as a template or a mold for making another chemical has long been considered in connection with genetics, but until the Watson-Crick model of DNA we had no detailed picture of *how* this could be done. Again, let us use the words of Watson and Crick from their second paper.

**Figure 17.10** Scale model of a DNA double helix. (Courtesy of Nova Research Group, Columbus, Ohio.)

"The phosphate-sugar backbone of our model is completely regular, but any sequence of the pairs of bases can fit into the structure. It follows that in a long molecule many different permutations are possible, and it therefore seems likely that the precise sequence of the bases is the code which carries the genetical information. If the actual order of the bases on one of the pair of chains were given, one could write down the exact order of the bases on the other one, because of the specific pairing. Thus one chain is, as it were, the complement of the other, and it is this feature which suggests how the deoxyribonucleic acid molecule might duplicate itself.

"Previous discussions of self-duplication have usually involved the concept of a template, or mould. Either the template was supposed to copy itself directly or it was to produce a 'negative,' which in its turn was to act as a template and produce the original 'positive' once again. In no case has it been explained in detail how it would do this in terms of atoms and molecules.

"Now our model for deoxyribonucleic acid is, in effect, a *pair* of templates, each of which is complementary to the other. We imagine that prior to duplication the hydrogen bonds are broken, and the two chains unwind and separate. Each chain then acts as a template for the formation on to itself of a new companion chain, so that eventually we shall have *two* pairs of chains, where we only had one before. Moreover, the sequence of the pairs of bases will have been duplicated exactly.

[The authors next discuss some of the uncertainties of their model as well as some of the evidence for it. The following quotation is from their closing paragraph.]

"Despite these uncertainties we feel that our proposed structure for deoxyribonucleic acid may help to solve one of the fundamental biological problems—the molecular basis of the template needed for genetic replication. The hypothesis we are suggesting is that the template is the pattern of bases formed by one chain of the deoxyribonucleic acid and that the gene contains a complementary pair of such templates."

It is as though genetic messages were written with a four-letter alphabet. Statistical theory enables us to calculate the number of isomers possible when four letters (four bases) are used many times. If there are, as indicated by genetic studies, about 1500 nucleotide pairs per gene, the possible number of isomers of such a gene would be $4^{1500}$. J. D. Watson estimates that this number greatly exceeds the number of different genes that have ever existed in all the chromosomes since the origin of life. A four-letter alphabet used enough times, is certainly sufficient for encoding any realistic amount of genetic information.

As for the amount of DNA in the human race and its responsibility for all the individuals in all their marvelous varieties, Theodosius Dobzhansky has made some interesting calculations. According to figures by Müller, the volume of DNA in a human gamete is about 4 cubic microns and the weight is a paltry $4 \times 10^{-12}$ g. The number of persons living is, very roughly, $3 \times 10^9$. They have come from twice that number of gametes, $6 \times 10^9$. The total volume of the physical carriers of genetic messages that the human species now alive has received from its ancestors is therefore about 2.4 mm$^3$, the total weight about 0.24 mg. This quantity, notes Dobzhansky, is about that of a raindrop.

Before cell division DNA molecules must be reproduced in duplicate; they must be *replicated*. According to the Watson-Crick theory, such replication consists of a temporary uncoiling, or partial uncoiling, of the two helices of a DNA double helix. This uncoiling occurs in a medium that contains building blocks for fresh DNA strands. From this mononucleotide pool each uncoiled DNA strand draws to itself the nucleotides that will fit with it via hydrogen bonds. Thus when an original A-T or G-C pair becomes unpaired, the old A becomes paired with a new T, the old T becomes paired with a new A, etc. Two new A-T pairs form, but on different double helices. Figure 17.11 describes this pairing in greater detail. Figure 17.12 shows in a schematic way how the growth of two new chains may occur as the parent double helix unwinds.

**The Meselson-Stahl Experiment.** The mechanism for DNA replication just described is called *semiconservative*. As indicated in Figure 17.13, there are three principal possibilities, conservative, semiconservative, and dispersive. In the conservative mechanism the strands of the parent double helix come back together again after serving as templates for two new strands. The dispersive mechanism implies that the original strands break up and then somehow recombine with new DNA.

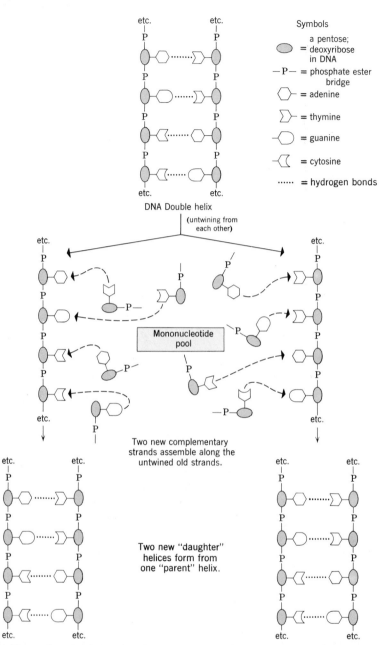

Figure 17.11 The replication of DNA. The complete untwining of the original double helix (top) implied here and suggested originally by Crick and Watson probably does not happen. Instead, as the long DNA double helix starts to unwind, replication may begin at the freshly exposed ends of the single strands as they appear (cf. Figure 17.12). New DNA is actually built up one nucleotide at a time, not (as implied here) all at once after the nucleotides have lined up. It is impractical to convey all relevant facts, and this figure stresses the *pairing* of bases as the most important feature in the replication of the genetic code. The actual mechanism of replication in the growing regions of DNA is a subject undergoing intensive research.

**Figure 17.12** According to one theory, replication of DNA occurs *while* the parent double helix unwinds rather than after it has become entirely unwound.

Meselson and Stahl in their experiment provided the most convincing evidence for the semiconservative model. They used the mass-15 isotope of nitrogen, combined in ammonia; we symbolize it as $^{15}NH_3$, the ordinary ammonia being $^{14}NH_3$. Working with the common colon bacillus, *Escherichia coli,* they grew some of this bacteria in a medium containing $^{15}NH_3$ as the only source of nitrogen. As a result, all the nitrogens of the bases in the DNA of freshly grown bacteria were of the heavier $^{15}N$. The density of this DNA was greater than that of the ordinary DNA of this species made with $^{14}N$. The analytical data were obtained in this experiment by measuring densities. Some of these bacteria grown in the $^{15}N$ medium were then transferred to a new medium where $^{14}NH_3$ was the only source of nitrogen. By the time a new generation had developed, the density of the new DNA indicated that each DNA molecule must consist of one strand of "heavy" ($^{15}N$) DNA and one strand of "light" ($^{14}N$) DNA; we may call this "isotopic hybrid DNA." In the second generation half the DNA molecules were of the isotopic hybrid type and half were light. Figure 17.13 illustrates this experiment, reported by M. Meselson and F. W. Stahl in 1958. Referring to Figure 17.13, we see that the conservative mechanism would not yield any isotopic hybrid DNA. The dispersive model would not, at least after just two generations, yield any wholly light DNA. The semiconservative model is consistent with the results. Other details of the mechanism of the biosynthesis of DNA have

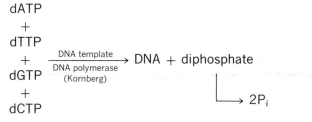

The four deoxyribonucleoside triphosphates

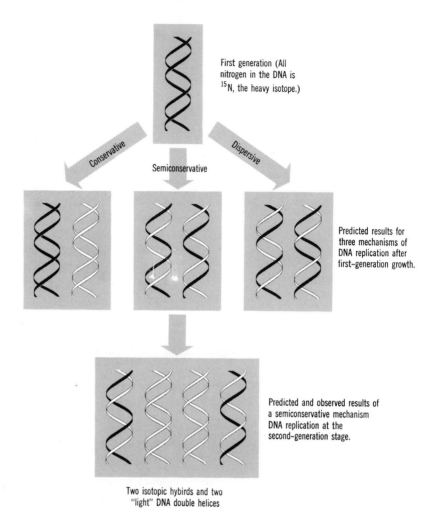

First generation (All
nitrogen in the DNA is
<sup>15</sup>N, the heavy isotope.)

Predicted results for
three mechanisms of
DNA replication after
first-generation growth.

Predicted and observed results of
a semiconservative mechanism
DNA replication at the
second-generation stage.

Two isotopic hybirds and two
"light" DNA double helices

**Figure 17.13**   The Meselson-Stahl experiment. The strands drawn with the heavy line represent
DNA in which all the nitrogen is in the form of the heavy isotope, <sup>15</sup>N. The lightly drawn strands
stand for DNA made of <sup>14</sup>N. By measuring densities of the DNA formed after the first and the
second generations, Meselson and Stahl (1958) found convincing evidence for the semiconservative mechanism.

been reported by Arthur Kornberg (Nobel prize, 1959). At Washington University
(St. Louis) a team of scientists led by him isolated an enzyme, *DNA polymerase,*
that would catalyze the formation of DNA in a test tube containing a small amount
of "primer" DNA and the triphosphate forms of the four nucleotides (i.e., dATP,
dTTP, dGTP and dCTP). (See bottom of page 555.)

The mechanism of DNA replication just described leaves many questions
unanswered. In higher plants and animals, for example, chromosomes have as much
as 50% protein. Just what does this protein have to do, if anything, with the chemistry

of heredity? How are the DNA molecules incorporated together with the protein to form a chromosome? We shall not suggest any theories to answer these questions, but it is possible to say a great deal about the most important function of DNA (aside from replication), its control over the synthesis of polypeptides and thence that of enzymes.

## DNA AND THE SYNTHESIS OF ENZYMES

The general scheme connecting DNA to polypeptides (enzymes, usually) is outlined in Figure 17.14; our discussion will give more details. The immediate task of chromosomal DNA is to participate in and determine the nucleotide sequence of RNA. The RNA molecules have all the features of DNA strands with two differences.

1. The —OH group at C-2′, absent in DNA, is of course present in RNA which is made from ribose rather than deoxyribose. This difference appears to be minor,

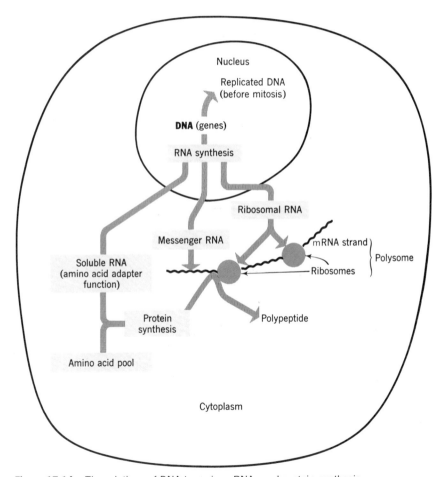

**Figure 17.14**   The relations of DNA to various RNAs and protein synthesis.

but it may be a factor in determining the final configuration of the RNA strand. Most RNA exists as single strands rather than as double helices, although such formations are known for RNA.

2. There is no thymine in RNA; its place is taken by uracil (Figure 17.3), which can pair with adenine just as well as thymine. Thus we can have A-T pairing and A-U pairing.

The three types of RNA known must be distinguished. All three types are synthesized under the control of DNA. (A fourth type of RNA associated with virus particles is also known.) For the discovery of an enzyme, *RNA polymerase,* which made possible *in vitro* synthesis of RNA, Severo Ochoa shared the 1959 Nobel prize in physiology and medicine with Arthur Kornberg.

**Ribosomal RNA (rRNA).** The main structural framework of a cell is the *endoplasmic reticulum,* a series of canaliculi and cisternae that interconnect and permeate most of the cytoplasm (see Figure 17.1). Studding portions of the endoplasmic reticulum are granules varying from 70 to 200 Å in diameter. Sometimes found free in the cytoplasm and made of RNA and protein, these granules may bind up about 75 to 80% of all the RNA of a cell. Ribosomes are the sites of protein synthesis, and yet ribosomal RNA itself does not appear to direct this work. Messenger RNA, the second type, directs protein synthesis. When we examine the details of polypeptide synthesis later in this chapter, we shall learn that ribosomal RNA may provide some stabilizing forces to a temporary complex between messenger RNA and the third type we study, soluble RNA. Ribosomes from a common source are known to be alike. They must therefore serve some nonspecific function that is common to the synthesis of all the protein (enzymes) made by that source.

**Messenger RNA (mRNA).** Messenger RNA molecules vary considerably in length and account for about 5 to 10% of the total RNA in a cell. As we shall see, mRNA molecules are the bearers of the genetic code after they have been synthesized under the direct supervision of DNA (Figure 17.15). Once made, mRNAs move out of the nucleus into the cytoplasm where they attach ribosomes to themselves at intervals along their chains. Such an assembly of many ribosomes along an mRNA chain is called a *polysome* (polyribosome; see Figure 17.16). Evidence indicates that mRNA can dissociate from ribosomes and that ribosomes can associate with different mRNA molecules of different molecular weight. When protein synthesis takes place, however, it is much more difficult for mRNA to dissociate from ribosomes. Additional evidence indicates that a ribosome moves along an mRNA chain while the synthesis of a polypeptide, directed by the mRNA, takes place. For such synthesis amino acids must be brought to the site in the order in which they are to appear in the final polypeptide. The cell uses the third type of RNA for this task.

**Soluble RNA (sRNA).** Because soluble RNA has a much lower formula weight than rRNA or mRNA, it is more soluble in media used to isolate the higher-formula-weight forms. Thus the designation soluble refers to a physical property and says nothing about function. When its function is to be stressed, various authors refer to sRNA as *transfer RNA* (tRNA) or as *amino acid adapter RNA* (aaRNA).

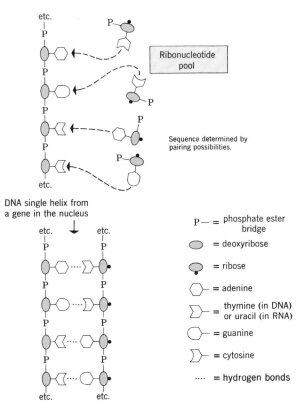

Sequence determined by
pairing possibilities.

DNA single helix from
a gene in the nucleus

P— = phosphate ester
bridge

= deoxyribose

= ribose

= adenine

= thymine (in DNA)
or uracil (in RNA)

= guanine

= cytosine

.... = hydrogen bonds

An RNA strand, complementary in terms of the
sequence of heterocyclic amines, assembles along the
DNA strand in the nucleus. A hybrid complex of this
type was observed (1960) by Alexander Rich of M.I.T.

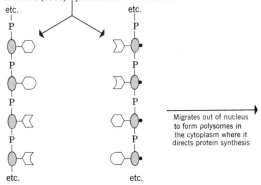

Migrates out of nucleus
to form polysomes in
the cytoplasm where it
directs protein synthesis

DNA strand in readiness
for synthesis of more RNA

Newly formed
mRNA strand

**Figure 17.15**  Transmission of a genetic code from DNA to mRNA in cell nuclei. The enzyme
*RNA polymerase,* itself a polypeptide that is dependent on a specific DNA molecule, controls
this synthesis along with the DNA shown here (ATP is also required). The ribonucleotide "pool"
consists of the triphosphate forms of the nucleotides, making ATP a member of the pool.

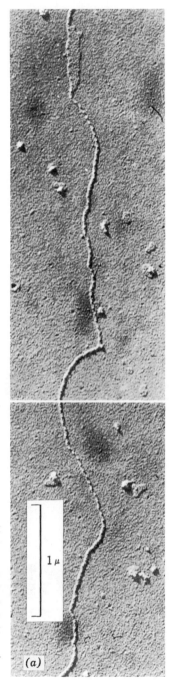

**Figure 17.16**  Polysomes. (*a*) Electron micrograph of a long polysome (polyribosome) isolated from rat skeletal muscle tissue by Lederle scientists. The threadlike material is mRNA, and what appear to be rather evenly spaced bulges are ribosome units. (Courtesy of Lederle Laboratories, a division of American Cyanamid Company.) (*b*) Electron micrograph of polysomes from the rabbit reticulocyte stained with uranyl acetate. Again, the dark thread running between ribosomes (large dark shapes) is mRNA. Its diameter is very close to 15 Å. (Courtesy of Professor Alexander Rich, Massachusetts Institute of Technology.)

We shall use sRNA for most of our purposes. The different species of sRNA molecules are believed to number at least twenty, one for each of the some twenty amino acids. The function of sRNA is to attach to itself the particular amino acid for which it is coded, carry it to a protein synthesis site on a polysome (specifically an mRNA site in contact with a ribosome), and there give it up to the growing end of a polypeptide chain at the particular moment called for by the genetic code. As illustrated in Figure 17.17, which is meant to convey only the most general aspects of the synthesis, the mRNA acts as a template. Its sequence of nucleotides was determined by a sequence on a DNA molecule, or a portion thereof (Figure 17.15). Although mRNAs cannot recognize individual amino acids, they do recognize individual sRNAs, for these assemble along an mRNA strand only in a sequence dictated by the requirements of base pairing. These requirements inevitably determine the order in which amino acids become lined up in a growing polypeptide chain.

Thus a sequence of nucleotides on a gene (DNA) molecule and a sequence of amino acids in a protein have a linear correlation. Since a protein synthesized under this genetic control is the apoenzyme portion of an enzyme, this theory offers a molecular explanation of the one-gene, one-enzyme idea first advanced with

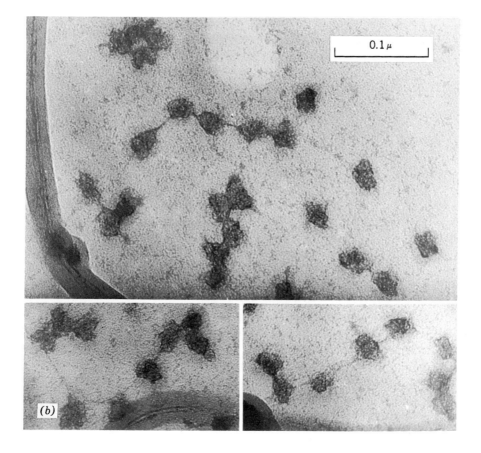

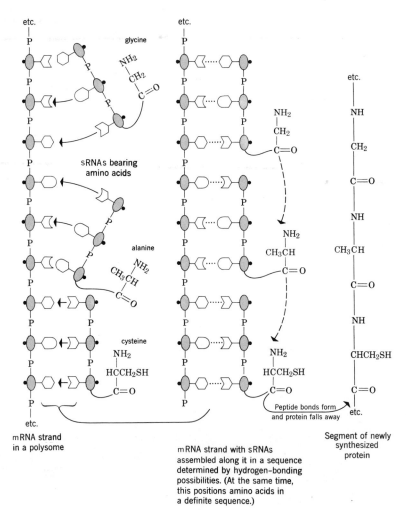

**Figure 17.17**   A generalized view of how sRNAs participate together with gene-coded mRNA in determining the order of amino acid units in a developing polypeptide. The figure emphasizes the importance of the pairing of complementary heterocyclic amines in this phase of gene-directed protein synthesis.

evidence in 1941 by George Beadle and Edward Tatum, who received the 1958 Nobel prize in physiology and medicine for this work. (Joshua Lederberg also shared the prize for work done on genetic recombination.) Since the work of Beadle and Tatum many enzymes have been found to consist of more than one polypeptide or protein strand. Because more than one gene is required to direct the synthesis of such enzymes, we now have a one-gene, one-polypeptide theory instead of a one-gene, one-enzyme theory, with the understanding that the polypeptide is usually an enzyme or is destined to become part of an enzyme.

 We now have a general idea of how the sequence of nucleotides on a gene molecule can be transcribed into a sequence on an mRNA molecule and then translated into a specific amino acid sequence. The importance of polymerizing enzymes, *DNA polymerase* and *RNA polymerase*, was indicated, and the need for triphosphates as sources of energy was noted. Many additional details are known, and those relating to how the various forms of RNA cooperate are described in the next sections. Although a few of the important experiments are mentioned, the reader should consult the reference list for further study. The names of several scientists have been and will be mentioned. Unfortunately, it is not practical to credit all the great scientists in the burgeoning field. In the opinion of some it is also unfortunate that more Nobel prizes cannot be given.

## TRANSLATING THE GENETIC CODE

**Codons.** If we start with one set of four different symbols and another set of twenty different symbols, it is obviously impossible to find a one-to-one correspondence of one set with the other. Yet this is the task we must perform if we insist on using a four-letter alphabet (the four nucleotides in DNA) to specify unique sequences based on a twenty-letter alphabet (amino acids for proteins). We could say, of course, that the *letters* of the four-letter alphabet are not themselves the code. Rather these letters form two-letter words and the *words* make up the code. Thus from A, T, C, and G we have the following "words":

| | | | |
|---|---|---|---|
| AA | AT | TC | CG |
| TT | TA | CT | GC |
| CC | AC | TG | |
| GG | CA | GT | |
| | AG | | |
| | GA | | |

This list totals sixteen, still not enough, still short of twenty. By using three-letter units, however, we can form sixty-four words, more than enough. Statistical considerations strongly suggest that the genetic code is made up of "words" rather than individual letters. The genetic "alphabet" is too short. Genetic evidence not only confirms this supposition but indicates that *groups of three nucleotides are the fundamental genetic units.* When expressed as sequences on mRNA, they are called *codons.* For example, a sequence of side chain bases on some section of a DNA molecule might be cytosine-thymine-thymine, or CTT. Transcription of this into mRNA will give a sequence of guanine-adenine-adenine, or GAA. The GAA sequence of the mRNA is the codon.

| | |
|---|---|
| Section of DNA strand | Sequence of bases in mRNA determined by complementary sequence on DNA |

**Sequence Recognition by Amino Acids.** Few of the amino acids have side chains to make direct pairing between them and bases on the side chains of mRNA even conceivable. There is no basis for a direct recognition by amino acids of the sequence they are to take up. The sRNA molecules mediate this difficulty by providing surfaces that can recognize pairing opportunities on an mRNA chain. To put an amino acid and an sRNA molecule together, the following events must occur. Wherever an enzyme is specified, its synthesis too is controlled by a gene (DNA).

1. The amino acid must be activated for eventual attachment to an sRNA molecule (this requires an activating enzyme and ATP for energy):

Activating enzyme + amino acid + ATP $\longrightarrow$

activating enzyme—amino acyl$\sim$AMP + diphosphate ion

An amino acyl group will have the form: $R{-}CH{-}\overset{\overset{\displaystyle O}{\|}}{C}{-}$ ; AMP is of course adenosine monophosphate.

$\underset{\displaystyle NH_2}{\big|}$

2. The amino acyl group is transferred to a specific sRNA molecule (the same enzyme serves):

Activating enzyme—amino acyl$\sim$AMP + sRNA $\longrightarrow$

amino acyl$\sim$sRNA + AMP + activating enzyme

$\qquad\qquad\qquad\downarrow\qquad\qquad\qquad\downarrow\qquad\qquad\qquad\downarrow$

| Will migrate to mRNA to incorporate the amino acid into a polypeptide | To be "recharged" to ATP via the respiratory chain | To be reused |

To get the right amino acid attached to the right sRNA molecule, the activating enzyme is crucial. It must be able to bind itself not only to just one specific amino acid but also to the sRNA adapter molecule. Thus some twenty specific enzymes (polypeptides) are needed, but it is not difficult to imagine this number each with secondary and tertiary structural features that can recognize on the one hand an amino acid and on the other hand, at some other part of the enzyme, an sRNA molecule. The lock and key theory of enzyme action serves well here (cf. p. 378).

In the sRNA portion of the amino acyl$\sim$sRNA complex we have the recognition site for a codon on an mRNA molecule. This recognition site must also be a sequence of three bases, and if the codon is GAA (example on p. 563), the recognition site must have the complementary sequence CUU, (not CTT; U replaces T in RNA). We call any sequence of three that is complementary to a codon an *anticodon*. (Some authors use the term *nodoc*.) The sRNA molecules, as noted earlier, are much longer than three nucleotides. The average molecular weight of sRNA is about 25,000, which corresponds to about eighty nucleotides linked together. The first structure determination of a specific RNA was for the sRNA molecule that attaches to itself the amino acid alanine. A team of scientists, associated with Cornell University and the U.S. Department of Agriculture and led by R. W. Holley,[3] reported in 1965 that *alanine transfer RNA* from yeast contains seventy-seven nucleotides, several of which are unusual (meaning that they are other than A, G, U, or C). In 1966 the structure of *tyrosine transfer RNA* was reported by J. Madison and G. Everett to contain seventy-eight nucleotides. Other sRNAs have been examined in enough detail to show that probably all of them have the CCA sequence of bases at one end of the chain. The other end (the C-5′ end) terminates in a G unit. These molecules are single-stranded, but by a hairpin fold most of the bases become paired; a double helix, as in DNA, makes up part of the configuration. Somewhere along this chain the anticodon sequence of three bases occurs, making it possible for the sRNA molecule to recognize a codon on an mRNA chain. Figure 17.18 shows how an sRNA molecule might look, both without and with an attached amino acyl group. The step-by-step growth of a polypeptide chain is pictured and discussed in Figures 17.19 and 17.20. According to a theory proposed by Alexander Rich (Massachusetts Institute of Technology), individual ribosomes move from one end of a mRNA chain to the other, making protein as they go. When the other end of the mRNA chain is reached, the ribosome drops off and the newly made polypeptide strand is released. As discussed earlier (p. 337), once the sequence of amino acids is put together, the protein automatically assumes its secondary or higher structural features. As one ribosome leaves the mRNA chain, a new empty one starts out at the other end according to Rich's theory. It must be borne in mind that this theory has by no means been proved. In fact, the way we have depicted the arrangement between mRNA and ribosomes in Figures 17.19 and 17.20 is still partly conjecture. The arrangement appears to be the easiest way to account for the facts, and it does help to rationalize the high rate of protein synthesis.

**Rate of Formation of Proteins.** The production rate of fresh hemoglobin can be estimated; on the average, one of its polypeptide chains must be made per ribosome every 90 seconds. Since there are about 150 amino acid units per polypeptide, the chain must grow at the rate of approximately two amino acid units per second.

**The Nirenberg-Matthaei Experiments and the Specific Code Words.** Marshall Nirenberg[3] and J. H. Matthaei of the National Institutes of Health reported on a series of experiments that gave support to the mechanisms we have just described and opened the way to assign specific three-letter code words, triplets, to specific amino acids. They worked with the colon bacillus *Escherichia coli* (not because these bacteria are more interesting than people, but because they are handier to use and because so many of their metabolic pathways are the same as those in people). By grinding and centrifuging these bacteria, Nirenberg and Matthaei were able to prepare a cell-free extract containing ribosomes, enzymes, and DNA.

---

[3] R. W. Holley, H. G. Khorana, and M. W. Nirenberg, who worked independently, shared the 1968 Nobel prize in physiology and medicine for their discoveries concerning the genetic code and its function in protein synthesis.

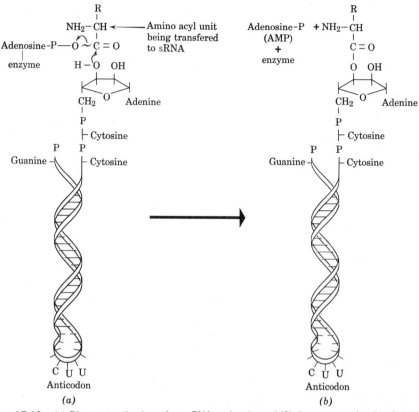

**Figure 17.18** (*a*) Diagrammatic view of an sRNA molecule and (*b*) the same molecule after it has been attached to an amino acyl group. The sRNA molecule is believed to be single-stranded but folded so that much of it has a double-helix configuration. Placing the triplet anticodon at the hairpin turn is for convenience only, as was the selection of the specific sequence cytosine-uracil-uracil (CUU) to illustrate an anticodon.

To this they added a mixture of the twenty amino acids that make up most proteins; they also supplied triphosphates for energy. This extract was able to synthesize protein. They then added an enzyme that catalyzes the destruction (by hydrolysis) of DNA. Yet for twenty more minutes the extract could make proteins. The implication was that during this time the synthesis of proteins already started was allowed to finish, but that to make new protein new mRNA would be needed. (It is known that mRNA is not particularly stable and must be replenished rather frequently.) But the DNA for directing the synthesis of the mRNA had been destroyed. Hence no new mRNA and no new protein. When they added RNA extracted from fresh batches of *E. coli,* protein synthesis recommenced. If at this point they added RNA not from the fresh batches of *E. coli* but rather from some entirely different organism such as yeast cells or tobacco mosaic virus, the extract still began to synthesize protein again. Apparently the ribosomes from *E. coli* (and presumably from any organism) are not themselves specific. It is not to the ribosome that we must

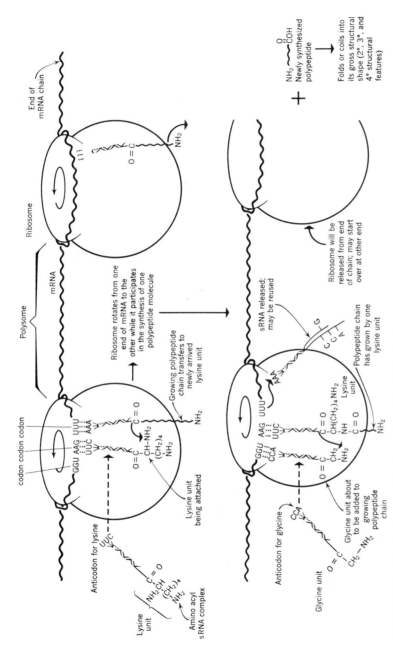

**Figure 17.19** Step-by-step growth of a polypeptide chain. Amino acyl sRNA complexes carry amino acyl groups to a polysome. Codon-anticodon pairing occurs between the sRNA and the mRNA at a ribosomal surface. The already partially grown polypeptide chain transfers to the amino group of the newly arrived amino acyl unit, and thus the chain grows. The ribosome moves along the mRNA strand as all this happens. At the end the ribosome drops off. This figure combines the ideas and results of many, especially those of Geoffrey Zubay (Brookhaven) and Alexander Rich (Massachusetts Institute of Technology).

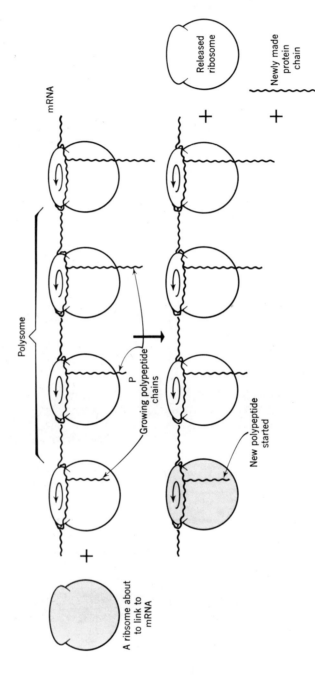

**Figure 17.20** Possible model for the dynamic function of a polysome, according to Alexander Rich. (From A. Rich, "Polyribosomes," *Scientific American*, December 1963, page 44.)

## Table 17.2 RNA Codon Assignments*

| Amino Acids | Codon Assignments | | | | | |
|---|---|---|---|---|---|---|
| Alanine | GCC | GCU | GCA | GGG | | |
| Arginine | CGC | AGA | CGU | CGG | AGG | |
| Asparagine | AAC | AAU | | | | |
| Aspartic acid | GAU | GAC | | | | |
| Cysteine | UGU | UGA | AGU | UGC | UGG | AGC |
| Glutamic acid | GAA | GAG | | | | |
| Glutamine | CAA | CAG | | | | |
| Glycine | GGU | GGA | GGC | GGG | | |
| Histidine | CAC | CAU | | | | |
| Isoleucine | AUU | AUC | | | | |
| Leucine | UUG | CUU | CUC | UUA | CUG | |
| Lysine | AAA | AAG | | | | |
| Methionine | AUG | AUA | | | | |
| Phenylalanine | UUU | UUC | | | | |
| Proline | CCC | CCU | CCA | CCG | | |
| Serine | UCU | UCC | UCG | AGU | UCA | AGC |
| Threonine | ACU | ACA | ACC | ACG | | |
| Tryptophan | UGG | UGA | | | | |
| Tyrosine | UAU | UAC | | | | |
| Valine | GUU | GUC | GUA | GUG | | |

* These assignments are for the colon bacillus *Escherichia coli*, but a large number of them are known to be codon assignments for several other species. Data from W. A. Groves and E. S. Kempner, *Science*, Vol. 156, April 21, 1967, page 389.

look for the genetic uniqueness of a species. The RNA (presumably including mRNA and sRNA) from other organisms is acceptable to the ribosomes of *E. coli* for the work of proton synthesis.

Finally, Nirenberg and Matthaei took RNA not from fresh *E. coli* or from yeasts or from a virus, but rather from a synthetic batch. They used a very simple polynucleotide, polyuridylic acid. In this RNA all the bases are uracils; it can be called poly-U. (Recall that uracil is found just in RNA, not in DNA. What they added, then, could serve only as RNA material.) When this poly-U was introduced as fresh RNA to the *E. coli* extract that had been treated to destroy its DNA but not its ribosomes or its enzymes, the extract proceeded to make polyphenylalanine. It made a protein in which the only amino acids were phenylalanine units. Evidently, the code for phenylalanine was at least one uracil unit, or a sequence of such units.

Ochoa's group at the New York University School of Medicine, H. G. Khorana's[4] group at the University of Wisconsin, and Crick's group at the Cavendish Laboratory, University of Cambridge, added their efforts to those of the Nirenberg-Matthaei (and P. Leder) group. The triplets of RNA bases that code for specific amino acids were quickly disclosed; Table 17.2 gives a summary. For most, more than one code word will do. Thus alanine can be recognized and transported by sRNA carrying any one of four coding triplets, GCC, GCU, GCA, or GGG. The code is therefore not a unique one; it is said to be *degenerate*. The significance of code degeneracy is not fully understood.

[4] Nobel prize, 1968. See footnote 3, page 565.

**Is the Genetic Code Universal? The Lipmann–von Ehrenstein Experiment.** The codons in Table 17.2 were determined for *E. coli*. Many of them have been found to apply to other organisms, and evidence for a universal genetic code, although by no means conclusive, is mounting. If the code is universal, whenever the cells of whatever species engage in protein synthesis and the next amino acid to be required is, say, alanine, one of the four code "words" of Table 17.2, GCC, GCU, GCA or GGG, will be the codon. It will be the codon for alanine whatever the species. F. Lipmann and G. von Ehrenstein (1961) reported one piece of dramatic evidence supporting this. From *E. coli* they prepared a mixture of amino acyl∼sRNAs charged with all the amino acids needed to make the globin of hemoglobin. One of the amino acyl groups, that of leucine, was made radioactive with carbon-14. In a cell-free system Lipmann and von Ehrenstein added to this mixture from *E. coli* purified ribosomes and polysomes obtained from the immature red cells of a rabbit. These ribosomes make globin for hemoglobin. Guanosine triphosphate (high-energy phosphate) and appropriate enzymes for globin synthesis were also added. These enzymes do not make globin without a supply of amino acyl∼sRNA. The question was whether the amino acyl∼sRNA from the bacterium *E. coli* could serve in the synthesis of the globin for the rabbit? Globin was indeed made, and about half the carbon-14 was found in it. The experiment not only hinted at the possibility of a universal genetic code but also gave evidence that the polysome component, presumably the mRNA thereof, has the information needed to control protein synthesis.

## HEREDITARY DEFECTS AND GENE-TO-ENZYME LINK

Several gene-to-enzyme-linked abnormalities in animal organisms illustrate how important are the accurate transcription and translation of the genetic code.

**Phenylketonuria, PKU.** Several of the pathways of phenylalanine and tyrosine metabolism are shown in Figure 17.21. Some babies in whom mental retardation develops are apparently born with a defective gene for making the enzyme needed to convert phenylalanine to tyrosine at ($A$) in Figure 17.21. Phenylalanine is therefore processed more frequently than normal by a pathway leading to phenylpyruvic acid. As this acid is formed in greater than normal concentrations, the kidneys remove some, and its appearance in urine is called *phenylketonuria* or PKU. Many people believe that the excess phenylpyruvic acid still circulating in the bloodstream causes the mental retardation often observed when PKU is detected. Unfortunately, the presence of phenylpyruvic acid in urine cannot be readily and reliably detected until an infant is six to eight weeks old. Efforts to reduce the dangers of PKU must probably be started earlier than this. Hence a simple blood test on a sample taken four or five days after birth was devised to detect excessive phenylalanine in circulation. About thirty-seven states have laws requiring that this test be made (as of 1968).

The treatment of PKU that initially showed the most promise in preventing the onset of mental retardation was the exclusion of as much phenylalanine from the diet as possible. Finding proteins low in phenylalanine is very difficult, and the diet is not only austere but also potentially dangerous—no milk, eggs, bread,

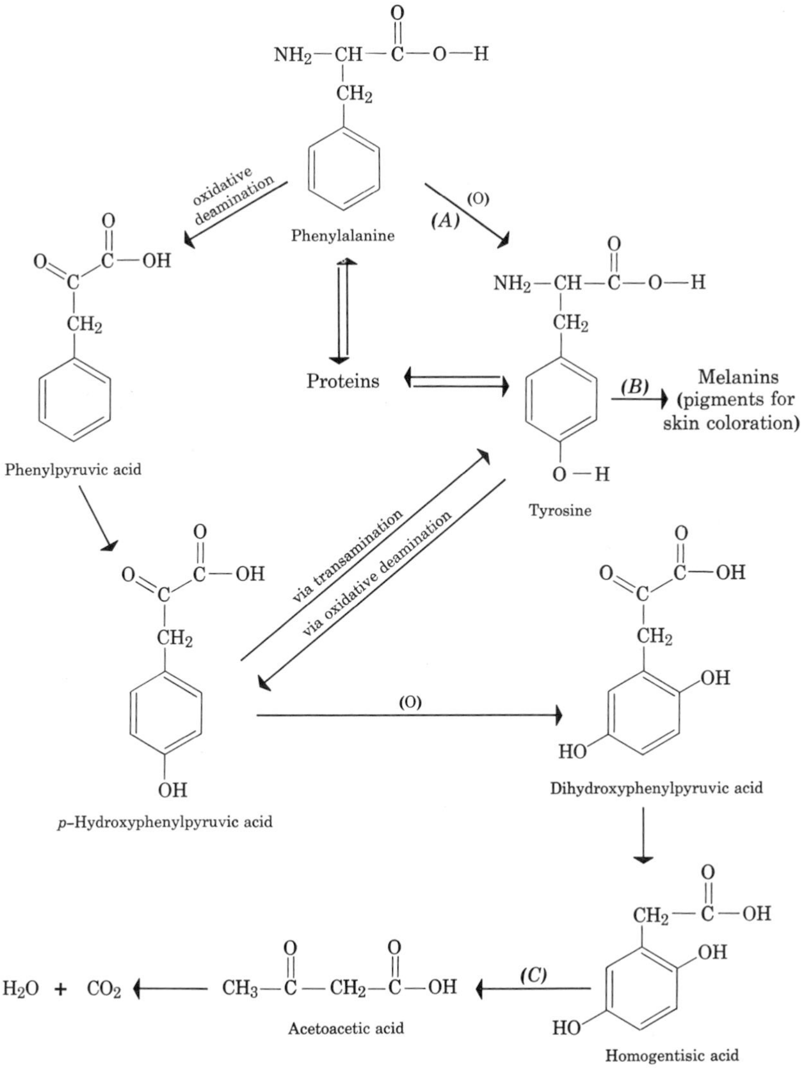

**Figure 17.21**  Pathways of phenylalanine and tyrosine metabolism. Letters in parentheses refer to textual discussion.

or meat. One slice of bread has all the phenylalanine a PKU baby is normally able to handle in a day. Special supplements are used, augmented by certain fruits, vegetables, and cereals.

Uncertainties in mass screenings for PKU among newborn infants, stemming in part from the unreliability of the blood test, together with the dangers of the PKU diet, have led the American Academy of Pediatrics to issue a formal statement of opposition to extension of PKU legislation.

**Albinism.** The pigments of hair, skin, and irises of eyes are polymeric sub-stances called *melanins* which form from tyrosine. At point (*B*) in Figure 17.21, a gene is needed to make the enzyme for this reaction. If the gene is not properly constructed, the enzyme fails to be made normally, and the melanin pigments do not form as they should. People suffering from this seemingly minor chemical defect are albinos.

**Sickle-Cell Anemia.** Largely through the work of V. M. Ingram (Massachusetts Institute of Technology), Pauling and Itano (California Institute of Technology), and many others, we have a better understanding of how faulty DNA causes production of the abnormal hemoglobin of sickle-cell anemia. Hemophilia, agammaglobulinemia, and Wilson's disease are other hereditary diseases believed to be caused by recessive genes that fail to produce effective enzymes.

## DNA AND MUTATIONS

We have learned that the genetic machinery of an organism must carry a code both faithful to the characteristics of the species and capable of accurate transmission from one generation to the next. At the same time it must also be capable of evolutionary change, which involves the appearance of mutants. Most mutants are too weak to survive; a few are improvements of the species. Whatever kind of mutation occurs must be transmitted to the next generation if it is to survive. Mutations, both those that survive and those that do not, are almost certainly related to changes in the structures of genetic molecules.

From what we have learned of the Watson-Crick theory, a change in the nucleotide sequence of a DNA molecule must necessarily affect the amino acid sequence of an enzyme. Of the important physical agencies for altering genetic material, atomic radiations and X-rays produce the most serious changes.

## CHEMICAL BASIS FOR RADIATION DAMAGE

Early workers with X-rays did not realize their danger, and an estimated 200 physicists, doctors, and radiologists have died over the years from excessive exposure to X-rays. In the 1920s girls painting radium watch dials in a New Jersey watch factory pointed their brushes by moistening them with their lips. Between 1921 and 1929 fifteen of the girls died after showing signs of severe jaw infection. Ionizing radiations of any type are dangerous and at the same time useful. This paradox has a chemical explanation, but before investigating it, we need to know more about the general symptoms of radiation sickness.

**Radiation Sickness. Gross Symptoms.** A drop in the level of the white cells in the blood; severe damage to the skin together with loss of hair and the appearance of ugly, nonhealing, ulcerating wounds; nausea, vomiting, diarrhea, and a feeling of weakness; internal bleeding—these are the principal symptoms of radiation illness. Even patients given deep X-ray or cobalt-ray treatment experience gastrointestinal disturbances and a falling white cell count.

These symptoms are so different in nature that it hardly seems possible one cause can have such diverse effects. Within the last three decades, however, scientists have done much to provide a chemical basis for understanding radiation sickness.

**Radiation Sickness. A Chemical Explanation.** When radiations penetrate living cells, they leave in their tracks a mixture of strange, unstable organic ions. As these seek stability, covalent bonds within them may break permanently. The ions may recombine either with themselves or with neighbors. New molecules, foreign to the cell, form.

Suppose that all these disruptions happen in a nucleic acid within a cell nucleus. These giant molecules simply cannot suffer structural changes and remain true to the genetic messages they bear. If a DNA molecule, a gene, is structurally altered, mutant daughter cells at the very least will form when the cell divides. At the worst, and this is common, the cell will not be able to divide at all and will be reproductively dead.

We could suppose that large enzymes also suffer similar structural damage. Although they sometimes do, the intensity of the radiation exposure needed to damage seriously an enzyme system is much greater than that needed to damage genes and chromosomes.

The best available evidence points to the conclusion that *the primary site of radiation damage is the genetic machinery in a cell's nucleus.* This damage occurs impartially in all cells exposed, but it is felt keenly only when they attempt to divide. If the exposure to radiation is intense, the changes in gene structure are likely to be so great that the cell is unable to divide when it comes time to do so. If the cells so damaged, as a result of deliberate exposure, are cancer cells, the cancer is halted. High doses of radiation aimed at cancerous cells during a reasonably short period stop the growth of cancer. On the other hand, if exposure is relatively mild and over a general area, damage to genes in cell nuclei is less likely to be profound. In fact, chromosomes and genes are capable of considerable self-repair. Minor changes that do remain are not likely to prevent cell division, but they may cause mutations. In future divisions these mutations may exhibit the wild, erratic behavior characteristic of cancer. Thus repeated low exposures to radiations are likely to induce cancerous growths in otherwise healthy tissue.

If the primary site of radiation damage is the genetic apparatus of a cell, the symptoms of overexposure will appear first among cells that most frequently undergo division. Cells in bone marrow are in this category. Since they are responsible for making white cells, it is not surprising that an early sign of radiation illness is a drop in the white cell count. Damage to other cells that divide less frequently becomes apparent as they reach their time for division. Whatever tissue they constitute will suffer. Of course, if the initial exposure is sufficiently severe, enough cells in enough parts of the body may be rendered reproductively dead to produce all effects of radiation illness quickly.

Damage to the contents of cell nuclei in *germ cells,* cells that produce sperm or eggs, can produce mutations in offspring. Whether or not these actually occur, and how important the changes may be, depends on the extent of the damage and on many other factors beyond the scope of this book. Any possibility of mutation is dangerous to the future of the species. So few mutations are in the direction of improving the species that claims for the desirability of an accelerated mutation rate must be considered unfounded. We should note, however, that man has always lived in the presence of radiations. Cosmic rays are ionizing radiations, and natural radioactivity has always been with us. The production of detrimental genes is unavoidable. Increases in the radiation level trouble all scientists and most informed people.

## POSTSCRIPT

The excitement created in scientific circles by the Watson-Crick theory has been great. The enthusiasm of some workers and students, however, has sometimes led to exaggerated claims and statements. Barry Commoner, one scientist who has argued for balance, has voiced some of the following concerns.

Described as a "living molecule," as a "self-duplicating molecule," DNA is neither. It cannot be duplicated apart from suitable enzymes, primer, and triphosphate forms of the nucleotides. *In vitro* syntheses of DNA have required these as well as the watchful, thinking mind and hands of the scientist. Biological evidence has long supported the conclusion that the least complex entity capable of self-duplication is the living cell. None of the recent biochemical evidence refutes this. The DNA molecule is not self-sufficient. Biological specificity is only partly due to it. The DNA from one cell cannot be introduced willy-nilly into *any* other kind of cell to make the cell from that point on do what the DNA tells it to do. Carefully selected host cells can have their genetic apparatus taken over by "alien" DNA, but such selection is necessary. The cell does not otherwise exhibit the expected changes. For genetic specificity not just the DNA but also the appropriate enzymes and the nucleotides must be present. For eventual enzyme synthesis all the needed sRNAs, the amino acids, and the needed enzymes must be available. Physicists have long hoped to uncover the secrets of the atomic nucleus by examining the debris from atom-smashing experiments. Recently, however, some physicists have suggested that the unique properties of the atomic nucleus are too intimately associated with its very complexity to be explained by data obtained from the fragments. Commoner argues that the living cell is no less complex and that this very complexity means, as it does for the physicists, that the *function* of DNA in the cell cannot be completely extricated from the dynamic *organization* of the cell. Commoner objects to the statement "DNA is the secret of life." He prefers "Life is the secret of DNA."

**The Kornberg-Goulian Experiment. Nucleic Acids and Viruses.** The viruses that have been studied are chemical aggregates made of two types of materials, protein and nucleic acid. The protein serves as an overcoat for the nucleic acid; in some viruses the nucleic acid is RNA, in others DNA. Each type of virus has its own peculiar host cell. The tobacco mosaic virus, for example, infects only the leaves of tobacco plants and causes no apparent harm to people who plant, cultivate, and harvest tobacco.

When a virus encounters its host cell, its nucleic acid core penetrates the cell. Once inside the living cell, the viral nucleic acid directs the making of more of itself and of protein overcoats for the new cores. When enough new viral particles are made, the host cell bursts, and neighboring host cells are invaded by the cores of the newly made virus particles. Thus the viral infection spreads.

One virus that has been extensively studied is labeled ΦX174 and has E. coli as its host cell. The nucleic acid in ΦX174, which is DNA, has 5500 nucleotide units and exists as a single-stranded closed ring. This DNA core material can be isolated from the virus and yet retain the natural infectivity of the virus.

One question that has bothered molecular biologists is whether DNA is made with only the four nucleotides and their four amines, A, T, G, and C. Perhaps the analytical techniques missed another unit or missed a nonnucleotide building block. If there were only one or two such units among several thousand nucleotides in a DNA molecule, they might escape detection. In 1967 Kornberg and Goulian reported powerful evidence that only the four nucleotides are required to make DNA. They took natural DNA cores from the ΦX174 virus and grew new cores in an artificial medium in which *only* the four different nucleotides were provided. Two enzymes were required, a DNA polymerase to join the nucleotides together in a string and a DNA-joining enzyme to catalyze the closure of the string into the ring form. No other building block molecules for DNA were needed. The synthetic DNA viral cores, when allowed to be in contact with their host cells, *E. coli,* were infectious. They entered the host cells, and many new virus particles were produced.

The Kornberg-Goulian experiment provided information about DNA, about viruses, and about the enzymes necessary for making DNA. One of the important scientific advances of the twentieth century, it brought closer the day when synthetic modifications of existing, but (humanly speaking) imperfect, DNA may be made and introduced into defective tissue to bring about its repair. It hastened the day when synthetic but noninfectious DNA may be used to take other "hitchhiker" molecules inside cells of defective tissue to act in a chemotherapeutic way. The experiment did not, however, constitute the creation of life in a test tube. Growing potatoes is not the same thing as creating potatoes. Growing viruses is not the same thing as creating viruses or creating life. Kornberg of course made no such extravagant claim, but his experiment did provide one more dramatic example of the molecular basis of life.

## REFERENCES AND ANNOTATED READING LIST

### BOOKS

G. Beadle and M. Beadle. *The Language of Life*. Doubleday and Company, Garden City, N.Y., 1966. Written especially for the scientific layman, this book is for him probably the best single introduction to what modern genetics is all about.

J. D. Watson. *Molecular Biology of the Gene*. W. A. Benjamin, New York, 1965. This is an outstanding treatment of molecular biology by one of the originators of the Watson-Crick theory. (Paperback.)

J. D. Watson. *The Double Helix*. Atheneum, New York, 1968. One who made history gives a personal account of exciting times in the development of a biological theory.

V. M. Ingram. *The Biosynthesis of Macromolecules*. W. A. Benjamin, New York, 1966. Written as a supplement to a biochemistry course, this book is also an excellent reference. (Paperback.)

P. E. Hartman and S. R. Suskind. *Gene Action*. Prentice-Hall, Englewood Cliffs, N.J., 1965. The relation of molecular biology to classical genetics is highlighted. (Paperback.)

Thomas P. Bennett and Earl Frieden. *Modern Topics in Biochemistry. Structure and Function of Biological Molecules*. The Macmillan Company, New York, 1966.

Robert F. Steiner and Harold Edelhoch. *Molecules and Life*. Van Nostrand Company, Princeton, N.J., 1965.

H. Stern and D. L. Nanney. *The Biology of Cells*. John Wiley and Sons, New York, 1965.

A. White, P. Handler, and E. L. Smith. *Principles of Biochemistry,* third edition. McGraw-Hill Book Company, New York, 1964.

B. Wallace and Th. Dobzhansky. *Radiation, Genes, and Man*. Holt, Rinehart, and Winston, New York, 1959. The book delineates the genetic aspects of radiation damage.

D. Grosch. *Biological Effects of Radiations*. Blaisdell Publishing Company, New York, 1965. After discussing the nature of radiations, their effects on molecules, cells, tissues, organs, organisms, and ecological communities are thoroughly described.

## ARTICLES

J. D. Watson and F. H. C. Crick. "A Structure for Deoxyribose Nucleic Acid." *Nature,* Vol. 171, 1953, page 737. "Genetical Implications of the Structure of Deoxyribonucleic Acid." *Nature,* Vol. 171, 1953, page 964.

F. H. C. Crick. "The Genetic Code." *Scientific American,* October 1962, page 66.

M. W. Nirenberg. "The Genetic Code: II." *Scientific American,* March 1963, page 80.

F. H. C. Crick. "The Genetic Code: III." *Scientific American,* October 1966, page 55.

R. T. Hinegardner and J. Engelberg. "Rationale for a Universal Genetic Code." *Science,* Vol. 142, 1963, page 1083. Neither abrupt nor gradual changes in the genetic code seem to be rational, argue the authors.

J. Hurwitz and J. J. Furth. "Messenger RNA." *Scientific American,* February 1962, page 41. The discovery of mRNA is described.

A. Rich. "Polyribosomes." *Scientific American,* December 1963, page 44. The author discusses the emergence of the idea that not single ribosomes acting in isolation but collections of ribosomes working together are the "factories" of protein synthesis.

G. Zubay. "Molecular Model for Protein Synthesis." *Science,* Vol. 140, 1963, page 1092. The author links structural and biochemical information into one coherent pattern and offers a stereochemically sound model for the template mechanism in protein synthesis.

M. F. Singer. "In Vitro Synthesis of DNA." *Science,* Vol. 158, 1967, page 1550. This is a report of the Kornberg-Goulian experiment.

A. Kornberg. "The Synthesis of DNA." *Scientific American,* October 1968, page 64. Kornberg describes the test tube synthesis of DNA that duplicates the viral activity of the DNA in the virus $\Phi$X174.

A. Kornberg. "Active Center of DNA Polymerase." *Science,* Vol. 163, 1969, page 1410. With considerable supporting evidence, Kornberg proposes what he calls a "speculative model for helix replication, in vivo," and he suggests how the enzyme DNA polymerase participates in achieving a sequential and practically simultaneous replication of both helical strands of DNA.

R. B. Merrifield. "The Automatic Synthesis of Proteins." *Scientific American,* March 1968, page 56. The *in vitro* synthesis of small proteins is described. Peptide chains are assembled one amino acid unit at a time by a remarkable solid phase method in which the growing chains are anchored on small beads of polystyrene.

J. P. Changeux. "The Control of Biochemical Reactions." *Scientific American,* April 1965, page 36. Ways in which feedback systems regulate the biosynthesis of cell products, including the work of repressor genes, are discussed.

B. F. C. Clark and K. A. Marcker. "How Proteins Start." *Scientific American*, January 1968, page 36. The authors describe how protein synthesis in bacteria is initiated.

B. Commoner. "DNA and the Chemistry of Inheritance." *American Scientist*, Vol. 52, 1964, page 365. "Life is the secret of DNA" would be a better guide for biological investigations than "DNA is the secret of life"—this is the argument of Commoner. See also "The Elusive Code of Life: Is DNA Really the Master Key to Heredity?" *Saturday Review*, October 1, 1966, page 71.

R. W. Holley. "The Nucleotide Sequence of a Nucleic Acid." *Scientific American*, February 1966, page 30. The work of determining the nucleotide sequence in alanine transfer RNA is described.

E. H. Davidson. "Hormones and Genes." *Scientific American*, June 1965, page 36. Evidence that some hormones act by controlling the activities of genes is described and discussed.

A. G. Bearn and J. L. German III. "Chromosomes and Disease." *Scientific American*, November 1961, page 66. Mongolism is linked to abnormalities in human chromosomes and to a defect in the genetic machinery for passing heredity material from parents to offspring.

A. G. Bearn. "The Chemistry of Hereditary Disease." *Scientific American*, December 1956, page 127. Several relatively rare hereditary diseases are examined in the light of the idea that a single gene may control the synthesis of a single enzyme.

H. Fraenkel-Conrat. "The Genetic Code of a Virus." *Scientific American*, October 1964, page 47. The virus that infects tobacco leaves consists of a coiled strand of RNA surrounded by a coat of protein molecules.

S. E. Stewart. "The Polyoma Virus." *Scientific American*, November 1960, page 63. Viruses, which contain nucleic acid, appear to operate by deranging the genetic machinery of the host cell.

L. Gorini. "Antibiotics and the Genetic Code." *Scientific American*, April 1966, page 102. Streptomycin and related drugs can change the meaning of the code that directs protein synthesis.

T. T. Puck. "Radiation and the Human Cell." *Scientific American*, April 1960, page 142. Remarkable photomicrographs of radiation-induced chromosome damage illustrate this article.

## PROBLEMS AND EXERCISES

1. Define each of the following terms.

   | | | |
   |---|---|---|
   | (*a*) zygote | (*b*) mitosis | (*c*) germ cells |
   | (*d*) somatic cells | (*e*) chromonemata | (*f*) chromosomes |
   | (*g*) genes | (*h*) replication | (*i*) nucleotide |
   | (*j*) nucleic acid | (*k*) DNA | (*l*) RNA |
   | (*m*) base pairing | (*n*) template | (*o*) rRNA |
   | (*p*) mRNA | (*q*) sRNA | (*r*) codon |
   | (*s*) anticodon | (*t*) genetic code | (*u*) phenylketonuria |
   | (*v*) melanins | | |

2. What structural features are common to all DNA molecules?
3. How do DNA molecules differ?
4. What structural features are common to all RNA molecules?
5. How do RNA molecules differ?
6. In terms of molecular structures, what specifically is meant by the genetic code?

7. By means of a "flow chart," discuss the one-gene, one-polypeptide theory.

8. If the sequences of heterocyclic bases on one section of a DNA molecule is A-T-C-G-G-T-T-A, what is the sequence in this region on the corresponding RNA molecule? (Remember that in RNA there is no thymine; uracil, U, takes its place.)

9. Discuss the particular contribution Watson and Crick made to our understanding of the relation between gene *structure* and gene *function*.

10. How does the Watson-Crick model correlate with Chargaff's data, A/T = 1 and G/C = 1, for the DNA of various species?

11. Discuss the relation of rRNA, mRNA, and sRNA to one another.

12. Discuss the relation of mRNA, ribosomes, and polysomes to one another.

13. What is believed to be the basis for mutations?

14. Liver cells are known to be particularly active in protein synthesis. They are also known to be particularly rich in RNA. How are these two facts correlated?

15. Explain in your own words how replication of DNA occurs.

16. How will small doses of radiations or X-rays over a long time possibly lead to cancer?

17. How does a well-focused "dose" of X-rays accomplish the arresting of cancer?

18. How is it that radiations can both cause cancer and cure it?

# The Formation
# of Acetyl Coenzyme A
# from Pyruvic Acid

The connecting link between glycolysis and the citric acid cycle is the oxidative decarboxylation of pyruvic acid and the formation of acetyl coenzyme A, or "active acetyl." Several steps and enough details of the reaction are known to enable the writing of a possible mechanism. We include this rather complicated

$$CH_3-\overset{O}{\overset{\|}{C}}-\overset{O}{\overset{\|}{C}}-O^- + CoA-SH + NAD^+ \longrightarrow CH_3-\overset{O}{\overset{\|}{C}}-S-CoA + CO_2 + NADH$$

Pyruvate ion        Coenzyme A                  Acetyl coenzyme A

$$\downarrow$$

Citric acid cycle

sequence of chemical events to illustrate what is apparently an important fact of physical life. Enzymes often work in teams. Some enzymes form systems, huge complexes of two, three, or more enzymes which are tightly bound into a structural matrix and which, being very close together, can work together. Thus the product made at one enzyme may, in a sense, be "handed" to a neighboring enzyme for further processing. Thus, if the substrate while at one enzyme is chemically changed to make it electrically attractive to some binding site at the neighboring enzyme, the newly altered substrate may move directly to it. The economy of such an arrangement is apparent; the new substrate need not drift away from the old enzyme surface, wander about, and then finally find attachment at the next enzyme surface. Whole sequences of reactions can occur swiftly.

In the mechanism that follows, several steps are presented as though they were isolated, but it must be understood that they are occurring at a multiple-enzyme complex. Four coenzymes are probably involved, lipoic acid, thiamine pyrophosphate (TPP), flavin adenine dinucleotide (FAD), and nicotine adenine dinucleotide (NAD). To simplify matters we shall define symbols to stand for the coenzymes before we discuss their action.

**Thiamine pyrophosphate**

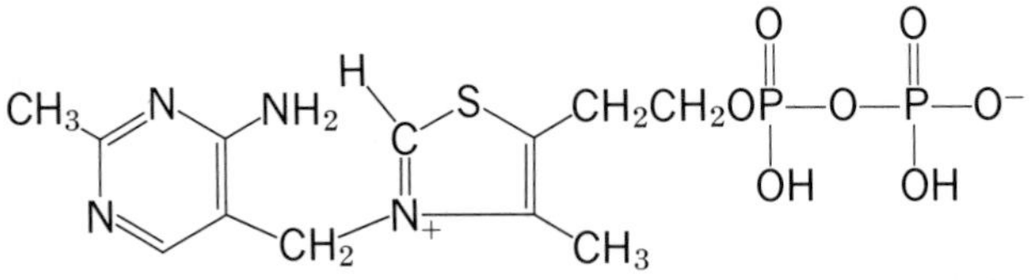

Thiamine pyrophosphate
(TPP)

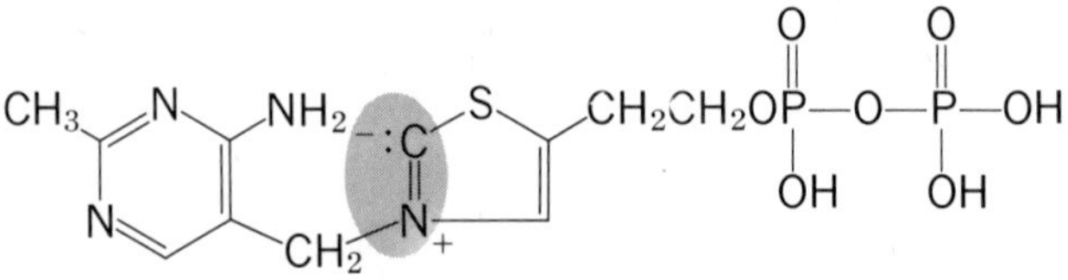

Presumed active form of TPP

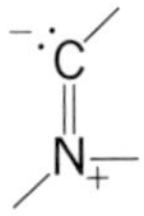

Symbol to be used for TPP

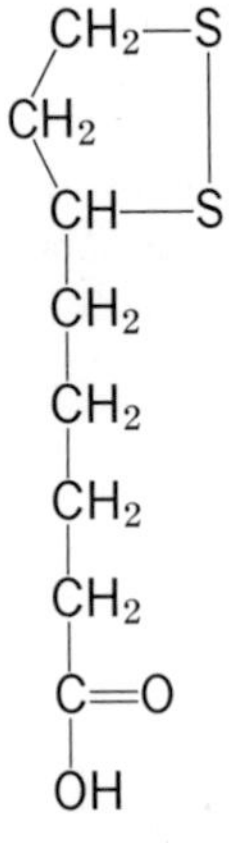

**Lipoic acid**

lipoyl unit

Lipoic acid

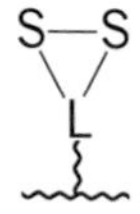

Symbol to be used for lipoyl unit on the enzyme

side chain of lysine, an amino acid, incorporated into the chain of the enzyme

protein backbone of enzyme

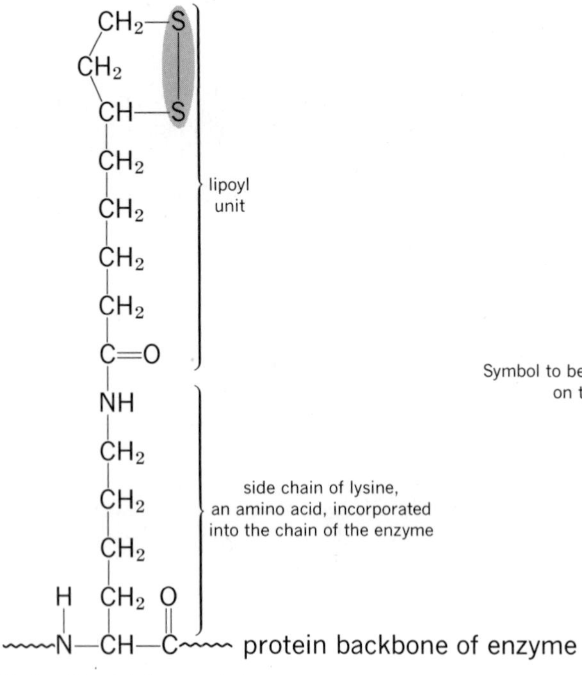

Lipoic acid as it is incorporated into the enzyme (protein)

**Coenzyme A**

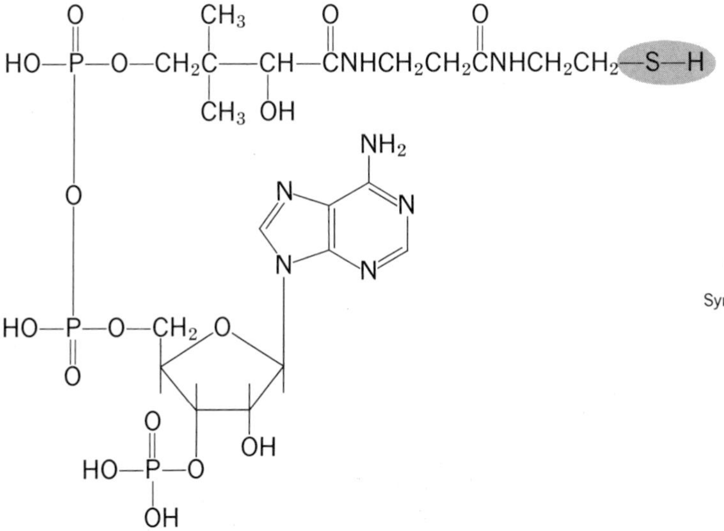

$$\text{HO}-\overset{\overset{\displaystyle O}{\|}}{\text{P}}-\text{O}-\text{CH}_2\overset{\overset{\displaystyle CH_3}{|}}{\underset{\underset{\displaystyle CH_3}{|}}{\text{C}}}-\overset{\overset{\phantom{O}}{|}}{\underset{\underset{\displaystyle OH}{|}}{\text{CH}}}-\overset{\overset{\displaystyle O}{\|}}{\text{C}}\text{NHCH}_2\text{CH}_2\overset{\overset{\displaystyle O}{\|}}{\text{C}}\text{NHCH}_2\text{CH}_2-\text{S}-\text{H}$$

CoA—S—H

Symbol to be used for
coenzyme A

Coenzyme A

What might seem a simple chemical change, the oxidative decarboxylation of pyruvic acid to acetyl coenzyme A, requires this team of coenzymes, each with a vitamin in its structure. The symbols for these coenzymes, to summarize them before we consider a possible mechanism for the oxidative decarboxylation of pyruvic acid, are as follows:

Thiamine pyrophosphate (TPP)

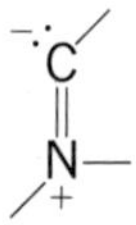

Lipoic acid

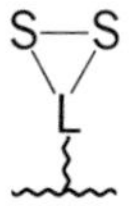

Coenzyme A                                CoA—S—H

A flavoprotein, $\alpha$-lipoyl-
dehydrogenase, containing FAD          FP        (oxidized form)
                                        $FPH_2$   (reduced form)

Nicotine adenine dinucleotide           $NAD^+$   (oxidized form)
                                        NADH      (reduced form)

1. TPP becomes temporarily bonded to the carbonyl carbon of the acetyl group that will be excised from pyruvic acid.

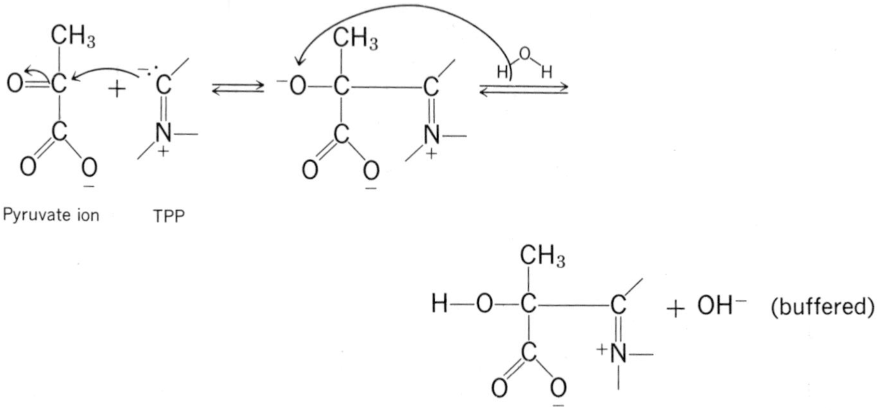

Pyruvate ion  TPP

$$H-O-\underset{\overset{|}{\underset{O}{C}}}{\overset{\overset{CH_3}{|}}{C}}-\overset{+N-}{C} + OH^- \quad \text{(buffered)}$$

2. Carbon dioxide is lost. This step is very much like the decarboxylation of a β-keto acid, if the $>$C$=$N is likened to a keto group,

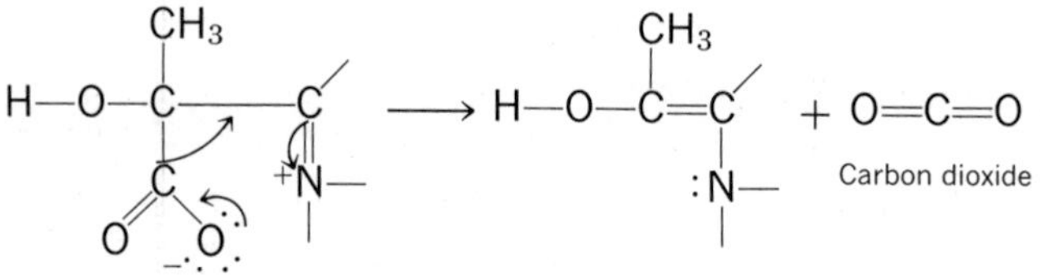

$$H-O-\overset{CH_3}{C}=C \quad + \quad O=C=O$$

Carbon dioxide

3. The acetyl unit is transferred to a lipoyl group. At this point a pair of electrons in the original pyruvate ion begins its departure; that is, oxidation is occurring here, and the disulfide unit begins to be reduced to two mercaptan groups.

Acetyl group is now attached to the lipoic acid system.    TPP recovered

4. The acetyl group is transferred to coenzyme A.

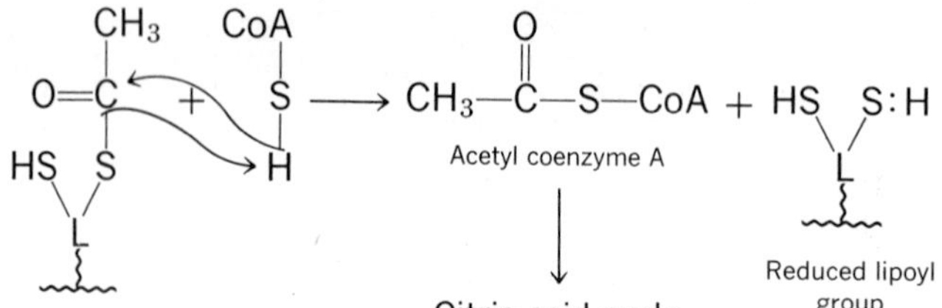

$$CH_3-\overset{\overset{O}{\|}}{C}-S-CoA + HS \quad S:H$$

Acetyl coenzyme A

Citric acid cycle

Reduced lipoyl group

5. Although acetyl coenzyme A ("active acetyl") has now been prepared, the enzyme system must be restored to its original form. Specifically, the two mercaptan groups on the reduced lipoyl unit must be reoxidized to the disulfide group of the original enzyme. For this to happen the elements of hydrogen are transferred somewhere, and the flavoprotein (FP), $\alpha$-lipoyldehydrogenase, is apparently the acceptor.

$$\text{L} \begin{array}{c} S-H \\ \\ S-H \end{array} + \text{FP} \longrightarrow \text{L} \begin{array}{c} S \\ | \\ S \end{array} + \text{FPH}_2$$

Restored disulfide form of the
lipoyl group

6. Finally, the $FPH_2$ is reoxidized, and we now are at the respiratory chain (cf. p. 449), the mechanism for transporting a pair of electrons and two hydrogen nuclei to the place where they will be combined with oxygen (while ATP is also synthesized).

$$FPH_2 + NAD^+ \longrightarrow FP + NADH + H^+$$

L. J. Reed[1] has proposed that the interactions in the enzyme complex occur by means of a "swinging arm mechanism." The long lipoyl group swings from one active site to another as it picks up the acetyl group (step 3), transfers it to coenzyme A (step 4), and gives off the elements of hydrogen to FP (step 5), as shown in Figure I.1.

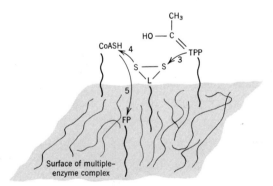

Figure I.1

[1] L. J. Reed, in *The Enzymes*, Vol. 3, edited by P. Boyer, Academic Press, New York, 1961, page 195.

# The Formation of Succinic Acid from $\alpha$-Ketoglutaric Acid in the Citric Acid Cycle

This reaction, probably the most complicated step in the citric acid cycle, may be written as

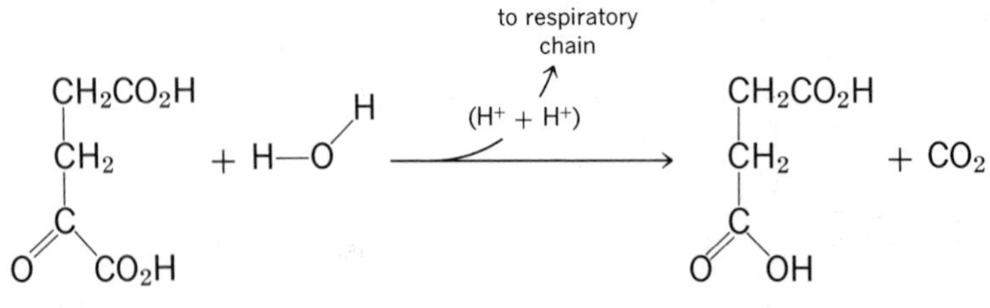

α-Ketoglutaric acid                  Succinic acid

Both dehydrogenation (oxidation) and decarboxylation occur. In Figure 13.9 we indicated that this step is accompanied by the production of ATP in a way not directly geared to events in the respiratory chain. This equation, obviously, is no more than a simplified statement of what happens, and what happens is very much like what occurs to pyruvic acid, as described in Appendix I.

$\alpha$-Ketoglutaric acid, on its way to becoming succinic acid, is first converted into the coenzyme A derivative of succinic acid.

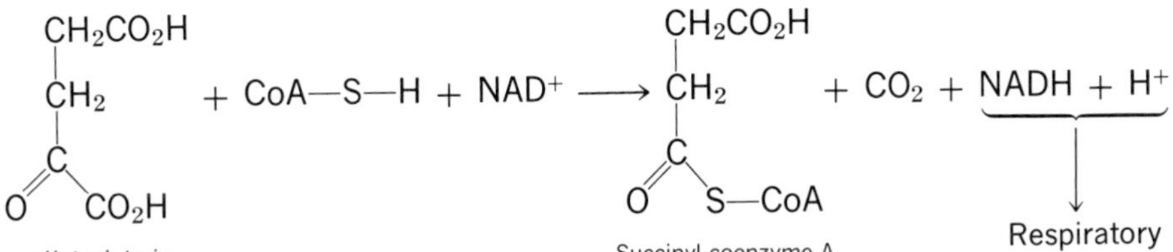
CH2CO2H
CH2
C
O
CO2H
+ CoA—S—H + NAD+
CH2CO2H
CH2
C
O
S—CoA
+ CO2 + NADH + H+
Respiratory chain

α-Ketoglutaric acid    Succinyl coenzyme A

The mechanism is apparently identical with that given in Appendix I, except that $HO_2CCH_2CH_2$— replaces $CH_3$— in the developing acetyl group.

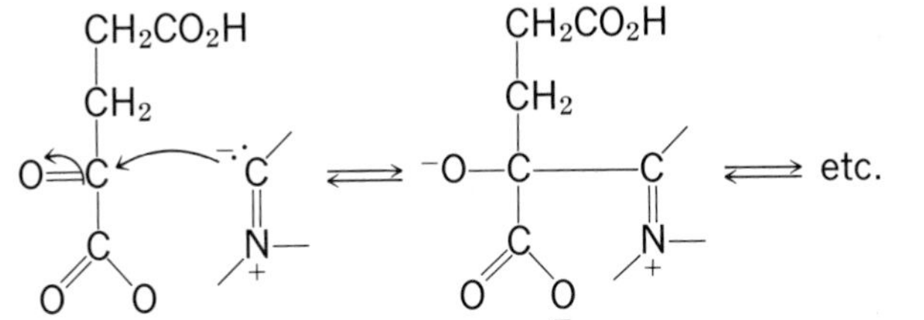
CH2CO2H
CH2
C
O
C
O
O
C
N—
+
CH2CO2H
CH2
-O—C
C
O
O
C
N—
+
etc.

Ketoglutarate ion    TPP

(succeeding steps analogous to those in Appendix I but leading to

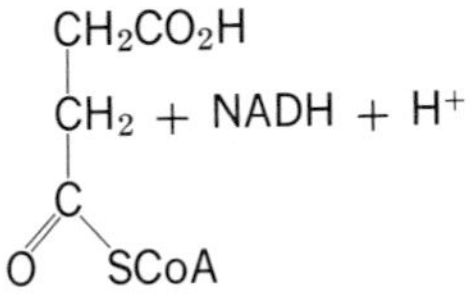
CH2CO2H
CH2 + NADH + H+
C
O
SCoA

Succinyl coenzyme A

Thus to get as far as succinyl coenzyme A, the respiratory chain is given the elements of hydrogen in the form of ($NADH + H^+$).

Succinyl coenzyme A is not, of course, succinic acid, but it does possess a "high-energy bond" that links the carbonyl group to sulfur. By a mechanism still not fully understood, succinyl coenzyme A reacts with guanosine diphosphate, GDP, a close relative of ADP, to form guanosine triphosphate (analogous to ATP). Then GTP transfers the phosphate unit to ADP, and a molecule of ATP forms. These reactions may be summarized by the following equations.

$$\text{Succinyl CoA} + \text{GDP} + \text{P}_i \longrightarrow \text{succinic acid} + \text{GTP} + \text{CoASH}$$

$$\text{GTP} + \text{ADP} \longrightarrow \text{GDP} + \text{ATP}$$

Thus ATP is synthesized at this step by a mechanism that does not use the respiratory chain directly.

# INDEX

Page numbers in italics refer to tables.

Ethyl propionate, synthesis of, 232
Ethyl p-tolyl ketone, synthesis of, 215
Ethyl valerate, 247
Extracellular fluid, 396

Famine, 520
Faraday, Michael, 72
α-Farnesene, 316
Farnesyl pyrophosphate, 507
Fasting level, glucose, 483
Fats, animal, 299, 300
Fatty acid cycle (spiral), 436, 501
Fatty acids, 300, 301
    in lipid catabolism, 500
    from lipogenesis, 495
Feces, 397
    pigments of, 534
Fehling's test, 199
    carbohydrates and, 278
Fermentation, alcohol, 477
Ferritin, 534
Fibrin, 348, 411, 423
Fibrinogen, 411, 423
Fibrosis, 423
Fish oil, fatty acids from, 300
Flavin adenine dinucleotide, 443
    in deaminations, 524
    in fatty acid cycle, 502
    reduced form of, 444
    in respiratory chain, 449
Flavin mononucleotide, 369, 443
    in lipogenesis, 498
Flavoproteins, 443, 581, 583
Flax, 293
Fleming, Alexander, 388
Fluoroacetic acid, 227
Fluorobenzene, 81
Folic acid, 372, 388
Foramide, 262
Formaldehyde, 197, 216
    Bakelite from, 217
    Grignard reaction of, 132
    hydrate of, 201, 216
    Melamine resin from, 218
Formalin, 216
Formic acid, 225
    from oxalic acid, 237
Free energy, Gibbs, 431
    from glucose, 437
    from palmitic acid, 437
    from pyruvic acid, 452, 454
Free radical, 25
Free rotation, 5

Freons, 155
Friedel, Charles, 84
Friedel-Crafts reactions, 75, 84, 214
Fructose, 198, 285
    absorption of, 480
    glucogenesis and, 482
    metabolism of, 462
    occurrence of in sucrose, 287
    osazone of, 279
Fructose 1,6-diphosphate, 465, 467, 469
    formation of, 213
    in glucogenesis, 480
Fructose 6-phosphate, 463, 465, 466, 468
    in glucogenesis, 482
Fuel, hydrocarbons as, 24
Fuel oil, 24
Fumarase, 376
Fumaric acid, 239, 256
    in citric acid cycle, 453, 458
    reaction of, with water, 131
Functional group, 4, 14
Functional isomers, 139
Furfural, 197

Galactosamine, 285
Galactose, 276, 284
    absorption of, 480
    glucogenesis and, 482
    metabolism of, 462
    occurrence of, in lactose, 287
    in lipids, 299, 311
Galactose 6-phosphate, 463
    in glucogenesis, 482
Gall bladder, 398, 405
Gammexane, 76
Gas oil, 24
Gasoline, 24
Gastric juice, 399
    analysis of, 402
Gastrin, 399
Gastritis, 402
Gelatin, 348
Gene, 539
    DNA and, 542; see also Deoxyribonucleic
      acid
    one-gene—enzyme theory, 562
Genetic gradients, principle of, 425
Genetotrophic principle, 426
"Geneva system," 17
Geometry and structure, 4, 32
Geranial, 318
Geraniol, 318
Geranyl pyrophosphate, 507